Geometry

PLANE AREA

1	Square	$A = s^2$	where s is the side
2	Rectangle	$A = lw$	where l is the length and w is the width
3	Parallelogram	$A = bh$	where b is the base and h is the height
4	Triangle	$A = \frac{1}{2}bh$	where b is the base and h is the height
5	Circle	$A = \pi r^2$	where r is the radius

SURFACE AREA

1	Cube	$S = 6s^2$	where s is the side
2	Rectangular Box	$S = 2(lw + wh + lh)$	where l is the length, w is the width, and h is the height
3	Sphere	$S = 4\pi r^2$	where r is the radius
4	Cylinder	$S = 2\pi rh + 2\pi r^2$	where r is the radius and h is the height
5	Cone	$S = \pi r^2 + \pi r\sqrt{r^2 + h^2}$	where r is the radius and h is the height

VOLUME

1	Cube	$V = s^3$	where s is the side
2	Rectangular Box	$V = lwh$	where l is the length, w is the width, and h is the height
3	Sphere	$V = \frac{4}{3}\pi r^3$	where r is the radius
4	Cylinder	$V = \pi r^2 h$	where r is the radius and h is the height
5	Cone	$V = \frac{1}{3}\pi r^2 h$	where r is the radius and h is the height

Precalculus

THIRD EDITION

Precalculus

MUSTAFA A. MUNEM

JAMES P. YIZZE

MACOMB COUNTY COMMUNITY COLLEGE

WORTH PUBLISHERS, INC.

PRECALCULUS, THIRD EDITION

PRINTED IN THE UNITED STATES OF AMERICA
LIBRARY OF CONGRESS CATALOG CARD NO. 77-81759
ISBN: 0-87901-086-X
FOURTH PRINTING, JULY 1981

DESIGN BY MALCOLM GREAR DESIGNERS

WORTH PUBLISHERS, INC.
444 PARK AVENUE SOUTH
NEW YORK, NEW YORK 10016

PREFACE

PURPOSE This third edition of our *Precalculus* textbook has been extensively rewritten and improved. The aims of the book, however, remain the same—to provide the preparation necessary for students who intend to study calculus or other courses in college mathematics. Our textbook also gives those students who will not take higher level mathematics courses an opportunity to investigate and understand this important area of college mathematics and the role it plays in their daily lives.

Our publisher conducted a survey of the many users of the previous edition, and substantial changes have been made in response to these findings. We were fortunate on this occasion to have received particularly constructive criticisms, not only of the second edition, but also of our proposed revisions. These suggestions in part provided us with a fresh vision of the book and its role as a tool for the student and for the teacher. As a result, there is hardly a paragraph in this third edition that has not been strengthened.

PREREQUISITES It is assumed that students who use this book have taken the equivalent of at least one year of high school plane geometry and one and one-half years of high school algebra, or that they have had a college course in intermediate algebra. However, a chapter-length review of the fundamentals of algebra has been included in this edition for students who need additional help.

OBJECTIVES Our book was written and revised with two goals in mind: first, that the book itself should provide all that is required for the student to learn the material; and second, that the book should enable students to *apply* the mathematical principles they have learned to the solution of specific problems. To help accomplish these objectives, the discussion of new concepts always includes numerous illustrative examples. Every attempt has been made to minimize the use of technical language and symbolism. However, those definitions, properties, and theorems that must be included in a book written at this level have been stated with care and precision. Theorems and properties are often preceded by a geometric discussion that provides the student with some insight into why the formal statements are written as they are. There is a reasonable balance between theory, on the one hand, and technique, drill, and application, on the other. The many problems in the new edition are intended to help students gain confidence in their understanding of the material covered, while at the same time indicating those areas in need of additional study.

FEATURES *Color* is again used to highlight theorems, properties, and definitions, important statements, phrases, and terms, as well as those parts of graphs that require emphasis. Since most of the explanations in the book build on intuition and geometric notions, we have included a large number of *illustrations* and *graphs*. In an effort to improve the graphs and illustrations, all of the artwork has been redrawn. Pertinent formulas from algebra, geometry, and trigonometry are listed inside the front and back covers of the book for convenient reference.

There are more than four hundred *examples* and over three thousand *problems* in the book. The examples are worked through in step-by-step detail, and the problems range from routine to challenging in order to satisfy all levels of student understanding. Numerous examples and *applied problems* from physics, engineering, economics, business, biology, ecology, medicine, and psychology have been added throughout the third edition.

The sets of problems at the end of each section have been organized to follow the examples in that section. As a rule, a text example is given for each type of problem in a set, and each group of a particular type of problem is graded in difficulty. The odd-numbered problems strive for the level of understanding desired for most students. Answers to these problems are provided in the back of the book. The even-numbered problems probe for a deeper, more conceptual understanding of the topics covered. This organization of problems will simplify assignment planning; one can thus choose a variety of computational and conceptual problems. At the end of each chapter, a set of *review problems* provides a final test of the student's grasp of the topics.

With the availability of *hand calculators,* the answers to many examples and problems can be computed more easily. In Chapters 4 and 5, for instance, alternative solutions that involve the use of a hand calculator have been included in some of the examples and in the appropriate problem sets. However, the availability of a calculator is *not* required for effective use of this book.

MAJOR CHANGES Chapter 1 now includes a review of topics from intermediate algebra, including polynomials, fractions, exponents, radicals, and factoring of polynomials. Individual topics can be covered briefly or omitted entirely without interrupting the flow of the material.

Chapter 2 has been expanded to include a detailed discussion of graphing techniques and a unique subsection on shifting the graphs of functions. We define a function initially as a correspondence and use this idea later to present the concept of a function as a set of ordered pairs.

Chapters 3 and 4 have been shortened and simplified to provide an earlier introduction to polynomial, exponential, and logarithmic functions, and applications of these kinds of functions have been added. Chapter 4 includes a section on common and natural logarithms.

Chapters 5 and 6 have been completely rewritten to give a more detailed exposition of trigonometric functions. The development and definitions of the trigonometric functions on real numbers and on angles are blended into one presentation in Chapter 5. Also, the basic properties and graphs of trigonometric functions are included in Chapter 5, although we postpone discussion of the inverse trigonometric functions until Chapter 6. In Chapter 6, we have included a new section on fundamental identities, as well as material on vector applications in trigonometry. The section on vectors has been placed at the end of the chapter so that it can be treated as optional material. Chapter 6 also includes trigonometric equations and triangle trigonometry, and the discussion of polar coordinates has been moved into this chapter.

Chapter 7 in this edition is essentially a rewritten version of Chapter 9 on analytic geometry from the second edition. The approach has been made less formal and a number of applications have been added.

The material on systems of linear equations is now found in Chapter 8. In this chapter, we have included new sections on systems of linear inequalities, systems of quadratic equations, and linear programming.

Chapter 9 has been reorganized to include complex numbers, mathematical induction, geometric series, and the binomial theorem.

PACE The book allows for considerable flexibility in the pace of the course, as well as the choice of topics. The following schedule, which was used for our four-credit, one-semester course, is meant only as a general guide:

Chapter 1: 4 lectures
Chapter 2: 10 lectures
Chapter 3: 10 lectures
Chapter 4: 8 lectures
Chapters 5 and 6: 20 lectures
Chapter 9: 4 lectures

Many other options are possible. For example, if the students are sufficiently prepared, Chapter 1 can be reviewed briefly in one lecture, so that the topics from Chapters 7 and 8 can be covered as time allows.

The book can also be used for separate, sequential courses in college algebra and college trigonometry. It can also be used for a trigonometry course that meets for 40 hours with the following schedule:

Chapter 1: 4 lectures
Chapter 2: 8 lectures
Chapters 5 and 6: 24 lectures
Chapter 8: 4 lectures

ADDITIONAL AIDS An accompanying third edition *Study Guide* is available for students who may need or desire additional drill or assistance. The *Study Guide* is written in a semi-programmed format, and its organization conforms with the arrangement of the topics in the book. It contains a great many carefully graded, fill-in statements and problems. Each topic in the *Study Guide* has been broken down into simpler units so that confidence can be built as each step in the learning process is taken. A test is included for each chapter, and all answers are provided to encourage self-testing.

ACKNOWLEDGMENTS Many users of the first two editions have contributed to the development of the third edition. We are especially grateful for critical and constructive reviews from the following professors: Peter G. Casazza of the University of Alabama, Huntsville; Charles S. Frady of Georgia State University; Mark P. Hale, Jr., of the University of Florida; Kay Hudspeth of Pennsylvania State University; Ronald P. Infante of Seton Hall University; Stanley M. Lukawecki of Clemson University; Paul McDougle of the University of Miami; George E. Mitchell of the University of Alabama; Rita V. Rodriguez of the University of Puerto Rico; and Tina H. Straley of Kennesaw College. Special thanks are in order for Professor Howard E. Taylor of West Georgia College, who read the entire third edition, both in manuscript and in the galley-proof stage, and Professor William Tschirhart, who read the entire book in the page-proof stage. We also thank our colleagues, Professors Wayne Hille and Douglas Marsh, for painstakingly working through every problem in the book. Finally, we wish to express our sincere gratitude to the entire staff of Worth Publishers, especially to Bob Andrews, and to our editor, Gordon Beckhorn, who contributed much to the quality of the third edition of *Precalculus* and who was a constant source of encouragement during its production.

Mustafa A. Munem
James P. Yizze

Warren, Michigan
January 1978

CONTENTS

CHAPTER 1

Fundamentals of Algebra

This chapter provides a review of the fundamentals of algebra that will be used throughout the remainder of the book. After introducing set terminology and describing special sets of real numbers, we briefly review polynomials, fractions, exponents, and radicals. Also included in the chapter are linear and absolute-value equations and inequalities.

1 Sets and Real Numbers

One objective of this section is to present basic set notation and terminology so that the language of sets can be used later to describe mathematical concepts. *Sets* are collections of objects. For example, we speak of the set of students in a particular course or the set of automobiles in a parking lot. In mathematics, we speak of the set of all counting numbers, or the set of prime numbers greater than 2 and less than 75. The objects in a set are normally called the *elements* of the set. Thus 1, 2, 3, and 4 are elements in the set of counting numbers.

The set that has no elements is called the *null set* or *empty set* and is denoted by $\varnothing$ or by $\{\ \ \}$. For example, the set of all women Presidents of the United States is an empty set, since no woman has yet been elected President of the United States.

A set is said to be *finite* if it is possible to list or enumerate *all* the elements of the set; a set that is neither finite nor empty is an *infinite* set. For example, if A is the set of all students in a particular class, A is a finite set since *all* its elements can be enumerated. On the other hand, if C is the set of all counting numbers, C is an infinite set since it is impossible to enumerate *all* the elements in this set.

Set descriptions are usually included between a pair of braces. For example, $A = \{a, b, c, d\}$ denotes that A is a finite set containing elements a, b, c, and d and no others. We use the notation $a \in A$ to indicate that a is an element of set A.

It is important to realize that $\varnothing$ is different from $\{0\}$, since $\{0\}$ is a set with one element, 0, whereas $\varnothing$ is a set that contains no elements.

Besides enumeration, another set description, *set builder notation,* takes the form $A = \{x \mid x \text{ has property } P\}$, which is read "$A$ is the set of all elements x such that x has property P." For example, $E = \{x \mid x$ is an even counting number} is read "E is the set of all elements x such that x is an even counting number." Notice that in this case, $2 \in E$, $4 \in E$, $6 \in E$, and so on. Since E is an infinite set, it is impossible to enumerate all of its elements; however, we can use the fact that the members of E form a generally known pattern to write E as $E = \{2, 4, 6, \ldots\}$, where the three dots mean the same as "and so on." We use the symbol $\notin$ to mean "is not an element of"; hence $1 \notin E$, $3 \notin E$, and $5 \notin E$.

1.1 Set Relations and Operations

Suppose that F is the set of all Ford automobiles and that M is the set of all motor vehicles. Clearly, all the elements of F are also found in M. The set F is considered to be a *subset* of M. The set of all girls in a biology class is a subset of the set of all students in that class. In general, set A is a *subset* of set B, written $A \subseteq B$, if every element of A is an element of B. *The empty set $\varnothing$ is considered to be a subset of every set.*

For example, if $C = \{1, 2, 3, 4, 5, 6\}$ and $D = \{2, 5, 3\}$, then $D \subseteq C$. Also, if $A = \{1, 2, 3\}$ and $B = \{x \mid x \text{ is a counting number less than } 4\}$, then $A \subseteq B$ and $B \subseteq A$. The latter example illustrates the definition of *equality of sets* because if $A \subseteq B$ and $B \subseteq A$, we consider sets A and B to be equal and we write $A = B$. In the first example, we have $D \subseteq C$, but $D \neq C$. Set D is an example of a *proper subset* of set C. In general, D is said to be a *proper subset* of C, written $D \subset C$ (notice that the horizontal bar is left off), if every element of D is in C and C has at least one element that is not in D; that is, $D \subseteq C$ but $D \neq C$.

EXAMPLE List all the subsets of $\{3, 5, 7\}$.

SOLUTION $\{3\}$, $\{5\}$, $\{7\}$, $\{3, 5\}$, $\{3, 7\}$, $\{5, 7\}$, $\{3, 5, 7\}$, and $\varnothing$ are the subsets of $\{3, 5, 7\}$. Note that all the subsets, with the exception of $\{3, 5, 7\}$ itself, are *proper* subsets of $\{3, 5, 7\}$.

When the selection of elements of subsets is limited to a fixed set, the limiting set is called a *universal set* or a *universe*. A universal set represents the complete set or the largest set from which the elements of all other sets in that same discussion can be chosen. The choice of the universal set is dependent on the situation being considered. For example, in one case it may be the set of all people in the United States, and in another it may be the set of all people in Michigan.

Consider a universal set $U = \{1, 2, 3, 4, 5, 6, 7, 8\}$. From U we can form $A = \{1, 2, 3, 4\}$ and $B = \{1, 3, 7\}$. How can sets A and B be used to form other sets? One way is simply to combine all the elements of A and B into one set to form $\{1, 2, 3, 4, 7\}$. The operation suggested by this example is that of *set union*.

A union B, written $A \cup B$, and represented by the entire shaded region in Figure 1a, is defined as

$$A \cup B = \{x \mid x \in A \text{ or } x \in B \text{ (or both)}\}$$

Figure 1

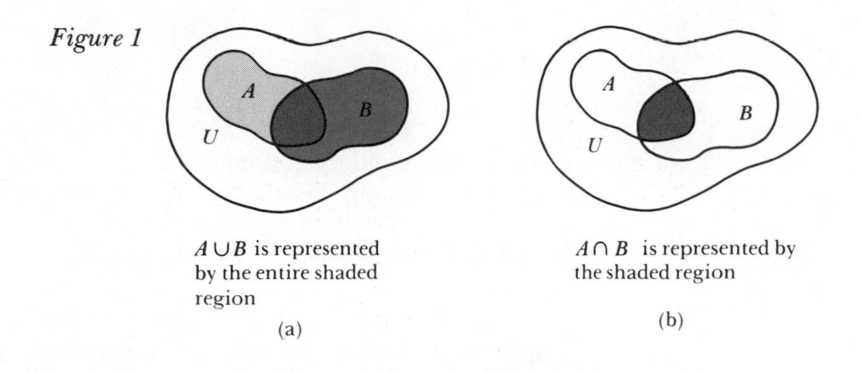

$A \cup B$ is represented by the entire shaded region

(a)

$A \cap B$ is represented by the shaded region

(b)

Hence, $\{1, 2, 3, 4\} \cup \{1, 3, 7\} = \{1, 2, 3, 4, 7\}$.

Another way to use $A = \{1, 2, 3, 4\}$ and $B = \{1, 3, 7\}$ to form another set is to form set $\{1, 3\}$, the set of all elements *common* to A and B. This operation is an example of *set intersection*.

A intersect B, written $A \cap B$, and represented by the shaded region in Figure 1b, is defined as

$$A \cap B = \{x \mid x \in A \text{ and (simultaneously) } x \in B\}$$

Thus, $\{1, 2, 3, 4\} \cap \{1, 3, 7\} = \{1, 3\}$.

If $A = \{1, 2, 3, 4\}$ and $B = \{5, 6, 7\}$, then $A \cap B = \varnothing$, and we say A and B are *disjoint sets.* In general, A and B are disjoint sets if $A \cap B = \varnothing$.

The union of two sets, then, is simply the set composed of all elements that are in at least one of the two sets; the intersection is the set of all elements common to the two sets. Note that when the union of two sets containing common elements is described, the common elements are *not* listed twice; hence $\{2, 3, 4\} \cup \{1, 4, 8\}$ is not written as $\{2, 3, 4, 1, 4, 8\}$ but rather as $\{2, 3, 4, 1, 8\}$, since the listing of 4 twice is superfluous.

EXAMPLES 1 Determine $A \cup B$ and $A \cap B$ if $A = \{1, 2, 3, 4, 5\}$ and $B = \{2, 5, 6, 7\}$.

SOLUTION $A \cup B = \{1, 2, 3, 4, 5, 6, 7\}$ and $A \cap B = \{2, 5\}$.

2 Let $A = \{x \mid x \text{ is a counting number}\}$ and let $B = \{x \mid x \text{ is an even counting number}\}$; that is, $B = \{2, 4, 6, 8, \ldots\}$. Find $A \cup B$ and $A \cap B$.

SOLUTION Note that $B \subset A$.

$A \cup B = \{x \mid x$ is a counting number or x is an even counting number$\}$. Therefore, $A \cup B = \{x \mid x \text{ is a counting number}\} = A$.

$A \cap B = \{x \mid x$ is a counting number and (simultaneously) x is an even counting number$\}$. Therefore, $A \cap B = \{x \mid x \text{ is an even counting number}\} = B$.

1.2 Real Number Sets

The language of sets is used to describe some of the number sets of algebra. By repeatedly adding 1 to itself, we can generate the set of counting numbers, also called the *set of positive integers,* $I_p = \{1, 2, 3, 4, 5, 6, 7, 8, 9, \ldots\}$.

The *set of negative integers*, $I_n = \{-1, -2, -3, \ldots\}$, consists of the negatives of the positive integers. The set

$$I_p \cup I_n \cup \{0\} = \{\ldots, -3, -2, -1, 0, 1, 2, 3, 4, \ldots\}$$

is called the *set of integers* and is denoted by I.

A *rational number* is any number that can be expressed in the form

$$\frac{a}{b} \qquad \left(a/b \text{ is an alternative way of writing the expression } \frac{a}{b}\right)$$

where a is an integer and b is a nonzero integer. For example, 3, $2\frac{1}{2}$, $-\frac{5}{7}$, and 53 percent are considered to be rational numbers, since they can be written as $\frac{3}{1}$, $\frac{5}{2}$, $\frac{-5}{7}$, and $\frac{53}{100}$, respectively. Q is generally used to denote the set of all rational numbers; hence

$$Q = \left\{q \,\middle|\, q = \frac{a}{b}, \quad \text{for } a \in I,\ b \in I, \text{ and } b \neq 0\right\}$$

We consider $a/1$ to be the same as a and hence we identify an integer as a rational number, so that $I \subseteq Q$. More precisely, since $I \neq Q$ (for instance, $\frac{1}{2} \in Q$, but $\frac{1}{2} \notin I$), I is a *proper subset* of Q. That is, all integers are rational numbers, but not all rational numbers are integers.

By using division, every rational number can be expressed as a decimal in which a block of one or more digits in the decimal repeats itself during the division process. For example, $\frac{2}{5} = 0.4000\ldots = 0.4\overline{0}$, $\frac{1}{3} = 0.333\ldots = 0.\overline{3}$, and $\frac{2}{11} = 0.18181818\ldots = 0.\overline{18}$. (In each example, the bar is used to indicate the block of digits that repeats itself.) Hence, it follows that every rational number can be represented by an eventually repeating decimal. The converse of this statement also holds; that is, *every eventually repeating decimal represents a rational number.* This latter statement means that any decimal number that eventually has a repeating block of digits in its decimal part can be represented by a ratio of two integers.

EXAMPLES 1 Express each rational number in decimal notation.

(a) $\frac{7}{8}$

(b) $\frac{1310}{99}$

SOLUTION By using division, we get

(a) $\frac{7}{8} = 0.875$

(b) $\frac{1310}{99} = 13.\overline{23}$

(Rational numbers, such as 2 or $\frac{7}{8}$, in which the repeating block is the digit 0, are sometimes called *terminating decimals.*)

2 Express each rational number as the ratio of two integers.

(a) 2.03 (b) $0.\overline{17}$

SOLUTION

(a) $2.03 = 2 + \frac{3}{100} = \frac{200}{100} + \frac{3}{100} = \frac{203}{100}$

(b) $0.17\overline{17}$ is another way of writing $0.\overline{17}$. Let $x = 0.17\overline{17}$. Multiplying both sides of this equation by 100 moves one of the repeating blocks of digits to the left of the decimal point, so that $100x = 17.\overline{17}$. Setting $x = 0.\overline{17}$ and subtracting the corresponding sides of these latter two equations results in $99x = 17$ or $x = \frac{17}{99}$, so that $0.\overline{17} = \frac{17}{99}$.

In summary, a rational number is a number that can be considered from two viewpoints: as a ratio of two integers or as a repeating decimal. (A more formal treatment of repeating decimals is given in Chapter 9, Section 6, where geometric series are considered.) If a rational number is represented in either of the two forms, it can be converted to the other form.

However, there are decimal numbers that do *not* repeat—for example, the decimal number 1.01001000100001. . . , where there is one more "0" after each "1" than there is before the "1." Another example of a nonrepeating decimal is $\pi = 3.14159265358. . . .$ (The equal sign is used here to mean "approximately equal to.") Also, $\sqrt{2}$ is not a rational number, since it can be shown that $\sqrt{2}$ has a nonrepeating decimal representation, $\sqrt{2} = 1.414213562. . . .$ The numbers $1 + \sqrt{2}$, $3 - \sqrt{2}$, $5\sqrt{2}$, $\sqrt{3}$, $\sqrt[3]{2}$, and $\sqrt{5}$ also have nonrepeating decimal forms. Such numbers are called *irrational numbers.*

The set of *real numbers* can be thought of as the set of all numbers that can be written as decimal numbers. Consequently, the set of real numbers can be written as the union of two disjoint sets of numbers: the rational numbers or repeating decimals, and the irrational numbers or nonrepeating decimals. If we use R to denote the set of real numbers and L to denote the set of irrational numbers, then we can express R as $R = Q \cup L$ (note that $Q \cap L = \varnothing$).

1.3 Real Number Properties

For convenience, we list the basic properties of real numbers. We assume that the operations of addition and multiplication on the set of real numbers are familiar to the reader.

PROPERTIES OF REAL NUMBERS

Let a, b, and c be real numbers.

1 *The Closure Properties*

(i) $a + b$ is a real number.
(ii) $a \cdot b$ is a real number.

2 *The Commutative Properties*

(i) $a + b = b + a$.
(ii) $a \cdot b = b \cdot a$.

3 *The Associative Properties*

(i) $(a + b) + c = a + (b + c)$.
(ii) $(a \cdot b) \cdot c = a \cdot (b \cdot c)$.

4 *The Distributive Properties*

(i) $a(b + c) = ab + ac$.
(ii) $(b + c)a = ba + ca$.

5 *The Identity Properties*

(i) $a + 0 = 0 + a = a$.
(ii) $a \cdot 1 = 1 \cdot a = a$.

6 *The Inverse Properties*

(i) For each real number a, there is a real number called the *additive inverse,* denoted by $-a$, such that

$$a + (-a) = (-a) + a = 0$$

(ii) For each real number $a \neq 0$, there is a real number called the *multiplicative inverse* or *reciprocal,* denoted by $1/a$ or a^{-1}, such that

$$a \cdot \frac{1}{a} = \frac{1}{a} \cdot a = 1$$

A set that satisfies the above properties is referred to as a *field.* For this reason, Properties 1–6 are sometimes called the *field properties* of real numbers.

Many other properties of the real number system can be derived from the above properties.

Let a, b, and c be real numbers. It is then possible to prove the following statements.

1 *The Cancellation Properties*

(i) If $a + c = b + c$, then $a = b$.
(ii) If $ac = bc$ and $c \neq 0$, then $a = b$.

2 *The Zero Property*

$a \cdot 0 = 0 \cdot a = 0$.

3 *Other Real Number Properties*

(i) $-(-a) = a$.
(ii) $(-a)(b) = a(-b) = -(ab)$.
(iii) $(-a)(-b) = ab$.
(iv) If $ab = 0$, then either $a = 0$ or $b = 0$.

EXAMPLE State the above property that justifies each of the following equalities.

(a) $\frac{5}{3} \cdot (-\frac{7}{2}) = (-\frac{7}{2}) \cdot \frac{5}{3}$
(b) $(2 \cdot 5) \cdot a = 2 \cdot (5a)$
(c) $14 \cdot (3 + y) = 14 \cdot 3 + 14y$
(d) $11 \cdot \frac{1}{11} = 1$
(e) $7 + (-7) = 0$
(f) $9 \cdot 0 = 0$
(g) $-(-4) = 4$
(h) $(-3)(-b) = 3b$
(i) If $3 + y = 3 + z$, then $y = z$.

SOLUTION

(a) $\frac{5}{3} \cdot (-\frac{7}{2}) = (-\frac{7}{2}) \cdot \frac{5}{3}$ (Commutative Property 2ii)
(b) $(2 \cdot 5) \cdot a = 2 \cdot (5a)$ (Associative Property 3ii)
(c) $14 \cdot (3 + y) = 14 \cdot 3 + 14y$ (Distributive Property 4i)
(d) $11 \cdot \frac{1}{11} = 1$ (Multiplicative Inverse Property 6ii)
(e) $7 + (-7) = 0$ (Additive Inverse Property 6i)
(f) $9 \cdot 0 = 0$ (Zero Property)
(g) $-(-4) = 4$ (Other Real Number Property 3i)
(h) $(-3)(-b) = 3b$ (Other Real Number Property 3iii)
(i) If $3 + y = 3 + z$, then $y = z$. (Cancellation Property 1i)

The operations of subtraction and division may be defined in terms of addition and multiplication, respectively. If a and b are real numbers, the *difference* between a and b, denoted by $a - b$, is defined by $a - b = a + (-b)$, and the *quotient* of a and b, denoted by $a \div b$ or a/b, is defined by the equation $a \div b = a \cdot (1/b)$, where $b \neq 0$. For example, $4 - (-5) = 4 + 5 = 9$, and $4 \div 5 = 4 \cdot \frac{1}{5} = \frac{4}{5}$. Division by zero is *not* defined in the real number system.

PROBLEM SET 1

In Problems 1–8, indicate which statements are true and which are false.

1 $2 \in \{2\}$ **2** $\{2, 3\} \in \{2, 3, 7, 8\}$ **3** $3 \in \{3, 4, 5\}$
4 $\{3\} \in \{3, 4, 5\}$ **5** $\{3\} \subseteq \{3, 4\}$ **6** $\varnothing \subseteq \{y, x\}$
7 $\{3, 4, 2\} = \{4, 3, 2\}$ **8** $\{\{3\}\} = \{3, \{3\}\}$

In Problems 9–11, use set builder notation, $\{x \mid x \text{ has property } P\}$, to describe each set. Also, describe the set by enumeration, if possible. Indicate which of the sets are finite and which are infinite.

9 A is the set of all even positive integers.
10 A is the set of all positive integers greater than 7 and less than 29.
11 A is the set of all positive integers divisible by 3, that is, all counting numbers that have a zero remainder when divided by 3.

12 Enumerate the elements of the following sets.
(a) $\{x \mid x \text{ is an even positive integer less than } 11\}$
(b) The set M of letters in the word "Massachusetts"

In Problems 13–17, list all the subsets of each set. Indicate which subsets are proper subsets. How many subsets does each set have?

13 $\{3\}$ **14** $\{4, 5\}$ **15** $\{6, 8, 10\}$ **16** $\{a, b, c\}$ **17** $\{5, 6, 7, 9\}$

18 Is it true that $\varnothing \subseteq \varnothing$? Why? How many elements are in the set $\varnothing$?

In Problems 19–28, use $A = \{1, 4, 7\}$, $B = \{1, 2, 5, 7\}$, $C = \{5, 6, 7, 8\}$, and $D = \{4, 5, 7\}$ to determine each set.

19 $A \cup C$ **20** $B \cap C$ **21** $A \cap B$
22 $B \cup D$ **23** $A \cup D$ **24** $B \cap \varnothing$
25 $A \cap (B \cup D)$ **26** $(A \cup B) \cap C$
27 $(A \cap D) \cup C$ **28** $(A \cap C) \cap D$

In Problems 29–30, find each set.

29 $\{0, 1, 2, 3, \ldots\} \cup \{1, 2, 3, 4, \ldots\}$

30 $\{\ldots, -3, -2, -1, 0\} \cap \{x \mid x \text{ is a nonnegative integer}\}$

In Problems 31–36, express each rational number in decimal notation.

31 $\frac{3}{5}$ **32** $-\frac{7}{4}$ **33** $-\frac{5}{6}$ **34** $\frac{5}{9}$

35 $-\frac{7}{3}$ **36** $-\frac{6}{7}$

In Problems 37–44, express each rational number as the ratio of two integers.

37 0.27 **38** 1.71 **39** -0.125 **40** -0.008

41 $0.\overline{5}$ **42** $0.\overline{46}$ **43** $-3.\overline{651}$ **44** $0.\overline{9}$

In Problems 45–60, state which property in Section 1.3 justifies each statement.

45 $5 + \sqrt{7}$ is a real number.

46 $5\sqrt{7}$ is a real number.

47 $\frac{7}{8} \cdot (-\frac{2}{3}) = (-\frac{2}{3}) \cdot \frac{7}{8}$

48 $\frac{9}{16} + (-\frac{8}{11}) = (-\frac{8}{11}) + \frac{9}{16}$

49 $a + (7 + 16) = (a + 7) + 16$

50 $(\frac{2}{7} \cdot c) \cdot d = \frac{2}{7} \cdot (c \cdot d)$

51 $8 \cdot (y + b) = 8 \cdot y + 8 \cdot b$

52 $1 \cdot c = c$

53 $(5 + 3) \cdot a = 5 \cdot a + 3 \cdot a$

54 $\sqrt{3} + (-\sqrt{3}) = 0$

55 $(-2) \cdot (-3) = 6$

56 If $5a = 0$, then $a = 0$.

57 If $3x = 3y$, then $x = y$.

58 If $3x + y = 3x + 5z$, then $y = 5z$.

59 $5 \cdot 0 = 0$

60 $-(-3) = 3$

2 Polynomials—Operations and Factoring

In this section, we present a brief review of the addition, subtraction, multiplication, and factorization of polynomials. In addition, we solve linear equations.

2.1 Positive Integer Exponents

Positive integer *exponents* provide a shorthand notation for representing repeated multiplication. The *exponential notation* a^n denotes a product in which a is used as a factor n times, that is,

$$a^n = \overbrace{a \cdot a \cdot a \cdots a}^{n \text{ factors}}$$

a^n is called the *nth power* of a; a is called the *base* and n is called the *exponent* of the expression a^n.

For example, $5^3 = 5 \cdot 5 \cdot 5$; 5 is the base, and 3 is the exponent. Similarly, $(x + y)^2 = (x + y)(x + y)$; $x + y$ is the base, and 2 is the exponent. Clearly,

$$x^5 \cdot x^3 = \overbrace{(x \cdot x \cdot x \cdot x \cdot x)}^{5 \text{ factors}} \cdot \overbrace{(x \cdot x \cdot x)}^{3 \text{ factors}} = x^8$$

In general, $x^m \cdot x^n = x^{m+n}$, where m and n represent positive integers.

2.2 Polynomial Terminology

Expressions such as $3x$, $5x + 3$, $y^2 + 7y + 2$, $3/x^2$, $(7x + 1)/(x + 2)$, and $z^5 - 3z^4 + 5z^2 + z + 3$ are called *algebraic expressions,* and the letters x, y, and z that represent real numbers are called *variables.* An algebraic expression that results from applying the operations of addition, subtraction, and multiplication *only* on a set of real numbers and variables is called a *polynomial.* Real numbers are also considered to be polynomials. Each of the expressions $5x + 2$, $3x^2 - 2x + 7$, and $4xy^4 - 3xy + 5x + 6y^2$ is an example of a polynomial, whereas $(7x + 1)/(x - 3)$ is *not* a polynomial because of the division by $x - 3$. The polynomial $5x + 2$ in one variable is the sum of the two *terms* $5x$ and 2. The number 5 in the term $5x$ is called its *coefficient* and the term 2, which contains no variable, is called the *constant* term. The polynomial $3x^2 - 2x + 7$ in one variable is the sum of three terms $3x^2$, $-2x$, and 7 whose coefficients are 3, −2, 7, respectively. The polynomial $4xy^4 - 3xy + 5x + 6y^2$ in two variables is the sum of four terms $4xy^4$, $-3xy$, $5x$, and $6y^2$ whose *numerical* coefficients are 4, −3, 5, and 6, respectively.

A polynomial containing one term is called a *monomial;* a polynomial of two terms is called a *binomial;* a polynomial of three terms is called a *trinomial.* For example, $3x^2 + x - 3$ is a trinomial, $\frac{1}{2}x^5 + 1$ is a binomial, and $x^3y^4 - 3xy + 5x$ is a trinomial in two variables. The *degree of a polynomial* in one variable is the highest exponent of that variable that appears. The polynomial $3x^2 + x - 3$ is of degree 2 in x, the polynomial $\frac{1}{2}x^5 + 1$ is of degree 5 in x, whereas the polynomial $x^3y^4 - 3xy + 5x$ has degree 4 in y and degree 3 in x. The polynomial $z^5 - 3z^4 + 5z^2 + z + 3$ has degree 5, and the coefficients are 1, −3, 5, 1, and 3, respectively.

2.3 Addition, Subtraction, and Multiplication of Polynomials

The addition, subtraction, and multiplication of polynomials are performed by using various properties of real numbers and simplifying the results. For example, the sum of $3x^3 + 2x^2 + 4$ and $7x^3 - 5x^2 + 8$ is obtained by adding the coefficients as follows:

$$\begin{aligned}(3x^3 + 2x^2 + 4) + (7x^3 - 5x^2 + 8) &= (3x^3 + 7x^3) + (2x^2 - 5x^2) + (4 + 8)\\ &= (3 + 7)x^3 + (2 - 5)x^2 + (4 + 8)\\ &= 10x^3 - 3x^2 + 12\end{aligned}$$

It is often more convenient to arrange the above work in the following "vertical scheme":

$$\begin{array}{r} 3x^3 + 2x^2 + 4 \\ (+)7x^3 - 5x^2 + 8 \\ \hline 10x^3 - 3x^2 + 12 \end{array}$$

EXAMPLE Perform the following operations.

(a) $(4x^3 + 7x - 13) + (-2x^3 + 5x + 17)$

(b) $(2x^3 + 3x^2 - 5x + 11) - (4x^3 - 5x^2 + 9)$

SOLUTION After arranging the polynomials in a vertical scheme, we have

(a)

$$\begin{array}{r} 4x^3 + 7x - 13 \\ (+)-2x^3 + 5x + 17 \\ \hline 2x^3 + 12x + 4 \end{array}$$

(b)

$$\begin{array}{r} 2x^3 + 3x^2 - 5x + 11 \\ (-)4x^3 - 5x^2 \qquad + 9 \\ \hline -2x^3 + 8x^2 - 5x + 2 \end{array}$$

To perform the multiplication of polynomials, we again use the various properties of the real number system. For example, to find the product $(2x)(5x^2)$, we apply the associative and commutative properties for multiplication to get $(2x)(5x^2) = 2 \cdot x \cdot 5 \cdot x^2 = 2 \cdot 5 \cdot x \cdot x^2 = 10x^3$. The multiplication of polynomials with more than one term requires the use of the distributive property. For example, $5(3x - 2) = 5(3x) + 5(-2) = 15x - 10$ and

$$\begin{aligned} (x + 4)(2x - 3) &= x(2x - 3) + 4(2x - 3) \\ &= 2x^2 - 3x + 8x - 12 \\ &= 2x^2 + 5x - 12 \end{aligned}$$

When more than two terms are involved, it is often easier to perform the multiplication of polynomials by using the vertical scheme as illustrated in the next example.

EXAMPLE Perform the multiplication $(x^2 - 2x + 1)(x^2 + x + 2)$.

SOLUTION

$$\begin{array}{lllll} x^2 & -\ 2x & +\ 1 & & \\ x^2 & +\ x & +\ 2 & & \\ \hline x^4 & -\ 2x^3 & +\ x^2 & & \\ & \quad x^3 & -\ 2x^2 & +\ x & \\ & & \quad 2x^2 & -\ 4x & +\ 2 \\ \hline x^4 & -\ x^3 & +\ x^2 & -\ 3x & +\ 2 \end{array}$$

Certain types of products of polynomials occur often enough in algebra to be worthy of special consideration. We list these types of products below.

SPECIAL PRODUCTS

Assume that a and b represent real numbers.

1 $(a + b)^2 = a^2 + 2ab + b^2$

2 $(a - b)^2 = a^2 - 2ab + b^2$

3 $(a - b)(a + b) = a^2 - b^2$

4 $(a + b)(a^2 - ab + b^2) = a^3 + b^3$

5 $(a - b)(a^2 + ab + b^2) = a^3 - b^3$

EXAMPLE Find the following products by using the special products above.

(a) $(x + 2y)^2$
(b) $(2x - y)^2$
(c) $(2x + 3y)(2x - 3y)$

SOLUTION

(a) Using Special Product 1, we have

$$(x + 2y)^2 = x^2 + 2(x)(2y) + (2y)^2 = x^2 + 4xy + 4y^2$$

(b) Using Special Product 2, we have

$$(2x - y)^2 = 4x^2 - 4xy + y^2$$

(c) Using Special Product 3, we have

$$(2x + 3y)(2x - 3y) = (2x)^2 - (3y)^2 = 4x^2 - 9y^2$$

2.4 Factoring Polynomials

If a polynomial is expressed as the product of two or more polynomials, then each polynomial in the product is called a *factor* of the original polynomial. This process is called *factoring* a polynomial. If a polynomial has no factors other than itself and 1, or its negative and -1, the polynomial is called *prime.* To factor a polynomial, we express it as the product of prime polynomials, or as the product of a monomial and prime polynomials. Once such a factorization is accomplished, we refer to the final result as a *complete factorization.* Unless otherwise specified, we accept as factors of any polynomial with integral coefficients only those prime factors that also contain integral coefficients.

The most familiar kind of factoring is removing a *common factor.* This is accomplished by using the distributive property as illustrated below.

EXAMPLE Factor each expression.

(a) $2x^2 + 5x$ (b) $3x^2y - 5xy^3$

SOLUTION

(a) $2x^2 + 5x = x(2x + 5)$
(b) $3x^2y - 5xy^3 = xy(3x) + xy(-5y^2) = xy(3x - 5y^2)$

Some factoring depends on recognizing polynomials that fit the forms of special products. The following example illustrates the use of these formulas.

EXAMPLE Factor each of the following.

(a) $16x^2 - 49y^2$ (b) $8x^3 + y^3$ (c) $x^3 - 27y^3$

SOLUTION

(a) Apply Special Product 3 to get

$$16x^2 - 49y^2 = (4x)^2 - (7y)^2 = (4x - 7y)(4x + 7y)$$

(b) By Special Product 4, we have

$$8x^3 + y^3 = (2x)^3 + y^3 = (2x + y)(4x^2 - 2xy + y^2)$$

(c) By Special Product 5, we have

$$x^3 - 27y^3 = x^3 - (3y)^3 = (x - 3y)(x^2 + 3xy + 9y^2)$$

Consider a second-degree polynomial $ax^2 + bx + c$, where a, b, and c are integers. If the binomials $(sx + p)$ and $(rx + q)$ are factors of the polynomial, then we can write

$$\begin{aligned} ax^2 + bx + c &= (sx + p)(rx + q) \\ &= srx^2 + sxq + prx + pq \\ &= rsx^2 + (sq + pr)x + pq \end{aligned}$$

for integers r, s, p, and q, such that

$$ax^2 + bx + c = rsx^2 + (sq + pr)x + pq$$

Therefore, to factor a polynomial $ax^2 + bx + c$, there are only limited choices for r, s, p, and q that satisfy the conditions $a = rs$, $b = sq + pr$, and $c = pq$, as illustrated in the following example.

EXAMPLE Factor each of the following trinomials.

(a) $x^2 + 7x + 10$ (b) $3x^2 + x - 2$

SOLUTION

(a) We attempt to find integers r, s, p, and q such that

$$x^2 + 7x + 10 = (sx + p)(rx + q)$$

That is, $sr = 1$, $pq = 10$, and $sq + pr = 7$. After trying various possibilities, we obtain $s = 1, r = 1, p = 2$, and $q = 5$. Thus

$$x^2 + 7x + 10 = (x + 2)(x + 5)$$

(b) We write $3x^2 + x - 2 = (sx + p)(rx + q)$, so that $sr = 3$, $pq = -2$, and $sq + pr = 1$. By trial and error, we obtain

$$3x^2 + x - 2 = (3x - 2)(x + 1)$$

2.5 Linear Equations

The most elementary type of equation *in one variable* is a *first-degree equation* or *linear equation.* Examples of such equations include $3x - 2 = 5$, $4x - 2 = 7x$, and $5t - 2 = 7t + 16$. A *solution* or a *root* of an equation is a real number that when substituted for the variable gives a true statement. All of the *solutions* of an equation form a set, and this set is called the *solution set* of the equation. For example, $3x - 4 = 8$, $3x = 12$, and $x = 4$ all have the same solution set, namely, $\{4\}$.

To solve a linear equation, we replace it by a chain of *equivalent* equations where each one is simpler than the preceding one. In the final equation the solution set will be obvious. This process of converting to equivalent equations is accomplished by using the following properties.

Assume that P, Q, and R represent polynomials. Then the following properties hold.

1 *Addition Property:* If $P = Q$, then $P + R = Q + R$.

2 *Multiplication Property:* If $R \neq 0$ and $P = Q$, then $PR = QR$.

Thus, if the same expression is added to (or subtracted from) both sides of an equation, Property 1 indicates that the resulting equation is equivalent to the original equation. Property 2 states that if both sides of an equation are multiplied (or divided) by a nonzero expression, the resulting equation is equivalent to the original equation.

The application of these properties to solving linear equations is illustrated in the following examples.

EXAMPLES **1** Find the solution set of $3x + 5 = 14$.

SOLUTION

$$\begin{aligned} 3x + 5 &= 14 \\ (3x + 5) - 5 &= 14 - 5 \\ 3x &= 9 \\ \tfrac{1}{3}(3x) &= \tfrac{1}{3}(9) \\ x &= 3 \end{aligned}$$

CHECK $3(3) + 5 = 14$. The solution set is $\{3\}$.

2 A taxpayer has a \$1200 exemption from income tax but pays 20 percent tax on the remainder of her income. If the total tax paid after the exemption is \$5440, find her total income.

SOLUTION Let x dollars be the total income. She pays taxes on $(x - 1200)$ dollars. Thus 20 percent of $(x - 1200)$ equals 5440, or

$$\begin{aligned} \tfrac{20}{100}(x - 1200) &= 5440 \\ \tfrac{1}{5}(x - 1200) &= 5440 \\ x - 1200 &= 27200 \\ x &= 28400 \end{aligned}$$

Therefore, her total income is \$28,400.

PROBLEM SET 2

In Problems 1–4, write each expression in equivalent exponential form.

1 $6 \cdot 6 \cdot 6 \cdot 6 \cdot 6 \cdot 6$

2 $x \cdot x \cdot x \cdot x \cdot x$

3 $2x \cdot x \cdot x \cdot y \cdot y - 3x \cdot x \cdot x$

4 $5x \cdot x \cdot x \cdot y \cdot y + 7y \cdot z \cdot z$

In Problems 5–8, identify each polynomial as a monomial, binomial, or trinomial. Also, find the degree of the polynomial and list its numerical coefficients.

5 $4x^2$ **6** $3x - 2$ **7** $5x^3 + 4$ **8** $-x^4 + 3x^2 + 13$

In Problems 9–22, perform the indicated operations.

9 $5xy^2 + (-3xy^2) + 2xy^2$
10 $(2x^2 + 3x + 1) + (5x^2 + 2x + 4)$
11 $(7x^2 + 4x + 3) + (-3x^2 + 2x - 5)$
12 $(4x^3 - 7x - 8) - (-2x^3 + 4x - 2)$
13 $(3x^2 + 5x + 7) - (x^2 - 3x + 21)$
14 $(3x^4 - 4x^3 + 6x^2 + x - 1) - (4 - x + 2x^2 - 3x^3 - x^4)$
15 $(3x - 1)(2x + 3)$
16 $(6x - 5y)(4x + 3y)$
17 $(9x + 7y)(5x - 4y)$
18 $(10x - 7)(5x + 8)$
19 $(2x + 1)(x^2 + 3x + 7)$
20 $(x^2 - 5)(x^2 + 8x - 11)$
21 $(x^2 - 5x + 6)(x^2 + 4x + 9)$
22 $(x^2 + 2xy + y^2)(x^2 - xy + 4y^2)$

In Problems 23–32, use the special products in Section 2.3 to perform each operation.

23 $(2x + y)^2$
24 $(3x - 5)^2$
25 $(8y - 8z)^2$
26 $(7t + 3)^2$
27 $(x - 3y)(x + 3y)$
28 $(3x - 5y)(3x + 5y)$
29 $(5r - 7s)(5r + 7s)$
30 $(1 - 10x)(1 + 10x)$
31 $(3 + y)(9 - 3y + y^2)$
32 $(3x - 4y)(9x^2 + 12xy + 16y^2)$

In Problems 33–60, factor each expression completely.

33 $9x^2 + 3x$
34 $17x^3y^2 - 34x^2y$
35 $12x^3y - 48x^2y^2$
36 $4xy^2z + x^2y^2z^2 - x^3y^3$
37 $m(x + y) + (x + y)$
38 $x(y - z) - (z - y)$
39 $1 - 9y^2$
40 $25 - 4a^2$
41 $16x^2 - 25y^2$
42 $144 - x^2z^2$
43 $x^4 - 81y^4$
44 $625a^4 - 81b^4$
45 $(x + y)^2 - (a - b)^2$
46 $(3x + 2y)^2 - 25b^2$
47 $64 + y^3$
48 $27x^3 - y^3$
49 $(y + 1)^3 + 8$
50 $64 - (x - 1)^3$
51 $x^2 - 16x + 63$
52 $x^2 - 9xy - 10y^2$
53 $12 - 4x - x^2$
54 $16 - x^2 - 6x$
55 $3x^2 + 5x - 2$
56 $10x^2 - 19x + 6$
57 $12x^2 + 17x - 5$
58 $42x^2 + x - 30$
59 $12 - 5x - 2x^2$
60 $6xy + 5y^2 - 8x^2$

In Problems 61–70, find the solution set of each equation.

61 $5x + 2 = x + 10$
62 $8x = 10 + 2(x + 1)$
63 $3x - 2(x + 1) = 2(x - 1)$
64 $7(x - 3) = 4(x + 5) - 47$
65 $5 + 8(x + 2) = 23 - 2(2x - 5)$
66 $11 - 7(1 - 2x) = 9(x + 1)$
67 $8 = 3x - 8(3 - 2x) - 63$
68 $8(5x - 1) + 36 = -3(x + 5)$
69 $34 - 3x = 8(7 - x) + 23$
70 $6(x - 10) + 3(2x - 7) = -45$

71 A movie theater charges \$3.25 admission for adults and \$1.00 for children. The receipts for a certain performance were \$903.00, and there were three times as many adults as children attending. How many tickets of each kind were sold?

72 A pay phone slot accepts quarters, dimes, and nickels. When opened, the phone box contained \$6.50 in coins. If there were four more dimes than quarters and three times as many nickels as dimes, find the number of coins of each kind.

73 A grocer has 100 pounds of candy selling at \$1.80 per pound. How many pounds of a different candy worth \$3.00 per pound should he mix with the 100 pounds in order to have a mixture worth \$2.40 per pound?

74 A gas station sells regular and premium gasoline. The regular gasoline sells for 61.9 cents per gallon, whereas the premium sells for 63.9 cents per gallon. If the gas station sells 14,000 gallons of gasoline in one week and takes in a total of \$8826, how many gallons of regular gasoline are sold in a week?

75 A woman has \$3900 invested, part of it at 6 percent and part at 7 percent. If the annual return on both investments is the same, how much does she have invested at each rate?

76 A nurse has a medicine containing 25 percent alcohol. How much medicine containing no alcohol should she add to 120 cubic centimeters of the 25 percent mixture so that the final mixture will contain only 20 percent alcohol?

3 Fractions

In Section 1.2 we indicated that a rational number can be represented in the form a/b, where a and b are integers and $b \neq 0$. The form a/b is also called a fraction whose *numerator* is a and whose *denominator* is b. If the numerator and the denominator of a fraction are polynomials, the fraction is called a *rational expression*. Examples of rational expressions are

$$\frac{1}{x}, \quad \frac{x+1}{5x+3}, \quad \frac{7}{t^2+11}, \quad \text{and} \quad \frac{3x^2+1}{7x^3+13}$$

Since division by 0 is *not* defined, it is always understood that the denominator of a rational expression cannot represent 0. Thus,

$$\text{for } \frac{x+2}{x-3}, \quad x \text{ cannot equal } 3$$

$$\text{for } \frac{x}{x^2-4}, \quad x \text{ cannot equal } 2 \text{ or } -2$$

Rational expressions, like rational numbers, can be written in *lowest terms* in which the numerator and the denominator have no common factor other than 1 and -1. Such a procedure is called *simplifying* or *reducing* the rational expression. The simplification technique is based on the *fundamental principle of rational expressions,* which states:

If P/Q is a rational expression, and if $K \neq 0$ is another rational expression, then

$$\frac{PK}{QK} = \frac{P}{Q}$$

EXAMPLE Reduce each of the following rational expressions.

(a) $\dfrac{12x^2y}{9xy^2}$ (b) $\dfrac{4x^2 - 1}{2x^2 + x}$ (c) $\dfrac{2x^2 + 5x - 3}{10x^2 + 9x - 7}$

SOLUTION

(a) $\dfrac{12x^2y}{9xy^2} = \dfrac{4x(3xy)}{3y(3xy)} = \dfrac{4x}{3y}$

(b) $\dfrac{4x^2 - 1}{2x^2 + x} = \dfrac{(2x - 1)(2x + 1)}{x(2x + 1)} = \dfrac{2x - 1}{x}$

(c) $\dfrac{2x^2 + 5x - 3}{10x^2 + 9x - 7} = \dfrac{(2x - 1)(x + 3)}{(2x - 1)(5x + 7)} = \dfrac{x + 3}{5x + 7}$

3.1 Multiplication and Division of Fractions

The multiplication and division of fractions are performed by using the following rules.

Let P/Q and R/S be rational expressions.

1 *Multiplication of Fractions*

$$\frac{P}{Q} \cdot \frac{R}{S} = \frac{P \cdot R}{Q \cdot S} = \frac{PR}{QS}$$

2 *Division of Fractions*

$$\frac{P}{Q} \div \frac{R}{S} = \frac{P}{Q} \cdot \frac{S}{R} = \frac{PS}{QR}, \quad \text{provided that } \frac{R}{S} \neq 0$$

EXAMPLE Perform the indicated operation and simplify the result for each of the following.

(a) $\dfrac{x^2 + 4x + 4}{2x^2 + 2x - 4} \cdot \dfrac{2x - 2}{x^2 + 2x}$ (b) $\dfrac{x^2 - 10x + 25}{x^2 - 100} \div \dfrac{x^2 - 7x + 10}{x^2 + 12x + 20}$

SOLUTION

(a) $\dfrac{x^2 + 4x + 4}{2x^2 + 2x - 4} \cdot \dfrac{2x - 2}{x^2 + 2x} = \dfrac{(x^2 + 4x + 4)(2x - 2)}{(2x^2 + 2x - 4)(x^2 + 2x)}$

$$= \frac{(x + 2)^2(2)(x - 1)}{(2)(x + 2)(x - 1)(x)(x + 2)} = \frac{1}{x}$$

(b) $$\frac{x^2-10x+25}{x^2-100} \div \frac{x^2-7x+10}{x^2+12x+20} = \frac{x^2-10x+25}{x^2-100}\cdot\frac{x^2+12x+20}{x^2-7x+10}$$
$$= \frac{(x^2-10x+25)(x^2+12x+20)}{(x^2-100)(x^2-7x+10)}$$
$$= \frac{(x-5)^2(x+2)(x+10)}{(x-10)(x+10)(x-2)(x-5)}$$
$$= \frac{(x-5)(x+2)}{(x-10)(x-2)}$$

3.2 Addition and Subtraction of Fractions

The addition and subtraction of fractions are performed by using the following rules.

Let P/Q and R/Q be rational expressions.

1 *Addition of Fractions*

$$\frac{P}{Q}+\frac{R}{Q}=\frac{P+R}{Q}$$

2 *Subtraction of Fractions*

$$\frac{P}{Q}-\frac{R}{Q}=\frac{P-R}{Q}$$

Thus, in order to be able to add or subtract fractions, it is necessary that the fractions have the same denominator. Normally, it is desirable to find the *least common denominator* (L.C.D.) of fractions when making the denominators the same. The L.C.D. is found by performing the following steps:

1 Factor each denominator completely.
2 List each factor with the largest exponent it has in any factored denominator. The product of the factors so listed is the L.C.D.

The process is illustrated in the following example.

EXAMPLE Perform the operation and simplify each of the following.

(a) $\dfrac{6}{x^2-2x-8}+\dfrac{x}{x+2}$ (b) $\dfrac{x}{x^2+6x+5}-\dfrac{2}{x^2+4x-5}$

SOLUTION

(a) The factorizations of the two denominators are $(x-4)(x+2)$ and $x+2$, respectively, so the L.C.D. is $(x-4)(x+2)$. Thus,

$$\frac{6}{x^2-2x-8}+\frac{x}{x+2}=\frac{6}{(x-4)(x+2)}+\frac{x}{x+2}$$
$$=\frac{6}{(x-4)(x+2)}+\frac{x(x-4)}{(x-4)(x+2)}$$
$$=\frac{6+x(x-4)}{(x-4)(x+2)}$$
$$=\frac{x^2-4x+6}{(x-4)(x+2)}$$

(b)
$$\frac{x}{x^2+6x+5}-\frac{2}{x^2+4x-5}=\frac{x}{(x+1)(x+5)}-\frac{2}{(x+5)(x-1)}$$
$$=\frac{x(x-1)}{(x+1)(x+5)(x-1)}-\frac{2(x+1)}{(x+1)(x+5)(x-1)}$$
$$=\frac{x(x-1)-2(x+1)}{(x+1)(x+5)(x-1)}$$
$$=\frac{x^2-x-2x-2}{(x+1)(x+5)(x-1)}$$
$$=\frac{x^2-3x-2}{(x+1)(x+5)(x-1)}$$

Sometimes the numerator or denominator (or both) of a fraction may themselves involve one or more fractions. The original fraction is then called a *complex fraction.* Examples of complex fractions are

$$\frac{\frac{1}{2}}{3},\qquad \frac{\frac{1}{x}}{\frac{3}{x+1}+7},\qquad \text{and}\qquad \frac{x-\frac{1}{x}}{\frac{2}{1+x}-\frac{x}{1-x}}$$

The following example illustrates a procedure for simplifying complex fractions based on the fact that P/Q means $P \div Q$.

EXAMPLE Simplify

$$\frac{\frac{1}{x-y}}{\frac{1}{x+y}+\frac{1}{x-y}}$$

SOLUTION

$$\frac{\frac{1}{x-y}}{\frac{1}{x+y}+\frac{1}{x-y}}=\frac{\frac{1}{x-y}}{\frac{x-y+x+y}{x^2-y^2}}=\frac{\frac{1}{x-y}}{\frac{2x}{x^2-y^2}}=\frac{1}{x-y}\div\frac{2x}{x^2-y^2}$$
$$=\frac{1}{x-y}\cdot\frac{x^2-y^2}{2x}=\frac{1}{x-y}\cdot\frac{(x-y)(x+y)}{2x}=\frac{x+y}{2x}$$

3.3 Equations Involving Fractions

Quite frequently, many applications result in equations that involve fractions. A technique for solving such equations is illustrated in the following examples.

EXAMPLES 1 Find the solution set of each equation.

(a) $\frac{3}{x}-\frac{1}{2x}=\frac{2}{x-1}$ (b) $\frac{5}{x-5}+6=\frac{x}{x-5}$

SOLUTION

(a) The L.C.D. of the fractions on both sides of the equation is $2x(x-1)$.

After multiplying both sides of the equation by $2x(x-1)$, we have

$$\begin{aligned} 2x(x-1)\left(\frac{3}{x}-\frac{1}{2x}\right) &= 2x(x-1)\left(\frac{2}{x-1}\right) \\ 6(x-1)-(x-1) &= 4x \\ 6x-6-x+1 &= 4x \\ 5x-5 &= 4x \\ x &= 5 \end{aligned}$$

CHECK

$$\frac{3}{5}-\frac{1}{10}=\frac{2}{4}$$

$$\frac{6-1}{10}=\frac{2}{4} \quad \text{or} \quad \frac{5}{10}=\frac{2}{4}$$

Therefore, the solution set is $\{5\}$.

(b) Multiply both sides of the equation by $x-5$ to get

$$\begin{aligned} (x-5)\left(\frac{5}{x-5}+6\right) &= (x-5)\left(\frac{x}{x-5}\right) \\ 5+6(x-5) &= x \\ 5+6x-30 &= x \end{aligned}$$

Thus $6x-25=x$ or $5x=25$ so $x=5$.

In checking the *proposed* solution, we encounter a zero in a denominator of a fraction in the equation. Therefore, there is no solution and the solution set is $\varnothing$. In this case, 5 is called an *extraneous root.* In retrospect, we see that we unknowingly multiplied both sides of the original equation by 0 (see Property 2 on page 13).

2 A computer can be used to do a biweekly school payroll in 8 hours. If a second computer is added, the two machines working together can do the payroll in 3 hours. If used alone, how long would it take the second computer to do the payroll?

SOLUTION Assume that the second computer can do the payroll in t hours. For each hour, $1/8$ of the payroll is done by the first computer, $1/t$ of the payroll is done by the second computer, and $1/3$ of the payroll is done by both computers. Thus,

$$\begin{aligned} \frac{1}{8}+\frac{1}{t} &= \frac{1}{3} \\ 24t\left(\frac{1}{8}+\frac{1}{t}\right) &= 24t\left(\frac{1}{3}\right) \\ 3t+24 &= 8t \end{aligned}$$

so that $5t=24$ or $t=\frac{24}{5}=4\frac{4}{5}$. (The student should check the proposed solution.) Therefore, it would take the second computer 4 hours and 48 minutes to do the payroll by itself.

PROBLEM SET 3

In Problems 1–8, reduce each rational expression.

1 $\dfrac{3x^3y(a-b)}{15x^4y^2(a-b)}$

2 $\dfrac{15x^2y^5(c-d)^3}{45x^3y^2(c-d)^2}$

3 $\dfrac{4x^2-9}{6x^2-9x}$

4 $\dfrac{x^2+4x}{x^2+6x+8}$

5 $\dfrac{7x^2-5xy}{49x^3-25xy^2}$

6 $\dfrac{x^2-1}{x^3+1}$

7 $\dfrac{x^2-4x-32}{x^2-10x+16}$

8 $\dfrac{x^2-8x-9}{x^2-14x+45}$

In Problems 9–26, perform the indicated operation and simplify the result.

9 $\dfrac{3x+6}{5x+5}\cdot\dfrac{x+1}{x^2+5x+6}$

10 $\dfrac{x+2}{x^2+8x-9}\cdot\dfrac{2x+18}{x^2-4}$

11 $\dfrac{a^2-1}{a+1}\cdot\dfrac{7a^2-5a-2}{a^2-2a+1}$

12 $\dfrac{a^2-9b^2}{a^2-b^2}\cdot\dfrac{5a-5b}{a^2+6ab+9b^2}$

13 $\dfrac{x^2-11x+10}{9x^2-25}\div\dfrac{x^2-8x-20}{12x^2+20x}$

14 $\dfrac{a^2+8a+16}{a^2-8a+16}\div\dfrac{a^3+4a^2}{a^2-16}$

15 $\dfrac{x-1}{x^2-1}\cdot\dfrac{2x+2}{x^2-4}\div\dfrac{3x+3}{x^2+4x+4}$

16 $\dfrac{x-3}{x^2+2x-3}\cdot\dfrac{x^2-5x+6}{x^2-2x-3}\div\dfrac{x^2-9}{x^2-1}$

17 $\dfrac{x}{x^2-25}+\dfrac{1}{x+5}$

18 $\dfrac{x}{x-1}-\dfrac{1}{x^2-x}$

19 $\dfrac{x}{x^2-9}-\dfrac{x-1}{x^2-5x+6}$

20 $\dfrac{x}{x^2+5x-6}+\dfrac{3}{x+6}$

21 $\dfrac{x-5}{x^2-5x-6}+\dfrac{x+4}{x^2-6x}$

22 $\dfrac{a}{a^2-2ab+b^2}-\dfrac{b}{a^2-b^2}$

23 $\dfrac{x}{x+2}-\dfrac{x}{x-2}-\dfrac{x^2}{x^2-4}$

24 $\dfrac{x-3}{x+3}-\dfrac{x+3}{3-x}+\dfrac{x^2}{9-x^2}$

25 $\dfrac{x+\dfrac{3}{x}}{1+\dfrac{3}{x^2}}$

26 $\dfrac{\dfrac{x-1}{x+1}-\dfrac{x+1}{x-1}}{\dfrac{x-1}{x+1}+\dfrac{x+1}{x-1}}$

In Problems 27–36, find the solution set of each equation.

27 $\dfrac{3x-2}{3}+\dfrac{x-3}{2}=\dfrac{5}{6}$

28 $\dfrac{x-14}{5}+4=\dfrac{x+16}{10}$

29 $\dfrac{1}{x}+\dfrac{2}{x}=3-\dfrac{3}{x}$

30 $\dfrac{2}{3x}+\dfrac{1}{6x}=\dfrac{1}{4}$

31 $\dfrac{2x}{x-2}=\dfrac{4}{x-2}-1$

32 $\dfrac{9}{5x-3}=\dfrac{5}{3x+7}$

33 $\dfrac{1}{x}+\dfrac{2}{x+a}=\dfrac{3}{x-a}$, a is a constant.

34 $\dfrac{3}{a-x}+\dfrac{a}{a+x}=\dfrac{1}{a^2-x^2}$, a is a constant.

35 $\dfrac{5}{2x-3} - \dfrac{3}{2} = \dfrac{1}{4x-6}$

36 $\dfrac{2}{x+3} = \dfrac{x-4}{x^2-9} + \dfrac{5}{x-3}$

37 An electronic computer can be used to solve a mathematical problem in 6 minutes. With the help of a newer computer the problem can be solved in 2 minutes. How long would it take the new computer to solve the problem alone?

38 One pipe can fill a tank in 18 minutes and another pipe can fill it in 24 minutes. A drain pipe can empty the tank in 15 minutes. With all three pipes open, how long will it take to fill the tank?

4 Exponents and Radicals

In Section 2 we defined the exponential notation a^n, where n is a positive integer, as

$$a^n = \overbrace{a \cdot a \cdot a \cdots a}^{n \text{ factors of } a}$$

In this section, we briefly study integer and rational number exponents as well as radicals.

A summary of the basic properties of exponents is given in the following theorem. Although the theorem is initially restricted to positive integer exponents, the properties hold for *all* real number exponents.

THEOREM 1 EXPONENT PROPERTIES

Suppose that m and n are positive integers and assume that a and b are real numbers. Then

(i) $a^n \cdot a^m = a^{n+m}$

(ii) $(a^m)^n = a^{mn}$

(iii) $(ab)^n = a^n b^n$

(iv) $\left(\dfrac{a}{b}\right)^n = \dfrac{a^n}{b^n}$, provided that $b \neq 0$

(v) $\dfrac{a^m}{a^n} = \begin{cases} a^{m-n} & \text{if } m \text{ is greater than } n \\ \dfrac{1}{a^{n-m}} & \text{if } n \text{ is greater than } m \\ 1 & \text{if } n = m \end{cases}$ provided that $a \neq 0$

The meaning of exponents can be extended to include the negative integers, together with zero, so that the properties of Theorem 1 continue to hold.

DEFINITION 1 a^0 AND a^{-n}

If a is a real number different from 0, then a^0 and a^{-n} are defined as follows:

(i) $a^0 = 1$

(ii) $a^{-n} = \dfrac{1}{a^n}$, where n is a positive integer

Because of Definition 1, Property (v) above can be restated as

$$\frac{a^m}{a^n} = a^{m-n}, \quad \text{provided that } a \neq 0$$

EXAMPLES Write each of the following expressions in a "simplified" form that has only positive integer exponents.

1 $\dfrac{3dc^{-2}}{c^3d^{-4}}$

SOLUTION

$$\frac{3dc^{-2}}{c^3d^{-4}} = 3d^{1-(-4)}c^{-2-3} = 3d^5c^{-5} = \frac{3d^5}{c^5}$$

2 $\left[\dfrac{(x^{-2})^{-1}(y^{-2})^0}{(x^{-1})^2(y^{-1})^2}\right]^2$

SOLUTION

$$\left[\frac{(x^{-2})^{-1}(y^{-2})^0}{(x^{-1})^2(y^{-1})^2}\right]^2 = \left(\frac{x^2 \cdot 1}{x^{-2}y^{-2}}\right)^2 = (x^4y^2)^2 = (x^4)^2(y^2)^2 = x^8y^4$$

3 $\dfrac{x^{-2} + y^{-2}}{(xy)^{-2}}$

SOLUTION

$$\frac{x^{-2} + y^{-2}}{(xy)^{-2}} = \frac{\dfrac{1}{x^2} + \dfrac{1}{y^2}}{\dfrac{1}{(xy)^2}} = \frac{\dfrac{y^2 + x^2}{x^2y^2}}{\dfrac{1}{x^2y^2}} = y^2 + x^2$$

Exponents can also be extended to include rational numbers so that the properties of Theorem 1 continue to hold. We begin by defining rational exponents of the form $1/n$, where n is a positive integer.

DEFINITION 2 PRINCIPAL ROOT

If a is a real number and n is a positive integer, then $a^{1/n}$, called the nth *principal root* of a, is defined to be the number x that satisfies $x^n = a$. If a is positive, $a^{1/n}$ is positive; and if a is negative and n odd, $a^{1/n}$ is negative. If a is negative and n is even, $a^{1/n}$ is not defined. The nth principal root of a, $a^{1/n}$, can also be expressed in the form $\sqrt[n]{a}$; the latter form is called a *radical* with index n.

For example, $4^{1/2} = 2$, since $2^2 = 4$ (note that $4^{1/2} = \sqrt{4} = 2$, not -2, because of Definition 2); $(-8)^{1/3} = -2$, since $(-2)^3 = -8$; and $(-9)^{1/2}$ is not defined, since there is no real number x for which $x^2 = -9$.

If p/q is a positive rational number, then in order for Property (ii) of Theorem 1 to hold, we define $a^{p/q}$ as follows.

DEFINITION 3 RATIONAL NUMBER EXPONENT

If p/q is a positive rational number and if a is a real number, then

$$a^{p/q} = (a^p)^{1/q} = (a^{1/q})^p$$

If p/q is a *negative* rational number and $a \neq 0$, then $a^{p/q} = 1/a^{-p/q}$ and the first part of the definition is applicable to $a^{-p/q}$, since $-p/q$ is a positive rational number.

For example, since

$$32^{2/5} = (32^{1/5})^2 = (\sqrt[5]{32})^2 = 2^2 = 4$$

it follows that

$$32^{-2/5} = \frac{1}{32^{2/5}} = \frac{1}{4}$$

Realizing that the exponent properties hold for rational number exponents, we have the following properties of radicals.

THEOREM 2 RADICAL PROPERTIES

Let m and n be positive integers, and assume that a and b are real numbers so that each of the roots exists. Then

(i) $\sqrt[n]{ab} = \sqrt[n]{a} \cdot \sqrt[n]{b}$

(ii) $\sqrt[n]{\dfrac{a}{b}} = \dfrac{\sqrt[n]{a}}{\sqrt[n]{b}}$, for $b \neq 0$

(iii) $\sqrt[n]{a^m} = (\sqrt[n]{a})^m$

These properties can be verified using the equivalent exponential notation, assuming that Theorem 1 holds for rational exponents (see Problem 36).

EXAMPLES 1 Simplify the following expressions. Assume that all variables represent positive real numbers.

(a) $x^{-1/2} \cdot x^{3/2} \cdot x^{5/9}$ (b) $\left(\dfrac{8x^{-9}}{y^{-6}}\right)^{-2/3}$ (c) $\sqrt{125x^2}$

(d) $\sqrt{\left(\dfrac{4}{25}\right)^3}$ (e) $\sqrt[3]{\sqrt{9} + \sqrt{25}}$ (f) $\sqrt{21} \cdot \sqrt{3}$

(g) $\sqrt{50x^3y^4}$ (h) $\sqrt[3]{54x^3y^4z^8}$ (i) $\sqrt[n]{x^n}$

SOLUTION

(a) $x^{-1/2} \cdot x^{3/2} \cdot x^{5/9} = x^{-1/2+3/2+5/9} = x^{1+5/9} = x^{14/9}$

(b) $\left(\dfrac{8x^{-9}}{y^{-6}}\right)^{-2/3} = \dfrac{(8x^{-9})^{-2/3}}{(y^{-6})^{-2/3}} = \dfrac{8^{-2/3}(x^{-9})^{-2/3}}{y^4} = \dfrac{(2^3)^{-2/3}x^6}{y^4} = \dfrac{2^{-2}x^6}{y^4} = \dfrac{x^6}{4y^4}$

(c) $\sqrt{125x^2} = \sqrt{25 \cdot 5 \cdot x^2} = \sqrt{25}\sqrt{5}\sqrt{x^2} = 5x\sqrt{5}$

(d) $\sqrt{\left(\frac{4}{25}\right)^3} = \left(\sqrt{\frac{4}{25}}\right)^3 = \left(\frac{\sqrt{4}}{\sqrt{25}}\right)^3 = \left(\frac{2}{5}\right)^3 = \frac{8}{125}$

(e) $\sqrt[3]{\sqrt{9} + \sqrt{25}} = \sqrt[3]{3 + 5} = \sqrt[3]{8} = 2$

(f) $\sqrt{21} \cdot \sqrt{3} = \sqrt{63} = \sqrt{9} \cdot \sqrt{7} = 3\sqrt{7}$

(g) $\sqrt{50x^3y^4} = \sqrt{25 \cdot 2 \cdot x^2 \cdot x \cdot y^4} = 5xy^2\sqrt{2x}$

(h) $\sqrt[3]{54x^3y^4z^8} = \sqrt[3]{27 \cdot 2 \cdot x^3 \cdot y^3 \cdot y \cdot z^6 \cdot z^2} = 3xyz^2\sqrt[3]{2yz^2}$

(i) $\sqrt[n]{x^n} = (x^n)^{1/n} = x$

2 Express each of the following fractions without radicals in the denominator. (This process is often referred to as *rationalizing the denominator.*)

(a) $\frac{1}{\sqrt{3}}$ (b) $\frac{1}{2 + \sqrt{3}}$ (c) $\frac{\sqrt{2}}{4 - \sqrt{14}}$

SOLUTION

(a) $\frac{1}{\sqrt{3}} = \frac{1}{\sqrt{3}} \cdot \frac{\sqrt{3}}{\sqrt{3}} = \frac{\sqrt{3}}{3}$

(b) $\frac{1}{2 + \sqrt{3}} = \frac{1}{2 + \sqrt{3}} \cdot \frac{2 - \sqrt{3}}{2 - \sqrt{3}} = \frac{2 - \sqrt{3}}{4 - 3} = 2 - \sqrt{3}$

(c) $$\begin{aligned}\frac{\sqrt{2}}{4 - \sqrt{14}} &= \frac{\sqrt{2}}{4 - \sqrt{14}} \cdot \frac{4 + \sqrt{14}}{4 + \sqrt{14}} \\ &= \frac{4\sqrt{2} + \sqrt{28}}{16 - 14} = \frac{4\sqrt{2} + 2\sqrt{7}}{2} \\ &= \frac{2(2\sqrt{2} + \sqrt{7})}{2} = 2\sqrt{2} + \sqrt{7}\end{aligned}$$

4.1 Equations Involving Radicals

Equations such as $\sqrt{x} = 5$, $\sqrt{5x + 1} = 4$, and $\sqrt[3]{5x + 2} = 3$ are called *radical equations.* To find the solution set of a radical equation such as $\sqrt{x} = 5$, we square both sides of the equation, so that $(\sqrt{x})^2 = 5^2$ or $x = 25$. Therefore, the solution set of the equation is $\{25\}$, for it is true that $\sqrt{25} = 5$.

In general, to find the solution set of an equation that involves square roots, we square both sides of the equation to eliminate the radical. It is important to realize that the process of squaring both sides of an equation to remove the radicals can introduce an *extraneous* root, that is, a number that does not satisfy the *original* equation. Therefore, it is necessary to check *all* solutions that result from this process to determine which of the proposed solutions are *accepted* or *rejected.* For example, to solve the equation $\sqrt{x} = -3$, we square both sides of the equation to get $(\sqrt{x})^2 = (-3)^2$ or $x = 9$. If we substitute $x = 9$ in the original equation, we have $\sqrt{9} = -3$, which is false. Therefore 9 is *not* a solution to the original equation. The number 9 is an *extraneous solution* of the above equation. The solution set is $\varnothing$.

EXAMPLES Find the solution set of each of the following equations. Check for extraneous roots.

1 $\sqrt{2x+5} = 3$

SOLUTION We square both sides of the equation to eliminate the radical, so that $(\sqrt{2x+5})^2 = 3^2$ or $2x + 5 = 9$. Solving the latter equation, we obtain $2x = 4$ or $x = 2$.

CHECK Substituting 2 for x in the original equation, we have

$$\sqrt{2(2)+5} = \sqrt{4+5} = \sqrt{9} = 3$$

Hence, $\{2\}$ is the solution set.

2 $\sqrt{4x^2 - 3} = 2x + 1$

SOLUTION Squaring both sides of the equation, we have

$$(\sqrt{4x^2 - 3})^2 = (2x + 1)^2$$

so that

$$4x^2 - 3 = 4x^2 + 4x + 1 \quad \text{or} \quad -3 = 4x + 1$$

Therefore,

$$4x = -4 \quad \text{or} \quad x = -1$$

CHECK Substituting -1 for x in the original equation, we have

$$\sqrt{4(-1)^2 - 3} = 2(-1) + 1$$

or

$$1 = -1$$

which is false. Hence, the solution set is $\varnothing$.

3 $\sqrt{x} + 1 = \sqrt{x+3}$

SOLUTION Squaring both sides of the equation, we get

$$(\sqrt{x} + 1)^2 = (\sqrt{x+3})^2$$

so that

$$x + 2\sqrt{x} + 1 = x + 3 \quad \text{or} \quad 2\sqrt{x} = 2 \quad \text{or} \quad \sqrt{x} = 1$$

We square both sides of the latter equation to get $x = 1$.

CHECK Substituting 1 for x in the original equation yields

$$\sqrt{1} + 1 = \sqrt{1+3}$$

or

$$2 = 2$$

Hence, the solution set is $\{1\}$.

PROBLEM SET 4

In Problems 1–24, write each expression in a form with positive integer exponents and simplify.

1 $\dfrac{3 \cdot 2^{-1} \cdot 4^{-1}}{2^2}$

2 $\dfrac{2^3 \cdot 2^4 \cdot 6^{-2}}{6^2}$

3 $\dfrac{2^{-3} \cdot 5^{-2}}{10^{-1} \cdot 16}$

4 $\dfrac{3^{-5} \cdot 4^{-2} \cdot 2^{-1}}{2^3 \cdot 3^{-2}}$

5 $16^{-2}[(2^{-1})(2)(2^5)]^4$

6 $[(5^{-1})(5^2 \cdot 5^{-3})]^{-1}$

7 $\left(\dfrac{3^2 \cdot 3^4 \cdot 9^{-1}}{4^3 \cdot 3^{-2} \cdot 5^0}\right)^{-1}$

8 $\left(\dfrac{3^0 \cdot 2^{-6}}{2^{-2} \cdot 4 \cdot 7^0}\right)^{-2}$

9 $\dfrac{x^{-4}y^2z^{-4}}{(xy)^{-2}(yz)^{-4}}$

10 $\left[\dfrac{(ab)^{-2}(bc)^{-3}}{(ac)^3(cd)^{-2}}\right]^{-2}$

11 $\dfrac{x^2 \cdot x^3 \cdot x^{-1} \cdot (x^{-2})^3}{x^{-3}}$

12 $\left[\dfrac{x^{-4}y^2z^{-3}}{x^3(yz)^{-2}}\right]^{-4}$

13 $\dfrac{a^{-1} - b^{-1}}{(a + b)^{-1}}$

14 $a^{-1}b + ab^{-1}$

15 $\dfrac{a^{-1} + b^{-1}}{a^{-1} - b^{-1}}$

16 $\dfrac{x^{-2} - y^{-2}}{x^{-1} - y^{-1}}$

17 $x^{2/3} \cdot x^{-1/2} \cdot x^{1/6}$

18 $(a^{7/9} \cdot a^{-1/3})^{-18}$

19 $\dfrac{a^{1/3} \cdot a^{3/8}}{a^{-7/2}}$

20 $\left(\dfrac{x^{-1} \cdot y^{-2/3}}{z^{-2}}\right)^{-3}$

21 $(x^{-3/8})^{-8/3} \cdot (y^{-1/3})^{3/2}$

22 $(x^{5/7} \cdot y^{3/14})^{-14} \cdot (x^{2/3})^6$

23 $\left(\dfrac{81x^{-12}}{y^{16}}\right)^{-1/4}$

24 $\left(\dfrac{x^{-1/4}y^{-5/2}}{x^3y^{-3}}\right)^4$

In Problems 25–35, simplify each expression. Assume that all variables represent positive real numbers.

25 $\sqrt{\sqrt{9} + \sqrt{16} + 2}$

26 $\sqrt{3} \cdot \sqrt[3]{27} \cdot \sqrt[4]{81}$

27 $\sqrt{2} \cdot \sqrt[3]{8} \cdot \sqrt[4]{4}$

28 $\sqrt{\sqrt[5]{1024}}$

29 $\sqrt{32x^3}$

30 $\sqrt[4]{x^2} \cdot \sqrt{x} \cdot \sqrt[3]{x}$

31 $2xy^2\sqrt[4]{x^7y^5}$

32 $\sqrt[3]{250x^4y^7}$

33 $\dfrac{\sqrt{3}\,\sqrt{6x}}{\sqrt{6}\sqrt{3x^3}}$

34 $\sqrt[4]{\dfrac{x^4y^3}{x^3y}} \cdot \dfrac{\sqrt[4]{x^5y}}{\sqrt[4]{xy^{-1}}}$

35 $\sqrt[n]{\dfrac{4^n \cdot 6}{4^{2n+1} + 2^{4n+1}}}$

36 Prove Theorem 2 by using Theorem 1, Definition 2, and Definition 3.

In Problems 37–44, rationalize the denominator in each expression. Assume that all variables represent positive real numbers.

37 $\dfrac{2}{\sqrt{2}}$

38 $\dfrac{10x}{3\sqrt{5x}}$

39 $\dfrac{5}{\sqrt{3} - \sqrt{2}}$

40 $\dfrac{5}{\sqrt{x} + 1}$

41 $\dfrac{x}{\sqrt{x} - \sqrt{y}}$

42 $\dfrac{\sqrt{x+1} + \sqrt{x-1}}{\sqrt{x+1} - \sqrt{x-1}}$

43 $\dfrac{\sqrt{3}}{3 - \sqrt{3}}$

44 $\dfrac{\sqrt{3} + 1}{\sqrt{2} + \sqrt{3} + \sqrt{5}}$
(*Hint:* It will take two steps to rationalize the denominator.)

In Problems 45–54, find the solution set of each equation. Check for extraneous roots.

45 $\sqrt{6x - 3} = 5$

46 $\sqrt{3x - 1} = 5$

47 $5 + \sqrt{x - 5} = 4$

48 $8 + \sqrt{x - 1} = 6$

49 $\sqrt{x^2 + 3x} = x + 1$

50 $5 + \sqrt{x^2 + 7} = x$

51 $\sqrt{x^2 - 3x + 3} = x + 1$

52 $\sqrt{x^2 + 5x + 2} = x$

53 $\sqrt{x + 12} = 2 + \sqrt{x}$

54 $\sqrt{x} = \sqrt{x + 16} - 2$

5 Inequalities

The set of real numbers can be represented geometrically as the set of all points along a line called a *number line* or a *real line* as follows. Suppose that an arbitrary point on a line is selected to represent the number 0 and another arbitrary point to the right of 0 is selected to represent the number 1. The point that represents 0 is called the *origin,* and the *line segment* determined by the point representing 0 and the point representing 1 defines the *scale unit* (Figure 1). By repeating the scale unit, moving from left to right, starting at 0, we can associate the set of positive integers $I_p = \{1, 2, 3, 4, \ldots\}$ with equispaced points on the line. Moving from right to left, starting at 0, we can associate the set of negative integers $I_n = \{-1, -2, -3, -4, \ldots\}$ with equispaced points on the line (Figure 1). The remaining real numbers can be "located" or "plotted" on the real line by using decimal representations.

Figure 1

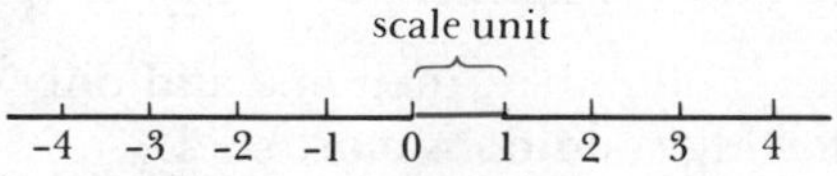

scale unit

-4 -3 -2 -1 0 1 2 3 4

EXAMPLE Locate the following numbers on the real line.

(a) 2.3 (b) $-\frac{17}{8}$ (c) $\sqrt{2}$

SOLUTION

(a) We can locate 2.3 by subdividing the portion of the number line between 2 and 3 into 10 equal parts; then, starting at 2, we move three parts to the right to 2.3 (Figure 2).

(b) $-\frac{17}{8} = -2.125$ (Figure 2).

(c) $\sqrt{2}$ is an irrational number, so $\sqrt{2}$ has a decimal representation that is a nonterminating, nonrepeating decimal ($\sqrt{2} = 1.41421\ldots$). If we were to attempt to locate $\sqrt{2}$ by using the decimal representation, we would become involved in an unending process in which we would "approach" but never actually locate the point. We can, however, locate $\sqrt{2}$ by using the geometry of the number line, together with the Pythagorean theorem (Figure 2).

Figure 2

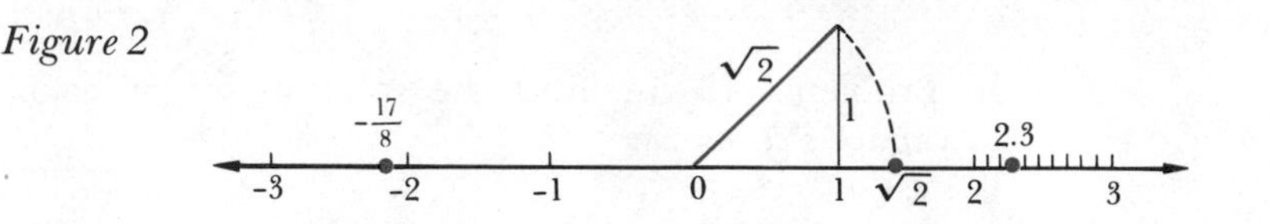

Notice that when a real number is "located" on the number line, the number is not the point, nor is the point the number. The point *represents* the number. It is customary, however, to use the words *real number* and *point* interchangeably. Thus we speak of the *point* $\frac{3}{4}$ rather than the *point corresponding to the real number* $\frac{3}{4}$.

5.1 Positive and Negative Numbers

All real numbers represented by points on the real line that lie to the "right" of the point 0 are *positive numbers,* whereas all real numbers "left" of the point 0 are *negative numbers.* Zero, then, is neither positive nor negative. Figure 3 illustrates the following principle:

Figure 3

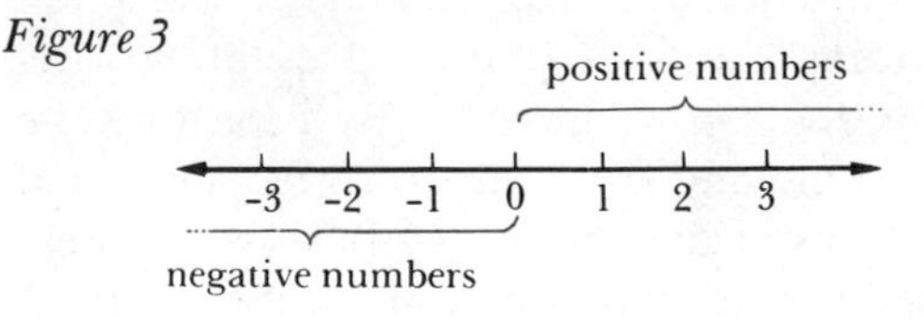

TRICHOTOMY PRINCIPLE

If a is a real number, then one and only one of the following conditions must hold:

(i) a is positive (a is to the right of 0)
(ii) a is negative (a is to the left of 0)
(iii) a is zero

The following axiom contains the basic assumptions that characterize the set of positive real numbers.

POSITIVE NUMBER AXIOM

The set of positive numbers is *closed under addition;* that is, the sum of any two positive numbers is always a positive number. The set of positive numbers is *closed under multiplication;* that is, the product of two positive numbers is always a positive number.

Using the notion of positive numbers, it is possible to give a precise "algebraic" characterization of the order of real numbers that is suggested by the "geometry" of the real line.

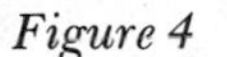

Figure 4

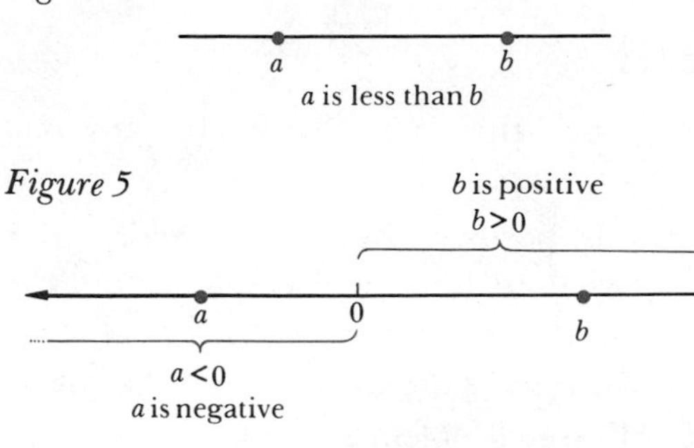

Figure 5

Geometrically, a real number a is less than a real number b if the point associated with a is to the "left" of the point associated with b on the real line (Figure 4).

DEFINITION 1 ORDER

Assume that a and b are real numbers. We say that *a is less than b*, written $a < b$, or, equivalently, *b is greater than a*, written $b > a$, if $b - a$ is a positive number.

Thus $2 < 3$ or $3 > 2$ because $3 - 2 = 1$ is a positive number; $-3 < -2$ because $-2 - (-3) = 1$; $-2 < 3$ since $3 - (-2) = 5$.

To say that $b > 0$ is to say that b is positive, and to say that $a < 0$ is to say that a is negative (Figure 5).

5.2 Properties of Inequalities

The positive number axiom and the definition of order can be used to prove some important properties of inequalities. If $a < b$ (a lies to the left of b) and if $b < c$ (b lies to the left of c), then a must lie to the left of c; that is, $a < c$ (Figure 6). This notion is generalized in the following theorem.

Figure 6

THEOREM 1 TRANSITIVE PROPERTY

If a, b, and c are real numbers such that $a < b$ and $b < c$, then $a < c$.

PROOF By Definition 1, $a < b$ means that $b - a = p$ is positive, and $b < c$ means that $c - b = q$ is positive, so that $(b - a) + (c - b) = p + q$, which is positive because the positive numbers are closed under addition. After simplifying, we get $c - a = p + q$, so that $c - a$ is positive, from which we can conclude, by Definition 1, that $a < c$.

For example, if $x < 3$ and $3 < y$, then $x < y$.

Let us consider the effect of adding any number to both sides of the inequality $a < b$. In this case, a is to the left of b on the number line, and if *any* number c is added to both a and b, then the result $a + c$ is to the left of $b + c$ on the number line. That is, $a + c < b + c$ (Figure 7). For example, if 3 is added to both sides of the inequality $x < 4$, we obtain $x + 3 < 4 + 3$ or $x + 3 < 7$.

Figure 7

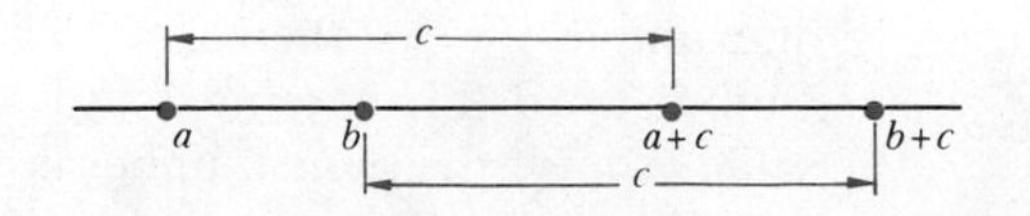

In general, we have the following theorem.

THEOREM 2 ADDITION PROPERTY

If a and b are real numbers with $a < b$, then $a + c < b + c$ for any real number c.

PROOF $a < b$ means that $b - a = p$ is positive. But $b - a = b - a + 0 = b - a + (c - c) = (b + c) - (a + c)$, so that $(b + c) - (a + c) = p$ is positive. Hence, by Definition 1, $a + c < b + c$.

For example, if $x < y$, then $x + 4 < y + 4$. Also $x - 5 < y - 5$, since $x - 5 = x + (-5)$ and $y - 5 = y + (-5)$ (see Problem 2a).

Figure 8 illustrates on the number line what happens to the inequality $2 < 3$ when each side is multiplied by a *positive* number, 2, and by a *negative* number, -2. If each side of $2 < 3$ is multiplied by $+2$, the resulting inequality, $4 < 6$, maintains the same order as the original inequality, whereas if each side of $2 < 3$ is multiplied by -2, the resulting inequality, $-4 > -6$, has its order *reversed* from that in the original inequality. In general, we have the following result.

Figure 8

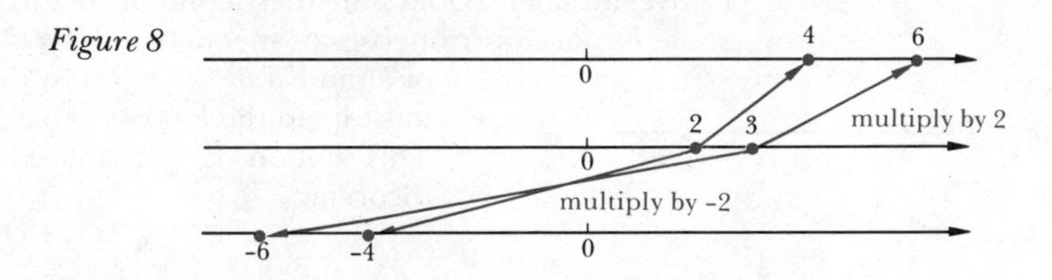

THEOREM 3 MULTIPLICATION PROPERTY

Let a, b, and c be real numbers.

(i) If $a < b$ and $c > 0$, then $ac < bc$.
(ii) If $a < b$ and $c < 0$, then $ac > bc$.

PROOF

(i) By Definition 1, $a < b$ means that $b - a = p$ is a positive number. But $c > 0$ and $p > 0$ implies that $pc > 0$, since the positive numbers are closed under multiplication. Hence $(b - a)c = pc$ is positive, so that $bc - ac$ is positive. Thus, by Definition 1, $ac < bc$.
(ii) Since $a < b$, $b - a = p$, which is a positive number. But c is negative, so that by the trichotomy principle, $-c$ is positive. Hence $p(-c)$ is a positive number because the positive numbers are closed under multiplication. Consequently, $p(-c) = (b - a)(-c) = ac - bc$ is positive and $bc < ac$.

For example, if $x < y$, then $5x < 5y$ and $-5x > -5y$.

In other words, Theorems 2 and 3 tell us that an inequality maintains the same order if the same number is added or subtracted (see Problem 2a) on both sides, or if the same *positive* number is multiplied or divided (see Problem 2b) on both sides; whereas if the same *negative* number is multiplied or divided (see Problem 2b) on both sides, the order of the inequality is *reversed.*

EXAMPLE Which of the above properties justifies the given statement?

(a) If $x < 2$, then $x + 3 < 2 + 3$.
(b) If $x < z$ and $z < 3$, then $x < 3$.
(c) If $x < -4$, then $-3x > 12$.
(d) If $y < -5$, then $7y < -35$.
(e) If $-5t < 35$, then $t > -7$.

SOLUTION

(a) Addition Property (Theorem 2)
(b) Transitive Property (Theorem 1)
(c) Multiplication Property (Theorem 3ii)
(d) Multiplication Property (Theorem 3i)
(e) Multiplication Property (Theorem 3ii)

By the trichotomy principle, if a is *not* less than b (written $a \nless b$), then either $a > b$ or else $a = b$. In this case, we say that a is *greater than or equal to* b or that b is *less than or equal to* a, and we write $a \geqslant b$ or $b \leqslant a$. Assertions of the form $a \leqslant b$ or $a \geqslant b$ are called *inequalities* in spite of the fact that they include the possibility of equality. The rules for manipulating such inequalities are the same as those given in the three theorems above. For example, if $-3x \leqslant 6$, then $x \geqslant -2$ because of Theorem 3ii.

5.3 Interval Notation

The notation $a < x < b$ means that $a < x$ and (simultaneously) that $x < b$ (Figure 9). Sets such as $\{x \mid a < x < b\}$ (Figure 9) are called *intervals.* The classification of intervals and the notation sometimes used to denote intervals are given below.

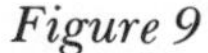
Figure 9

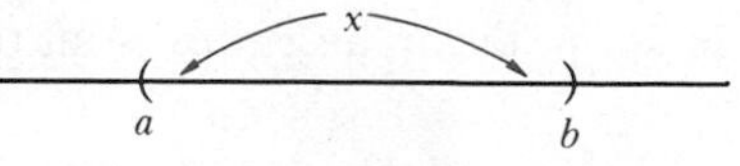

Figure 10

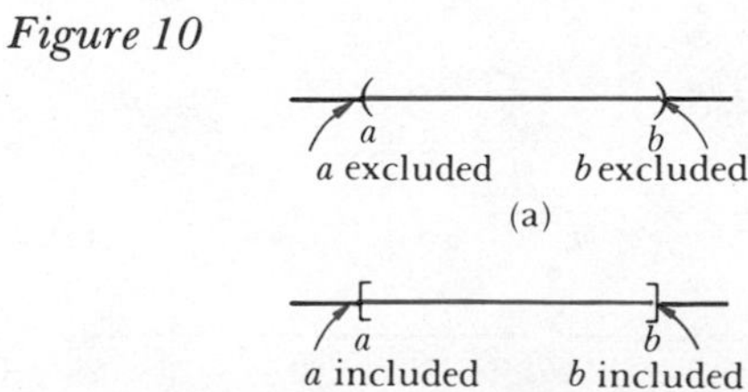

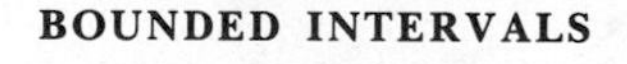
BOUNDED INTERVALS

Let a and b be real numbers with $a < b$.

1 The *open interval* from a to b, denoted by (a, b), is defined by $(a, b) = \{x \mid a < x < b\}$.

Notice that the end points a and b do *not* belong to the interval (Figure 10a).

2 The *closed interval* from a to b, denoted by $[a, b]$, is defined by $[a, b] = \{x \mid a \leqslant x \leqslant b\}$.

Notice that the closed interval $[a, b]$ contains *both* of its end points (Figure 10b).

Figure 11

Figure 12

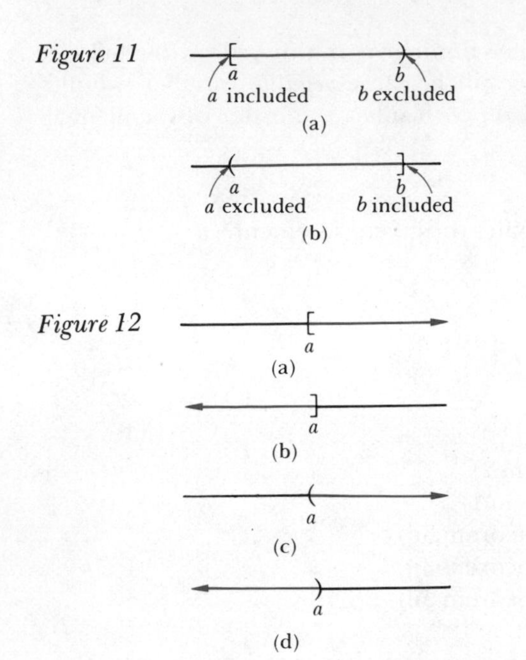

3 An interval from a to b including one end point but excluding the other end point is written as $[a, b) = \{x | a \leq x < b\}$ (Figure 11a) or $(a, b] = \{x | a < x \leq b\}$ (Figure 11b).

UNBOUNDED INTERVALS

We use the symbols ∞ ("infinity") and $-\infty$ (∞ and $-\infty$ are just convenient symbols and are *not* real numbers) to describe *unbounded* intervals as follows. If a is a real number, then

1 $[a, \infty) = \{x | x \geq a\}$ (Figure 12a)

2 $(-\infty, a] = \{x | x \leq a\}$ (Figure 12b)

3 $(a, \infty) = \{x | x > a\}$ (Figure 12c)

4 $(-\infty, a) = \{x | x < a\}$ (Figure 12d)

Finally, we use the symbol $(-\infty, \infty)$ to denote the "interval" consisting of the set of all real numbers R.

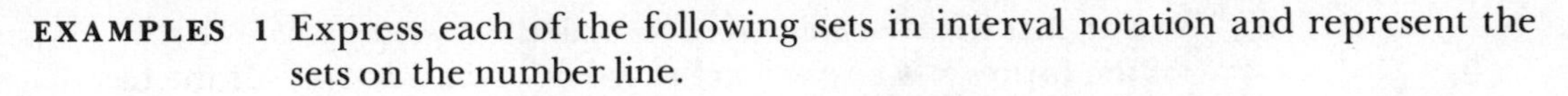

EXAMPLES 1 Express each of the following sets in interval notation and represent the sets on the number line.

Figure 13

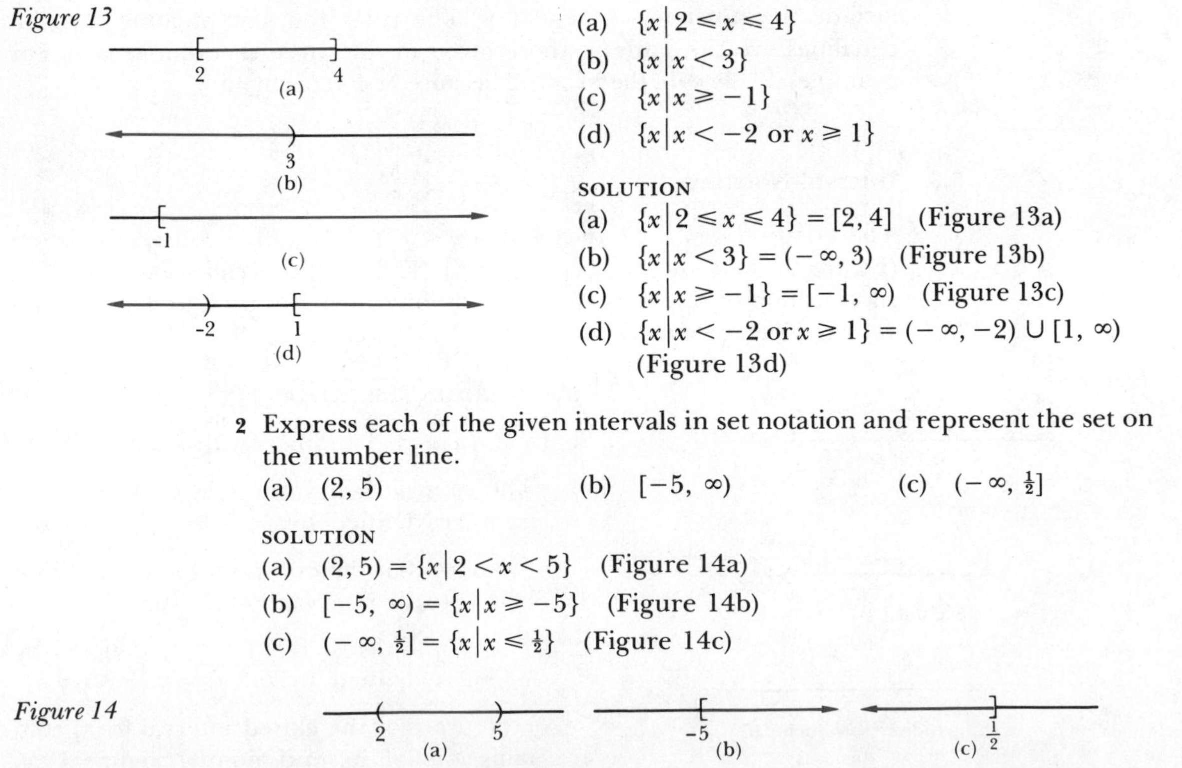

(a) $\{x | 2 \leq x \leq 4\}$

(b) $\{x | x < 3\}$

(c) $\{x | x \geq -1\}$

(d) $\{x | x < -2 \text{ or } x \geq 1\}$

SOLUTION

(a) $\{x | 2 \leq x \leq 4\} = [2, 4]$ (Figure 13a)

(b) $\{x | x < 3\} = (-\infty, 3)$ (Figure 13b)

(c) $\{x | x \geq -1\} = [-1, \infty)$ (Figure 13c)

(d) $\{x | x < -2 \text{ or } x \geq 1\} = (-\infty, -2) \cup [1, \infty)$ (Figure 13d)

2 Express each of the given intervals in set notation and represent the set on the number line.

(a) $(2, 5)$ (b) $[-5, \infty)$ (c) $(-\infty, \frac{1}{2}]$

SOLUTION

(a) $(2, 5) = \{x | 2 < x < 5\}$ (Figure 14a)

(b) $[-5, \infty) = \{x | x \geq -5\}$ (Figure 14b)

(c) $(-\infty, \frac{1}{2}] = \{x | x \leq \frac{1}{2}\}$ (Figure 14c)

Figure 14

3 The *complement* A^c of a set A relative to the set of real numbers is the set of all real numbers that are *not* contained in A. Thus, if A is a set of real numbers, then $A^c = \{x \mid x \notin A\}$. Find the complement of each of the following sets, which are described by interval notation.

(a) $(3, \infty)$ (b) $(8, 9)$

SOLUTION

(a) $(3, \infty)^c = (-\infty, 3]$ (Figure 15a)

(b) $(8, 9)^c = (-\infty, 8] \cup [9, \infty)$ (Figure 15b)

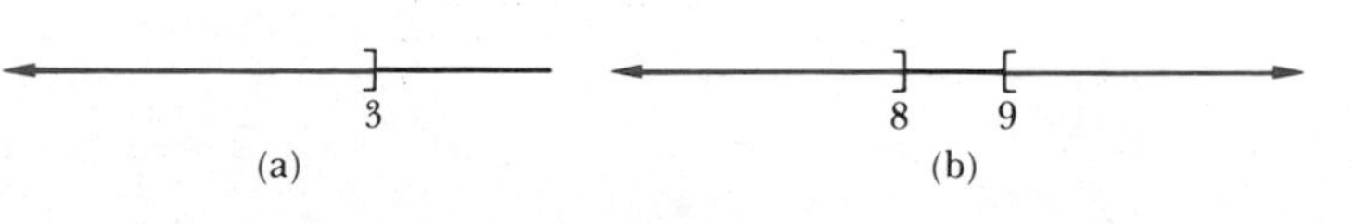

Figure 15

5.4 Linear Inequalities in One Variable

Inequalities such as $3x + 2 \geq 5$ and $(-3x + 5)/2 < 3$ are called *linear inequalities.* The solution set of a linear inequality in one variable, namely, the set of all numbers that satisfy the given inequality, can be determined in much the same manner as the solution set of a linear equation. We replace an inequality by an equivalent inequality with a solution set that is obvious. As with equations, the properties of inequalities enable us to convert a given inequality to an equivalent one. A few examples will help to clarify the method for solving linear inequalities in one variable.

EXAMPLES Find the solution set of each of the following inequalities and represent the solution set on the number line.

1 $3x - 2 < 7$

SOLUTION Adding 2 to both sides of the inequality (Theorem 2), we get $3x - 2 + 2 < 7 + 2$ or $3x < 9$. We next multiply both sides by $\frac{1}{3}$ (Theorem 3) to obtain $\frac{1}{3}(3x) < \frac{1}{3}(9)$, and this is equivalent to $x < 3$. Hence the solution set is $\{x \mid x < 3\} = (-\infty, 3)$ (Figure 16).

Figure 16 *Figure 17*

2 $x + 2 \geq 7x - 1$

SOLUTION After adding $-7x$ to both sides of the inequality, we obtain the expression $-6x + 2 \geq -1$. Subtracting 2 from each side gives us $-6x \geq -3$. After multiplying each side by $-\frac{1}{6}$, we get $(-\frac{1}{6})(-6x) \leq (-\frac{1}{6})(-3)$, which is the same as $x \leq \frac{1}{2}$. The solution set is $\{x \mid x \leq \frac{1}{2}\} = (-\infty, \frac{1}{2}]$ (Figure 17).

3 $\frac{3}{4} - \frac{7x}{5} \leq -\frac{9}{20}$

SOLUTION After multiplying both sides by 20, we have

$$20\left(\frac{3}{4} - \frac{7x}{5}\right) \leq 20\left(\frac{-9}{20}\right)$$

so that $15 - 28x \leq -9$. Adding -15 to each side results in $-28x \leq -24$. After multiplying each side by $-\frac{1}{28}$, we have $x \geq \frac{6}{7}$.

Figure 18

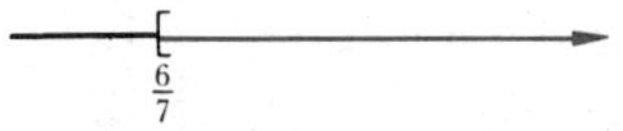

$\frac{6}{7}$

The solution set is

$$\left\{x \mid x \geq \frac{6}{7}\right\} = \left[\frac{6}{7}, \infty\right) \qquad \text{(Figure 18)}$$

4 A pocket calculator is sold at a profit; that is, the revenue obtained from the sale of a calculator is greater than the cost of its production. If it costs the company \$50 for the parts of each calculator and an additional \$7500 per day to run the production line, how many calculators must be sold each day at \$125 per unit in order for the company to show a profit?

SOLUTION Let x represent the number of calculators sold each day. Then the revenue in dollars obtained from the sale is $125x$. The cost of producing the calculators in dollars is $50x + 7500$ each day. The problem is to determine a positive integer x so that $125x > 50x + 7500$. But this latter inequality is equivalent to $75x > 7500$ or $x > 100$. Hence, the company will show a profit if it sells more than 100 calculators each day.

PROBLEM SET 5

1 Locate each of the following real numbers on the number line. Use decimal approximations in parts (c) and (d).
(a) $-2\frac{1}{5}$ (b) 3.7 (c) $3\sqrt{2}$ (d) $\pi/2$

2 (a) Theorem 2, together with the fact that $a - c = a + (-c)$, implies that if the same number is subtracted from both sides of an equality, the order of the inequality remains the same; that is, if $a < b$, then $a - c < b - c$ for any real number c. Prove this implication and give two examples.
(b) Using the fact that $a \div c = a(1/c)$ for $c \neq 0$, restate Theorem 3 for division. Prove each statement and give two examples of each.

In Problems 3–16, indicate the appropriate property or theorem that justifies each statement. Assume that all variables represent real numbers.

3 If $x < 5$, then $x + 7 < 5 + 7$.
4 If $x > -3$, then $x + 4 > -3 + 4$.
5 If $y \geq -4$, then $y + 5 \geq -4 + 5$.
6 If $t \leq 6$, then $t + 10 \leq 6 + 10$.
7 If $a < 4$ and $b > 4$, then $a < b$.
8 If $c \geq 5$ and $d \leq 5$, then $c \geq d$.
9 If $x \geq 7$, then $4x \geq 28$.
10 If $t \leq 3$, then $4t \leq 12$.
11 If $a < b$, then $-7a > -7b$.
12 If $x < y$, then $x/(-10) > y/(-10)$.
13 If $-3p \leq 27$, then $p \geq -9$.
14 If $-8r > 40$, then $r < -5$.
15 If $b > a$ and $a > 1$, then $b > 1$.
16 If $-r \leq -7$, then $r \geq 7$.

In Problems 17–22, express each set in interval notation and represent the set on the number line.

17 $\{x \mid 2 < x < 5\}$ **18** $\{x \mid -4 \leqslant x < 5\}$ **19** $\{x \mid x \leqslant -2\}$

20 $\{x \mid x \geqslant 2 \text{ or } x \leqslant -1\}$ **21** $\{x \mid x \geqslant -7\}$ **22** $\{x \mid x < -5 \text{ or } x \geqslant 3\}$

In Problems 23–26, express the given interval in set notation and represent the set on the number line.

23 $(1, 5)$ **24** $[3, 8]$ **25** $(-4, \infty)$ **26** $(-\infty, 7]$

In Problems 27–28, find the complement of each set relative to the set of real numbers.

27 $A = (-\infty, 3)$ **28** $B = [3, 4)$

In Problems 29–33, use interval notation and set operations to represent each of the sets.

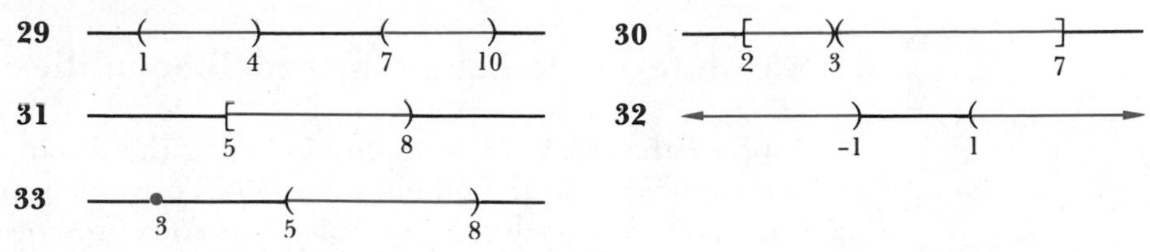

34 If $0 < x < y$, show that $1/x > 1/y$.

In Problems 35–54, express the solution set of each inequality in both set notation and interval notation and then represent the solution set on the number line.

35 $3x < 9$ **36** $-21w \leqslant -63$ **37** $-6x \geqslant -18$

38 $-15t > -75$ **39** $4x + 3 \geqslant 12$ **40** $3x - 2 > 7$

41 $2t - 5 \geqslant 3$ **42** $-8t - 4 \leqslant -16$ **43** $-9x - 2 < 16$

44 $6 < x - 2$ **45** $3(x + 2) - 5x \leqslant 4x$

46 $-(8 + x) - 5 + 4x \geqslant 1$ **47** $x + 6 \leqslant 4 - 3x$

48 $5 - x < -x + 3$ **49** $6 < x + 2 < 8$

50 $-4 \leqslant x + 3 \leqslant 6$ **51** $-2 \leqslant 3t - 1 \leqslant 7$

52 $-8 \leqslant x - 5 \leqslant 3$ **53** $\dfrac{x}{3} + 2 \leqslant \dfrac{x}{4} - 2x$

54 $\dfrac{3x - 7}{6} - 13 \geqslant 1 - \dfrac{x}{2}$

In Problems 55–58, let a, b, c, and d be real numbers. Prove each inequality.

55 If $a < b$, then $a < (b + a)/2 < b$. **56** If $0 < a < 1$, then $a^2 < a$.

57 If $a < b$ and $c < d$, then $a + c < b + d$.

58 If $a > 0$ and $b > 0$, then $(a + b)/2 \geqslant \sqrt{ab}$.

59 A woman has stock worth \$60 per share which pays her an 11 percent dividend every full year. In how many full years will each stock have paid her more than \$58 in dividends?

60 If one of the dimensions of a rectangular room is 13 feet and its area is less than 432 square feet, what can be concluded about the other dimension of the room?

61 A professor saves $100 monthly in a teachers' credit union. How many months will it take her to save at least $3865 in order to pay cash for a new car?

62 As part of a payroll savings plan, an employee asks for a $7 deduction from his weekly salary. How many weeks will it take to save up $437 to pay cash for a microwave oven?

63 A tire manufacturing company shows a profit (the revenue obtained from the sales of its tires is greater than the cost of producing them). If each tire costs $12 in materials, and if it costs $1500 a week to produce the tires, how many tires must be sold each week at $30 each in order for the company to show a profit?

6 Absolute-Value Equations and Inequalities

Suppose that we are interested in finding the distance between 0 and any real number x on the number line. For convenience, we use the notation $|x|$, which is read the *absolute value of* x, to represent the distance between x and 0. Then $|3| = 3$, $|0| = 0$, and $|-5| = 5$.

In fact, if $x > 0$, we have $|x| = x - 0 = x$ (Figure 1a). If $x < 0$, then $|x| = 0 - x = -x$ (Figure 1b). Finally, if $x = 0$, $|x| = 0$.

Figure 1

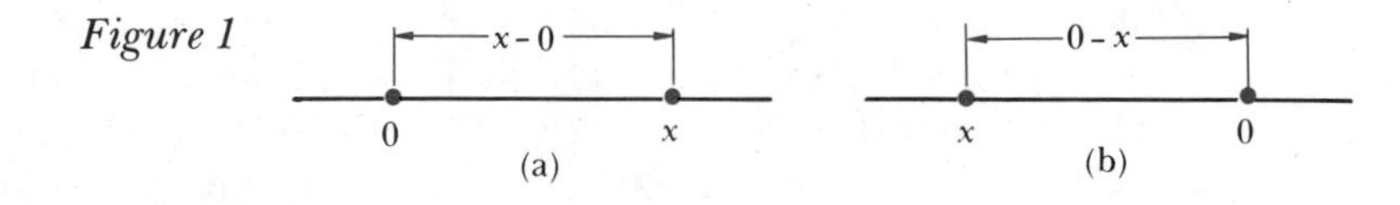

Thus we have the following definition.

DEFINITION 1 ABSOLUTE VALUE

If x is a real number, the *absolute value of* x, denoted by $|x|$, is defined as follows:

$$|x| = \begin{cases} x & \text{if } x \geq 0 \\ -x & \text{if } x < 0 \end{cases}$$

Thus, it follows from Definition 1 that $|3| = 3$ because $3 > 0$; $|0| = 0$; and $|-5| = -(-5) = 5$ because $-5 < 0$.

EXAMPLES 1 Use $x = 4$ and $y = -7$ to compute the value of each of the following expressions.

(a) $|x + 2y|$ (b) $|x| + |2y|$ (c) $|xy|$ (d) $|x| \cdot |y|$

(e) $|x - y|$ (f) $|x| - |y|$ (g) $\left|\frac{x}{y}\right|$ (h) $\frac{|x|}{|y|}$

(i) $|x|^2$ (j) $|y|^2$

SOLUTION

(a) $|x + 2y| = |4 + 2(-7)| = |4 - 14| = |-10| = 10$

(b) $|x| + |2y| = |4| + |2(-7)| = |4| + |-14| = 4 + 14 = 18$

(c) $|xy| = |4(-7)| = |-28| = 28$

(d) $|x| \cdot |y| = |4| \cdot |-7| = (4)(7) = 28$

(e) $|x - y| = |4 - (-7)| = |11| = 11$

(f) $|x| - |y| = |4| - |-7| = 4 - 7 = -3$

(g) $\left|\frac{x}{y}\right| = \left|\frac{4}{-7}\right| = \frac{4}{7}$ (h) $\frac{|x|}{|y|} = \frac{|4|}{|-7|} = \frac{4}{7}$

(i) $|x|^2 = |4|^2 = 4^2 = 16$ (j) $|y|^2 = |-7|^2 = 7^2 = 49$

2 Simplify $\frac{x}{|x|}$ if x is a nonzero real number.

SOLUTION If $x > 0$, $|x| = x$, so that

$$\frac{x}{|x|} = \frac{x}{x} = 1$$

If $x < 0$, $|x| = -x$, so that

$$\frac{x}{|x|} = \frac{x}{-x} = -1$$

Consequently, $\frac{x}{|x|} = \begin{cases} 1 & \text{if } x > 0 \\ -1 & \text{if } x < 0. \end{cases}$

6.1 Absolute-Value Equations

Given any two real numbers x and y such that $x \leqslant y$, the distance d between x and y is considered to be the *nonnegative* number $y - x$. Thus, $d = y - x$ for $x \leqslant y$ (Figure 2). For example, to find the distance d between -1 and 3, we note that $-1 < 3$, so that $d = 3 - (-1) = 4$. Also, the distance between -5 and -2 is given by $d = (-2) - (-5) = 3$ since $-5 < -2$ (Figure 3). Using absolute values, we can also express the distance d between x and y, without regard to the order relation between x and y, as $d = |x - y|$. The distance d between -1 and 3 is given by $d = |-1 - 3| = |-4| = 4$ or by $d = |3 - (-1)| = |4| = 4$.

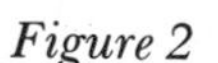
Figure 2

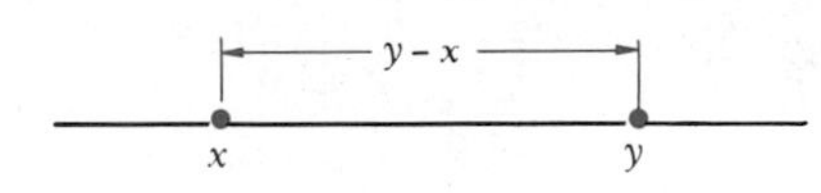

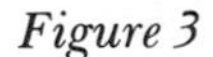
Figure 3

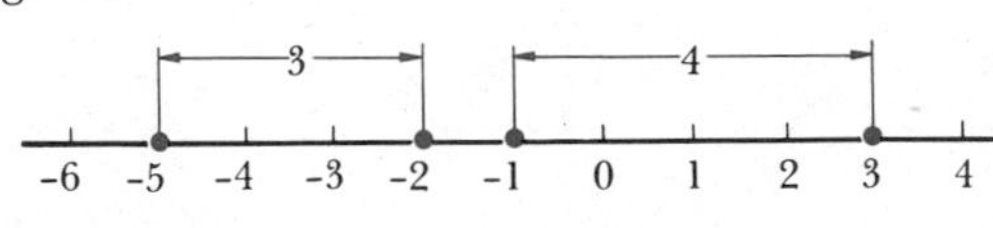

We can use Definition 1 of absolute value to prove the following properties of absolute value.

THEOREM 1

$|x| \geqslant 0$ for any real number x.

PROOF Using Definition 1 of absolute value, if $x \geqslant 0$, then $|x| = x \geqslant 0$ and if $x < 0$, then $|x| = -x > 0$.

THEOREM 2

$|-x| = |x|$ for any real number x.

PROOF If $x \geqslant 0$, then $-x \leqslant 0$, so that $|x| = x$ and $|-x| = -(-x) = x$. On the other hand, if $x < 0$, then $-x > 0$, so that $|x| = -x$ and $|-x| = -x$. In any case, $|x| = |-x|$.

Geometrically, $|-x| = |x|$ is interpreted to mean that the distance between 0 and x is the same as the distance between 0 and $-x$. For example, $|-7| = |7| = 7$.

THEOREM 3

$|x|^2 = x^2$ for any real number x.

PROOF For $x \geqslant 0$, $|x| = x$, so that $|x|^2 = (x)^2 = x^2$. For $x < 0$, $|x| = -x$, so that $|x|^2 = (-x)^2 = x^2$. Hence $|x|^2 = x^2$ for all possible values of x.

The above properties, together with Definition 1 of absolute value, can be used to solve absolute value equations of the form $|ax + b| = c$, where a, b, and c are real numbers with $c \geqslant 0$.

EXAMPLES Find the solution set of each equation.

1 $|x| = 3$

SOLUTION Since $|3| = |-3| = 3$, then $x = -3$ or $x = 3$. Therefore, the solution set is $\{-3, 3\}$.

2 $|x - 3| = 4$

SOLUTION Since $|x - 3| = 4$, it follows that

$$x - 3 = 4 \quad \text{or} \quad x - 3 = -4$$

so $x = 7$ or $x = -1$. Therefore, the solution set is $\{-1, 7\}$.

3 $|3x - 4| = 5$

SOLUTION Since $|3x - 4| = 5$, it follows that

$$3x - 4 = 5 \quad \text{or} \quad 3x - 4 = -5$$

so $x = 3$ or $x = -\frac{1}{3}$. Therefore, the solution set is $\{-\frac{1}{3}, 3\}$.

4 $|x + 2| = |x - 7|$

SOLUTION The equation will be satisfied if $x + 2$ and $x - 7$ are equal, or if $x + 2$ and $x - 7$ are negatives of each other. Thus,

$$\begin{aligned} x + 2 &= x - 7 \quad \text{or} \quad & x + 2 &= -(x - 7) \\ 2 &= -7 & x + 2 &= -x + 7 \\ & & 2x &= 5 \\ & & x &= \tfrac{5}{2} \end{aligned}$$

The first equation, $x + 2 = x - 7$, leads to the false statement $2 = -7$. Thus, the only solution of the equation $|x + 2| = |x - 7|$ is the number $\frac{5}{2}$. The solution set is $\{\frac{5}{2}\}$.

6.2 Absolute-Value Inequalities

Figure 4

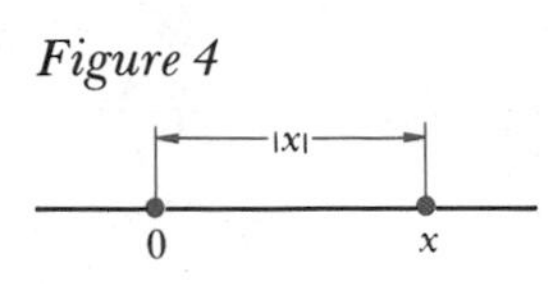

We know that $|x|$ represents the distance between 0 and x as shown in Figure 4, where x is illustrated as a positive number. Now we can use this geometric interpretation, together with the results above, to get a clear understanding of absolute-value inequalities of the forms $|x| < a$ or $|x| > a$, where a is a positive number.

THEOREM 4

If $|x| < a$, where $a > 0$, then $-a < x < a$.

GEOMETRIC INTERPRETATION Quite simply, $|x| < a$ means that the distance between 0 and x is less than a units; or, equivalently, x is within a units of 0 (Figure 5). Using inequalities, this means that $-a < x < a$.

Figure 5

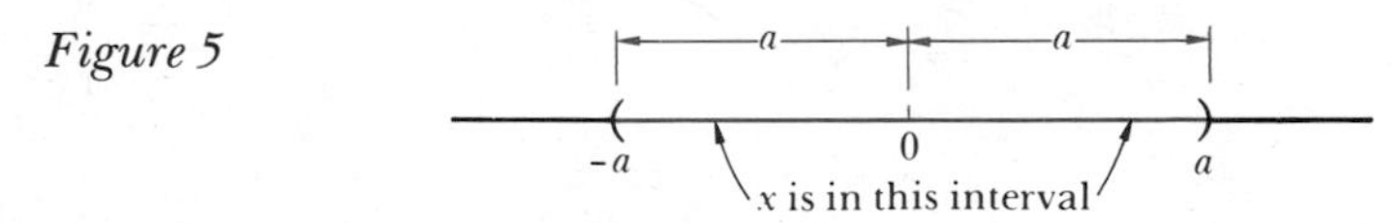

PROOF OF THEOREM If $|x| < a$, then $-a < -|x|$ (why?). By Definition 1 of absolute value, $|x| = x$ or $|x| = -x$, so that $-a < -|x| \leq x \leq |x| < a$ (see Problem 54a). By the transitive property of inequalities, $-a < x < a$.

THEOREM 5

If $|x| > a$, where $a > 0$, then $x < -a$ or $x > a$.

GEOMETRIC INTERPRETATION The expression $|x| > a$ means that the distance between 0 and x is more than a units; or, equivalently, x is more than a units from 0 (Figure 6). Using inequalities, this means that $x < -a$ or $x > a$.

Figure 6

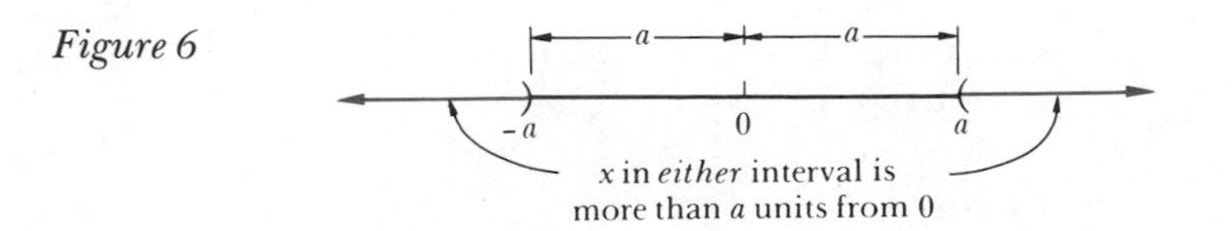

PROOF OF THEOREM By Definition 1 of absolute value, either $|x| = x$ or $|x| = -x$. Hence $|x| = x > a$ or $|x| = -x > a$. That is, $x > a$ or $-x > a$. But $-x > a$ implies that $x < -a$ (why?), so $x < -a$ or $x > a$.

EXAMPLES Find the solution set of each inequality and represent the solution on the number line.

1 $|x| < 3$

SOLUTION By Theorem 4, the inequality can be written as $-3 < x < 3$. The solution set is $\{x \mid -3 < x < 3\}$ (Figure 7).

Figure 7

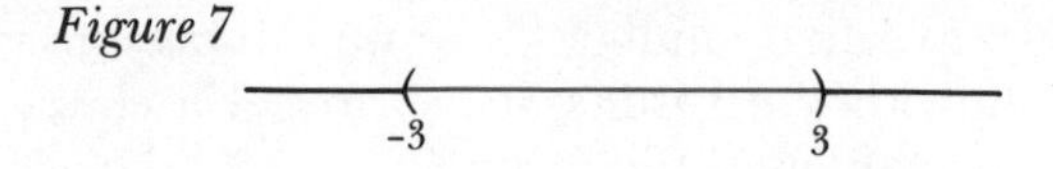

2 $|3x - 2| \leqslant 8$

SOLUTION The inequality can also be written as

$$-8 \leqslant 3x - 2 \leqslant 8$$
$$-6 \leqslant 3x \leqslant 10$$
$$-2 \leqslant x \leqslant \tfrac{10}{3}$$

Thus the solution set for this inequality is $\{x \mid -2 \leqslant x \leqslant \frac{10}{3}\}$ (Figure 8).

Figure 8

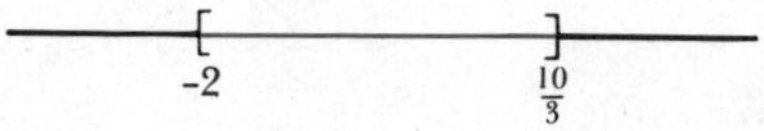

3 $|x| - 3 > 4$

SOLUTION Add 3 to both sides of the inequality to get $|x| > 7$. By Theorem 5, we have $x > 7$ or $x < -7$. Therefore, the solution set, as shown in Figure 9, is $\{x \mid x < -7 \text{ or } x > 7\}$.

4 $|2x - 3| \geqslant 5$

SOLUTION $|2x - 3| \geqslant 5$ implies

$$2x - 3 \leqslant -5 \quad \text{or} \quad 2x - 3 \geqslant 5$$
$$2x \leqslant -2 \qquad\qquad 2x \geqslant 8$$
$$x \leqslant -1 \qquad\qquad x \geqslant 4$$

Therefore, the solution set is $\{x \mid x \leqslant -1 \text{ or } x \geqslant 4\}$ (Figure 10).

Figure 9

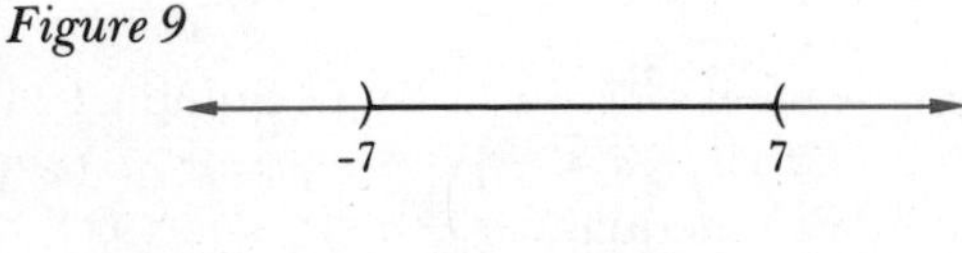

Figure 10

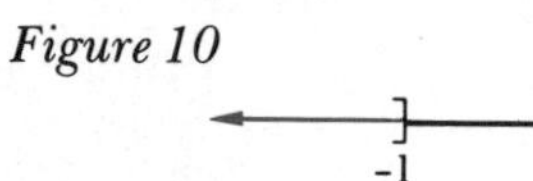

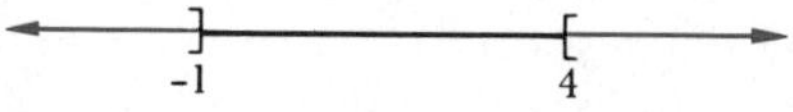

Definition 1 of absolute value, together with Theorems 4 and 5, can be used to verify the *triangle inequality:* $|a + b| \leqslant |a| + |b|$ for any real numbers a and b. (See Problem 54b.)

EXAMPLE Show that if $|x| < 3$ and $|y| < 1$, then $|x + y| < 4$.

SOLUTION By the triangle inequality, $|x + y| \leqslant |x| + |y| < 3 + 1$, so that, by transitivity, $|x + y| < 4$.

PROBLEM SET 6

In Problems 1–8, use $x = 3$ and $y = -4$ to compute the value of each expression.

1	$\lvert x\rvert + \lvert y\rvert$	**2**	$\lvert x + y\rvert$	**3**	$\lvert x - y\rvert$	**4**	$\lvert x\rvert - \lvert y\rvert$
5	$\lvert xy\rvert$	**6**	$\lvert x/y\rvert$	**7**	$3\lvert x\rvert + \lvert -4y\rvert$	**8**	$\lvert 3x\rvert - 4\lvert y\rvert$

9 Simplify the expression

$$\frac{x+1}{|x+1|}, \quad \text{where } x \neq -1$$

10 (a) Under what conditions does $|x+y| = |x| + |y|$?
(b) Under what conditions does $|x| = |y|$?

In Problems 11–30, find the solution set of each equation.

11 $|x| = 4$ **12** $|t| - 3 = 5$ **13** $|y| + 2 = 4$
14 $|x| + 7 = 7$ **15** $|3x| = 15$ **16** $|-3x| = 12$
17 $|x| + |-3| = 3$ **18** $|x| = |-5|$ **19** $|t+2| = 5$
20 $|y-2| = 6$ **21** $|p-1| = -9$ **22** $|x-4| = -3$
23 $|3x+2| = 5$ **24** $|2t+6| = 18$ **25** $|\frac{2}{5}x - 1| = 3$
26 $|\frac{1}{2}t + 7| = 0$ **27** $|x-3| = |x+5|$ **28** $|x-7| = |3x+1|$
29 $|y-2| = |y+3|$ **30** $|2x-9| = |5x-3|$

In Problems 31–52, find the solution set of each absolute value inequality and represent the solution set on the number line.

31 $|x| < 2$ **32** $|x| + 3 \leq 7$ **33** $|2x| - 1 \leq 5$
34 $|3x| - 1 < 5$ **35** $|x-1| \leq 3$ **36** $|3-2x| \leq 5$
37 $|2x+3| < 1$ **38** $|4x-2| \leq 4$ **39** $|2t-5| \leq 11$
40 $|3x-6| \leq 0$ **41** $|x| > 1$ **42** $|3y| \geq 24$
43 $|5p| - 1 \geq 14$ **44** $|6x| - 3 > 9$ **45** $|x+2| \geq 5$
46 $|t+1| > 7$ **47** $|4x-3| > 9$ **48** $|5x-3| - 2 \geq 10$
49 $|2x-4| > 0$ **50** $|x+5| < |x+1|$ **51** $|6y-3| > 7$
52 $|5-3x| \geq -2$

53 Show that if $|x-a| < \frac{1}{10}$ and $|a-y| < \frac{1}{10}$, then $|x-y| < \frac{1}{5}$.

54 (a) Prove that $-|x| \leq x \leq |x|$ for every real number x.
(b) Use the inequality in part (a) to prove the triangle inequality, $|a+b| \leq |a| + |b|$, where a and b are any two real numbers.

REVIEW PROBLEM SET

In Problems 1–8, let A be a set containing five elements and let B be a set containing three elements. Indicate which of the statements are true and which are false.

1 B is a subset of A.
2 $A \cap B$ is a subset of A.
3 $A \cap B$ contains exactly five elements.
4 $A \cup B$ contains exactly five elements.
5 If $A \cap B = \varnothing$, then $A \cup B = \varnothing$.
6 If $x \in A$ and $x \in B$, then $A \cap B$ is not an empty set.
7 If $A \cap B$ contains three elements, then B is a subset of A.
8 $A \cap B$ is a subset of $A \cup B$.

9 List all the subsets of the set $\{5, 10, 11\}$. Indicate which subsets are proper subsets.

10 Let $U = \{x, y, u, v\}$. If $A = \{x, y\}$, $A \cap B = \{x\}$, and $A \cup B = U$, find the set B.

In Problems 11–16, let $A = \{a, b, c\}$, $B = \{c, d, e, f\}$, and $C = \{a, c, d, g\}$. Form each of the following sets.

11 $A \cap B$ **12** $B \cap C$ **13** $C \cup B$
14 $B \cup A$ **15** $A \cap (B \cup C)$ **16** $A \cup (C \cap B)$

In Problems 17–18, describe $A \cap B$ in each case, where x is assumed to be a positive integer. (A positive integer m is said to be a *multiple* of a positive integer n if there exists a positive integer k such that $m = kn$. For example, 18 is a multiple of 6, since $18 = 3 \cdot 6$.)

17 $A = \{x \mid x \text{ is a multiple of } 3\}$ and $B = \{x \mid x \text{ is a multiple of } 5\}$
18 $A = \{x \mid x \text{ is a multiple of } 3\}$ and $B = \{x \mid x \text{ is a multiple of } 6\}$

In Problems 19–22, express each rational number in decimal form.

19 $\frac{7}{40}$ **20** $-\frac{3}{11}$ **21** $-\frac{11}{22}$ **22** $\frac{7}{6}$

In Problems 23–26, express each rational number as the ratio of two integers.

23 0.16 **24** -0.0035 **25** $-0.0\overline{31}$ **26** $0.\overline{629}$

In Problems 27–36, justify each statement by giving the appropriate property. Assume that all variables represent real numbers.

27 $x + 5$ is a real number **28** $a + (b + 3) = (a + b) + 3$
29 $3x = x \cdot 3$ **30** $2(xy) = (2x)y$
31 $1 \cdot 4 = 4$ **32** $5 + 0 = 5$
33 If $x = y$, then $2x = 2y$ **34** $x \cdot (1/x) = 1$, for $x \neq 0$
35 If $7a = 0$, then $a = 0$ **36** If $x + 9 = y + 9$, then $x = y$

In Problems 37–48, perform the indicated operations.

37 $(2x^2 + 3x - 4) + (x^2 - 5x + 7)$
38 $(3x^2 + 7x + 8) - (2x^2 + 3x + 2)$
39 $(5x^2 - 3x + 2) + (2x^2 + 5x - 7) - (3x^2 - 4x - 1)$
40 $(x - y)(x^2 - 2xy + y^2)$ **41** $(x^2 - x + 1)(2x^2 - 3x + 2)$
42 $(2x - 3)(x + 5)$ **43** $(7 - 5xy)(4 + 3xy)$
44 $(3x - 2)^2$ **45** $(x^2 + 4)(x^2 - 4)$
46 $(1 + 7y)^2$ **47** $(x^2 - y)(x^4 + xy + y^2)$
48 $(2x + 7)(4x^2 - 14x + 49)$

In Problems 49–58, factor each polynomial completely.

49 $26x^3y^2 + 39x^5y^4 - 52x^2y^3$ **50** $7xy - 7yz + 14y^2z - 14xy^2$
51 $y^2 - 121$ **52** $y^3 - 216$
53 $25y^2 - 81z^2$ **54** $x^8 - 256$
55 $x^3 + 64$ **56** $x^4 - 10x^2 + 9$
57 $x^2 - x - 56$ **58** $6x^2 - 29x + 35$

59 Find the solution set of each equation.
(a) $12(x - 2) + 8 = 5(x - 1) + 2x$
(b) $x - 7(4 + x) = 5x - 6(3 - 4x)$

60 A monthly phone bill includes a charge of \$4.80 for local calls plus an additional charge of \$2.60 for each long distance call placed within a certain area. Suppose that a tax of 9 percent of all charges is added to the total bill and assume that all the long distance calls were within the \$2.60 area. How many long distance calls were made if the total bill, including taxes, is \$25.07?

In Problems 61–68, perform the indicated operations and simplify the result.

61 $\dfrac{x^2 - 16}{x^2 - 4x} \cdot \dfrac{x - 4}{x + 4}$

62 $\dfrac{x^2 + 5x - 6}{x^2 + x - 2} \cdot \dfrac{x^2 + 3x - 4}{x^2 + 7x + 12}$

63 $\dfrac{x^2 - x - 2}{x^2 - x - 6} \div \dfrac{x^2 - 2x}{2x + x^2}$

64 $\dfrac{y^3 + 1}{x^2 - 4y^2} \div \dfrac{y^2 - y + 1}{x - 2y}$

65 $\dfrac{3}{x^2 - 7x + 12} + \dfrac{2}{x^2 - 5x + 4}$

66 $\dfrac{5}{x^2 + 8x + 15} - \dfrac{4}{x^2 + 2x - 3}$

67 $\dfrac{1 - \dfrac{1}{x}}{x - 2 + \dfrac{1}{x}}$

68 $\dfrac{\dfrac{y}{y^2 - 1} - \dfrac{1}{y + 1}}{\dfrac{y}{y - 1} + \dfrac{1}{y + 1}}$

69 Find the solution set of each equation.
(a) $\dfrac{2}{x} + \dfrac{x - 1}{3x} = \dfrac{2}{5}$
(b) $\dfrac{10 - x}{x} + \dfrac{3x + 3}{3x} = 3$

70 The manager of a parking garage agreed to pay a troop of Boy Scouts a fixed amount for washing some cars, enough to give each boy in the troop \$1.00. When 25 boys failed to show up, those already present agreed to do the work, which meant that each boy who worked got \$1.50. How many boys are in the troop?

In Problems 71–80, simplify each expression. Assume that all variables represent positive real numbers.

71 $(x^{-2}y^{-3}z^0)^{-3}$

72 $\dfrac{(xy^{-1}z)^{-2}}{(xy^{-1}z)^{-6}}$

73 $\dfrac{(x^{-1}y^{-2/3})^{-3}}{(x^{-1}y^{-2/3})^{-5}}$

74 $\dfrac{x^{-2} - y^{-2}}{x^{-1} + y^{-1}}$

75 $\sqrt{32x^5y^{10}z^{15}}$

76 $\sqrt[3]{81x^4z^5w^6}$

77 $\sqrt[6]{128x^{13}y^{25}}$

78 $\sqrt{98(x + 2y)^2}$

79 $\dfrac{\sqrt{7} - \sqrt{6}}{\sqrt{7} + \sqrt{6}}$

80 $\dfrac{\sqrt{x}}{3 - \sqrt{y}}$

81 Find the solution set of each equation.
(a) $\sqrt{2x + 5} - 7 = -4$
(b) $2\sqrt{2x - 3} + 4 = 1$

82 Indicate which of the following statements are true for *all* real numbers a and b with $a < b$.
(a) $2a > -(-2)b$ (b) $5/a < 5/b$
(c) $a - c < b - c$ (d) $a + 3 < b + 4$

In Problems 83–86, indicate the appropriate property or theorem that justifies each statement. Assume that all variables represent real numbers.

83 If $x > y$ and $z > 0$, then $x/z > y/z$.
84 If $x < y$, then $x/(-3) > y/(-3)$.
85 If $-7y < 35$, then $y > -5$.
86 If $a < 5$ and $5 < d$, then $a < d$.

87 Express each set in interval notation and represent the set on the number line.
(a) $\{x \mid -3 \leq x \leq 2\}$ (b) $\{x \mid x \geq 2\}$

88 Express each interval in set notation and represent the set on the number line.
(a) $[2, 7]$ (b) $(-\infty, -2)$ (c) $(3, \infty)$

In Problems 89–92, find the solution set of each inequality and represent the solution set on the number line.

89 $5x - 9 > 2x + 3$
90 $\frac{2x}{3} + \frac{1}{5} > \frac{7}{15} + \frac{4x}{5}$
91 $5x - 1 \leq 8x - 5$
92 $5x - 2 \geq 6x + 5$

93 In a fund-raising gathering, a woman sold five times as many tickets as a man, but the woman could not have sold more than 35 tickets. How many tickets might the man have sold?

94 Compute $\left|\frac{x}{|x|}\right|$ if x is a nonzero real number.

In Problems 95–104, find the solution set of each equation or inequality. Show the solution set on the number line.

95 $|3x + 4| = 12$
96 $|5 - 4x| = 11$
97 $|x - 1| = 2x$
98 $|x - 2| = |2 - x|$
99 $|3x + 1| > 8$
100 $|x - 5| \geq \frac{3}{2}$
101 $|2x + 5| \leq 9$
102 $|3t - 7| + 4 < 5$
103 $|x - 2| > -5$
104 $|x + 5| < |x - 1|$

CHAPTER 2

Functions and Graphs

In Chapter 1, we reviewed the fundamentals of algebra. The remainder of this book assumes a knowledge of these fundamentals. In this chapter, we present the concept of a *function*—a concept that is basic not only to this course, but also to the later study of mathematics or any area of science. The algebra of functions and the composition of functions are introduced, as well as inverse functions. Through the use of graphs, we investigate some of the properties that functions can possess. Because of this emphasis on graphs, it is appropriate to begin the chapter with a discussion of the cartesian coordinate system.

1 Cartesian Coordinate System and Distance Formula

We have seen in Chapter 1 that the number line provides us with a geometric representation of the real numbers as *points on a line.* This geometric representation was used in investigating the order of the real numbers and the notion of distance between points on a line. In this section we consider a method of representing "ordered pairs" of real numbers as *points in a plane.* Then this geometric representation is used to determine a method of finding the distance between two points in a plane.

The elements in the set $\{5, 6\}$ do not have to be listed in any particular order. This set could be written either as $\{5, 6\}$ or as $\{6, 5\}$; in other words, $\{5, 6\} = \{6, 5\}$. By contrast, $(5, 6)$ is an *ordered pair* consisting of *first, the element* 5, and *second, the element* 6. [Although this notation is the same as that used to denote open intervals, the context will be sufficiently clear to indicate whether (a, b) represents an interval or an ordered pair.]

Two ordered pairs are considered to be *equal* only when they have equal *first elements* (members) *and* equal *second elements* (members). For example, $(1, 2) \neq (2, 1)$, even though each pair contains the same elements. Likewise, $(4, 3) \neq (4, 4)$, whereas $(9, y) = (x, 8)$ if and only if $x = 9$ and $y = 8$.

1.1 Cartesian Coordinate System

The set of all ordered pairs of real numbers can be represented as the set of all points in a plane by using a *cartesian coordinate system,* which is constructed as follows.

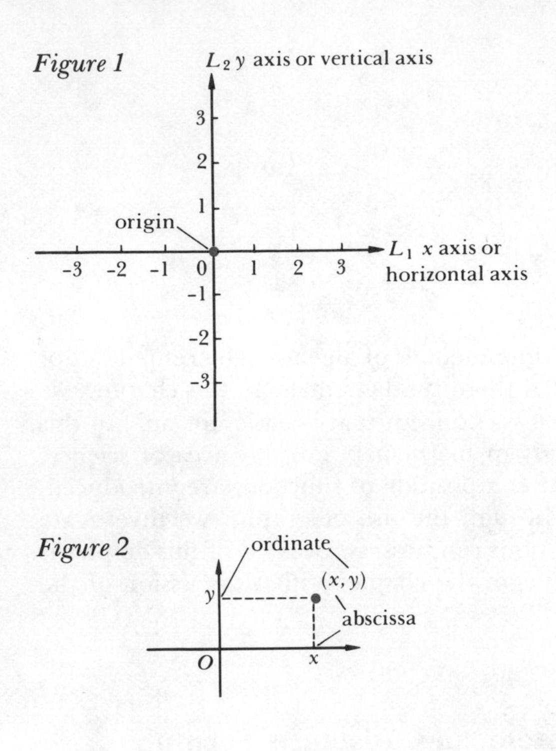

Figure 1

Figure 2

First, two mutually perpendicular number lines, L_1 and L_2, are constructed and scaled as illustrated in Figure 1. The point of intersection of the two number lines is called the *origin.* The two number lines are called the *coordinate axes.* Because of the locations of the lines, L_1 and L_2 are normally referred to as the *horizontal axis* (or x axis) and the *vertical axis* (or y axis), respectively.

Given an ordered pair of real numbers (x, y), we can use the cartesian coordinate system to represent or locate (x, y) as a point in the plane as follows. (The first member of the pair, x, is called the *abscissa;* the second member of the pair, y, is called the *ordinate;* x and y are called the *coordinates.*) The abscissa x is located on the horizontal axis. Then a line is drawn perpendicular to this axis at point x. The ordinate y is located on the vertical axis and a second line is drawn perpendicular to this axis at point y. The intersection of these two perpendicular lines is the point in the plane that represents the ordered pair (x, y) (Figure 2).

For example, the ordered pair (1, 1) is located by moving 1 unit to the right of 0 on the x axis, then 1 unit up from the x axis. Similarly, (7, 5) is located by moving 7 units to the right of 0 and 5 units up; $(-\frac{1}{2}, 0)$ is located by moving $\frac{1}{2}$ unit to the left of 0 and 0 units up from the x axis (Figure 3).

The coordinate axes divide the plane into four disjoint regions called *quadrants,* which are described in Figure 4. Notice that the coordinate axes have no point in common with the four quadrants.

Figure 3

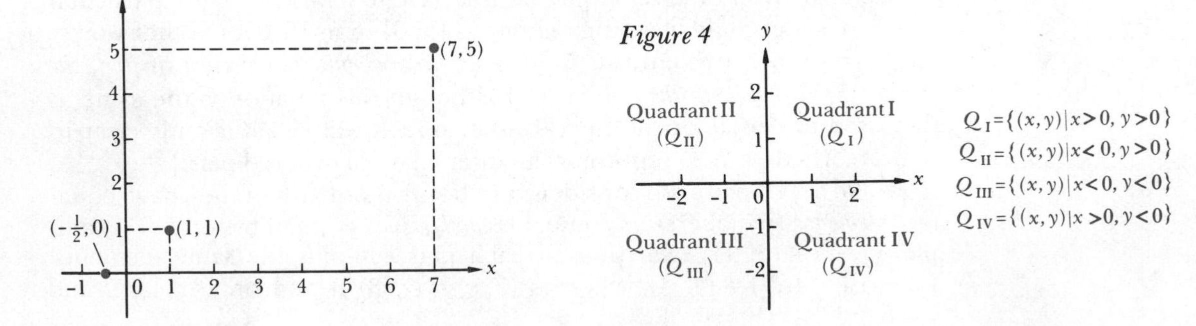

Figure 4

EXAMPLES 1 Locate each of the following points and indicate which quadrant, if any, contains the point.

(a) $\left(\frac{1}{2}, -\frac{1}{2}\right)$ (b) $\left(\pi, \frac{1}{\pi}\right)$ (c) $(-2, -6)$

(d) $(-3, 4)$ (e) $(-\sqrt{2}, 0)$ (f) $(0, -5)$

SOLUTION After locating the points (Figure 5), we see that
(a) $(\frac{1}{2}, -\frac{1}{2})$ lies in quadrant IV.
(b) $\left(\pi, \frac{1}{\pi}\right)$ lies in quadrant I.
(c) $(-2, -6)$ lies in quadrant III.
(d) $(-3, 4)$ lies in quadrant II.
(e) $(-\sqrt{2}, 0)$ is on the x axis. It is not in any quadrant.
(f) $(0, -5)$ is on the y axis. It is not in any quadrant.

2 Locate the point $P = (-2, 3)$ and give the coordinates of point Q if the line segment $\overline{PQ}$ is perpendicular to the y axis and bisected by it.

SOLUTION Because of the location of P (Figure 6), the coordinates of Q are given by $(2, 3)$ if the y axis is to bisect line segment $\overline{PQ}$.

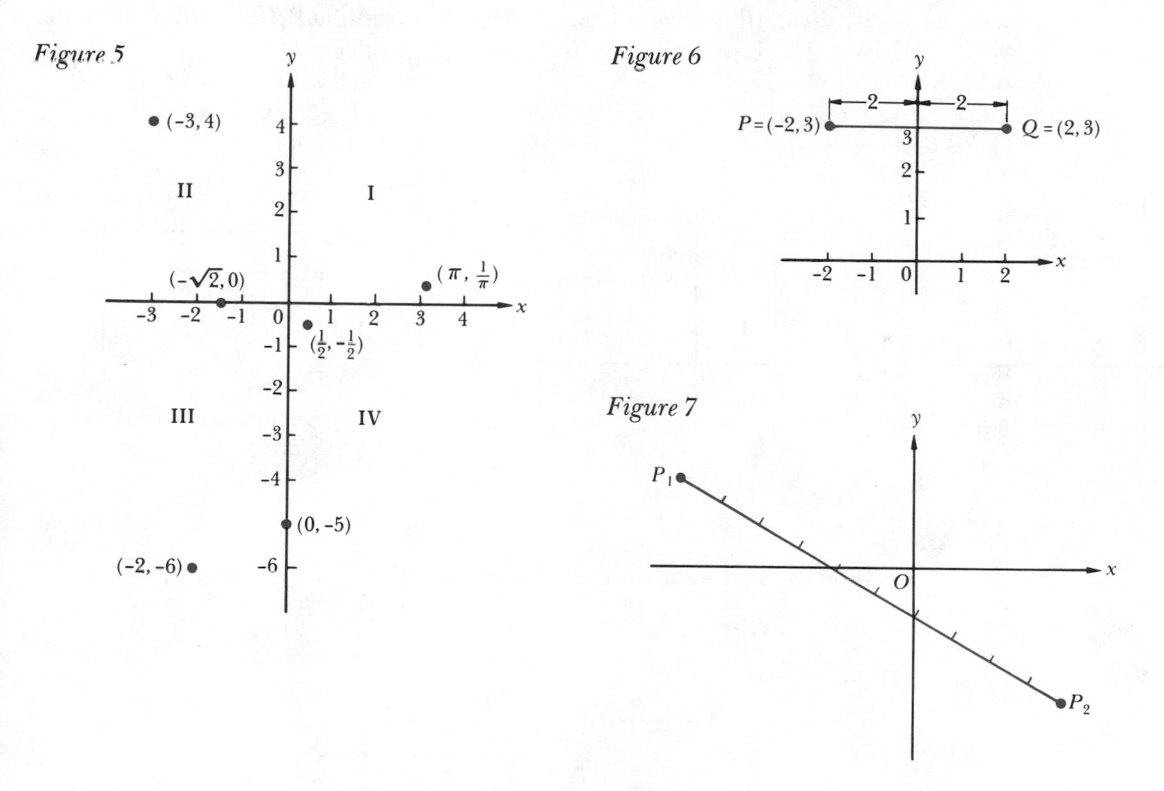

Figure 5

Figure 6

Figure 7

1.2 Distance Between Two Points

Suppose that a cartesian coordinate system is established using the same scale units for both the x and y axes. Then the distance between any two points, say, P_1 and P_2, is the length of the line segment determined by the two points (Figure 7). The following theorem establishes a formula for the distance between points P_1 and P_2 in terms of their cartesian coordinates.

THEOREM 1 DISTANCE FORMULA

Given any two points P_1 and P_2 with coordinates (x_1, y_1) and (x_2, y_2), respectively, the distance d between P_1 and P_2 is given by the formula

$$d = \sqrt{(x_1 - x_2)^2 + (y_1 - y_2)^2}$$

PROOF The distance d between P_1 and P_2 in terms of coordinates x_1, y_1, x_2, and y_2 can be derived by considering three cases:

(i) If the two points lie on the same vertical line, that is, $x_1 = x_2$, then $d = |y_1 - y_2|$ (Figure 8a).

(ii) If the two points lie on the same horizontal line, that is, $y_1 = y_2$, then $d = |x_1 - x_2|$ (Figure 8b).

(iii) If the two points lie on a line that is neither horizontal nor vertical, then a right triangle, that is, a triangle P_1PP_2 with a 90° angle, can be constructed as shown in Figure 8c.

Figure 8

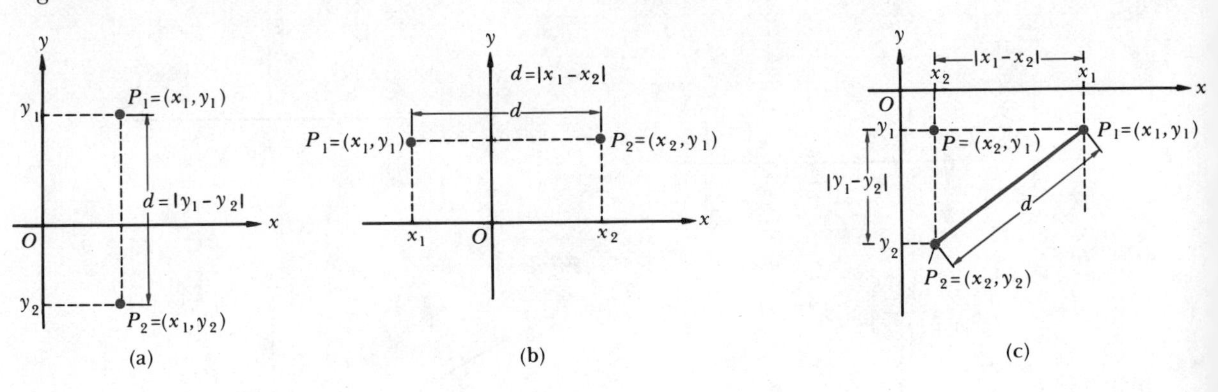

By using the Pythagorean theorem we get

$$d^2 = |\overline{PP}_1|^2 + |\overline{PP}_2|^2$$

so that

$$d^2 = |x_1 - x_2|^2 + |y_1 - y_2|^2$$

or

$$d^2 = (x_1 - x_2)^2 + (y_1 - y_2)^2$$

Hence,

$$d = \sqrt{(x_1 - x_2)^2 + (y_1 - y_2)^2}$$

Notice that this latter formula is also applicable in the special cases where P_1 and P_2 are on the same vertical line or same horizontal line (why?). Since $(a - b)^2 = (b - a)^2$, the "order" of subtracting the abscissas or the ordinates is irrelevant.

EXAMPLES 1 Use the distance formula to find the distance between $(-1, -2)$ and $(3, -4)$.

SOLUTION The distance d is given by

$$d = \sqrt{[3-(-1)]^2 + [-4-(-2)]^2} = \sqrt{4^2 + (-2)^2} = \sqrt{20} = 2\sqrt{5}$$

2 Use the distance formula to show that the triangle whose vertices are $(-2, 6)$, $(-1, -1)$, and $(1, 0)$ is a right triangle.

SOLUTION Let $A = (-2, 6)$, $B = (-1, -1)$, and $C = (1, 0)$ be the vertices of the triangle. Then

$$|\overline{AB}| = \sqrt{(-1+2)^2 + (-1-6)^2} = \sqrt{1+49} = \sqrt{50}$$
$$|\overline{BC}| = \sqrt{(1+1)^2 + (0+1)^2} = \sqrt{4+1} = \sqrt{5}$$
$$|\overline{AC}| = \sqrt{(1+2)^2 + (0-6)^2} = \sqrt{9+36} = \sqrt{45}$$

so that $|\overline{AB}|^2 = 50$, $|\overline{BC}|^2 = 5$, and $|\overline{AC}|^2 = 45$. Thus,

$$|\overline{AB}|^2 = |\overline{BC}|^2 + |\overline{AC}|^2$$

and the triangle is a right triangle (Figure 9) because of the converse of the Pythagorean theorem.

Figure 9

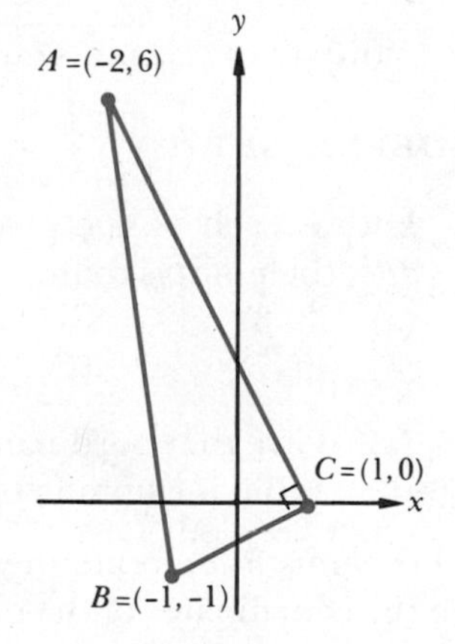

3 (a) Derive a formula for the distance between the origin and any point (x, y) in the plane.

(b) The *unit circle* is the circle with center at the origin and radius 1. Use the distance formula to express the relationship between x and y if (x, y) is a point on the unit circle.

SOLUTION

(a) The distance d between $(0, 0)$ and (x, y) is given by

$$d = \sqrt{(x-0)^2 + (y-0)^2}$$

so that the formula is

$$d = \sqrt{x^2 + y^2} \qquad \text{(Figure 10a)}$$

Figure 10

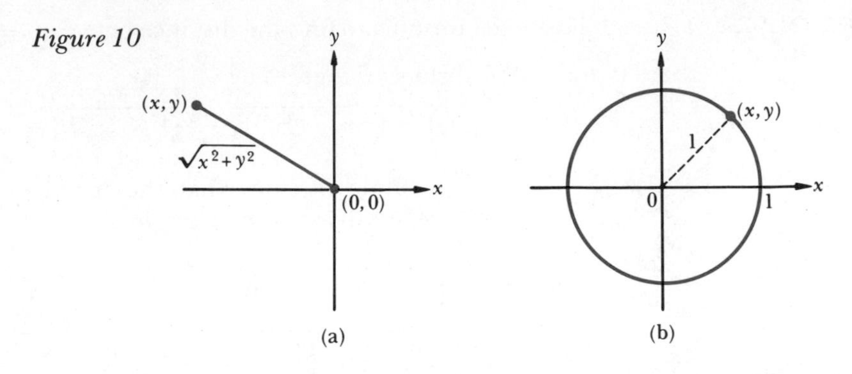

(b) Suppose that (x, y) is any point on the unit circle (Figure 10b); then, by the distance formula,

$$1 = \sqrt{(x-0)^2 + (y-0)^2}$$

so that

$$1 = x^2 + y^2$$

In fact,

$$\{(x, y) \mid x^2 + y^2 = 1\}$$

is the set of *all* points on the unit circle.

PROBLEM SET 1

1 Locate each of the given points and indicate in which quadrant, if any, the point is found.
(a) $(3, 3)$ (b) $(-2, 4)$ (c) $(5, -1)$ (d) $(3, -2)$
(e) $(0, 7)$ (f) $(0, -3)$ (g) $(-3, 0)$ (h) $(-1, -5)$

2 (a) Give the coordinates of any five points on the x axis.
(b) What is common to the coordinates of all points on the x axis?

In Problems 3–8, locate the point P on a cartesian coordinate system and give the coordinates of the points Q, R, and S such that:
(a) The line segment $\overline{PQ}$ is perpendicular to the x axis and is bisected by it.
(b) The line segment $\overline{PR}$ is perpendicular to the y axis and is bisected by it.
(c) The line segment $\overline{PS}$ is bisected by the origin.

3 $P = (1, 4)$ **4** $P = (-4, -2)$ **5** $P = (-3, 2)$
6 $P = (4, -1)$ **7** $P = (2, -3)$ **8** $P = (2, 0)$

In Problems 9–18, use the distance formula to find the distance between the two given points.

9 $(1, 2)$ and $(7, 10)$ **10** $(-3, -4)$ and $(-5, -7)$
11 $(1, 1)$ and $(-3, 2)$ **12** $(-2, 5)$ and $(3, -1)$
13 $(5, 0)$ and $(-7, 3)$ **14** $(t, 8)$ and $(t, 7)$
15 $(2, 3)$ and $(-\frac{1}{2}, 1)$ **16** $(t, u+1)$ and $(t+1, u)$
17 $(5, -t)$ and $(7, t)$ **18** $(2t-1, (t-1)^2)$ and $(t^2, 0)$

In Problems 19–22, use the distance formula to show that the triangle with the given vertices is a right triangle.

19 (0, 0), (−3, 0), and (−3, 4) **20** (−3, 1), (3, 1), and (3, 10)

21 (1, 1), (5, 1), and (5, 7) **22** (−2, −2), (0, 0), and (3, −3)

23 Find a point $(x, 1)$ in the first quadrant for which the triangle whose vertices are $A = (1,1)$, $B = (4,7)$, and $C = (x,1)$ is an isosceles triangle, with $|\overline{AB}| = |\overline{AC}|$.

24 Find all real numbers x for which the distance between (4, 2) and $(4, x)$ is 5 units.

25 Given points $P_1 = (-3, -2)$, $P_2 = (1, 2)$, and $P_3 = (3, 4)$, do the following:
(a) Locate the points.
(b) Find the lengths of segments $\overline{P_1P_2}$, $\overline{P_2P_3}$, and $\overline{P_1P_3}$.
(c) Three points are said to be *collinear* if they all lie on the same straight line. Are P_1, P_2, and P_3 collincar? Explain.

26 Suppose that $P_1 = (a, b)$, $P_2 = (c, d)$, and $P_3 = ((a + c)/2, (b + d)/2)$ are collinear points; then do the following:
(a) Find the lengths of $\overline{P_1P_3}$ and $\overline{P_2P_3}$ in terms of a, b, c, and d.
(b) How do the lengths of these two line segments compare?
(c) What can you conclude about the geometric position of P_3 with respect to P_1 and P_2?
(d) Give a specific example to illustrate this situation.

27 Use the distance formula to express the relationship between x and y if (x, y) is a point on a circle of radius 1 with center at P_0.
(a) $P_0 = (1, 0)$ (b) $P_0 = (-1, 0)$ (c) $P_0 = (0, 1)$
(d) $P_0 = (0, -1)$ (e) $P_0 = (2, 3)$

28 Use the fact that $\left(\frac{a+c}{2}, \frac{b+d}{2}\right)$ represents the midpoint of the line segment with end points (a, b) and (c, d) to answer the following:
(a) What is the midpoint of the line segment with end points (1, −3) and (5, 8)?
(b) A line segment has (−2, −4) as one end point and (1, −2) as the midpoint. What are the coordinates of the other end point?

2 Relations and Their Graphs

The word "relation" or "relationship" at times is used in mathematics in much the same way that it is commonly used. In everyday language, we speak of the relation between miles traveled by a car and the number of gallons of gas consumed, or the relation between the cost of a trip and the distance traveled, or the relation between demand and the price of a product. Similarly, we can speak of the relation between x and y expressed by the formula $y = 2x$. That is, a *relation* suggests a correspondence or an

Table 1

$x \to y$
$1 \to 2$
$2 \to 4$
$3 \to 6$
$4 \to 8$
$5 \to 10$
....
$x \to 2x$

association between the elements of two sets. For example, Table 1 displays a relation between the numbers in the x column and the numbers in the y column. Here the correspondence between the numbers in the x column and the numbers in the y column is given by the formula $y = 2x$, where x is a positive integer.

In the relation illustrated in Table 1, there are three main ingredients: a first set, a second set, and a correspondence between the members of the two sets.

In order to define a relation so that the corresponding members of the two sets are clearly identified, ordered-pair notation is sometimes used.

DEFINITION 1 RELATION

A *relation* is a set of ordered pairs. The *domain* of a relation is the set of all first members of the ordered pairs, and the *range* of a relation is the set of all second members of the ordered pairs.

For instance, the domain of the relation $R = \{(1, -1), (3, 2), (4, 7)\}$ is $\{1, 3, 4\}$, and the range is $\{-1, 2, 7\}$.

EXAMPLE List the members and identify the domain and range of the relation $B = \{(x, y) | y = -3x; x \in \{-2, -1, 0, 1, 2\}\}$.

SOLUTION If we replace x by -2, -1, 0, 1, and 2 in the equation $y = -3x$, we have $y = -3(-2) = 6$, $y = -3(-1) = 3$, $y = -3(0) = 0$, $y = -3(1) = -3$, and $y = -3(2) = -6$, respectively; so the relation is described by enumeration as $B = \{(-2, 6), (-1, 3), (0, 0), (1, -3), (2, -6)\}$. The domain of B is $\{-2, -1, 0, 1, 2\}$ and its range is $\{6, 3, 0, -3, -6\}$.

The *graph* of a relation is the set of all points in the cartesian plane that represent the ordered pairs in the relation. In other words, if (x, y) is a member of a relation, we can consider the real numbers x and y from two viewpoints (Figure 1). On the one hand, x is a member of the domain and y is the corresponding member of the range of the relation. On the other hand, x represents the abscissa and y the corresponding ordinate of a point on the graph of the relation. On the graph, then, the members of the domain are the abscissas, and the members of the range are the ordinates.

Figure 1

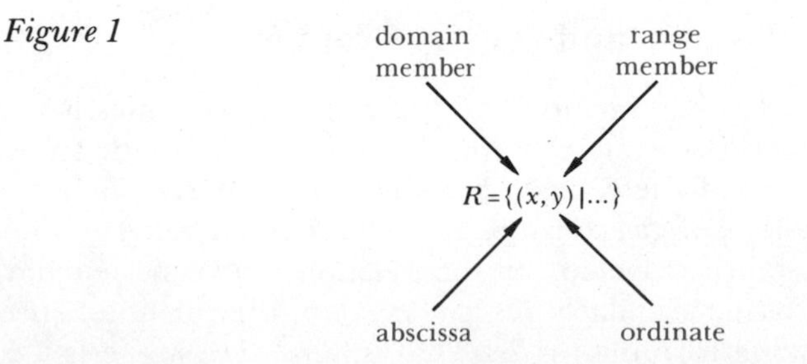

Any restriction on the domain is a restriction on "the horizontal position of the points of the graph," and any restriction on the range is a restriction on "the vertical position of the points of the graph."

Unless otherwise stated, we assume that the universal set is R, the set of all real numbers, when defining the domain and range of a relation.

EXAMPLES 1 Assume that the graph of a relation is given in Figure 2. Use the graph to identify the domain and the range of the relation in interval notation.

Figure 2

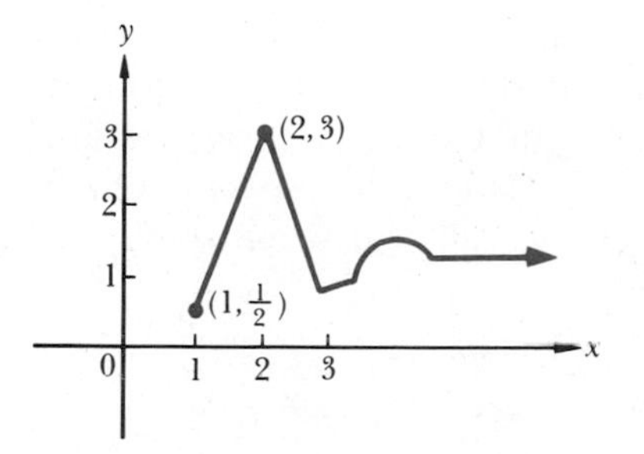

SOLUTION It can be seen from the graph in Figure 2 that each of the abscissas x satisfies $x \geq 1$ and each of the ordinates y satisfies $\frac{1}{2} \leq y \leq 3$; consequently, the domain is the set $[1, \infty)$ and the range is the set $[\frac{1}{2}, 3]$.

2 Graph the relation $R_1 = \{(x, y) \mid x > 1 \text{ and } y \leq -2\}$.

SOLUTION The graph of R_1 is the shaded region in Figure 3. Note that the solid line is part of the region, whereas the dashed line is not part of the region.

Figure 3

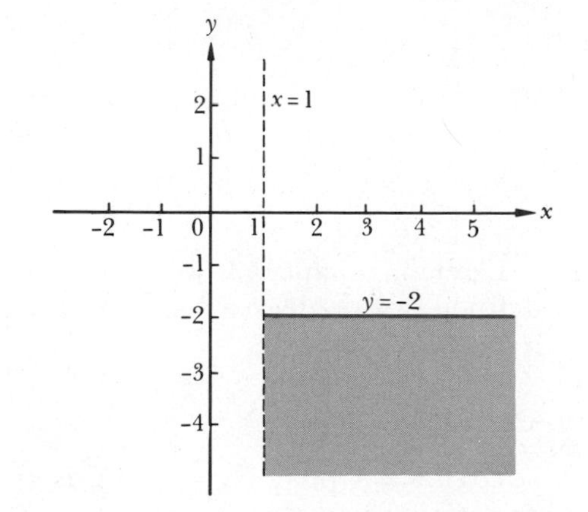

Quite often set notation is *not* used to describe a relation; instead, the relation is defined by an equation or an inequality. For example, assuming that x represents a member of the domain and y represents a member of the range, $x < y$ defines the relation $\{(x, y) | x < y\}$ and $y = 2x$ defines the relation $\{(x, y) | y = 2x\}$.

EXAMPLES Graph each relation. Assume that x represents a member of the domain and y represents a member of the range.

1 $y = 5x$

SOLUTION It is impossible here to list all the members of this relation. The best we can do is to plot some of the points to discover the "pattern" of the graph. Thus (0, 0), (1, 5), ($\frac{1}{5}$, 1), ($-\frac{1}{5}$, -1), and (-1, -5) are recorded in the table next to Figure 4. If we were to continue to plot members of this relation, the points would appear to lie on a straight line. In fact, the graph actually turns out to be a straight line (Figure 4).

Figure 4

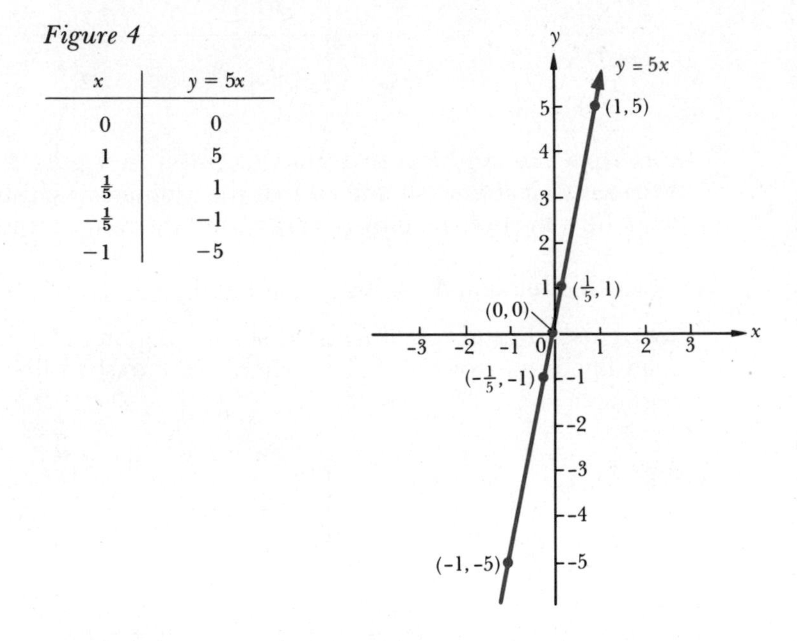

x	$y = 5x$
0	0
1	5
$\frac{1}{5}$	1
$-\frac{1}{5}$	-1
-1	-5

Later, in Chapter 3, we shall see that the graph of any relation that is defined by first-degree equations such as $y = 5x$, $y = 1 - 8x$, and $y = 3x + 7$ is a straight line.

2 $y^2 = 25x^2$

SOLUTION A point (x, y) is on this graph if and only if $y^2 - 25x^2 = 0$. This equation can be written as $(y - 5x)(y + 5x) = 0$. Thus, $y - 5x = 0$ or $y + 5x = 0$ so that $y = 5x$ or $y = -5x$. The set of points for which $y = 5x$ is the line L_1, and the set of points for which $y = -5x$ is the line L_2, so the graph of $y^2 = 25x^2$ is the union of L_1 and L_2 (Figure 5).

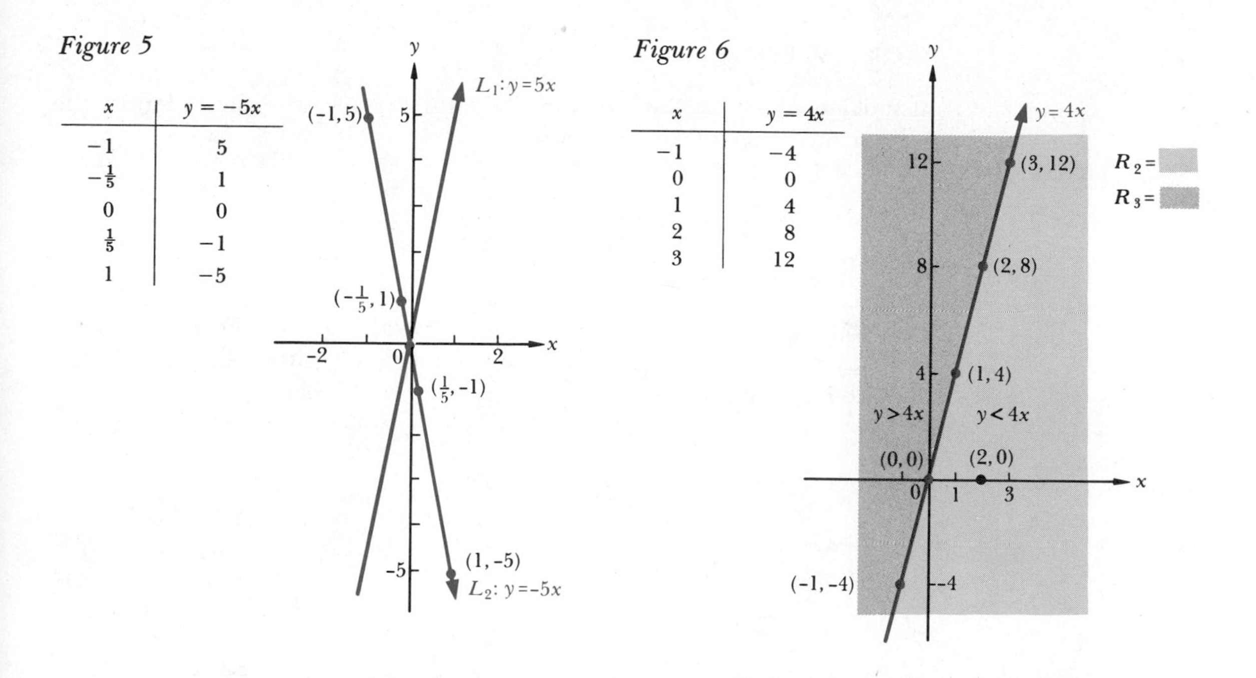

Figure 5

x	$y = -5x$
-1	5
$-\frac{1}{5}$	1
0	0
$\frac{1}{5}$	-1
1	-5

Figure 6

x	$y = 4x$
-1	-4
0	0
1	4
2	8
3	12

3 (a) $y = 4x$, (b) $y < 4x$, and (c) $y > 4x$, on the same coordinate system

SOLUTION

(a) We list a few points and locate them to determine the graph of $y = 4x$ (Figure 6). The line defined by $y = 4x$ divides the plane into two disjoint regions so that all points in one of the regions satisfy $y > 4x$ and all points in the other region satisfy $y < 4x$. To determine which region contains which points, it is enough to test *one* point in either region. If we select the point (2, 0), then $x = 2$ and $y = 0$ satisfy $y < 4x$, since $0 < 8$. Thus we have the following results.

(b) The graph of $y < 4x$ is the shaded region R_2 in Figure 6.

(c) The graph of $y > 4x$ is the shaded region R_3 in Figure 6.

Figure 7

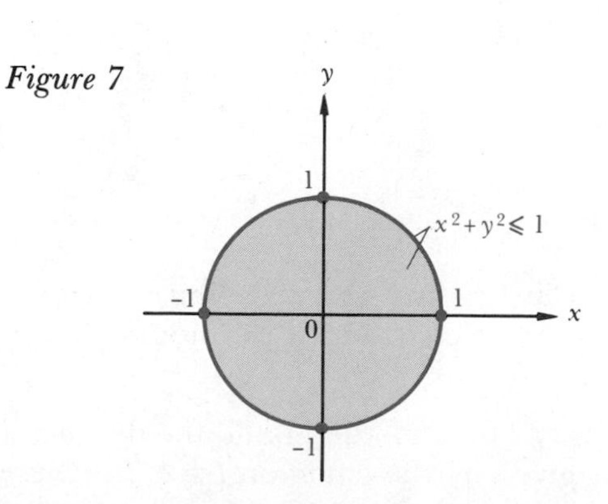

4 $x^2 + y^2 \leqslant 1$

SOLUTION Earlier we determined that the graph of $x^2 + y^2 = 1$ is the unit circle. By selecting points *outside* the circle, we find that the coordinates satisfy $x^2 + y^2 > 1$. Thus the graph of the relation $x^2 + y^2 \leqslant 1$ is the shaded region that includes the interior as well as the boundary of the circle in Figure 7.

PROBLEM SET 2

In Problems 1–4, list the members, when necessary, and then identify the domain and range of each relation.

1 $R_1 = \{(1, 2), (3, 4), (2, -1)\}$

2 $R_2 = \{(x, y) | y = -1; x \in \{1, 2, 3\}\}$

3 $R_3 = \{(x, y) | y = 4x; x \in \{-3, -1, 2\}\}$

4 $R_4 = \{(x, y) | y = 4x^2; x \in \{-3, -1, 2\}\}$

In Problems 5–6, assume that a relation has the graph shown in the accompanying figure. Indicate the domain and range of the relation.

5 See Figure 8. **6** See Figure 9.

Figure 8

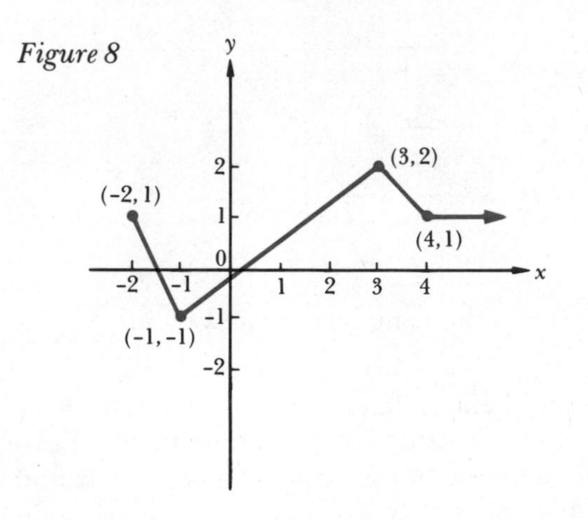

Figure 9

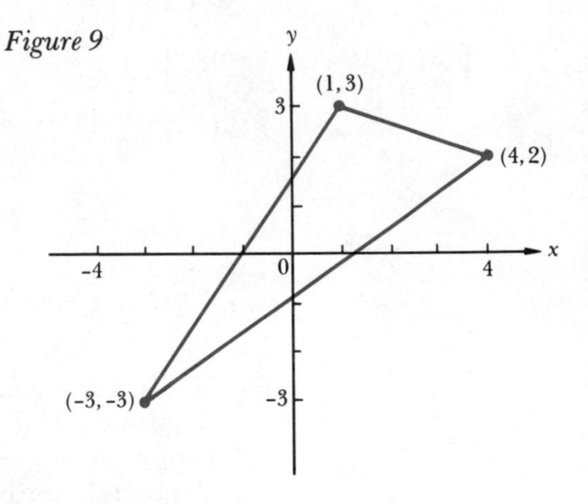

In Problems 7–24, graph each relation. Assume that x represents a member of the domain and y represents a member of the range.

7 $\{(x, y) | x \geq 1 \text{ and } y \leq -1\}$

8 $\{(x, y) | |x| \leq 2 \text{ and } |y| \leq 1\}$

9 $\{(x, y) | |x| \leq 3 \text{ and } |y| \geq 2\}$

10 $\{(x, y) | |x| \leq 1 \text{ or } |y| > 2\}$

11 $y = 3x$

12 $y = 2x + \frac{5}{3}$

13 $y^2 = 9x^2$

14 $y = -3x$

15 $y \geq 3x$

16 $y < 2x + 1$

17 $y < 3x$

18 $y \geq 3x - 2$

19 $y = -2$

20 $x = 3$

21 $x^2 + y^2 \geq 1$

22 $y = |x|$

23 $y \geq -3x + 1$

24 $|x| + |y| = 1$

25 A truck driver leaves New York City heading west traveling at 55 miles per hour. Express the distance d traveled in terms of t hours. Sketch the graph of the relation.

26 The relation between the intensity I of a floodlight and the distance d from the (source) floodlight is given by the equation $I = k/d^2$, where k is a constant. Sketch the graph of the relation for $k = 2$.

3 Functions

The idea of function is encountered quite often in everyday living. For example, if it is known that an automobile averages 20 miles per gallon, then the number of gallons of gasoline required for any automobile trip is a function of the number of miles traveled; the number of books to be ordered for a course is a function of the number of students enrolled in the course; the tuition charge for a student is a function of the number of credit hours taken. Intuitively, the word *function* suggests some kind of *correspondence.* In each of the examples above, there is an established correspondence between numbers—the number of gallons corresponds to the miles traveled; the number of books corresponds to the number of students; the tuition charge corresponds to the number of credit hours.

In mathematics, the general idea of a function is simple. Suppose that one variable quantity, say, y, depends in a definite way on another variable quantity, say, x. Then for each particular value of x, there is *one* corresponding value of y. Such a correspondence defines a *function,* and we say that (the variable) y is a function of (the variable) x. For example, if x is used to denote the radius of a circle and y is used to denote the area of this circle, then y depends on x in a definite way, namely, $y = \pi x^2$. Thus, we say that the area y of a circle is a function of its radius x. In a sense, the value of y depends on the value assigned to x. For this reason, we sometimes refer to x as the *independent variable* and y as the *dependent variable.* If $x = 5$, $y = 25\pi$; if $x = 7$, $y = 49\pi$; if $x = 10$, $y = 100\pi$. More formally, we have the following definition.

DEFINITION 1 FUNCTION AS A CORRESPONDENCE

A *function* is a correspondence that assigns to each member in a certain set, called the *domain* of the function, one and only one member in a second set, called the *range* of the function. The *independent variable* of the function can take on any value in the domain of the function. The set of all possible corresponding values that the *dependent variable* assumes is the range of the function.

In the example above, where $y = \pi x^2$, the domain of the function is the set of all positive real numbers, since the radius of any circle is a positive number. The corresponding values for y are positive numbers; for this reason, the range of the function is the set of positive real numbers.

Often a function is defined by a formula or an equation. If the domain of the function is not specified, then it is understood to be the subset of real numbers for which the formula or the equation is meaningful. For instance, the domain of the function defined by the equation $y = 1/x$, where x represents the independent variable, does *not* include zero, since division by zero is not defined. The domain of the function given by the equation $y = -5x + 3$ is the set of all real numbers, since $-5x + 3$ is defined for any real number x.

EXAMPLE Find the domain of the function defined by $y = \sqrt{4 - x}$, where x represents the independent variable.

SOLUTION The expression $\sqrt{4 - x}$ is a real number when $4 - x \geq 0$, so that the domain of the function is the set of all real numbers for which $4 - x \geq 0$, that is, $\{x \mid x \leq 4\}$.

Although the domain of a function defined by an equation may often be determined by inspection, the range is more difficult to find and can be determined more readily from the graph of the function. Later in the book, as more functions are studied, various techniques for determining the domain and range of a function will be explored.

3.1 Function Notation

Suppose that a function is defined by the equation $3r + 5t = 3$. We need to identify the independent variable and the dependent variable. Function notation gives us a way of providing such information. Thus, if r represents the independent variable for the function defined by $3r + 5t = 3$, we say that t is a function of r and describe it by the function notation $t = f(r)$. In other words, $t = f(r)$, which is read "t equals f of r," means that t is a function of r, where r represents the independent variable and t represents the dependent variable. Solving for t in terms of r results in $t = -\frac{3}{5}r + \frac{3}{5}$, so $t = f(r) = -\frac{3}{5}r + \frac{3}{5}$.

On the other hand, if $r = f(t)$ for $3r + 5t = 3$, we know that t represents the independent variable and r the dependent variable. Solving for r in terms of t results in $r = -\frac{5}{3}t + 1$, so $r = f(t) = -\frac{5}{3}t + 1$.

EXAMPLE Assume that $y = f(x)$ is defined by the equation $13x - 5y = 65$. Find the expression for $f(x)$.

SOLUTION Solving the equation $13x - 5y = 65$ for y in terms of x results in $y = \frac{13}{5}x - 13$, so $y = f(x) = \frac{13}{5}x - 13$.

It is important to know that letters other than f are often used to denote functions. For example, g and h, as well as F, G, and H, are favorites for this purpose. If f is a function and x represents a member of the domain, then $f(x)$ represents the corresponding member of the range. Note that $f(x)$ is *not* the function f. In the interest of brevity, however, we frequently use the phrase "the function $y = f(x)$." There is no great harm in this practice as long as it is understood that the phrase "the function $y = f(x)$" actually means "f is a function defined by the equation $y = f(x)$." Although we shall avoid this practice when absolute precision is desired, we shall indulge in it whenever it seems convenient. Thus $g(t) = t^2$, $h(x) = x + 7$, $V(r) = \frac{4}{3}\pi r^3$, and $F(s) = \sqrt{s}$ all denote functions with independent variables t, x, r, and s, respectively. If A is the function that gives the area of a circle in terms of its radius r, then we write $A(r) = \pi r^2$.

When function notation is used, sometimes it is helpful to think of the variable that represents the members of the domain as a "blank." For example, $g(t) = t^2$ can be thought of as $g(\ ) = (\ )^2$; hence if any expression (representing a real number) is used to represent a member of the domain, it is easy to see where this same expression is to be substituted in the equation describing the function.

Using $g(t) = t^2$ again, $g(x + h)$ can be determined by first writing the function as $g(\ ) = (\ )^2$ so that, after substituting the expression $x + h$ into the blank, we get

$$g(x + h) = (x + h)^2 = x^2 + 2xh + h^2$$

Similarly, we know that $A(r) = \pi r^2$ denotes a function that gives the area A of a circle in terms of its radius r. Suppose that the diameter of the circle is represented by d. Then $r = d/2$. Substituting $d/2$ into the function A yields

$$A\left(\frac{d}{2}\right) = \pi\left(\frac{d}{2}\right)^2 = \frac{\pi d^2}{4}$$

EXAMPLES 1 Assume that $f(x) = 7x + 2$. Determine the following values.

(a) $f(1)$ (b) $f(2)$ (c) $f(3)$ (d) $f(4)$ (e) $\sqrt{f(2)}$ (f) $[f(4)]^2$

SOLUTION Since $f(x) = 7x + 2$, we have

(a) $f(1) = 7(1) + 2 = 7 + 2 = 9$
(b) $f(2) = 7(2) + 2 = 14 + 2 = 16$
(c) $f(3) = 7(3) + 2 = 21 + 2 = 23$
(d) $f(4) = 7(4) + 2 = 28 + 2 = 30$
(e) $\sqrt{f(2)} = \sqrt{7(2) + 2} = \sqrt{16} = 4$
(f) $[f(4)]^2 = [7(4) + 2]^2 = (30)^2 = 900$

2 Given that $g(x) = x^2 - 1$, determine each of the following values.

(a) $g(3) + g(5)$ (b) $g(x + 1)$ (c) $g(2a + 4)$
(d) $2g(b - 2)$ (e) $g(-a)$ (f) $[g(5c)]^2$

SOLUTION

(a) $g(3) = 3^2 - 1 = 8$ and $g(5) = 5^2 - 1 = 24$, so $g(3) + g(5) = 8 + 24 = 32$
(b) $g(x + 1) = (x + 1)^2 - 1 = x^2 + 2x$
(c) $g(2a + 4) = (2a + 4)^2 - 1 = 4a^2 + 16a + 15$
(d) $2g(b - 2) = 2[(b - 2)^2 - 1] = 2(b^2 - 4b + 3) = 2b^2 - 8b + 6$
(e) $g(-a) = (-a)^2 - 1 = a^2 - 1$
(f) $[g(5c)]^2 = [(5c)^2 - 1]^2 = (25c^2 - 1)^2 = 625c^4 - 50c^2 + 1$

3 The *difference quotient* of a function $y = f(x)$ is defined as

$$\frac{f(x + h) - f(x)}{h}, \quad \text{for } h \neq 0$$

Compute the difference quotient for each of the following functions.

(a) $f(x) = 2x + 1$ (b) $f(x) = x^2$

SOLUTION

(a) $f(x + h) = 2(x + h) + 1$, so

$$\frac{f(x+h) - f(x)}{h} = \frac{[2(x+h) + 1] - (2x + 1)}{h}$$
$$= \frac{2x + 2h + 1 - 2x - 1}{h} = \frac{2h}{h} = 2$$

(b) $f(x + h) = (x + h)^2 = x^2 + 2xh + h^2$ implies that

$$\frac{f(x+h) - f(x)}{h} = \frac{x^2 + 2xh + h^2 - x^2}{h}$$
$$= \frac{h(2x + h)}{h} = 2x + h$$

Figure 1

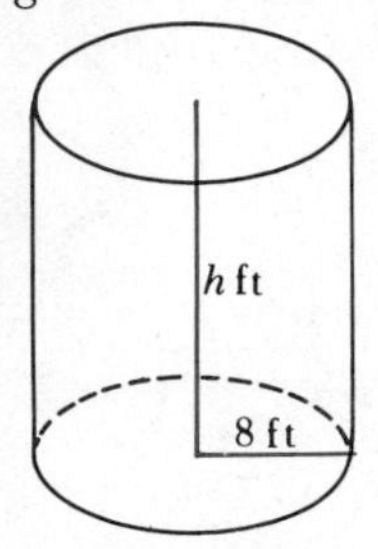

4 Water is being pumped into a cylindrical tank of radius $r = 8$ feet and height h feet (Figure 1). Express the volume V of the water in the tank as a function of the height h. Find the volume when $h = 10$ feet.

SOLUTION Let V cubic feet be the volume of the tank whose radius is 8 feet and height is h feet. The volume V of the cylinder is given by the formula $V = \pi r^2 h = \pi 8^2 h = 64\pi h$, so that if $h = 10$, then $V = 64\pi(10) = 640\pi$. Therefore, the volume of the tank is 640π cubic feet.

3.2 Functions as Relations

Since a function establishes a pairing of each member of the domain with a corresponding member of the range, it seems natural that we can think of a function in terms of ordered pairs. If x is a member of the domain of a function and y is the corresponding member of the range, we can represent the correspondence between x and y as the ordered pair (x, y). In fact, we can say that the function is the set of all such ordered pairs. For example, the function f defined by $f(x) = 2x - 1$ could be represented as $f = \{(x, y) \mid y = 2x - 1\}$.

Recalling that a relation is a set of ordered pairs, we have the following alternative definition of a function.

DEFINITION 2 FUNCTION AS A RELATION

A *function* is a relation in which no two different ordered pairs have the same first member. The set of all first members (of the ordered pairs) is called the *domain* of the function. The set of all second members (of the ordered pairs) is called the *range* of the function. For each ordered pair (x, y) of a function we say that y corresponds to x under the function.

From this definition of a function we conclude that all functions are relations; however, *not all relations are functions.* For example, $\{(1, 1), (2, 2), (3, 7), (3, 5)\}$ is a relation with domain $\{1, 2, 3\}$ and range $\{1, 2, 7, 5\}$, but it is *not* a function because $(3, 7)$ and $(3, 5)$ have the same first members. By contrast, $\{(1, 2), (3, 4), (4, 4)\}$ is a relation that is a function with domain

$\{1, 3, 4\}$ and range $\{2, 4\}$. Note that in the latter example, two pairs have the same *second* member; this does not violate the definition of a function.

Hence, if at least two different ordered pairs of a relation have the same first member, the relation is not a function. Geometrically, this means that if the graph of a relation has more than one point with the same abscissa, the relation is not a function. Thus, if a vertical line intersects the graph of a relation at more than one point, the relation is *not* a function.

EXAMPLE Examine the graphs of each of the relations given in Figures 2a and 2b below to decide whether or not the relation is a function.

SOLUTION Since no vertical line intersects the graph of the relation in Figure 2a at more than one point, we conclude that the relation is a function. By comparison, the relation that is graphed in Figure 2b is *not* a function because the vertical line drawn through the point (2, 0) intersects the graph at two points.

Assume that $y = f(x)$ is defined by $y = 2x - 1$ with domain $\{0, 1, 2\}$. Since $f(0) = -1$, $f(1) = 1$, and $f(2) = 3$, the range of f is $\{-1, 1, 3\}$. Using ordered-pair notation, f could also be described as $f = \{(0, -1), (1, 1), (2, 3)\}$. Both function notation and ordered-pair notation indicate precisely which member of the domain is associated with each member of the range. Another way of denoting this association is to consider the function as a *mapping* of members of the domain onto corresponding members in the range as described in Table 1 below:

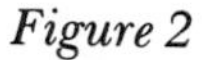

Figure 2

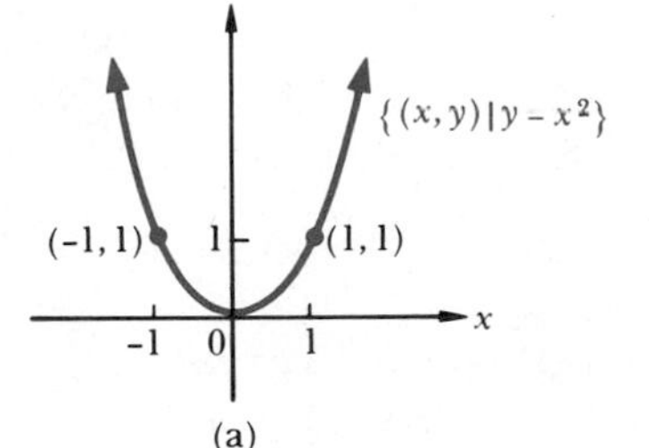

(a)

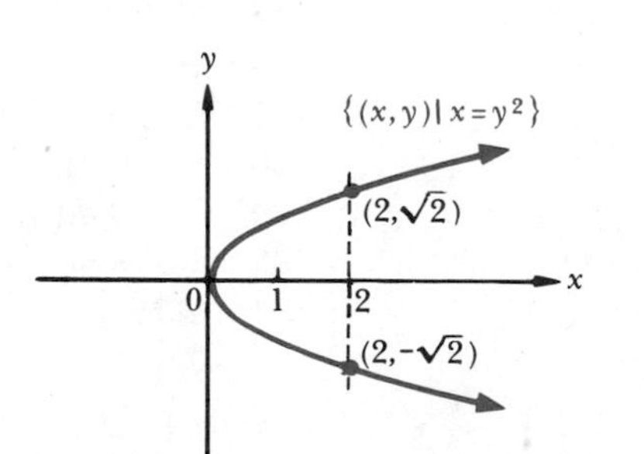

(b)

Table 1

Function Notation	*Mapping Representation*
$f(0) = -1$	0 is mapped onto -1 or -1 is the image of 0 under f
$f(1) = 1$	1 is mapped onto 1 or 1 is the image of 1 under f
$f(2) = 3$	2 is mapped onto 3 or 3 is the image of 2 under f

This mapping representation is symbolized in the following ways:

$$\begin{array}{lclcl} f{:}0 \to -1 & \text{or} & 0 \xrightarrow{f} -1 & \text{or} & 0 \to f(0) = -1 \\ f{:}1 \to 1 & & 1 \xrightarrow{f} 1 & & 1 \to f(1) = 1 \\ f{:}2 \to 3 & & 2 \xrightarrow{f} 3 & & 2 \to f(2) = 3 \end{array}$$

In general, if $y = f(x)$, we say that *f maps x onto y* or *x is mapped onto y by f* or *y is the image of x under f.*

$$f{:}\, x \to y \quad \text{or} \quad x \xrightarrow{f} y \quad \text{or} \quad x \to f(x)$$

EXAMPLE Assume that g is defined by $g(x) = x^3 - x$. Find the values $g(-1)$, $g(1)$, $g(2)$, and $g(3)$. Represent these function values using function notation, mapping notation, and ordered-pair notation.

SOLUTION Since

$$\begin{aligned} g(-1) &= (-1)^3 - (-1) = -1 + 1 = 0 \\ g(1) &= 1^3 - 1 = 0 \\ g(2) &= 2^3 - 2 = 8 - 2 = 6 \\ g(3) &= 3^3 - 3 = 27 - 3 = 24 \end{aligned}$$

we have these representations:

Function Notation $y = g(x)$	*Mapping* $g\colon x \to y$	*Ordered Pairs* $(x, y) \in g$
$g(-1) = 0$	$g\colon -1 \to 0$	$(-1, 0)$
$g(1) = 0$	$g\colon 1 \to 0$	$(1, 0)$
$g(2) = 6$	$g\colon 2 \to 6$	$(2, 6)$
$g(3) = 24$	$g\colon 3 \to 24$	$(3, 24)$

PROBLEM SET 3

In Problems 1–8, assume that $y = f(x)$ is defined by the given equation. Find the expression for $f(x)$.

1 $x + y = 7$
2 $3y - 5 = 0$
3 $y - 3x^2 + 2 = 0$
4 $xy = 5$
5 $xy + 3y = 4$
6 $2y - x^2y + 1 = 0$
7 $x^2 + y^2 = 9$, for $y \geqslant 0$
8 $x^2 + y^2 = 9$, for $y \leqslant 0$

In Problems 9–50, let $f(x) = 2x + 5$, $g(x) = x^2 - 2x + 8$, $h(x) = \dfrac{x+3}{x-2}$, and $k(x) = 1 + \sqrt{x}$. Find each of the following values.

9 $f(0)$
10 $k(0)$
11 $h(0)$
12 $f(3)$
13 $g(3)$
14 $h(3)$
15 $\sqrt{f(-2)}$
16 $\sqrt{g(-2)}$
17 $\sqrt{h(6)}$
18 $k(81)$
19 $k(1)$
20 $k(4)$
21 $f(a) + 2$
22 $k(a) + 2$
23 $h(a) + 2$
24 $f(a + 2)$
25 $k(a + 2)$
26 $h(a + 2)$
27 $f(a) + f(2)$
28 $k(a) + k(2)$
29 $h(a) + h(3)$
30 $f(a + b) - f(a)$
31 $g(a + b) - g(a)$
32 $h(a + b) - h(a)$
33 $f(7x)$
34 $g(7x)$
35 $h(7x)$
36 $7f(x)$
37 $7g(x)$
38 $7h(x)$
39 $f\left(\dfrac{1}{x+a}\right)$
40 $g\left(\dfrac{1}{x+a}\right)$
41 $h\left(\dfrac{1}{x+a}\right)$
42 $f(-x)$
43 $g(-x)$
44 $h(-x)$
45 $f(x^2)$
46 $g(x^2)$
47 $h(x^2)$
48 $[f(x)]^2$
49 $[g(x)]^2$
50 $[h(x)]^2$

In Problems 51–58, find the domain of each function.

51 $f(x) = 5x^3 + 1$
52 $g(x) = \dfrac{1}{x-1}$
53 $g(x) = \sqrt{2x - 1}$

54 $h(x) = \sqrt{x} + 5$ **55** $f(x) = \dfrac{5}{x-2}$ **56** $f(x) = \dfrac{4}{x^3}$

57 $g(x) = \dfrac{1}{\sqrt{3x+2}}$ **58** $h(x) = \dfrac{2}{\sqrt{x^2+1}}$

In Problems 59–63, form the difference quotient (see Example 3, page 59), and then simplify the resulting expression.

59 $f(x) = 2$ **60** $f(x) = 3x + 5$ **61** $f(x) = -5x + 1$

62 $f(x) = x^2 + 3$ **63** $f(x) = 1/x$

64 Let $f(x) = x + 2$, $g(x) = 2x - 1$, and $h(x) = x - 1$. Determine which of the following equations hold for all values of x.

(a) $f\left(\dfrac{1}{x}\right) = \dfrac{1}{f(x)}$ (b) $g(x) \cdot \left[\dfrac{1}{g(x)}\right] = 1$

(c) $h(x-1) = h(x) - h(1)$ (d) $f(x+2) = f(x) + f(2)$

(e) $g(2x) = 2g(x)$

65 Let $f(x) = 3x^2 - x$.

(a) Determine each of the values $f(-2)$, $f(-1)$, $f(0)$, $f(1)$, and $f(2)$.

(b) Represent the function values using function notation, mapping notation, and ordered-pair notation.

66 Let $g(t) = 2t - 4$, for any real number t. Find all the values of t such that $t \xrightarrow{g} 12$.

67 Which of the graphs in Figure 3 represent functions?

Figure 3

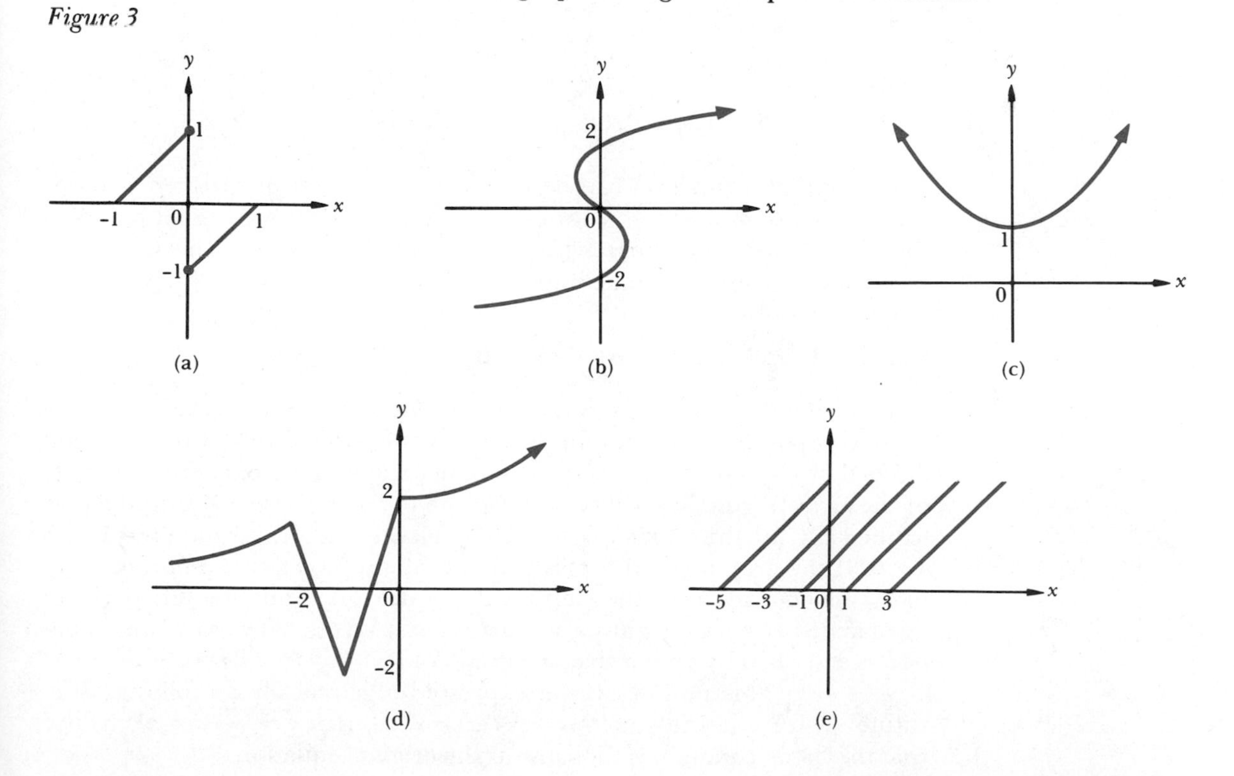

68 A closed box with a square base of x centimeters has a volume of 400 cubic centimeters. Express the total surface area A of the exterior of the box as a function of x.

69 Write an equation that expresses the radius r of a circle as a function of its circumference C.

70 Find an expression for the length of the diagonal D of a rectangular box as a function of the width s if the length is twice the width and the height is three times the width.

71 Equal squares are cut from the four corners of a rectangular piece of cardboard that is 10 inches by 14 inches. An open box is then formed by folding up the flaps. Express the volume V of the box as a function of x, where x is the length of the side of the squares removed.

72 A manufacturer's sales representative receives a weekly salary of \$125 plus a commission of 10 percent of the sales made that week. Express the weekly salary S of the sales representative as a function of the amount of the weekly sales x, where x is in dollars.

73 The function C for converting temperature from Fahrenheit degrees F to Celsius (centigrade) degrees is given by $C(F) = \frac{5}{9}(F - 32)$. Convert each of the following from Fahrenheit to Celsius.
(a) 32°F (b) 86°F (c) 0°F (d) −13°F

4 Graphs and Properties of Functions

Recalling that a function can be represented as a set of ordered pairs of numbers, we consider the *graph* of a function f to be the set of all points in the plane whose coordinates (x, y) satisfy $y = f(x)$. We use the graphs of functions to investigate some important function properties.

4.1 Even and Odd Functions: Symmetry

In sketching the graph of a function, it is often helpful to make use of any symmetry possessed by the graph. For instance, the graph of the function defined by $f(x) = x^4$ can be divided symmetrically into a part to the "right" of the y axis (Figure 1a), which is the graph of $f(x) = x^4$ for $x \geq 0$, and a part to the "left" of the y axis (Figure 1b), which is the graph of $f(x) = x^4$ for $x < 0$. If the graph of $f(x) = x^4$ is "folded" along the y axis, then the "right part" will coincide with the "left part" (Figure 1c). Thus the left part can be obtained from the right part by *reflecting* it across the y axis. This situation is described by saying that the graph of $f(x) = x^4$ is *symmetric with respect to the y axis.* This notion can be expressed algebraically as follows: Substitute $-x$ for x in the equation $y = f(x) = x^4$, so that $y = (-x)^4 = x^4$. Notice that the "new" equation is the same as the original equation.

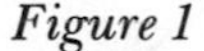
Figure 1

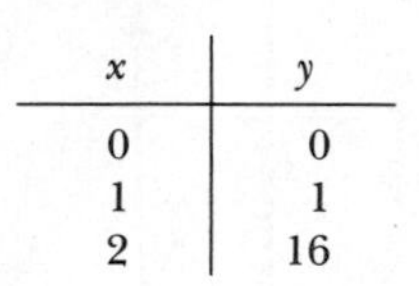

x	y
0	0
1	1
2	16

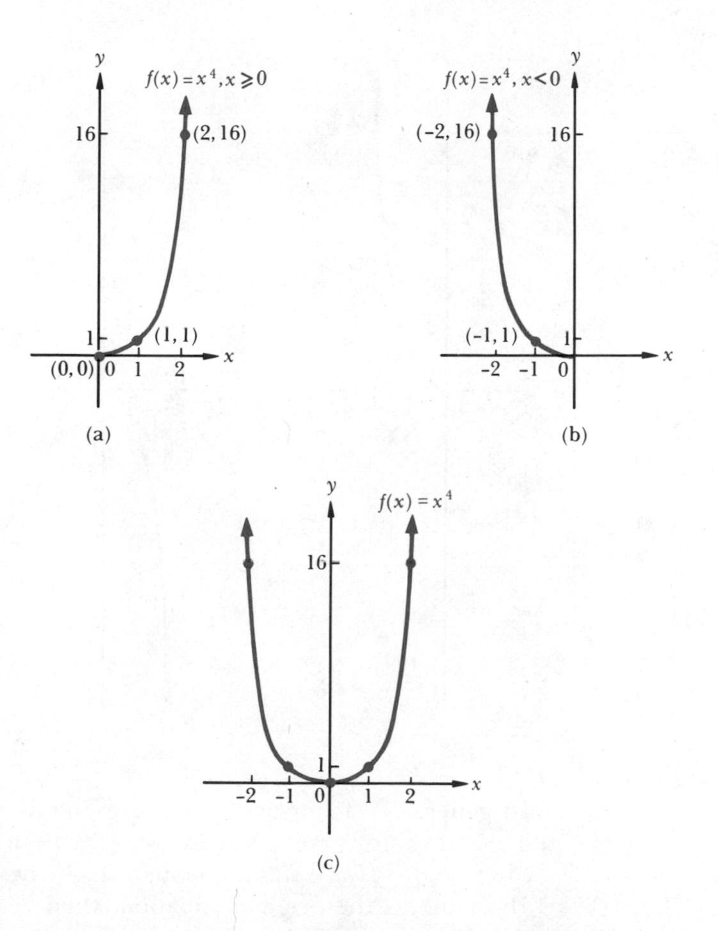

In general, if a function f is represented by an equation, we can test for symmetry of the graph of f with respect to the y axis by substituting $-x$ for x. If the new equation, when simplified, is the same as the original equation, then both (x, y) and $(-x, y)$ are on the graph of f; that is, $f(-x) = f(x)$, and the graph of the function f is symmetric with respect to the y axis. A function f possessing the property that $f(-x) = f(x)$ for all x in the domain of f is called an *even function.*

For example, the function f defined by $f(x) = x^4$ (Figure 1c) is an even function, since $f(-x) = (-x)^4 = x^4 = f(x)$. Its graph is symmetric with respect to the y axis.

Now consider the graph of $g(x) = x^3$. The graph of g can be divided symmetrically into a part in quadrant I (Figure 2a), which is the graph of $g(x) = x^3$ for $x \geqslant 0$, and another part in quadrant III (Figure 2b), which is the graph of $g(x) = x^3$ for $x < 0$. The part in quadrant III can be obtained from the part in quadrant I by reflecting it pointwise across the origin (Figure 2c). This situation is described by saying the graph of $g(x) = x^3$ is *symmetric with respect to the origin.* To express the notion algebraically, we substitute $-x$ for x and $-y$ for y simultaneously into the equation $y = g(x) = x^3$ to get $-y = (-x)^3 = -x^3$. The "new" equation is equivalent to the given equation.

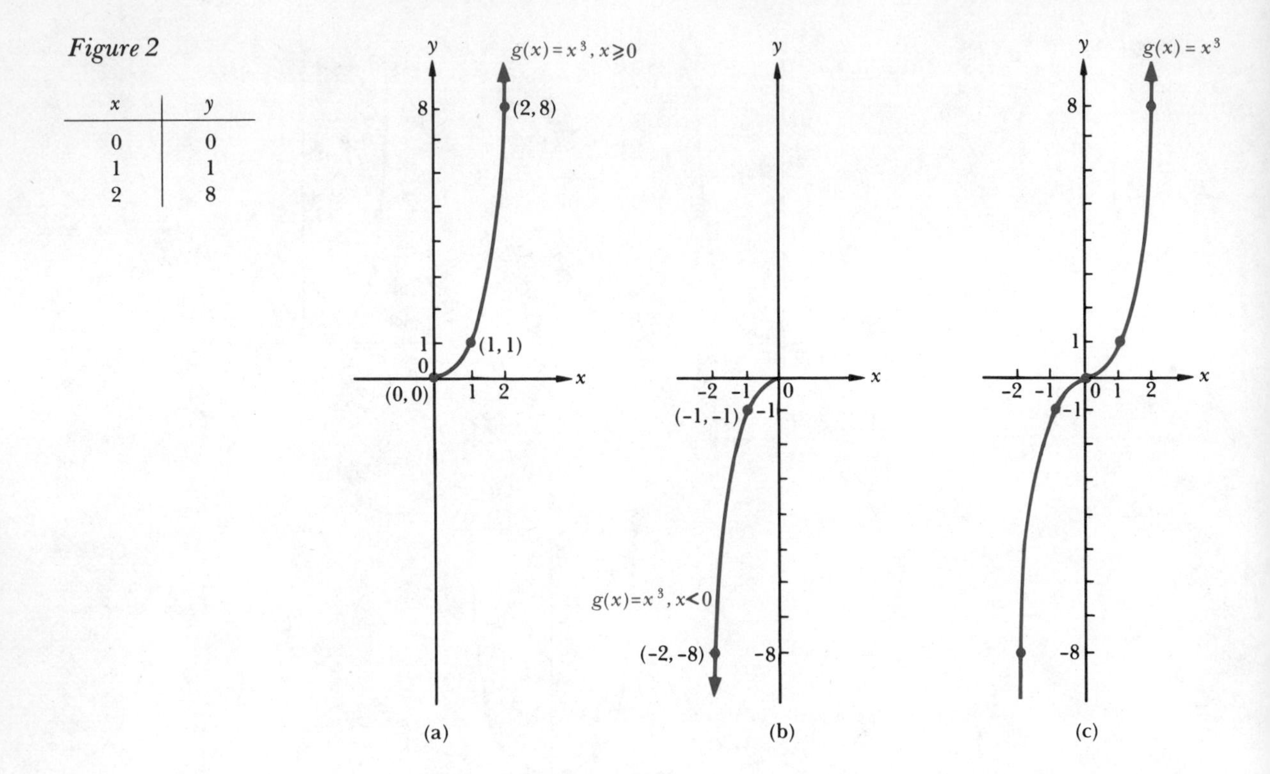

Figure 2

x	y
0	0
1	1
2	8

In general, if a function g is represented by an equation, then we can test the graph of g for symmetry with respect to the origin by substituting $-x$ for x and $-y$ for y, simultaneously. If the new equation, when simplified, is the same as the original equation, then $(-x, -y)$ is on the graph of g whenever (x, y) is on the graph of g; that is, $g(-x) = -g(x)$, and the graph is symmetric with respect to the origin. A function g possessing the property that $g(-x) = -g(x)$ for all x in the domain of g is called an *odd function.*

For instance, the function g defined by $g(x) = x^3$ (Figure 2) is an odd function, since $g(-x) = (-x)^3 = -x^3 = -g(x)$. The graph is symmetric with respect to the origin.

EXAMPLES Test each of the given functions for symmetry with respect to the y axis and the origin by determining if the function is even or odd. Use the symmetry to graph the function.

1 $f(x) = 4x^2 + 1$

SOLUTION Substituting $-x$ for x in $f(x) = 4x^2 + 1$ results in the equation $f(-x) = 4(-x)^2 + 1 = 4x^2 + 1 = f(x)$. Thus f is an even function and the graph of f is symmetric with respect to the y axis. Because of this symmetry, we can obtain the graph of f as follows:

First, graph f for $x \geq 0$ by plotting a few points (Figure 3a). Next, use the symmetry to obtain the remainder of the graph by reflecting the graph in Figure 3a across the y axis to complete the graph of f (Figure 3b).

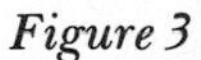

Figure 3

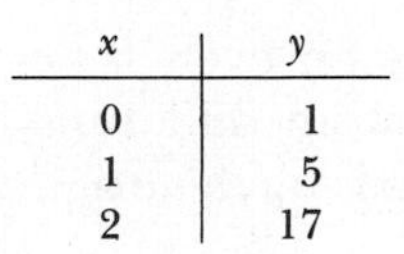

x	y
0	1
1	5
2	17

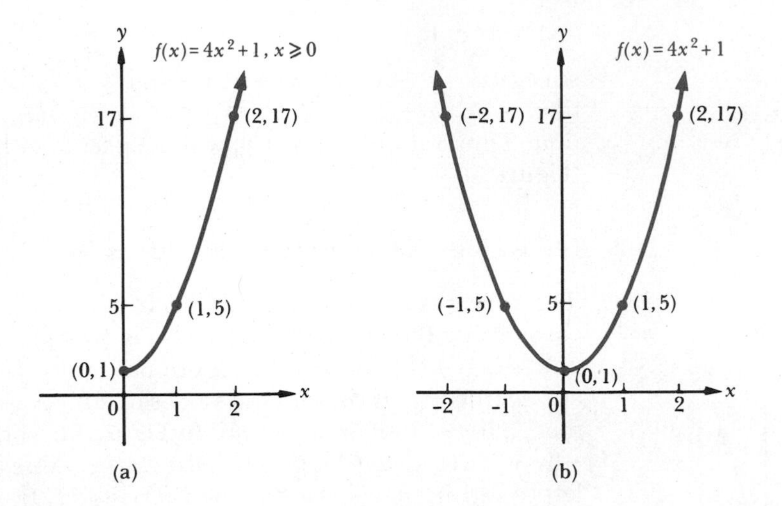

2 $g(x) = 2x^3$

SOLUTION Since $g(-x) = 2(-x)^3 = -2x^3 = -g(x)$, g is an odd function and the graph of g is symmetric with respect to the origin. This symmetry can be used to simplify the graphing of g as follows:

First, graph g for $x \geqslant 0$ by plotting a few points (Figure 4a). The graph of g is then obtained by reflecting the graph in Figure 4a across the origin, which results in the complete graph of g (Figure 4b).

Figure 4

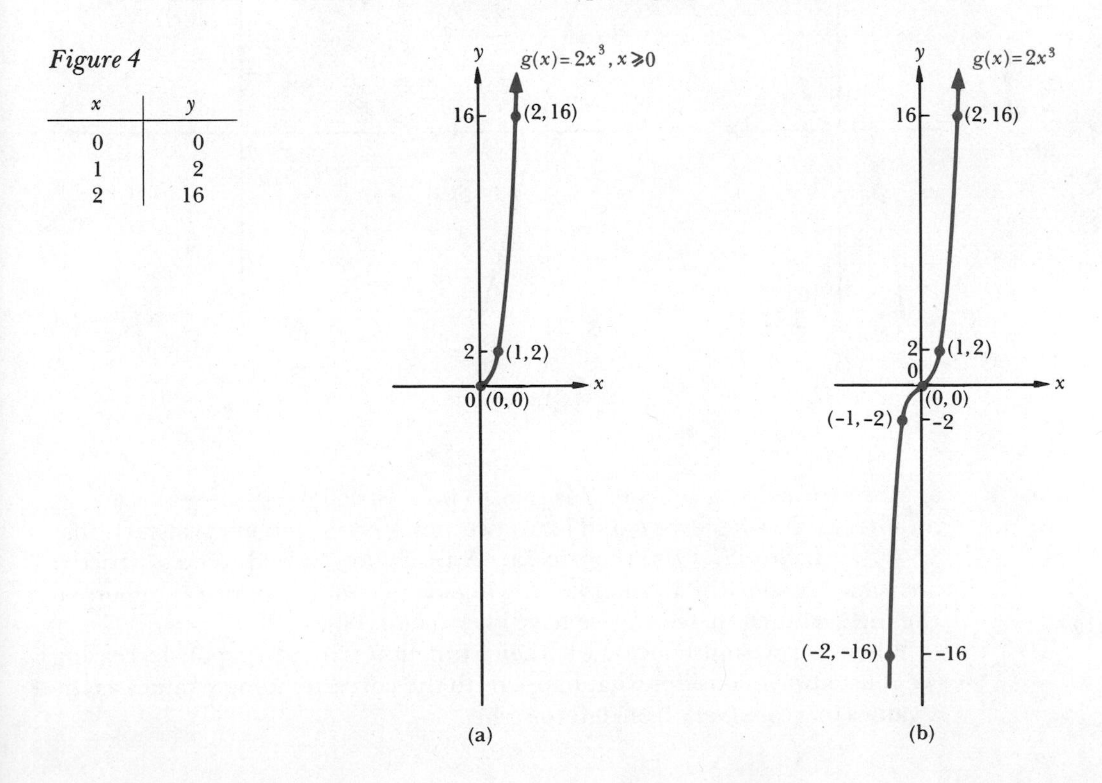

x	y
0	0
1	2
2	16

3 $h(x) = 3x + 1$

SOLUTION $h(-x) = 3(-x) + 1 = -3x + 1 \neq h(x)$, so h is *not* an even function. Also, $h(-x) = -3x + 1 \neq -3x - 1 = -h(x)$, so h is *not* an odd function. Thus h is not symmetric with respect to either the y axis or the origin (Figure 5).

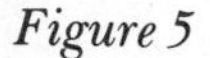

Figure 5

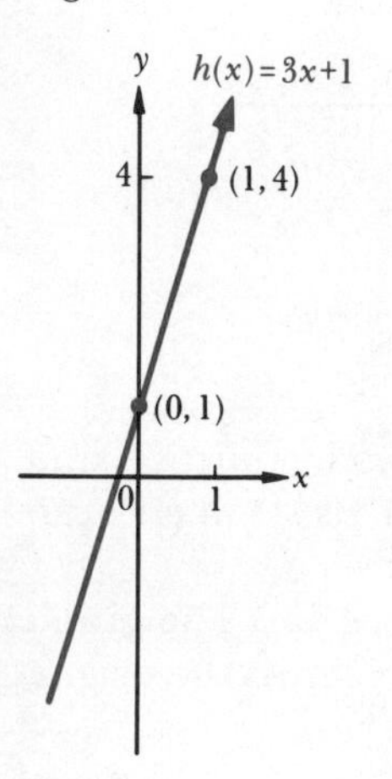

4.2 Increasing and Decreasing Functions

The concepts of increasing and decreasing functions can be developed by considering the graphs of $f(x) = x^3$ (Figure 6a), $g(x) = -3x + 1$ (Figure 6b), and $h(x) = 4$ (Figure 6c). In Figure 6a, the function values of $f(x) = x^3$ increase (the graph rises) as the values of x increase (vary from left to right); that is, if $a < b$, then $f(a) < f(b)$. In Figure 6b, the function values of $g(x) = -3x + 1$ decrease (the graph falls) as the values of x increase (vary from left to right); that is, if $a < b$, then $g(a) > g(b)$. In Figure 6c, we see that as x increases, $f(x) = 4$ neither increases nor decreases.

Figure 6

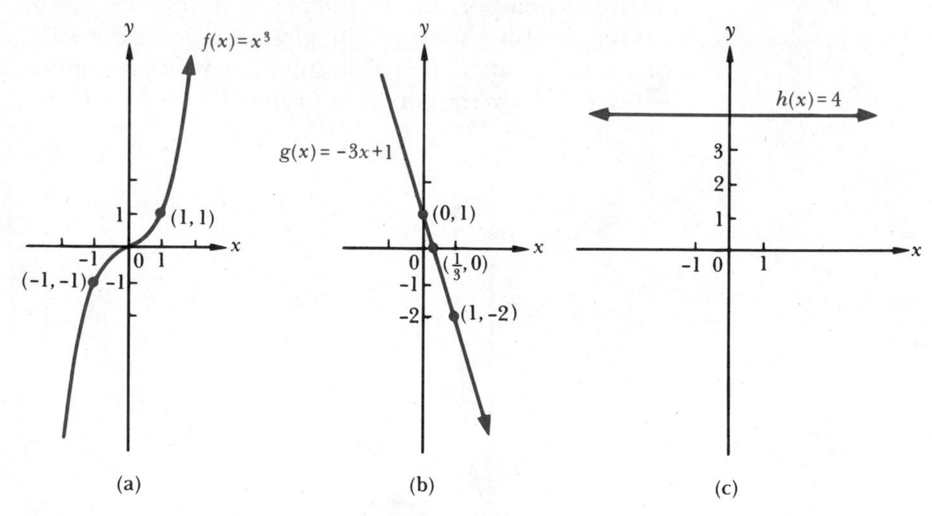

More formally, a function f is said to be a (strictly) *increasing function* in an interval if, whenever a and b are two numbers in the interval such that $a < b$, we have $f(a) < f(b)$ (Figure 7a). A function g is said to be a (strictly) *decreasing function* in an interval if, whenever a and b are two numbers in the interval such that $a < b$, we have $g(a) > g(b)$ (Figure 7b).

Whenever we must decide whether a function is increasing or decreasing we must always consider what happens to the corresponding y values as the x values increase (vary from left to right).

Figure 7

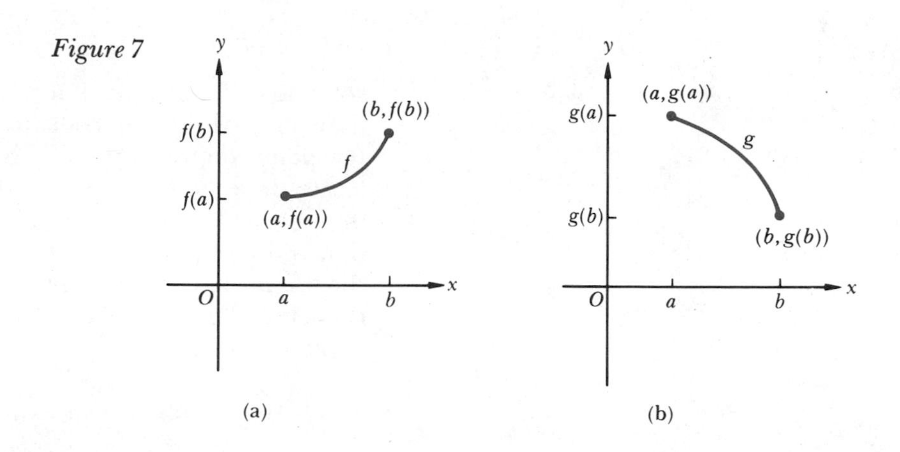

EXAMPLE Given the graph in Figure 8, indicate the intervals for which f is increasing, the intervals for which f is decreasing, and the intervals for which f is neither increasing nor decreasing.

Figure 8

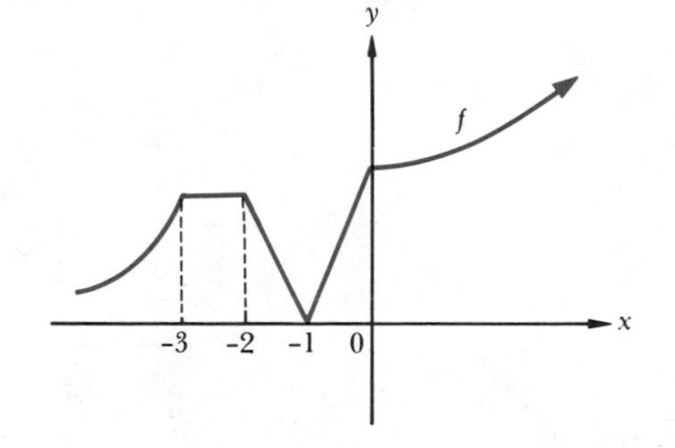

SOLUTION The graph of f indicates that the function is increasing in the intervals $(-\infty, -3]$ and $[-1, \infty)$. It is decreasing in the interval $[-2, -1]$. The function is neither increasing nor decreasing in the interval $[-3, -2]$.

4.3 Graphs of Functions

The graph of a function usually conveys at a glance whether the function is increasing or decreasing and whether the function is even (symmetric with respect to the y axis) or odd (symmetric with respect to the origin). The graph also displays the domain and range of the function. A few examples will help to demonstrate the relationship between function properties and the graph of the function.

EXAMPLES In each of the following examples, find the domain of f and sketch its graph. Is f even or odd? Does the graph of f have symmetry? From the graph of f, determine the range of the function. Also indicate the intervals in which f is increasing or decreasing.

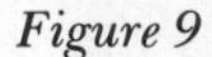
Figure 9

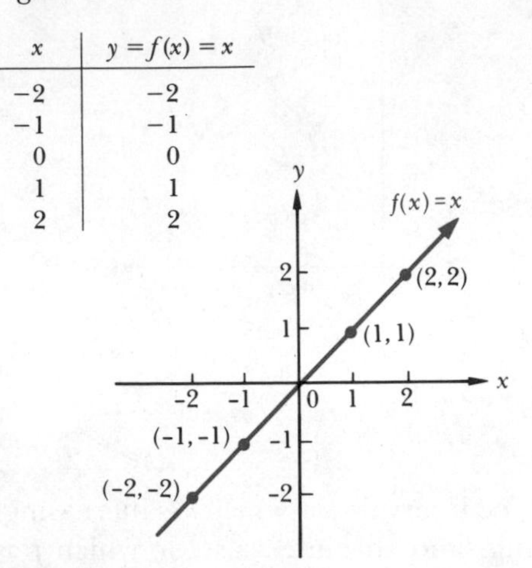

x	$y=f(x)=x$
-2	-2
-1	-1
0	0
1	1
2	2

1 $f(x)=x$ (identity function)

SOLUTION The domain of f is the set of all real numbers. We list the coordinates $(x, f(x))$ of a few points on the graph of f in a table and then locate these points to obtain the graph (Figure 9). Since $f(-x)=-x=-f(x)$, f is an odd function and the graph is symmetric with respect to the origin. The graph shows the range of f to be the set of all real numbers, and f is increasing in R (Figure 9).

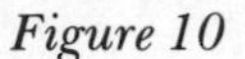
Figure 10

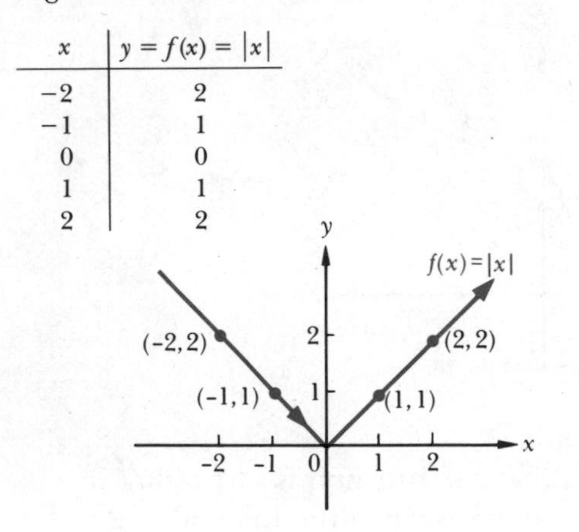

x	$y=f(x)=\|x\|$
-2	2
-1	1
0	0
1	1
2	2

2 $f(x)=|x|$ (absolute value function)

SOLUTION The domain of f is the set of all real numbers. The graph is obtained by locating a few points (Figure 10). Since $f(-x)=|-x|=|x|=f(x)$, f is an even function and the graph is symmetric with respect to the y axis. From the graph we see that the range of f is the set of nonnegative real numbers, that is, $\{y|y \geq 0\}$. The graph also indicates that f is increasing in the interval $[0, \infty)$ and decreasing in the interval $(-\infty, 0]$ (Figure 10).

3 $f(x)=3$ (constant function)

SOLUTION Functions of the form $f(x)=c$, where c is a constant, are called *constant functions*. The domain of f is the set of all real numbers; the graph is given in Figure 11. Since $f(-x)=3=f(x)$, f is an even function and the graph is symmetric with respect to the y axis. Clearly, the graph shows that the range of f is the set $\{3\}$ and that f is neither increasing nor decreasing (Figure 11).

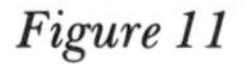
Figure 11

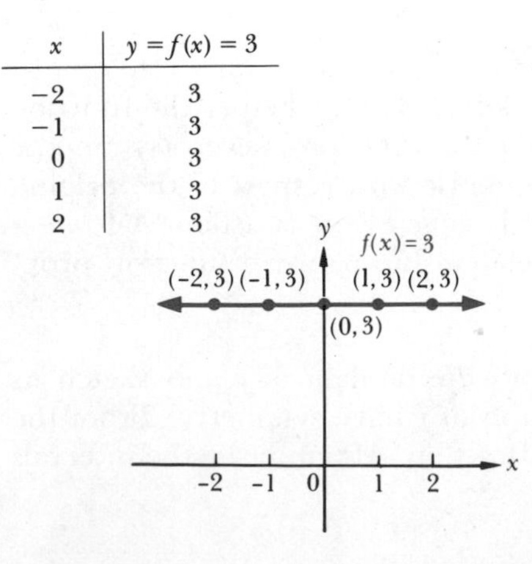

x	$y=f(x)=3$
-2	3
-1	3
0	3
1	3
2	3

4 $f(x)=[\![x]\!]$ (greatest integer function)

SOLUTION The *greatest integer of a real number* x, written $[\![x]\!]$, is the integer n that satisfies $n \leq x < n+1$. In other words, $[\![x]\!]$ is the "nearest" integer less than or equal to x. Thus $[\![5\frac{1}{4}]\!]=5$, $[\![-2\frac{1}{2}]\!]=-3$, $[\![-\frac{1}{2}]\!]=-1$, and $[\![\sqrt{2}]\!]=1$. The domain of f is the set of all real numbers, and the graph is given in Figure 12. The function is neither even nor odd, and the graph is not symmetric with respect to either the y axis or the origin. The graph reflects the fact that

Figure 12

x	$y = f(x) = [\![x]\!]$
-4	-4
$-\frac{7}{2}$	-4
-3	-3
$-\frac{5}{2}$	-3
-2	-2
$-\frac{3}{2}$	-2
-1	-1
$-\frac{1}{2}$	-1
0	0
$\frac{1}{2}$	0
1	1
$\frac{3}{2}$	1
2	2
3	3
4	4

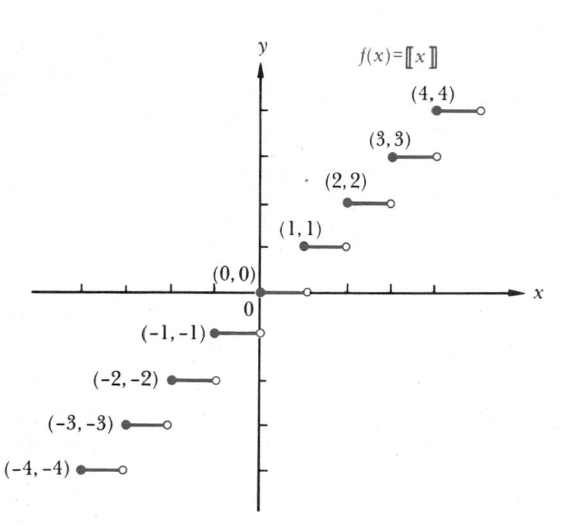

Figure 13

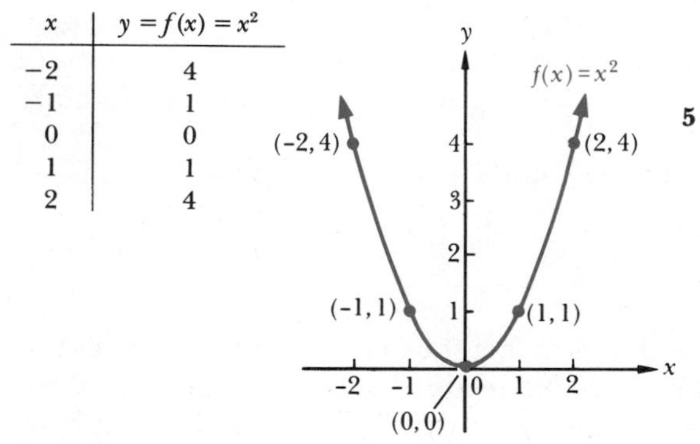

x	$y = f(x) = x^2$
-2	4
-1	1
0	0
1	1
2	4

Figure 14

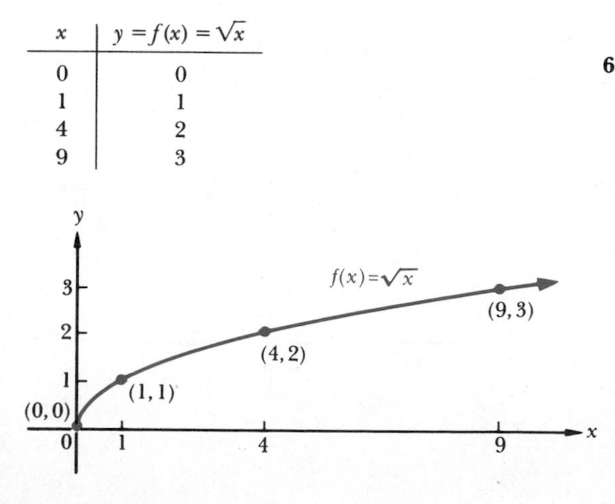

x	$y = f(x) = \sqrt{x}$
0	0
1	1
4	2
9	3

the range of f is the set of all integers and that f is neither increasing nor decreasing in R (Figure 12).

5 $f(x) = x^2$ (square function)

SOLUTION The domain of f is the set of all real numbers, and the graph is given in Figure 13. Since $f(-x) = (-x)^2 = x^2 = f(x)$, f is an even function and the graph is symmetric with respect to the y axis. The graph indicates that the range of f is the set of all nonnegative real numbers, that is, $\{y \mid y \geq 0\}$, and that f is increasing in the interval $[0, \infty)$ and decreasing in the interval $(-\infty, 0]$ (Figure 13).

6 $f(x) = \sqrt{x}$ (square root function)

SOLUTION The domain of f is the set of nonnegative real numbers, that is, $\{x \mid x \geq 0\}$, since the square root of a negative number is not defined in the real number system. The graph is given in Figure 14. The function is neither even nor odd, so the graph is not symmetric with respect to either the y axis or the origin. From the graph of the function, we see that the range of f is the set of all nonnegative real numbers, that is, $\{y \mid y \geq 0\}$, and that f is an increasing function in the interval $[0, \infty)$ (Figure 14).

7 $f(x) = \begin{cases} x & \text{if } x \geq 1 \\ x^2 & \text{if } x < 1 \end{cases}$

Figure 15

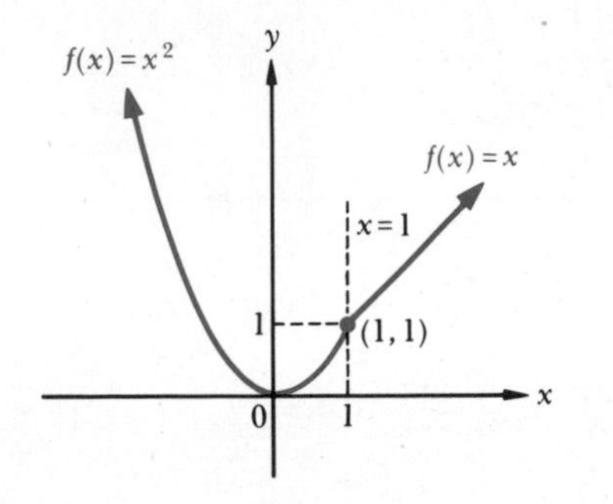

SOLUTION The independent variable x can take on any value, so the domain of f is the set of all real numbers. In sketching the graph, we consider the portion to the right of the vertical line $x = 1$ separately from the portion to the left of this line (see Examples 1 and 5). Since f is neither even nor odd, the graph of f is not symmetric with respect to either the y axis or the origin (Figure 15). The graph also indicates that the range of f is the set of all nonnegative real numbers, that is, $\{y \mid y \geq 0\}$, and that f is an increasing function in the interval $[0, \infty)$ and a decreasing function in the interval $(-\infty, 0]$ (Figure 15).

4.4 Graphing Techniques

At times it is possible to use known graphs of functions, together with geometric shifting, stretching, shrinking, and/or reflecting, to obtain graphs of other functions. The following examples clarify these graphing techniques.

EXAMPLES Use the graphs of $f(x) = |x|$, $g(x) = x^2$, and $h(x) = x^3$ to sketch the graphs of the following functions.

1 $F(x) = |x| + 2$

SOLUTION Since $F(x) = f(x) + 2$, we can get the graph of F from the graph of f by adding 2 units to every y value obtained for the graph of f. Geometrically, this can be accomplished by vertically shifting the graph of $f(x) = |x|$ (Figure 16a) 2 units upward (Figure 16b).

Figure 16

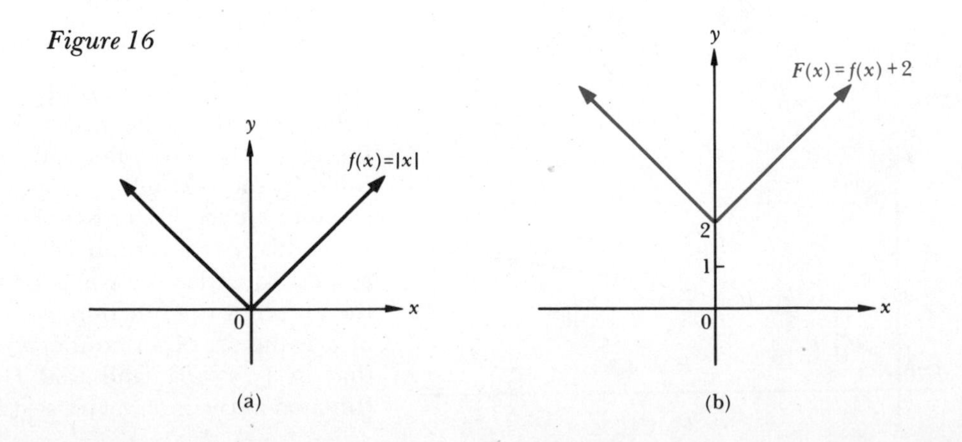

2 $G(x) = 3x^2$

SOLUTION $G(x) = 3g(x)$. Consequently, the graph of G can be determined from the graph of g by multiplying each ordinate in Figure 17a by 3 to obtain the graph in Figure 17b. Geometrically, this is equivalent to vertically stretching the graph of g by a factor of 3.

Figure 17

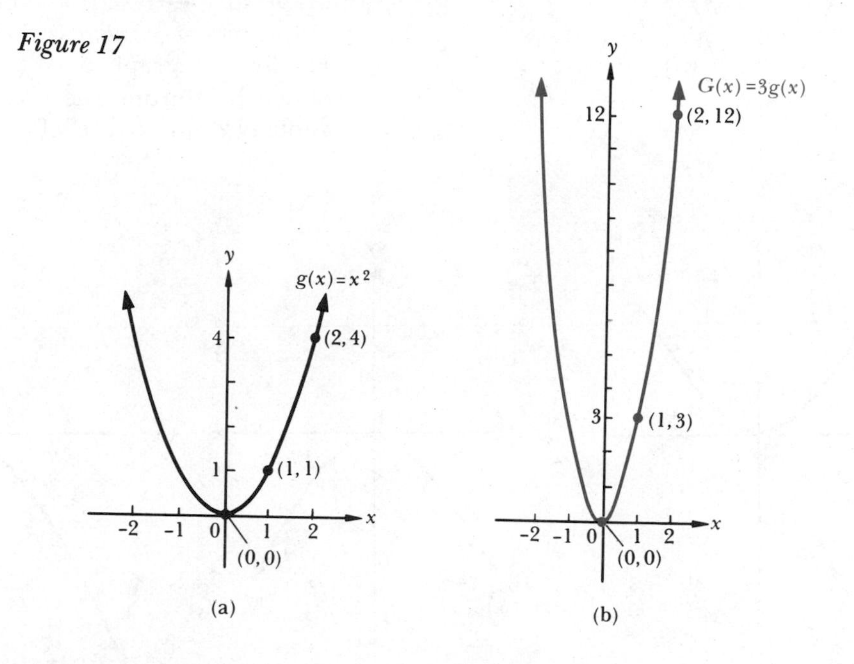

3 $H(x) = (x - 1)^3$

SOLUTION Since $H(x) = h(x - 1)$, the graph of H is obtained geometrically by shifting the graph of $h(x) = x^3$ (Figure 18a) horizontally 1 unit to the right (Figure 18b).

Figure 18

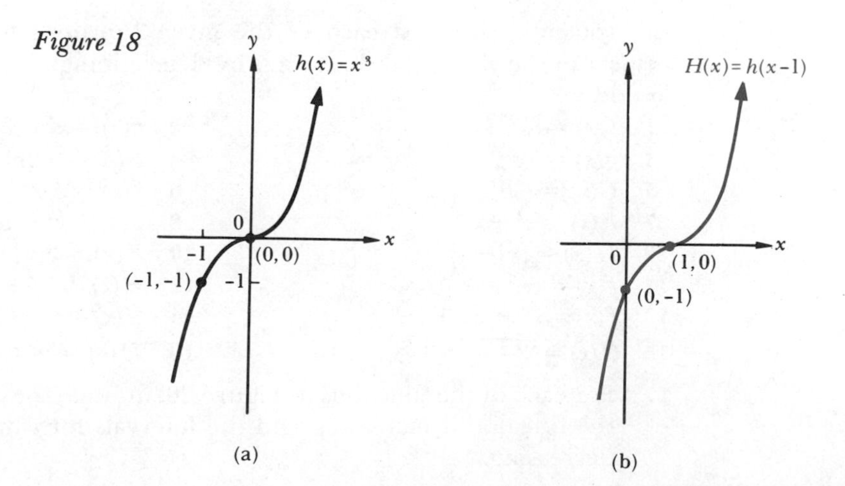

Figure 19

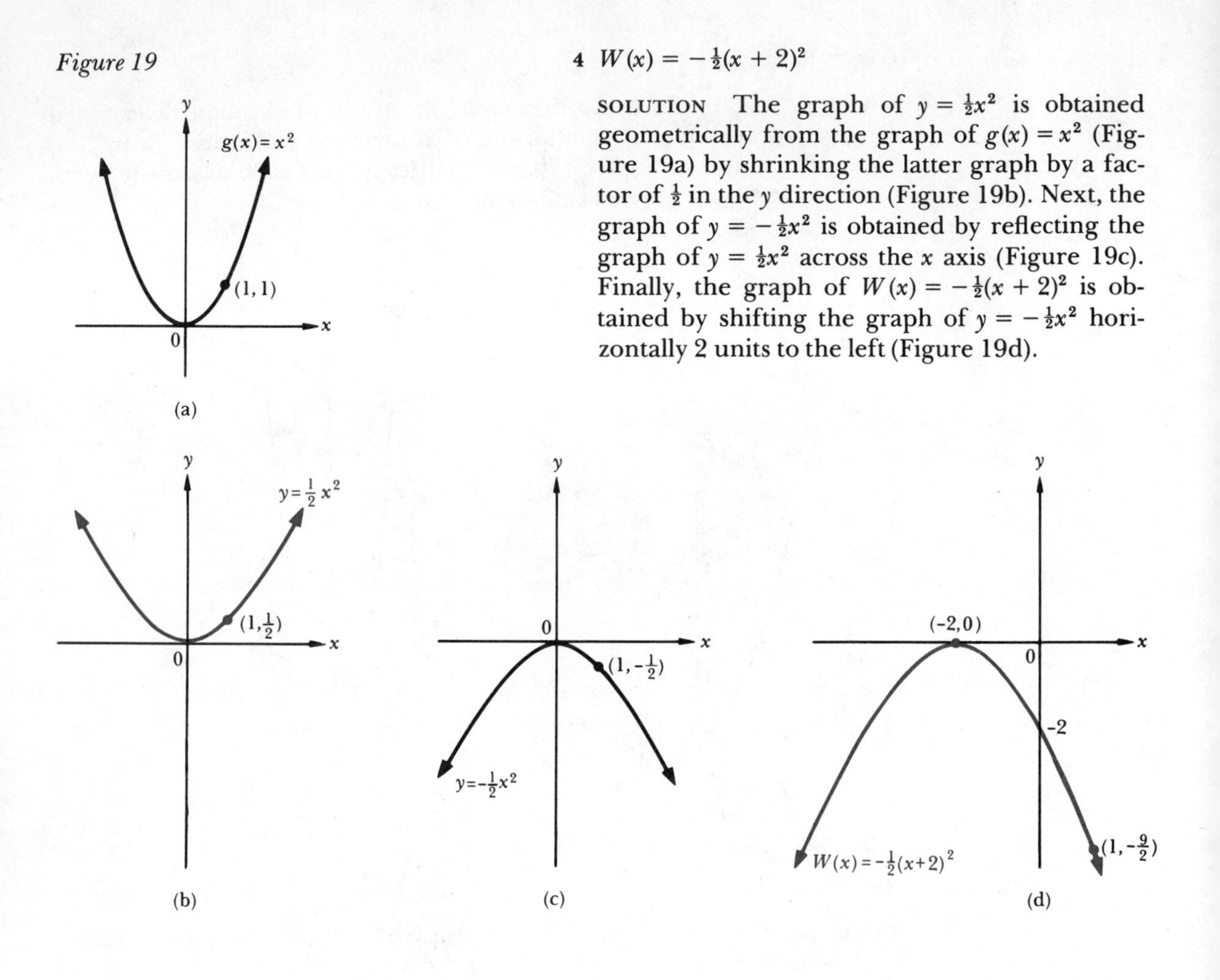

4 $W(x) = -\frac{1}{2}(x + 2)^2$

SOLUTION The graph of $y = \frac{1}{2}x^2$ is obtained geometrically from the graph of $g(x) = x^2$ (Figure 19a) by shrinking the latter graph by a factor of $\frac{1}{2}$ in the y direction (Figure 19b). Next, the graph of $y = -\frac{1}{2}x^2$ is obtained by reflecting the graph of $y = \frac{1}{2}x^2$ across the x axis (Figure 19c). Finally, the graph of $W(x) = -\frac{1}{2}(x + 2)^2$ is obtained by shifting the graph of $y = -\frac{1}{2}x^2$ horizontally 2 units to the left (Figure 19d).

PROBLEM SET 4

In Problems 1–16, test each of the given functions for symmetry with respect to the y axis and the origin by determining if the function is even or odd.

1 $f(x) = 8x^2 + 5$

2 $f(x) = x^4 + 3$

3 $g(x) = 7x^3$

4 $F(x) = -4x^3 + 1$

5 $G(x) = -6$

6 $f(x) = 5$

7 $F(x) = x^2 + x$

8 $g(x) = x^2 - 5x$

9 $H(x) = |x| - x$

10 $F(x) = 3|x| - 5$

11 $f(x) = 5x^3 - x$

12 $H(x) = (x^2 + 1)^3$

13 $f(x) = x^3 - 2x + (1/x)$

14 $G(x) = x^4 - 3x^2 + 10$

15 $G(x) = \sqrt{4 - x^2}$

16 $H(x) = -\sqrt{9 - x}$

17 For each of the functions in Figure 20, indicate the intervals for which the function is increasing and the intervals for which the function is decreasing.

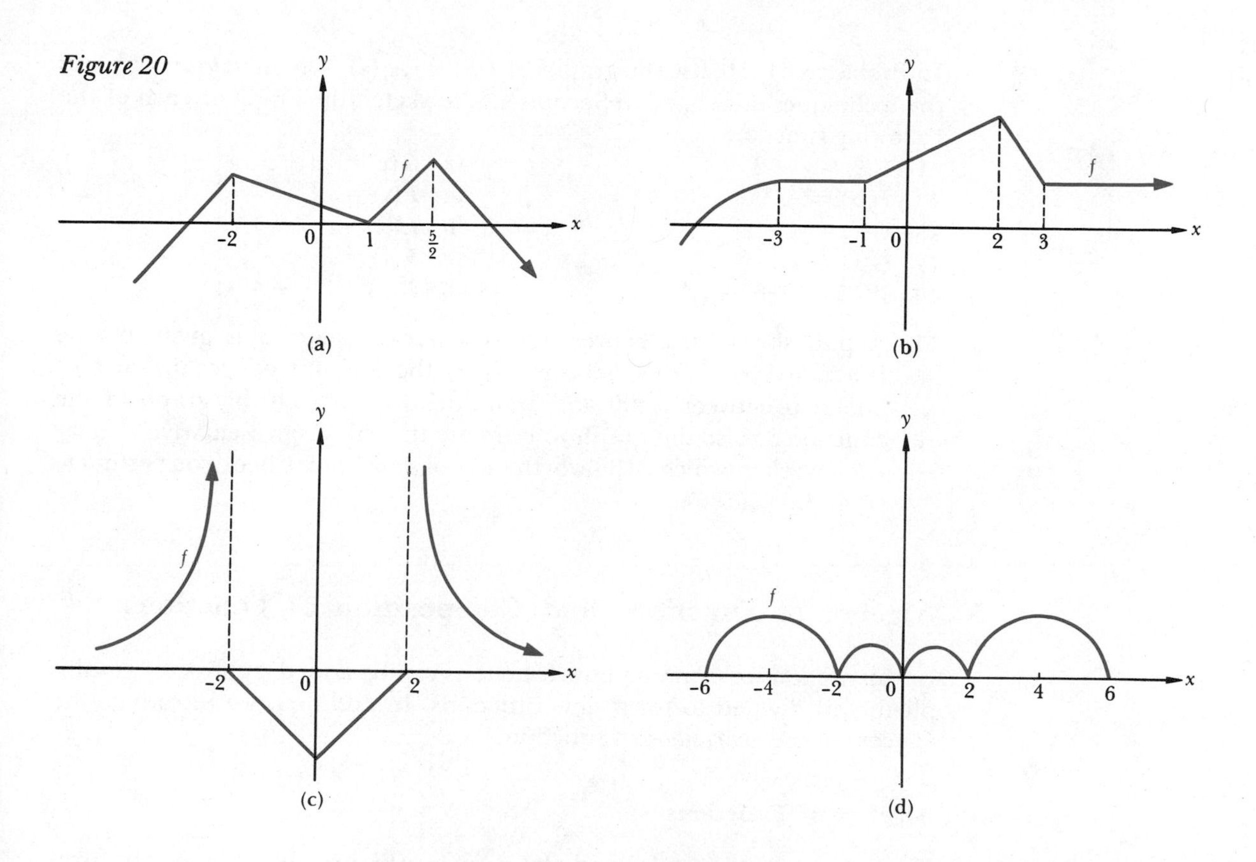

Figure 20

18 If $y = f(x)$ is an increasing function, indicate whether each of the following functions is increasing or decreasing.
(a) $g(x) = 2f(x)$ (b) $K(x) = -\frac{1}{2}f(x)$ (c) $h(x) = f(x) + 2$

In Problems 19–40, find the domain of the function and sketch its graph. Is the function even or odd? Does the graph of the function have symmetry? From the graph of the function, find the range of the function and indicate the intervals where the function is increasing and the intervals where the function is decreasing.

19 $f(x) = 5x + 2$
20 $f(x) = -5x + 3$
21 $f(x) = |x - 1|$
22 $f(x) = |4x - 1|$
23 $f(x) = 2$
24 $f(x) = -3$
25 $f(x) = \sqrt{x - 1}$
26 $f(x) = \sqrt{x + 2}$
27 $f(x) = -5x^2$
28 $f(x) = -1 - x^2$
29 $f(x) = x^3 + 1$
30 $f(x) = -\sqrt[3]{x}$
31 $f(x) = |x| + 4$
32 $f(x) = |x| - x$
33 $f(x) = -5\sqrt{x}$
34 $f(x) = \sqrt{-5x}$
35 $f(x) = [\![5x]\!]$
36 $f(x) = [\![-3x]\!]$
37 $f(x) = x + |x - 2|$
38 $f(x) = |x| + |x - 2|$
39 $f(x) = \begin{cases} x + 3 & \text{if } x \leq 1 \\ 4x^2 & \text{if } x > 1 \end{cases}$
40 $f(x) = \begin{cases} 2x + 1 & \text{if } x < 3 \\ 10 - x & \text{if } x \geq 3 \end{cases}$

In Problems 41–50, use the graphs of $f(x) = x$, $g(x) = |x|$, and $h(x) = x^3$, and the techniques illustrated in Section 4.4, to sketch the graph of each of the following functions.

41 $F(x) = x - 2$

42 $G(x) = |x| + 3$

43 $H(x) = -x^3$

44 $F(x) = 2x$

45 $G(x) = |x - 2|$

46 $H(x) = -2x^3 + 1$

47 $G(x) = |x/2|$

48 $F(x) = -4x + 1$

49 $W(x) = -\frac{1}{3}(x + 1)^3$

50 $R(x) = 3|x - 4|$

51 A publisher's sales representative's weekly salary S is given by the function $S(x) = 120 + 0.3x$, where x, the amount of weekly sales of books, is between \$100 and \$300 inclusive. Sketch the graph of the function S. Use the graph to estimate the sales representative's salary for a week in which \$180 worth of books are sold. Check your estimate by finding $S(180)$.

5 Algebra of Functions and Composition of Functions

In this section we examine how functions can be added, subtracted, multiplied, and divided to form new functions. In addition, we introduce the concept of the *composition* of functions.

5.1 Algebra of Functions

If $f(x) = x^2 - 1$ (Figure 1a) and $g(x) = 2x + 1$ (Figure 1b), then we can form a new function simply by adding $f(x)$ and $g(x)$ to get

$$h(x) = f(x) + g(x) = (x^2 - 1) + (2x + 1) = x^2 + 2x$$

Figure 1

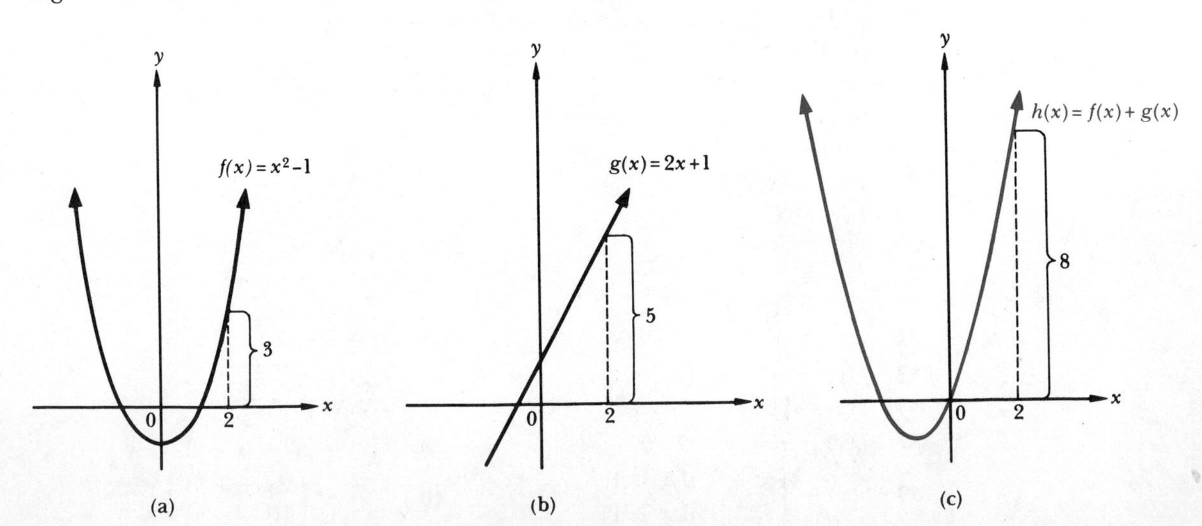

We refer to the function h as the *sum* of the functions f and g and we write $h = f + g$. Note that the graph of h is obtained from the graphs of f and g by adding ordinates; for instance, $h(2) = f(2) + g(2) = 3 + 5 = 8$ (Figure 1c).

It is also possible to subtract, multiply, and divide functions as specified in the following definition.

DEFINITION 1 SUM, DIFFERENCE, PRODUCT, AND QUOTIENT FUNCTIONS

Let f and g be any two functions. We define the functions $f + g$, $f - g$, $f \cdot g$, and f/g as follows:

(i) *Sum function:* $(f + g)(x) = f(x) + g(x)$

(ii) *Difference function:* $(f - g)(x) = f(x) - g(x)$

(iii) *Product function:* $(f \cdot g)(x) = f(x) \cdot g(x)$

(iv) *Quotient function:* $\left(\frac{f}{g}\right)(x) = \frac{f(x)}{g(x)}$, for $g(x) \neq 0$

In each case, the domain of the defined function consists of all values of x *common* to the domains of f and g, except for the quotient function, in which case the values of x for which $g(x) = 0$ are excluded.

EXAMPLES Find the expression that defines each of the following four functions for the pair of given functions.

(a) $(f + g)(x)$ (b) $(f - g)(x)$ (c) $(f \cdot g)(x)$ (d) $\left(\frac{f}{g}\right)(x)$

Also, indicate the domain of f/g.

1 $f(x) = x^2$ and $g(x) = x - 1$

SOLUTION

(a) $(f + g)(x) = f(x) + g(x) = x^2 + x - 1$

(b) $(f - g)(x) = f(x) - g(x) = x^2 - (x - 1) = x^2 - x + 1$

(c) $(f \cdot g)(x) = f(x) \cdot g(x) = x^2(x - 1) = x^3 - x^2$

(d) $\left(\frac{f}{g}\right)(x) = \frac{f(x)}{g(x)} = \frac{x^2}{x - 1}$

Since $x - 1 = 0$ if $x = 1$ and division by 0 is not defined, the domain of f/g is the set of all real numbers except 1.

2 $f(x) = 3x^3 + 7$ and $g(x) = x^2 - 1$

SOLUTION

(a) $(f + g)(x) = f(x) + g(x) = (3x^3 + 7) + (x^2 - 1) = 3x^3 + x^2 + 6$

(b) $(f - g)(x) = f(x) - g(x) = (3x^3 + 7) - (x^2 - 1) = 3x^3 - x^2 + 8$

(c) $(f \cdot g)(x) = f(x) \cdot g(x) = (3x^3 + 7)(x^2 - 1) = 3x^5 - 3x^3 + 7x^2 - 7$

(d) $\left(\frac{f}{g}\right)(x) = \frac{f(x)}{g(x)} = \frac{3x^2 + 7}{x^2 - 1}$

Since $x^2 - 1 = 0$ if $x = 1$ or $x = -1$, the domain of f/g includes all real numbers except 1 and -1.

5.2 Composition of Functions

The basic idea of the *composition* of two functions is that of a kind of "chain reaction" in which the functions occur one after the other. Let us consider a specific example.

We know from solid geometry that the volume V of a sphere of radius r is given by the formula $V = \frac{4}{3}\pi r^3$. Now, suppose that air is being pumped into a spherical balloon so that at the end of t seconds, the radius r satisfies the equation $r = t^2 + 1$. Given the equations $V = \frac{4}{3}\pi r^3$ and $r = t^2 + 1$, we can express V in terms of t by substituting the expression for r from the second equation into the first equation to obtain $V = \frac{4}{3}\pi(t^2 + 1)^3$.

The equation $V = \frac{4}{3}\pi r^3$ defines a function $f(r) = \frac{4}{3}\pi r^3$, whereas the equation $r = t^2 + 1$ defines a function $g(t) = t^2 + 1$. Thus, the original first and second equations can be represented as the functions $V = f(r)$ and $r = g(t)$, respectively. Again we can express V in terms of t by using substitution to obtain $V = f(g(t))$. In order to prevent a "pile up" of parentheses, we often replace the outer parentheses in the latter equation by square brackets and write $V = f[g(t)]$. The equation $V = f[g(t)]$ defines a new function $h(t) = f[g(t)]$. The function h, which was obtained by "chaining" g followed by f, is called the *composition* of g by f and is sometimes written as $h = f \circ g$. This idea of composition of functions is expressed precisely in the following definition.

DEFINITION 2 COMPOSITION OF FUNCTIONS

Let f and g be two functions satisfying the condition that at least one number in the range of g belongs to the domain of f. Then the *composition* of g by f is the function $f \circ g$ defined by the equation $(f \circ g)(x) = f[g(x)]$.

The domain of the composite function $f \circ g$ is the set of all values x in the domain of g such that $g(x)$ belongs to the domain of f. The range of $f \circ g$ is the set of all possible values of $f[g(x)]$. Schematically, $f \circ g$ is shown as a mapping of g followed by f in Figure 2. Note that the symbol $\circ$ is used for composition function $f \circ g$ so as not to confuse it with the product $f \cdot g$. We shall use the notation $f[g(x)]$ interchangeably with the notation $(f \circ g)(x)$ to denote the composition of g by f.

Figure 2

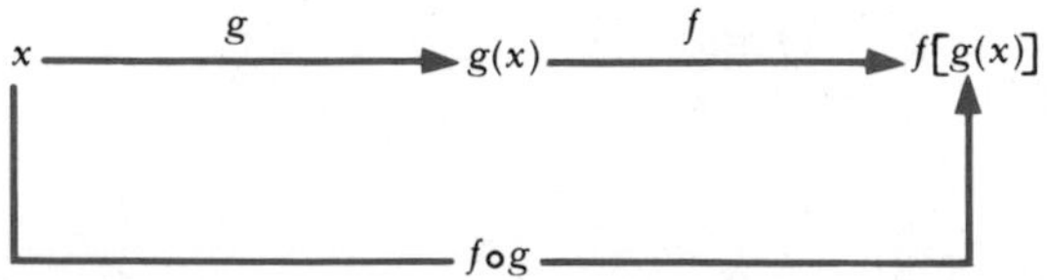

EXAMPLES 1 Let $f(x) = x^2$ and $g(x) = 2x - 3$. Find $(f \circ g)(2)$ and $(g \circ f)(2)$; note whether or not $(f \circ g)(2)$ and $(g \circ f)(2)$ are equal.

SOLUTION Since

$$(f \circ g)(2) = f[g(2)] = f(1) = 1 \quad \text{and} \quad (g \circ f)(2) = g[f(2)] = g(4) = 5$$

we have $(f \circ g)(2) \neq (g \circ f)(2)$.

2 Let $f(x) = 2x^2$ and $g(x) = 4x + 1$. Find the expression that defines each of the following composite functions.

(a) $f[g(x)]$ (b) $g[f(x)]$ (c) $f[f(x)]$ (d) $g[g(x)]$

SOLUTION

(a) $f[g(x)] = f(4x + 1) = 2(4x + 1)^2$
$= 32x^2 + 16x + 2$

(b) $g[f(x)] = g(2x^2) = 4(2x^2) + 1$
$= 8x^2 + 1$

(c) $f[f(x)] = f(2x^2) = 2(2x^2)^2$
$= 2(4x^4) = 8x^4$

(d) $g[g(x)] = g(4x + 1) = 4(4x + 1) + 1$
$= 16x + 5$

3 Let $f(x) = x - 1$ and $g(x) = \sqrt{x}$. Determine the expression and the domain for the following composite functions.

(a) $f[g(x)]$ (b) $g[f(x)]$

SOLUTION

(a) $f[g(x)] = f(\sqrt{x}) = \sqrt{x} - 1$. The domain of $f \circ g$ is the set of all nonnegative real numbers, that is, $\{x | x \geq 0\}$.

(b) $g[f(x)] = g(x - 1) = \sqrt{x - 1}$. The domain of $g \circ f$ is the set of all real numbers x such that $x - 1 \geq 0$, that is, $\{x | x \geq 1\}$. Notice that the domain of $g \circ f$ is different from the domain of f.

4 Find two functions, f and g, that will produce the composite function $h(x) = (2x + 1)^3$ if $h = f \circ g$.

SOLUTION If we let $f(u) = u^3$ and $u = g(x) = 2x + 1$, then h is defined by $h(x) = f[g(x)] = f(2x + 1) = (2x + 1)^3$.

5 A damaged oil tanker leaks oil into the Atlantic Ocean, and its oil spill is spreading on the ocean surface in a form of a circle of radius r (in kilometers). Suppose that r is expressed as a function of time t (in hours) by the equation $r(t) = 0.75 + 2.25t$, where $t = 0$ corresponds to the time at which the radius of the spill is 0.75 kilometers.

(a) Express the area A of the oil spill as a function of t; that is, find $(A \circ r)(t)$ explicitly.

(b) What is the area of the spill at time $t = 3$ hours?

SOLUTION

(a) The area A of the oil spill is given by $A = \pi r^2$ and r is given by the equation $r(t) = 0.75 + 2.25t$. Thus, the expression for the composite function $A \circ r$ is given by

$$\begin{aligned}(A \circ r)(t) &= A[r(t)] = A(0.75 + 2.25t)\\ &= \pi(0.75 + 2.25t)^2\\ &= \pi(0.5625 + 3.375t + 5.0625t^2)\end{aligned}$$

(b) If $t = 3$ hours, and 3.14 is used as an approximate value for π, then

$$\begin{aligned}A(3) &= A[r(3)] = \pi[0.5625 + 3(3.375) + 9(5.0625)]\\ &= 3.14\,(0.5625 + 10.1250 + 45.5625)\\ &= 3.14(56.25) = 176.625 \text{ square kilometers}\end{aligned}$$

PROBLEM SET 5

In Problems 1–10, find the expression that defines each of the following four functions for the pair of given functions.

(a) $(f+g)(x)$ (b) $(f-g)(x)$ (c) $(f \cdot g)(x)$ (d) $(f/g)(x)$

What is the domain of f/g?

1 $f(x) = 3x + 1$ and $g(x) = 3x - 7$
2 $f(x) = x$ and $g(x) = -5$
3 $f(x) = 4x - 5$ and $g(x) = x + 6$
4 $f(x) = -5$ and $g(x) = 7$
5 $f(x) = x^2 + 5$ and $g(x) = 2x - 1$
6 $f(x) = x^2 - 3x$ and $g(x) = 4x + 1$
7 $f(x) = 7x + 1$ and $g(x) = -3x + 8$
8 $f(x) = x^3$ and $g(x) = 2x^2$
9 $f(x) = x^2 + 5$ and $g(x) = -x^2 + x$
10 $f = \{(1, 2), (2, 2), (3, 5), (4, 6)\}$ and $g = \{(2, 2), (-1, 3), (3, 7)\}$

In Problems 11–20, use $f(x) = 2x^2 + 6$ and $g(x) = 7x + 2$ to find each of the following values.

11 $(f \circ g)(2)$ **12** $(g \circ f)(2)$ **13** $(f \circ f)(2)$ **14** $(g \circ g)(2)$
15 $g[f(4)]$ **16** $f[g(3)]$ **17** $f[g(5)]$ **18** $g[f(5)]$
19 $f[f(-1)]$ **20** $g[g(-1)]$

In Problems 21–28, find

(a) $f[g(x)]$ and the domain of $f \circ g$ (b) $g[f(x)]$ and the domain of $g \circ f$

21 $f(x) = 2x$ and $g(x) = 5x - 3$
22 $f(x) = 2x^2 + 5$ and $g(x) = 7x$
23 $f(x) = 3x$ and $g(x) = -3x$
24 $f(x) = x^3 + 1$ and $g(x) = \sqrt[3]{x - 1}$
25 $f(x) = 11x + 2$ and $g(x) = \dfrac{x}{11} - \dfrac{2}{11}$
26 $f(x) = x^2 + 3$ and $g(x) = 5$
27 $f(x) = 6$ and $g(x) = 9$
28 $f(x) = \sqrt{2x + 1}$ and $g(x) = x^2 + 9$

In Problems 29–34, find two functions, f and g, that will produce the given composite function $h = f \circ g$.

29 $h(x) = (5x - 3)^3$ **30** $h(x) = (x^2 + 5x)^4$
31 $h(t) = (t^2 - 2)^{-2}$ **32** $h(s) = \left(\dfrac{s+1}{s-1}\right)^3$
33 $h(x) = (x + x^{-1})^5$ **34** $h(x) = (x^2 + 30)^{-6}$

35 Let $F(x) = 3x - 7$ and $G(x) = 2x + k$. Determine the value of k so that $F[G(x)] = G[F(x)]$.

36 Let $H(x) = (2x^3 + 7)^3$. Find a function P such that $H[P(x)] = P[H(x)]$.

37 Let $f(x) = x^2 + 1$ and $g(x) = (x + 1)^2$. Form $f \circ g$ and $g \circ f$.

38 Let $F(x) = x$ and $G(x) = 1/x$. Find the domains of $F \circ G$ and $G \circ F$.

39 Let $f(x) = 3x + 1$ and $g(x) = -5x + 2$.
(a) Find $f[g(x)]$.
(b) For what values of x does $f[g(x)] = 2$?

40 Verify that $f[f[f(x)]] = x$ if $f(x) = 1 - (1/x)$

41 Let $f(x) = x^3 + 2$ and $g(x) = \sqrt[3]{x + 7}$.
(a) Find $f[g(x)]$.
(b) For what values of x does $f[g(x)] = 13$?

42 Suppose that $f(x) = x^{1/3}$, $g(x) = (x^9 + x^6)^{1/2}$, and $h(x) = x(x + 1)^{1/2}$. Show that $g[f(x)] = h(x)$.

43 Let $f(x) = ax + 1$. Find the value of a such that $f[f(x)] = x$.

In Problems 44–46, let $f(x) = x$, $g(x) = |x|$, and $h(x) = [\![x]\!]$. Answer "true" if the given equation is true for *all* real values; answer "false" if the equation is false for *some* real value or values. If your answer is "false," give an example to support your claim.

44 (a) $g(x^3) = [g(x)]^3$ (b) $g(x + y) = g(x) + g(y)$
(c) $g(xy) = g(x) \cdot g(y)$

45 (a) $f(x^3) = [f(x)]^3$ (b) $f(x + y) = f(x) + f(y)$
(c) $f(xy) = f(x) \cdot f(y)$

46 (a) $h(x^3) = [h(x)]^3$ (b) $h(x + y) = h(x) + h(y)$
(c) $h(xy) = h(x) \cdot h(y)$

47 Suppose a cylindrical vessel has a circular base of radius 4 inches.
(a) Express the volume V of the vessel as a function of the height h.
(b) Express the height h as a function of time if, after t seconds, the height is $2t + 4$.
(c) Use parts (a) and (b) to construct a function that expresses the volume of the vessel as a function of time.

48 A baseball diamond is a square, 90 feet long on each side. A ball is hit down the third-base line at the rate of 50 feet per second. Let y denote the distance in feet between the ball and first base, let x denote its distance in feet from home plate, and let t denote the elapsed time in seconds after the ball was hit. Express y as a function f of x and x as a function g of t. Find $(f \circ g)(t)$ explicitly.

49 A manufacturer of electric appliances finds that the production cost C (in dollars per unit) for its deluxe model can opener is a function of the number x of can openers produced, defined by

$$C = \frac{x^2 + 120x + 8000}{20x}$$

The selling price S of each can opener, which is a function of the production cost C per unit, is given by $S(C) = 1.02C$. Express the selling price as a function of the number of can openers produced; that is, find $(S \circ C)(x)$ explicitly. What is the selling price per unit if 1000 can openers are produced?

6 Inverse Functions

Suppose that the function $f = \{(1, 2), (3, 4), (-1, 0)\}$. If we were to form another set of ordered pairs from f by interchanging the members within each ordered pair, the result would yield $g = \{(2, 1), (4, 3), (0, -1)\}$, which is also a function. In this case, the functions f and g are said to be *invertible,* and g is considered to be the inverse of function f. By contrast, if we began with the function $h = \{(2, 5), (-1, 5), (-3, 4)\}$, the relation resulting from interchanging the members of each pair, $\{(5, 2), (5, -1), (4, -3)\}$, would *not* be a function because 5 is the first member of more than one pair. The function h is said to be *not invertible.*

Let us compare the graph of f (Figure 1a) with the graph of the inverse function g (Figure 1b). Notice that the graph of the function g can be obtained by reflecting the graph of the function f across the line $y = x$.

Figure 1

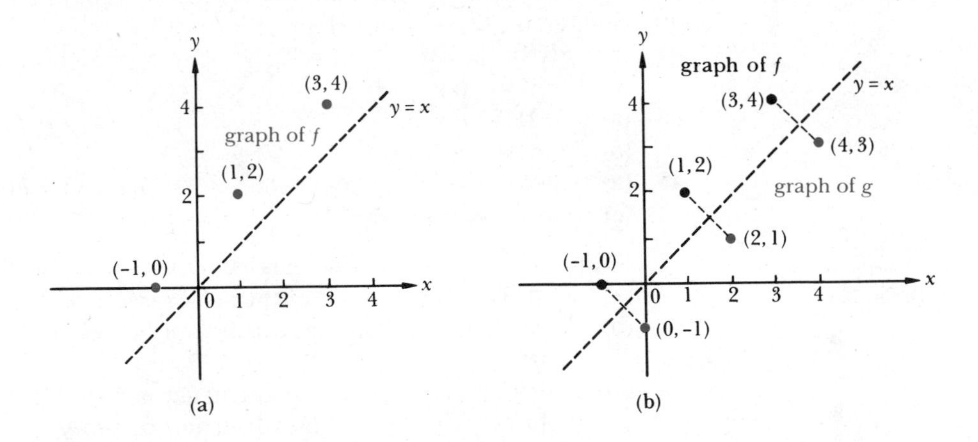

In general, a function is said to be *invertible* if the relation formed by interchanging the members of each ordered pair of the given function is also a function. If a function is *invertible,* the function formed by interchanging the members of the ordered pairs of f is called the *inverse* of f and is denoted by f^{-1}.

Suppose, for instance, that f is an invertible function such that (a, b) is an ordered pair of f. Then (b, a) is an ordered pair of f^{-1}. Using function notation, this means that $b = f(a)$ and $a = f^{-1}(b)$ so that, by substitution, $b = f(a) = f[f^{-1}(b)]$ and $a = f^{-1}(b) = f^{-1}[f(a)]$.

In general, $f^{-1}[f(x)] = x$ for every x in the domain of f and $f[f^{-1}(x)] = x$ for every x in the domain of f^{-1}.

EXAMPLES Suppose that f and $f^{-1} = g$ are as specified below. Show that f and g satisfy (a) $f[g(x)] = x$ and (b) $g[f(x)] = x$.

1 $f(x) = 5x$ and $g(x) = \dfrac{x}{5}$

SOLUTION

(a) $f[g(x)] = f\left(\frac{x}{5}\right) = 5\left(\frac{x}{5}\right) = x$

(b) $g[f(x)] = g(5x) = \frac{1}{5}(5x) = x$

2 $f(x) = 3x + 7$ and $g(x) = \frac{x-7}{3}$

SOLUTION

(a) $f[g(x)] = f\left(\frac{x-7}{3}\right) = 3\left(\frac{x-7}{3}\right) + 7 = x$

(b) $g[f(x)] = g(3x+7) = \frac{(3x+7)-7}{3} = x$

6.1 Existence of the Inverse Function

If $y = f(x)$, we know from the definition of a function that for each x there is one and only one y; hence each of all possible *vertical* lines (representing all possible values of x) intersects the graph no more than once. We indicated earlier that if f is invertible, the relation formed by interchanging the members of any ordered pair (x, y) of f must also be a function. Consequently, in order for f to be invertible, it must also be true that for each y in the range of f there is one and only one x in the domain of f. This property holds for f if each of all possible *horizontal* lines (representing all possible values of y) intersects the graph of $y = f(x)$ no more than once. This means that we can use the graph of a function to discover whether or not the function has an inverse.

EXAMPLES Use the graph of the given function to decide whether or not the function is invertible.

1 $f(x) = x^2$

SOLUTION From the graph of f (Figure 2), we see that any horizontal line above the x axis intersects the graph twice. Hence the function is *not* invertible.

Figure 2

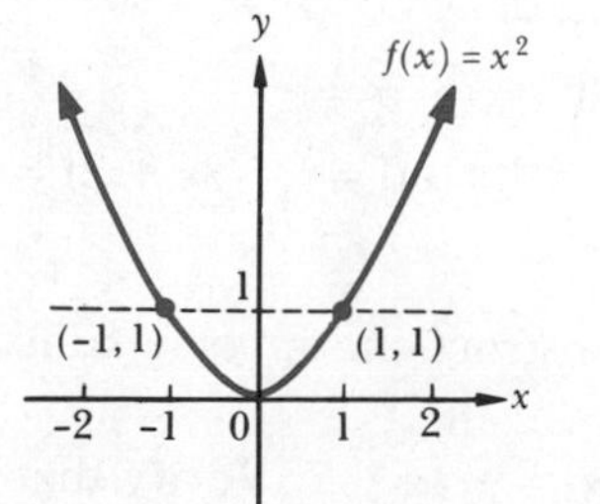

2 $f(x) = 3x - 5$

SOLUTION From the graph of f (Figure 3), it can be seen that no horizontal line intersects the graph more than once. Hence for each range number there is one and only one corresponding domain number, and f is invertible.

Figure 3

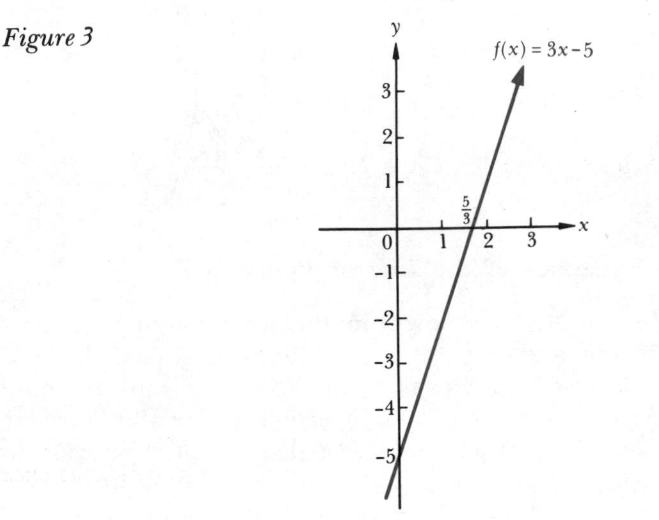

6.2 Construction of the Inverse Function and Its Graph

If f is invertible and is defined by the equation $y = f(x)$, then the equation that defines the inverse function can be constructed by expressing x in terms of y.

For example, suppose that $y = f(x)$ is given by $y = 3x + 2$. Solving for x in terms of y, we obtain

$$x = \tfrac{1}{3}y - \tfrac{2}{3}$$

Now this equation defines the inverse function f^{-1}. Thus $f^{-1}(y) = \frac{1}{3}y - \frac{2}{3}$. However, when using function notation, we normally use x rather than y to represent members of the domain. Consequently, we write the inverse function as $f^{-1}(x) = \frac{1}{3}x - \frac{2}{3}$. Note that

$$f[f^{-1}(x)] = f(\tfrac{1}{3}x - \tfrac{2}{3}) = 3(\tfrac{1}{3}x - \tfrac{2}{3}) + 2 = x - 2 + 2 = x$$

and

$$f^{-1}[f(x)] = f^{-1}(3x + 2) = \tfrac{1}{3}(3x + 2) - \tfrac{2}{3} = x + \tfrac{2}{3} - \tfrac{2}{3} = x$$

EXAMPLES Construct the inverse function of each of the given invertible functions.

1 $f(x) = -\frac{1}{2}x + 7$. Verify that $f[f^{-1}(x)] = f^{-1}[f(x)] = x$.

SOLUTION Let $y = -\frac{1}{2}x + 7$. Solving for x in terms of y, we get

$$\tfrac{1}{2}x = -y + 7 \quad \text{or} \quad x = -2y + 14$$

so

$$f^{-1}(y) = -2y + 14$$

After changing notation, we have

$$f^{-1}(x) = -2x + 14$$

Note that

$$\begin{aligned} f[f^{-1}(x)] &= f(-2x + 14) = -\tfrac{1}{2}(-2x + 14) + 7 \\ &= x - 7 + 7 = x \end{aligned}$$

and

$$\begin{aligned} f^{-1}[f(x)] &= f^{-1}(-\tfrac{1}{2}x + 7) = -2(-\tfrac{1}{2}x + 7) + 14 \\ &= x - 14 + 14 = x \end{aligned}$$

2 $g(x) = \dfrac{1}{x + 1}$. Verify that $g[g^{-1}(x)] = g^{-1}[g(x)] = x$.

SOLUTION Let

$$y = \frac{1}{x + 1}$$

Then

$$x + 1 = \frac{1}{y} \quad \text{or} \quad x = \frac{1}{y} - 1 = \frac{1 - y}{y}$$

so

$$g^{-1}(y) = \frac{1 - y}{y}$$

After changing notation, we get

$$g^{-1}(x) = \frac{1 - x}{x}$$

Note that

$$g[g^{-1}(x)] = g\left(\frac{1 - x}{x}\right) = \frac{1}{\dfrac{1 - x}{x} + 1} = \frac{x}{1 - x + x} = x$$

and

$$g^{-1}[g(x)] = g^{-1}\left(\frac{1}{x + 1}\right) = \frac{1 - \dfrac{1}{x + 1}}{\dfrac{1}{x + 1}} = \frac{x + 1 - 1}{x + 1} \cdot \frac{(x + 1)}{1} = x$$

The graph of f^{-1} can be obtained geometrically from the graph of f by reflecting the graph of f across the line $y = x$.

For example, the reflection of the graph of $f(x) = 3x + 2$ across the line $y = x$ is the graph of $f^{-1}(x) = \frac{1}{3}x - \frac{2}{3}$. Figure 4 illustrates the reflections of the points (1, 5), (0, 2), (2, 8), and (−2, −4) on the graph of f, onto the points (5, 1), (2, 0), (8, 2), and (−4, −2), respectively, of the graph of f^{-1}.

Figure 4

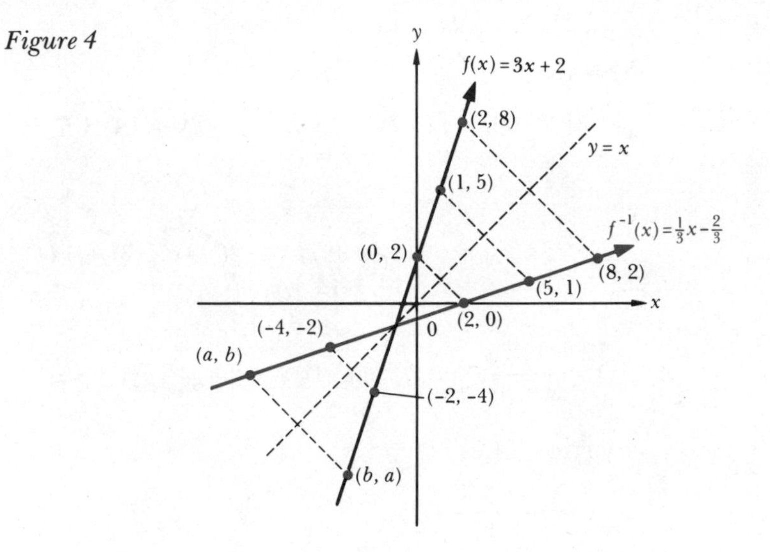

EXAMPLES Construct the inverse of the given invertible function. Sketch the graph of f and f^{-1} on the same coordinate system.

1 $f(x) = 2x - 3$

SOLUTION Let $y = 2x - 3$. Then

$$x = \frac{y + 3}{2}$$

results from solving for x in terms of y; therefore, after changing the notation, we have

$$f^{-1}(x) = \frac{x + 3}{2}$$

After graphing

$$f(x) = 2x - 3 \quad \text{and} \quad f^{-1}(x) = \frac{x + 3}{2}$$

on the same coordinate system (Figure 5), we observe that the graph of the inverse function f^{-1} is a reflection of the graph of f across the line $y = x$.

Figure 5

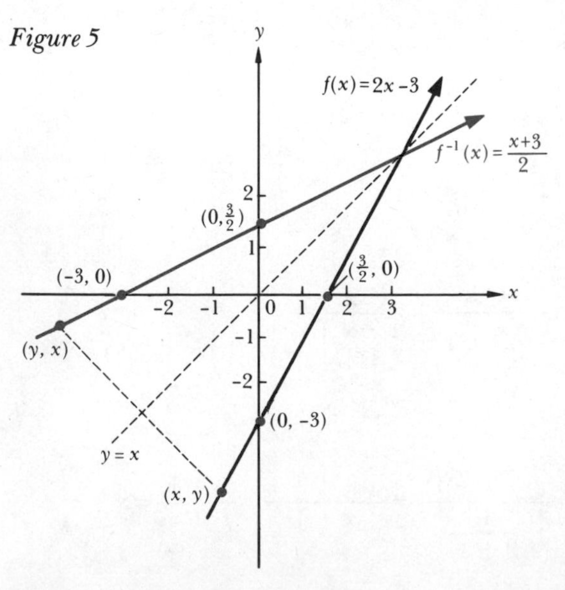

2 $f(x) = x^3$

SOLUTION Let $y = x^3$. Solving for x in terms of y, we get $x = \sqrt[3]{y}$ so that, after changing notation, we have $f^{-1}(x) = \sqrt[3]{x}$. When both $f(x) = x^3$ and $f^{-1}(x) = \sqrt[3]{x}$ are graphed on the same coordinate system (Figure 6), we see that the graph of f^{-1} is a reflection of the graph of f across the line $y = x$.

Figure 6

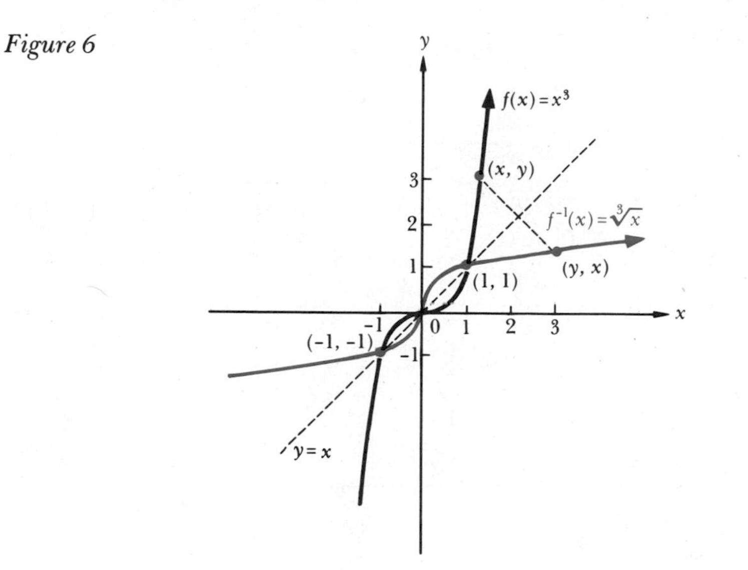

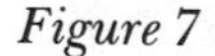

3 $f(x) = \dfrac{1}{x}$

SOLUTION Let $y = 1/x$. Solving for x in terms of y results in $x = 1/y$. This implies that $f^{-1}(x) = 1/x$. Notice that this function is its own inverse. Again the graph of f^{-1} is a reflection of the graph of f across $y = x$ (Figure 7). The graphs of f and f^{-1} coincide because the graph of f is already symmetric with respect to the line $y = x$.

Figure 7

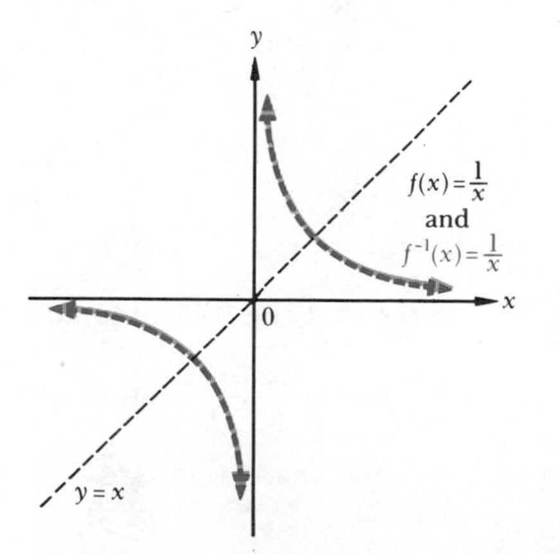

PROBLEM SET 6

In Problems 1–4, for the given functions f and g, verify that $f[g(x)] = x$ and $g[f(x)] = x$ and thus that $g = f^{-1}$.

1 $f(x) = 7x - 2$ and $g(x) = \dfrac{x}{7} + \dfrac{2}{7}$

2 $f(x) = 1 - 5x$ and $g(x) = \dfrac{1}{5} - \dfrac{x}{5}$

3 $f(x) = x^4$, where $x \geqslant 0$, and $g(x) = \sqrt[4]{x}$

4 $f(x) = mx + b$, where m and b are constants and $m \neq 0$, and $g(x) = \dfrac{1}{m}x - \dfrac{b}{m}$

5 Decide whether each function whose graph is given in Figure 8 has an inverse.

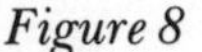

Figure 8

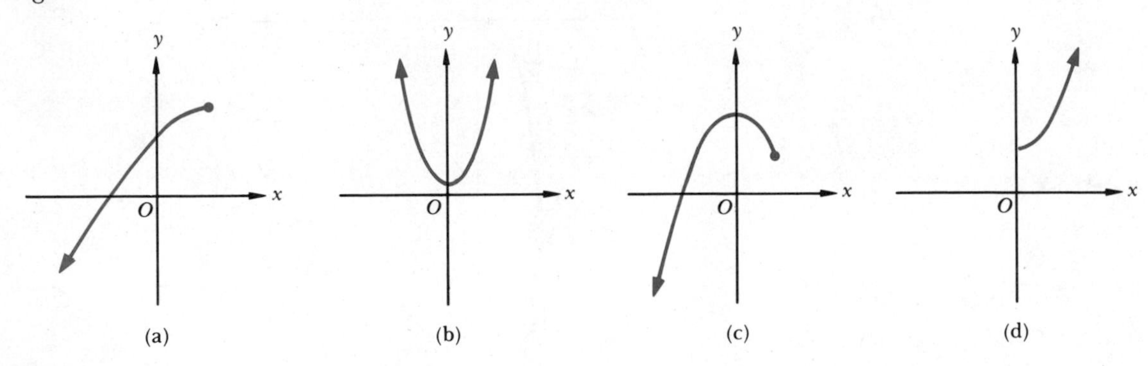

6 Graph each pair of functions in Problems 1 and 2 on the same coordinate system and note that the graph of f^{-1} is a reflection of the graph of f across the line $y = x$.

In Problems 7–16, each function has an inverse. Find f^{-1} and verify that $f[f^{-1}(x)] = x$ and $f^{-1}[f(x)] = x$.

7 $f(x) = 3x - 7$

8 $f(x) = 5 - 11x$

9 $f(x) = \frac{3}{4}x + 5$

10 $f(x) = x^2 - 3$, for $x \geqslant 0$

11 $f(x) = -\dfrac{5}{x}$

12 $f(x) = -x^2$, for $x \geqslant 0$

13 $f(x) = 8x^3$

14 $f(x) = \dfrac{1}{x+1}$

15 $f(x) = x^2$, for $x \leqslant 0$

16 $f = \{(1, 4), (4, 1), (6, 3), (3, 6)\}$

17 Show that the following functions are inverses of each other.

$$f(x) = \frac{3x - 7}{x + 1} \quad \text{and} \quad g(x) = \frac{7 + x}{3 - x}$$

18 Assume that $f(x) = \sqrt{4 - x^2}$ for $0 \leqslant x \leqslant 2$. Show that f is its own inverse.

19 (a) Sketch the graph of $f(x) = |x + 1|$ and indicate if f^{-1} exists.
(b) Sketch the graph of $g(x) = 3$ and indicate if g^{-1} exists.

20 (a) Use the graph of $f(x) = -3x^2 + 1$ to show that f^{-1} does not exist.
(b) Show that $f(x) = -3x^2 + 1$ for $x \leqslant 0$ has an inverse function and find f^{-1}.
(c) Explain the difference between the two functions in parts (a) and (b).

In Problems 21–28, examine the graph of the given function to determine whether or not f^{-1} exists. If f^{-1} exists, find it and use the graph of f to graph f^{-1} on the same coordinate system.

21 $f(x) = 7x + 5$
22 $f(x) = x^2 - 4,\quad \text{for } x \geqslant 0$
23 $f(x) = 1 - 3x$
24 $f(x) = -2|x|$
25 $f(x) = 3/x$
26 $f(x) = x^3 + 5$
27 $f(x) = 3$
28 $f(x) = [\![2x]\!]$

REVIEW PROBLEM SET

1 Locate each of the given points and indicate in which quadrant, if any, the point is found.
(a) $(1, 2)$ (b) $(-1, 1)$ (c) $(2, -1)$ (d) $(-1, -1)$
(e) $(1, -\frac{1}{2})$ (f) $(4, 0)$ (g) $(-2, 0)$ (h) $(0, -4)$

2 Locate the point $P = (-1, 4)$ on a cartesian coordinate system and then give the coordinates of Q and S if
(a) Line segment $\overline{PQ}$ is perpendicular to and bisected by the x axis.
(b) Line segment $\overline{PS}$ is bisected by the origin.

In Problems 3–6, use the distance formula to find the distance between each given pair of points.

3 $(2, 1)$ and $(4, -5)$
4 $(-3, 2)$ and $(6, -1)$
5 $(-6, -3)$ and $(2, 1)$
6 (a, b) and $(b, 3)$

7 Show that the triangle whose vertices are the points $P_1 = (-2, 4)$, $P_2 = (-5, 1)$, and $P_3 = (-6, 5)$ is an isosceles triangle.

8 The abscissa of a point P is 3 and its distance from the point $(3, 4)$ is $2\sqrt{13}$. Find the possible ordinates of P.

9 List the members and identify the domain and range of the relation $R = \{(x, y) | y = 5x - 2; x \in \{-2, -1, 0, 1, 2\}\}$.

10 Graph the relation $\{(x, y) | \, |x| \leqslant 1 \text{ and } |y| \leqslant 3\}$.

In Problems 11–12, graph each relation. Assume that x represents a member of the domain and y represents a member of the range.

11 $y \geqslant 3x + 2$
12 $|x| - |y| = 1$

In Problems 13–14, assume that $y = g(x)$ is defined by the given equation. Find an expression for $g(x)$.

13 $5xy - 2y = 7$
14 $3y - 5x^2y + 2 = 0$

In Problems 15–26, assume that $f(x) = x^3$, $g(x) = 2x + 2$, and $h(x) = x^2 - 5x$. Find each value.

15 $f(-1)$ **16** $g(-1)$ **17** $h(-5)$

18 $f(\sqrt{2})$ **19** $f(\sqrt[3]{2})$ **20** $h(\sqrt{2})$

21 $f(\frac{1}{3})$ **22** $g(a + b) - g(a)$ **23** $\dfrac{g(a + b) - g(a)}{b}$

24 $f(b^2)$ **25** $[f(\sqrt{x})]^2$ **26** $g\left(\dfrac{1}{x + a}\right)$

In Problems 27–28, find the domain of each function.

27 $f(x) = \dfrac{1}{(x - 2)^2}$ **28** $g(x) = \sqrt{x^2 - 9}$

29 Figure 1 shows the graphs of four relations. Which graphs represent functions?

Figure 1

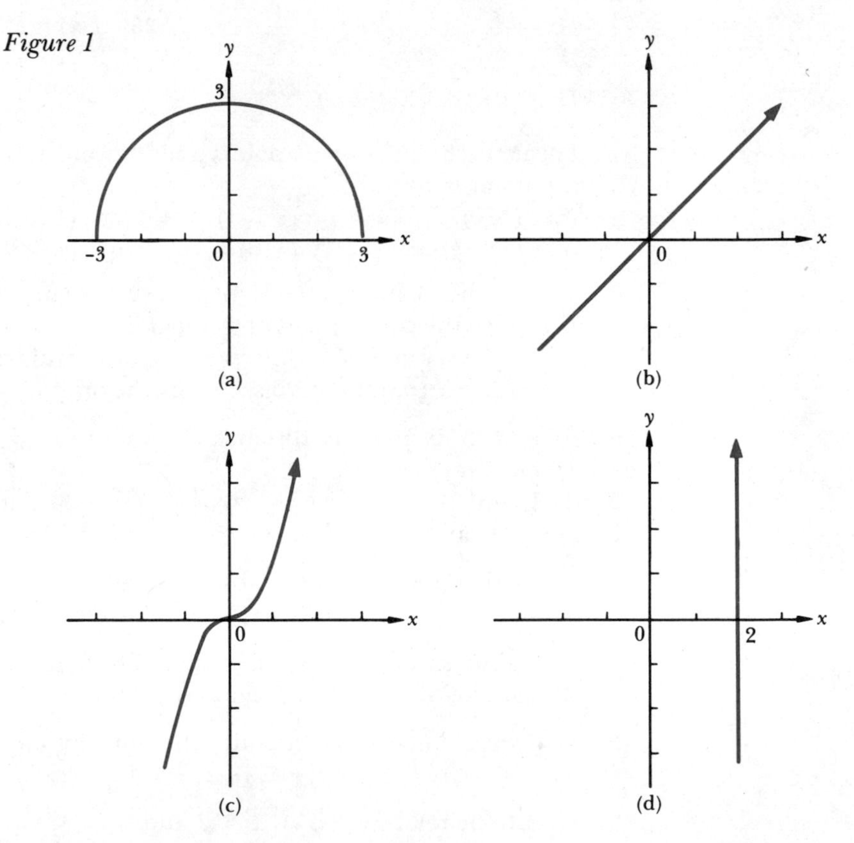

30 Determine which of these points lie on the graph of $f(x) = |5x - 1|$: $(-1, 6)$, $(0, 2)$, $(3, 14)$, $(-\frac{1}{2}, -\frac{7}{2})$, and $(\frac{1}{3}, \frac{2}{3})$. What is the domain of f?

31 Let $f(x) = \sqrt{x^2 + 5}$.

(a) Determine each of the values $f(-2)$, $f(2)$, and $f(0)$.

(b) Describe the function values using function notation, mapping notation, and ordered-pair notation.

32 The demand for a certain kind of candy bar is such that the product of the demand d (in thousands of cartons) and the price p (per carton in dollars) is always equal to 250,000, so long as the price is not less than \$2.00 nor more than \$3.50. Express the demand d as a function of the price p. Compute the demand when the price is
(a) \$2.00 (b) \$2.50 (c) \$3.00 (d) \$3.50

33 A rectangle of sides $2x$ and $2y$ units is inscribed in a circle of radius 3 units and whose center is at the origin. Express y as a function of x. Express the area A of the rectangle as a function of x.

34 Let $g(x) = x^4 + 5x^2$. Which of the following holds for all real numbers?
(a) $g(-x) = g(x)$ (b) $g(-x) = -g(x)$

In Problems 35–44, find the domain of the function and sketch its graph. Is the function even or odd? Does the graph of the function have symmetry? From the graph of the function, find the range of the function and indicate the intervals where the function is increasing and the intervals where the function is decreasing.

35 $f(x) = -2x + 3$

36 $f(x) = 2x^2 + 5$

37 $g(x) = |2x| - 2x$

38 $f(x) = |3x| + 3x$

39 $h(x) = -1$

40 $h(x) = 2\sqrt[3]{x}$

41 $H(x) = \begin{cases} -x^2 + 2 & \text{if } x \geq 1 \\ 2x - 1 & \text{if } x < 1 \end{cases}$

42 $F(x) = \begin{cases} x^4 + 1 & \text{if } x \geq -1 \\ 5 + 3x & \text{if } x < -1 \end{cases}$

43 $F(x) = -2x^3$

44 $f(x) = 5$

45 Use the graph of $g(x) = x^2$ and the techniques in Section 4.4 to sketch the graph of each of the following functions.
(a) $h(x) = x^2 + 4$ (b) $w(x) = (x - \frac{1}{2})^2$ (c) $f(x) = -2(x - \frac{1}{2})^2$

46 Figure 2 shows the graph of a function f. Use the graph of f to sketch the graph of g.
(a) $g(x) = f(x) - 3$ (b) $g(x) = f(x - 3)$ (c) $g(x) = -3f(x)$

Figure 2

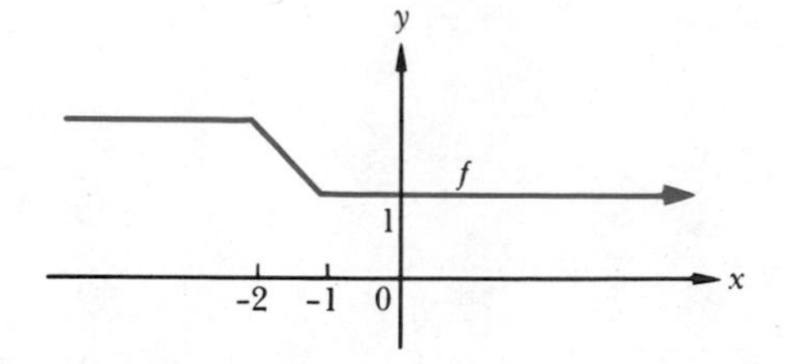

In Problems 47–52, find expressions for $(f + g)(x)$, $(f - g)(x)$, $(f \cdot g)(x)$, and $(f/g)(x)$. Indicate the domain of f/g.

47 $f(x) = x + 2$ and $g(x) = x - 1$

48 $f(x) = x^3$ and $g(x) = -x$

49 $f(x) = x^2$ and $g(x) = 2x + 1$

50 $f(x) = 2x^2 - 1$ and $g(x) = 2x^2 + 1$

51 $f(x) = -9$ and $g(x) = -2$

52 $f(x) = 4$ and $g(x) = -7$

In Problems 53–62, let $f(x) = 7 - x^2$ and $g(x) = 1 + 5x$. Find each of the following.

53 $(f \circ g)(5)$ **54** $(g \circ f)(3)$ **55** $f[g(x)]$ **56** $g[f(x)]$

57 $f[f(2)]$ **58** $g[g(-3)]$ **59** $g[g(x)]$ **60** $f[f(x)]$

61 $f\left[\frac{1}{g(x)}\right]$ **62** $g\left[\frac{1}{f(x)}\right]$

63 Express the given function h as the composition of two other functions f and g so that $h = f \circ g$.
(a) $h(x) = (7x + 2)^5$ (b) $h(t) = \sqrt{t^2 + 17}$

64 A city estimates that its population during the next 6 years will be approximated by the function $p(t) = 100t^2 + 20{,}000$, where t is the time in years and $t = 0$ corresponds to the present year. The city administration also estimates that its average daily pollution index I is a function of the population p of the city and is given by the function $I(p) = 20 + p/1000$. Express the pollution index I as a function of t explicitly; that is, find $I[p(t)]$.

In Problems 65–66, each function has an inverse. Find f^{-1} and verify that $f[f^{-1}(x)] = x$ and $f^{-1}[f(x)] = x$.

65 $f(x) = 7 - 13x$ **66** $f(x) = \frac{1}{3}x + \frac{7}{5}$

In Problems 67–70, examine the graph of the given function to determine if the function has an inverse f^{-1}. If f^{-1} exists, find it and use the graph of f to sketch f^{-1} on the same coordinate system.

67 $f(x) = \frac{1}{2}x + 2$ **68** $f(x) = \sqrt{x - 3}$

69 $f(x) = -3|x|$ **70** $f(x) = -3x^4$

CHAPTER 3

Polynomial and Rational Functions

In this chapter we study two particular types of functions—*polynomial functions* and *rational functions*—that have important applications in mathematics, business, and the sciences. Quadratic equations and inequalities and synthetic division are incorporated into our discussion of these functions.

The identity function $f(x) = x$ and the constant function $f(x) = c$, which were introduced in the preceding chapter (see Examples 1 and 3 on page 70), are specific examples of a general type of polynomial function called a *linear function.*

1 Linear Functions

Suppose that we have two thermometers that are used simultaneously to measure temperature—one graduated according to the Fahrenheit scale, the other graduated according to the Celsius (centigrade) scale. If x represents the Celsius reading and y represents the corresponding Fahrenheit reading, a functional relationship between x and y can be found as follows.

The freezing point of water is 0° Celsius or 32° Fahrenheit, whereas the boiling point of water is 100° Celsius or 212° Fahrenheit. The ratio of Fahrenheit degrees to Celsius degrees is equivalent to $\frac{212-32}{100-0} = \frac{180}{100} = \frac{9}{5}$. Now, if x is the Celsius reading, x represents the "directed" number (x may be negative) of Celsius degrees, so $\frac{9}{5}x$ represents the corresponding *directed number* of Fahrenheit degrees. Since the "starting point" of the Fahrenheit scale is 32, as compared with 0 on the Celsius scale, it follows that a Fahrenheit reading of y corresponding to a Celsius reading of x is given by the equation $y = \frac{9}{5}x + 32$ (Figure 1).

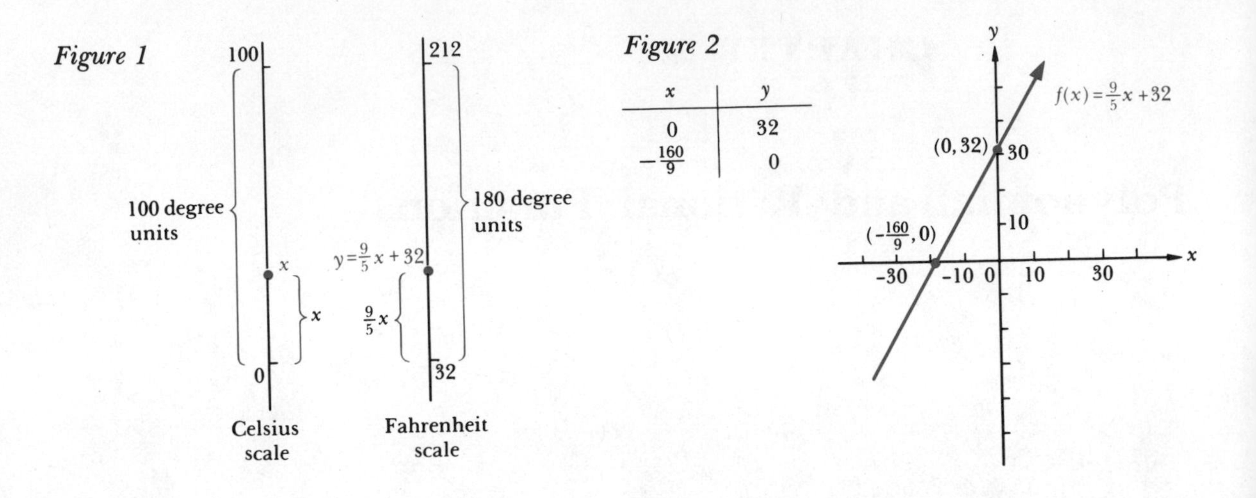

If x is assumed to represent the independent variable, then $y = \frac{9}{5}x + 32$ defines the function $f(x) = \frac{9}{5}x + 32$ (Figure 2). The function f is an example of a linear function.

In general, a linear function is defined as follows:

DEFINITION 1 LINEAR FUNCTION

A function of the form $f(x) = mx + b$, where m and b are constants, is called a *linear function.*

Such functions are called *linear* functions because their graphs are straight lines. The following example illustrates this fact.

EXAMPLE Show that the three points on the graph of the linear function $f(x) = 2x + 1$ with abscissas $x = 0$, 1, and 2 lie on the same straight line; that is, show that these three points are *collinear.*

SOLUTION Since $f(0) = 1$, $f(1) = 3$, and $f(2) = 5$, the three points that belong to the graph are $P_1 = (0, 1)$, $P_2 = (1, 3)$, and $P_3 = (2, 5)$ (Figure 3). The three points P_1, P_2, and P_3 are collinear if $|\overline{P_1P_2}| + |\overline{P_2P_3}| = |\overline{P_1P_3}|$, since the shortest distance between two points is a straight line.

By the distance formula,

$$|\overline{P_1P_2}| = \sqrt{(1-0)^2 + (3-1)^2} = \sqrt{1+4} = \sqrt{5}$$
$$|\overline{P_2P_3}| = \sqrt{(2-1)^2 + (5-3)^2} = \sqrt{1+4} = \sqrt{5}$$
$$|\overline{P_1P_3}| = \sqrt{(2-0)^2 + (5-1)^2} = \sqrt{4+16} = \sqrt{20} = 2\sqrt{5}$$

Since $\sqrt{5} + \sqrt{5} = 2\sqrt{5}$, $|\overline{P_1P_2}| + |\overline{P_2P_3}| = |\overline{P_1P_3}|$, and the three points are collinear.

Figure 3

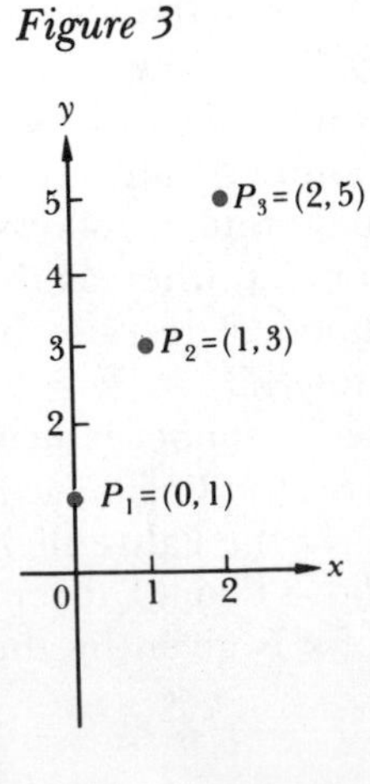

The argument used in the above example can be generalized to show that any three points on the graph of a linear function $f(x) = mx + b$ are collinear; that is, the graph of a linear function is always a straight line.

Thus, if $P_1 = (x_1, y_1)$, $P_2 = (x_2, y_2)$, and $P_3 = (x_3, y_3)$ are on the graph of $f(x) = mx + b$, then P_1, P_2, and P_3 are collinear (Figure 4).

Figure 4

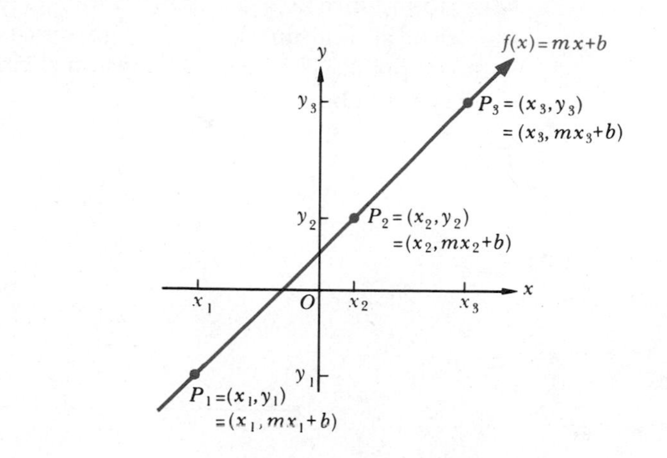

Since the graph of a linear function is a straight line, and a straight line is determined uniquely by any two of its points, it is enough to locate two points in order to determine the graph of a linear function.

In general, the graph of an equation of the form $ax + by + c = 0$, where a, b, and c are constants such that a and b are not both zero, is a straight line. Equations of this form are called *linear equations* (see Problem 26).

EXAMPLES 1 Explain why the constant function $f(x) = 3$ is a linear function.

SOLUTION The function $f(x) = 3$ can be expressed as

$$f(x) = 0 \cdot x + 3$$

Hence f is a linear function with $m = 0$ and $b = 3$. The ordinates of the graph of f equal 3 for *all* values of x in the set of real numbers, which is the domain of f. Consequently, the graph of f contains all points of the form $(x, 3)$, where x is any real number. These points form a horizontal line 3 units above the x axis (Figure 5).

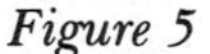

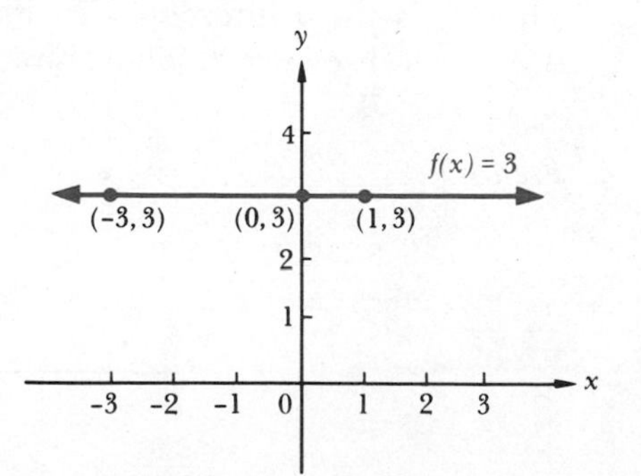

2 Graph the linear function $f(x) = 2x + 5$ and determine the domain and range.

SOLUTION Since $2x + 5$ is defined for *any* real number x, the domain of f is the set of all real numbers. The graph can be determined by locating at least two points (Figure 6). The graph clearly shows the range to be the set of all real numbers.

Figure 6

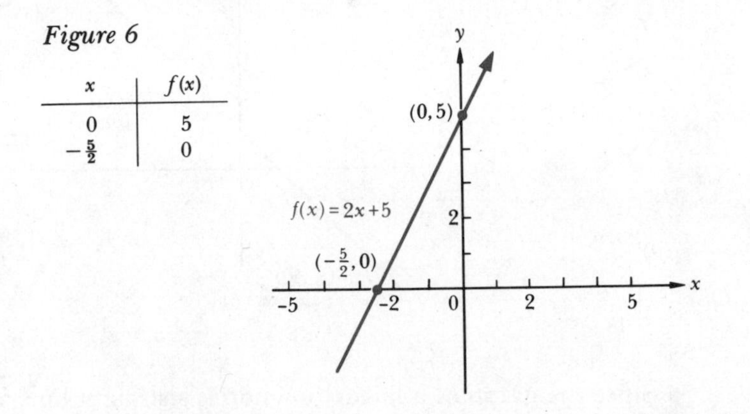

x	$f(x)$
0	5
$-\frac{5}{2}$	0

In Example 2 above, the function f is defined by the equation $y = 2x + 5$. If we set $x = 0$ in the equation $y = 2x + 5$, we obtain $y = 5$. On the other hand, after substituting $y = 0$ into the equation we get $2x + 5 = 0$, so $x = -\frac{5}{2}$. Hence $(-\frac{5}{2}, 0)$ and $(0, 5)$ are the points where the graph of f intersects the x and y axes, respectively (Figure 6). We call the number $-\frac{5}{2}$ the *x intercept* and the number 5 the *y intercept* of the graph of f.

In general, the *x intercept* of a linear function f defined by $y = mx + b$, where $m \neq 0$, is determined by substituting $y = 0$ into the equation and solving for x to obtain $x = -b/m$. The *y intercept* is found by setting $x = 0$ in the equation to obtain $y = b$. Thus the graph of f intersects the x axis at $(-b/m, 0)$ and the y axis at $(0, b)$.

EXAMPLES 1 Sketch the graph of the linear function $y = -\frac{1}{2}x + 1$ by locating the x and y intercepts.

SOLUTION To find the y intercept we set $x = 0$ to get $y = -\frac{1}{2} \cdot 0 + 1 = 1$. Thus, 1 is the y intercept. To find the x intercept, we set $y = 0$ so that $0 = -\frac{1}{2}x + 1$, or $x = 2$. Therefore, 2 is the x intercept (Figure 7).

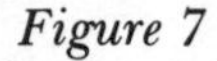

Figure 7

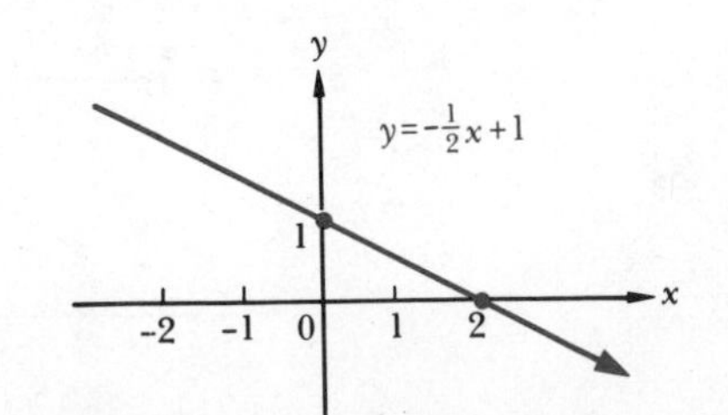

2 Find an expression that defines the linear function $y = f(x)$ if -3 and 5 are the x and y intercepts of the graph, respectively.

SOLUTION One form of the linear function is $y = mx + b$. Since $(-3, 0)$ and $(0, 5)$ both lie on the graph, the two ordered pairs of numbers must satisfy the equation $y = mx + b$ simultaneously. Hence $0 = -3m + b$ and $5 = 0 \cdot m + b$, so $b = 5$ and $0 = -3m + 5$. Thus, $m = \frac{5}{3}$ and the function is defined by $y = \frac{5}{3}x + 5$.

1.1 Slope of a Line

Figure 8 displays the graphs of linear functions defined by the equations $y = x$, $y = 2x$, and $y = 5x$. Note that each of these functions is "increasing" at a different rate. For example, as x increases from 1 to 2, $y = x$ increases from 1 to 2 (a 1-unit increase), $y = 2x$ increases from 2 to 4 (a 2-unit increase), and $y = 5x$ increases from 5 to 10 (a 5-unit increase). What we would like to do next is find some way of measuring the "inclination" or "rate of change" of lines and then relate such a measure to the equation of the line. One way to do this is to determine what is called the *slope of a line*.

Figure 8

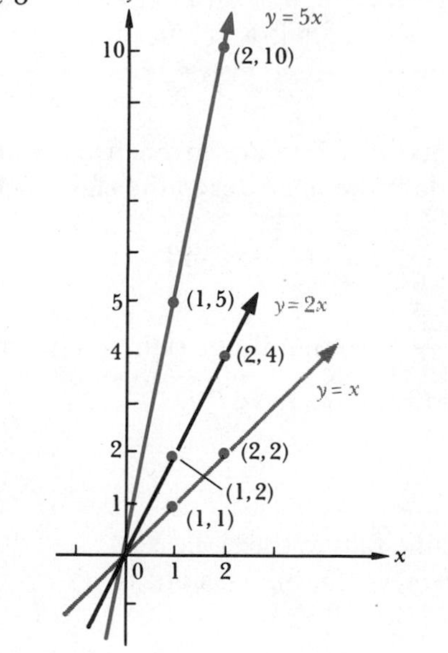

DEFINITION 2 SLOPE OF A LINE

Suppose that (x_1, y_1) and (x_2, y_2) are any two points of a line such that $x_1 \neq x_2$ (Figure 9). The number s that is defined by the equation

$$s = \frac{y_2 - y_1}{x_2 - x_1} \qquad x_1 \neq x_2$$

is called the *slope* of the line.

At times the ratio used to define the slope is described as the ratio of the rise over the run, or more briefly as $\frac{\text{rise}}{\text{run}}$, where the rise is the vertical "directed" distance and the run is the horizontal "directed" distance (Figure 9).

Figure 9

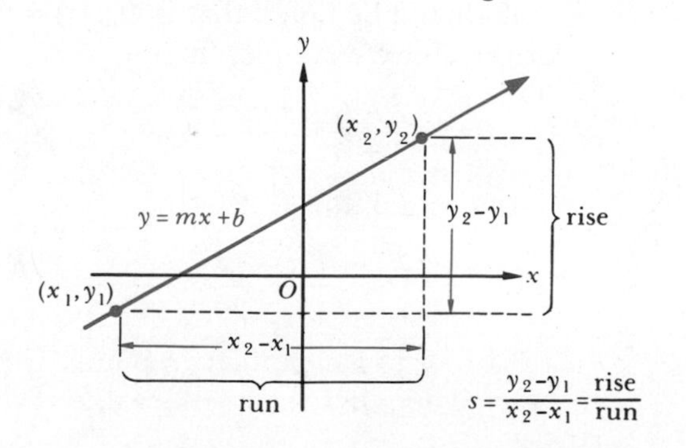

Figure 10

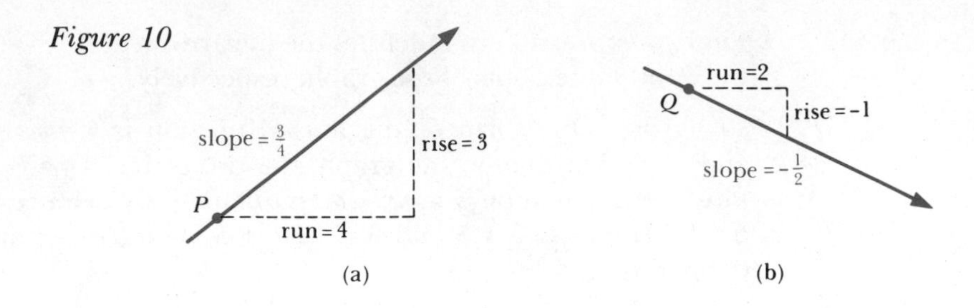

From the geometric viewpoint, a line that contains a point P and has a slope of $\frac{3}{4}$ is "rising" to the right (Figure 10a), whereas a line that contains a point Q and has a slope of $-\frac{1}{2}$ is "falling" to the right (Figure 10b).

In general, if the slope of a line is positive, the line rises to the right and the corresponding linear function that has as its graph the given line is increasing. If the slope of a line is negative, the line falls to the right and the corresponding linear function that has as its graph the given line is decreasing.

EXAMPLE In each of the following, two points of a line are given. Determine the slope, then, without graphing, decide if the line rises to the right or falls to the right.

(a) (6, 2) and (3, 7) (b) (3, −2) and (5, 6)

SOLUTION

(a) The slope is defined by $s = \dfrac{y_2 - y_1}{x_2 - x_1}$, so that if one point $(x_1, y_1) = (6, 2)$ and the second point $(x_2, y_2) = (3, 7)$, we have

$$s = \frac{7-2}{3-6} = \frac{5}{-3} = -\frac{5}{3}$$

Since the slope is negative, the line falls to the right.

(b) Letting $(x_1, y_1) = (3, -2)$ and $(x_2, y_2) = (5, 6)$, the slope is

$$s = \frac{6-(-2)}{5-3} = \frac{8}{2} = 4$$

Since the slope is positive, the line rises to the right.

It should be noted that if $(x_1, y_1) = (5, 6)$ and $(x_2, y_2) = (3, -2)$ in part (b) of the above example, then

$$s = \frac{-2-6}{3-5} = \frac{-8}{-2} = 4$$

In general, since

$$\frac{y_1 - y_2}{x_1 - x_2} = \frac{-(y_1 - y_2)}{-(x_1 - x_2)} = \frac{y_2 - y_1}{x_2 - x_1}$$

the order in which the two points are taken when the coordinates are subtracted does not change the value of the slope.

Notice that the values of the slopes of the lines in the example above were calculated using two particular points. Will the value of the slope of a line be the same no matter what two points on the line are used to compute the slope? This question is answered by the following theorem.

THEOREM 1

The slope of the line that is the graph of $f(x) = mx + b$ is m. In other words, no matter which points are selected to compute the slope, the result will be the same value, m, the coefficient of x.

PROOF Consider any x_1 and x_2 such that $x_1 \neq x_2$. Then $(x_1, f(x_1))$ and $(x_2, f(x_2))$ are two points of the line, so that the slope s can be computed as follows:

$$s = \frac{f(x_2) - f(x_1)}{x_2 - x_1} = \frac{(mx_2 + b) - (mx_1 + b)}{x_2 - x_1}$$

$$= \frac{mx_2 - mx_1}{x_2 - x_1} = \frac{m(x_2 - x_1)}{x_2 - x_1} = m$$

For example, the slope of the line defined by $y = 2x + 1$ is 2; the slope of the line defined by $y = -\frac{1}{2}x$ is $-\frac{1}{2}$.

Because of Theorem 1, we usually use the letter m to represent the slope of a straight line.

If $x_1 = x_2$ for all points on a line, then the formula for the slope of the line would give

$$m = \frac{y_2 - y_1}{x_2 - x_1} = \frac{y_2 - y_1}{0}$$

so the slope is not defined because of the division by 0. This situation occurs whenever the abscissas of all the points on the line are the same. Such a line is parallel to the y axis or coincides with the y axis, and the relation that has as its graph the given line is *not* a function (see Problem 2). For example, the graph of $x = 4$ is a line parallel to the y axis (Figure 11); the slope of this line is not defined.

Figure 11

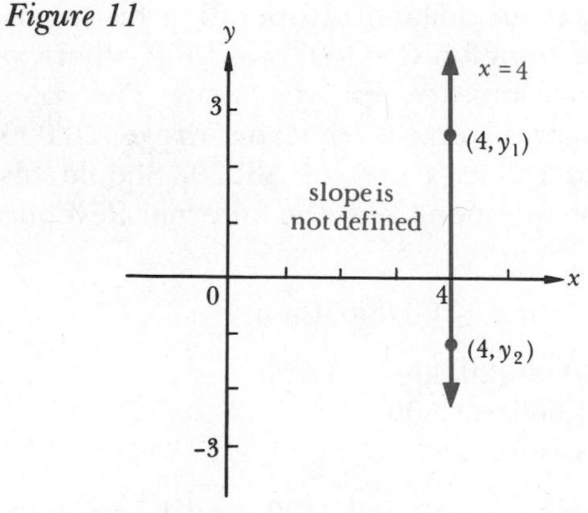

If the slope of a line is zero, then the linear function $f(x) = mx + b$ that defines the line assumes the form $f(x) = b$. That is, the function is constant and its graph is a horizontal line. An example of such a function is $f(x) = 3$ (Figure 5 on page 95).

EXAMPLES In Examples 1–3, determine the slope of the line defined by the given function, where x represents a member of the domain and y represents a member of the range. Use the slope to decide whether the function is increasing, decreasing, or constant and then graph the function. Label the intercepts.

Figure 12

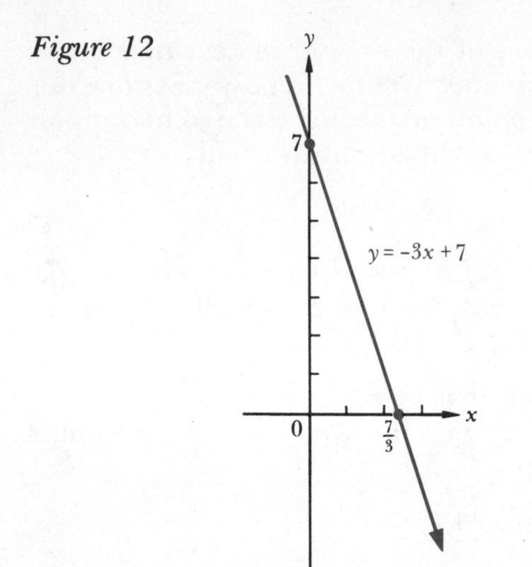

1 $y + 3x = 7$

SOLUTION Solving the equation for y in terms of x, we get $y = -3x + 7$. Hence the slope is -3. Since the slope is negative, the function is decreasing. Setting $x = 0$, we find $y = 7$, so the y intercept is 7. Letting $y = 0$, we have $x = \frac{7}{3}$, so the x intercept is $\frac{7}{3}$ (Figure 12).

2 $3y - 2x = 6$

SOLUTION $3y - 2x = 6$ is equivalent to $y = \frac{2}{3}x + 2$, so the slope is $\frac{2}{3}$ and the function defined by the equation is increasing. The y intercept is 2, and the x intercept is -3 (Figure 13).

Figure 13

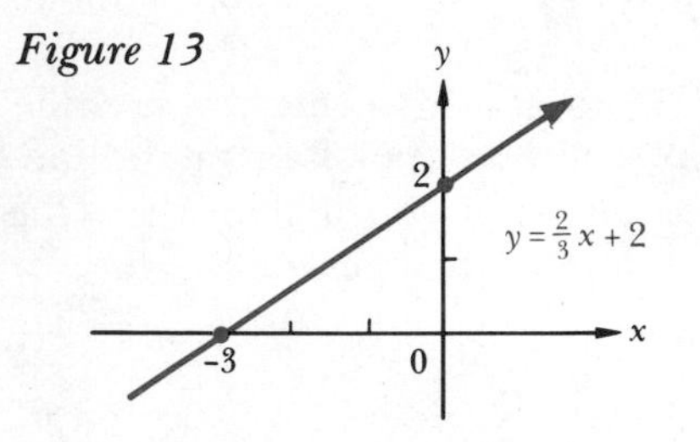

3 $y = -2$

SOLUTION Since $y = -2 = 0 \cdot x - 2$, the slope is 0. The y intercept is -2. The function neither increases nor decreases. There is no x intercept. The graph is a horizontal line (Figure 14).

Figure 14

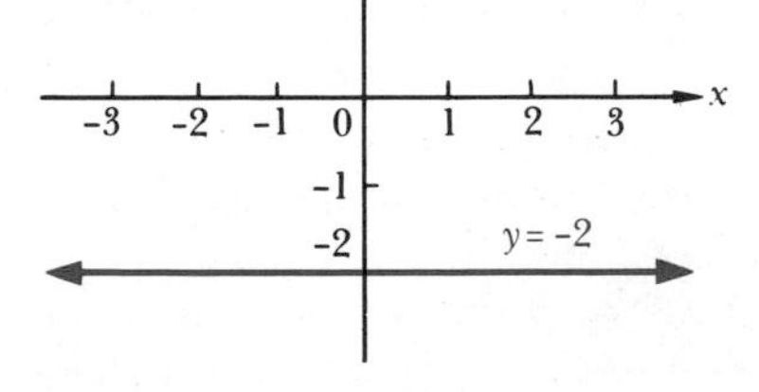

4 When figuring the deductions for his income taxes, a sales representative estimates the annual cost C (in dollars) of operating his car by using the equation $C = 0.08m + 1500$, where m is his total mileage for the year. The sales representative reports that he drove 20,000 miles and claims a cost of \$3250. Should his figures be questioned by the Internal Revenue Service?

SOLUTION If $m = 20{,}000$, then

$$\begin{aligned} C &= (0.08)(20{,}000) + 1{,}500 \\ &= 1{,}600 + 1{,}500 \\ &= 3{,}100 \end{aligned}$$

Since his actual cost is \$3100, and he claims a cost of \$3250, his figures should be questioned.

1.2 Forms of Equations of Lines

SLOPE-INTERCEPT FORM

Given the linear function defined by the equation $y = mx + b$, we have established the fact that m is the slope and b is the y intercept. Since the equation $y = mx + b$ displays the slope and the y intercept, it is called the *slope-intercept form* for the equation of the line.

POINT-SLOPE FORM

Suppose that the slope of a line is m and that (x_1, y_1) is a point on the line. If (x, y) is used to represent any point on the line and $x \neq x_1$, then

$$m = \frac{y - y_1}{x - x_1}$$

Figure 15

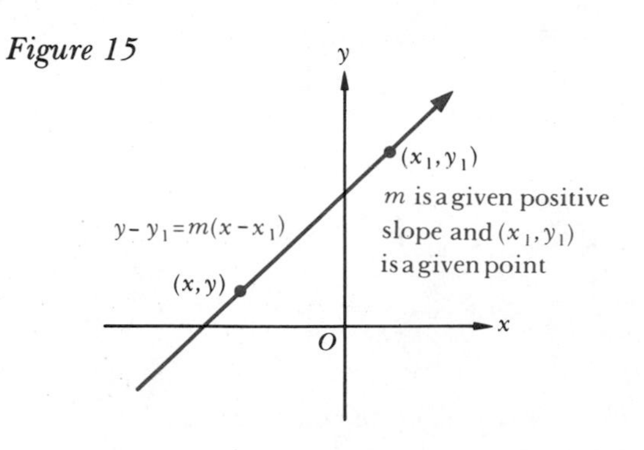

and $y - y_1 = m(x - x_1)$ (Figure 15). Notice that (x_1, y_1) also satisfies the equation $y - y_1 = m(x - x_1)$. The form of the equation of the line

$$y - y_1 = m(x - x_1)$$

where m is the slope and (x_1, y_1) is *any* point on the line, is called the *point-slope form.*

For example, the point-slope form for the equation of the line that contains the point $(-1, 2)$ with slope 3 is $y - 2 = 3[x - (-1)]$ or $y - 2 = 3(x + 1)$.

EXAMPLES In Examples 1 and 2, find both the slope-intercept form and point-slope form for the equations of the given line.

1 The slope is 5 and the line contains the point (2, 3).

SOLUTION Using the point-slope form, that is, $y - y_1 = m(x - x_1)$ with $m = 5$ and $(x_1, y_1) = (2, 3)$, we have $y - 3 = 5(x - 2)$; or, equivalently, $y = 5x - 7$, which is the slope-intercept form for the equation of the line.

2 The line contains the points $(-2, 5)$ and $(3, -4)$.

SOLUTION The slope of the line is given by

$$\frac{5 - (-4)}{-2 - 3} = -\frac{9}{5}$$

The point-slope form for the equation of the line is $y - 5 = -\frac{9}{5}(x + 2)$; equivalently, the slope-intercept form, obtained by solving for y in terms of x, is $y = -\frac{9}{5}x + \frac{7}{5}$.

3 Find the slope-intercept form for the equation of the line defined by the linear function f if the slope is -2 and $f(3) = 1$.

SOLUTION $f(x) = mx + b$ with $m = -2$, so $f(3) = -2(3) + b = 1$; hence, $b = 7$ and $f(x) = -2x + 7$. The slope-intercept form is $y = -2x + 7$.

4 A supplier can sell 26,000 bricks if he charges 15 cents per brick and 8000 bricks if he charges 19 cents per brick. Assume that the relationship between the number of bricks y that the supplier can sell and the price x of each brick is linear. Find an equation that expresses y in terms of x. How many bricks can he sell if he charges 18 cents per brick?

SOLUTION Here $x = 15$ when $y = 26{,}000$ and $x = 19$ when $y = 8{,}000$.
The slope of the line containing the points (15, 26000) and (19, 8000) is

$$\frac{8{,}000 - 26{,}000}{19 - 15} = -\frac{18{,}000}{4} = -4{,}500$$

Substituting into the equation $y - y_1 = m(x - x_1)$, where $(x_1, y_1) = (15, 26000)$ and $m = -4{,}500$, we get

$$y - 26{,}000 = -4{,}500(x - 15)$$

If $x = 18$, then

$$\begin{aligned} y &= -4{,}500(18 - 15) + 26{,}000 \\ &= (-4{,}500)(3) + 26{,}000 \\ &= -13{,}500 + 26{,}000 \\ &= 12{,}500 \end{aligned}$$

The supplier can sell 12,500 bricks if he charges 18 cents per brick.

1.3 Geometry of Two Lines

If two *different* lines are graphed on the same coordinate system, either they are parallel or they intersect at one point. We can use the slopes of two lines to determine whether they are parallel or intersect. For instance, the graphs of the linear functions defined by $y = 3x - 1$ and $y = 3x + 2$ suggest that the two lines are parallel (Figure 16a). Both lines have the same slope, 3. By contrast, the graphs of the linear functions defined by $y = 3x - 1$ and $y = -\frac{1}{3}x + 2$ show that the two lines intersect (Figure 16b). [The point of intersection is $(\frac{9}{10}, \frac{17}{10})$. The point at which two nonparallel lines intersect can be found by solving the equations of the two lines simultaneously. Methods for solving systems of linear equations will be studied in Chapter 8.] In this case the two lines have different slopes, 3 and $-\frac{1}{3}$. More specifi-

Figure 16

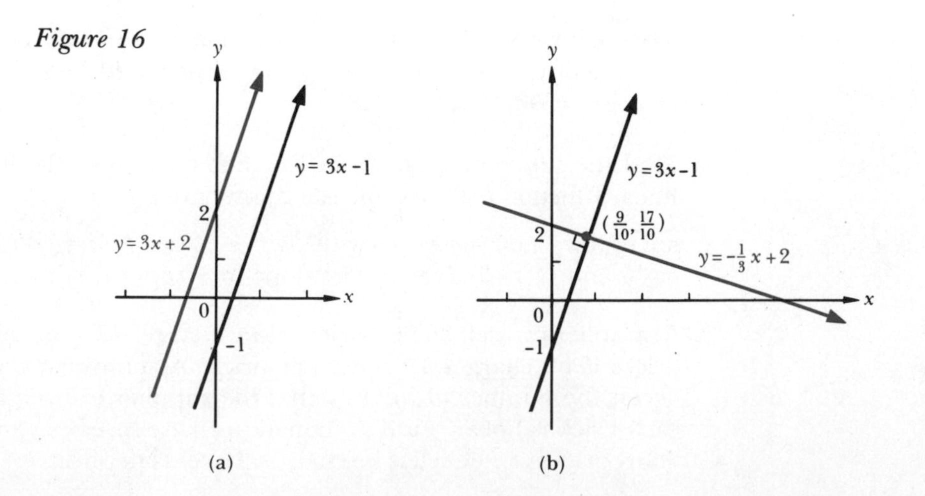

cally, the two lines are actually perpendicular. Observe that the product of the slopes 3 and $-\frac{1}{3}$ is $(3)(-\frac{1}{3}) = -1$. We have the following property, which is presented without proof.

PROPERTY OF SLOPES

(i) Two *different* lines with slopes m_1 and m_2 intersect if and only if $m_1 \neq m_2$ and are parallel if and only if $m_1 = m_2$.

(ii) Two lines with slopes m_1 and m_2 are perpendicular if and only if $m_1 m_2 = -1$.

Property (ii) does not apply to horizontal or vertical lines. Recall that if a line is vertical its slope is undefined; if it is horizontal its slope is zero.

EXAMPLES In Examples 1–3, given a pair of lines, determine whether the lines are parallel or intersect. If the lines intersect, find if they are perpendicular.

1 $x - 2y = 4$; $3x + 2y = 4$

SOLUTION The equation $x - 2y = 4$ can be expressed in the slope-intercept form by solving for y in terms of x to get $y = \frac{1}{2}x - 2$. Similarly, $3x + 2y = 4$ is equivalent to $y = -\frac{3}{2}x + 2$. Hence the slopes of the two lines are, respectively, $\frac{1}{2}$ and $-\frac{3}{2}$. Since the slopes are different, the lines intersect (Figure 17). Also, $(\frac{1}{2})(-\frac{3}{2}) = -\frac{3}{4} \neq -1$, so the lines are *not* perpendicular.

2 The line containing (3, 3) and (5, 6); the line containing (−1, 1) and (1, 4)

SOLUTION The line containing the points $P_1 = (3, 3)$ and $P_2 = (5, 6)$ is parallel to the line containing the points $P_3 = (-1, 1)$ and $P_4 = (1, 4)$, since their slopes are, respectively,

$$m_1 = \frac{6-3}{5-3} = \frac{3}{2} \quad \text{and} \quad m_2 = \frac{4-1}{1+1} = \frac{3}{2} \qquad \text{(Figure 18)}$$

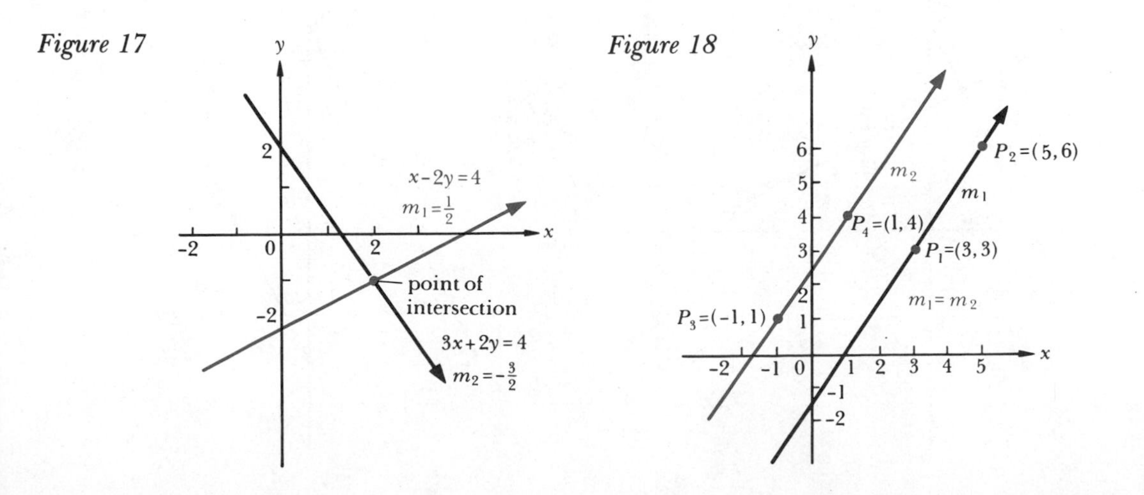

Figure 19

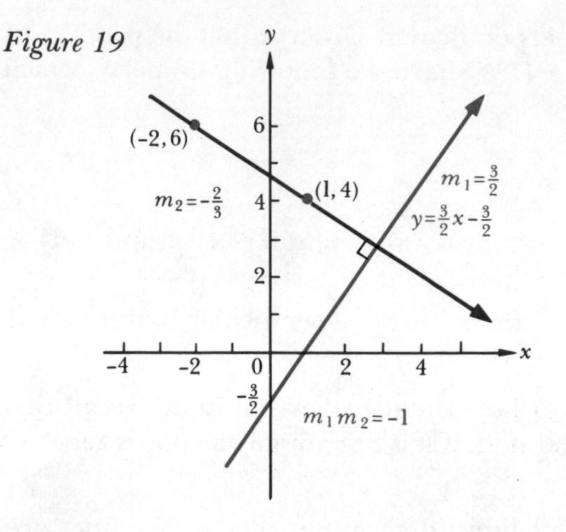

3 $y = \frac{3}{2}x - \frac{3}{2}$; the line containing (1, 4) and (−2, 6)

SOLUTION The line $y = \frac{3}{2}x - \frac{3}{2}$ has slope $m_1 = \frac{3}{2}$. The line containing the points (1, 4) and (−2, 6) has slope

$$m_2 = \frac{6-4}{-2-1} = -\frac{2}{3}$$

Since $m_1 m_2 = (\frac{3}{2})(-\frac{2}{3}) = -1$, the two lines are perpendicular (Figure 19).

4 Given a line L containing the points (−4, −2) and (−2, 2), find the equation of the line that is (a) perpendicular to L and contains the point (3, 2), and (b) parallel to L and contains the point (3, 2).

SOLUTION

(a) The slope of the line L that contains points (−4, −2) and (−2, 2) is

$$\frac{2-(-2)}{-2-(-4)} = \frac{4}{2} = 2$$

so the slope of the perpendicular line is $-\frac{1}{2}$. Hence the equation of the line perpendicular to the given line L and containing (3, 2) is given by $y - 2 = -\frac{1}{2}(x - 3)$, or, equivalently, $y = -\frac{1}{2}x + \frac{7}{2}$ (Figure 20a).

(b) Since the unknown line is parallel to the given line L, its slope must be equal to 2. Thus the equation of the unknown line is $y - 2 = 2(x - 3)$ or, equivalently, $y = 2x - 4$ (Figure 20b).

Figure 20

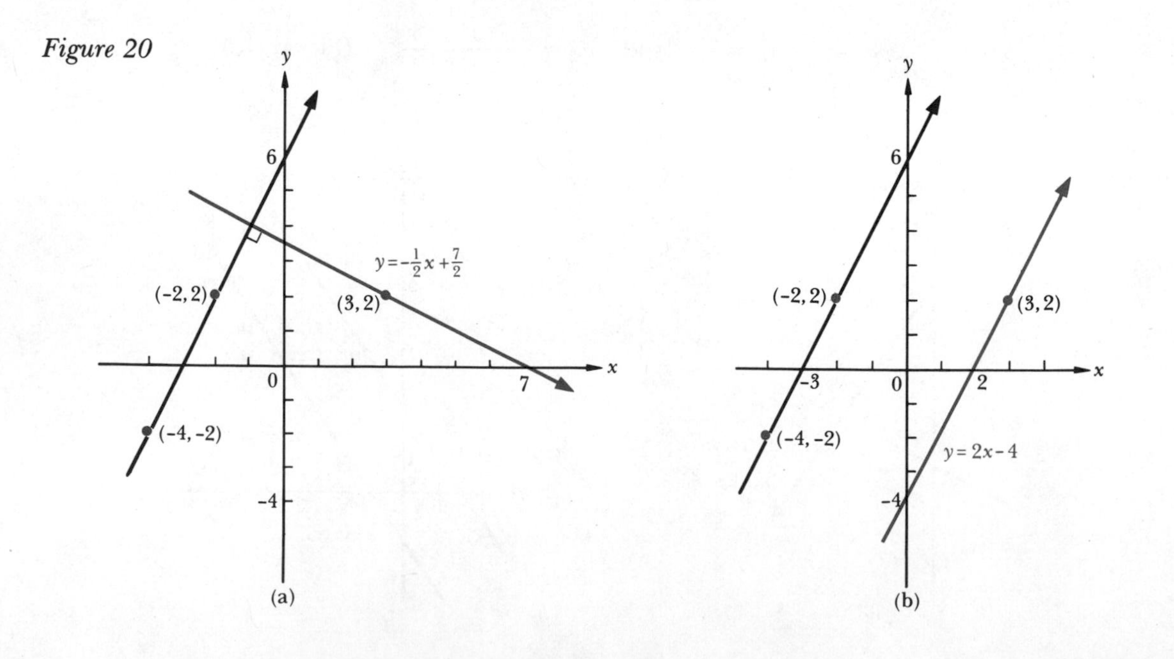

PROBLEM SET 1

1 (a) Use the distance formula to determine whether or not (2, 5), (4, 3), and (3, 4) are collinear.
(b) Use slopes to decide whether or not (1, 1), (2, 4), and (3, 2) are collinear.

2 Let k be a constant. Discuss the relation $\{(x, y) \mid x = k\}$. What is the domain and range? Graph the relation. Is the relation a function?

In Problems 3–11, find the slope of the graph of the linear function defined by the given equation. Use the slope to decide if the function is increasing or decreasing. Graph the function and locate the x and y intercepts. Assume that $y = f(x)$.

3 $y = 5x - 7$ **4** $x + y = 1$ **5** $3x - 5y = 0$

6 $y + 1 = 2(x - 1)$ **7** $y = 5 - 3x$ **8** $\frac{y}{2} - \frac{3x}{5} = 1$

9 $3x - y + 5 = 0$ **10** $y = -2x + 3$ **11** $y - 8 = 0$

12 Graph on the same coordinate system all the linear functions defined by $y = mx$, where $m \in \{-10, -5, -2, -1, 0, 1, 2, 5, 10\}$. What happens to the location of the line as the slope increases from -10 to 10?

In Problems 13–17, find the slope of the line containing the given pair of points. Then, without graphing, decide if the line rises to the right or falls to the right.

13 (0, 0) and (3, 7) **14** (0, 1) and (−1, 3)
15 (−2, 3) and (−3, 5) **16** (−2, −2) and (−3, −5)
17 (−2, 7) and (2, 7)

18 (a) Show that an equation of a line with y intercept $b \neq 0$ and with x intercept $a \neq 0$ can be written in the form

$$\frac{x}{a} + \frac{y}{b} = 1$$

This equation is called the *intercept form* of the equation of a line.
(b) Use the result in part (a) to write in intercept form an equation of the line that contains the points (4, 0) and (0, −6).

In Problems 19–25, find an equation that defines the linear function satisfying the given conditions, and then graph the line. Indicate the form for the equation of the line used.

19 The slope is 3 and the line contains the point (1, 1).
20 The line contains the points (−2, 5) and (2, −3).
21 The slope is 0 and the line contains the point (−3, −2).
22 $f(1) = 2$ and the slope is -3.
23 $f(3) = 3$ and $f(1) = -2$.
24 The line is parallel to the line $3x + 2y + 2 = 0$ and contains the point (−3, −1).
25 The line is perpendicular to the line $3x + 2y + 2 = 0$ and contains the point (−3, −5).

26 Recall that if a, b, and c are constants such that a and b are not both zero, then the graph of $ax + by + c = 0$ is a straight line. This equation is called the *general form* of a linear equation.

(a) Find the slope and the y intercept of the line whose equation is $2x - 3y + 5 = 0$.

(b) Show that $ax + by + c = 0$, where $a \neq 0$ and $b \neq 0$, determines a line that intersects the x and y axes at the points $(-c/a, 0)$ and $(0, -c/b)$, respectively.

In Problems 27–29, for each pair of given lines, determine whether the lines are parallel or intersect. If the lines intersect, indicate if they are perpendicular.

27 $3x - 2y = 7$; $9x - 6y = -1$

28 $y = 5x - 7$; the line containing $(7, 6)$ and $(2, 7)$

29 $y = 3x + 1$; $x - y = 7$

30 Prove that if the product of the slopes of two lines is -1, then the two lines are perpendicular. (*Hint:* Use the distance formula.)

31 Use slopes to prove that the triangles with the following vertices are right triangles. Identify which point is the vertex of the right angle.

(a) $(-4, -2)$, $(2, -8)$, and $(4, 6)$

(b) $(2, 3)$, $(6, 0)$, and $(5, 7)$

32 A projectile fired straight up attains a velocity of v feet per second after t seconds of flight, and the relation between the numbers v and t is linear. If the projectile is fired at a velocity of 200 feet per second and reaches a velocity of 90 feet per second after 2 seconds of flight, express v as a function of t. How soon after it is fired does the projectile reach its highest point?

33 An apartment building was built in 1969 at a cost of \$350,000. What is its value (for tax purposes) in 1990, if it is being *depreciated linearly* over 40 years according to the formula $v = c - (c/N)n$, where c (in dollars) is the original cost of the property, N is the fixed number of years over which the property is depreciated, and v is the value of the undepreciated balance at the end of n years?

34 Assume that the total amount of money spent annually on radio advertising increases at a constant rate. Given the information that the actual 1966 and 1976 expenditures for radio advertising were \$545 million and \$889 million, respectively, find the linear equation of the *trend line* that contains these two points, and use it to predict the amount that will be spent on radio advertising in the year 1986.

35 Let the temperature k meters above the surface of the earth be t degrees Celsius and assume that the function that relates t and k is linear. If the temperature on the surface of the earth is 30° Celsius and the temperature at 2500 meters is 16° Celsius, what is the temperature at 5000 meters?

36 On the 1977 federal tax schedule, a taxpayer earning an income I of at least \$100,000 must pay a tax T of \$45,180 plus 62 percent of the excess over \$100,000.

(a) Assuming that the relationship between I and T is linear, express T as a function of I, where $I \geq 100,000$.

(b) Sketch the graph of the function determined in part (a).

2 Quadratic Functions

Several practical applications from physics and other sciences lead us to investigate other types of functions. For instance, we know from physics that if a ball is thrown straight up from 65 feet above the ground with a speed of 64 feet per second, then its height h in feet above the ground t seconds later is given by the function $h(t) = -16t^2 + 64t + 65$. This is an example of a quadratic function. More formally, we have the following definition.

DEFINITION 1 QUADRATIC FUNCTION

A function f defined by $f(x) = ax^2 + bx + c$, where a, b, and c are real numbers and $a \neq 0$, is called a *quadratic function.*

Examples of quadratic functions include $h(x) = x^2$, $f(x) = 3x^2 - 1$, and $g(x) = \frac{1}{2}x^2 - x$. As with a linear function, the y and x intercepts of the graph of a quadratic function defined by $y = f(x) = ax^2 + bx + c$, with $a \neq 0$, are obtained by setting $x = 0$ and $y = 0$, respectively. To locate the *y intercept,* substitute $x = 0$ into the equation to get $y = c$. The *x intercepts* are located by setting $y = 0$ and then solving the quadratic equation $ax^2 + bx + c = 0$.

2.1 Quadratic Equations

An equation that is equivalent to an equation of the form $ax^2 + bx + c = 0$, where a, b, and c are real numbers and $a \neq 0$, is called a *quadratic equation in x.* For example, $3x^2 = 7x + 3$ and $7x^2 = 5$ are quadratic equations because the first can be expressed as $3x^2 + (-7)x + (-3) = 0$, and the second as $7x^2 + 0x + (-5) = 0$.

Quadratic equations can be solved by one of the following methods. Recall that numbers that satisfy a given equation are referred to as *roots* of that equation.

FACTOR METHOD

We know from the algebra of real numbers that certain quadratic expressions can be represented in *factored form,* that is, as the product of two linear expressions containing real numbers (see Chapter 1, page 12). This factorization can be used to solve quadratic equations by applying the following property:

$$ab = 0 \quad \text{if and only if} \quad a = 0 \text{ or } b = 0$$

For example, the equation $x^2 - 5x + 4 = 0$ may be solved in the following manner:

$$x^2 - 5x + 4 = (x - 1)(x - 4) = 0$$

so

$$x - 1 = 0 \quad \text{or} \quad x - 4 = 0$$
$$x = 1 \qquad\qquad x = 4$$

Hence the roots are 1 and 4, and the solution set is $\{1, 4\}$.

EXAMPLES Find the solution set of each of the following quadratic equations by the factor method.

1 $x^2 - 5x + 6 = 0$

SOLUTION

$$x^2 - 5x + 6 = (x - 2)(x - 3) = 0$$

so

$$x - 2 = 0 \quad \text{or} \quad x - 3 = 0$$
$$x = 2 \qquad\qquad x = 3$$

Thus the roots are 2 and 3, and the solution set is $\{2, 3\}$.

2 $2x^2 - 5x + 2 = 0$

SOLUTION

$$2x^2 - 5x + 2 = (2x - 1)(x - 2) = 0$$

so

$$2x - 1 = 0 \quad \text{or} \quad x - 2 = 0$$
$$x = \tfrac{1}{2} \qquad\qquad x = 2$$

Thus, the roots are $\frac{1}{2}$ and 2, and the solution set is $\{\frac{1}{2}, 2\}$.

COMPLETING-THE-SQUARE METHOD

Suppose that we are to solve a quadratic equation that is not readily factorable. Let us say, for example, that we are to solve $3x^2 - 2x - 2 = 0$. This quadratic equation can be solved by a process known as *completing the square,* which proceeds as follows.

First, "isolate" the x terms of

$$3x^2 - 2x - 2 = 0 \quad \text{to get} \quad 3x^2 - 2x = 2$$

Next, change the resulting equation to an equivalent equation that has 1 as the coefficient of the x^2 term by dividing both sides by 3 to get

$$x^2 - \tfrac{2}{3}x = \tfrac{2}{3}$$

Finally, make the left-hand side a "perfect square" by adding the appropriate number. In order to form a perfect square on the left side, take one-half the coefficient of x, square it, and then add the result to *both* sides of the equation to get

$$\begin{aligned} x^2 - \tfrac{2}{3}x + (\tfrac{1}{3})^2 &= \tfrac{2}{3} + (\tfrac{1}{3})^2 \\ x^2 - \tfrac{2}{3}x + \tfrac{1}{9} &= \tfrac{7}{9} \\ (x - \tfrac{1}{3})^2 &= \tfrac{7}{9} \end{aligned}$$

This last equation implies, then, that

$$x - \frac{1}{3} = \sqrt{\frac{7}{9}} = \frac{\sqrt{7}}{3} \quad \text{or} \quad x - \frac{1}{3} = -\sqrt{\frac{7}{9}} = -\frac{\sqrt{7}}{3}$$

Hence

$$x = \frac{1}{3} + \frac{\sqrt{7}}{3} = \frac{1 + \sqrt{7}}{3} \quad \text{or} \quad x = \frac{1}{3} - \frac{\sqrt{7}}{3} = \frac{1 - \sqrt{7}}{3}$$

and the solution set of the equation is

$$\left\{\frac{1 + \sqrt{7}}{3}, \frac{1 - \sqrt{7}}{3}\right\}$$

EXAMPLES Find the solution set of each of the following quadratic equations by the method of completing the square.

1 $x^2 + 4x + 2 = 0$

SOLUTION The equation

$$x^2 + 4x + 2 = 0$$

is equivalent to

$$x^2 + 4x = -2$$

If the square of one-half the coefficient of the x term, namely, $[\frac{1}{2}(4)]^2 = 4$, is then added to each side of the latter equation, we have

$$x^2 + 4x + 4 = -2 + 4 \quad \text{or} \quad x^2 + 4x + 4 = 2$$

The left-hand side of the equation is a perfect square, so the equation can be written as

$$(x + 2)^2 = 2$$

which implies that $x + 2 = \sqrt{2}$ or $x + 2 = -\sqrt{2}$. Thus

$$x = -2 + \sqrt{2} \quad \text{or} \quad x = -2 - \sqrt{2}$$

and the solution set is $\{-2 + \sqrt{2}, -2 - \sqrt{2}\}$.

2 $2x^2 - 2x - 1 = 0$

SOLUTION

$$2x^2 - 2x = 1 \quad \text{(Isolate } x \text{ terms)}$$
$$x^2 - x = \tfrac{1}{2} \quad \text{(Make coefficient of } x^2 \text{ equal to 1)}$$
$$x^2 - x + \tfrac{1}{4} = \tfrac{1}{2} + \tfrac{1}{4} \quad \text{(Complete the square)}$$
$$(x - \tfrac{1}{2})^2 = \tfrac{3}{4} \quad \text{(Factor and simplify)}$$

Thus

$$x - \frac{1}{2} = \sqrt{\frac{3}{4}} = \frac{\sqrt{3}}{2} \quad \text{or} \quad x - \frac{1}{2} = -\sqrt{\frac{3}{4}} = -\frac{\sqrt{3}}{2}$$

so

$$x = \frac{1}{2} + \frac{\sqrt{3}}{2} = \frac{1+\sqrt{3}}{2} \quad \text{or} \quad x = \frac{1}{2} - \frac{\sqrt{3}}{2} = \frac{1-\sqrt{3}}{2}$$

The solution set is

$$\left\{\frac{1+\sqrt{3}}{2}, \frac{1-\sqrt{3}}{2}\right\}$$

Instead of repeating the process of completing the square for each quadratic equation that is not easily factorable, we can generalize the method of completing the square to arrive at a formula that enables us to solve quadratic equations with relative ease in the real number system. (Later, in Chapter 9, we shall see that the formula can also be used if the quadratic equation does *not* have real number roots.)

THEOREM 1 QUADRATIC FORMULA

If $ax^2 + bx + c = 0$, with a, b, and c real numbers and $a \neq 0$, then the roots of the equation can be determined by the formula

$$x = \frac{-b \pm \sqrt{b^2 - 4ac}}{2a}$$

PROOF Following the method for completing the square given in Example 2 above, we have

$$ax^2 + bx + c = 0$$
$$ax^2 + bx = -c$$
$$x^2 + \frac{bx}{a} = \frac{-c}{a}$$
$$x^2 + \frac{bx}{a} + \left(\frac{b}{2a}\right)^2 = \frac{-c}{a} + \left(\frac{b}{2a}\right)^2$$
$$\left(x + \frac{b}{2a}\right)^2 = \frac{b^2}{4a^2} + \frac{-c}{a} = \frac{b^2 - 4ac}{4a^2}$$

Hence

$$x + \frac{b}{2a} = \frac{\sqrt{b^2 - 4ac}}{2a} \quad \text{or} \quad x + \frac{b}{2a} = -\frac{\sqrt{b^2 - 4ac}}{2a}$$

From this we get

$$x = \frac{-b}{2a} + \frac{\sqrt{b^2 - 4ac}}{2a} \quad \text{or} \quad x = \frac{-b}{2a} - \frac{\sqrt{b^2 - 4ac}}{2a}$$

However,

$$x = \frac{-b \pm \sqrt{b^2 - 4ac}}{2a}$$

is usually used as an abbreviated way of writing these latter two equations.

EXAMPLES Find the solution set of each of the following quadratic equations by the quadratic formula.

1 $x^2 + x - 3 = 0$

SOLUTION Here $a = 1$, $b = 1$, and $c = -3$. Substituting into the formula

$$x = \frac{-b \pm \sqrt{b^2 - 4ac}}{2a}$$

we get $x = \dfrac{-1 \pm \sqrt{1 - 4(1)(-3)}}{2} = \dfrac{-1 \pm \sqrt{1 + 12}}{2} = \dfrac{-1 \pm \sqrt{13}}{2}$

so that $x = \dfrac{-1 - \sqrt{13}}{2}$ or $x = \dfrac{-1 + \sqrt{13}}{2}$.

The solution is $\left\{ \dfrac{-1 - \sqrt{13}}{2}, \dfrac{-1 + \sqrt{13}}{2} \right\}$.

2 $2x^2 - 8x + 3 = 0$

SOLUTION Here $a = 2, b = -8$, and $c = 3$, so

$$\begin{aligned} x &= \frac{-b \pm \sqrt{b^2 - 4ac}}{2a} \\ &= \frac{-(-8) \pm \sqrt{64 - 4(2)(3)}}{2(2)} = \frac{8 \pm \sqrt{64 - 24}}{4} \\ &= \frac{8 \pm \sqrt{40}}{4} = \frac{8 \pm 2\sqrt{10}}{4} \\ &= \frac{2(4 \pm \sqrt{10})}{4} = \frac{4 \pm \sqrt{10}}{2} \end{aligned}$$

so $x = 2 + \dfrac{\sqrt{10}}{2}$ or $x = 2 - \dfrac{\sqrt{10}}{2}$.

The solution set is $\left\{ 2 - \dfrac{\sqrt{10}}{2}, 2 + \dfrac{\sqrt{10}}{2} \right\}$.

3 $5x^2 + x + 1 = 0$

SOLUTION $a = 5, b = 1$, and $c = 1$, so

$$x = \frac{-1 \pm \sqrt{1 - 4(5)(1)}}{10} = \frac{-1 \pm \sqrt{-19}}{10}$$

Since $\sqrt{-19}$ is not defined in the real number system, we conclude that the equation does not have any real number roots and the solution set is $\varnothing$.

The number $b^2 - 4ac$ is called the *discriminant.* The discriminant of a quadratic equation indicates the *type of roots* that a quadratic equation has. If it is zero, there is only one real number root, a double root. (Why?) If it is positive, there are two distinct real number roots. (Why?) If it is negative, there are no real number roots. (Why?)

EXAMPLE Determine the type of roots of each quadratic equation by using the discriminant.

(a) $2x^2 - 4x + 1 = 0$
(b) $x^2 - 4x + 4 = 0$
(c) $2x^2 + 3x + 2 = 0$

SOLUTION

(a) Since $a = 2, b = -4$, and $c = 1$, we have $b^2 - 4ac = (-4)^2 - 4(2)(1) = 8$. Therefore, the equation has two distinct real number roots.
(b) Since $a = 1$, $b = -4$, and $c = 4$, then $b^2 - 4ac = (-4)^2 - 4(1)(4) = 0$, and the equation has one real number root, a double root.
(c) Since $a = 2$, $b = 3$, and $c = 2$, then $b^2 - 4ac = 3^2 - 4(2)(2) = -7$, and the equation has no real number roots.

Some equations that are not quadratic can still be solved by the methods of this section. These equations are equivalent in form to $au^2 + bu + c = 0$, where u is an expression involving another variable. For example, the equations $x^4 - 5x^2 + 6 = 0$, $x^{-2} + x^{-1} - 6 = 0$, and $(x^2 - 1)^2 + 2(x^2 - 1) - 3 = 0$ are quadratic in form. The following example illustrates a technique for solving such equations.

EXAMPLE Find the solution set of

$$\left(x - \frac{8}{x}\right)^2 + \left(x - \frac{8}{x}\right) - 42 = 0$$

SOLUTION Let $u = x - \dfrac{8}{x}$ and substitute u in the given equation to obtain

$$u^2 + u - 42 = (u + 7)(u - 6) = 0$$

Then

$$u + 7 = 0 \quad \text{or} \quad u - 6 = 0$$
$$u = -7 \qquad\qquad u = 6$$

Thus

$$x - \frac{8}{x} = -7 \quad \text{or} \quad x - \frac{8}{x} = 6$$

so that

$$x^2 + 7x - 8 = 0 \quad \text{or} \quad x^2 - 6x - 8 = 0$$

Since -8 and 1 are the roots of the equation $x^2 + 7x - 8 = 0$ (why?), and $3 - \sqrt{17}$ and $3 + \sqrt{17}$ are the roots of the equation $x^2 - 6x - 8 = 0$ (why?), it follows that the solution set is $\{-8, 1, 3 - \sqrt{17}, 3 + \sqrt{17}\}$.

2.2 Properties of Quadratic Functions

Figure 1

The graph of any quadratic function f defined by $f(x) = ax^2 + bx + c$ will have the same general shape as the graph of $f(x) = x^2$ (Figure 1), although the location of the graph and whether it opens upward or downward will vary depending on the specific values of a, b, and c.

Several examples of graphs of quadratic functions are shown in Figure 2. These graphs are called *parabolas.* (A detailed discussion of parabolas is given in Chapter 7, Section 4.) Note that parabolas may open upward and have low points or *minimum points* (Figures 2a and 2c); other parabolas open downward and have high points or *maximum points* (Figures 2b and 2d).

Figure 2

(a)

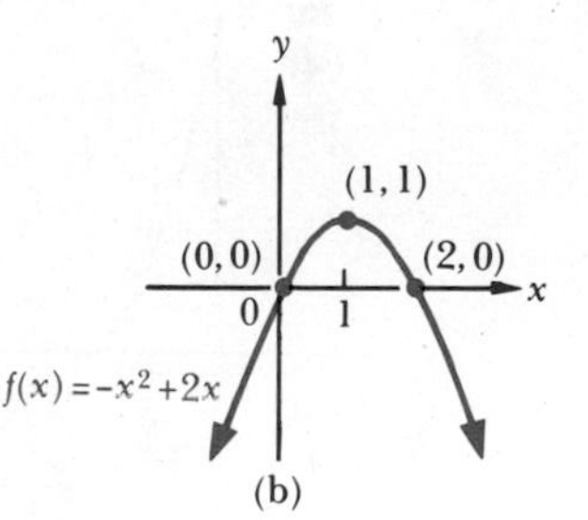

(b)

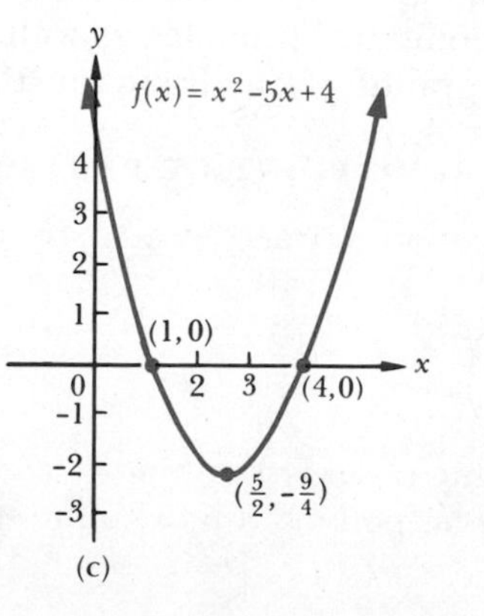

(c)

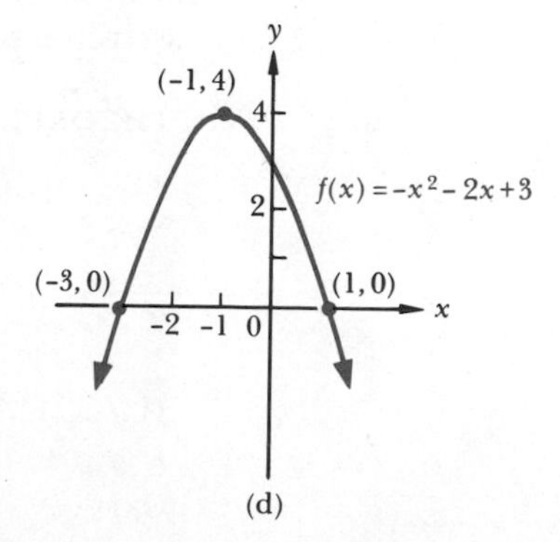

(d)

If the graph of a function has a maximum or a minimum point, the function is said to have an *extreme point.*

In graphing most quadratic functions, rather than plot a number of points, we locate the x and y intercepts and the extreme point of the parabola and then use these points to get a rough sketch of the graph.

EXAMPLE Sketch the graph of the function defined by $y = f(x) = 3x^2 - 5x + 2$ by locating the intercepts and the extreme point of the parabola. Use the graph to determine the range of f.

SOLUTION To find the y intercept, let $x = 0$. Then $y = 2$, so the y intercept is 2. In order to find the x intercepts, we let $y = 0$ to obtain the quadratic equation

$$3x^2 - 5x + 2 = (3x - 2)(x - 1) = 0$$

so that $x = \frac{2}{3}$ or $x = 1$. Thus, the x intercepts are $\frac{2}{3}$ and 1. To find the extreme point (the process used here to determine the extreme point will be replaced with an easier method later), first factor out 3 from the two terms involving x in the given quadratic function $y = 3x^2 - 5x + 2$ to get

$$y = 3(x^2 - \tfrac{5}{3}x) + 2$$

Next, complete the square of the expression $x^2 - \frac{5}{3}x$ to obtain

$$y = 3(x^2 - \tfrac{5}{3}x + \tfrac{25}{36}) + 2 - \tfrac{25}{12} = 3(x - \tfrac{5}{6})^2 - \tfrac{1}{12}$$

Figure 3

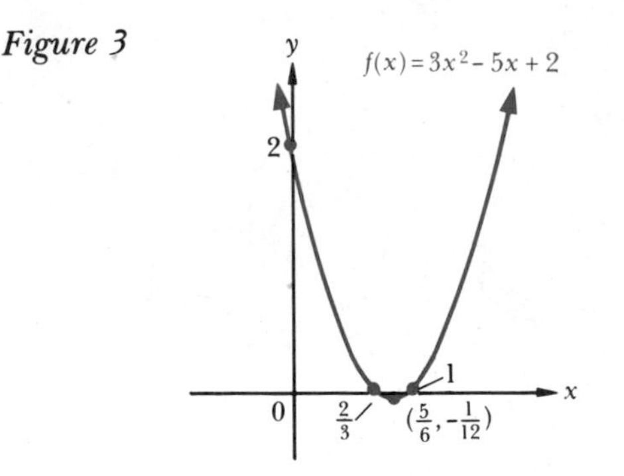

Since $3(x - \frac{5}{6})^2 \geqslant 0$, it follows that $y \geqslant -\frac{1}{12}$. This means that y is always greater than or equal to $-\frac{1}{12}$, so $-\frac{1}{12}$ is a *minimum value* and the parabola opens upward. Notice that the minimum value for y occurs when $x - \frac{5}{6} = 0$, that is, when $x = \frac{5}{6}$, so that $(\frac{5}{6}, -\frac{1}{12})$ is the minimum point. Next, the graph of f is obtained by locating the intercepts and minimum point and then sketching a parabola that contains the points (Figure 3). The graph shows that the range of f is $\{y \mid y \geqslant -\frac{1}{12}\}$.

The next theorem generalizes the process used in the above example to find the extreme point and provides us with a simple way of locating the extreme points of graphs of quadratic functions.

THEOREM 2 EXTREME POINT OF PARABOLA

The quadratic function defined by $y = f(x) = ax^2 + bx + c$ has an extreme point

$$\left(\frac{-b}{2a}, f\left(\frac{-b}{2a}\right)\right)$$

If $a > 0$, the extreme point is a minimum and the parabola opens upward. If $a < 0$, the extreme point is a maximum and the parabola opens downward.

PROOF First, we factor out a from the two terms involving x in the equation $y = ax^2 + bx + c$ to get

$$y = a\left(x^2 + \frac{b}{a}x\right) + c$$

Next we complete the square of the expression $x^2 + (b/a)x$ to obtain

$$\begin{aligned} y &= a\left(x^2 + \frac{b}{a}x + \frac{b^2}{4a^2}\right) + c - \frac{b^2}{4a} \\ &= a\left(x + \frac{b}{2a}\right)^2 + c - \frac{b^2}{4a} \end{aligned}$$

If $a > 0$, then

$$a\left(x + \frac{b}{2a}\right)^2 \geqslant 0 \quad \text{since} \quad \left(x + \frac{b}{2a}\right)^2 \geqslant 0$$

Hence

Figure 4

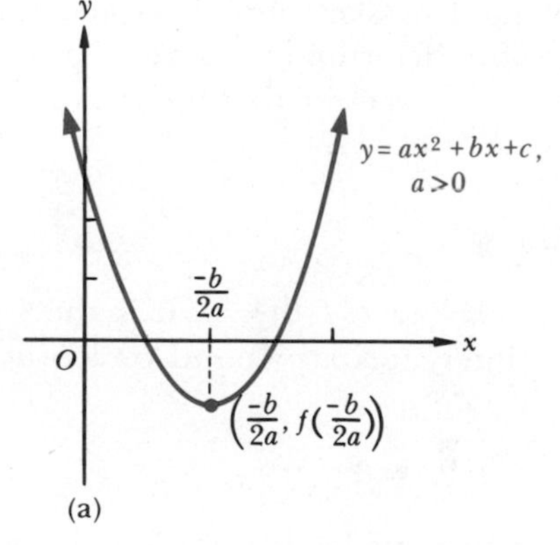

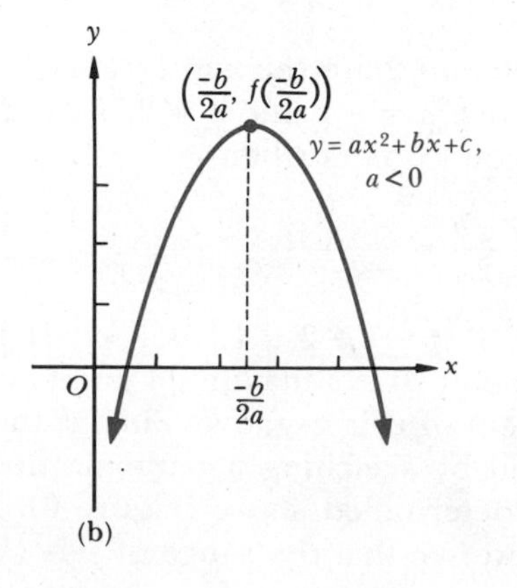

(i) $y = \left[a\left(x + \frac{b}{2a}\right)^2\right] + c - \frac{b^2}{4a} \geqslant c - \frac{b^2}{4a}$

if $a > 0$ (Figure 4a).

(ii) $y \leqslant c - \frac{b^2}{4a}$

if $a < 0$ (Figure 4b).

For verification of (ii), see Problem 26.

Now, if we next substitute $x = -b/2a$ into the given quadratic function $y = ax^2 + bx + c$, we get

$$\begin{aligned} y &= a\left(\frac{b^2}{4a^2}\right) + b\left(\frac{-b}{2a}\right) + c \\ &= c - \frac{b^2}{4a} \end{aligned}$$

which is the same as the right-hand expressions in both inequalities (i) and (ii) above. Consequently, if $a > 0$, $f(-b/2a)$ is the *minimum* value of y; the graph of f, a parabola, opens upward as in Figure 4a. If $a < 0$, $f(-b/2a)$ is the *maximum* value of y; the graph of f, a parabola, opens downward as in Figure 4b. In either case, the value of the function $f(x) = ax^2 + bx + c$ at $x = -b/2a$ gives the *extreme* value of $y = f(x)$. Thus the *extreme point* of the graph of a quadratic function $y = f(x)$ is

$$\left(\frac{-b}{2a}, f\left(\frac{-b}{2a}\right)\right)$$

EXAMPLES In Examples 1 and 2, sketch the graph of the quadratic function by locating the x and y intercepts and the extreme point of the parabola. Use the graph to determine the range of the function.

1 $y = x^2 - 3x + 2$, where $y = f(x)$

SOLUTION If $x = 0$, then $y = 2$, so the y intercept is 2. The x intercepts occur when $y = 0$, that is, when

$$x^2 - 3x + 2 = (x - 2)(x - 1) = 0$$

so that 2 and 1 are the x intercepts.

Using Theorem 2 above, the extreme point occurs when

$$x = \frac{-b}{2a} = \frac{-(-3)}{2} = \frac{3}{2}$$

Since

$$f(\tfrac{3}{2}) = (\tfrac{3}{2})^2 - 3(\tfrac{3}{2}) + 2 = \tfrac{9}{4} - \tfrac{9}{2} + 2 = -\tfrac{1}{4}$$

$(\frac{3}{2}, -\frac{1}{4})$ is the extreme point. Since the coefficient of x^2 is positive, the extreme point is a minimum. A rough sketch of the graph of $y = x^2 - 3x + 2$ is obtained by sketching a parabola containing the points determined above (Figure 5). From the graph, it is clear that the range of the function is $\{y \mid y \geq -\frac{1}{4}\}$.

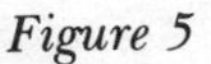

Figure 5

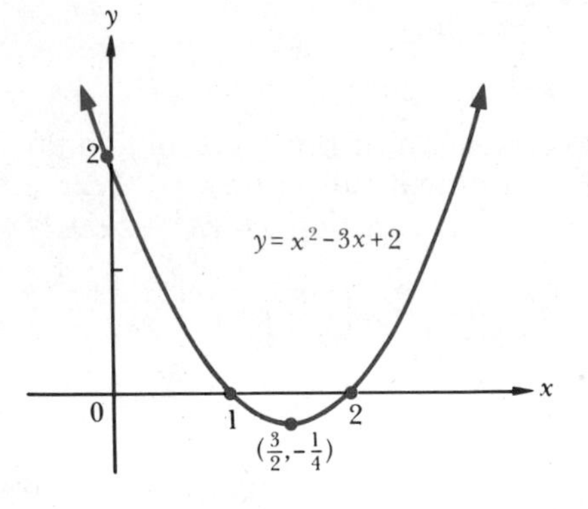

2 $f(x) = -x^2 + 2x$

SOLUTION Since $f(0) = 0$, 0 is the y intercept. The x intercepts are found by letting $f(x) = 0$, so that

$$-x^2 + 2x = 0$$

Solving this quadratic equation yields

$$-x^2 + 2x = x(-x + 2) = 0$$

so that 0 and 2 are the x intercepts.

We have $a = -1$ and $b = 2$, so that the extreme point occurs when

$$x = \frac{-b}{2a} = \frac{-2}{-2} = 1$$

Since $f(1) = -1 + 2 = 1$, then $(1, 1)$ is the extreme point. It is a maximum point because the coefficient of x^2 is negative. Finally, the graph is obtained by sketching a parabola through the points determined above (Figure 6). From the graph we see that the range of f is $\{y \mid y \leq 1\}$.

Figure 6

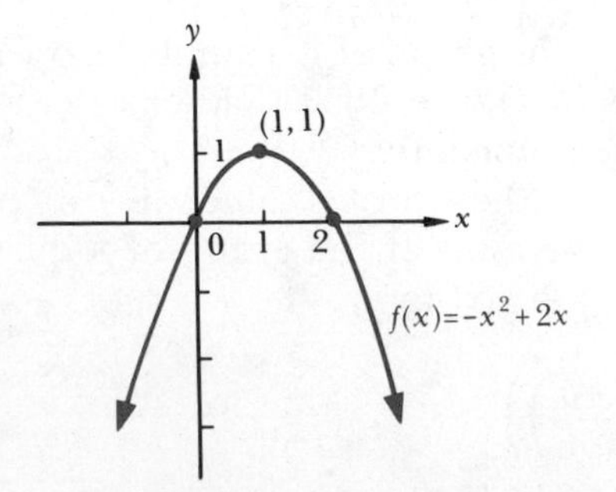

3 A projectile is fired from the ground in such a way that it is h feet above ground level t seconds after the firing. If $h = 96t - 16t^2$, find (a) h when $t = 1$, (b) the maximum height reached by the projectile, and (c) the graph of the equation.

Figure 7

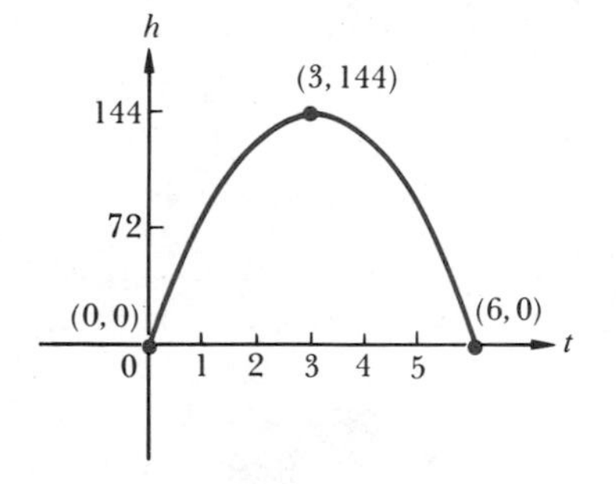

SOLUTION

(a) At $t = 1$, $h = 96(1) - 16(1)^2 = 80$.

(b) $h = -16t^2 + 96t$, so the maximum value of h occurs when

$$t = \frac{-b}{2a} = \frac{-96}{-32} = 3$$

Substituting $t = 3$ into $h = 96t - 16t^2$, then

$$h = 96(3) - 16(9) = 144$$

Hence, h has a maximum value of 144 feet when $t = 3$ seconds.

(c) The t intercepts are found by setting $h = 0$. Thus,

$$96t - 16t^2 = -16t(t - 6) = 0$$

so $t = 0$ or $t = 6$. Notice that the h intercept is 0. The graph is determined by constructing the parabola that contains (0, 0), (6, 0), and the maximum point, (3, 144) (Figure 7). The graph does not extend beyond the interval [0, 6] on the t axis. (Why?)

4 A woman wants to design a rectangular vegetable garden with an ornamental fence around it. The fencing for three sides of the garden costs \$2 per foot. The fencing for the fourth side costs \$3 per foot. She is spending \$120 on the fence. What dimensions should she give the garden to maximize its area?

SOLUTION Let x feet be the length of the side costing \$3 per foot. Let y feet be the width of the garden (Figure 8). Since the cost of the fence is \$120 and it costs \$2 per foot for the other three sides, we have the equation $120 = 2(x + 2y) + 3x$ or, equivalently,

Figure 8

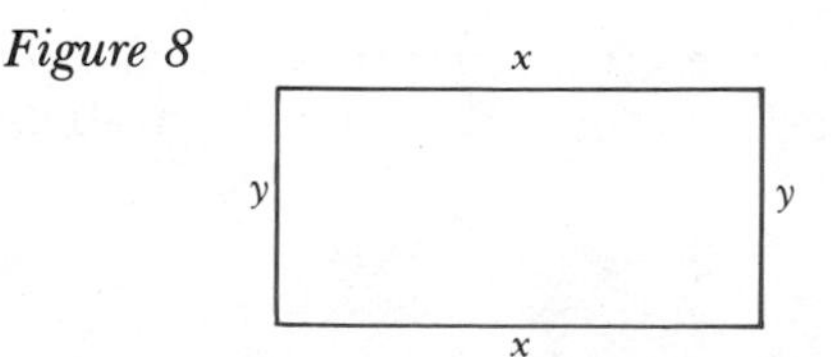

$$y = \frac{120 - 5x}{4}$$

The area A of the garden is given by $A = xy$, so

$$A(x) = x\left(\frac{120 - 5x}{4}\right) = \frac{-5x^2}{4} + 30x$$

Thus A is a quadratic function with $a = -\frac{5}{4}$ and $b = 30$. Hence the maximum area is obtained when

$$x = \frac{-b}{2a} = 12 \quad \text{so} \quad y = \frac{120 - 60}{4} = 15$$

Therefore, the dimensions of the garden that give a maximum area are 12 feet by 15 feet.

PROBLEM SET 2

In Problems 1–6, use the factor method to find the solution set of the given quadratic equation.

1 $x^2 + 5x = 0$
2 $x^2 - 49 = 0$
3 $x^2 - 6x + 8 = 0$
4 $3x^2 - 7x = 0$
5 $-6x^2 + 5x - 1 = 0$
6 $-2x^2 + 19x + 33 = 0$

In Problems 7–12, use the method of completing the square to find the solution set of the quadratic equation.

7 $x^2 - 2x - 2 = 0$
8 $x^2 + 10x + 3 = 0$
9 $x^2 + 3x - 1 = 0$
10 $x^2 - x - 1 = 0$
11 $-3x^2 + 8x - 2 = 0$
12 $-2x^2 + 4x + 1 = 0$

In Problems 13–18, determine the discriminant and then find the solution set of the quadratic equation by using the quadratic formula.

13 $2x^2 + x - 1 = 0$
14 $2x^2 - 5x + 1 = 0$
15 $x^2 - 12x + 4 = 0$
16 $6x^2 - x - 5 = 0$
17 $-6x^2 + 5x + 2 = 0$
18 $x^2 - 9x + 4 = 0$

In Problems 19–25, find the solution set of each equation. Reduce the given equation to a quadratic equation by using the suggested substitution for u.

19 $x^4 - 13x^2 + 36 = 0, u = x^2$
20 $x^3 - 9x^{3/2} + 8 = 0, u = x^{3/2}$
21 $(x^2 + 1)^2 - 3(x^2 + 1) + 2 = 0, u = x^2 + 1$
22 $(2x^2 + 7x)^2 - 3(2x^2 + 7x) = 10, u = 2x^2 + 7x$
23 $\left(3x - \frac{2}{x}\right)^2 + 6\left(3x - \frac{2}{x}\right) + 5 = 0, u = 3x - \frac{2}{x}$
24 $\left(x - \frac{5}{x}\right)^2 - 2x + \frac{10}{x} = 8, u = x - \frac{5}{x}$
25 $\frac{x+1}{x} - 3\left(\frac{x}{x+1}\right) + 2 = 0, u = \frac{x}{x+1}$

26 (a) Prove that if $a < 0$ and $y = ax^2 + bx + c$, then $y \leq c - (b^2/4a)$.
(b) Explain why quadratic functions do *not* have inverse functions.

In Problems 27–38, sketch the graph of the given quadratic function by locating the x and y intercepts and the extreme point of the parabola. Use the graph to determine the range.

27 $f(x) = x^2 - 9$
28 $g(x) = x^2 + 2x$
29 $g(x) = x^2 + 5x - 14$
30 $f(x) = x^2 + 3x - 9$
31 $f(x) = -x^2 - 4x - 4$
32 $h(x) = -x^2 - 1$
33 $f(x) = 6 - x - x^2$
34 $g(x) = -x^2 - x + 12$
35 $g(x) = 2x^2 + 4x - 3$
36 $f(x) = 3x^2 - x - 3$
37 $f(x) = -2x^2 + 3x - 1$
38 $f(x) = x^2 + x + 5$

39 (a) Graph on the same coordinate system the equation $y = ax^2$, where $a \in \{-10, -5, -2, -1, 0, 1, 2, 5, 10\}$. Compare the graphs to the different values of a. What do you notice?

(b) Sketch the graph of each of the following four functions on the same coordinate system by using the graph of $y = x^2$.

$$f(x) = x^2 - 2 \qquad f(x) = x^2 - 1$$
$$f(x) = x^2 + 1 \qquad f(x) = x^2 + 2$$

In general, how do graphs of functions of the form $f(x) = x^2 + k$, where k is a constant, compare for different values of k?

40 From the graph of $f(x) = x^2$, obtain the graph of each function by an appropriate horizontal shift.
(a) $g(x) = (x - 1)^2$ (b) $h(x) = (x + 2)^2$

41 A projectile is fired from a balloon in such a way that it is h feet above the ground t seconds after the firing. If $h = -16t^2 + 96t + 256$, find
(a) h when $t = 0$
(b) The maximum height reached by the projectile
(c) The graph of the equation

42 A guided missile follows a parabolic path according to the equation $P(t) = -t^2 + t + 1$, where P, in feet, represents the height and t, in seconds, represents the time elapsed. Determine the value of t at which the missile reaches its highest point and at which the missile strikes the ground.

43 A rectangular field is adjacent to a river and is to have fencing on three sides, the side on the river requiring no fencing. What is the largest area that can be enclosed if 50 yards of fencing is used?

44 A travel agency advertises all-expenses-paid trips to the World Series for special groups. Transportation is by charter bus, which seats 48 passengers, and the charge per person is \$80 plus an additional \$2 for each empty seat. (Thus, if there are four empty seats each person has to pay \$88; if there are six empty seats each person has to pay \$92; and so on.) If there are x empty seats, how many passengers are there on the bus? How much does each passenger have to pay? Determine and then graph the function that relates the travel agency's total receipts to the number of empty seats.

45 In chemistry, a catalyst is defined as a substance that alters the rate of a chemical reaction without itself undergoing a change; this phenomenon is called *catalysis.* If, in a chemical reaction, the product of the reaction serves as a catalyst for the reaction, then the process is called *autocatalysis.* Suppose that in an autocatalytic process we start with an amount A of a given substance. Let x be the amount of the product. It is assumed that the rate of reaction R depends on both the amount of the product x and the amount of remaining substance $(A - x)$. If this rate is given by the equation $R = kx(A - x)$, where A and k are known positive constants, for what amount of x is the rate of reaction greatest?

3 Quadratic Inequalities

The graphs of quadratic functions discussed in the previous section can be used to solve inequalities in any one of these forms: $ax^2 + bx + c > 0$, $ax^2 + bx + c \geq 0$, $ax^2 + bx + c < 0$, or $ax^2 + bx + c \leq 0$, where $a \neq 0$. Such inequalities are called *quadratic inequalities.*

Two such quadratic inequalities are $x^2 + 5x + 6 < 0$ and $x^2 + 5x + 6 > 0$. To solve these two inequalities, consider the graph of the quadratic function $y = f(x) = x^2 + 5x + 6$ (Figure 1). Observe that the graph of f intersects the x axis at -3 and -2. Notice also that if $x < -3$ or if $x > -2$, then $f(x) > 0$; if $-3 < x < -2$, then $f(x) < 0$. Hence, $x^2 + 5x + 6 > 0$ whenever $x < -3$ or $x > -2$, and $x^2 + 5x + 6 < 0$ when $-3 < x < -2$.

Figure 1

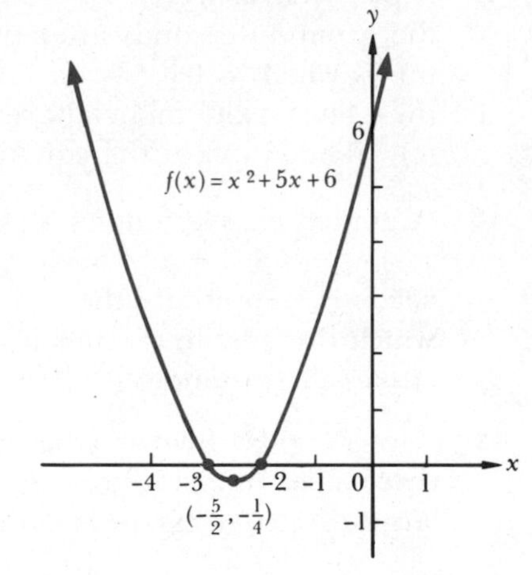

EXAMPLES 1 Use the graph of $f(x) = 2x^2 + x - 6$ to find the solution set of the inequality $2x^2 + x - 6 < 0$.

SOLUTION From the graph of f in Figure 2, we see that the given inequality is satisfied by all values of x such that $f(x) < 0$. Thus, $2x^2 + x - 6 < 0$ if $-2 < x < \frac{3}{2}$, and the solution set is $\{x \mid -2 < x < \frac{3}{2}\}$.

Figure 2

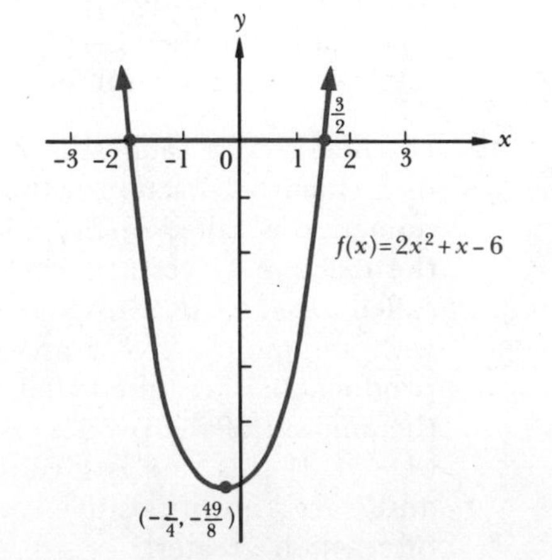

2 Use the graph of $f(x) = x^2 - 2x - 3$ to find the solution set of the inequality $x^2 - 2x - 3 \geq 0$.

SOLUTION The graph of $f(x) = x^2 - 2x - 3$ (Figure 3) indicates that the inequality is satisfied by all values of x such that $f(x) \geq 0$. Hence, the inequality $x^2 - 2x - 3 \geq 0$ is true if $x \leq -1$ or $x \geq 3$, and the solution set is $\{x | x \leq -1 \text{ or } x \geq 3\}$.

Figure 3

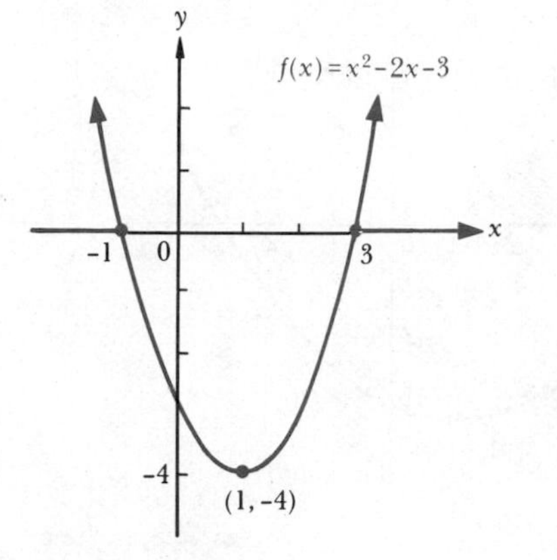

3 Find the solution set of $x^2 < -1$.

SOLUTION The inequality $x^2 < -1$ is equivalent to $x^2 + 1 < 0$. Let us examine the graph of the associated function $f(x) = x^2 + 1$ (Figure 4). Since $x^2 + 1 > 0$ for *all* real numbers, $x^2 < -1$ has no real number solution and the solution set is $\varnothing$.

Figure 4

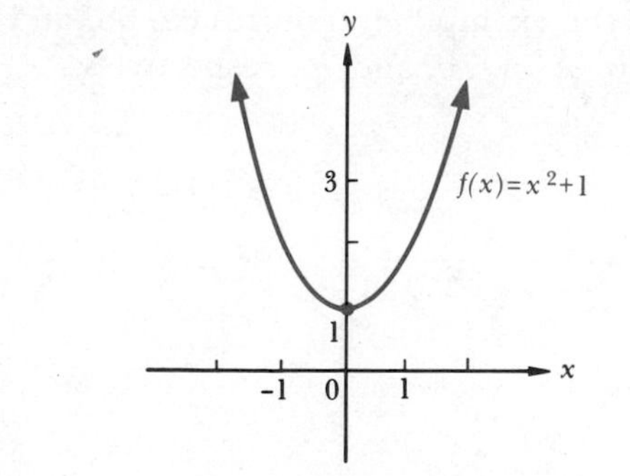

3.1 Cut-Point Method

We have seen how to solve quadratic inequalities by using the graphs of the associated functions. Although this method is easy to follow, another algebraic method, which is based on the ideas presented above, can be used. We shall refer to this method as the *cut-point method.* The *cut points* of a quadratic inequality are merely the x intercepts of the graph of the associated quadratic function.

For example, the cut points of $x^2 + 2x - 15 < 0$ are the x intercepts of the graph of $f(x) = x^2 + 2x - 15$, namely, -5 and 3 (Figure 5a). These points divide the number line into three intervals: $(-\infty, -5)$, $(-5, 3)$, and $(3, \infty)$ (Figure 5b). Notice that $x^2 + 2x - 15 > 0$ if $x < -5$ or $x > 3$, and $x^2 + 2x - 15 < 0$ if $-5 < x < 3$.

Figure 5

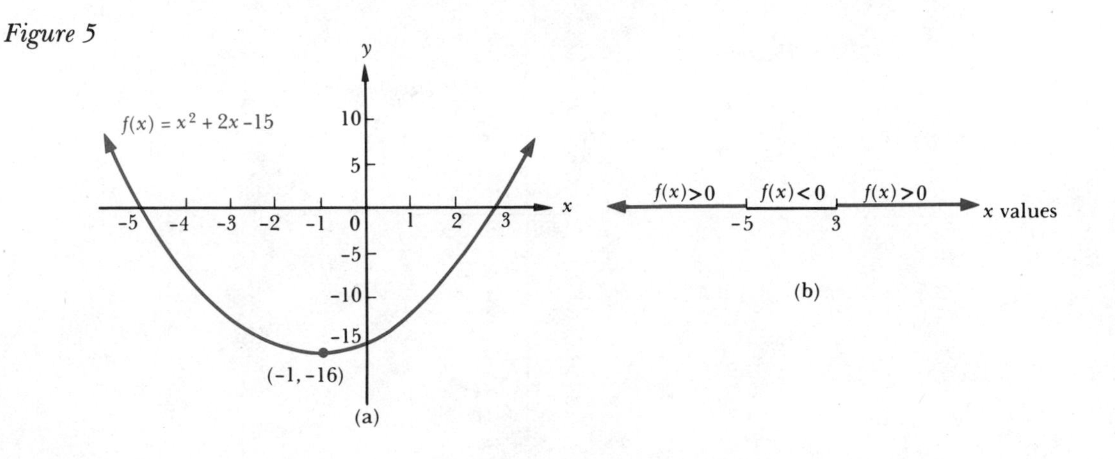

Since the graphs of quadratic functions are parabolas, it follows that there are only three possible cases in determining cut points: Either there is no cut point (Figure 6a), there is one cut point (Figure 6b), or there are two cut points (Figure 6c).

In order to determine the intervals in which $f(x)$ is positive and the intervals in which $f(x)$ is negative, it is enough to know the cut points; hence the examples in Figures 6a, 6b, and 6c result in the situations given in Figures 7a, 7b, and 7c, respectively.

Figure 6

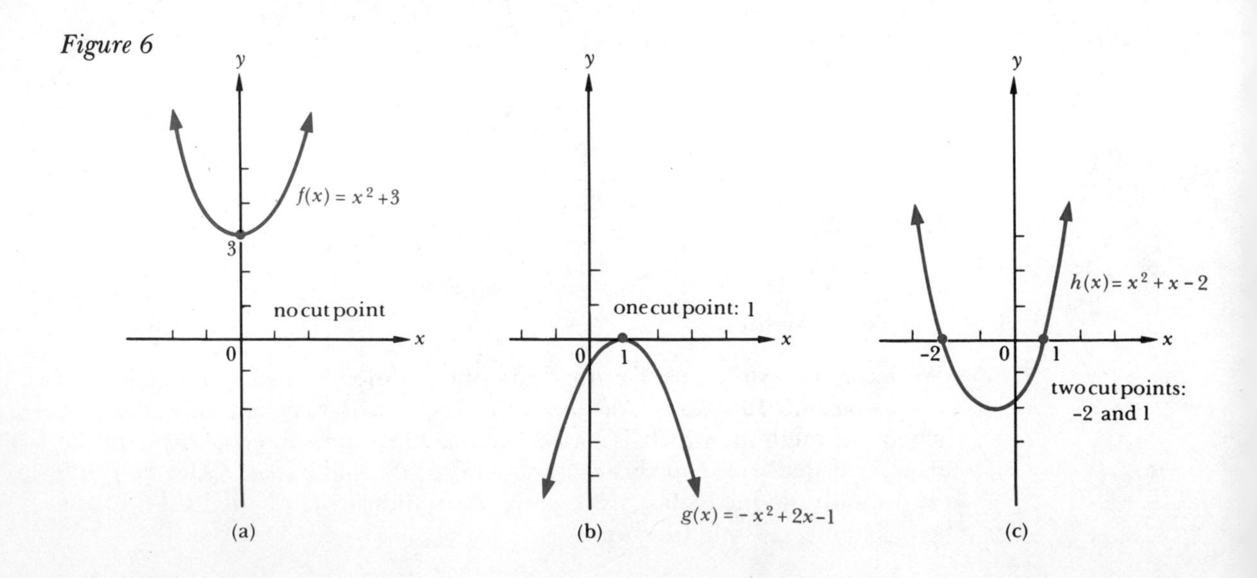

Figure 7

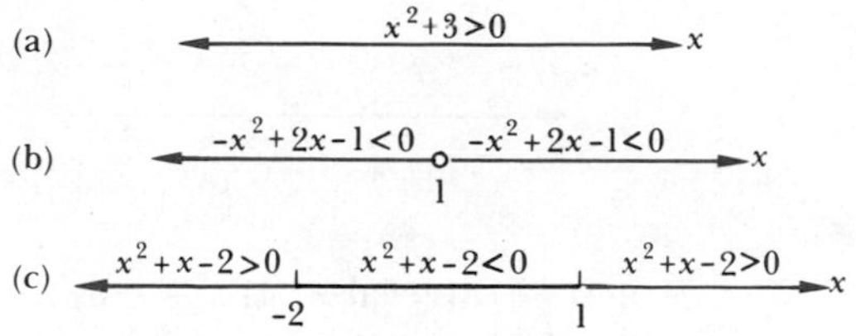

(a) $x^2+3>0$

(b) $-x^2+2x-1<0$ ○ $-x^2+2x-1<0$; 1

(c) $x^2+x-2>0$ | $x^2+x-2<0$ | $x^2+x-2>0$; −2, 1

In general, once the cut points are determined, the solution set of a quadratic inequality is obtained by testing *one value* in each of the intervals determined by the cut points, since the sign of the value of the associated quadratic function is the same for all values in any one of these intervals.

For example, to find the solution set of the inequality $x^2 - 2x - 3 > 0$, we notice that the cut points occur when $x^2 - 2x - 3 = 0$. Solving this equation, we obtain the cut points -1 and 3, which *partition* (separate) the real numbers into three intervals denoted by A, B, and C as shown in Figure 8a.

Since we know that the quadratic expression is always positive or always negative on each of these intervals, it is enough to test *any one value* in each interval to decide whether *all values* in the interval yield positive or negative values for the quadratic. This procedure can be simplified by adopting the following steps:

Step 1 Interval A: $x < -1$ ($x = -2$ is used to test the interval). Since $(-2)^2 - 2(-2) - 3 = 5 > 0$, all $x < -1$ will satisfy $x^2 - 2x - 3 > 0$.

Step 2 Interval B: $-1 < x < 3$ ($x = 0$ is used to test the interval). Since $0^2 - 2 \cdot 0 - 3 = -3 \not> 0$, any value x such that $-1 < x < 3$ will *not* satisfy $x^2 - 2x - 3 > 0$.

Step 3 Interval C: $x > 3$ ($x = 4$ is used to test the interval). Since $4^2 - 2(4) - 3 = 5 > 0$, all $x > 3$ will satisfy $x^2 - 2x - 3 > 0$.

Thus, $x^2 - 2x - 3 > 0$ whenever $x < -1$ or $x > 3$ (Figure 8b), and the solution set is $\{x \mid x < -1 \text{ or } x > 3\}$.

Figure 8

EXAMPLES Use the cut-point method to find the solution set of each quadratic inequality.

1 $2x^2 + 13x - 7 \geq 0$

SOLUTION First the cut points are determined by solving $2x^2 + 13x - 7 = 0$. Since $2x^2 + 13x - 7 = (2x - 1)(x + 7) = 0$, the cut points are -7 and $\frac{1}{2}$. The numbers -7 and $\frac{1}{2}$ partition the number line into three intervals, indicated by A, B, and C (Figure 9a). We select numbers from each interval and check them using the following steps:

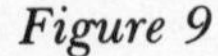
Figure 9

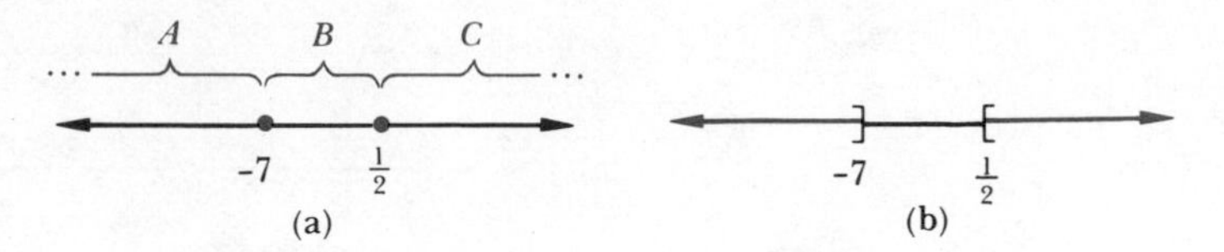

Step 1 Interval A: If $x = -8$, then $2(-8)^2 + 13(-8) - 7 = 17 > 0$.
Step 2 Interval B: If $x = 0$, then $2(0)^2 + 13(0) - 7 = -7 \not> 0$.
Step 3 Interval C: If $x = 1$, then $2(1)^2 + 13(1) - 7 = 8 > 0$.

Therefore, the solution set includes all the numbers in intervals A and C and also includes -7 and $\frac{1}{2}$ (Figure 9b). The solution set of the inequality is $\{x \mid x \leq -7 \text{ or } x \geq \frac{1}{2}\}$.

2 $4x^2 - 12x + 9 < 0$

SOLUTION Since $4x^2 - 12x + 9 = (2x - 3)^2 = 0$, then $\frac{3}{2}$ is the only cut point. The number $\frac{3}{2}$ partitions the number line into two intervals indicated by A and B in Figure 10. We select numbers from each interval and check them as follows:

Step 1 Interval A: If $x = 0$, then $4(0) - 12(0) + 9 = 9 \not< 0$.
Step 2 Interval B: If $x = 2$, then $4(4) - 12(2) + 9 = 1 \not< 0$.

Figure 10

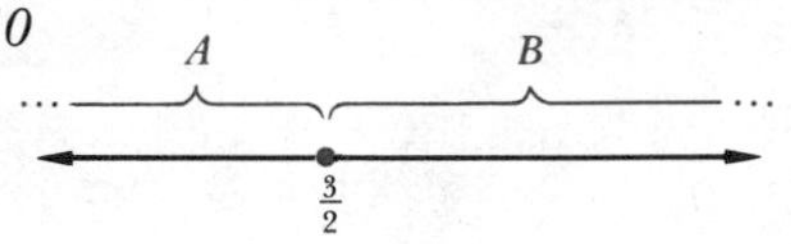

Therefore, the solution set does *not* include any real numbers in intervals A and B and the solution set is $\varnothing$.

Although the following inequality is not quadratic, it can be solved by a method similar to the one used in the preceding examples.

EXAMPLE Find the solution set of the inequality

$$\frac{3x - 12}{x + 2} < 0$$

SOLUTION We multiply both sides of the inequality by $(x + 2)^2$. Since $(x + 2)^2$ is always positive for $x \neq -2$, we have

$$(x + 2)^2\left(\frac{3x - 12}{x + 2}\right) < (x + 2)^2 \cdot 0 \quad \text{or} \quad (x + 2)(3x - 12) < 0$$

The cut points are determined by solving $(x + 2)(3x - 12) = 0$, to get -2 and 4. The numbers -2 and 4 partition the number line into the three intervals denoted by A, B, and C in Figure 11a. We select numbers from each interval and check them using the following steps:

Step 1 Interval A: If $x = -3$, then $(-3 + 2)(-9 - 12) = 21 \not< 0$.
Step 2 Interval B: If $x = 0$, then $(0 + 2)(0 - 12) = -24 < 0$.
Step 3 Interval C: If $x = 5$, then $(5 + 2)(15 - 12) = 21 \not< 0$.

Figure 11

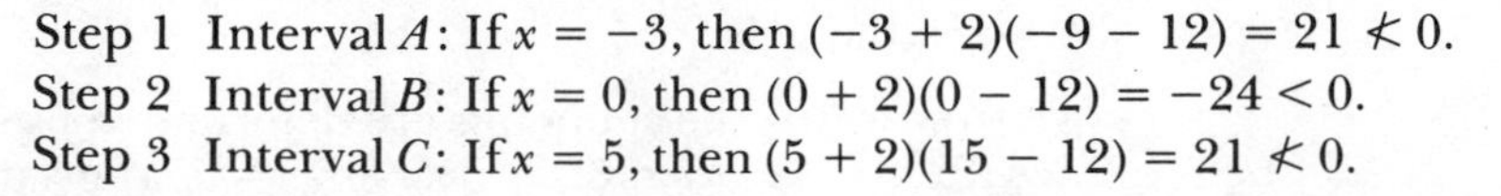

Therefore, the solution set includes the numbers in interval B (Figure 11b). Thus the solution set is $\{x \mid -2 < x < 4\}$.

Certain applied problems involve quadratic inequalities. The following example illustrates one such application.

EXAMPLE A man has a swimming pool 20 feet by 30 feet. He wishes to pour a strip of cement of uniform width around the edge of the pool. He has enough mixed cement to cover an area greater than 336 square feet. What is the possible width of the strip of cement?

SOLUTION Figure 12 illustrates this situation. Let x feet represent the width of the cement strip. The width of the large rectangle is given by $20 + 2x$, and the length of the large rectangle is given by $30 + 2x$. The area of the large rectangle is $(20 + 2x)(30 + 2x)$, and the area of the pool is $(20)(30) = 600$ square feet. Thus the area of the cement strip must satisfy the inequality

Figure 12

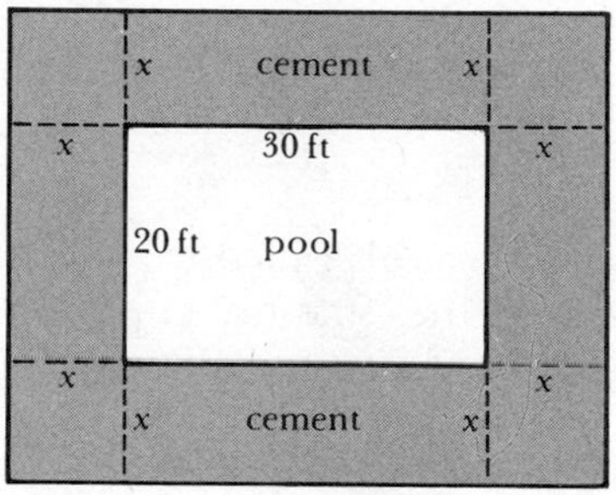

$$(20 + 2x)(30 + 2x) - 600 > 336$$

or

$$600 + 100x + 4x^2 - 600 > 336$$

so

$$4x^2 + 100x - 336 > 0$$

We solve this inequality by the cut-point method. Since

$$4x^2 + 100x - 336 = 4(x + 28)(x - 3) = 0$$

then 3 and -28 are the cut points. The numbers 3 and -28 partition the number line into three intervals. After selecting numbers from each interval and checking them, we find that $x < -28$ or $x > 3$. Since the width of the strip cannot be negative, $x > 3$. Obviously, x would have to be restricted further according to the total amount of cement available.

PROBLEM SET 3

In Problems 1–10, graph the quadratic function determined by the given equation, and then use the graph to find the solution set of the corresponding inequality. Display the solution set on the number line.

1 $y = x^2 - 9x + 14;\ x^2 - 9x + 14 > 0$

2 $y = -x^2 + x + 20;\ -x^2 + x + 20 \geq 0$

3 $y = -2x^2 + 5x - 3;\ -2x^2 + 5x - 3 \leq 0$

4 $y = 2x^2 - x - 1;\ 2x^2 - x - 1 > 0$

5 $y = (x - 1)^2;\ x^2 - 2x < -1$

6 $y = x^2 + 6x + 8;\ -x^2 - 6x - 8 \geq 0$

7 $y = -x^2 - 2;\ x^2 + 2 > 0$

8 $y = 2x^2 - 7x + 6$; $2x^2 - 7x + 6 > 0$

9 $y = 3x^2 - 5x + 1$; $3x^2 - 5x + 1 < 0$

10 $y = \frac{1}{3}x^2 - \frac{1}{2}x - \frac{3}{2}$; $\frac{1}{3}x^2 - \frac{1}{2}x - \frac{3}{2} < 0$

In Problems 11–22, use the cut-point method to find the solution set of each quadratic inequality. Display the solution set on the number line.

11 $x^2 - 9 < 0$ **12** $x^2 - 7x + 6 \geq 0$

13 $x^2 + 2x - 3 \leq 0$ **14** $x^2 - 6x + 8 < 0$

15 $3x^2 - 7x + 2 > 0$ **16** $4x^2 + 11x - 3 \geq 0$

17 $2x^2 + 3x - 9 \leq 0$ **18** $6x^2 + 5x - 14 \leq 0$

19 $4x^2 + 4x + 1 < 0$ **20** $x^2 + 4 > 0$

21 $-6x^2 - 13x + 5 \geq 0$ **22** $-3x^2 + 48 > 0$

In Problems 23–28, find the solution set of each inequality. Display the solution set on the number line.

23 $\dfrac{x-1}{x-4} < 0$ **24** $\dfrac{2x-1}{x+2} < 0$ **25** $\dfrac{5x-1}{x-2} > 0$

26 $\dfrac{x-2}{3x-1} > 0$ **27** $\dfrac{1-x}{x} < 1$ **28** $\dfrac{1}{x+2} > 3$

29 The length of a rectangle is 7 centimeters more than twice the width. Find the possible widths such that the area of the rectangle is greater than 60 square centimeters.

30 One aspect of economics deals with what is called the *supply equation*. It relates the price per item p and the quantity supplied q by the equation $q = ap + b$, where a and b are constants depending on the particular situation. The total profit from selling q items equals the profit per item times q. Suppose that it costs a wheat farmer \$4 per bushel to grow wheat. The supply equation is $q = 8p + 90$. What price should the farmer charge per bushel to realize a profit of at least \$315?

4 Polynomial Functions of Degree Greater Than 2

Up to now, we have encountered special kinds of polynomial functions—the linear and the quadratic functions. In this section, we briefly survey polynomial functions of degree greater than 2.

A function defined by the equation

$$f(x) = a_n x^n + a_{n-1}x^{n-1} + a_{n-2}x^{n-2} + \cdots + a_1 x + a_0$$

where n is a positive integer and $a_n, a_{n-1}, a_{n-2}, \ldots, a_1$ and a_0 are real numbers with $a_n \neq 0$, is called a *polynomial function of degree n in x*. The numbers $a_n, a_{n-1}, a_{n-2}, \ldots, a_1$, and a_0 are called the *coefficients* of the polynomial function. If $n = 0$, then $f(x) = a_0$, where $a_0 \neq 0$, is called a *zero-degree* polynomial function. The function $f(x) = 0$ is called the *zero* polynomial function, and no degree is assigned to it.

For example, $f(x) = 2x^3 - 5x^2 + 3$ is a polynomial function of degree 3, since it can be written in the form

$$f(x) = 2x^3 - 5x^2 + 0x + 3$$

The coefficients of this polynomial function are 2, −5, 0, and 3, where $a_3 = 2, a_2 = -5, a_1 = 0$, and $a_0 = 3$.

The function $g(x) = 4$ is an example of a zero-degree polynomial function with one coefficient, $a_0 = 4$, whereas $h(x) = (2x^3 + 1)/x^2 = 2x + x^{-2}$ is not a polynomial function in x because of the -2 exponent.

In graphing a polynomial function, it is helpful to determine the x intercepts. Clearly, as with the linear and the quadratic function, the x intercepts of a graph of a polynomial function occur at those values of x that have zero as the corresponding range value. These values of x are called the *zeros* of the function. For example, the x intercepts of the graph of f defined by $f(x) = x^3 - x^2$ are determined by solving the equation $x^3 - x^2 = 0$. Since $x^3 - x^2 = x^2(x - 1) = 0$, then $x = 0$ or $x = 1$, so 0 and 1 are the x intercepts of the graph of f, or 0 and 1 are the zeros of the polynominal function $f(x) = x^3 - x^2$.

The zeros of the polynomial function $g(x) = 2x^2 + x - 1$ can be determined by solving the equation $2x^2 + x - 1 = 0$ to get -1 and $\frac{1}{2}$; the zeros of $g(x) = 4x^3 - x^2$ are 0 and $\frac{1}{4}$; the function $h(x) = x^2 + 1$ has no real number zeros because $x^2 + 1 = 0$ has no solution in the real number system.

The graphing of a polynomial function also requires plotting a few points. This means, of course, that specific members of the domain must be substituted into the function to determine corresponding range members. For certain situations the computation of the range members is not difficult. For example, if $f(x) = x^3 + 1$, then $f(-1) = 0$, $f(0) = 1$, $f(1) = 2$, and $f(2) = 9$ can easily be determined by substitution. In other situations the computation becomes rather tedious. For instance, if g is the function given by $g(x) = x^5 - 3x^4 + 7x - 8$, the determination of $g(3)$, $g(-2)$, and $g(7)$ becomes quite involved.

A shorthand division process, called *synthetic division*, can be used to simplify computations of the latter type. However, before examining synthetic division, it is necessary to comment briefly on the division of polynomials.

4.1 Division of Polynomials—Synthetic Division

The polynomial $x^3 + 2x^2 - 2x + 3$ is *divisible* by $x + 3$, since

$$x^3 + 2x^2 - 2x + 3 = (x^2 - x + 1)(x + 3)$$

whereas the division of $x^2 + 3x + 7$ by $x + 1$ gives a quotient of $x + 2$ and a remainder of 5, so that

$$x^2 + 3x + 7 = (x + 2)(x + 1) + 5$$

In either case, the division of polynomials is based on a property called the *division algorithm*, which is stated as follows:

PROPERTY 1 DIVISION ALGORITHM

If $f(x)$ and $D(x)$ are nonconstant polynomials such that the degree of $f(x)$ is greater than or equal to the degree of $D(x)$ with $D(x) \neq 0$, then there exist unique polynomials $Q(x)$ and $R(x)$ such that $f(x) = D(x) \cdot Q(x) + R(x)$, where the degree of $R(x)$ is less than the degree of $D(x)$ or $R(x) = 0$. The expression $D(x)$ is called the *divisor*, $f(x)$ is the *dividend*, $Q(x)$ is the *quotient*, and $R(x)$ is the *remainder*.

EXAMPLE Suppose that $f(x) = 3x^3 - x^2 + 2x - 1$, and $D(x) = x - 3$. Find $Q(x)$ and $R(x)$ such that $3x^3 - x^2 + 2x - 1 = (x - 3)Q(x) + R(x)$.

SOLUTION We arrange this division in the following manner:

$$
\begin{array}{r|l}
 & 3x^2 + 8x + 26 \\ \hline
x - 3 & 3x^3 - x^2 + 2x - 1 \\
 & \underline{3x^3 - 9x^2} \\
 & \quad 8x^2 + 2x \\
 & \quad \underline{8x^2 - 24x} \\
 & \qquad 26x - 1 \\
 & \qquad \underline{26x - 78} \\
 & \qquad\qquad 77
\end{array}
$$

Hence $Q(x) = 3x^2 + 8x + 26$, and $R(x) = 77$.

Synthetic division is merely a *shorthand method* for performing the division of any polynomial by a polynomial of the form $x - c$. In the example above, the division can be displayed as follows:

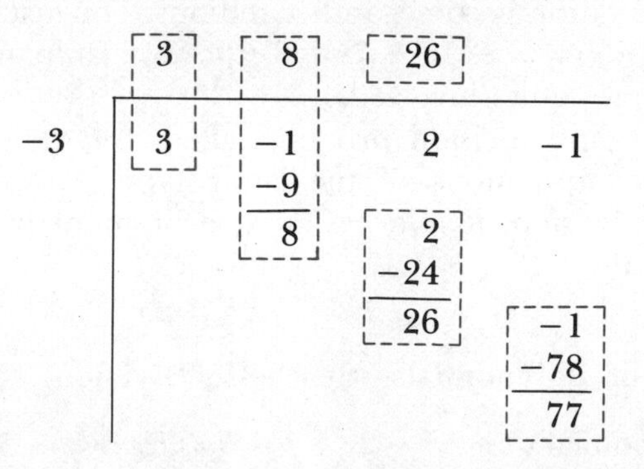

Here the variables are implied rather than indicated explicitly. This latter form could also be rearranged as follows:

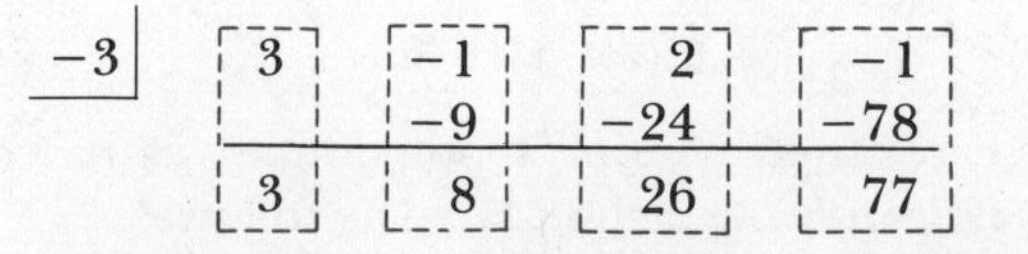

Note that in this representation, the coefficients of the quotient appear in the last row, together with the remainder in the last position. Finally, the last form can be abbreviated as

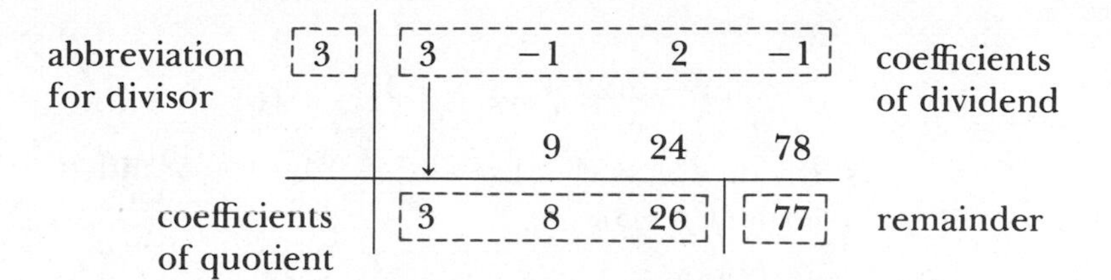

The last form of the division is called *synthetic division.*

Since the sign of the number to the left of the bar (−3 in this example) is changed, it is not necessary to perform the subtraction as we do in the long division method. All we need do is to *add* algebraically. Hence,

$$3x^3 - x^2 + 2x - 1 = (x - 3)(3x^2 + 8x + 26) + 77$$

EXAMPLES Use synthetic division to work the following examples.

1 Find the quotient $Q(x)$ and the remainder $R(x)$ if the dividend is given by $f(x) = 3x^3 - 8x + 1$ and the divisor is $D(x) = x + 2$.

SOLUTION First, write the divisor $x + 2$ in the form $x - c$ as $x - (-2)$. Then use -2 as the "divisor" in the synthetic division as follows:

−2	3	0	−8	1
	↓	−6	12	−8
	3	−6	4	−7

Notice that 0 is used as the coefficient of the "missing" x^2 term. Also notice the pattern in the synthetic division: We bring down the first coefficient, 3, multiply 3 by −2 to get −6, and then add 0 to −6 to get −6. Next, we multiply −6 by −2 to get 12, and then add −8 to 12 to get 4. Finally, we multiply 4 by −2 to get −8, and then add 1 to −8 to get the remainder −7. Consequently,

$$3x^3 - 8x + 1 = (3x^2 - 6x + 4)(x + 2) + (-7)$$

so

$$Q(x) = 3x^2 - 6x + 4 \quad \text{and} \quad R(x) = -7$$

2 Divide $3x^3 - 2x^2 + 1$ by $x - 2$.

SOLUTION

2	3	−2	0	1
		6	8	16
	3	4	8	17

Hence,

$$3x^3 - 2x^2 + 1 = (3x^2 + 4x + 8)(x - 2) + 17$$

Notice that if $x = 2$ is substituted into each side of the latter equation, we get

$$3(2)^3 - 2(2)^2 + 1 = 0 + 17$$

3 Divide $f(x) = x^5 - 4x^3 - 6x^2 - 9$ by $x + 3$ and then use the result to determine $f(-3)$.

SOLUTION

-3	1	0	-4	-6	0	-9
		-3	9	-15	63	-189
	1	-3	5	-21	63	-198

Hence,

$$\begin{aligned} f(x) &= x^5 - 4x^3 - 6x^2 - 9 \\ &= (x^4 - 3x^3 + 5x^2 - 21x + 63)(x + 3) + (-198) \end{aligned}$$

so $f(-3) = 0 + (-198) = -198$. Notice that the function value $f(-3)$ is the same as the remainder we get when dividing $f(x)$ by $x - (-3)$.

This last example suggests two theorems.

THEOREM 1 REMAINDER THEOREM

If a polynomial $f(x)$ of degree $n > 0$ is divided by $x - c$, the remainder R is a constant and is equal to the value of the polynomial when c is substituted for x; that is, $f(c) = R$.

PROOF Let $Q(x)$ be the quotient. By the division algorithm,

$$f(x) = (x - c)Q(x) + R(x)$$

Since the remainder $R(x)$ is of degree less than the divisor $x - c$, $R(x)$ must be constant and we denote it as R. The equation

$$f(x) = (x - c)Q(x) + R$$

holds for all x, and if we set $x = c$, we find that

$$f(c) = (c - c)Q(c) + R = 0 \cdot Q(c) + R = R$$

so that $f(c) = R$.

THEOREM 2 FACTOR THEOREM

If $f(x)$ is a polynomial of degree $n > 0$ and $f(c) = 0$, then $x - c$ is a factor of the polynomial $f(x)$; conversely, if $x - c$ is a factor, then c is a zero of f. (The factor theorem is a *corollary* of the remainder theorem. A corollary is a theorem that follows directly from another theorem.)

PROOF Let $f(x) = (x - c)Q(x) + R$. If $f(c) = 0$, then, by Theorem 1, $R = 0$, so that $f(x) = Q(x)(x - c)$ and $(x - c)$ is a factor of $f(x)$. Conversely, if $x - c$ is a factor of $f(x)$, then $f(x) = Q(x)(x - c)$, so $f(c) = Q(c) \cdot 0 = 0$ and c is a zero of f.

Theorem 1 tells us what was hinted in the second and third examples above. The value of a polynomial $f(c)$ is the same as the remainder we get when the polynomial is divided by $x - c$. But this type of operation can be performed by synthetic division. Hence the values of a polynomial function can be determined by synthetic division.

For example, $f(3)$ for $f(x) = x^5 - 3x^4 + 7x - 8$ can be determined as follows:

3	1	−3	0	0	7	−8
		3	0	0	0	21
	1	0	0	0	7	13

so that $f(3) = 13$. (Notice that in many cases it is just as simple to calculate the value directly.)

EXAMPLES Use synthetic division and Theorems 1 and 2 to work the following examples.

1 Find the quotient $Q(x)$ and the remainder R if $f(x) = 3x^5 - 5x^3 + 1$ is divided by $x - 2$. Use your result to determine $f(2)$.

SOLUTION

2	3	0	−5	0	0	1
		6	12	14	28	56
	3	6	7	14	28	57

Hence, $Q(x) = 3x^4 + 6x^3 + 7x^2 + 14x + 28$ and $R = 57$. By Theorem 1, $f(2) = 57$.

2 If $g(x) = x^3 - 2x^2 + 5x - 4$, evaluate

(a) $g(1)$ (b) $g(-2)$ (c) $g(3)$

SOLUTION

(a)

1	1	−2	5	−4
		1	−1	4
	1	−1	4	0

Hence, $g(1) = 0$. In other words, $x - 1$ is a factor of $g(x)$.

(b)

−2	1	−2	5	−4
		−2	8	−26
	1	−4	13	−30

Hence, $g(-2) = -30$.

(c)

3	1	−2	5	−4
		3	3	24
	1	1	8	20

Hence, $g(3) = 20$.

3 Show that $x + 2$ is a factor of $f(x) = x^4 - 2x^2 + 3x - 2$ and find $Q(x)$. Express $f(x)$ in factored form.

SOLUTION

-2	1	0	-2	3	-2
		-2	4	-4	2
	1	-2	2	-1	0

Hence, $f(-2) = 0$, and $Q(x) = x^3 - 2x^2 + 2x - 1$, so $f(x)$ can be written as $f(x) = (x + 2)(x^3 - 2x^2 + 2x - 1)$.

4 Find the value of k to make $x - 3$ a factor of $f(x) = 3x^3 - 4x^2 - kx - 33$.

SOLUTION

3	3	-4	$-k$	-33
		9	15	$-3k + 45$
	3	5	$-k + 15$	$-3k + 12$

Since $x - 3$ is a factor of $f(x)$, it follows from Theorem 2 that $-3k + 12 = 0$. Hence, $k = 4$.

4.2 Rational Zeros

The zeros of polynomial functions are used to determine the x intercepts of their graphs. If the polynomial function is of first or second degree, it is easy to determine the zeros, but as the degree increases, the problem of determining the zeros becomes much more difficult. The general method to be taken up here enables us to determine rational zeros (if they exist) of polynomial functions with integer or rational number coefficients. Before stating the result, it is necessary to recall one of the basic properties of the system of integers: a positive integer has a unique factorization; that is, it can be expressed uniquely as a product of prime numbers.

For example, $24 = 2^3 \cdot 3$, so that 2 and 3 are the only prime factors of the number 24.

THEOREM 3 RATIONAL ZERO THEOREM

Assume that $f(x) = a_n x^n + a_{n-1}x^{n-1} + \cdots + a_1 x + a_0$, $a_n \neq 0$ and n is a positive integer. If the coefficients are integers and p/q is a rational zero in lowest terms, then p is a divisor of a_0 and q is a divisor of a_n.

PROOF Since

$$a_n\left(\frac{p}{q}\right)^n + a_{n-1}\left(\frac{p}{q}\right)^{n-1} + \cdots + a_1\left(\frac{p}{q}\right) + a_0 = 0$$

it follows, after multiplying both sides by q^n, that

$$a_n p^n + a_{n-1}p^{n-1}q + \cdots + a_1 pq^{n-1} + a_0 q^n = 0$$

Thus

(i) $a_n p^n + a_{n-1} p^{n-1} q + \cdots + a_1 p q^{n-1} = -a_0 q^n$
(ii) $a_{n-1} p^{n-1} q + \cdots + a_1 p q^{n-1} + a_0 q^n = -a_n p^n$

Since both sides of Equation (i) are integers and p is a divisor of the left side, then, by the unique factorization property, p is also a divisor of the right side. But p and q have no common factors, since p/q is in lowest terms. Hence, p is a divisor of a_0.

Similarly, in Equation (ii), q is a divisor of the left side, and hence of the right side as well. As before, q has no factors in common with p, so q must be a divisor of a_n.

EXAMPLES Use Theorem 3 and synthetic division to find all the possible rational zeros of each of the following polynomial functions.

1 $f(x) = x^3 - 2x^2 - x + 2$

SOLUTION Assume that p/q is a rational zero of f. By Theorem 3, p is a divisor of 2 and q is a divisor of 1. Hence, p can be any of the integers -1, 1, -2, or 2, and q can be either -1 or 1. Therefore, the *possible values* of p/q are -1, 1, -2, and 2. Of these four possibilities, not more than three can be zeros of f. Using synthetic division to find $f(2)$, we have

2	1	−2	−1	2
		2	0	−2
	1	0	−1	0

so $f(2) = 0$ by the remainder theorem. Hence, 2 is a zero of f. Also

$$x^3 - 2x^2 - x + 2 = (x - 2)(x^2 - 1)$$

The remaining zeros of f are the zeros of $Q(x) = x^2 - 1 = (x - 1)(x + 1)$, namely, 1 and -1. Thus the rational zeros of f are -1, 1, and 2.

2 $f(x) = x^3 + 2x^2 - 4x - 8$

SOLUTION Assume that p/q is a rational zero of f. By Theorem 3, p is a divisor of -8 and q is a divisor of 1, so p can assume any of the values 1, -1, 2, -2, 4, -4, 8, or -8, and q can assume values 1 or -1; therefore, the *possible* rational zeros p/q are given by 1, -1, 2, -2, 4, -4, 8, or -8. After testing these rational zeros by synthetic division, we find that 2 and -2 are zeros of f; in fact, $x^3 + 2x^2 - 4x - 8 = (x + 2)^2(x - 2)$. Thus 2 and -2 are the only zeros.

It is important to realize that the rational zero theorem has only limited usefulness in determining roots of polynomial equations. For example, the roots of the equation $x^2 - 3x + 1 = 0$ cannot be determined by the method used in the above examples, because the roots of this equation are not rational numbers.

4.3 Graphs of Polynomial Functions of Degree Greater Than 2

The graphing of polynomial functions of degree greater than 2 is more difficult than graphing polynomial functions of degree less than or equal to 2. It can be shown (in calculus) that all polynomial functions are *continuous functions.* The term "continuous" is used here to describe functions that have graphs without breaks or jumps from one point on the graph to another. This statement is not very precise and, in fact, is not correct in all cases. But for the functions considered in this book, it will be sufficient.

Figure 1

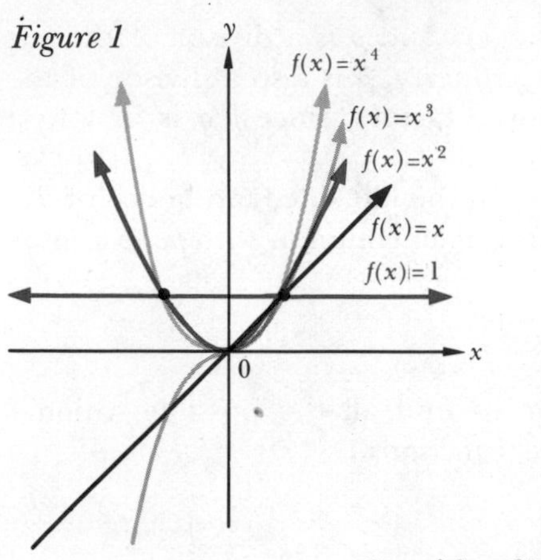

We use continuity to graph polynomial functions. For example, we graph $f(x) = 1$, $f(x) = x$, $f(x) = x^2$, $f(x) = x^3$, and $f(x) = x^4$ by plotting a few points and assuming continuity between these points for our sketches (Figure 1).

Consider $f(x) = x^3 + \frac{3}{2}x^2 - 6x - 2$. First, we can prepare a table of values of x and $f(x)$. (Here synthetic division could be used in the evaluations.) Next the points can be located. From the table we observe that the points $(x, f(x))$ to be located are $(-3, 2.5)$, $(-2, 8)$, $(-1, 4.5)$, $(0, -2)$, $(1, -5.5)$, and $(2, 0)$ (Figure 2).

The question is whether or not the points we have already plotted for $f(x) = x^3 - \frac{3}{2}x^2 - 6x - 2$ are sufficient to give us a fairly accurate sketch of the graph. Are there hidden peaks not shown thus far? We are *not* in a position to answer this question, but if we plot more points in between those already located, we can get a rough sketch of the graph (Figure 3).

Figure 2 *Figure 3*

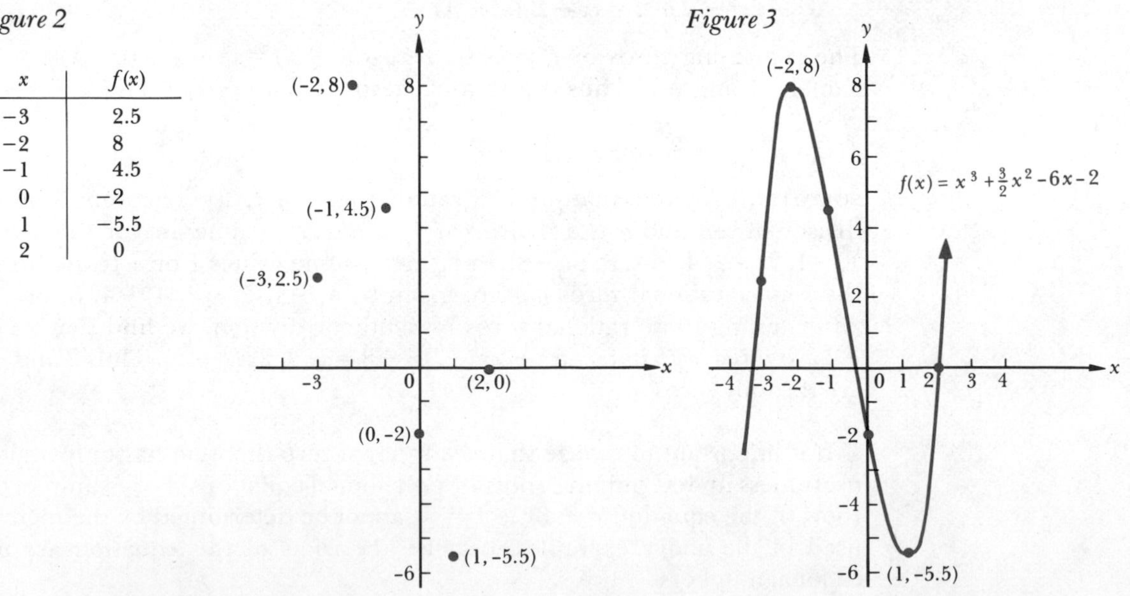

x	$f(x)$
−3	2.5
−2	8
−1	4.5
0	−2
1	−5.5
2	0

EXAMPLES Sketch the graph of each function by locating a few points. Identify the x and y intercepts, and then use the graph to find the solution set of the corresponding inequality.

1 $f(x) = (x + 1)(x - 2)(x - 3)$; $(x + 1)(x - 2)(x - 3) \leq 0$

Figure 4

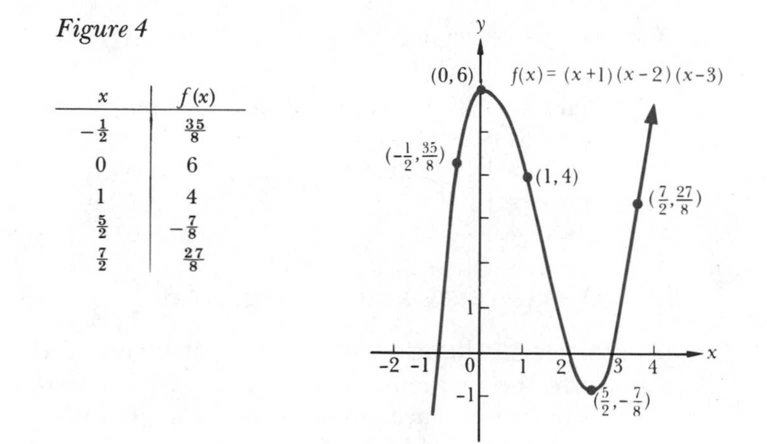

x	$f(x)$
$-\frac{1}{2}$	$\frac{35}{8}$
0	6
1	4
$\frac{5}{2}$	$-\frac{7}{8}$
$\frac{7}{2}$	$\frac{27}{8}$

SOLUTION The graph of f intersects the x axis only at -1, 2, and 3. Some additional points on the graph of f are given in the table accompanying Figure 4. Notice that the portion of the graph below or on the x axis suggests the values of x that satisfy the inequality. Thus a solution of the inequality $(x + 1)(x - 2)(x - 3) \leq 0$ is any x such that $x \leq -1$ or $2 \leq x \leq 3$. The solution set is $\{x \mid x \leq -1 \text{ or } 2 \leq x \leq 3\}$.

Figure 5

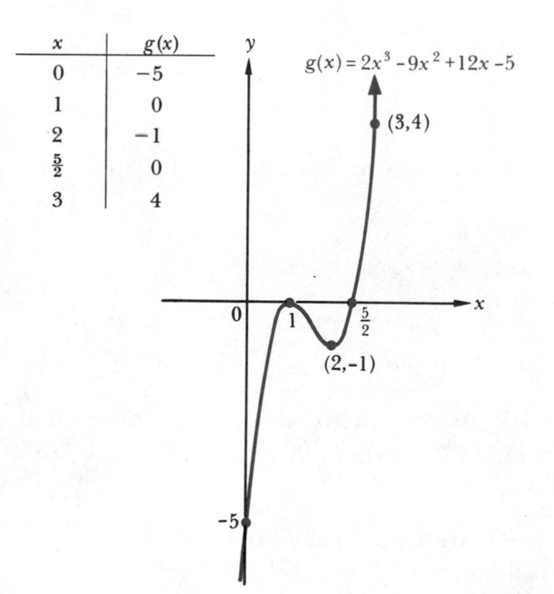

x	$g(x)$
0	-5
1	0
2	-1
$\frac{5}{2}$	0
3	4

2 $g(x) = 2x^3 - 9x^2 + 12x - 5$; $2x^3 - 9x^2 + 12x - 5 > 0$

SOLUTION By using the rational zero theorem, together with the following synthetic division,

$$\begin{array}{r|rrrr} 1 & 2 & -9 & 12 & -5 \\ & & 2 & -7 & 5 \\ \hline & 2 & -7 & 5 & \multicolumn{1}{|r}{0} \end{array}$$

we find that 1 is a zero of g and $g(x) = (2x^2 - 7x + 5)(x - 1)$. However, $2x^2 - 7x + 5 = (2x - 5)(x - 1)$, so $g(x) = (2x - 5)(x - 1)^2$. Thus, $\frac{5}{2}$ and 1 are the zeros of g, and $\frac{5}{2}$ and 1 are the x intercepts. The y intercept is -5. Finally, synthetic division is used to form the table of values for the graph of g (Figure 5). From the graph of g, we see that a solution of the inequality $2x^3 - 9x^2 + 12x - 5 > 0$ is any x such that $x > \frac{5}{2}$. Thus the solution set is $\{x \mid x > \frac{5}{2}\}$.

PROBLEM SET 4

In Problems 1–4, indicate whether or not the given function is a polynomial function. Specify the degree and identify the coefficients of each polynomial function.

1 $f(x) = 5^{-1}$

2 $f(x) = x^{-2} + x$

3 $f(x) = \dfrac{1}{x^3} + 2x^2 + x - 2$

4 $f(x) = 2x - 3^{1/2}x^2 + 5x^3 - 7$

In Problems 5–14, find all zeros of the given polynomial function.

5 $f(x) = -3x + 2$

6 $g(x) = 3x^2 + x - 2$

7 $h(x) = (x - 1)(x^2 - 3x + 2)$

8 $f(x) = (x - 1)(x - 2)(x - 5)$

9 $h(x) = x^4 - 16$

10 $g(x) = (x - 1)^2 - 9$

11 $f(x) = x^3 - 9x$

12 $f(x) = x^4 - x^3$

13 $g(x) = (x + 1)(x - 2)(x^2 - 25)$

14 $h(x) = (2x - 1)(3x + 1)(x^2 - x - 2)$

In Problems 15–20, use both (a) long division and (b) synthetic division to perform the divisions. Assume that $f(x)$ represents each polynomial expression; express each result in the form $f(x) = D(x) \cdot Q(x) + R$.

15 $5x^3 - 2x^2 + 3x - 4$ by $x - 3$

16 $2x^4 + 3x^3 - 5x^2 + 2x - 1$ by $x + 1$

17 $5x^5 - 3x^4 + 2x^3 + x^2 - 7x + 3$ by $x - 2$

18 $2x^4 - 3x^3 + 5x^2 + 6x - 3$ by $x + 2$

19 $-4x^6 - 5x^3 + 3x^2 + x + 7$ by $x - 1$

20 $2x^4 + 3x^3 - 3x^2 + x - 1$ by $x + 4$

In Problems 21–25, use synthetic division to find $Q(x)$ and $f(c)$ so that $f(x) = (x - c)Q(x) + f(c)$.

21 $f(x) = 3x^3 + 6x^2 - 10x + 7$ and $c = 2$

22 $f(x) = 3x^3 + 4x^2 - 7x + 16$ and $c = -1$

23 $f(x) = 2x^3 - 5x^2 + 5x + 11$ and $c = \frac{1}{2}$

24 $f(x) = -2x^4 + 3x^3 + 5x - 13$ and $c = 3$

25 $f(x) = -3x^4 - 3x^3 + 3x^2 + 2x - 4$ and $c = -2$

26 If $f(x) = x^3 + 2x^2 - 13x + 10$, use synthetic division to determine $f(-5)$, $f(-4)$, $f(-3)$, $f(-1)$, $f(0)$, $f(1)$, $f(2)$, $f(3)$, $f(4)$, and $f(5)$. What are the factors of $f(x)$?

27 (a) If $f(x) = 2x^3 - 6x^2 + x + k$, find k so that $f(3) = -2$.
(b) Find k so that $x - 2$ is a factor of $f(x) = 3x^3 + 4x^2 + kx - 20$.

28 (a) Show that $x - 1$ is a factor of $f(x) = 14x^{99} - 65x^{56} + 51$.
(b) Show that $x + 4$ is a factor of $f(x) = 2x^2 + 3x - 20$.

In Problems 29–37, write down all rational numbers that *might* be zeros of the given polynomial function. Use synthetic division and the remainder theorem to test these possibilities to determine which of them are zeros.

29 $f(x) = x^3 - x^2 - 4x + 4$

30 $g(x) = x^3 + 2x - 12$

31 $h(x) = 5x^3 - 12x^2 + 17x - 10$

32 $f(x) = x^3 - x^2 - 14x + 24$

33 $f(x) = 3x^3 - 7x^2 + 8x - 2$
34 $h(x) = 4x^3 - 13x + 6$
35 $g(x) = 10x^3 + x^2 - 7x + 2$
36 $f(x) = 4x^4 - 15x^2 + 5x + 6$
37 $h(x) = 4x^4 - 4x^3 - 7x^2 + 4x + 3$

38 An open box with a rectangular base is to be constructed from a rectangular piece of cardboard 16 centimeters wide and 21 centimeters long by cutting out a square whose side has length x from each corner and then bending up the sides. Find the size of the corner when the volume of the box is to equal 266 cubic centimeters.

In Problems 39–44, sketch the graph of the given function by locating the x and y intercepts and plotting a few points.

39 $g(x) = x^3 - 2x^2 - 5x + 6$
40 $f(x) = (2x + 1)^4$
41 $h(x) = x(x - 1)(x + 2)$
42 $f(x) = 2x^3 - 3x^2 - 3x + 2$
43 $f(x) = x^3 - 3x^2 + 4$
44 $g(x) = 6x^3 - x^2 - 5x + 2$

In Problems 45–50, use the graphs from Problems 39–44 to find the solution set of the given inequality.

45 $x^3 - 2x^2 - 5x + 6 \leq 0$
46 $(2x + 1)^4 < 0$
47 $x(x - 1)(x + 2) > 0$
48 $2x^3 - 3x^2 - 3x + 2 > 0$
49 $x^3 - 3x^2 + 4 \leq 0$
50 $6x^3 - x^2 - 5x + 2 \geq 0$

5 Rational Functions

In this section we study functions that are quotients of polynomials, which are called *rational functions.* For example,

$$k(x) = \frac{x+1}{x+2}, \quad h(x) = \frac{5}{x}, \quad \text{and} \quad g(x) = \frac{3x^3 + 5x}{x^2 + 2}$$

are rational functions.

It should be noted that the domain of a rational function $S(x) = \frac{f(x)}{g(x)}$ does not contain the zeros of the polynomial function g, since division by zero is not defined.

If the real number a is a zero of both f and g in the rational function $S(x) = f(x)/g(x)$, that is, if $f(a) = 0$ and $g(a) = 0$ for $S(x) = f(x)/g(x)$, it follows from the factor theorem that

$$\begin{aligned} S(x) = \frac{f(x)}{g(x)} &= \frac{f_1(x)(x-a)}{g_1(x)(x-a)} \\ &= \frac{f_1(x)}{g_1(x)}, \quad \text{for } x \neq a \end{aligned}$$

For example, the rational function

$$\begin{aligned} S(x) = \frac{x^2 - 4}{x - 2} &= \frac{(x+2)(x-2)}{x-2} \\ &= x + 2, \quad \text{for } x \neq 2 \end{aligned}$$

Note that $S(2)$ is *not* defined. The graph in Figure 1 displays the fact that 2 is *not* in the domain of S.

Figure 1

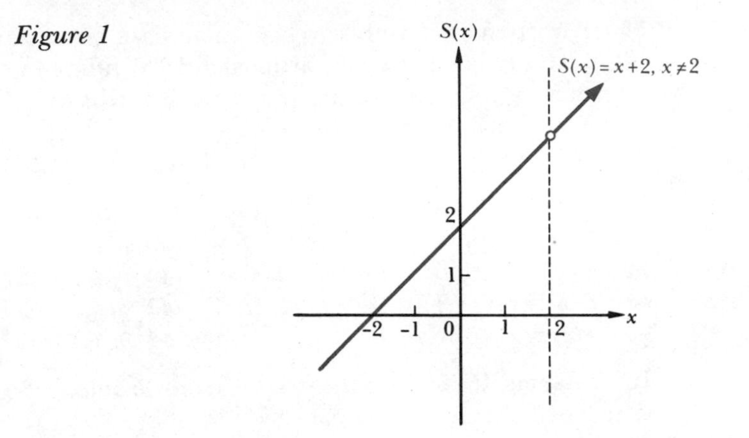

Next let us consider an example of a rational function $T(x) = f(x)/g(x)$, where $g(a) = 0$ and $f(a) \neq 0$, to see what happens to the graph near a. Suppose that

$$T(x) = \frac{2x + 1}{x - 1}$$

T has a y intercept at -1 and an x intercept at $-\frac{1}{2}$. The domain of T is the set of all real numbers *except* 1, that is, $\{x \mid x \neq 1\}$. Upon examining the behavior of the values in Table 1, we see that as x gets closer to 1 "from the right," the corresponding values of $T(x)$ become very large.

Table 1

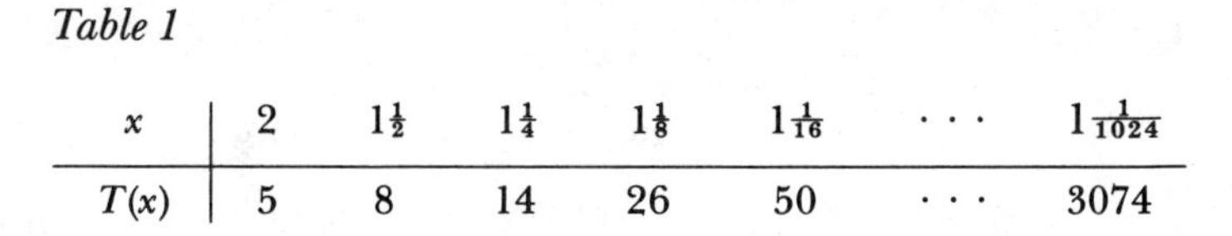

x	2	$1\frac{1}{2}$	$1\frac{1}{4}$	$1\frac{1}{8}$	$1\frac{1}{16}$	$\cdots$	$1\frac{1}{1024}$
$T(x)$	5	8	14	26	50	$\cdots$	3074

This situation results from the fact that the denominator of the fraction, $x - 1$, is becoming very close to zero, while the numerator of the fraction, $2x + 1$, is getting closer to 3 in value.

Table 2 shows that as x gets closer to 1 "from the left," the corresponding values of $T(x)$, which are negative numbers, become very large in absolute value.

Table 2

x	$\frac{1}{2}$	$\frac{3}{4}$	$\frac{7}{8}$	$\frac{9}{10}$	$\cdots$	$\frac{1999}{2002}$
$T(x)$	-4	-10	-22	-28	$\cdots$	-2000

Figure 2 displays the behavior of $T(x)$ near $x = 1$. Note that there is no point on the graph corresponding to $x = 1$. Furthermore, the values of $T(x)$ become larger in absolute value as x gets closer to 1. This situation is described by saying that the graph of the function T is getting closer to the line $x = 1$ "asymptotically," and the line $x = 1$ is called a *vertical asymptote.*

Figure 2

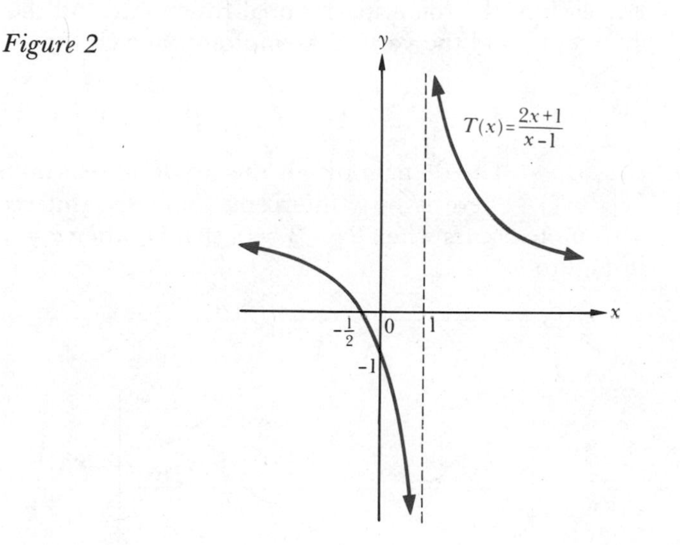

It is important to compare

$$S(x) = \frac{x^2 - 4}{x - 2} \quad \text{and} \quad T(x) = \frac{2x + 1}{x - 1}$$

Each of these two functions is undefined when the denominator is zero. On the one hand, for

$$S(x) = \frac{x^2 - 4}{x - 2}$$

the restriction on the denominator results in a graph with "one missing point," since both the numerator and denominator are simultaneously zero when $x = 2$ (Figure 1). On the other hand, for

$$T(x) = \frac{2x + 1}{x - 1}$$

the restriction on the denominator results in a graph that not only has a "missing point" when the denominator is zero, but whose points become *infinitely* distant from the x axis as x gets closer to the number that results in the denominator becoming zero (Figure 2). The latter situation occurs because *only* the denominator is zero when $x = 1$, whereas the numerator is not zero at $x = 1$. Thus, for the functions defined above, $x = 2$ is *not* an asymptote for S, whereas $x = 1$ is an asymptote for T.

In general, $S(x) = \dfrac{f(x)}{g(x)}$ has a *vertical asymptote* at $x = a$ if $g(a) = 0$ and if $\left|\dfrac{f(x)}{g(x)}\right|$ becomes infinitely large as x gets closer to a. This situation occurs when $g(a) = 0$ and $f(a) \neq 0$.

EXAMPLES For each of the following rational functions, find the domain, the x and y intercepts, and the vertical asymptotes. Sketch the graph.

1 $f(x) = \dfrac{4}{2x - 3}$

SOLUTION The domain of f is the set of all real numbers *except* $\frac{3}{2}$, that is, $\{x \mid x \neq \frac{3}{2}\}$. There is no x intercept, and the y intercept is $-\frac{4}{3}$. A vertical asymptote occurs when $2x - 3 = 0$, that is, when $x = \frac{3}{2}$. The graph is given in Figure 3.

Figure 3

x	$f(x)$
-2	$-\frac{4}{7}$
-1	$-\frac{4}{5}$
0	$-\frac{4}{3}$
1	-4
$1\frac{1}{4}$	-8
$1\frac{2}{5}$	-20
$1\frac{49}{100}$	-200

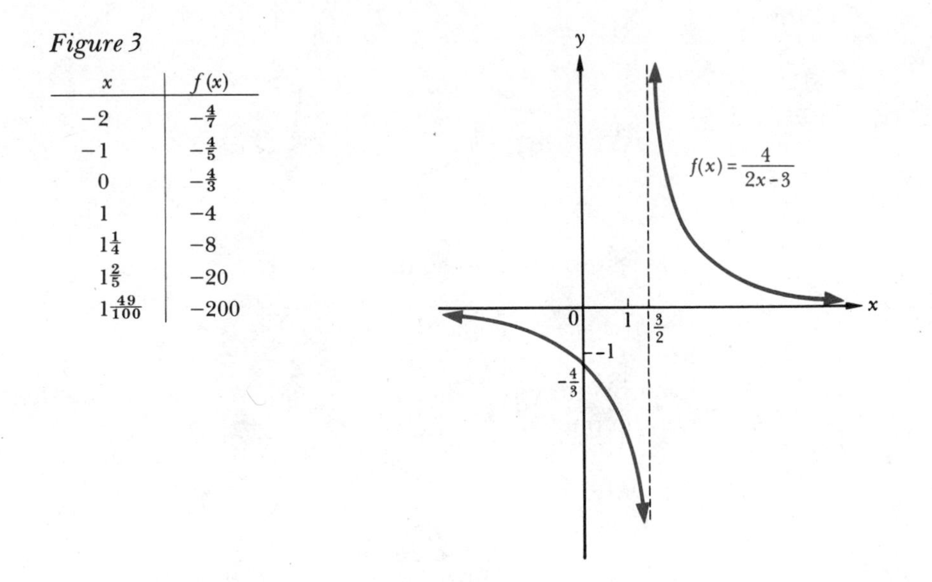

Notice that as x approaches $\frac{3}{2}$, $|f(x)|$ becomes very large. More precisely, as x gets closer to $\frac{3}{2}$ "from the left," the values of $f(x)$ become "very large" negative numbers, and as x gets closer to $\frac{3}{2}$ "from the right," the values of $f(x)$ become very large positive numbers.

2 $f(x) = \dfrac{x - \frac{1}{2}}{x^2 - 1}$

SOLUTION The domain of f is the set of all real numbers *except* 1 and -1, that is, $\{x \mid x \neq 1 \text{ and } x \neq -1\}$. The graph of f intersects the y axis at $\frac{1}{2}$ and the x axis at $\frac{1}{2}$; the lines $x = 1$ and $x = -1$ are vertical asymptotes (Figure 4).

Figure 4

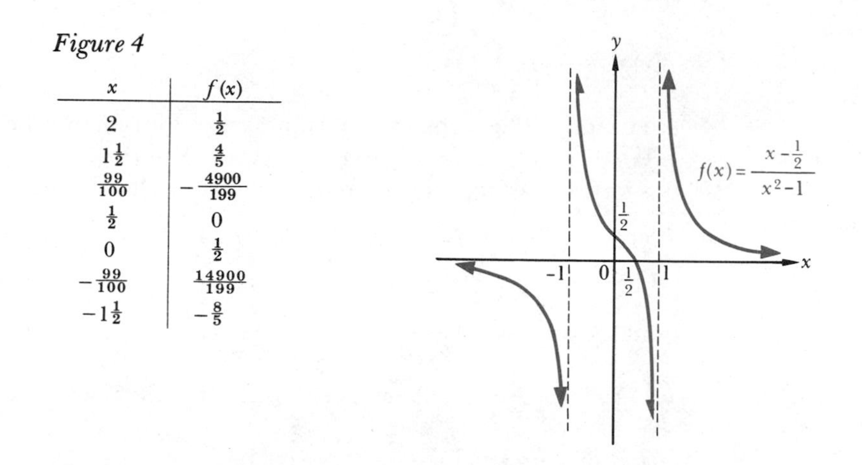

x	$f(x)$
2	$\frac{1}{2}$
$1\frac{1}{2}$	$\frac{4}{5}$
$\frac{99}{100}$	$-\frac{4900}{199}$
$\frac{1}{2}$	0
0	$\frac{1}{2}$
$-\frac{99}{100}$	$\frac{14900}{199}$
$-1\frac{1}{2}$	$-\frac{8}{5}$

The behavior of f near the asymptotes is examined in these steps:

(a) As x gets closer to -1 from the left (for $x < -1$), then $x - \frac{1}{2} < 0$ and $x^2 - 1 > 0$, and $x^2 - 1$ is close to 0. Thus, $f(x)$ is "becoming negatively large"; that is, $f(x)$ is negative and its absolute value is increasing.

(b) As x approaches -1 from the right (for $x > -1$), then $x - \frac{1}{2} < 0$ and $x^2 - 1 < 0$, and $x^2 - 1$ is close to 0. Thus, $f(x)$ is "becoming positively large"; that is, $f(x)$ is positive and its absolute value is increasing.

(c) As x approaches 1 from the left (for $x < 1$), then $x - \frac{1}{2} > 0$ and $x^2 - 1 < 0$, and $x^2 - 1$ is close to 0. Thus, $f(x)$ is "becoming negatively large."

(d) As x approaches 1 from the right (for $x > 1$), then $x - \frac{1}{2} > 0$ and $x^2 - 1 > 0$, and $x^2 - 1$ is close to 0. Thus $f(x)$ is "becoming positively large."

3 $f(x) = \dfrac{5}{x^2 + 4}$

SOLUTION The domain of f is the set of all real numbers. There is no x intercept and the y intercept is $\frac{5}{4}$. There is no vertical asymptote, since the equation $x^2 + 4 = 0$ has no real roots. Thus $f(x) > 0$ for all real numbers x. The graph of f is symmetric with respect to the y axis because $f(-x) = f(x)$. The symmetry of the graph of f is shown in Figure 5.

Figure 5

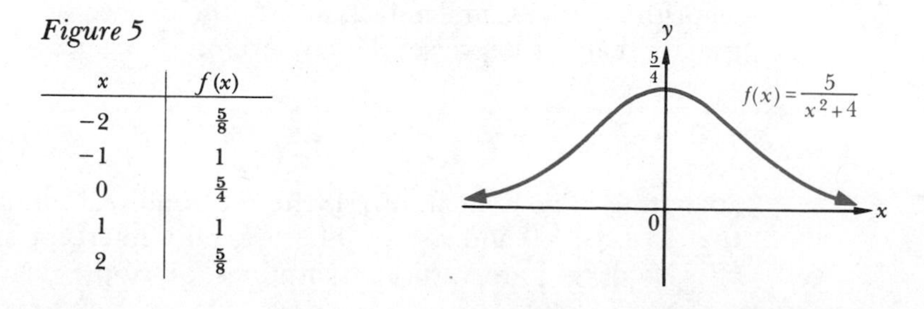

x	$f(x)$
-2	$\frac{5}{8}$
-1	1
0	$\frac{5}{4}$
1	1
2	$\frac{5}{8}$

4 $f(x) = \dfrac{3x}{x-1}$

SOLUTION The domain of f is the set of all real numbers *except* 1, that is, $\{x \mid x \neq 1\}$. The x and y intercepts are 0 and 0, respectively. Also $x = 1$ is a vertical asymptote. Using the division algorithm, we can rewrite f as $f(x) = 3 + \dfrac{3}{x-1}$. Now, as x becomes larger in absolute value, that is, as x becomes very large in the positive sense or "negatively large," the rational expression $\dfrac{3}{x-1}$ approaches zero, so that if x is very large, $f(x) = 3 + \dfrac{3}{x-1}$ gets very close to 3. In other words, as x becomes very large (or very small), $\dfrac{3}{x-1}$ approaches zero so that $f(x)$ approaches 3 "asymptotically." The line $y = 3$ is a *horizontal asymptote.* Notice that if $x > 1$, then $f(x) > 3$, and if $x < 1$, then $f(x) < 3$. The graph of f illustrates this behavior (Figure 6).

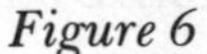
Figure 6

x	$f(x)$
-2	2
$-\frac{3}{2}$	$\frac{9}{5}$
-1	$\frac{3}{2}$
0	0
$\frac{1}{2}$	-3
$\frac{3}{2}$	9

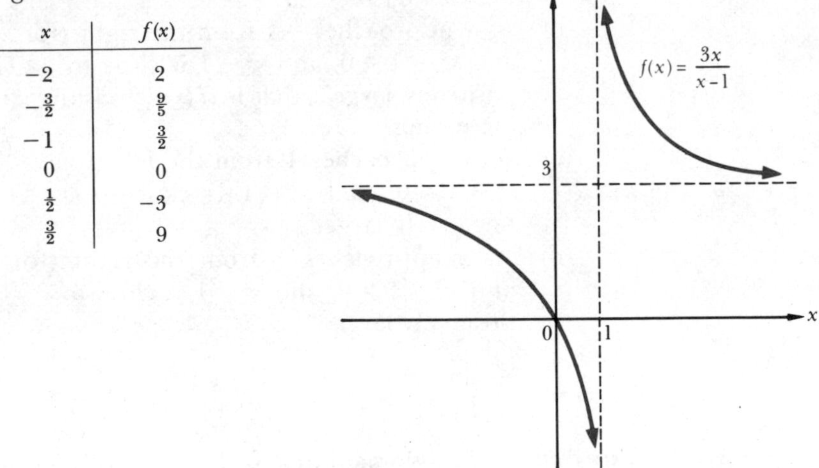

Notice that in this example the degree of the numerator is at least as large as the degree of the denominator. This comparison of the degrees of the numerator and the denominator is what prompts the use of the division algorithm. (In Examples 1, 2, and 3, the degree of the denominator was greater than the degree of the numerator.)

5 $f(x) = \dfrac{2x^2 + 1}{2x^2 - 3x}$

SOLUTION The domain of f is the set of all real numbers *except* 0 and $\frac{3}{2}$, that is, $\{x \mid x \neq 0 \text{ and } x \neq \frac{3}{2}\}$. There is no x intercept and no y intercept; $x = 0$ and $x = \frac{3}{2}$ are vertical asymptotes. Since the degree of the numerator is the same as the degree of the denominator, we use the division algorithm (as in Example 4) to get

$$f(x) = \frac{2x^2 + 1}{2x^2 - 3x} = 1 + \frac{3x + 1}{2x^2 - 3x}$$

Then, upon rewriting

$$\frac{3x + 1}{2x^2 - 3x} \quad \text{as} \quad \frac{\frac{3}{x} + \frac{1}{x^2}}{2 - \frac{3}{x}}$$

by dividing both the numerator and the denominator by x^2, we notice that as $|x|$ becomes very large, $(3x + 1)/(2x^2 - 3x)$ approaches 0 so that $f(x)$ approaches 1 asymptotically; that is, $y = 1$ is a horizontal asymptote (Figure 7).

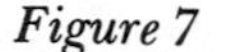
Figure 7

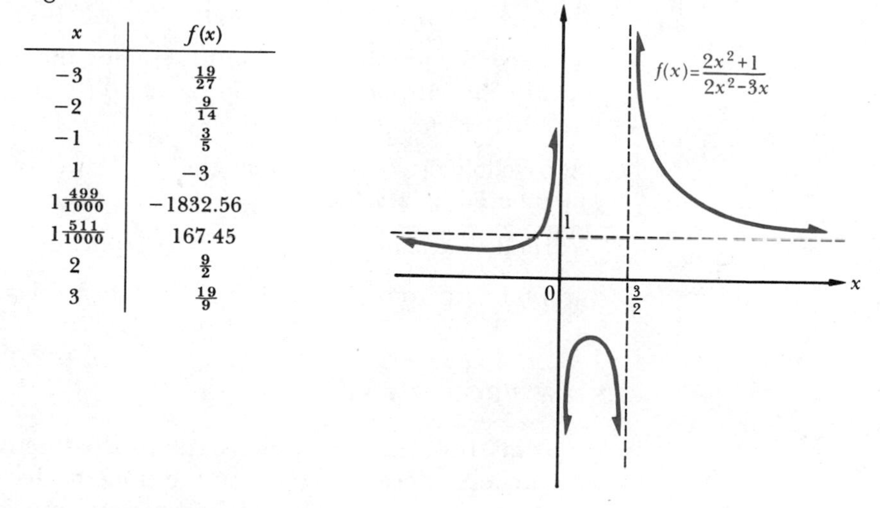

x	$f(x)$
-3	$\frac{19}{27}$
-2	$\frac{9}{14}$
-1	$\frac{3}{5}$
1	-3
$1\frac{499}{1000}$	-1832.56
$1\frac{511}{1000}$	167.45
2	$\frac{9}{2}$
3	$\frac{19}{9}$

PROBLEM SET 5

In Problems 1–15, for each rational function, determine the domain, the asymptotes, and the x and y intercepts. Sketch the graph. Use the division algorithm to examine those functions where the degree of the numerator is greater than or equal to the degree of the denominator.

1 $f(x) = \dfrac{9x^2 - 4}{3x - 2}$

2 $f(x) = \dfrac{x + 3}{x^2 + x - 6}$

3 $f(x) = \dfrac{1}{x}$

4 $f(x) = \dfrac{-2}{x^3}$

5 $f(x) = \dfrac{5}{3x + 2}$

6 $f(x) = \dfrac{6x}{x^2 - 4}$

7 $f(x) = \dfrac{8}{3x^2}$

8 $f(x) = \dfrac{x^2 + 1}{x^2 - 1}$

9 $f(x) = \dfrac{4}{x^2 - 9}$

10 $f(x) = \dfrac{x^2}{(x-1)(x-3)}$

11 $f(x) = \dfrac{x}{x^2 - 4}$

12 $f(x) = \dfrac{x(x-1)}{(x^2-1)(x-3)}$

13 $f(x) = \dfrac{3(x-2)}{x^2 - 3x + 2}$

14 $f(x) = \dfrac{x^2+1}{3(x^2+3)}$

15 $f(x) = \dfrac{x^3}{x^3+1}$

16 Living neural tissue can be excited by an electric current only if the current across the tissue reaches or exceeds a certain threshold, which we denote by I. The threshold I depends on the duration t of current flow. Weiss' law expresses I as a function of t by the equation

$$I = \frac{a}{t} + b, \quad \text{where } a \text{ and } b \text{ are positive constants}$$

(a) Graph the rational function $g(t) = \dfrac{4}{t}$ for $t > 0$.

(b) Use the graph of g to graph $I = \dfrac{4}{t} + 3$ for $t > 0$.

(c) Use the graph of part (b) to describe the behavior of the graph of the threshold I as t approaches 0 (from the right of 0) and as $|t|$ becomes very large.

17 A psychologist estimates that the percent P of certain material remembered by a subject in an experiment t months after the subject learned that material is given by the function $P = \dfrac{100}{1+t}$. Sketch the graph of P; from the graph interpret the values of P as t gets larger.

REVIEW PROBLEM SET

In Problems 1–6, determine the slope of the line defined by the given linear function where $y = f(x)$. Use the slope to decide if the function is increasing or decreasing. Graph the function and locate the intercepts.

1 $y = \dfrac{3}{5}x + \dfrac{7}{3}$

2 $\dfrac{x}{3} + \dfrac{y}{5} = 1$

3 $7x + 5y - 13 = 0$

4 $y = 5$

5 $x = -3 - y$

6 $y - 2 = -\frac{3}{4}(x + 1)$

In Problems 7–14, find an equation of each line and sketch its graph.

7 Slope $m = 3$; contains the point $(4, 5)$

8 Slope $m = 0$; contains the point $(-2, 3)$

9 Contains the points $(3, -2)$ and $(5, 6)$

10 Contains the points $(0, 2)$ and $(3, 0)$

11 Parallel to the line $3x + y = 5$ and containing the point $(-2, 3)$

12 Containing the point $(-3, -5)$ and parallel to the line containing points $(5, 8)$ and $(-1, 3)$

13 Perpendicular to the line $8x + 7y = -3$ and containing the point $(2, -4)$

14 Perpendicular to the line containing the points $(-3, 7)$ and $(-1, -1)$ and with x intercept -2

15 The three vertices of a right triangle are $(9, 3)$, $(5, 9)$, and $(-7, 1)$. Identify the vertex of the right angle.

16 Two opposite vertices of a rectangle are $(6, 2)$ and $(-5, 4)$. Two sides of the rectangle are parallel to the line $8x - 6y + 5 = 0$. Find equations of the lines containing the two sides.

17 A duplicating machine produces y sheets in x hours. The relationship between y and x is linear. If the machine has produced 120 sheets of paper after running for $\frac{1}{2}$ hour and 470 sheets of paper after running for $1\frac{1}{2}$ hours, find a function f such that $y = f(x)$.

18 A projectile fired upward attains a speed of v feet per second after t seconds of flight, and the relationship between the two numbers v and t is linear. If the projectile is fired at a speed of 2000 feet per second and reaches a speed of 90 feet per second after 2 seconds of flight, express v as a function of t.

In Problems 19–24, find the solution set of the given quadratic equation.

19 $(x + 1)^2 = 7$

20 $x^2 + 2x - 3 = 0$

21 $-2x^2 + x + 1 = 0$

22 $3x^2 - x - 1 = 0$

23 $4x^2 + 6x - 1 = 0$

24 $6x^2 - 7x - 5 = 0$

In Problems 25–29, graph each of the quadratic functions by locating the x and y intercepts and the extreme point. Then use the graph to determine the range of f and to solve the given associated inequality.

25 $f(x) = -x^2 + 2;\ -x^2 + 2 \geq 0$

26 $g(x) = 2x^2 - x - 1;\ 2x^2 - x - 1 < 0$

27 $f(x) = 6x^2 - 5x - 4;\ 6x^2 - 5x - 4 \leq 0$

28 $h(x) = -3 + 10x - 8x^2;\ -3 + 10x - 8x^2 < 0$

29 $g(x) = -x^2 + 2x - 1;\ -x^2 + 2x - 1 < 0$

30 A projectile is fired from the ground in such a way that it is h feet above the ground t seconds after the firing, where $h = 88t - 16t^2$.
(a) Find t when $h = 0$.
(b) Find the maximum height reached by the projectile.

31 A manufacturer can sell x units of a product at a price of p cents per unit, where $p = 300 - 0.25x$. What number of units should be sold to achieve maximum income?

32 An apple orchard now has 30 trees per acre, and the average yield is 400 apples per tree. For each additional tree planted per acre, the average yield is reduced by approximately 10 apples. How many trees per acre will give the largest crop of apples?

In Problems 33–38, find all zeros of the given polynomial function.

33 $h(x) = \frac{1}{2}x + \frac{1}{3}$

34 $g(x) = x^4 - x$

35 $f(x) = (x - 1)(x^2 - 5x + 4)$

36 $h(x) = x^2 - 5x$

37 $F(x) = (x^2 - 4)(x^2 - 3x + 2)$
38 $G(x) = (2x^2 + 5x - 3)(6x^2 - 11x - 10)$

In Problems 39–42, use synthetic division to determine $Q(x)$ and $f(c)$ so that $f(x) = (x - c)Q(x) + f(c)$.

39 $f(x) = 3x^3 + 5x^2 + 7x - 3; c = 2$
40 $f(x) = 5x^4 - 2x^3 + 11x^2 + 5x + 36; c = 1$
41 $f(x) = 2x^3 + 4x^2 + 3x - 18; c = -1$
42 $f(x) = x^7 - 5; c = 2$

43 Let $f(x) = x^5 + x^4 - 3x^3 - 5x^2 + x + 7$. Use synthetic division to find each of the following values.
(a) $f(2)$ (b) $f(3)$ (c) $f(-1)$ (d) $f(-2)$ (e) $f(\frac{1}{2})$

44 Use synthetic division to find k so that -2 is a zero of the function $g(x) = 5x^3 - 2x^2 + 4x + k$.

In Problems 45–48, find all rational number zeros.

45 $f(x) = x^3 - 4x^2 + x + 6$
46 $g(x) = x^4 - 7x^3 + 18x^2 - 20x + 8$
47 $h(x) = 2x^4 + x^3 - 19x^2 - 9x + 9$
48 $f(x) = 4x^4 + 4x^3 - 3x^2 - 2x + 1$

49 Use the graph of $f(x) = (x + 1)(x - 2)(x + 3)$ to find the solution set of $(x + 1)(x - 2)(x + 3) \geqslant 0$.

50 Explain why the graphs of $f(x) = x - 4$ and $f(x) = \dfrac{x^2 - 16}{x + 4}$ are different.

In Problems 51–55, determine the domain, the x and y intercepts, and the asymptotes, then sketch the graph of each of the rational functions.

51 $f(x) = \dfrac{2}{x - 5}$

52 $g(x) = \dfrac{x}{x + 3}$

53 $h(x) = \dfrac{x - 2}{x^2 - 9}$

54 $f(x) = \dfrac{x^2}{x^2 + 1}$

55 $G(x) = \dfrac{x^2 + 2}{x^2 + 3}$

56 According to Boyle's law, the pressure p and the volume v of a certain container of gas satisfy the equation $p = 3000/v$. The pressure is measured in pounds per square inch and the volume in cubic inches. Sketch the graph of the equation $p = 3000/v$ for $v > 0$.

CHAPTER 4

Exponential and Logarithmic Functions

In Chapters 2 and 3 we studied polynomial and rational functions. In this chapter, we introduce two more important types of functions—*exponential functions* and *logarithmic functions.* These functions are indispensable in mathematics and in many of its practical applications. The concept of logarithms and their properties are covered extensively in this chapter because logarithms are needed to solve certain types of equations that arise from formulas used in business, engineering, and the sciences.

1 Exponential Functions

In this section, we introduce functions that satisfy the basic laws of exponents considered in Section 4 of Chapter 1. Such functions are appropriate for modeling population growth and similar applications.

Suppose, for instance, that a biologist is investigating a colony of a certain kind of bacteria. As part of her investigation she wishes to discover how the number of bacteria in the culture changes with time. She discovers that the time required for the number of bacteria to triple does not depend on the number of bacteria that are present initially. To be specific, assume that on a given day there are x_0 bacteria present and that the number triples each day. There are $3x_0$ present after 1 day, $3(3x_0) = 3^2x_0$ after 2 days, $3(3^2x_0) = 3^3x_0$ after 3 days, and so on, so that after n days there are 3^nx_0 bacteria present. This phenomenon of bacteria growth can be represented by the function $f(t) = 3^tx_0$, where t represents the number of days after the experiment begins and x_0 represents the number of bacteria present initially.

Now, although it is true that the actual experiment supplies data only when t is a positive integer, we can assume that the function $f(t) = 3^tx_0$ is continuous, so that $f(t)$ indicates the number of bacteria present after t days, where t is *any* positive number. For example, after 2 days, 4 hours ($2\frac{1}{6}$ or $\frac{13}{6}$ days), there would be $3^{13/6}x_0$ bacteria present.

It is our purpose here to investigate the properties of functions of the type $f(t) = 3^tx_0$.

DEFINITION 1 EXPONENTIAL FUNCTION

If b is a positive number, then the function $f(x) = b^x$ is called an *exponential function* with *base b.*

For example, $f(x) = 3^x$ is an exponential function with base 3; $h(t) = (\frac{1}{3})^t$ is an exponential function with base $\frac{1}{3}$; $g(x) = 2^{-x} = (2^{-1})^x = (\frac{1}{2})^x$ is an exponential function with base $\frac{1}{2}$.

To sketch the graph of an exponential function, we first calculate function values for some domain values. For instance, to graph $f(x) = 2^x$, we first find the values of $f(x)$ for $x = -4, -3, -2, -1, 0, 1, 2, 3$, and 4.

$$f(-4) = 2^{-4} = \frac{1}{2^4} = \frac{1}{16} \qquad f(1) = 2^1 = 2$$

$$f(-3) = 2^{-3} = \frac{1}{2^3} = \frac{1}{8} \qquad f(2) = 2^2 = 4$$

$$f(-2) = 2^{-2} = \frac{1}{2^2} = \frac{1}{4} \qquad f(3) = 2^3 = 8$$

$$f(-1) = 2^{-1} = \frac{1}{2} \qquad f(4) = 2^4 = 16$$

$$f(0) = 2^0 = 1$$

1.1 Exponential Function Properties

As with polynomial functions, it can be shown through the use of more advanced techniques of calculus that every exponential function is continuous everywhere in its domain. Thus, after tabulating and then plotting the points from the calculations done above, we obtain the graph of $f(x) = 2^x$ by connecting these points to form a smooth curve (Figure 1).

Figure 1

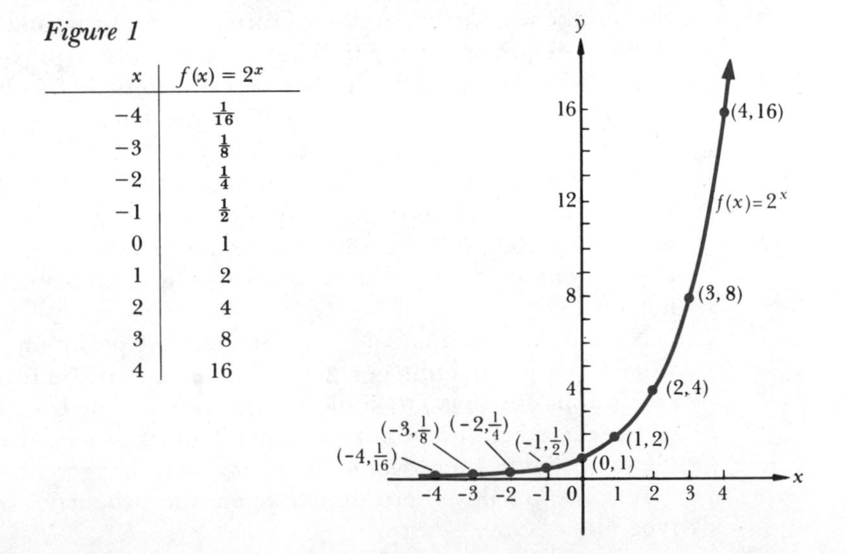

x	$f(x) = 2^x$
-4	$\frac{1}{16}$
-3	$\frac{1}{8}$
-2	$\frac{1}{4}$
-1	$\frac{1}{2}$
0	1
1	2
2	4
3	8
4	16

Using the extended properties of real number exponents (see Section 4 of Chapter 1, page 21), we conclude that the domain of f is the set of all real numbers. In fact, the domain of all exponential functions is the set of real numbers. The graph of f (Figure 1) reveals that as x increases through *positive values,* the function f is increasing very rapidly; and as x gets smaller through *negative values,* the graph of f gets closer and closer to the x axis. That is, the x axis is a horizontal asymptote for $f(x) = 2^x$. It is also evident from the graph that the range of f is the set of positive real numbers.

EXAMPLES Sketch the graph of each function in Examples 1 and 2. Indicate the domain and range of the function. Also, specify whether the function is increasing or decreasing.

Figure 2

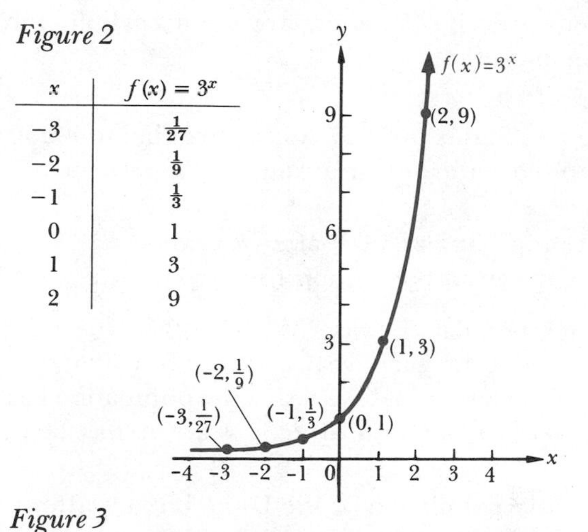

x	$f(x) = 3^x$
-3	$\frac{1}{27}$
-2	$\frac{1}{9}$
-1	$\frac{1}{3}$
0	1
1	3
2	9

1 $f(x) = 3^x$

SOLUTION The function $f(x) = 3^x$ has base 3. The domain of f is the set of all real numbers. After plotting a few points, we use continuity to obtain the graph (Figure 2). The graph of f shows that f is an increasing function; the range of f is the set of all positive real numbers, that is, $\{y \mid y > 0\}$. Notice also that the x axis is a horizontal asymptote and that f does have an inverse.

2 $g(x) = (\frac{1}{4})^x$

SOLUTION The function $g(x) = (\frac{1}{4})^x$ has base $\frac{1}{4}$; the domain of g is the set of all real numbers. Here again we use the continuity of exponential functions to obtain the graph given in Figure 3. The function g is decreasing and has an inverse. Also, the x axis is a horizontal asymptote. The range of g is the set of positive real numbers, that is, $\{y \mid y > 0\}$.

Figure 3

x	$g(x) = (\frac{1}{4})^x$
-1	4
0	1
1	$\frac{1}{4}$
2	$\frac{1}{16}$

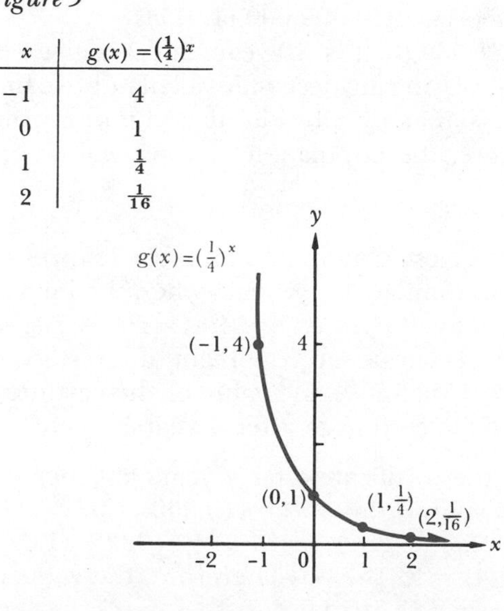

3 If the graph of an exponential function f contains the point (2, 25), what is the base?

SOLUTION If the function is exponential, it takes the form $f(x) = b^x$. Since (2, 25) is on the graph of f, it follows that $25 = b^2$, so $b = 5$ is the base. Note that although -5 satisfies $b^2 = 25$, -5 is *not* the base because of the definition of an exponential function.

1.2 Exponential Functions—Applications

Exponential functions occur in many applications. One such example is the natural growth of bacteria in a culture. Other examples involve population growth, nuclear decay, chemical decomposition, cell division, decay of current in an electrical circuit, and compound interest. In performing the computations that are encountered in these applications we *can* use logarithms, which enable us to raise numbers to certain powers with comparative ease. (These and other uses of the logarithmic tables will be explored in Section 5.) However, with the development of modern electronic calculators and with the consistent decreases in prices of such calculators, the logarithmic tables and other traditional computing devices are becoming obsolete. A calculator can perform a complicated computation with incredible speed and accuracy. In this textbook, we recommend the use of calculators in working *some* problems, but if calculators are not accessible, the problems can be done by using logarithms.

EXAMPLES 1 The population P of the United States in 1977 was approximately 230 million. Suppose that the population grows at a rate of 1.7 percent per year.

(a) Find a function that gives the population P after n years.
(b) What is the approximate projected population for 1987?

SOLUTION If the United States population was 230,000,000 in the year 1977, and it increases by 1.7 percent each year, we can say that the population is multiplied by 1.017 each year. After 2 years, the population has been multiplied by $(1.017)^2$, and so on until after n years it has been multiplied by $(1.017)^n$.

(a) The function P that gives the population of the United States after n years is, therefore, given by $P(n) = (230{,}000{,}000)(1.017)^n$.
(b) The projected population for 1987 (that is, 10 years later) is given by $P(10) = (230{,}000{,}000)(1.017)^{10}$. Using an electronic calculator, we find that $(1.017)^{10} = 1.183612463$, so that, by substituting and multiplying, $P(10) = 272{,}230{,}867$. Therefore, the population in 1987 will be approximately 272,000,000.

2 If P dollars represents the amount invested at an annual rate r (expressed as a decimal), then the amount A accumulated in n years when the interest is compounded annually is given by the function $A(n) = P(1 + r)^n$. If \$10,000 is invested in a savings certificate at 7 percent interest compounded annually, find a function A that gives the value of the certificate after n years. What is the value of the certificate after 4 years?

SOLUTION If $A(n)$ is the value of the certificate after n years at 7 percent compounded annually, then $A(n)$ is given by $A(n) = (10{,}000)(1.07)^n$. The value of the certificate after 4 years is given by $A(4) = (10{,}000)(1.07)^4$. Using a calculator, we find that $A(4) = 13{,}107.96$. Therefore, the value of the certificate at the end of 4 years is \$13,107.96.

1.3 Special Exponential Equations

In the two examples above, one might ask the questions: When will the population of the United States reach a certain number, say, 460 million? When will the investment in the savings certificate be worth a certain number of dollars, say, $33,000? Such questions are answered by solving equations involving exponents called *exponential equations.* For example, $460{,}000{,}000 = (230{,}000{,}000)(1.017)^n$ and $33{,}000 = (10{,}000)(1.07)^n$ are exponential equations. Later, in Section 5, we shall solve such equations by using logarithms. However, we can deal with some *special situations* now.

For example, suppose that the number of bacteria in a certain culture after t hours is given by the function $Q(t) = 2000(3^t)$. One might ask how long it will take to grow 486,000 bacteria in the culture; that is, when is $Q(t) = 486{,}000$? This question is answered by solving the equation $2{,}000(3^t) = 486{,}000$ or $3^t = 243 = 3^5$. In order for 3^t to equal 3^5, t and 5 must be the same number. Therefore, $t = 5$. That is, it will take 5 hours to grow 486,000 bacteria in the culture.

In general, to solve exponential equations, we use the property

$$a^x = a^y \quad \text{if and only if} \quad x = y \text{ for } a > 0 \text{ and } a \neq 1$$

EXAMPLES Find the solution set of each equation.

1 $5^x = 125$

SOLUTION $5^x = 125 = 5^3$, so $x = 3$. Therefore, the solution set is $\{3\}$.

2 $27^x = 81$

SOLUTION $27^x = (3^3)^x = 3^{3x}$. Also, $81 = 3^4$. Thus, $3^{3x} = 3^4$, so $3x = 4$ or $x = \frac{4}{3}$. Therefore, the solution set is $\{\frac{4}{3}\}$.

3 $4^{2x-1} = 16$

SOLUTION $4^{2x-1} = (2^2)^{2x-1} = 2^{4x-2}$. Also, $16 = 2^4$. Thus, $2^{4x-2} = 2^4$, so that $4x - 2 = 4$ or $x = \frac{3}{2}$. Therefore, the solution set is $\{\frac{3}{2}\}$.

4 $2^{2x+2} - 9(2^x) + 2 = 0$

SOLUTION The equation $2^{2x+2} - 9(2^x) + 2 = 0$ can be written in the form $(2^x)^2 2^2 - 9(2^x) + 2 = 0$. If we let $y = 2^x$, we get $4y^2 - 9y + 2 = 0$, that is, $(4y - 1)(y - 2) = 0$, so that $y = \frac{1}{4}$ or $y = 2$, and $2^x = \frac{1}{4}$ or $2^x = 2$. Hence $x = -2$ or $x = 1$. Therefore, the solution set is $\{-2, 1\}$.

PROBLEM SET 1

In Problems 1–12, use $f(x) = (\frac{2}{3})^x$, $g(x) = 4^{-x/2}$, and $h(x) = 2^{-(1+x)^2}$ to find each of the following values.

1 $f(-1)$	**2** $g(-1)$	**3** $h(-1)$	**4** $f(-2)$
5 $g(-2)$	**6** $h(-3)$	**7** $f(2)$	**8** $g(2)$
9 $h(0)$	**10** $f(3)$	**11** $g(3)$	**12** $h(3)$

In Problems 13–23, graph the function. Indicate the domain and range. Also specify whether the function is increasing or decreasing in R.

13 $f(x) = 5^x$ **14** $h(x) = 1^x$ **15** $f(x) = 3^{x+1}$
16 $f(x) = (\sqrt{2})^x$ **17** $g(x) = (\frac{4}{7})^x$ **18** $f(x) = 2^{-x} + 1$
19 $h(x) = (\frac{1}{3})^{-x}$ **20** $g(x) = (0.1)^x$ **21** $f(x) = (5)(3^x)$
22 $h(x) = -4^x$ **23** $g(x) = (0.7)^x$

24 Use the graph of $f(x) = 3^x$ to approximate each of the following numbers.
(a) $3^{1/2}$ (b) $3^{5/2}$ (c) 3^{π} (d) $3^{\sqrt{2}}$ (e) $3^{-1.2}$

In Problems 25–29, find the base of the exponential function $f(x) = b^x$ when its graph contains the given point.

25 $(2, 9)$ **26** $(-2, 16)$ **27** $(3, 27)$
28 $(\frac{1}{2}, \sqrt{10})$ **29** $(0, 1)$

30 Let $f(x) = b^x$.
(a) Find $f[f(x)]$.
(b) Show that $f(c + d) = f(c) \cdot f(d)$.

31 (a) Given an exponential function $f(x) = b^x$, indicate whether f is increasing or decreasing for $b = 1$; for $0 < b < 1$; and for $b > 1$. Give examples to support your conclusions.
(b) Can an exponential function f be even? Odd? Explain.

32 Graph $f(x) = 5^x$ and $g(x) = -5^x$ on the same coordinate system. Compare the two graphs. Can you generalize your results so that you can use the graph of $f(x) = b^x$ to determine the graph of $g(x) = -b^x$?

33 Graph $f(x) = 10^x$.
(a) Use the graph to explain why f has an inverse.
(b) Use the graph of $f(x) = 10^x$, together with reflection across the line $y = x$, to obtain the graph of f^{-1}.
(c) What are the domain and range of f?
(d) What are the domain and range of f^{-1}?

34 Graph $g(x) = (\frac{2}{3})^x$. Answer questions (a)–(d) from Problem 33 above.

In Problems 35–44, find the solution set of the given exponential equation.

35 $3^{3x} = 27$ **36** $6^{2x-1} = 216$ **37** $2^{x+1} = \frac{1}{8}$
38 $4^{3x-1} = 16$ **39** $16^{3x} = 8^{2x-1}$ **40** $7^{5x+4} = 1$
41 $36^{x-6} = 6$ **42** $3^{-8x+6} = 27^{-x-8}$ **43** $2^{2x+2} + 2^{x+2} = 3$
44 $5^{x^2+3x} = 625$

45 Suppose that a biologist grows a culture of a certain kind of bacteria. Assume that in an experiment, $N = x_0 3^n$ represents the number of bacteria present at the end of n days, where x_0 is the number of bacteria that are present initially. Suppose that there are 333,000 bacteria present at the end of 2 days.
(a) Find the number of bacteria at the end of 4 days.
(b) In how many days were there 111,000 bacteria present?

46 A department store's annual profit from the sale of a certain toy is given by the equation $y = 8{,}000 + 30{,}000(2^{-0.4x})$, where y is the annual profit in dollars and x denotes the number of years the toy has been on the market.

(a) Calculate the store's annual profit for $x = 1, 2, 3, 5, 10$, and 15.

(b) Use the results of part (a) to sketch the graph of this equation.

47 A nation had a population of 20,000,000 in 1977. If the population grows at the rate of 2 percent per year, write a function that gives the population P after n years. How many people will there be in 1982? [Use $(1.02)^5 = 1.1041$.]

48 Medical researchers estimate that the number of cancer cells present in a certain tissue sample after t days is given by the function $N(t) = 200(2^{0.12t})$. Find the number of cancer cells present in the tissue sample after 8 days. [Use $2^{0.96} = 1.9453$.]

49 A savings bond certificate pays 11 percent interest compounded annually for the period of 7 years. Find a function f that gives the value of the certificate including interest after x years if the certificate is originally worth \$5000. What is the value of the certificate after 7 years? [Use $(1.11)^7 = 2.0762$.]

2 Logarithmic Functions

Figures 1, 2, and 3 of Section 1 suggest that any exponential function given by $y = b^x$, where $b \neq 1$, is either an increasing or decreasing function (see Problem Set 1, Problem 31a). Consequently, this kind of exponential function is invertible, and the inverse function can be constructed by using the technique presented in Chapter 2.

We start with the invertible exponential function $y = b^x$, where $b > 0$ and $b \neq 1$. Next, we "switch" the x and y to get $x = b^y$. In Chapter 2, we completed the process of constructing the inverse function by expressing y in terms of x. The difficulty here is that we do not have the algebraic tools needed to solve $x = b^y$ for y in terms of x. In order to maintain the convention of expressing the dependent variable y in terms of the independent variable x, we introduce some new terminology.

DEFINITION 1 LOGARITHM

$y = \log_b x$ (read as logarithm of x to base b) is equivalent to $x = b^y$, where $b > 0$ and $b \neq 1$.

For example, the equation $3 = \log_2 8$ is equivalent to $8 = 2^3$, and the equation $x = \log_3 64$ is equivalent to $64 = 3^x$.

In view of Definition 1, we can rewrite $x = b^y$ as $y = \log_b x$ so that the inverse of the exponential function $f(x) = b^x$, where $b > 0$ and $b \neq 1$, is the *logarithmic function* $f^{-1}(x) = \log_b x$.

For example, if $f(x) = 10^x$, then $f^{-1}(x) = \log_{10} x$.

To reinforce the idea presented here, let us work a few examples.

EXAMPLES 1 Write each of the following exponential equations as an equivalent logarithmetic equation.

(a) $3^2 = 9$ (b) $10^0 = 1$ (c) $\sqrt[3]{27} = 3$

SOLUTION By referring above to Definition 1 of a logarithm, we know that $b^y = x$ is equivalent to $y = \log_b x$. Hence

(a) $3^2 = 9$ is equivalent to $2 = \log_3 9$.
(b) $10^0 = 1$ is equivalent to $0 = \log_{10} 1$.
(c) $\sqrt[3]{27} = 27^{1/3} = 3$ is equivalent to $\frac{1}{3} = \log_{27} 3$.

2 Write each of the following logarithmic equations as an equivalent exponential equation.

(a) $\log_{10} 10 = 1$ (b) $\log_{1/2} 4 = -2$ (c) $\log_{1/2} \frac{1}{4} = 2$

SOLUTION By reference again to Definition 1 of a logarithm, we know that $y = \log_b x$ is equivalent to $b^y = x$. Hence

(a) $\log_{10} 10 = 1$ is equivalent to $10^1 = 10$.
(b) $\log_{1/2} 4 = -2$ is equivalent to $(\frac{1}{2})^{-2} = 4$.
(c) $\log_{1/2} \frac{1}{4} = 2$ is equivalent to $(\frac{1}{2})^2 = \frac{1}{4}$.

3 Let $f(x) = \log_5 x$. Find

(a) $f(25)$ (b) $f(5)$ (c) $f(\frac{1}{5})$ (d) $f(\frac{1}{25})$ (e) $f(1)$

SOLUTION

(a) $f(25) = \log_5 25$. Setting $t = \log_5 25$, we get $5^t = 25 = 5^2$ so that $t = 2$. Therefore, $f(25) = \log_5 25 = 2$.
(b) $f(5) = \log_5 5$. Let $t = \log_5 5$, then $5^t = 5$ so that $t = 1$. Therefore, $f(5) = \log_5 5 = 1$.
(c) $f(\frac{1}{5}) = \log_5 \frac{1}{5}$. Assume that $t = \log_5 \frac{1}{5}$, so that $5^t = \frac{1}{5} = 5^{-1}$ or $t = -1$. Therefore, $f(\frac{1}{5}) = \log_5 \frac{1}{5} = -1$.
(d) $f(\frac{1}{25}) = \log_5 \frac{1}{25}$. Setting $t = \log_5 \frac{1}{25}$, we get

$$5^t = \frac{1}{25} = \frac{1}{5^2} = 5^{-2} \quad \text{or} \quad t = -2$$

Therefore, $f(\frac{1}{25}) = \log_5 \frac{1}{25} = -2$.
(e) $f(1) = \log_5 1$. Let $t = \log_5 1$, then $5^t = 1$ so that $t = 0$. Therefore, $f(1) = \log_5 1 = 0$.

4 Evaluate

(a) $\log_b b$ (b) $\log_b 1$

SOLUTION

(a) If $t = \log_b b$, then $b^t = b$, so $t = 1$. Hence $\log_b b = 1$.
(b) If $t = \log_b 1$, then $b^t = 1$, so $t = 0$. Thus $\log_b 1 = 0$.

5 Evaluate $\dfrac{\log_{49} 7 - \log_8 64}{\log_9 27 + \log_{10} 100}$.

SOLUTION Since $\log_{49} 7 = \frac{1}{2}$, $\log_8 64 = 2$, $\log_9 27 = \frac{3}{2}$, and $\log_{10} 100 = 2$

$$\frac{\log_{49} 7 - \log_8 64}{\log_9 27 + \log_{10} 100} = \frac{\frac{1}{2} - 2}{\frac{3}{2} + 2} = \frac{-\frac{3}{2}}{\frac{7}{2}} = -\frac{3}{7}$$

6 If the graph of a logarithmic function defined by $f(x) = \log_b x$ contains the point (216, 3), what is the base?

SOLUTION Since the point (216, 3) is on the graph of f, then $\log_b 216 = 3$, so $b^3 = 216 = 6^3$ and $b = 6$. Hence the base is 6.

2.1 Logarithmic Function Properties

Since the graph of f^{-1} is a reflection of the graph of f across the line $y = x$, the graph of $f^{-1}(x) = \log_b x$ is a reflection of the graph of $f(x) = b^x$, where $b \neq 1$, across the line $y = x$. For example, $y = g(x) = \log_2 x$ is the inverse of $f(x) = 2^x$. Figure 1 shows that the graph of $y = g(x) = \log_2 x$ is the reflection of the graph of $f(x) = 2^x$.

Figure 1 reveals that the domain of f is the range of g, and the range of f is the domain of g. The domain of g is the set of all positive real numbers, whereas the range of g is the set of all real numbers. Figure 1 also shows that the function g is continuous and increasing.

In general, the domain of f^{-1} is the range of f. Consequently, since the range of f defined by $f(x) = b^x$ is the set of all positive real numbers whenever $b \neq 1$, the domain of g defined by $g(x) = \log_b x$ is also the set of all positive real numbers, that is, $\{x \mid x > 0\}$. As a result, it is only possible to compute the logarithm of a positive number.

Figure 1

$x = 2^y$	y
1	0
2	1
4	2
8	3

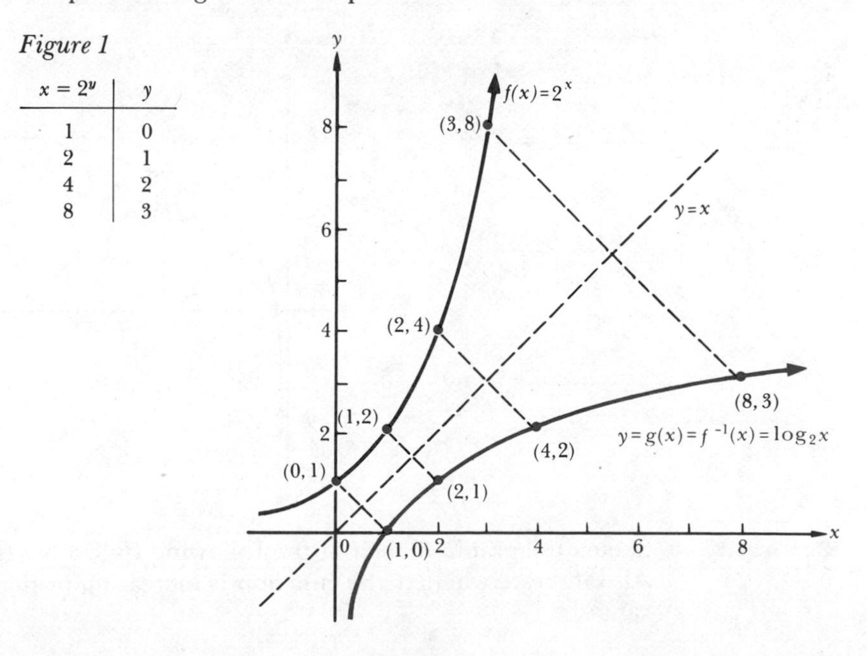

EXAMPLES Find the domain of each of the following functions.

1 $y = \log_5 x$

SOLUTION The domain consists of all values of x such that x is positive, that is, $\{x \mid x > 0\}$.

2 $f(x) = \log_8(2x - 1)$

SOLUTION Since the logarithm is defined only for positive numbers, $2x - 1 > 0$, so $x > \frac{1}{2}$. Hence, the domain of f is $\{x \mid x > \frac{1}{2}\}$.

3 $g(x) = \log_{10}(x^2 - 4)$

SOLUTION $x^2 - 4$ must be positive in order for $g(x)$ to be defined. Solving the quadratic inequality $x^2 - 4 > 0$ results in $x < -2$ or $x > 2$. Hence, the domain of g is $\{x \mid x < -2 \text{ or } x > 2\}$.

2.2 Graphing Logarithmic Functions

A more direct way to graph a logarithmic function $y = \log_b x$, other than to reflect the graph of $y = b^x$, is to use the equivalent equation $x = b^y$ to locate a few points and then use continuity to connect the points to form the graph.

Thus, to graph $y = \log_3 x$, we can use the equivalent equation $x = 3^y$. For $y = 0, x = 3^0 = 1$; for $y = 1, x = 3^1 = 3$; for $y = 2, x = 3^2 = 9$; and so on. One should note that we are *reversing* the usual technique in finding points on the graph. Here we have *first* selected a value of $y = f(x)$ and then determined the corresponding value of x (Figure 2).

Figure 2

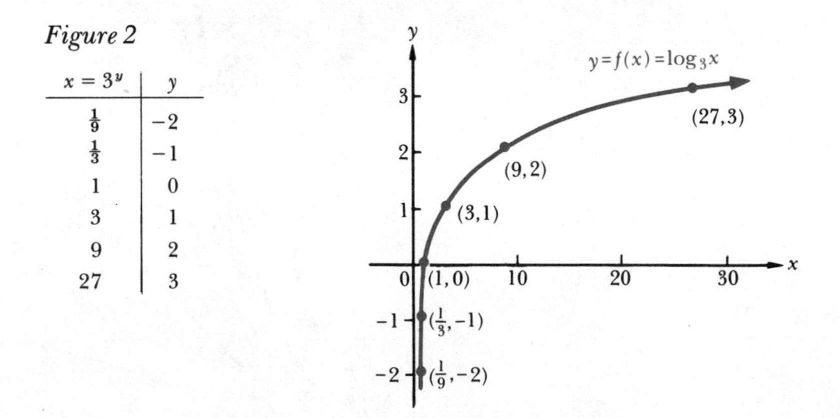

$x = 3^y$	y
$\frac{1}{9}$	-2
$\frac{1}{3}$	-1
1	0
3	1
9	2
27	3

EXAMPLES Sketch the graph of each of the following functions. Indicate the domain. Also indicate whether the function is increasing or decreasing and find its range.

1 $f(x) = \log_4 x$

SOLUTION The domain of f is the set of all positive real numbers, that is, $\{x \mid x > 0\}$. We know that $y = f(x) = \log_4 x$ is equivalent to $4^y = x$. A table of values can be determined for this exponential equation. The graph of $f(x) = \log_4 x$ (Figure 3) shows that f is an increasing function. Notice that the range is the set of all real numbers.

Figure 3

$x = 4^y$	y
$\frac{1}{4}$	-1
1	0
4	1
16	2
64	3

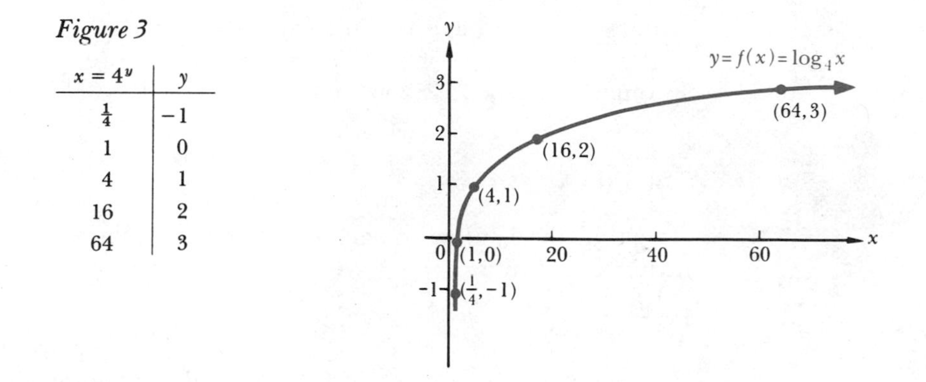

2 $g(x) = \log_{1/4} x$

SOLUTION The domain of g is the set of all positive real numbers, that is, $\{x \mid x > 0\}$. The graph can be determined by considering $y = g(x) = \log_{1/4} x$ as $(\frac{1}{4})^y = x$ to locate a few points (Figure 4). Thus, $g(x) = \log_{1/4} x$ is a decreasing function. The range is the set of all real numbers.

3 $h(x) = \log_2(-x)$, for $x < 0$

SOLUTION Since $-x$ must be positive, the domain of h is the set of all negative real numbers, that is, $\{x \mid x < 0\}$. The graph of h can be determined by considering $y = h(x) = \log_2(-x)$ as $2^y = -x$ to locate a few points (Figure 5). Thus h is decreasing and the range of h is the set of all real numbers.

Figure 4

$x = (\frac{1}{4})^y$	y
$\frac{1}{2}$	$\frac{1}{2}$
1	0
2	$-\frac{1}{2}$

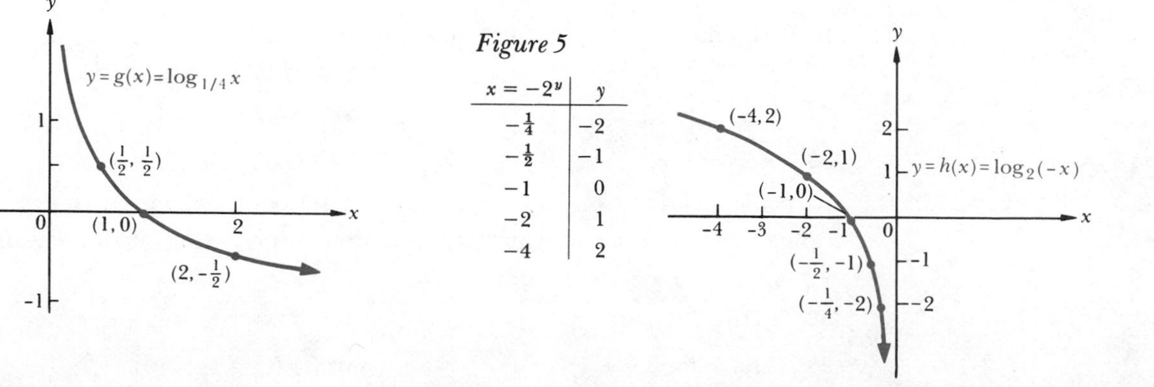

Figure 5

$x = -2^y$	y
$-\frac{1}{4}$	-2
$-\frac{1}{2}$	-1
-1	0
-2	1
-4	2

Logarithmic functions occur frequently in applications from business and the sciences, as illustrated in the following example.

EXAMPLE A company finds that the weekly profit P in dollars can be determined by the function $P(x) = 70 \log_2 x$, where x is the amount (in dollars) spent on TV commercials and $x > 2$. What is the company's weekly profit when it spends $256 on TV commercials?

SOLUTION We have to find the value of $P(x)$ when $x = 256$, so that $P(256) = 70 \log_2 256 = 70(8) = 560$. Therefore, the company's weekly profit when it spends $256 on TV commercials is $560.

PROBLEM SET 2

In Problems 1–6, write each exponential equation as an equivalent logarithmic equation.

1 $5^3 = 125$ **2** $4^{-2} = \frac{1}{16}$ **3** $\sqrt{9} = 3$
4 $(\frac{1}{3})^{-2} = 9$ **5** $\sqrt[5]{32} = 2$ **6** $\pi^t = w$

In Problems 7–12, write each logarithmic equation as an equivalent exponential equation.

7 $\log_9 81 = 2$ **8** $\log_{10} 0.0001 = -4$ **9** $\log_{1/3} 9 = -2$
10 $\log_{10} \frac{1}{10} = -1$ **11** $\log_{36} 216 = \frac{3}{2}$ **12** $\log_w z = s$

In Problems 13–20, use $f(x) = \log_7 x$, $g(x) = \log_8 x$, or $h(x) = \log_{1/6} x$ to find each value.

13 $f(49)$ **14** $g(64)$ **15** $h(216)$ **16** $f(1)$
17 $g(\frac{1}{64})$ **18** $h(\frac{1}{36})$ **19** $h(1)$ **20** $f(\frac{1}{7})$

In Problems 21–22, simplify each expression.

21 $\dfrac{\log_2 \sqrt{\frac{1}{8}} - \log_2 16^{1/3}}{\log_8 64 \cdot \log_3 3^{1/6}}$ **22** $\dfrac{\log_3 243}{\log_2 \sqrt[4]{64} + \log_8 8^{-10}}$

In Problems 23–28, find the base of the logarithmic function $f(x) = \log_b x$ if its graph contains the given point.

23 $(2, \frac{1}{2})$ **24** $(25, 2)$ **25** $(8, 3)$
26 $(\frac{1}{64}, -3)$ **27** $(\frac{1}{27}, -3)$ **28** $(\frac{1}{216}, -3)$

In Problems 29–34, determine the domain of each function.

29 $f(x) = \log_{10}(3x + 2)$ **30** $f(x) = \log_a |x + 1|$
31 $g(x) = \log_3(x^2 - 5x + 4)$ **32** $h(x) = \log_5 \left(\dfrac{1}{x}\right)$
33 $F(x) = \log_6(1 - x)$ **34** $G(x) = \log_7 |-x|$

In Problems 35–40, sketch the graph of the function. Find the domain, range, and intervals in which the function is increasing or decreasing.

35 $f(x) = \log_6 x$ **36** $g(x) = 3 \log_5 2x$
37 $g(x) = \log_{1/2} x^2$ **38** $f(x) = \dfrac{1}{2} \log_{1/3} \dfrac{1}{x}$
39 $h(x) = \log_\pi (x + 1)$ **40** $g(x) = \log_3 |x|$

In Problems 41–43, determine $f[g(x)]$ and $g[f(x)]$.

41 $f(x) = \log_{10} x$ and $g(x) = 3x$ **42** $f(x) = \log_5 x$ and $g(x) = \sqrt{x}$

43 $f(x) = \log_2 x$ and $g(x) = 5x - 1$

44 Use the graph of $f(x) = \log_{10} x$ to approximate the following values.
(a) $\log_{10} \frac{1}{2}$ (b) $\log_{10} 2$ (c) $\log_{10} \pi$
(d) $\log_{10}\sqrt{2}$ (e) $\log_{10} 0.1$

45 Graph $f(x) = \log_b x$, where $b = 2, 3, 4$, and 5, on the same coordinate system. How do the graphs compare?

46 Graph $f(x) = \log_5 x$ and $g(x) = 5^x$ on the same coordinate system. Explain the symmetry with respect to the line $y = x$.

47 Sketch the graphs of $f(x) = \log_3 x$ and $g(x) = \log_{1/3} x$. Compare the properties of f and g. In general, how do the logarithmic functions with base less than 1 compare to the logarithmic functions with base greater than 1?

48 Let $f(x) = 10^x$ and $g(x) = \log_{10} x$. Find $f[g(x)]$ and $g[f(x)]$. Simplify your answer by using the fact that $g(x) = \log_{10} x$ is equivalent to $10^{g(x)} = x$.

49 Graph $y = \log_{10} x$.
(a) For what values of x is $y < 0$?
(b) For what values of x is $y > 0$?
(c) For what value of x is $y = 0$?
(d) If $x_1 < x_2$, how does $\log_{10} x_1$ compare to $\log_{10} x_2$?

50 Let $f(x) = b^x$ and $g(x) = \log_b x$.
(a) Show that $b^{\log_b M} = M$.
(b) Use the result in part (a) to simplify $f[g(x)]$ and $g[f(x)]$.

51 A company's weekly profit P in dollars is estimated by the function $P(x) = 914 \log_9 x$, where x is the number of parts produced each minute. Find the company's weekly profit if it produced 729 parts per minute.

52 The pH of a solution is a shorthand method of expressing the acidity or alkalinity of a solution. The pH P is defined as $P = -\log_{10} h$, where h represents the hydrogen ion concentration. Find the pH if the hydrogen ion concentration is
(a) 10^{-3} mole (b) 0.00001 mole (c) 10^{-7} mole (d) 100 moles

3 Properties of Logarithms

Since logarithms are exponents, their properties are derived from the properties of exponents (see Chapter 1, Section 4) and Definition 1 in Section 2 of the present chapter. These properties are used in computations involving logarithms.

THEOREM 1 LOGARITHM PROPERTIES

Suppose that M, N, and b are positive real numbers, where $b \neq 1$, and r is any real number. Then

(i) $\log_b MN = \log_b M + \log_b N$

(ii) $\log_b \frac{M}{N} = \log_b M - \log_b N$

(iii) $\log_b N^r = r \log_b N$

PROOF

(i) Since $M = b^{\log_b M}$ and $N = b^{\log_b N}$ (see Problem Set 2, Problem 50a),

$$MN = b^{\log_b M} \cdot b^{\log_b N} = b^{\log_b M + \log_b N}$$

Thus, by Definition 1 of a logarithm given in Section 2,

$$\log_b MN = \log_b M + \log_b N$$

(ii) Since $M = \left(\frac{M}{N}\right)N$, we can use Property (i) to get

$$\log_b M = \log_b \frac{M}{N} + \log_b N$$

so that

$$\log_b \frac{M}{N} = \log_b M - \log_b N$$

(iii) Since $N = b^{\log_b N}$, then

$$N^r = (b^{\log_b N})^r = b^{r \log_b N}$$

so

$$\log_b N^r = r \log_b N$$

EXAMPLES 1 Use the properties in Theorem 1 to write each of the following expressions as a sum, difference, or multiple of logarithms.

(a) $\log_8 \frac{17}{5}$ (b) $\log_a (x^7 y^{11})$, for $x > 0$ and $y > 0$

SOLUTION

(a) $\log_8 \frac{17}{5} = \log_8 17 - \log_8 5$ [Theorem 1 (ii)]

(b) $\log_a (x^7 y^{11}) = \log_a x^7 + \log_a y^{11}$ [Theorem 1 (i)]
$= 7 \log_a x + 11 \log_a y$ [Theorem 1 (iii)]

2 Assume that $\log_b 2 = 0.35$, $\log_b 3 = 0.55$, and $\log_b 5 = 0.82$. Use Theorem 1 to find each of the following values.

(a) $\log_b \frac{2}{3}$ (b) $\log_b 2^3$ (c) $\frac{\log_b 2}{\log_b 3}$ (d) $(\log_b 2)^3$

(e) $\log_b 24$ (f) $\log_b \sqrt{\frac{2}{3}}$ (g) $\log_b \left(\frac{60}{b}\right)$ (h) $\log_b 0.6$

SOLUTION

(a) $\log_b \frac{2}{3} = \log_b 2 - \log_b 3 = 0.35 - 0.55 = -0.20$

(b) $\log_b 2^3 = 3 \log_b 2 = 3(0.35) = 1.05$

(c) $\dfrac{\log_b 2}{\log_b 3} = \dfrac{0.35}{0.55} = \dfrac{7}{11}$

(d) $(\log_b 2)^3 = (0.35)^3 = 0.042875$

(e) $\log_b 24 = \log_b(2^3 \cdot 3) = \log_b 2^3 + \log_b 3$
$= 3 \log_b 2 + \log_b 3 = 3(0.35) + 0.55 = 1.60$

(f) $\log_b(\frac{2}{3})^{1/2} = \frac{1}{2}\log_b(\frac{2}{3}) = \frac{1}{2}(\log_b 2 - \log_b 3) = \frac{1}{2}(0.35 - 0.55) = -0.10$

(g) $\log_b \left(\dfrac{60}{b}\right) = \log_b 60 - \log_b b = \log_b(2^2 \cdot 3 \cdot 5) - \log_b b$
$= 2 \log_b 2 + \log_b 3 + \log_b 5 - \log_b b$
$= 2(0.35) + 0.55 + 0.82 - 1 = 0.70 + 0.55 + 0.82 - 1 = 1.07$

(h) $\log_b 0.6 = \log_b \frac{3}{5} = \log_b 3 - \log_b 5 = 0.55 - 0.82 = -0.27$

3 Use the properties of logarithms to write each of the following expressions as a single logarithm.

(a) $\log_2 \frac{13}{5} + 2 \log_2 \frac{5}{2} - \log_2 \frac{169}{8}$ (b) $\log_a(c^2 - cd) - \log_a(2c - 2d)$

SOLUTION Using Theorem 1, we have

(a) $\log_2 \frac{13}{5} + 2 \log_2 \frac{5}{2} - \log_2 \frac{169}{8} = \log_2 \frac{13}{5} + \log_2(\frac{5}{2})^2 - \log_2 \frac{169}{8}$
$= \log_2 \frac{13}{5} + \log_2 \frac{25}{4} - \log_2 \frac{169}{8}$
$= \log_2(\frac{13}{5} \cdot \frac{25}{4}) - \log_2 \frac{169}{8}$
$= \log_2 \left(\dfrac{\frac{13}{5} \cdot \frac{25}{4}}{\frac{169}{8}}\right) = \log_2 \frac{10}{13}$

(b) $\log_a(c^2 - cd) - \log_a(2c - 2d) = \log_a \left(\dfrac{c^2 - cd}{2c - 2d}\right) = \log_a \dfrac{c(c-d)}{2(c-d)} = \log_a \dfrac{c}{2}$

The basic properties of logarithms, along with Definition 1 of Section 2, are used to solve logarithmic equations as illustrated below.

EXAMPLES Find the solution set of each equation.

1 $\log_4(3x - 2) = 2$

SOLUTION The equation $\log_4(3x - 2) = 2$ is equivalent to $4^2 = 3x - 2$, so $16 = 3x - 2$ or $18 = 3x$. Thus, $x = 6$, and the solution set is $\{6\}$.

2 $\log_3(x + 1) + \log_3(x + 3) = 1$

SOLUTION We see that $\log_3(x + 1) + \log_3(x + 3) = \log_3[(x + 1)(x + 3)] = 1$. Hence, $x^2 + 4x + 3 = 3$ or $x^2 + 4x = 0$, so $x = 0$ or $x = -4$. However, -4 does not satisfy the *original* equation because the logarithm of a negative number is not defined. Hence, $x = 0$ is the only solution, and the solution set is $\{0\}$.

3 $\log_4(x + 3) - \log_4 x = 1$

SOLUTION Notice that $\log_4(x + 3) - \log_4 x = \log_4[(x + 3)/x] = 1$. Therefore, $(x + 3)/x = 4^1 = 4$, from which it follows that $x = 1$. Thus, the solution set is $\{1\}$.

PROBLEM SET 3

In Problems 1–8, evaluate each expression.

1 $\log_3 3^5$ **2** $\log_{10} \frac{1}{10^7}$ **3** $\log_2 16^{1/5}$

4 $\log_5\left(\frac{25^{2/3}}{5^{1/2}}\right)$ **5** $\log_7\left(\frac{49^{2/3}}{7^{3/2}}\right)$ **6** $3^{\log_3 7}$

7 $7^{\log_7 5}$ **8** $\log_2(16^{1/5} \cdot 64^{1/4})$

In Problems 9–14, write each of the expressions as a sum, difference, or multiple of logarithms. Assume that all variables represent positive real numbers.

9 $\log_b(x^2y^3)$ **10** $\log_b\left(\frac{x^9}{y^7}\right)$ **11** $\log_b \sqrt[6]{\frac{x^5}{y^3}}$

12 $\log_b\sqrt{x\sqrt{xy}}$ **13** $\log_b\sqrt[3]{x\sqrt[3]{y}}$ **14** $\log_b\left(\frac{\sqrt[5]{x} \cdot \sqrt[3]{y}}{z^2}\right)$

In Problems 15–25, assume that $\log_{10} 2 = 0.3010$ and $\log_{10} 3 = 0.4771$. Use the properties in Theorem 1 to evaluate the given logarithm.

15 $\log_{10} 4$ **16** $\log_{10} 18$ **17** $\log_{10}\sqrt{3}$

18 $\log_{10}\sqrt[5]{2}$ **19** $\log_{10} 3000$ **20** $\log_{10} 5$

21 $\log_{10} 60$ **22** $\log_{10} 0.5$ **23** $\log_{10} \frac{1}{3}$

24 $\log_{10} \frac{3}{4}$ **25** $\log_{10} 0.24$

26 Use $x_1 = 10000$, $x_2 = 10$, $b = 10$, and $p = 3$ to show that each statement is *false*.

(a) $\log_b\left(\frac{x_1}{x_2}\right) = \frac{\log_b x_1}{\log_b x_2}$

(b) $\frac{\log_b x_1}{\log_b x_2} = \log_b x_1 - \log_b x_2$

(c) $\log_b x_1 \cdot \log_b x_2 = \log_b x_1 + \log_b x_2$

(d) $\log_b(x_1x_2) = \log_b x_1 \cdot \log_b x_2$

(e) $\log_b(x_1^p) = (\log_b x_1)^p$

(f) $(\log_b x_1)^p = p \log_b x_1$

In Problems 27–34, use Theorem 1 to write each expression as a single logarithm. Assume that all variables are positive real numbers different from 1.

27 $\log_5 \frac{5}{7} + \log_5 \frac{40}{25}$ **28** $\log_2 \frac{32}{11} + \log_2 \frac{121}{16} - \log_2 \frac{4}{5}$

29 $\log_3 \frac{3}{4} - \log_3 \frac{3}{8}$ **30** $\log_x\left(a + \frac{a}{b}\right) - \log_x\left(c + \frac{c}{b}\right)$

31 $\log_c(a^2 - ab) - \log_c(7a - 7b)$

32 $\log_7\left(\frac{1}{4} - \frac{1}{x^2}\right) - \log_7\left(\frac{1}{2} - \frac{1}{x}\right)$ **33** $\log_a\left(\frac{a}{\sqrt[3]{x}}\right) - \log_a\left(\frac{\sqrt[3]{x}}{a}\right)$

34 $3\log_e\left(\frac{a^2b}{c^2}\right) + 2\log_e\left(\frac{bc^2}{a^4}\right) + 2\log_e\left(\frac{abc}{2}\right)$

In Problems 35–51, find the solution set.

35 $\log_3 x = 2$ **36** $\log_{27} x = \frac{1}{3}$ **37** $\log_{16} x^2 = -\frac{1}{2}$

38 $\log_x 4 = \frac{2}{5}$ **39** $\log_3(x + 1) < 2$ **40** $x^{\log_5 3} = 3$

41 $\log_7(2x - 7) = 2$

42 $\log_2(3x - 1) > 3$

43 $\log_5(5x - 1) = -2$

44 $\log_8 \sqrt{\dfrac{3x+4}{x}} = 0$

45 $\log_2(x^2 + 3x + 4) = 1$

46 $\log_2 |4x - 3| < 1$

47 $\log_2(x^2 - 9) - \log_2(x + 3) = 2$

48 $\log_{10}(x + 1) - \log_{10} x = 1$

49 $\log_4 x + \log_4(6x + 10) = 1$

50 $\log_3 x + \log_3(x - 6) = \log_3 7$

51 $\log_{10}(x^2 - 144) - \log_{10}(x + 12) = 1$

52 Prove that $\log_b(xyz) = \log_b x + \log_b y + \log_b z$, where b, x, y, and z are positive numbers and $b \neq 1$. [*Hint:* Use Theorem 1.]

53 Let $\log_b 2 = A$, $\log_b 3 = B$, and $\log_b 5 = C$. Express $\log_b(0.012)$ in terms of A, B, and C.

4 Common and Natural Logarithms

The properties of logarithms established in the preceding section provide us with the mathematical tools for manipulating logarithmic expressions. In this section we study two specific logarithmic bases.

The two logarithmic bases that are most widely used are 10 and e, where e is an irrational number, which is approximately equal to 2.71828. . . . The choice of 10 hardly needs an explanation, since we are accustomed to working with powers of 10. The alternative choice of the irrational number e is based on the fact that several formulas used in business, engineering, statistics, and science use e as a base.

4.1 Logarithms—Base 10

Base 10 is called the *common base.* By convention, we usually do not write the index 10 when using logarithmic notation; $\log x$ is the abbreviated way of writing $\log_{10} x$. For certain values of x it is easy to determine $\log x$. For example, $\log 10 = 1$ (why?) and $\log 100 = \log 10^2 = 2$ (why?); in fact, $\log 10^n = n$.

For other values of x it is *not* so easy to determine $\log x$. Let us examine two such values to see how far we can get in computing $\log x$.

Suppose that $x = 5340$. Clearly, we could represent x as $5.34 \cdot 10^3$, so that

$$\begin{aligned}\log 5340 &= \log(5.34 \cdot 10^3)\\ &= \log 5.34 + \log 10^3\\ &= \log 5.34 + 3\end{aligned}$$

Hence the problem has been reduced to finding $\log 5.34$.

Suppose that $x = 0.000234$. Then

$$\begin{aligned}\log 0.000234 &= \log(2.34 \cdot 10^{-4})\\ &= \log 2.34 + \log(10^{-4})\\ &= \log 2.34 + (-4)\end{aligned}$$

Here the problem is reduced to finding $\log 2.34$.

In both cases, the computation of the logarithm of the given number has been reduced to the computation of the logarithm of a number between 1 and 10. The procedure that is suggested by these two examples can be generalized.

Any positive number x can be represented as $x = s \cdot 10^n$, where $1 \leq s < 10$ and n is an integer. This form is called the *scientific notation* for x, and n is called the *characteristic* of $\log x$. Now

$$\begin{aligned} \log x &= \log(s \cdot 10^n) \\ &= \log s + \log 10^n \\ &= (\log s) + n, \qquad \text{where } s \text{ satisfies } 1 \leq s < 10 \end{aligned}$$

The latter form is called the *standard form* of $\log x$, and $\log s$ is called the *mantissa* of $\log x$. Note that since $y = \log x$ is increasing and $1 \leq s < 10$, it follows that $\log 1 \leq \log s < \log 10$, that is, $0 \leq \log s < 1$. In other words, the mantissa is always a number between 0 and 1, with the possibility of being equal to 0.

Hence the task of determining $\log x$ can always be reduced to determining $\log s$, where s is always between 1 and 10, and $\log s$ can be determined from the *common log table* (Appendix Table I), which gives us *approximations* of the values of the logarithms.

EXAMPLES In each of the following, determine the common logarithm value. Indicate the characteristic and the mantissa.

1 53,900

SOLUTION Since $53{,}900 = 5.39 \cdot 10^4$, the characteristic of log 53,900 is 4. Referring to Appendix Table I, we find that the mantissa of log 5.39 is 0.7316, so that

$$\begin{aligned} \log 53{,}900 &= \log 5.39 + 4 \\ &= 0.7316 + 4 = 4.7316 \end{aligned}$$

2 0.0035

SOLUTION Since $0.0035 = 3.5 \cdot 10^{-3}$, the characteristic of log 0.0035 is -3. The mantissa of $\log 3.5 = 0.5441$. Hence

$$\begin{aligned} \log 0.0035 &= \log 3.5 - 3 \\ &= 0.5441 - 3 = -2.4559 \end{aligned}$$

Note that although $\log 0.0035 = -2.4559$, the mantissa of log 0.0035 is 0.5441 and *not* 0.4559 or even -0.4559. Thus, we write the equivalent form of -2.4559 as $0.5441 - 3$ to overcome the difficulty of negative mantissas, since they do not appear in Appendix Table I.

4.2 Antilogarithms—Base 10

The process of determining the logarithm of a given number can also be "reversed." In other words, given the logarithm value, we can find what number has the given value as its logarithm. For example, suppose that $\log x = 3$. Then $x = 1000$. This relationship between x and 3 is sometimes written as $x = \text{antilog } 3 = 1000$.

In general, $x = \text{antilog } r$ is equivalent to $\log x = r$. As with logarithms, it is easy to determine antilogarithms of certain numbers. For example, $x = \text{antilog}(-3)$ is equivalent to $\log x = -3$, so that $x = 10^{-3} = 0.001$ and $\text{antilog}(-3) = 0.001$. Similarly, $w = \text{antilog } 5$ is equivalent to $\log w = 5$, so that $w = 10^5 = 100{,}000$ and $\text{antilog } 5 = 100{,}000$. However, it is *not* as easy to determine the value of antilog 4.4969.

Antilog 4.4969, or the solution of $\log x = 4.4969$, can be determined by using Appendix Table I if we reverse the process of determining logarithms in the following manner:

$$\begin{aligned}\log x &= 4.4969 \\ &= 0.4969 + 4 && \text{(Standard Form: mantissa plus characteristic)} \\ &= \log 3.14 + 4 && \text{(Appendix Table I)} \\ &= \log 3.14 + \log 10^4 \\ &= \log(3.14 \cdot 10^4) = \log 31400\end{aligned}$$

so that $x = 31{,}400$.

EXAMPLES Find the value of each of the following antilogarithms.

1 antilog 0.5740

SOLUTION Assume that antilog $0.5740 = x$. Then $\log x = 0.5740$. Using Appendix Table I, we find that $x = 3.75$, so antilog $0.5740 = 3.75$.

2 antilog 2.7210

SOLUTION Assume that antilog $2.7210 = x$, so

$$\begin{aligned}\log x &= 2.7210 \\ &= 0.7210 + 2 && \text{(Standard Form: mantissa plus characteristic)} \\ &= \log 5.26 + 2 && \text{(Appendix Table I)} \\ &= \log 5.26 + \log 100 \\ &= \log(5.26 \cdot 10^2) = \log 526\end{aligned}$$

Thus, $x = 526$ and antilog $2.7210 = 526$.

3 antilog(−2.0804)

SOLUTION Assume that $\text{antilog}(-2.0804) = t$. Then

$$\begin{aligned}\log t &= -2.0804 \\ &= (-2.0804 + 3) - 3 \\ &= 0.9196 + (-3) && \text{(Standard Form: mantissa plus characteristic)} \\ &= \log 8.31 + (-3) && \text{(Appendix Table I)} \\ &= \log(8.31 \cdot 10^{-3}) \\ &= \log 0.00831\end{aligned}$$

Thus $t = 0.00831$ and $\text{antilog}(-2.0804) = 0.00831$.

4.3 Linear Interpolation

The logarithms and antilogarithms that we computed up to now were special in the sense that we were able to find the necessary numbers in Appendix Table I. But what if this were not the case? Suppose, for example, that we wanted to find log 1.234, or suppose that we wanted to find antilog 0.2217. We would not be able to find log 1.234 or mantissa 0.2217 in Appendix Table I. This problem can be resolved by using an approximation method called *linear interpolation.*

Let us investigate the meaning of linear interpolation by determining log 1.234. From Appendix Table I we find that $\log 1.24 = 0.0934$ and $\log 1.23 = 0.0899$. Notice that $1.23 < 1.234 < 1.24$.

Next, examine that part of the graph of $y = \log x$ where $1.23 \leq x \leq 1.24$ (Figure 1a). Note that $\log 1.234 = 0.0899 + \bar{d}$, where $\bar{d}$ is the number that we are to approximate. First, we "replace" the arc of the log graph with a line segment (Figure 1b). Next, we assume that $\bar{d}$ is approximately the same as d in Figure 1b. (The amount of "curvature" of $y = \log x$ has been exaggerated in Figure 1b for illustrative purposes.) Finally, d can be determined by using the proportionality of the sides of the similar right triangles that have been formed. Thus

$$\frac{d}{0.0035} = \frac{0.004}{0.01}$$

so $d = 0.0014$.

Consequently, the approximate value of log 1.234 is given by

$$\begin{aligned}\log 1.234 &= \log 1.23 + (0.0014)\\ &= 0.0899 + 0.0014\\ &= 0.0913\end{aligned}$$

Figure 1

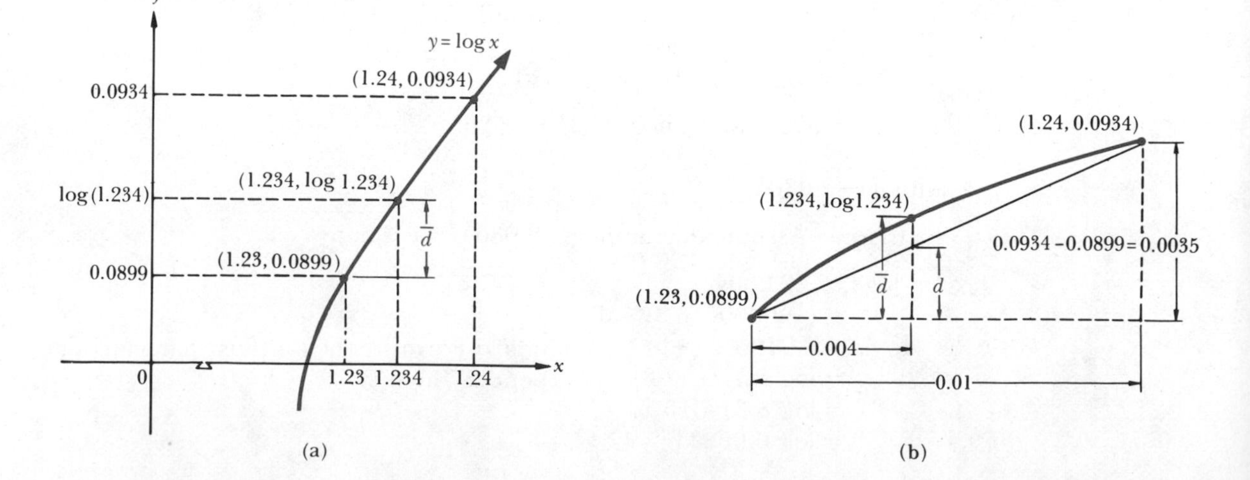

EXAMPLE Use Appendix Table I, together with linear interpolation, to determine antilog 0.2217.

SOLUTION Assume that $x = \text{antilog } 0.2217$. Then $\log x = 0.2217$. From Appendix Table I we find the following values:

$$\log 1.67 = 0.2227$$
$$\log x = 0.2217$$
$$\log 1.66 = 0.2201$$

As before, we examine that portion of the graph of $y = \log x$ where $1.66 \leq x \leq 1.67$ (Figure 2).

Figure 2

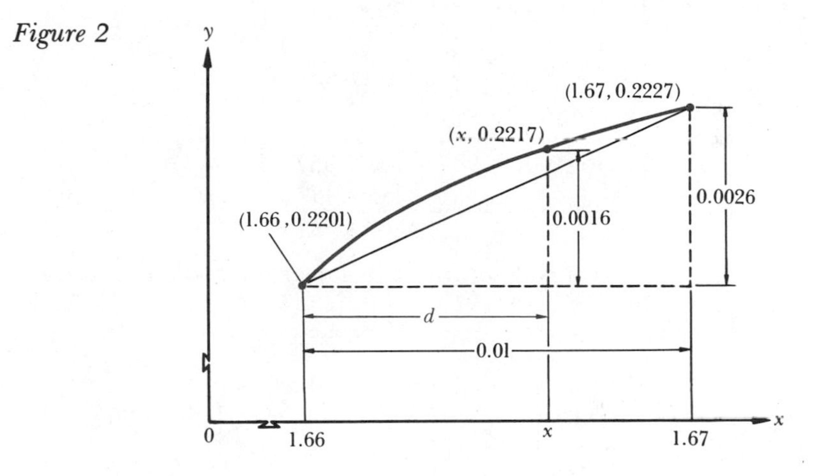

Hence d is *approximated* by

$$\frac{d}{0.01} = \frac{0.0016}{0.0026}$$

so $d = 0.006$. Thus,

$$x = 1.66 + 0.006 = 1.666$$

and

$$\text{antilog } 0.2217 = 1.666$$

Essentially, then, linear interpolation is an approximation method that replaces an arc of a curve with a straight line segment; the accuracy of this method of approximation depends on the "straightness" of the curve between the end points.

The evaluation of log 1.234 that was explained in detail above can also be organized in a schematic arrangement to display the mechanics involved in linear interpolation. This scheme is illustrated for log 1.234 in the first of the following examples.

EXAMPLES Use linear interpolation to determine each value.

1 log 1.234

SOLUTION From Appendix Table I, we have

$$0.01\left[\begin{array}{ll} & \begin{array}{cc} t & \log t \\ 1.24 & 0.0934 \end{array} \\ 0.004\left[\begin{array}{l}1.234\\1.23\end{array}\right. & \left.\begin{array}{c}m\\0.0899\end{array}\right]d\end{array}\right]0.0035 \qquad \frac{d}{0.0035}=\frac{0.004}{0.01}$$

so $d = 0.0014$. Thus, $m = 0.0899 + d = 0.0899 + 0.0014 = 0.0913$. Therefore, $\log 1.234 = 0.0913$.

2 log 0.0007957

SOLUTION

$$\begin{aligned}\log 0.0007957 &= \log(7.957 \cdot 10^{-4})\\ &= \log 7.957 + (-4)\\ &= m + (-4).\end{aligned}$$

Next m is determined by linear interpolation. From Appendix Table I, we have

$$0.01\left[\begin{array}{ll} & \begin{array}{cc} t & \log t \\ 7.96 & 0.9009 \end{array} \\ 0.007\left[\begin{array}{l}7.957\\7.95\end{array}\right. & \left.\begin{array}{c}m\\0.9004\end{array}\right]d\end{array}\right]0.0005 \qquad \frac{d}{0.0005}=\frac{0.007}{0.01}$$

so the approximate value of d is given by $d = 0.0004$. Consequently,

$$m = 0.9004 + 0.0004 = 0.9008$$

Thus, $\log 0.0007957 = 0.9008 + (-4) = -3.0992$.

3 antilog 4.5544

SOLUTION If $x = \text{antilog } 4.5544$, then $\log x = 4.5544$. Thus

$$\log x = 4.5544 = 0.5544 + 4 = \log s + 4$$

Next, s is determined by linear interpolation. From Appendix Table I, we have

$$0.01\left[\begin{array}{ll} & \begin{array}{cc} t & \log t \\ 3.59 & 0.5551 \end{array} \\ d\left[\begin{array}{l}s\\3.58\end{array}\right. & \left.\begin{array}{c}0.5544\\0.5539\end{array}\right]0.0005\end{array}\right]0.0012 \qquad \frac{d}{0.01}=\frac{0.0005}{0.0012}$$

so d is approximated by $d = 0.004$. Thus,

$$s = 3.58 + 0.004 = 3.584$$

Consequently,

$$\begin{aligned}\log x &= \log 3.584 + 4\\ &= \log(3.584 \cdot 10^4)\end{aligned}$$

so that $x = 35{,}840$ and antilog $4.5544 = 35{,}840$.

4.4 Base e Computations

Base e occurs in many of the formulas that are used in applied mathematics. Base e is called the *natural base,* and the *natural logarithm,* $\log_e x$, is usually written $\ln x$. In order to compute natural logarithms, we use Appendix Table II, together with the general properties of logarithms. In Appendix Table I (common logarithms) it was necessary to write down only the values of $\log x$ for $1 \leq x < 10$. Logarithms of other numbers were obtained by using the characteristics, the logarithms of the integral powers of 10. Appendix Table II contains values of $\ln x$ for $1 \leq x < 10$ as in Appendix Table I, but now it is more difficult to find logarithms of other numbers, because the logarithms of the integer powers of 10 are not themselves integers. For example, $\ln 3.28 = 1.1878$ can be found directly in Appendix Table II, whereas $\ln 32.8$ is found as follows:

$$\begin{aligned}\ln 32.8 &= \ln[(3.28)(10)] = \ln 3.28 + \ln 10 = \ln 3.28 + \ln[(2)(5)]\\ &= \ln 3.28 + \ln 2 + \ln 5\\ &= 1.1878 + 0.6931 + 1.6094\\ &= 3.4903\end{aligned}$$

EXAMPLES 1 Use Appendix Table II to determine the following values.
(a) $\ln 7.47$ (b) $\ln 17.4$

SOLUTION
(a) Using Appendix Table II, we have $\ln 7.47 = 2.0109$.
(b) 17.4 can be written as $(1.74)(10)$, so

$$\begin{aligned}\ln 17.4 &= \ln 1.74 + \ln 10 = \ln 1.74 + \ln[(2)(5)]\\ &= \ln 1.74 + \ln 2 + \ln 5\\ &= 0.5539 + 0.6931 + 1.6094\\ &= 2.8564\end{aligned}$$

2 Use Appendix Table II, together with linear interpolation, to determine $\ln 7.115$.

SOLUTION From Appendix Table II, we have

$$0.01\left[\begin{array}{cc} t & \ln t\\ 7.12 & 1.9629\\ 0.005\left[\begin{array}{l}7.115\\7.11\end{array}\right. & \left.\begin{array}{r}?\\1.9615\end{array}\right]d \end{array}\right]0.0014 \qquad \frac{d}{0.0014} = \frac{0.005}{0.01}$$

so $d = 0.0007$. Thus $\ln 7.115 = 1.9615 + 0.0007 = 1.9622$.

The next theorem enables us to relate natural and common logarithms.

THEOREM 1

$$\log_b x = \frac{\log_a x}{\log_a b}$$

where a and b are positive real numbers different from 1.

PROOF We know that $b^{\log_b x} = x$, so $\log_a(b^{\log_b x}) = \log_a x$. However, recall that $\log_a(b^{\log_b x}) = \log_b x \cdot \log_a b$, so $\log_b x \cdot \log_a b = \log_a x$, from which it follows that

$$\log_b x = \frac{\log_a x}{\log_a b}$$

EXAMPLES In Examples 1 and 2, use Theorem 1 above to express each logarithm in terms of common logarithms.

1 $\ln x$

SOLUTION

$$\ln x = \log_e x = \frac{\log x}{\log e}$$

2 $\log_2 10$

SOLUTION

$$\log_2 10 = \frac{\log 10}{\log 2} = \frac{1}{\log 2}$$

3 Let a and b be positive numbers other than 1; show that $\log_a b \cdot \log_b a = 1$.

SOLUTION

$$\log_a b = \frac{\log_b b}{\log_b a} = \frac{1}{\log_b a}$$

so $\log_a b \cdot \log_b a = 1$.

PROBLEM SET 4

In Problems 1–10, use Appendix Table I to find the value of the given common logarithm.

1 log 8.7
2 log 575
3 log 47.3
4 log 3
5 log 13,600
6 log 550,000
7 log 0.175
8 log 0.00343
9 log 0.000023
10 log 0.0000000714

In Problems 11–18, use Appendix Table I, together with linear interpolation, to find the value of the given common logarithm.

11 log 6.137
12 log 555.5
13 log 12.35

14 log 75.56 **15** log 0.5473 **16** log 1111
17 log 472,800 **18** log 0.0001766

In Problems 19–28, use Appendix Table I to find the value of the given antilogarithm.

19 antilog 0.9138 **20** antilog 0.6021
21 antilog 1.7419 **22** antilog 2.3139
23 antilog 2.1959 **24** antilog(−0.4473)
25 antilog(−1.2549) **26** antilog(−9.8125)
27 antilog(−3.3979) **28** antilog(−4.9957)

In Problems 29–36, use Appendix Table I, together with linear interpolation, to find the value of the given antilogarithm.

29 antilog 0.5627 **30** antilog 3.8665
31 antilog 1.1979 **32** antilog 2.0020
33 antilog(−0.7777) **34** antilog(−1.341)
35 antilog(−2.1234) **36** antilog(−5.7)

In Problems 37–42, use Appendix Table II to find the value of the given natural logarithm.

37 ln 5 **38** ln 3.88 **39** ln 12
40 ln 100 **41** ln 456 **42** ln 0.035

In Problems 43–46, use Appendix Table II, together with linear interpolation, to find the value of the given natural logarithm.

43 ln 4.253 **44** ln 17.18 **45** ln 7523
46 ln 0.09174

In Problems 47–50, use Theorem 1 on page 170 to approximate the given logarithm by converting to common logarithms.

47 $\log_2 5$ **48** ln 3 **49** $\log_7 97$ **50** $\log_{1/3}(\frac{4}{3})$

5 Applications of Logarithms

In this section, we examine three applications: computations, solving exponential equations, and solving applied problems from science and business.

5.1 Computations Using Logarithms

In the past, tedious computations such as $(1.07)^{10}$, $[(134)^5(0.35)^8]/(49)^3$, and $\sqrt[5]{17}$ were performed by the use of logarithms, as will be illustrated in the examples below. But, as we indicated in Section 1, these types of calculations are now frequently accomplished with the aid of calculators. However, since calculators are not always available, it is worthwhile to be familiar with the use of logarithms for performing such computations.

EXAMPLES Use logarithms to *approximate* the value of each expression.

1 $(1.07)^{10}$

SOLUTION Let $x = (1.07)^{10}$. Then

$$\begin{aligned}\log x &= \log[(1.07)^{10}] = 10 \log(1.07)\\ &= 10(0.0294) = 0.2940\end{aligned}$$

so $x = \text{antilog } 0.2940$. By using linear interpolation we find that $x = 1.968$. Thus $(1.07)^{10}$ is approximately 1.968.

2 $\dfrac{(134)^5(0.35)^8}{(49)^3}$

SOLUTION Let

$$x = \frac{(134)^5(0.35)^8}{(49)^3}$$

Then

$$\begin{aligned}\log x &= \log\left[\frac{(134)^5(0.35)^8}{(49)^3}\right]\\ &= \log(134)^5 + \log(0.35)^8 - \log(49)^3\\ &= 5 \log 134 + 8 \log 0.35 - 3 \log 49\\ &= 5(2.1271) + 8(-0.4559) - 3(1.6902)\\ &= 10.6355 - 3.6472 - 5.0706\\ &= 1.9177\end{aligned}$$

so $x = \text{antilog } 1.9177$. By using linear interpolation we get $x = 82.74$, so $\dfrac{(134)^5(0.35)^8}{(49)^3}$ is approximately 82.74.

3 $\sqrt[5]{17}$

SOLUTION Let $x = \sqrt[5]{17}$. Then $x = 17^{1/5}$, so

$$\begin{aligned}\log x &= \tfrac{1}{5} \log 17\\ &= \tfrac{1}{5}(1.2304)\\ &= 0.2461\end{aligned}$$

Using linear interpolation, we have $x = \text{antilog } 0.2461 = 1.762$ so that $\sqrt[5]{17}$ is approximately 1.762.

5.2 Exponential Equations

We indicated in Section 1.3 that *special* types of exponential equations are solved by applying the properties of exponents, together with the techniques we have already studied for solving equations. However, in solving equations such as $5^x = 7$, $e^{-3t} = 0.5$, and $41^{2x-1} = 3^x$, we need to use logarithms. The following examples illustrate the process.

EXAMPLES Use logarithms to find the solution set of each of the following exponential equations.

1 $5^x = 7$

SOLUTION Since $5^x = 7$,

$$\log 5^x = \log 7$$
$$x \log 5 = \log 7$$
$$x = \frac{\log 7}{\log 5} = \frac{0.8451}{0.6990} = 1.21$$

Hence, the solution set is $\{1.21\}$.

2 $e^{-3t} = 0.5$

SOLUTION $\ln e^{-3t} = \ln 0.5$, so

$$-3t = -0.6931 \quad \text{and} \quad t = \frac{0.6931}{3} = 0.23$$

Therefore, the solution set is $\{0.23\}$.

3 $41^{2x-1} = 3^x$

SOLUTION Since $41^{2x-1} = 3^x$, we have

$$\log 41^{2x-1} = \log 3^x$$
$$(2x - 1)\log 41 = x \log 3$$
$$2x \log 41 - \log 41 = x \log 3$$
$$2x \log 41 - x \log 3 = \log 41$$
$$x(2 \log 41 - \log 3) = \log 41$$
$$x = \frac{\log 41}{2 \log 41 - \log 3} = \frac{1.6128}{3.2256 - 0.4771} = 0.59$$

Thus, the solution set is $\{0.59\}$.

5.3 Applied Problems from Science and Business

The population growth and the compound interest problems that we considered in Section 1 led us to exponential equations of bases different than 10 or e. Thus they were more easily solved by the use of calculators. It turns out that *any* exponential equation, regardless of the base, can be converted by using logarithms to another equation containing base 10 or base e. Here, we present some applications and demonstrate how logarithms are used to solve the resulting exponential equations.

EXAMPLES 1 Boyle's law for expansion of air at a constant temperature is given by the equation $PV^{1.4} = C$, where P is the pressure, V is the volume, and C is a constant. At a certain instant, the volume of air in a container is 75.2 cubic inches and $C = 12{,}600$. Find the pressure P at that moment.

SOLUTION $PV^{1.4} = C$, so $P = C/V^{1.4} = CV^{-1.4}$. Thus, $P = (12{,}600)(75.2)^{-1.4}$. Using logarithms to calculate P, we have

$$\begin{aligned}\log P &= \log[(12{,}600) \cdot (75.2)^{-1.4}] \\ &= \log 12{,}600 + \log(75.2)^{-1.4} \\ &= \log 12{,}600 - 1.4 \log 75.2 \\ &= 4.1004 - 1.4(1.8762) \\ &= 1.4737\end{aligned}$$

Therefore, $P = \text{antilog}(1.4737) = 29.76$. Thus, the pressure P is 29.76 pounds per square inch.

2 If P dollars represents the amount invested at an annual interest rate r (expressed as a decimal), then the amount A_n accumulated in n years when interest is compounded t times a year is given by the formula

$$A_n = P\left(1 + \frac{r}{t}\right)^{nt}$$

Suppose that \$1000 is invested at a yearly interest rate of 6 percent compounded every 4 months.
(a) How much money is accumulated after 4 years?
(b) In how many years will the money double at this rate?

SOLUTION $P = 1000$, $r = 0.06$, and $t = 3$, so

$$A_n = 1000\left(1 + \frac{0.06}{3}\right)^{3n} \quad \text{or} \quad A_n = 1000(1.02)^{3n}$$

(a) If $n = 4$, $A_4 = 1000(1.02)^{12}$. Now we can use logarithms to get

$$\begin{aligned}\log A_4 &= \log[1000(1.02)^{12}] \\ &= \log 1000 + \log(1.02)^{12} \\ &= 3 + 12 \log 1.02 \\ &= 3 + 12(0.0086) \\ &= 3.1032\end{aligned}$$

Using linear interpolation, we have, $A_4 = \text{antilog } 3.1032 = 1268$. Therefore, the money accumulated after 4 years is \$1268.

(b) We must solve the equation $2000 = 1000(1.02)^{3n}$ or $(1.02)^{3n} = 2$, so

$$\log(1.02)^{3n} = \log 2$$

or

$$3n \log 1.02 = \log 2$$

and

$$n = \frac{\log 2}{3 \log 1.02} = 11.7$$

Therefore, at this rate of interest, the original \$1000 investment would double in approximately 11.7 years.

3 Suppose that the number of bacteria N present in a culture after t hours is given by $N = Pe^{2t}$, where P represents the number of bacteria present initially. How long will it take for the number of bacteria to triple?

SOLUTION We want $N = 3P$, so $3P = Pe^{2t}$, that is, $3 = e^{2t}$, from which it follows that

$$\ln 3 = \ln e^{2t} = 2t \ln e$$

or, equivalently,

$$t = \frac{\ln 3}{2} = \frac{1.0986}{2} = 0.55$$

Therefore, it will take approximately 0.55 hours for the number of bacteria to triple.

4 In a problem involving decomposition of radioactive material, suppose that the amount of radium R present after t years is given by $R = Pe^{-kt}$, where P represents the number of grams present initially and k is a constant. Find k if 30 percent of the radium disappears in 100 years.

SOLUTION When $t = 100$, $R = P - 0.3P = 0.7P$. Hence

$$0.7P = Pe^{-100k}$$

or

$$e^{-100k} = 0.7$$

so

$$-100k \ln e = \ln 0.7$$

that is,

$$k = \frac{\ln 0.7}{-100} = \frac{-0.3566}{-100} = 0.0036$$

Hence, k is approximately 0.0036.

An *annuity* is a series of regular payments due at equal intervals of time. The interval of time between successive payments is called the *payment period.* As the word "annuity" implies, one year was the original basis for the payment period; that is, payments were made on an annual basis. The term annuity is now applied to such familiar payment schemes as insurance premiums, pension funds, home mortgages, and savings accounts, where the fixed payment interval may be annual, semiannual, quarterly, or monthly.

One common type of annuity problem is to find the total value to which the series of payments will accumulate at the end of the term of the annuity, assuming that these payments earn interest at a prescribed rate. An annuity is said to be *simple* if the interval between successive payments coincides with the *conversion period,* the time at which interest earned is paid. If the size of each payment is P dollars and the interest rate per conversion period is r (expressed as a decimal), then the accumulated value S of all payments at the time of the nth payment is given by the formula

$$S = P\left[\frac{(1+r)^n - 1}{r}\right]$$

EXAMPLE At the end of each month a teacher deposits \$100 in a credit union. If the annual interest rate earned on these deposits is 6 percent, converted monthly, find the amount in the account after 4 years.

SOLUTION The annuity consists of $n = 12(4) = 48$ payments. The interest rate per conversion period is $r = \frac{1}{12}(0.06) = 0.005$, and the periodic payment $P = 100$. Then

$$\begin{aligned} S &= 100\left[\frac{(1+0.005)^{48} - 1}{0.005}\right] \\ &= 20{,}000[(1.005)^{48} - 1] \\ &= 20{,}000(1.005)^{48} - 20{,}000 \end{aligned}$$

To evaluate the expression $20{,}000(1.005)^{48}$, we make use of logarithms. Let $x = 20{,}000(1.005)^{48}$. Then

$$\begin{aligned} \log x &= \log[20{,}000(1.005)^{48}] = \log 20{,}000 + \log(1.005)^{48} \\ &= \log 20{,}000 + 48\log(1.005) = 4.3010 + 48(0.0022) \\ &= 4.3010 + 0.1056 = 4.4066 \end{aligned}$$

Thus, $x = \text{antilog } 4.4066 = 25{,}500$, so $S = 25{,}500 - 20{,}000 = 5500$. The amount in the account after 4 years is approximately \$5500.

The *present value* of an annuity is the total amount paid into the annuity, *discounted* to the beginning of the first payment period. In other words, it is the amount of the annuity after all the payments have been made, less the total of the interest paid. The present value A is given by the formula

$$A = P\left[\frac{1 - (1+r)^{-n}}{r}\right]$$

where P is the amount of each payment, r is the interest rate per conversion period, and n is the number of payments. This formula can also be used, for instance, to determine the total amount paid for a mortgaged home, minus the interest charges, as is illustrated in the following example.

EXAMPLE A young couple bought a new home by making a down payment of \$8000 and contracting for monthly payments of \$275 for 25 years. [Obviously, the total cost to the couple of their new home will be 8,000 + (275)(12)(25), or \$90,500, which represents the cost of the home itself, plus the mortgage.] If the annual interest rate on the mortgage is 9 percent, what price will the couple pay for the home itself, discounting the interest on the mortgage? (Assume that the first mortgage payment occurs 1 month after the down payment has been made.)

SOLUTION Let A be the amount of the mortgage on the home. Thus, we have $n = (25)(12) = 300$ and $r = \frac{1}{12}(0.09) = 0.0075$. Also, $P = 275$. Then

$$\begin{aligned} A = P\left[\frac{1-(1+r)^{-n}}{r}\right] &= 275\left[\frac{1-(1+0.0075)^{-300}}{0.0075}\right] \\ &= 36{,}666.67[1-(1.0075)^{-300}] \\ &= 36{,}666.67 - (36{,}666.67)(1.0075)^{-300} \end{aligned}$$

To evaluate the expression $36{,}666.67(1.0075)^{-300}$, we use a calculator to compute the required logarithmic values, since they cannot be found in our Appendix Table I. Let $x = (36{,}666.67)(1.0075)^{-300}$; then

$$\begin{aligned} \log x &= \log[(36{,}666.67)(1.0075)^{-300}] = \log(36{,}666.67) + \log(1.0075)^{-300} \\ &= \log(36{,}666.67) - 300\log(1.0075) = 4.5643 - 300(0.0032) \\ &= 4.5643 - 0.9600 = 3.6043 \end{aligned}$$

Therefore, $x = \text{antilog } 3.6043 = 4{,}020.68$, and $A = 36{,}666.67 - 4{,}020.68 = 32{,}645.99$. The cost of the home is \$32,645.99 + \$8,000.00 = \$40,645.99. [The amount of interest the couple paid on their mortgage can be found by subtraction, that is, \$90,500.00 − \$40,645.99 = \$49,854.01.]

PROBLEM SET 5

In Problems 1–9, use logarithms to approximate the value of the given expression to the nearest hundredth place.

1 $(3.11)(3.41)(5.32)$ **2** $(5.3)^7$ **3** $(0.213)^3(2.6)^5$

4 $\dfrac{(0.515)^3(4.7)^5}{(97)^2}$ **5** $\sqrt[3]{731}$ **6** $(8.11)^{3/5}$

7 $\dfrac{\sqrt{12}(3.14)^2}{4^3}$ **8** $\sqrt[7]{32.46}$ **9** $\dfrac{(1.07)^5(2.72)^{-2}}{\sqrt[3]{100}}$

10 Use a calculator to evaluate Problems 1–9.

In Problems 11–20, use logarithms to find the solution set of the given exponential equation.

11 $6^x = 12$ **12** $3^{5x} = 2$ **13** $7^{3x-1} = 5$

14 $0.2^{x+1} = 0.5$ **15** $3^{x+1} = 17^{2x}$ **16** $43^{2x-1} = 5^x$

17 $e^x = 5$ **18** $3e^{2x} = 17$ **19** $e^{-5x} = 4$

20 $e^x = 10^{x+1}$

21 An experimental rocket is launched vertically from the ground and reaches a height of h miles in t minutes, where h is given by the equation $h = 150(e^{0.2t} - 1)$. Find the height after 3 minutes.

22 The diameter D of a tree (in centimeters) increases according to the formula $D = D_0 \cdot 10^{0.05t}$, where D_0 is the initial diameter and t represents the number of years of growth. How long will it take for the diameter to double?

23 Assume that it is known that the number N of bacteria present in a culture after t minutes is given the equation $N = 1000e^{0.25t}$.

(a) How many bacteria are present after $\frac{1}{2}$ hour?

(b) How long will it take to grow 50,000 bacteria?

24 The half-life of radium is defined to be the time required for a given amount of radium to decrease by one-half due to radioactive decay. Assume that the amount R of radium present after t years is given by the formula $R = Pe^{-kt}$, where P represents the number of grams present initially and k is a constant.
(a) Find k if the half-life is 50 years.
(b) Find the half-life if $k = 0.005$.

25 Suppose that \$1500 is put into a savings plan that yields a yearly interest rate of $5\frac{1}{2}$ percent. In how many years would the money triple if the interest is compounded
(a) Annually? (b) Semiannually?
(c) Quarterly? (d) Monthly?

26 Suppose that \$500 is invested at an interest rate of 6 percent compounded annually.
(a) How much money is accumulated after 7 years?
(b) In how many years would the money double at this rate?

27 If interest is compounded continuously, that is, if the number of conversions within any year becomes "infinite," the continuously compounded interest formula $A = Pe^{nr}$ is used. In this formula, A represents the total accumulation after n years, P represents the principal, r represents the rate of interest per year expressed as a decimal, and e is the natural logarithm base. Use this formula to answer the following questions.

Suppose that \$1000 is invested at a yearly interest rate of 6 percent compounded continuously.
(a) How much money is accumulated after 4 years?
(b) In how many years would the \$1000 double at this rate?

28 Suppose that, due to inflation, prices are rising at a rate of 6 percent per year compounded continuously. How long does it take a car initially costing \$3125 to reach a cost of \$5315 due to the inflationary factor alone?

29 For \$10,000, a retired couple bought a treasury note that paid 4 percent annual interest compounded monthly. They redeemed the note after 21 months. Find the total amount they received on their investment.

30 On the day she was born, a girl's parents invested \$2000 in a bank to save for her college education. At 6 percent interest compounded annually, how much money will have accumulated by her eighteenth birthday?

31 A woman deposits \$20 in a bank at the end of each month. If the interest rate is 5 percent converted monthly, find the amount in her account after 10 years.

32 If \$8000 is deposited in a savings account at the end of every quarter and the interest rate is 6 percent converted quarterly, how much money is on deposit after 10 years?

33 A man buys a house by making a down payment of \$10,000 and a monthly payment of \$250 for 30 years. If the interest rate on the mortgage is 9 percent, what price will the man pay for the house itself? (Assume that the first mortgage payment occurs 1 month after the down payment has been made.)

34 A football player will receive a royalty payment of \$1000 semiannually for the next 5 years for a commercial he made for television. Determine the present value of these payments if the money is invested at 8 percent converted semiannually.

REVIEW PROBLEM SET

In Problems 1–5, graph the function. Indicate the domain, the range, and the intervals in which the function is increasing or decreasing.

1 $f(x) = (\frac{1}{2})^x$ **2** $h(x) = (\frac{3}{5})^{-x}$ **3** $g(x) = 4^x$

4 $g(x) = \pi^x$ **5** $f(x) = 3^{x^2}$

In Problems 6–13, find the solution set of each equation without using logarithms.

6 $2^x = 8^{x-1}$ **7** $4^{-x} = 8^{x+2}$

8 $3^{2x+2} - 18(3^x) + 9 = 0$ **9** $7^{(1/2)x+3} = 1$

10 $(\frac{1}{2})^{x-1/3} = 8$ **11** $(\frac{2}{3})^x = \frac{81}{16}$

12 $2^{x^2+4x} = 32$ **13** $10^{3x^2-2x} = 100{,}000$

14 Let $f(x) = cb^x$, $f(0) = 7$, and $f(1) = 14$. Find b and c.

15 Let $f(x) = \log_{16} x$. Find the domain and the range of f; also find each of the following values.
(a) $f(256)$ (b) $f(64)$ (c) $f(32)$ (d) $f(\sqrt[5]{2})$

In Problems 16–19, determine the domain of the function.

16 $f(x) = \log_5(2x - 1)$ **17** $g(x) = \log|x|$

18 $h(x) = \ln|x + 2|$ **19** $f(x) = \ln e^x$

In Problems 20–25, graph the function and determine the domain and range.

20 $f(x) = \log_7 x$ **21** $h(x) = \log_{1/5} x$

22 $g(x) = \ln x$ **23** $g(x) = \log(x - 1)$

24 $h(x) = \log_3|-2x|$ **25** $F(x) = \log_5(2 + x)$

In Problems 26–37, find the value of each logarithm without using the Appendix tables.

26 $\log_3 9$ **27** $\log_4 8$ **28** $\log_6 1$

29 $\log_5 0.04$ **30** $\log_{100} 0.001$ **31** $\log_9 \frac{1}{3}$

32 $\log_{0.36} 0.6$ **33** $\log_2 \frac{1}{256}$ **34** $\log_5 \frac{1}{625}$

35 $\log_2 \frac{1}{1024}$ **36** $\log_4 8\sqrt{2}$ **37** $\log_{6/5} \frac{25}{36}$

In Problems 38–43, use the properties of logarithms to write each expression as a sum, difference, or multiple of logarithms. Assume that all variables represent positive real numbers and that $a \neq 1$.

38 $\log_2(3^6 \cdot 5^7)$ **39** $\log_8\left(\frac{5^7}{9^3}\right)$ **40** $\log_2(xy^5)$

41 $\log_2 \sqrt[7]{5^3 \cdot 5^6}$ **42** $\log_a\left(\frac{x^2}{y^4}\right)$ **43** $\log_5(2^6 \cdot 3^7 \cdot 5^2)$

In Problems 44–49, use the properties of logarithms to write each expression as a single logarithm. Assume that all variables represent positive real numbers and that $a \neq 1$.

44 $\log_2 \frac{3}{7} + \log_2 \frac{14}{27}$ **45** $\log_3 \frac{5}{13} + \log_3 \frac{4}{15}$

46 $\log_5 \frac{6}{7} - \log_5 \frac{27}{4} + \log_5 \frac{21}{16}$ **47** $\log_9 \frac{11}{5} + \log_9 \frac{14}{5} - \log_9 \frac{22}{15}$

48 $5 \log_a x - 3 \log_a y$ **49** $2 \log_a x^3 + \log_a\left(\frac{2}{x}\right) - \log_a\left(\frac{2}{x^4}\right)$

In Problems 50–57, let $\log_a 2 = 0.69$, $\log_a 3 = 1.10$, $\log_a 5 = 1.62$, and $\log_a 7 = 1.94$. Find each of the values.

50 $\log_a(3^5 \cdot 3^7)$ **51** $\log_a \sqrt[5]{5^3 \cdot 7^4}$

52 $\log_a \sqrt[3]{16}$ **53** $\sqrt[3]{\log_a 16}$

54 $\log_a\left(\frac{2^4}{3^4}\right)$ **55** $\log_a \sqrt[3]{\frac{3}{2}}$

56 $\log_a\left(\frac{60}{a}\right)$ **57** $\log_a \frac{25}{27}$

In Problems 58–67, find the solution set.

58 $\log_4(x + 3) = -1$ **59** $\log(x^2 - 4) = 0$

60 $\log_7(2x - 1) > 2$ **61** $\log_5|3x + 7| < 0$

62 $\log_4 3x = \log_4 3 + \log_4 5$ **63** $\log_7 2x = \log_7 8 - \log_7 2$

64 $\log_5(2x - 1) + \log_5(2x + 1) = 1$

65 $\log_3(x - 1) + \log_3(x - 2) = \log_3 6$

66 $\log_{1/2}(4x^2 - 1) - \log_{1/2}(2x + 1) = 1$

67 $\log_3(x^2 - 1) - \log_3(x - 1) = 4$

68 If $\log_b x = 3$, determine each of the following values.

(a) $\log_{1/b} x$ (b) $\log_b\left(\frac{1}{x}\right)$

69 Let $\log_b 2 = A$, $\log_b 3 = B$, and $\log_b 5 = C$. Express $\log_b 0.006$ in terms of A, B, and C.

In Problems 70–79, evaluate the expression by using the appropriate log table in the Appendix. Interpolate if necessary.

70 log 27,300 **71** log 74.47

72 antilog 2.9557 **73** antilog(−1.5467)

74 $\ln 8.86$ **75** $\ln 10$

76 $\ln 123.4$ **77** $\log(\ln 7.3)$

78 $\log_5 8$ **79** $\log_3 12$

In Problems 80–83, use logarithms to approximate the value. Use a calculator to evaluate and then compare the answer from each approach.

80 $(1.25)^5$ **81** $\dfrac{(834)^3(1.34)^2}{(75)^4}$

82 $\sqrt[3]{17.1}$ **83** $1000(1.05)^{15}$

In Problems 84–89, use logarithms to find the solution set of each equation.

84 $(2.06)^x = 300$ **85** $10^x = 6$ **86** $5^{2x-1} = 2^{x+3}$

87 $e^{x^2} = 9$ **88** $15e^x = 5$ **89** $e^{-x} = 3^{x+1}$

90 (a) What amount must have been deposited 20 years ago to be worth \$10,000 now at the yearly interest rate of 3 percent compounded continuously?

(b) Find the accumulated amount from an initial \$1000 investment at the end of 4 years if the annual interest rate is 5 percent, compounded continuously.

91 In the study of "probability theory," the normal distribution curve is described by the equation $y = ae^{-cx^2}$. Find the value of c when $a = 2.2$ if $x = 0.5$ and $y = 0.345$.

92 Suppose that \$1000 is put into a savings plan that yields a yearly interest rate of $6\frac{1}{2}$ percent. How much money is accumulated after 8 years if the interest is compounded

(a) Annually? (b) Semiannually?

(c) Quarterly? (d) Monthly?

93 The charge Q on a condenser (in coulombs) is described by the formula

$$Q = C \cdot E[1 - e^{-t/(C \cdot R)}]$$

where C is the capacity of the condenser in farads, E is the applied voltage, R is the resistance in ohms, and t is the time in seconds after the voltage is applied. Find Q if $C = 7 \times 10^{-6}$ farad, $R = 600$ ohms, $E = 130$ volts, and $t = 0.025$ seconds.

94 A sum of \$5000 is invested in a stock whose growth-rate average is 15 percent compounded annually. Assuming that the rate of growth continues, find the value of the stock after 6 years.

95 A man initially deposits \$1500 in a bank. At the end of each month for the next 18 months, he deposits an additional \$150. The annual interest that the bank pays is 6 percent, converted monthly. How much is in the account after 18 months?

CHAPTER 5

Trigonometric Functions

The polynomial, exponential, and logarithmic functions discussed in the preceding chapters are not sufficient for some applications of mathematics to physics and engineering. In this chapter, we study other types of functions, called the *trigonometric functions,* which play an important role in science and technology. These functions arose from the fields of astronomy and navigation, and from the surveying techniques of the ancient Egyptians and Greeks.

Originally, the principal application of trigonometry was restricted to the problem of solving for missing parts of triangles. Today, however, this use has been extended to important applications of trigonometric functions in solving problems involving periodic phenomena, such as sound waves, alternating current, and business cycles.

We begin by introducing the *wrapping function,* which is used to construct the trigonometric functions defined on real numbers. Next, we investigate the properties and graphs of the trigonometric functions. Finally, we turn our attention to the trigonometric functions defined on angles and to the evaluations of such functions.

1 The Wrapping Function

Recall from Chapter 2, Section 1, that the unit circle is the circle of radius 1 with center at (0, 0) and with the equation $x^2 + y^2 = 1$ (Figure 1).

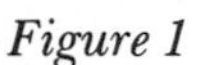

Figure 1

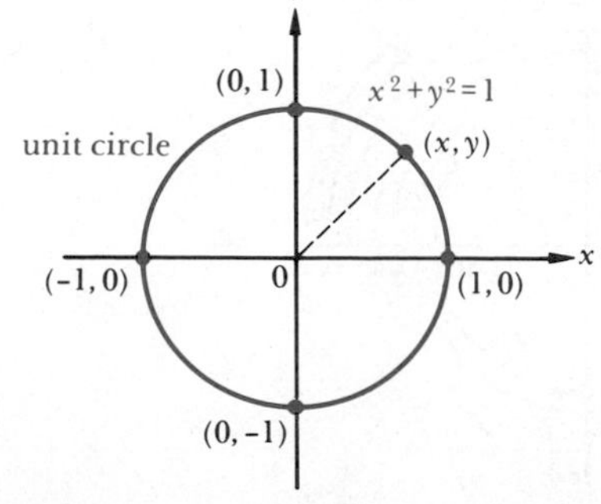

It is possible to associate each real number with the coordinates of a point on the unit circle by using the following scheme. First, assume that a number line has the same scale unit as the one used for the unit circle. Next, the number 0 on the number line is placed so as to coincide with the point (1, 0) on the unit circle. The *positive* part of the real line is "wrapped around" the circle in a *counterclockwise* sense and the *negative* part of the real line is "wrapped around" in a *clockwise* sense. Thus, if $t_1 > 0$, it is associated with a point on the unit circle by moving t_1 units counterclockwise along the circumference of the circle, starting at the point (1, 0) and ending at (x_1, y_1) (Figure 2a). If $t_2 < 0$, it is associated with a point on the circle by moving $|t_2|$ units clockwise along the circumference of the circle, starting at the point (1, 0) and ending at (x_2, y_2) (Figure 2b).

Figure 2

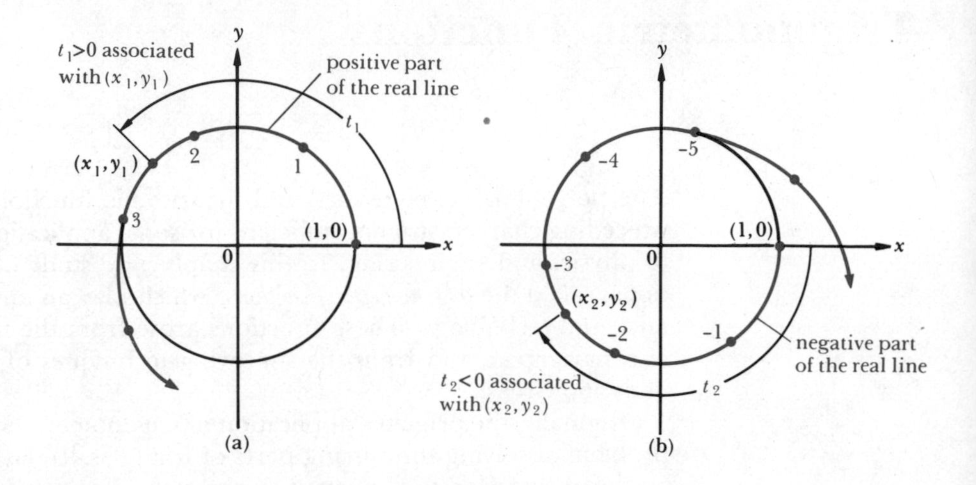

Since the circumference C of any circle is given by $C = 2\pi r$, where r represents the measure of the radius of the circle, it follows that the circumference of the unit circle is 2π.

Using the wrapping scheme described above, together with Figure 3, it becomes clear that:

0	is associated with the point (1, 0)
$\frac{\pi}{2}$	is associated with the point (0, 1)
π	is associated with the point (−1, 0)
$\frac{3\pi}{2}$	is associated with the point (0, −1)
2π	is associated with the point (1, 0)

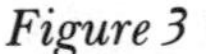
Figure 3

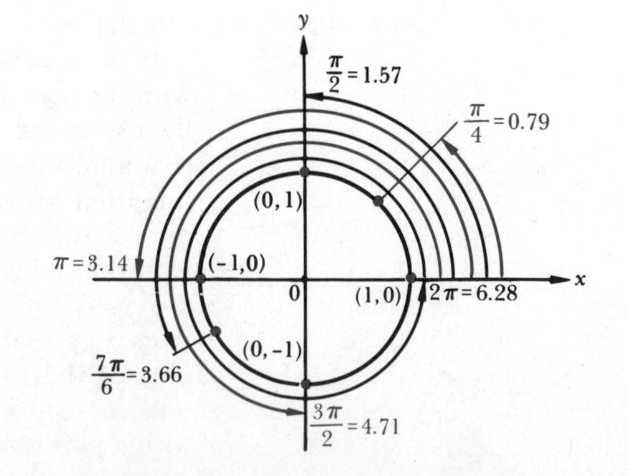

Table 1, together with Figure 3, displays the location of some points along the circumference of the unit circle. (If 3.14 is used as an approximation for π, the circumference of the unit circle is approximately 6.28 units.)

Table 1

Start at (1, 0) *and Move Counterclockwise*	*Actual Distance*	*Approximate Distance* ($\pi = 3.14$)
One-eighth of the way around the circumference	$\frac{1}{8}(2\pi) = \frac{\pi}{4}$	0.79
One-fourth of the way around the circumference	$\frac{1}{4}(2\pi) = \frac{\pi}{2}$	1.57
One-half of the way around the circumference	$\frac{1}{2}(2\pi) = \pi$	3.14
Seven-twelfths of the way around the circumference	$\frac{7}{12}(2\pi) = \frac{7\pi}{6}$	3.66
Three-fourths of the way around the circumference	$\frac{3}{4}(2\pi) = \frac{3\pi}{2}$	4.71
One complete revolution around the circumference	$1(2\pi) = 2\pi$	6.28

The *scheme* that has been described above defines a function, called the *wrapping function P,* which associates each real number t with a point (x, y) on the unit circle (Figure 4).

Figure 4

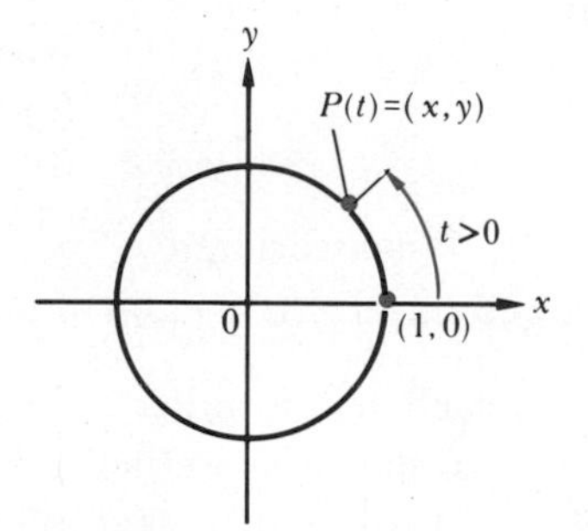

Thus $P(0) = (1, 0)$; $P(\pi/2) = (0, 1)$; $P(\pi) = (-1, 0)$; $P(3\pi/2) = (0, -1)$; and $P(2\pi) = (1, 0)$.

EXAMPLES 1 Display the location of each of the following points on the unit circle. Indicate the quadrant in which $P(t)$ lies.

(a) $P\left(\frac{\pi}{6}\right)$

(b) $P(2)$

(c) $P\left(-\frac{3\pi}{4}\right)$

(d) $P(-1)$

Figure 5

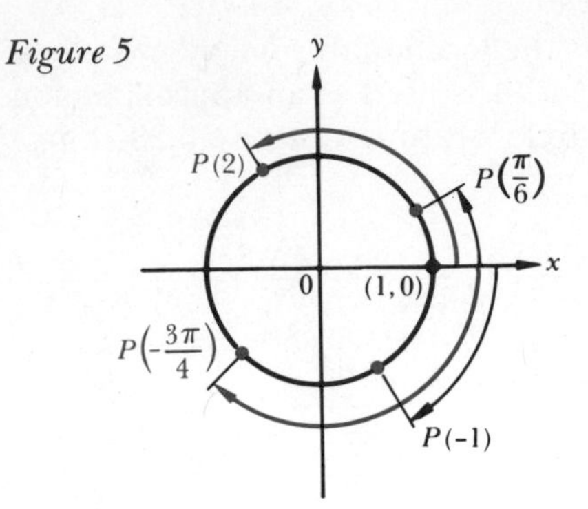

SOLUTION

(a) Since

$$0 < \frac{\pi}{6} < \frac{\pi}{2}$$

it follows that $P(\pi/6)$ lies in quadrant I (Figure 5).

(b) Since

$$\frac{\pi}{2} < 2 < \pi$$

$P(2)$ is in quadrant II (Figure 5).

(c) Since

$$-\pi < -\frac{3\pi}{4} < -\frac{\pi}{2}$$

$P(-3\pi/4)$ lies in quadrant III (Figure 5).

(d) Since

$$-\frac{\pi}{2} < -1 < 0$$

$P(-1)$ is in quadrant IV (Figure 5).

Figure 6

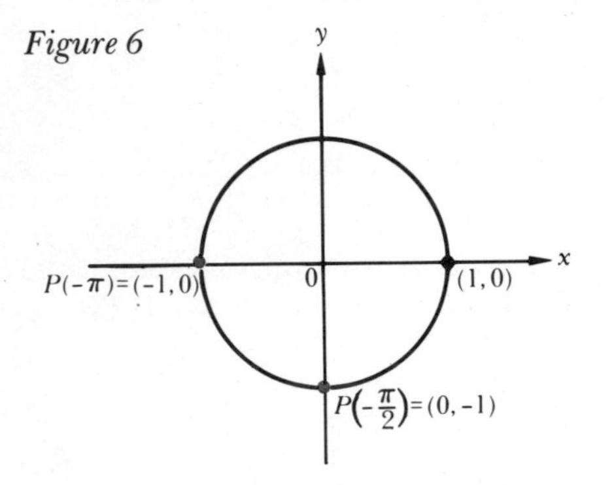

2 Find the coordinates of the following points.

(a) $P(-\pi)$ (b) $P(-\pi/2)$

SOLUTION

(a) $P(-\pi) = (-1, 0)$ (Figure 6).

(b) $P(-\pi/2) = (0, -1)$ (Figure 6).

1.1 Trigonometric (Circular) Functions

With each real number t, the wrapping function P associates a point (x, y) on the unit circle, so that $P(t) = (x, y)$, where $x^2 + y^2 = 1$. That is, the *range* of the wrapping function can be thought of as a set of ordered pairs of real numbers. Now we use the wrapping function P to construct other functions, called *circular functions,* or, more commonly, the *trigonometric functions defined on real numbers.*

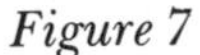
Figure 7

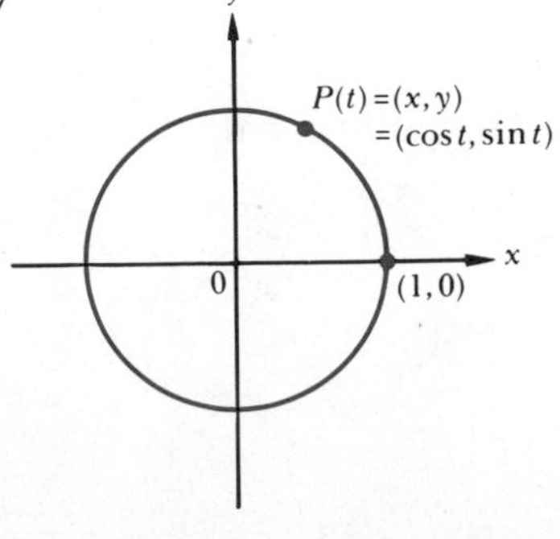

DEFINITION 1 TRIGONOMETRIC FUNCTIONS OF REAL NUMBERS

Let t be a real number and $P(t) = (x, y)$ (Figure 7). Then the *trigonometric* (circular) *functions* are named and defined on the real number t as follows:

Name of Function	*Designation and Defined Value*
sine of t	$\sin t = y$
cosine of t	$\cos t = x$
tangent of t	$\tan t = \dfrac{y}{x}$, if $x \neq 0$
cosecant of t	$\csc t = \dfrac{1}{y}$, if $y \neq 0$
secant of t	$\sec t = \dfrac{1}{x}$, if $x \neq 0$
cotangent of t	$\cot t = \dfrac{x}{y}$, if $y \neq 0$

The next few examples help to clarify the meaning of Definition 1.

EXAMPLES Use Definition 1 to determine the values of the six trigonometric functions for the given value of t.

1 (a) $t = 0$ (b) $t = -3\pi/2$

SOLUTION

(a) Since $P(0) = (1, 0)$ (Figure 8), the ordinate y of $P(0)$ is 0, so it follows from Definition 1 that the cosecant and the cotangent functions are undefined. The remaining functions are evaluated as follows:

$$\cos 0 = x = 1 \qquad \sin 0 = y = 0$$

$$\tan 0 = \frac{y}{x} = \frac{0}{1} = 0 \qquad \sec 0 = \frac{1}{x} = \frac{1}{1} = 1$$

Figure 8

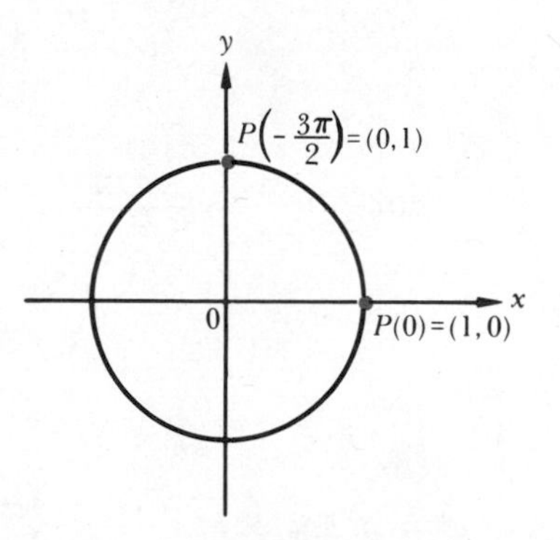

(b) Since $P(-3\pi/2) = (0, 1)$ (Figure 8), the abscissa x of $P(\pi/2)$ is 0, so from Definition 1 it follows that the tangent and secant functions are undefined and

$$\cos\left(-\frac{3\pi}{2}\right) = x = 0 \qquad \sin\left(-\frac{3\pi}{2}\right) = y = 1$$

$$\cot\left(-\frac{3\pi}{2}\right) = \frac{x}{y} = \frac{0}{1} = 0 \qquad \csc\left(-\frac{3\pi}{2}\right) = \frac{1}{y} = \frac{1}{1} = 1$$

2 $t = \dfrac{\pi}{4}$

SOLUTION Let $P(\pi/4) = (x, y)$. Since

$$P\left(\frac{\pi}{4}\right) = P\left(\frac{1}{2} \cdot \frac{\pi}{2}\right)$$

$P(\pi/4)$ is the midpoint of the arc joining the points (1, 0) and (0, 1) on the unit circle (Figure 9). Using the symmetry of the circle, we have $x = y$. Since $x^2 + y^2 = 1$, it follows that $x^2 + x^2 = 2x^2 = 1$; that is,

Figure 9

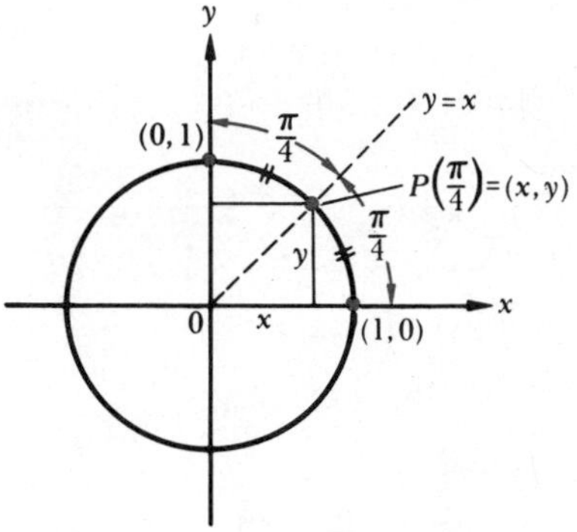

$$x = \frac{1}{\sqrt{2}} = \frac{\sqrt{2}}{2}$$

since x is positive in quadrant I. Hence

$$x = y = \frac{\sqrt{2}}{2}$$

Consequently,

$$P\left(\frac{\pi}{4}\right) = \left(\frac{\sqrt{2}}{2}, \frac{\sqrt{2}}{2}\right)$$

Using Definition 1, we have

$$\sin\frac{\pi}{4} = y = \frac{\sqrt{2}}{2} \qquad \csc\frac{\pi}{4} = \frac{1}{y} = \frac{1}{\sqrt{2}/2} = \frac{2}{\sqrt{2}} = \sqrt{2}$$

$$\cos\frac{\pi}{4} = x = \frac{\sqrt{2}}{2} \qquad \sec\frac{\pi}{4} = \frac{1}{x} = \frac{1}{\sqrt{2}/2} = \frac{2}{\sqrt{2}} = \sqrt{2}$$

$$\tan\frac{\pi}{4} = \frac{y}{x} = \frac{\sqrt{2}/2}{\sqrt{2}/2} = 1 \qquad \cot\frac{\pi}{4} = \frac{x}{y} = \frac{\sqrt{2}/2}{\sqrt{2}/2} = 1$$

3 $t = \dfrac{\pi}{3}$

Figure 10

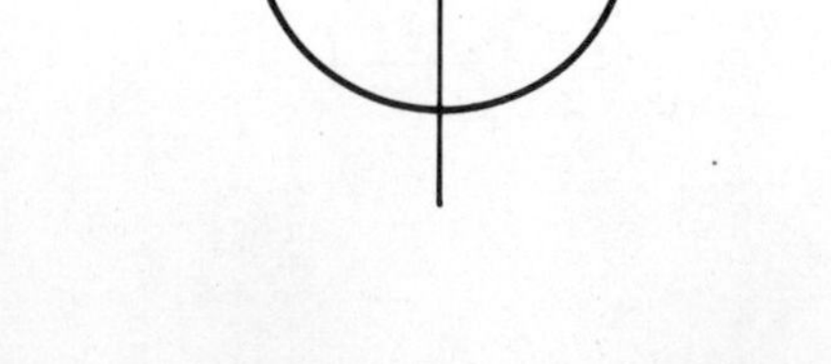

SOLUTION Assume that $P(\pi/3) = (x, y)$ and let $A = (1, 0)$, $B = (x, y)$, and $C = (-x, y)$. Since the lengths of arcs $\widehat{AB}$ and $\widehat{BC}$ are equal (each is of length $\pi/3$) (Figure 10), it follows from geometry that the chords $\overline{AB}$ and $\overline{BC}$ are equal in length, so that

$$|\overline{AB}| = |\overline{BC}|$$

and thus

$$|\overline{AB}|^2 = |\overline{BC}|^2$$

Using the distance formula, we get

$$(x-1)^2+y^2=4x^2$$

But, $x^2+y^2=1$, so $y^2=1-x^2$. Then, by substitution into the above equation, we have

$$(x-1)^2+(1-x^2)=4x^2$$

so that

$$4x^2+2x-2=0$$

or, equivalently,

$$2(2x-1)(x+1)=0$$

from which it follows that $\frac{1}{2}$ is the value of x. [Notice that -1 is also a root, but since the point $P(\pi/3)$ is in quadrant I, we choose only $x=\frac{1}{2}$.] Hence

$$y^2=1-\frac{1}{4}=\frac{3}{4} \quad \text{and} \quad y=\frac{\sqrt{3}}{2} \qquad \text{(Why?)}$$

Consequently,

$$P\left(\frac{\pi}{3}\right)=\left(\frac{1}{2},\frac{\sqrt{3}}{2}\right)$$

Using Definition 1, we have

$$\sin\frac{\pi}{3}=y=\frac{\sqrt{3}}{2} \qquad \csc\frac{\pi}{3}=\frac{1}{y}=\frac{1}{\sqrt{3}/2}=\frac{2}{\sqrt{3}}=\frac{2\sqrt{3}}{3}$$

$$\cos\frac{\pi}{3}=x=\frac{1}{2} \qquad \sec\frac{\pi}{3}=\frac{1}{x}=\frac{1}{1/2}=2$$

$$\tan\frac{\pi}{3}=\frac{y}{x}=\frac{\sqrt{3}/2}{1/2}=\sqrt{3} \qquad \cot\frac{\pi}{3}=\frac{x}{y}=\frac{1/2}{\sqrt{3}/2}=\frac{1}{\sqrt{3}}=\frac{\sqrt{3}}{3}$$

Examples 2 and 3 above establish the values of the trigonometric functions for *special values* of t that appear frequently, namely, $t=\pi/4$ and $t=\pi/3$. Another important value is $\pi/6$. It can be shown that

$$P\left(\frac{\pi}{6}\right)=\left(\frac{\sqrt{3}}{2},\frac{1}{2}\right)$$

(see Problem 28) so that, by Definition 1, we have

$$\sin\frac{\pi}{6}=y=\frac{1}{2} \qquad \csc\frac{\pi}{6}=\frac{1}{y}=\frac{1}{1/2}=2$$

$$\cos\frac{\pi}{6}=x=\frac{\sqrt{3}}{2} \qquad \sec\frac{\pi}{6}=\frac{1}{x}=\frac{1}{\sqrt{3}/2}=\frac{2}{\sqrt{3}}=\frac{2\sqrt{3}}{3}$$

$$\tan\frac{\pi}{6}=\frac{y}{x}=\frac{1/2}{\sqrt{3}/2}=\frac{\sqrt{3}}{3} \qquad \cot\frac{\pi}{6}=\frac{x}{y}=\frac{\sqrt{3}/2}{1/2}=\sqrt{3}$$

1.2 Values of $P(t)$ in Quadrants II, III, and IV

The symmetry of the unit circle enables us to use known values of $P(t)$ in quadrant I to evaluate $P(t)$ in quadrants II, III, and IV. This procedure is illustrated in the following examples.

EXAMPLES Use the symmetry of the unit circle to determine the coordinates of the following points.

1 $P\left(\frac{3\pi}{4}\right)$, $P\left(\frac{5\pi}{4}\right)$, $P\left(\frac{7\pi}{4}\right)$, and $P\left(-\frac{\pi}{4}\right)$ using

$$P\left(\frac{\pi}{4}\right) = \left(\frac{\sqrt{2}}{2}, \frac{\sqrt{2}}{2}\right)$$

SOLUTION

$$P\left(\frac{\pi}{4}\right) = \left(\frac{\sqrt{2}}{2}, \frac{\sqrt{2}}{2}\right)$$

is the midpoint of the arc of the unit circle in quadrant I and $P(3\pi/4)$ is the midpoint of the arc of the unit circle in quadrant II. Using the symmetry of the circle, we conclude that

$$P\left(\frac{3\pi}{4}\right) = \left(-\frac{\sqrt{2}}{2}, \frac{\sqrt{2}}{2}\right) \qquad \text{(Figure 11)}$$

Similarly,

$$P\left(\frac{5\pi}{4}\right) = \left(-\frac{\sqrt{2}}{2}, -\frac{\sqrt{2}}{2}\right)$$

and

$$P\left(\frac{7\pi}{4}\right) = P\left(-\frac{\pi}{4}\right) = \left(\frac{\sqrt{2}}{2}, -\frac{\sqrt{2}}{2}\right)$$

Figure 11

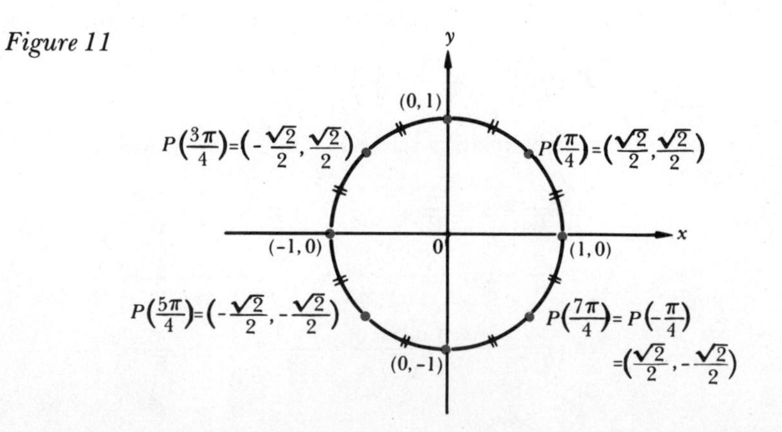

Figure 12

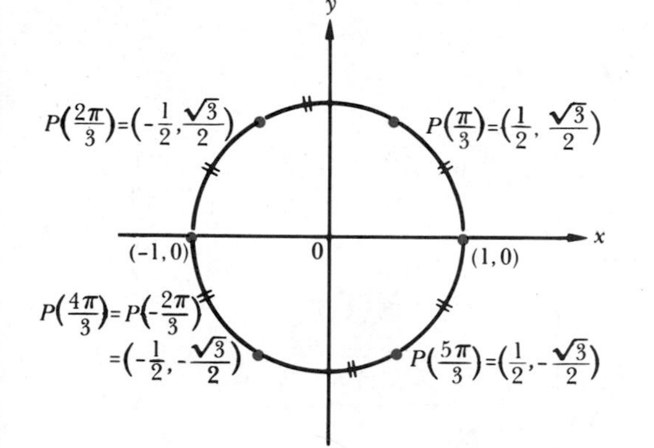

2 $P(2\pi/3), P(4\pi/3), P(5\pi/3)$, and $P(-2\pi/3)$ using

$$P\left(\frac{\pi}{3}\right)=\left(\frac{1}{2},\frac{\sqrt{3}}{2}\right)$$

SOLUTION Using the symmetry of the unit circle, together with the fact that

$$P\left(\frac{\pi}{3}\right)=\left(\frac{1}{2},\frac{\sqrt{3}}{2}\right) \qquad \text{(Figure 12)}$$

we get

$$P\left(\frac{2\pi}{3}\right)=\left(-\frac{1}{2},\frac{\sqrt{3}}{2}\right)$$
$$P\left(\frac{4\pi}{3}\right)=P\left(-\frac{2\pi}{3}\right)=\left(-\frac{1}{2},-\frac{\sqrt{3}}{2}\right)$$
$$P\left(\frac{5\pi}{3}\right)=\left(\frac{1}{2},-\frac{\sqrt{3}}{2}\right)$$

3 $P(5\pi/6), P(7\pi/6), P(11\pi/6)$, and $P(-7\pi/6)$ using

$$P\left(\frac{\pi}{6}\right)=\left(\frac{\sqrt{3}}{2},\frac{1}{2}\right)$$

SOLUTION The symmetry of the unit circle, together with the fact that

$$P\left(\frac{\pi}{6}\right)=\left(\frac{\sqrt{3}}{2},\frac{1}{2}\right) \qquad \text{(Figure 13)}$$

is used to conclude that

$$P\left(\frac{5\pi}{6}\right)=P\left(-\frac{7\pi}{6}\right)=\left(-\frac{\sqrt{3}}{2},\frac{1}{2}\right)$$
$$P\left(\frac{7\pi}{6}\right)=\left(-\frac{\sqrt{3}}{2},-\frac{1}{2}\right)$$
$$P\left(\frac{11\pi}{6}\right)=\left(\frac{\sqrt{3}}{2},-\frac{1}{2}\right)$$

Figure 13

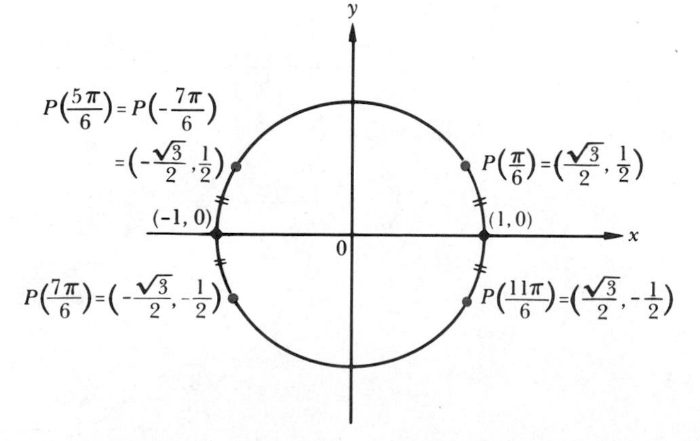

For convenience, Figure 14 summarizes the coordinates of $P(t)$ for the special values of t that we derived above.

Figure 14

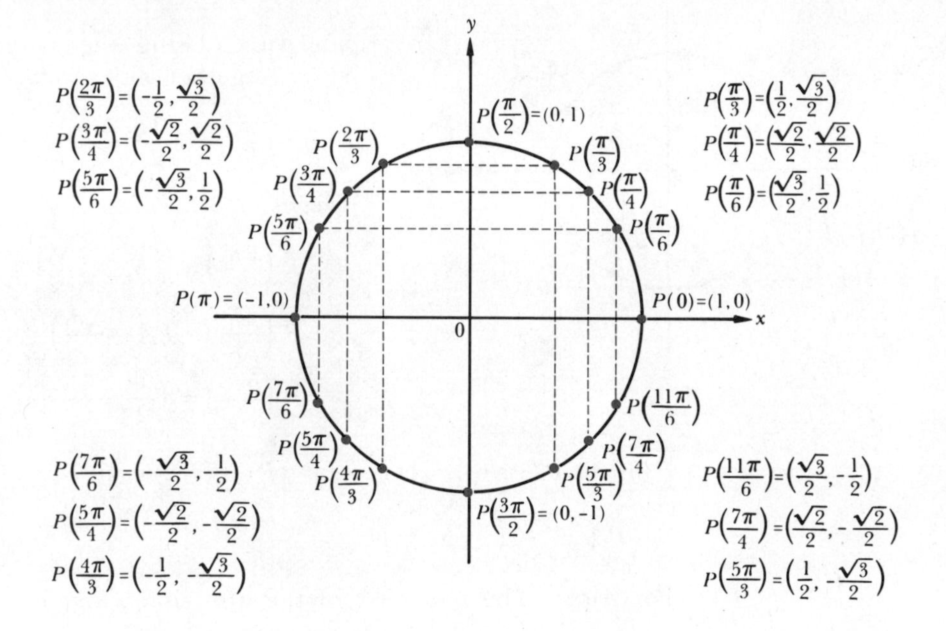

EXAMPLE Find each of the following values by using Definition 1 and Figure 14.

(a) $\sin 3\pi/4$ (b) $\cos(-\pi/4)$ (c) $\tan 5\pi/3$
(d) $\cot(-2\pi/3)$ (e) $\sec 11\pi/6$ (f) $\csc(-7\pi/6)$

SOLUTION [Refer to Figure 14 for the values of $P(t)$ below.]

(a) $$P\left(\frac{3\pi}{4}\right) = \left(-\frac{\sqrt{2}}{2}, \frac{\sqrt{2}}{2}\right)$$

so

$$\sin \frac{3\pi}{4} = y = \frac{\sqrt{2}}{2}$$

(b) $$P\left(-\frac{\pi}{4}\right) = P\left(\frac{7\pi}{4}\right) = \left(\frac{\sqrt{2}}{2}, -\frac{\sqrt{2}}{2}\right)$$

implies that

$$\cos\left(-\frac{\pi}{4}\right) = x = \frac{\sqrt{2}}{2}$$

(c) $$P\left(\frac{5\pi}{3}\right) = \left(\frac{1}{2}, -\frac{\sqrt{3}}{2}\right)$$

so

$$\tan \frac{5\pi}{3} = \frac{y}{x} = \frac{-\sqrt{3}/2}{1/2} = -\sqrt{3}$$

(d) $$P\left(-\frac{2\pi}{3}\right) = P\left(\frac{4\pi}{3}\right) = \left(-\frac{1}{2}, -\frac{\sqrt{3}}{2}\right)$$

so

$$\cot\left(-\frac{2\pi}{3}\right) = \frac{x}{y} = \frac{-1/2}{-\sqrt{3}/2} = \frac{1}{\sqrt{3}} = \frac{\sqrt{3}}{3}$$

(e) $$P\left(\frac{11\pi}{6}\right) = \left(\frac{\sqrt{3}}{2}, -\frac{1}{2}\right)$$

so we have

$$\sec\frac{11\pi}{6} = \frac{1}{x} = \frac{1}{\sqrt{3}/2} = \frac{2}{\sqrt{3}} = \frac{2\sqrt{3}}{3}$$

(f) $$P\left(-\frac{7\pi}{6}\right) = P\left(\frac{5\pi}{6}\right) = \left(-\frac{\sqrt{3}}{2}, \frac{1}{2}\right)$$

implies that

$$\csc\left(-\frac{7\pi}{6}\right) = \frac{1}{y} = \frac{1}{1/2} = 2$$

The following theorem shows how to express the tangent, cotangent, secant, and cosecant functions in terms of the sine and cosine functions.

THEOREM 1

Let t be a real number. Then

(i) $\tan t = \dfrac{\sin t}{\cos t}$, if $\cos t \neq 0$ (ii) $\cot t = \dfrac{\cos t}{\sin t}$, if $\sin t \neq 0$

(iii) $\sec t = \dfrac{1}{\cos t}$, if $\cos t \neq 0$ (iv) $\csc t = \dfrac{1}{\sin t}$, if $\sin t \neq 0$

PROOF Let $P(t) = (x, y)$. By Definition 1, we have

(i) $\tan t = \dfrac{y}{x} = \dfrac{\sin t}{\cos t}$, if $x \neq 0$ (ii) $\cot t = \dfrac{x}{y} = \dfrac{\cos t}{\sin t}$, if $y \neq 0$

(iii) $\sec t = \dfrac{1}{x} = \dfrac{1}{\cos t}$, if $x \neq 0$ (iv) $\csc t = \dfrac{1}{y} = \dfrac{1}{\sin t}$, if $y \neq 0$

EXAMPLES 1 Use Theorem 1 to find the values of the six trigonometric functions at t if the coordinates of $P(t)$ are given by $(3/5, 4/5)$.

SOLUTION Notice that $(3/5, 4/5)$ lies on the unit circle because

$$\left(\frac{3}{5}\right)^2 + \left(\frac{4}{5}\right)^2 = \frac{9}{25} + \frac{16}{25} = 1$$

Since $P(t) = \left(\frac{3}{5}, \frac{4}{5}\right)$, it follows from Definition 1 that

$$\cos t = x = \frac{3}{5} \quad \text{and} \quad \sin t = y = \frac{4}{5}$$

Thus

$$\tan t = \frac{\sin t}{\cos t} = \frac{4/5}{3/5} = \frac{4}{3} \qquad \cot t = \frac{\cos t}{\sin t} = \frac{3/5}{4/5} = \frac{3}{4}$$

$$\sec t = \frac{1}{\cos t} = \frac{1}{3/5} = \frac{5}{3} \qquad \csc t = \frac{1}{\sin t} = \frac{1}{4/5} = \frac{5}{4}$$

2 Use Theorem 1 to prove that

$$\cot t = \frac{1}{\tan t}$$

PROOF By Theorem 1, we have

$$\tan t = \frac{\sin t}{\cos t} \quad \text{and} \quad \cot t = \frac{\cos t}{\sin t}$$

so

$$\cot t = \frac{\cos t}{\sin t} = \frac{1}{\sin t/\cos t} = \frac{1}{\tan t}$$

PROBLEM SET 1

In Problems 1–4, let A, B, C, and D be points located on the unit circle as shown in Figure 15. Find the smallest positive value of t in terms of π for each given point $P(t)$ on the unit circle. (Assume that a counterclockwise motion is used in moving from one point to the next.)

Figure 15

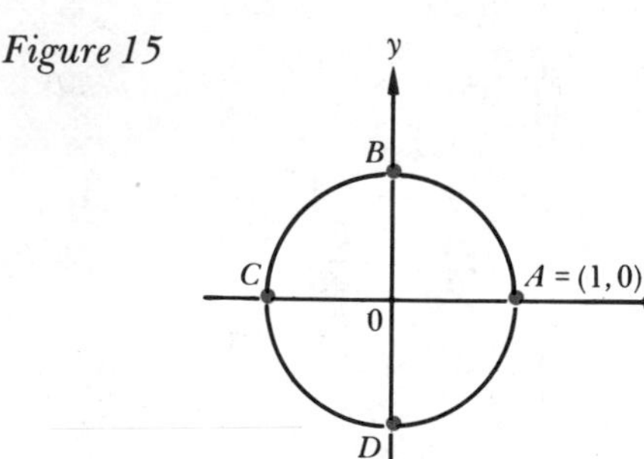

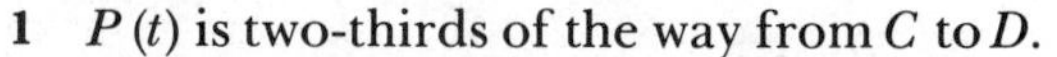

1 $P(t)$ is two-thirds of the way from C to D.
2 $P(t)$ is one-fourth of the way from B to A.
3 $P(t)$ is five-eighths of the way from C to B.
4 $P(t)$ is one-sixth of the way from D to A.

In Problems 5–8, find the value of t if a given point is located by starting at point (1, 0) and moving around the circumference of the unit circle as follows:

5 One-third of the way, clockwise
6 Three-eighths of the way, clockwise
7 Five-fourths of the way, counterclockwise
8 One-sixth of the way, counterclockwise

In Problems 9–27, display each point on the unit circle and determine the quadrant (if any) in which each point lies. (Use $\pi = 3.14$.)

9 $P(-2\pi)$ **10** $P\left(-\frac{5\pi}{3}\right)$ **11** $P\left(\frac{3\pi}{4}\right)$

12 $P\left(\frac{5\pi}{6}\right)$ **13** $P\left(-\frac{7\pi}{6}\right)$ **14** $P\left(-\frac{\pi}{3}\right)$

15 $P\left(-\frac{5\pi}{4}\right)$ **16** $P\left(-\frac{11\pi}{6}\right)$ **17** $P\left(-\frac{\pi}{6}\right)$

18 $P(0.8)$ **19** $P(3.6)$ **20** $P(4.5)$
21 $P(6)$ **22** $P(1.4)$ **23** $P(-1.3)$
24 $P(-2.7)$ **25** $P(-3.7)$ **26** $P(-5.7)$
27 $P(-4)$

28 Prove that

$$P\left(\frac{\pi}{6}\right) = \left(\frac{\sqrt{3}}{2}, \frac{1}{2}\right)$$

In Problems 29–36, use Figure 14 on page 192 to determine the coordinates of the given point $P(t)$ of the unit circle.

29 $P\left(-\frac{3\pi}{4}\right)$ **30** $P\left(-\frac{5\pi}{4}\right)$ **31** $P\left(-\frac{7\pi}{4}\right)$

32 $P\left(-\frac{4\pi}{3}\right)$ **33** $P\left(-\frac{5\pi}{3}\right)$ **34** $P\left(-\frac{\pi}{3}\right)$

35 $P\left(-\frac{\pi}{6}\right)$ **36** $P\left(-\frac{5\pi}{6}\right)$

In Problems 37–40, show that each point lies on the unit circle, and determine one positive and one negative value of t for which $P(t)$ is equal to the given ordered pair. Refer to Figure 14 if necessary.

37 $P(t) = \left(\frac{1}{2}, \frac{\sqrt{3}}{2}\right)$ **38** $P(t) = (-1, 0)$

39 $P(t) = \left(-\frac{\sqrt{2}}{2}, -\frac{\sqrt{2}}{2}\right)$ **40** $P(t) = \left(-\frac{1}{2}, \frac{\sqrt{3}}{2}\right)$

In Problems 41–60, use Definition 1 and Figure 14, if necessary, to find the value of the indicated trigonometric function.

41 $\sin \frac{\pi}{2}$ **42** $\cos 2\pi$ **43** $\tan \pi$

44 $\sec(-\pi)$ **45** $\csc(-2\pi)$ **46** $\cot(-\pi)$

47 $\tan\left(-\frac{\pi}{4}\right)$ **48** $\cot \frac{7\pi}{6}$ **49** $\sec \frac{3\pi}{4}$

50 $\sin \frac{7\pi}{6}$ **51** $\cos\left(-\frac{7\pi}{6}\right)$ **52** $\cot\left(-\frac{11\pi}{6}\right)$

53 $\sec\left(-\frac{5\pi}{4}\right)$ **54** $\csc\left(-\frac{7\pi}{4}\right)$ **55** $\cos \frac{11\pi}{6}$

56 $\cot \frac{5\pi}{3}$ **57** $\sin\left(-\frac{5\pi}{3}\right)$ **58** $\cot\left(-\frac{4\pi}{3}\right)$

59 $\cot\left(-\frac{7\pi}{4}\right)$ **60** $\sec\left(-\frac{2\pi}{3}\right)$

In Problems 61–66, show that the point lies on the unit circle, then find the values of $\sin t$ and $\cos t$. Use Theorem 1 to evaluate $\tan t$, $\cot t$, $\sec t$, and $\csc t$.

61 $P(t) = \left(-\frac{\sqrt{3}}{2}, \frac{1}{2}\right)$ **62** $P(t) = \left(-\frac{4}{5}, \frac{3}{5}\right)$

63 $P(t) = \left(\frac{5}{13}, -\frac{12}{13}\right)$

64 $P(t) = \left(\frac{1}{\sqrt{10}}, \frac{3}{\sqrt{10}}\right)$

65 $P(t) = \left(-\frac{3}{\sqrt{13}}, \frac{2}{\sqrt{13}}\right)$

66 $P(t) = \left(-\frac{4}{\sqrt{17}}, -\frac{1}{\sqrt{17}}\right)$

67 Observe the behavior of $P(t)$ on the unit circle as specified below; then determine whether the values of the cos t and sin t increase or decrease.
(a) As t increases from 0 to $\pi/2$
(b) As t increases from $\pi/2$ to π
(c) As t increases from π to $3\pi/2$
(d) As t increases from $3\pi/2$ to 2π

68 Use Figure 14 on page 192 to complete Table 2.

Table 2

t	cos t	sin t	tan t	cot t	sec t	csc t
0	1	0	0	undefined	1	undefined
$\frac{\pi}{6}$						
$\frac{\pi}{4}$						
$\frac{\pi}{3}$						
$\frac{\pi}{2}$						
$\frac{2\pi}{3}$						
$\frac{3\pi}{4}$						
$\frac{5\pi}{6}$						
π						
$\frac{7\pi}{6}$						
$\frac{5\pi}{4}$						
$\frac{4\pi}{3}$						
$\frac{3\pi}{2}$						
$\frac{5\pi}{3}$						
$\frac{7\pi}{4}$						
$\frac{11\pi}{6}$						

2 Properties of the Trigonometric Functions

So far, we have defined the trigonometric functions on real numbers and computed their values for some special numbers. In this section, we discover that the values of the trigonometric functions repeat every 2π units. Also, we study the domains, the signs of the values, and some of the relationships of the trigonometric functions to each other.

2.1 Periodicity of the Trigonometric Functions

Since the circumference of the unit circle is 2π, we have

$$P(t) = P(t + 2\pi)$$

for any real number t (Figure 1).

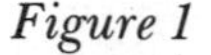

Figure 1

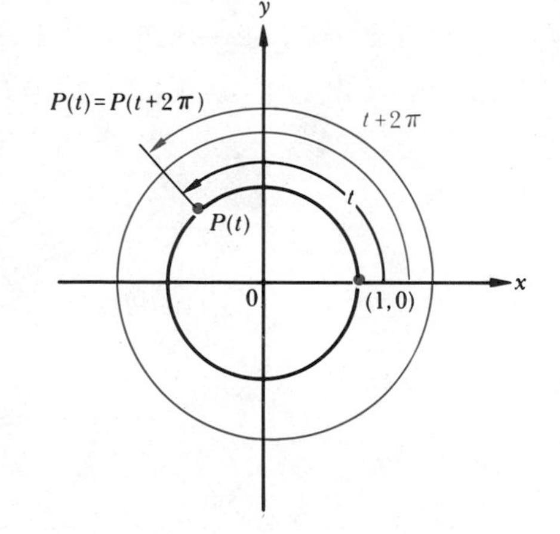

For example,

$$P(0) = P(2\pi) = P(4\pi) = P(6\pi) = (1, 0)$$

The function values of P repeat every 2π units. That is,

$$P(t) = P(t + 2\pi) = P(t + 4\pi) = P(t + 6\pi)$$

and so forth. Functions with this repetitive property are called *periodic functions.* More formally, we have the following definition.

DEFINITION 1 PERIODIC FUNCTION

A function f is said to be a *periodic function of period a,* where a is a nonzero constant, if, for all x in the domain of f, $x + a$ is also in the domain of f, and

$$f(x + a) = f(x)$$

The smallest *positive* period of f is called the *fundamental period of f.*

Thus, the wrapping function P is a periodic function of periods $\pm 2\pi$, $\pm 4\pi$, $\pm 6\pi$, or any multiple of 2π. Its fundamental period is 2π. In other words, if t is any real number, then

$$P(t) = P(t + 2\pi n), \quad \text{for } any \text{ integer } n$$

EXAMPLES **1** Use the periodicity to find a value for t, where $0 \leq t < 2\pi$, so that $P(t)$ coincides with the given point. Locate each point on the unit circle.

(a) $P(7)$ (b) $P(15)$ (c) $P(-2)$ (d) $P(-45)$

SOLUTION (Note that 2π is approximately 6.28 if we use $\pi = 3.14$.)

(a) Since $7 = 0.72 + 6.28$, it follows, from the fact that P is a periodic function of period 2π (or 6.28), that

$$P(7) = P(0.72 + 6.28) = P(0.72) \qquad \text{(Figure 2)}$$

(b) Since $15 = 2.44 + 2(6.28)$, we get

$$P(15) = P[2.44 + 2(6.28)] = P(2.44) \qquad \text{(Figure 2)}$$

(c) $P(-2) = P(-2 + 6.28) = P(4.28)$ (Figure 2)

(d) $P(-45) = P[-45 + 8(6.28)] = P(5.24)$ (Figure 2)

Figure 2

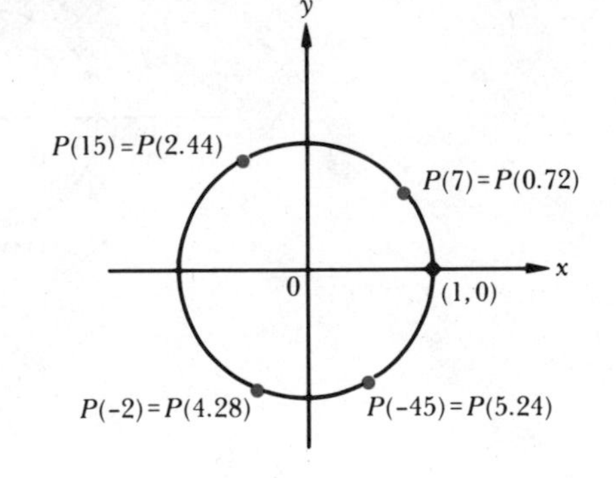

2 Use periodicity to find the coordinates of the following points. Refer to Figure 14 on page 192.

(a) $P\left(\frac{11\pi}{2}\right)$ (b) $P\left(\frac{25\pi}{6}\right)$ (c) $P\left(-\frac{2\pi}{3}\right)$ (d) $P\left(-\frac{27\pi}{4}\right)$

SOLUTION Using the periodicity of P and Figure 14, we have

(a) $P\left(\frac{11\pi}{2}\right) = P\left(\frac{3\pi}{2} + 2 \cdot 2\pi\right) = P\left(\frac{3\pi}{2}\right) = (0, -1)$

(b) $P\left(\frac{25\pi}{6}\right) = P\left(\frac{\pi}{6} + 2 \cdot 2\pi\right) = P\left(\frac{\pi}{6}\right) = \left(\frac{\sqrt{3}}{2}, \frac{1}{2}\right)$

(c) $P\left(-\frac{2\pi}{3}\right) = P\left(-\frac{2\pi}{3} + 2\pi\right) = P\left(\frac{4\pi}{3}\right) = \left(-\frac{1}{2}, -\frac{\sqrt{3}}{2}\right)$

(d) $P\left(-\frac{27\pi}{4}\right) = P\left(-\frac{27\pi}{4} + 4 \cdot 2\pi\right) = P\left(\frac{5\pi}{4}\right) = \left(-\frac{\sqrt{2}}{2}, -\frac{\sqrt{2}}{2}\right)$

Using the fact that $P(t) = (x, y) = P(t + 2\pi)$ from Page 197, it follows that $\cos t = x = \cos(t + 2\pi)$ and $\sin t = y = \sin(t + 2\pi)$. Hence, the *cosine* and *sine* are periodic functions, each of fundamental period 2π. Thus, if we can determine the values of the $\cos t$ and $\sin t$ for $0 \leq t < 2\pi$, we can use these values to determine $\cos t$ and $\sin t$ for any real number t. We shall use this property later to graph the cosine and sine functions.

EXAMPLES 1 Use the results of Section 1, together with the periodicity of the sine and cosine functions, to find the value of $\cos t$ and $\sin t$ if

(a) $t = \dfrac{25\pi}{6}$ (b) $t = -\dfrac{27\pi}{4}$

SOLUTION (Refer to Figure 14 on page 192.)

(a) $$\cos\frac{25\pi}{6} = \cos\left(\frac{\pi}{6} + 2 \cdot 2\pi\right) = \cos\frac{\pi}{6} = \frac{\sqrt{3}}{2}$$

and

$$\sin\frac{25\pi}{6} = \sin\left(\frac{\pi}{6} + 2 \cdot 2\pi\right) = \sin\frac{\pi}{6} = \frac{1}{2}$$

(b) $$\cos\left(-\frac{27\pi}{4}\right) = \cos\left(-\frac{27\pi}{4} + 4 \cdot 2\pi\right) = \cos\left(\frac{5\pi}{4}\right) = -\frac{\sqrt{2}}{2}$$

and

$$\sin\left(-\frac{27\pi}{4}\right) = \sin\left(-\frac{27\pi}{4} + 4 \cdot 2\pi\right) = \sin\left(\frac{5\pi}{4}\right) = -\frac{\sqrt{2}}{2}$$

2 Use the fact that the sine and cosine are periodic functions of period 2π to determine the period of each of the following functions.

(a) $f(t) = \sin 2t$ (b) $g(t) = \cos\dfrac{t}{5}$

SOLUTION

(a) $$\begin{aligned} f(t) &= \sin 2t \\ &= \sin(2t + 2\pi) \qquad \text{(The sine function has period } 2\pi.) \\ &= \sin[2(t + \pi)] \\ &= f(t + \pi) \end{aligned}$$

so f is a periodic function of period π.

(b) $$\begin{aligned} g(t) &= \cos\frac{t}{5} = \cos\left(\frac{t}{5} + 2\pi\right) \\ &= \cos\left[\frac{1}{5}(t + 10\pi)\right] = g(t + 10\pi) \end{aligned}$$

so g is a periodic function of period 10π.

The following theorem shows how the periodicity of the sine and cosine functions are used to determine that the other trigonometric functions are periodic.

THEOREM 1

The tangent, cotangent, secant, and cosecant are periodic functions of period 2π. (Later, in Chapter 6, pages 261 and 266, we shall see that the *fundamental period* of the tangent and cotangent is actually π.)

PROOF We prove the tangent and the secant to be periodic functions of period 2π, and leave the cotangent and cosecant as an exercise (see Problem 20). Using Theorem 1 of Section 1, we have

$$\tan(t + 2\pi) = \frac{\sin(t + 2\pi)}{\cos(t + 2\pi)} = \frac{\sin t}{\cos t} = \tan t$$

and

$$\sec(t + 2\pi) = \frac{1}{\cos(t + 2\pi)} = \frac{1}{\cos t} = \sec t$$

EXAMPLE Use the periodicity of the trigonometric functions and Theorem 1 of Section 1 to find each of the following values.

(a) $\tan \frac{25\pi}{6}$ (b) $\csc \frac{25\pi}{6}$

(c) $\cot \left(-\frac{27\pi}{4}\right)$ (d) $\sec \left(-\frac{27\pi}{4}\right)$

SOLUTION (Refer to Figure 14 on page 192.)

(a)
$$\tan \frac{25\pi}{6} = \tan \left(\frac{\pi}{6} + 2 \cdot 2\pi\right) = \tan \frac{\pi}{6} = \frac{\sin(\pi/6)}{\cos(\pi/6)} = \frac{1/2}{\sqrt{3}/2} = \frac{1}{\sqrt{3}} = \frac{\sqrt{3}}{3}$$

(b)
$$\csc \frac{25\pi}{6} = \csc \left(\frac{\pi}{6} + 2 \cdot 2\pi\right) = \csc \frac{\pi}{6} = \frac{1}{\sin(\pi/6)} = \frac{1}{1/2} = 2$$

(c)
$$\cot \left(-\frac{27\pi}{4}\right) = \cot \left(-\frac{27\pi}{4} + 4 \cdot 2\pi\right) = \cot \frac{5\pi}{4} = \frac{\cos(5\pi/4)}{\sin(5\pi/4)} = \frac{-\sqrt{2}/2}{-\sqrt{2}/2} = 1$$

(d)
$$\sec \left(-\frac{27\pi}{4}\right) = \sec \left(-\frac{27\pi}{4} + 4 \cdot 2\pi\right) = \sec \frac{5\pi}{4} = \frac{1}{\cos(5\pi/4)} = \frac{1}{-\sqrt{2}/2} = \frac{-2}{\sqrt{2}} = -\sqrt{2}$$

2.2 Signs of the Values of the Trigonometric Functions

It is important to note that the signs of the values of the trigonometric functions depend on the quadrant in which $P(t)$ is located (Figure 3). Thus, if $P(t_1)$ is in quadrant I, then *all* functional values are positive since x_1 and y_1 are positive. If $P(t_2)$ is in quadrant II, then $x_2 < 0$ and $y_2 > 0$, and consequently, *only* the sine and cosecant values of t_2 are positive. If $P(t_3)$ is in quadrant III, then $x_3 < 0$ and $y_3 < 0$, and so *only* the tangent and the cotangent values of t_3 are positive. Finally, if $P(t_4)$ is located in quadrant IV, then $x_4 > 0$ and $y_4 < 0$, and *only* the cosine and secant values of t_4 are positive.

EXAMPLE Indicate the quadrant in which $P(t)$ lies if $\sin t > 0$ and $\cot t < 0$.

SOLUTION From Figure 3, we see that $\sin t > 0$ and $\cot t < 0$ if $P(t)$ is located in quadrant II.

Figure 3

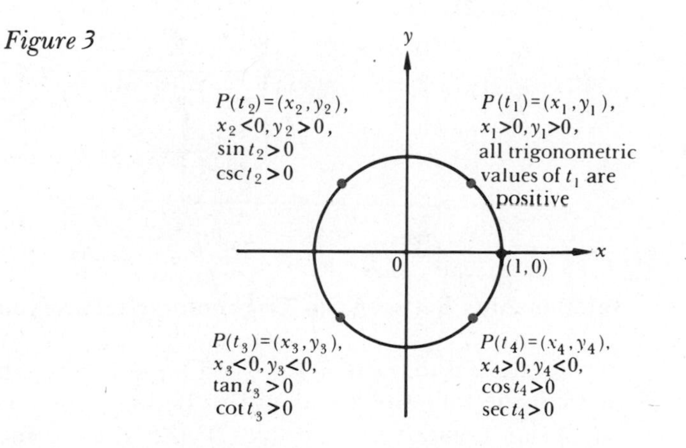

2.3 Domains and Ranges of the Trigonometric Functions

The domain of each trigonometric function can be determined as follows. Since $P(t) = (x, y)$ is defined for any real number t, $\cos t$ and $\sin t$ are defined for any real number; that is, the domain for both the cosine and sine functions is the set of all real numbers.

Since

$$\tan t = \frac{\sin t}{\cos t} \quad \text{and} \quad \sec t = \frac{1}{\cos t}$$

the domain for both the tangent and the secant function consists of all real numbers for which $\cos t \neq 0$; that is, the domain is

$$\left\{t \,\middle|\, t \neq \frac{\pi}{2} + n\pi, \quad \text{where } n \text{ is any integer}\right\}$$

Similarly, the domain of each of the cotangent and cosecant functions consists of all real numbers for which $\sin t \neq 0$, since

$$\cot t = \frac{\cos t}{\sin t} \quad \text{and} \quad \csc t = \frac{1}{\sin t}$$

Thus, the domain of the cotangent and the cosecant functions is

$$\{t \mid t \neq n\pi, \quad \text{where } n \text{ is any integer}\}$$

Because of the fact that any point $P(t) = (x, y)$ on the unit circle satisfies $-1 \leq x \leq 1$ and $-1 \leq y \leq 1$, $\cos t = x$ has as its range the set $[-1, 1]$ and $\sin t = y$ has as its range the set $[-1, 1]$ (Figure 4). The ranges of the remaining four trigonometric functions will be determined by using their graphs in Section 3.

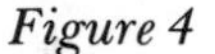

Figure 4

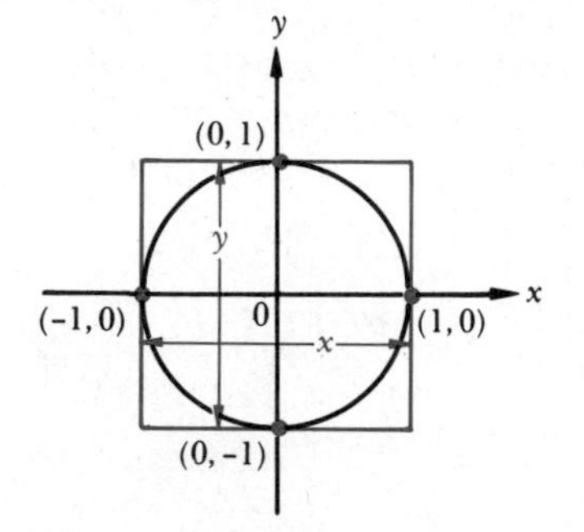

2.4 Relationships Between the Trigonometric Functions

We use the fact that $P(t) = (x, y)$ is a point on the unit circle $x^2 + y^2 = 1$ to establish the following relationships between the trigonometric functions. The power notation used for $(\cos t)^n$ is $\cos^n t$ and for $(\sin t)^n$ is $\sin^n t$. Note that $\cos^n t \neq \cos(t^n)$ and $\sin^n t \neq \sin(t^n)$.

THEOREM 2

Let $P(t) = (x, y)$. Then, $\cos t = x$ and $\sin t = y$, where $x^2 + y^2 = 1$.

(i) $\cos^2 t + \sin^2 t = 1$

(ii) $\tan^2 t + 1 = \sec^2 t$

(iii) $1 + \cot^2 t = \csc^2 t$

PROOF

(i) Let $P(t) = (x, y)$. Then, $\cos t = x$ and $\sin t = y$, where $x^2 + y^2 = 1$. Substituting $\cos t$ for x and $\sin t$ for y, we have $\cos^2 t + \sin^2 t = 1$.

(ii) $\tan^2 t + 1 = \dfrac{\sin^2 t}{\cos^2 t} + 1 = \dfrac{\sin^2 t + \cos^2 t}{\cos^2 t} = \dfrac{1}{\cos^2 t} = \sec^2 t$

(iii) $1 + \cot^2 t = 1 + \dfrac{\cos^2 t}{\sin^2 t} = \dfrac{\sin^2 t + \cos^2 t}{\sin^2 t} = \dfrac{1}{\sin^2 t} = \csc^2 t$

EXAMPLES Use Theorem 2 and the signs of the values of the trigonometric functions to solve the following examples.

1 If $\sin t = 3/10$ and $P(t)$ is in quadrant I, find $\cos t$ and $\cot t$.

SOLUTION Using $\cos^2 t + \sin^2 t = 1$, we have $\cos^2 t = 1 - \sin^2 t$. Therefore, $\cos t = \pm\sqrt{1 - \sin^2 t}$. Since $P(t)$ lies in quadrant I, $\cos t > 0$ and

$$\cos t = \sqrt{1 - \sin^2 t} = \sqrt{1 - \left(\frac{3}{10}\right)^2} = \sqrt{1 - \frac{9}{100}} = \sqrt{\frac{91}{100}} = \frac{\sqrt{91}}{10}$$

$$\cot t = \frac{\cos t}{\sin t} = \frac{\sqrt{91}/10}{3/10} = \frac{\sqrt{91}}{3}$$

2 If $\cos t = \sqrt{2}/2$, and $P(t)$ is in quadrant IV, find the remaining trigonometric function values of t.

SOLUTION Since $\sin t < 0$ for $P(t)$ in quadrant IV, we have

$$\sin t = -\sqrt{1 - \cos^2 t} = -\sqrt{1 - \left(\frac{\sqrt{2}}{2}\right)^2} = -\sqrt{1 - \frac{1}{2}} = -\frac{\sqrt{2}}{2}$$

$$\tan t = \frac{\sin t}{\cos t} = \frac{-\sqrt{2}/2}{\sqrt{2}/2} = -1$$

$$\cot t = \frac{\cos t}{\sin t} = \frac{\sqrt{2}/2}{-\sqrt{2}/2} = -1$$

$$\sec t = \frac{1}{\cos t} = \frac{1}{\sqrt{2}/2} = \frac{2}{\sqrt{2}} = \sqrt{2}$$

$$\csc t = \frac{1}{\sin t} = \frac{1}{-\sqrt{2}/2} = -\frac{2}{\sqrt{2}} = -\sqrt{2}$$

3 If $\tan t = -5/12$ and $P(t)$ is in quadrant II, find the remaining trigonometric function values of t.

SOLUTION Since $\sec t < 0$ for $P(t)$ in quadrant II, and since it has been established that $\sec^2 t = \tan^2 t + 1$, we have

$$\sec t = -\sqrt{\tan^2 t + 1} = -\sqrt{\left(-\frac{5}{12}\right)^2 + 1}$$

$$= -\sqrt{\frac{25 + 144}{144}} = -\sqrt{\frac{169}{144}} = -\frac{13}{12}$$

Also

$$\cot t = \frac{1}{\tan t} = \frac{1}{-5/12} = -\frac{12}{5}$$

Since $\csc t > 0$ in quadrant II, and since $\csc^2 t = 1 + \cot^2 t$, we have

$$\csc t = \sqrt{1 + \cot^2 t} = \sqrt{1 + \left(-\frac{12}{5}\right)^2} = \sqrt{\frac{25 + 144}{25}} = \frac{13}{5}$$

Finally,

$$\cos t = \frac{1}{\sec t} = \frac{1}{-13/12} = -\frac{12}{13}$$

and

$$\sin t = \frac{1}{\csc t} = \frac{1}{13/5} = \frac{5}{13}$$

2.5 Even and Odd Trigonometric Functions

Now we establish which of the trigonometric functions are even and which are odd by the following theorem.

THEOREM 3

Let t be any real number in the domain of the given function. Then

(i) $\cos(-t) = \cos t$ (The cosine is an even function.)

(ii) $\sin(-t) = -\sin t$ (The sine is an odd function.)

(iii) $\tan(-t) = -\tan t$ (The tangent is an odd function.)

(iv) $\sec(-t) = \sec t$ (The secant is an even function.)

(v) $\csc(-t) = -\csc t$ (The cosecant is an odd function.)

(vi) $\cot(-t) = -\cot t$ (The cotangent is an odd function.)

Figure 5

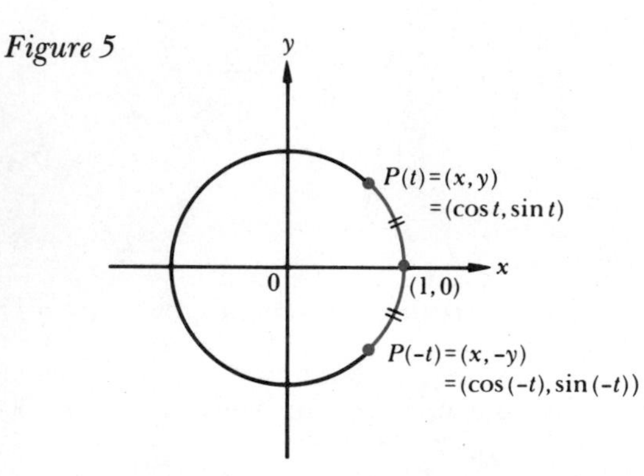

PROOF

(i–ii) Although we first illustrate this result for a value of t that displays $P(t)$ in quadrant I, it is important to remember that the argument can be used for any real number t. Assume that $P(t) = (x, y)$. Then, by Definition 1 of Section 1, together with the symmetry of the circle, we have, as shown in Figure 5,

$$P(t) = (x, y) = (\cos t, \sin t)$$

and

$$P(-t) = (x, -y) = (\cos(-t), \sin(-t))$$

Hence $\cos t = x = \cos(-t)$ and $\sin(-t) = -y = -\sin t$.

(iii) $\tan(-t) = \dfrac{\sin(-t)}{\cos(-t)} = \dfrac{-\sin t}{\cos t} = -\tan t$

(iv–vi) One can easily show that the secant is an even function, whereas the cotangent and the cosecant functions are odd (see Problem 60).

EXAMPLE Use Theorem 3 and Figure 14 on page 192 to find each of the following values.

(a) $\sin\left(-\dfrac{3\pi}{4}\right)$ (b) $\cos\left(-\dfrac{3\pi}{4}\right)$ (c) $\tan\left(-\dfrac{3\pi}{4}\right)$

SOLUTION

(a) $\sin\left(-\dfrac{3\pi}{4}\right) = -\sin\dfrac{3\pi}{4} = -\dfrac{\sqrt{2}}{2}$

(b) $\cos\left(-\dfrac{3\pi}{4}\right) = \cos\dfrac{3\pi}{4} = -\dfrac{\sqrt{2}}{2}$

(c) $\tan\left(-\dfrac{3\pi}{4}\right) = -\tan\dfrac{3\pi}{4} = -\left(\dfrac{\sqrt{2}/2}{-\sqrt{2}/2}\right) = 1$

PROBLEM SET 2

In Problems 1–6, use the periodicity of the wrapping function P to find a value for t, where $0 \leq t < 2\pi$, so that the point $P(t)$ coincides with the given point. Locate each given point on the unit circle. (Use $\pi = 3.14$.)

1 $P(0.8 + 6.28)$ **2** $P(0.7 - 6.28)$ **3** $P(10)$
4 $P(-23)$ **5** $P(-7.8)$ **6** $P(29.9)$

In Problems 7–19, use the periodicity of the wrapping function, together with Figure 14 on page 192, to determine the coordinates of each point.

7 $P\left(\frac{\pi}{2} + 2\pi\right)$ **8** $P\left(\frac{\pi}{6} - 4\pi\right)$ **9** $P\left(-\frac{\pi}{4} - 4\pi\right)$
10 $P\left(\frac{\pi}{3} + 6\pi\right)$ **11** $P(5\pi)$ **12** $P\left(\frac{9\pi}{2}\right)$
13 $P\left(-\frac{7\pi}{2}\right)$ **14** $P\left(-\frac{13\pi}{4}\right)$ **15** $P\left(\frac{59\pi}{6}\right)$
16 $P\left(\frac{41\pi}{3}\right)$ **17** $P\left(-\frac{22\pi}{3}\right)$ **18** $P\left(-\frac{71\pi}{3}\right)$
19 $P\left(\frac{23\pi}{4}\right)$

20 Prove that the cotangent and cosecant functions are periodic functions of period 2π.

In Problems 21–42, use periodicity and Figure 14 on page 192 to determine each of the given values.

21 $\sin\left(\frac{\pi}{6} + 2\pi\right)$ **22** $\cos\left(\frac{\pi}{6} - 2\pi\right)$ **23** $\tan\left(-\frac{\pi}{6} + 2\pi\right)$
24 $\cot\left(-\frac{\pi}{6} - 2\pi\right)$ **25** $\cos\left(\frac{\pi}{4} + 4\pi\right)$ **26** $\sec\left(-\frac{\pi}{4} + 4\pi\right)$
27 $\csc\frac{23\pi}{4}$ **28** $\cot\frac{25\pi}{3}$ **29** $\sin\left(-\frac{25\pi}{3}\right)$
30 $\cos\frac{29\pi}{3}$ **31** $\sin\frac{9\pi}{2}$ **32** $\cos\left(-\frac{13\pi}{4}\right)$
33 $\tan\frac{17\pi}{6}$ **34** $\cot\frac{22\pi}{3}$ **35** $\cos\frac{41\pi}{6}$
36 $\sec\frac{71\pi}{3}$ **37** $\cot\left(-\frac{31\pi}{4}\right)$ **38** $\sin(-25\pi)$
39 $\sec\frac{107\pi}{3}$ **40** $\cos\left(-\frac{77\pi}{4}\right)$ **41** $\sin\left(-\frac{91\pi}{6}\right)$
42 $\cot\left(-\frac{129\pi}{6}\right)$

In Problems 43–50, use the fact that the trigonometric functions are periodic functions of period 2π to find a period of the given function.

43 $f(t) = \sin 4t$ **44** $g(t) = \csc 4t$ **45** $h(t) = \cos 3t$
46 $f(t) = \tan 2t$ **47** $F(t) = \sin(2t + 1)$ **48** $G(t) = \sec 5t$
49 $g(t) = \cos\frac{t}{2}$ **50** $F(t) = \cot\left(\frac{7t}{2} - 1\right)$

51 In Table 1, the signs of two of the trigonometric function values of t are given. Indicate the quadrant in which $P(t)$ lies, and the signs of the other trigonometric function values of t.

Table 1

Quadrant										
$\sin t$	+	+				−				
$\cos t$			+						+	−
$\tan t$		+	−	+	+		+			+
$\cot t$									+	
$\sec t$					−	+		+		
$\csc t$	+			−			+	−		

52 Find a and the six trigonometric functions of t if $0 < t < \frac{\pi}{2}$ and

$$P(t) = \left(\frac{3}{a}, \frac{4}{a}\right).$$

In Problems 53–58, find the trigonometric function values of t in each case.

53 $\sin t$ and $\csc t$ if $\cos t = -\sqrt{3}/2$ and $P(t)$ is in quadrant III.
54 $\cos t$ and $\sec t$ if $\sin t = 12/13$ and $P(t)$ is in quadrant II.
55 $\tan t$ and $\cot t$ if $\sin t = -5/13$ and $P(t)$ is in quadrant IV.
56 $\sec t$ and $\cot t$ if $\tan t = 4/3$ and $P(t)$ is in quadrant I.
57 $\csc t$ and $\tan t$ if $\cot t = -12/5$ and $P(t)$ is in quadrant IV.
58 $\cos t$ and $\cot t$ if $\sec t = -2$ and $P(t)$ is in quadrant II.

59 Given $P(t) = (\cos t, \sin t)$, explain why $P(-t) = (\cos t, -\sin t)$.

60 Prove that the cotangent and the cosecant functions are odd functions, and that the secant function is an even function.

61 Use Theorem 3 and Figure 14 on page 192 to find the value of

(a) $\sin\left(-\frac{11\pi}{6}\right)$

(b) $\cot\left(-\frac{11\pi}{6}\right)$

(c) $\sec\left(-\frac{11\pi}{6}\right)$

62 If $\sin t = \frac{4}{5}$ and $\cos t < 0$, find

(a) $\sin(-t)$ (b) $\cos(-t)$

In Problems 63–64, find $f[g(t)]$ and $g[f(t)]$ and indicate which of the composite functions is even, which is odd, and which is neither.

63 $f(t) = \sin t$ and $g(t) = t^2$
64 $f(t) = \cos t$ and $g(t) = t^3 + 5$

3 Graphs of the Trigonometric Functions

The properties and evaluations of the trigonometric functions explored in the preceding sections are used here to sketch their graphs. We first obtain a rough sketch for one period by plotting a few points and drawing a smooth, continuous curve containing them. Then, using periodicity, we complete the sketch by extending the graph over one period in both directions (to the left and right).

In constructing these graphs, we also use the techniques for graphing functions that were covered in Chapter 2. Hereafter we ordinarily use x instead of t to denote domain elements. In this usage, x can be thought of as representing an arc length along the unit circle, just as t did.

3.1 Graphs of Sine and Cosine Functions

We begin by investigating the graphs of $f(x) = \sin x$ and $g(x) = \cos x$. Since $f(x) = \sin x$ is a periodic function of period 2π, it is sufficient to restrict our attention to the values of x for which $0 \leq x < 2\pi$. Then the graph of f defined by $f(x) = \sin x$ for any real number x can be extended as far as we like, since the graph repeats every 2π units (Figure 1a).

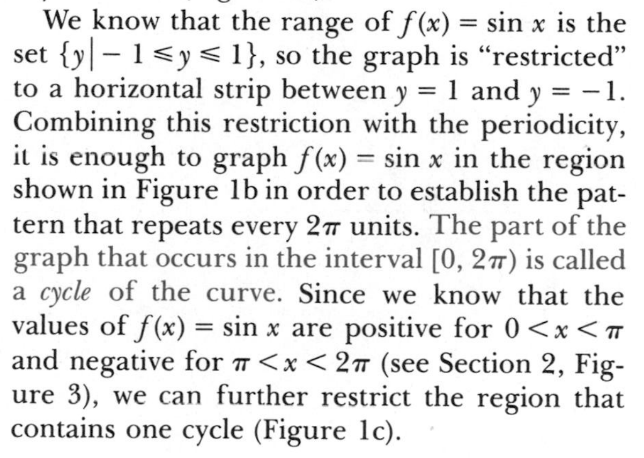

We know that the range of $f(x) = \sin x$ is the set $\{y | -1 \leq y \leq 1\}$, so the graph is "restricted" to a horizontal strip between $y = 1$ and $y = -1$. Combining this restriction with the periodicity, it is enough to graph $f(x) = \sin x$ in the region shown in Figure 1b in order to establish the pattern that repeats every 2π units. The part of the graph that occurs in the interval $[0, 2\pi)$ is called a *cycle* of the curve. Since we know that the values of $f(x) = \sin x$ are positive for $0 < x < \pi$ and negative for $\pi < x < 2\pi$ (see Section 2, Figure 3), we can further restrict the region that contains one cycle (Figure 1c).

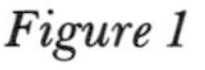

Figure 1

(a)

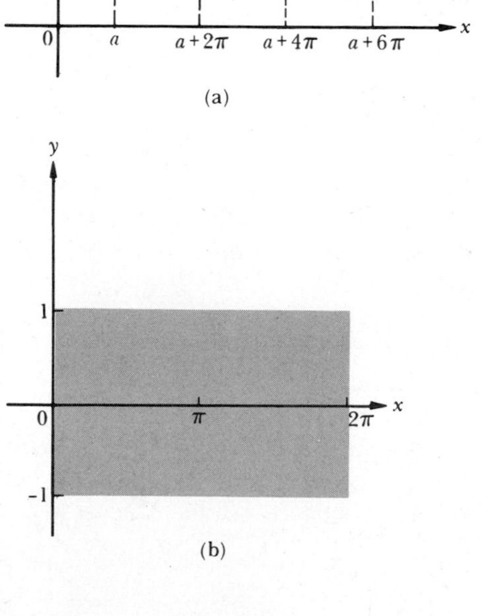

(b)

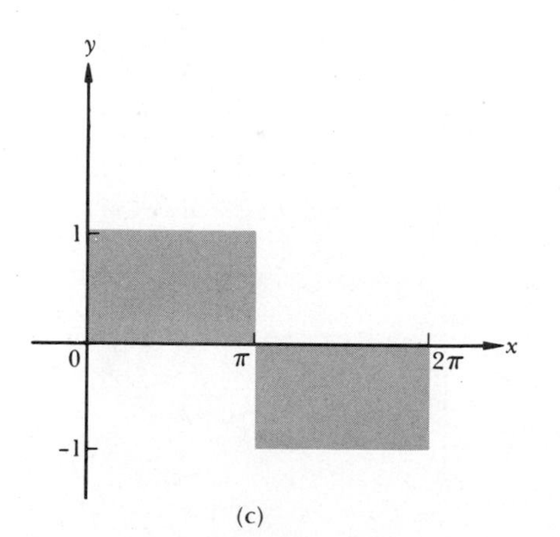

(c)

Figure 2

x	$\sin x$
0	0
$\pi/6$	$1/2$
$\pi/4$	$\sqrt{2}/2$
$\pi/3$	$\sqrt{3}/2$
$\pi/2$	1
π	0
$7\pi/6$	$-1/2$
$3\pi/2$	-1
$11\pi/6$	$-1/2$

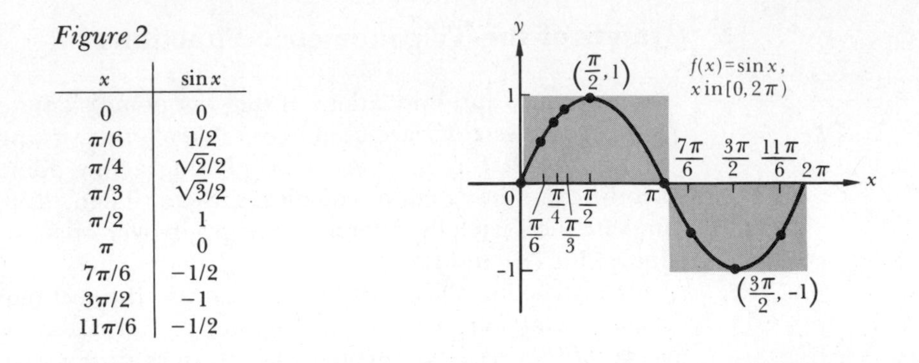

Finally, some known ordered pairs $(x, \sin x)$ can be used, together with the fact that f is increasing in the interval $(0, \pi/2)$, decreasing in the interval $(\pi/2, 3\pi/2)$, and again is increasing in the interval $(3\pi/2, 2\pi)$ (see Problem Set 1, Problem 67), to obtain the graph on the interval $[0, 2\pi)$ contained in the shaded region (Figure 2). This is the graph of one *cycle* of the sine function.

Figure 3

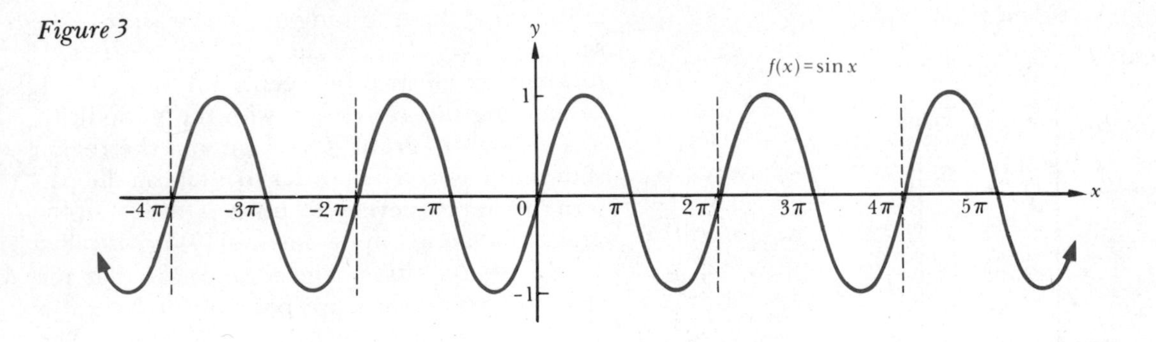

Using the periodicity of the sine function, the complete graph of $f(x) = \sin x$ is obtained by repeating the cycle every 2π units (Figure 3). Notice that the graph displays the fact that the sine is an odd function (its graph is symmetric with respect to the origin) and that it is not invertible. (Why?)

The graph of $g(x) = \cos x$ can be determined in the same way that the graph of the sine function was. Since the cosine is a periodic function of period 2π, it is enough to graph it on the interval $[0, 2\pi)$. As with the sine function, the range of the cosine function is $\{y \mid -1 \leq y \leq 1\}$, so we need only consider the graph in the region of Figure 4a.

Figure 4a

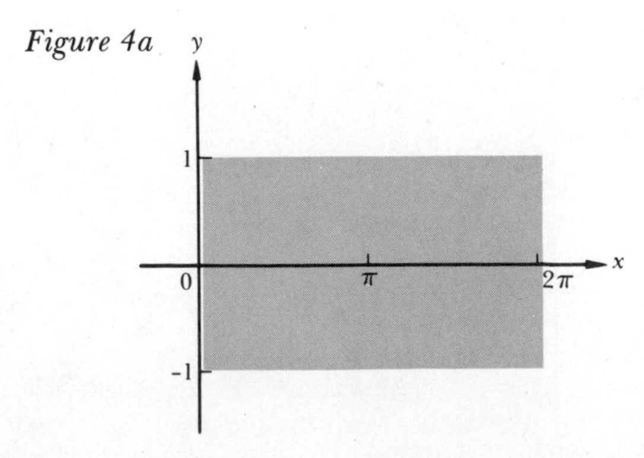

Figure 4b

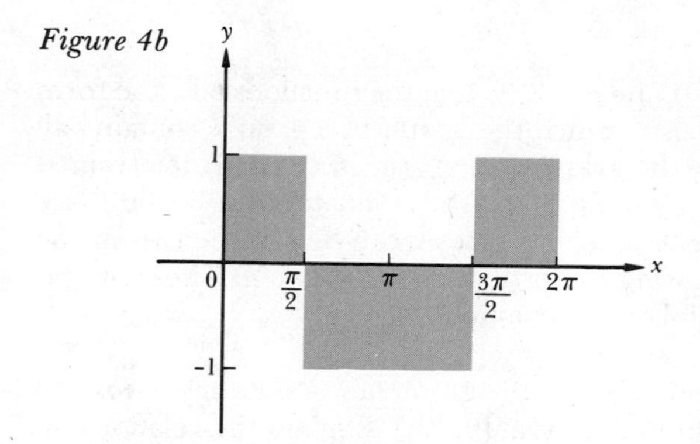

We can further restrict the region that contains the graph by using the fact that $g(x) = \cos x$ is positive for x in the interval $(0, \pi/2)$ and in the interval $(3\pi/2, 2\pi)$, and negative for x in the interval $(\pi/2, 3\pi/2)$ (see Figure 3 on Page 201) (Figure 4b).

Finally, by using some function values of the cosine function, together with the known behavior of the cosine (see Problem Set 1, Problem 67), we get the graph in the interval $[0, 2\pi)$, a cycle of the cosine curve (Figure 5).

Figure 5

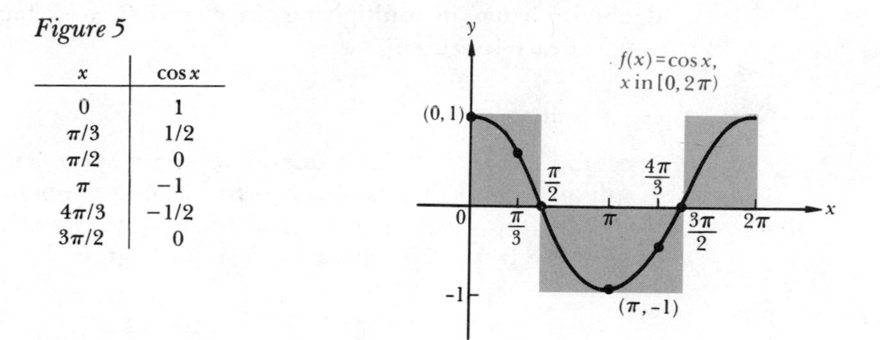

x	$\cos x$
0	1
$\pi/3$	$1/2$
$\pi/2$	0
π	-1
$4\pi/3$	$-1/2$
$3\pi/2$	0

By repeating the cycle every 2π units, we get the complete graph of $g(x) = \cos x$ (Figure 6). Notice that the graph displays both the fact that $g(x) = \cos x$ is not invertible and that it is an even function (the graph is symmetric with respect to the y axis).

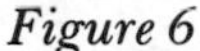

Figure 6

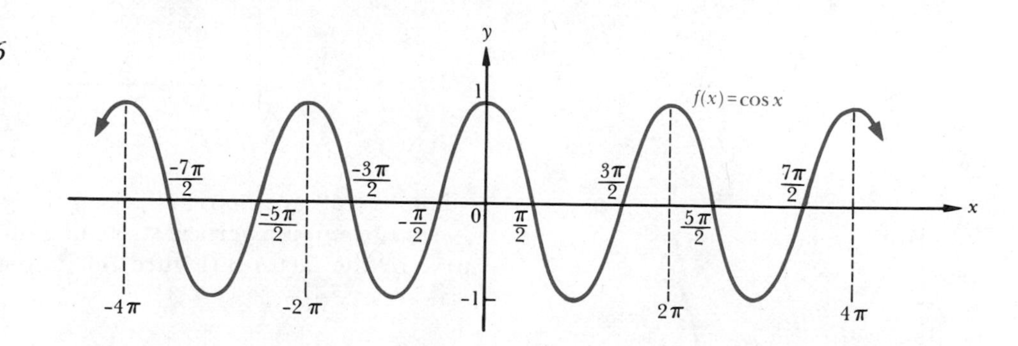

The graphs of functions of the form $F(x) = a \sin(kx + b)$ or of the form $G(x) = a \cos(kx + b)$ are obtained from the known graphs of $f(x) = \sin x$ and $g(x) = \cos x$ by using the graphing techniques studies in Chapter 2. To determine how to do this, we examine the effects of the constants a, k, and b on the graphs of f and g.

Table 1

x	$\sin x$	$a \sin x$
.	.	.
.	.	.
.	.	.
x_0	y_0	ay_0
.	.	.
.	.	.
.	.	.

AMPLITUDE

First, let us suppose that $k = 1$ and $b = 0$, so that the functions take the form $F(x) = a \sin x$ or $G(x) = a \cos x$. Since the graph of $y = \sin x$ contains all points of the form $(x, \sin x)$, the graph of $F(x) = a \sin x$ can be determined from the graph of $y = \sin x$ by multiplying the ordinates by a factor a (see Table 1). Similarly, the graph of $G(x) = a \cos x$ can be obtained from the graph of $y = \cos x$ by multiplying the ordinates by a factor a. These operations are illustrated in the following examples.

EXAMPLES In each of the following examples, use the graph of $y = \sin x$ or $y = \cos x$ to graph the given function on the interval $[0, 2\pi)$. Explain the relationship between the geometric change of the graph of $y = \sin x$ or $y = \cos x$ and the algebraic change of multiplying $\sin x$ or $\cos x$ by a factor. Also, find the range of each function.

1 $f(x) = 3 \sin x$

SOLUTION First, we graph one cycle of $y = \sin x$ (Figure 7a). Next, the graph of $f(x) = 3 \sin x$ is obtained by locating the points $(x, 3 \sin x)$. Geometrically, this means that each ordinate of $(x, \sin x)$ is "stretched" vertically by the factor 3 to obtain $(x, 3 \sin x)$ (Figure 7b).

Figure 7

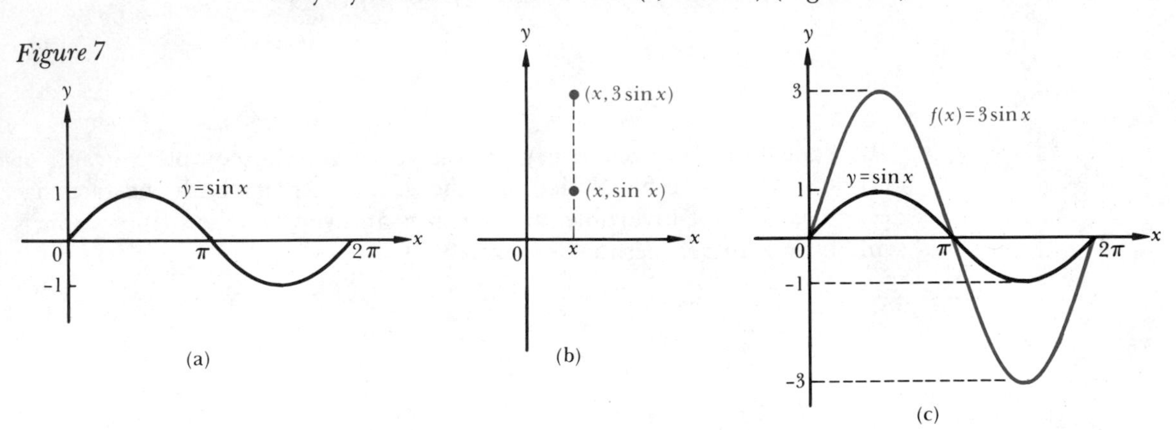

Figure 8

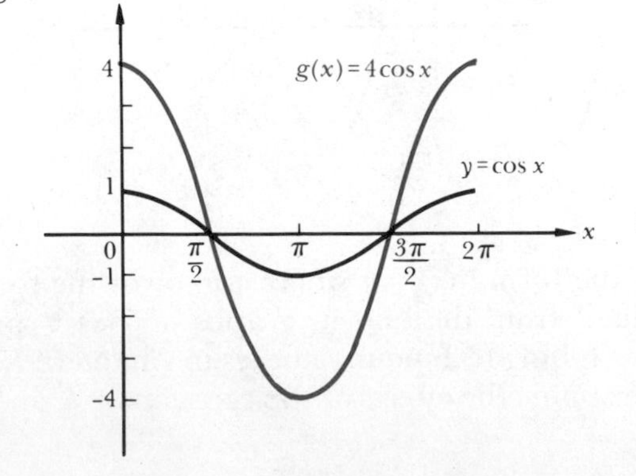

The overall geometric effect on the graph of $y = \sin x$, then, is a vertical stretching of the sine curve by the factor 3 (Figure 7c). The range of f is the set $\{y \mid -3 \leq y \leq 3\}$.

2 $g(x) = 4 \cos x$

SOLUTION The graph of $g(x) = 4 \cos x$ is similar to the graph of $y = \cos x$. Here, the graph of $y = \cos x$ is stretched vertically by a factor 4 to obtain the graph of g (Figure 8). The range of g is the set $\{y \mid -4 \leq y \leq 4\}$.

3 $h(x) = (-1/5)\sin x$

SOLUTION In this case each ordinate of $y = \sin x$ is multiplied by $-1/5$ in order to obtain the graph of $h(x) = (-1/5)\sin x$. This multiplication has the geometric effect of reflecting the graph of $y = \sin x$ across the x axis (because of the negative value) and also of vertically shrinking the sine curve by the factor 1/5 (Figure 9). The range of h is the set $\{y \mid -1/5 \leq y \leq 1/5\}$.

Figure 9

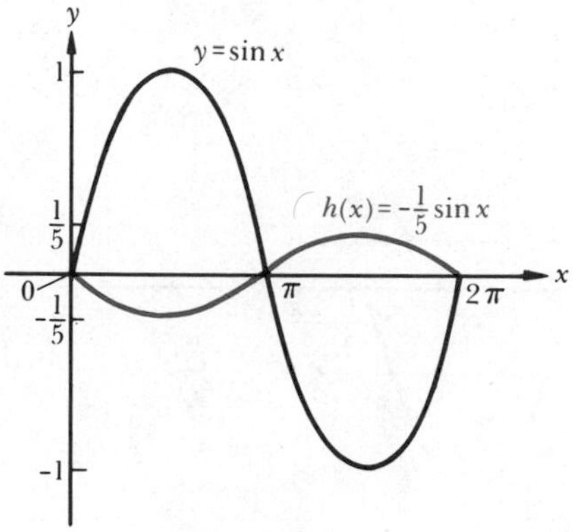

In each of the examples above, notice that the factor can be used to determine the range of the function. This factor is used to compute the *amplitude* of the function. The *amplitude* of the sine and cosine functions is defined to be $\frac{1}{2}(M - m)$, where M is the largest value of the function and m is the smallest value of the function. Hence, the amplitude of $f(x) = 3 \sin x$ is $\frac{1}{2}[3 - (-3)] = 3$; the amplitude of $g(x) = 4 \cos x$ is $\frac{1}{2}[4 - (-4)] = 4$; and the amplitude of $h(x) = -\frac{1}{5}\sin x$ is $\frac{1}{2}[\frac{1}{5} - (-\frac{1}{5})] = \frac{1}{5}$.

In general, if $F(x) = a \sin x$ or $G(x) = a \cos x$, the amplitude is $|a|$ and the range is the set $\{y \mid -|a| \leq y \leq |a|\}$.

PHASE SHIFT AND PERIODICITY

To graph functions of the form $F(x) = \sin(kx + b)$ and $G(x) = \cos(kx + b)$, we use the known graph of $y = \sin x$ and $y = \cos x$, respectively, together with periodicity. Since both the sine and the cosine are periodic functions of period 2π, it follows that the function F and the function G generate one cycle of the sine graph and the cosine graph, respectively, when $kx + b$ varies from 0 to 2π, that is, when $0 \leq kx + b < 2\pi$.

Suppose, for example, that $f(x) = \sin(3x - 1)$. Then, f generates one cycle of the sine graph when

$3x - 1$ varies from 0 to 2π $\quad (0 \leq 3x - 1 < 2\pi)$

that is, when

$3x$ varies from 1 to $2\pi + 1$ $\quad (1 \leq 3x < 2\pi + 1)$

or

x varies from $\frac{1}{3}$ to $\frac{2\pi}{3} + \frac{1}{3}$ $\quad \left(\frac{1}{3} \leq x < \frac{2\pi}{3} + \frac{1}{3}\right)$

Hence, if x varies from $1/3$ to $(1/3) + (2\pi/3)$, then $3x - 1$ varies from 0 to 2π, so that, in turn, $f(x) = \sin(3x - 1)$ generates one cycle of the sine curve (Figure 10a). The curve begins at $1/3$ (the *phase shift*) and covers an interval of length $2\pi/3$, the period of f. Using the periodicity, we obtain the complete graph of $f(x) = \sin(3x - 1)$ (Figure 10b).

Figure 10

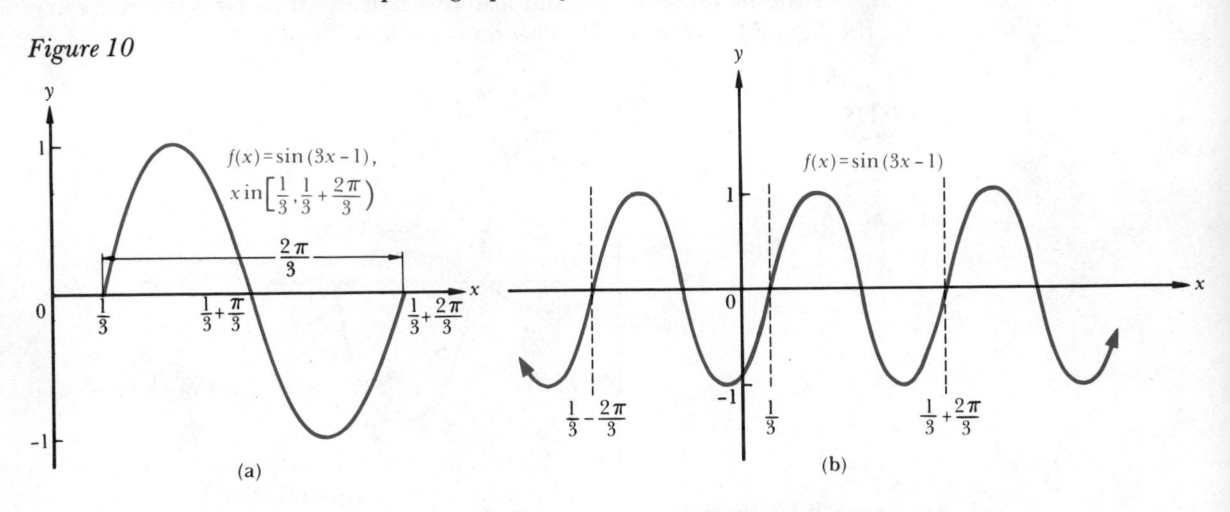

EXAMPLES Use the graph of $y = \sin x$ or $y = \cos x$ to graph *one cycle* of each of the following functions. Indicate the amplitude, period, and phase shift.

1 $f(x) = \sin \frac{1}{2}x$

SOLUTION The amplitude of f is 1. The function $f(x) = \sin \frac{1}{2}x$ generates one cycle if

$$\tfrac{1}{2}x \text{ varies from } 0 \text{ to } 2\pi \qquad (0 \leq \tfrac{1}{2}x < 2\pi)$$

that is, if

$$x \text{ varies from } 0 \text{ to } 4\pi \qquad (0 \leq x < 4\pi)$$

Consequently, the period is 4π and the phase shift is 0, since the graph "begins" at 0 and covers an interval of length 4π (Figure 11).

Figure 11

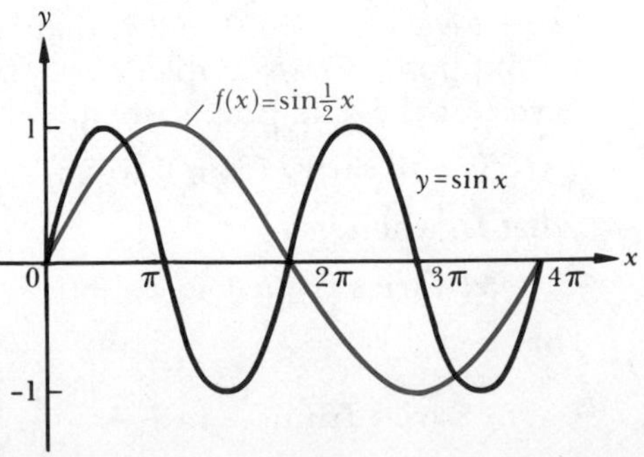

2 $g(x) = \cos 3x$

SOLUTION The amplitude of g is 1. The function $g(x) = \cos 3x$ generates one cycle if

$$3x \text{ varies from 0 to } 2\pi \qquad (0 \leq 3x < 2\pi)$$

that is, if

$$x \text{ varies from 0 to } \frac{2\pi}{3} \qquad \left(0 \leq x < \frac{2\pi}{3}\right)$$

Consequently, the period is $2\pi/3$ and the phase shift is 0 (Figure 12).

Figure 12

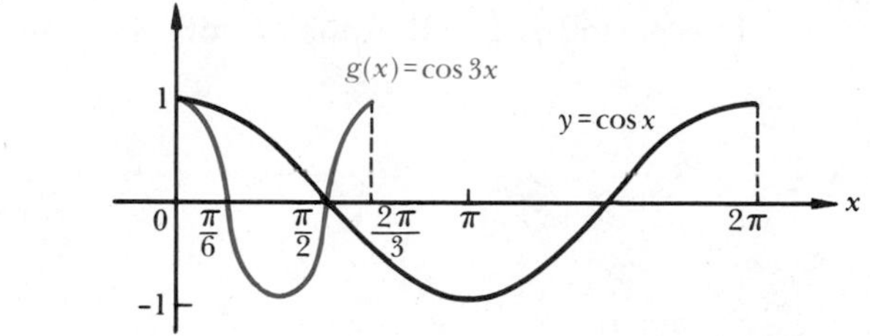

3 $h(x) = \frac{3}{2}\sin(2x - \pi)$

SOLUTION Here, the amplitude is $\frac{3}{2}$; the function h generates a cycle of the sine curve if

$$2x - \pi \text{ varies from 0 to } 2\pi \qquad (0 \leq 2x - \pi < 2\pi)$$

that is, if

$$2x \text{ varies from } \pi \text{ to } 3\pi \qquad (\pi \leq 2x < 3\pi)$$

or

$$x \text{ varies from } \frac{\pi}{2} \text{ to } \frac{3\pi}{2} \qquad \left(\frac{\pi}{2} \leq x < \frac{3\pi}{2}\right)$$

Hence, the period is π and the phase shift is $\frac{\pi}{2}$ (Figure 13).

Figure 13

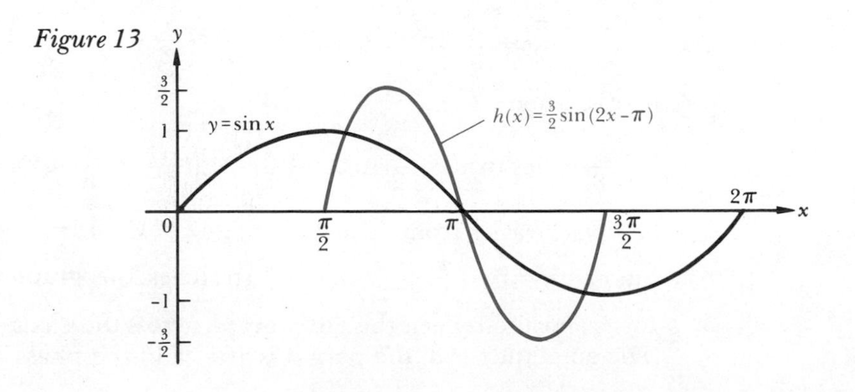

4 $F(x) = \sin(-x + 1)$

SOLUTION Since the sine function is an odd function,

$$F(x) = \sin(-x + 1) = \sin[-(x - 1)] = -\sin(x - 1)$$

First, we graph $y = \sin(x - 1)$, which generates a cycle of the sine curve if

$$x - 1 \text{ varies from } 0 \text{ to } 2\pi \qquad (0 \leq x - 1 < 2\pi)$$

that is, if

$$x \text{ varies from } 1 \text{ to } 2\pi + 1 \qquad (1 \leq x < 2\pi + 1) \quad \text{(Figure 14a)}$$

Then, we can obtain the graph of $F(x) = -\sin(x - 1)$ from the graph of $y = \sin(x - 1)$ by reflecting the latter graph across the x axis (Figure 14b). The period is 2π, the phase shift is 1, and the amplitude is 1.

Figure 14

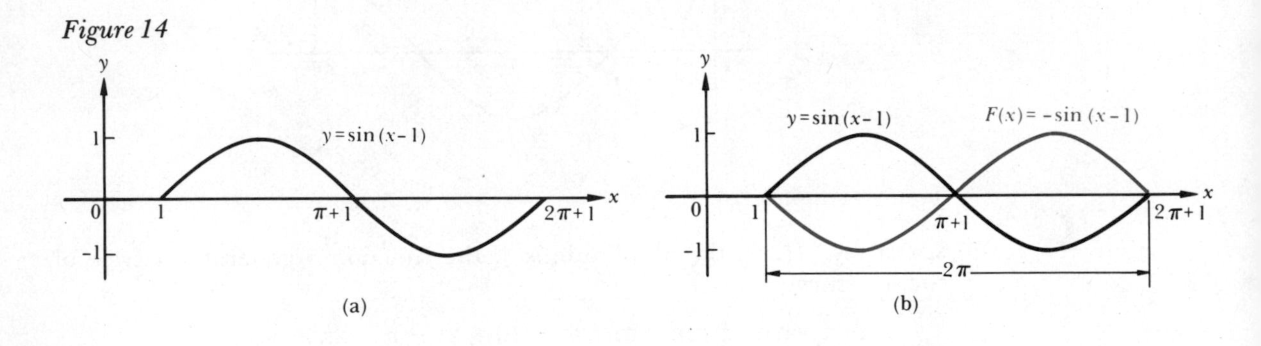

5 $G(x) = -3\cos\left(\frac{x}{8} + \frac{\pi}{2}\right)$

SOLUTION First, we investigate the graph of

$$y = \cos\left(\frac{x}{8} + \frac{\pi}{2}\right)$$

This function generates a cycle of the cosine curve when

$$\frac{x}{8} + \frac{\pi}{2} \text{ varies from } 0 \text{ to } 2\pi \qquad \left(0 \leq \frac{x}{8} + \frac{\pi}{2} < 2\pi\right)$$

that is, when

$$\frac{x}{8} \text{ varies from } -\frac{\pi}{2} \text{ to } 2\pi - \frac{\pi}{2} \qquad \left(-\frac{\pi}{2} \leq \frac{x}{8} < 2\pi - \frac{\pi}{2}\right)$$

or when x varies from -4π to 12π $(-4\pi \leq x < 12\pi)$ (Figure 15a). Multiplication of $y = \cos\left(\frac{\pi}{8} + \frac{\pi}{2}\right)$ by -3 stretches the graph vertically by the factor 3, and also reflects this latter graph across the x axis (why?) (Figure 15b). The amplitude is 3, the period is 16π, and the phase shift is -4π.

Figure 15

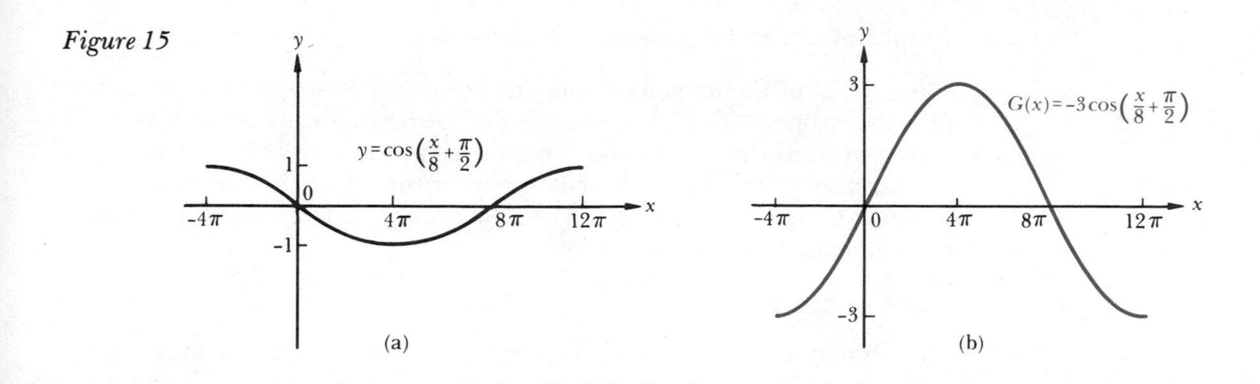

Functions defined by equations of the form

$$y = a\ \cos(kt + b) \quad \text{or} \quad y = a\ \sin(kt + b)$$

where t represents time and a, k, and b are constants, describe periodic motion. Problems involving periodic motion often occur in physics, biology, and other sciences. These equations describe a *pure wave* or, as it is sometimes called, a *simple harmonic motion.* If t is measured in seconds, then the fundamental period is $2\pi/k$ seconds and, in an interval of one second, the graph of the function will contain $k/2\pi$ cycles. The number $k/2\pi$ is called the *frequency* of the motion.

EXAMPLE Consider a condenser with a capacitance of C farads and containing a charge of Q_0 coulombs. When connected in series with a coil of negligible resistance and having an inductance of L henrys, the charge Q on the condenser after t seconds is given by the function

$$Q(t) = Q_0 \sin\left(\frac{t}{\sqrt{LC}} - \frac{\pi}{2}\right)$$

where L and C are constants. If $L = 0.4$ henry and $C = 10^{-5}$ farad, find the period and frequency of this circuit.

SOLUTION Here, $L = 0.4$ henry and $C = 10^{-5}$ farad, so

$$\frac{1}{\sqrt{LC}} = \frac{1}{\sqrt{(0.4)(10^{-5})}} = \frac{1}{\sqrt{4 \times 10^{-6}}} = \frac{1000}{2} = 500$$

The above equation describes a simple harmonic motion, whose period is

$$\frac{2\pi}{\frac{1}{\sqrt{LC}}} = 2\pi\sqrt{LC} = \frac{\pi}{250} = 0.004\pi = 0.01256 \quad \text{(approximately)}$$

The frequency of the circuit is

$$\frac{\frac{1}{\sqrt{LC}}}{2\pi} = \frac{500}{2\pi} = \frac{250}{\pi} = 79.62 \text{ cycles per second} \quad \text{(approximately)}$$

3.2 Graphs of Other Trigonometric Functions

The graphs of the tangent, cotangent, secant, and cosecant functions display the properties of being periodic; of being even or odd, that is, symmetric with respect to either the y axis or the origin; and of being increasing or decreasing. We discussed the domains of these functions in Section 2.3. Here we sketch their graphs and use the graphs to determine the range of each function.

THE TANGENT

The domain of $f(x) = \tan x$ is $\{x \mid x \neq (\pi/2) + n\pi$, where n is an integer$\}$, so the graph of the tangent function *cannot* intersect any line of the form $x = (\pi/2) + n\pi$, where n is an integer (Figure 16). In fact, lines of the form $x = (\pi/2) + n\pi$ are vertical asymptotes of the tangent function.

Figure 16

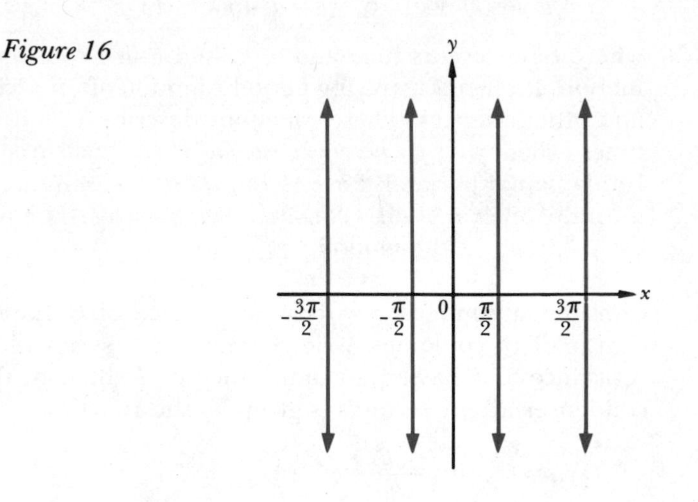

Since $\tan x = \sin x/\cos x$, $f(x) = \tan x$ is increasing in the interval $[0, \pi/2)$ because the sine function is increasing and the cosine function is decreasing in this interval. This behavior of f can be observed by examining the values in Table 2.

Table 2

x	$\sin x$	$\cos x$	$\tan x = \sin x/\cos x$
0	0	1.000	0
$\pi/6$	0.500	0.866	0.577
$\pi/4$	0.707	0.707	1.000
$\pi/3$	0.866	0.500	1.732
$\pi/2$	1.000	0	undefined

As x increases from 0 toward $\pi/2$, $\sin x$ increases from 0 toward 1 and $\cos x$ decreases from 1 toward 0. Hence, the quotient $\sin x/\cos x$ becomes increasingly large as x gets "closer" to $\pi/2$. That is, the values of $\tan x$ increase indefinitely as x increases from 0 toward $\pi/2$, and the line $x = \pi/2$ is a vertical asymptote of the graph of $f(x) = \tan x$ (Figure 17a).

Since $f(x) = \tan x$ is an odd function, we can use the fact that the graph of f is symmetric with respect to the origin to obtain the graph of f in the interval $(-\pi/2, 0]$ from the graph of f in the interval $[0, \pi/2)$ (Figure 17b).

Figure 17

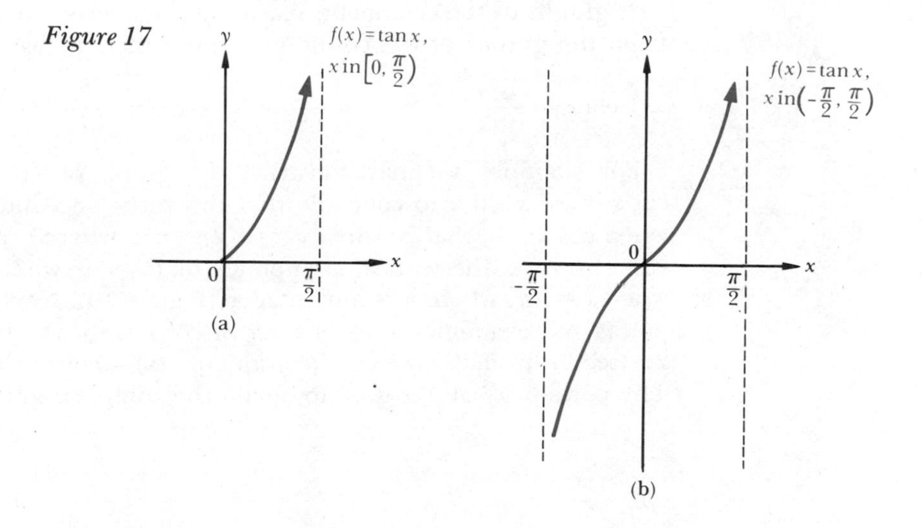

It will be established in Chapter 6, on page 261, that the *fundamental period* of $f(x) = \tan x$ is π. Consequently, the graph in Figure 17b can be repeated every π units in either direction to obtain the complete graph of $f(x) = \tan x$ (Figure 18).

Figure 18

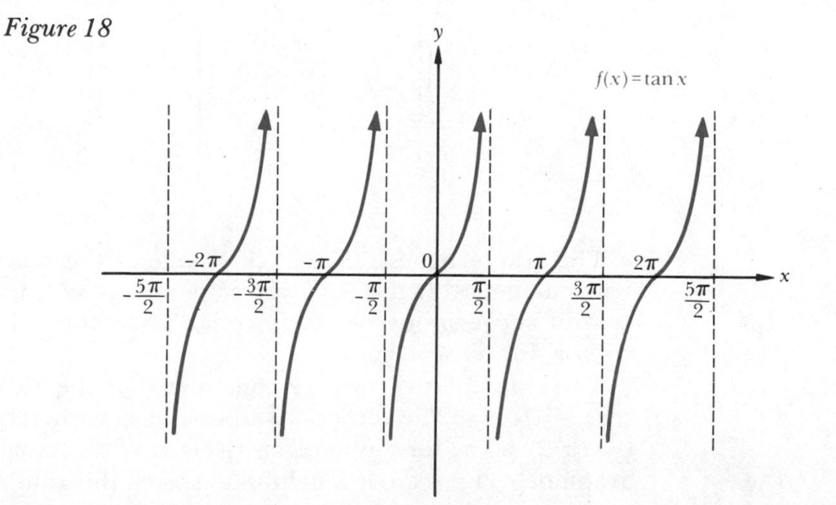

The graph of $f(x) = \tan x$ suggests that (1) the range of the tangent is the set of all real numbers; (2) the zeros of $f(x) = \tan x$ are given by $x = n\pi$, where n is any integer; and (3) the function increases between any two consecutive asymptotes, which occur whenever $x = \dfrac{\pi}{2} + n\pi$, for any integer n.

THE COTANGENT, SECANT, AND COSECANT

The graphs of the cotangent, secant, and cosecant functions are obtained from the graphs of the cosine and sine functions by using the relations

$$\cot x = \frac{\cos x}{\sin x} \qquad \sec x = \frac{1}{\cos x} \qquad \csc x = \frac{1}{\sin x}$$

For example, we may construct the graph of $f(x) = \cot x$ by using $\cot x = \cos x/\sin x$ to conclude that the zeros or x intercepts of f occur when $\cos x = 0$, that is, when $x = (\pi/2) + n\pi$, where n is any integer (Figure 19). Also, the vertical asymptotes of f occur when $\sin x = 0$, that is, when $x = n\pi$, where n is any integer (Figure 19). Next, by locating some points and examining the behavior of $\cos x/\sin x$ as x increases from 0 to 2π (see Problem 25), we get the graph of $f(x) = \cot x$ in the interval $(0, 2\pi)$. The periodicity of f is used to obtain the complete graph (Figure 19).

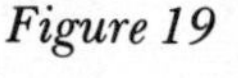
Figure 19

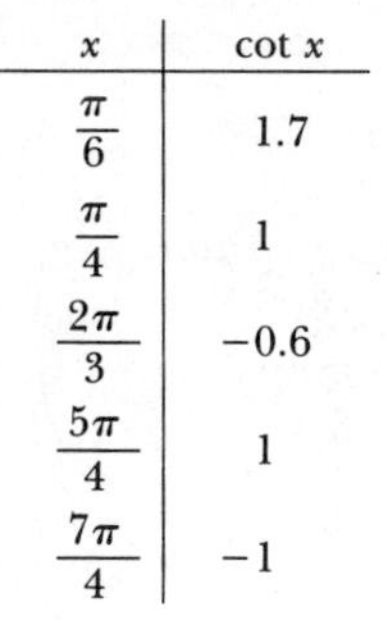

x	$\cot x$
$\frac{\pi}{6}$	1.7
$\frac{\pi}{4}$	1
$\frac{2\pi}{3}$	−0.6
$\frac{5\pi}{4}$	1
$\frac{7\pi}{4}$	−1

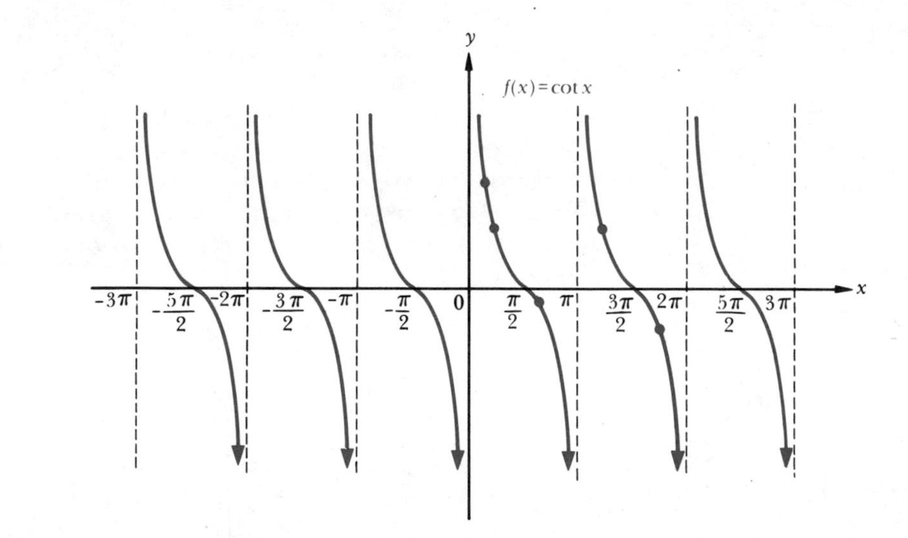

The graph suggests that (1) the range of the function f is the set of all real numbers; (2) the *fundamental* period is π; and (3) the function decreases between any two consecutive asymptotes, which occur whenever $x = n\pi$, for any integer n.

The graph of $g(x) = \sec x$ is constructed by the use of the known relation $\sec x = 1/\cos x$. The vertical asymptotes occur when $\cos x = 0$, that is, when $x = (\pi/2) + n\pi$, for any integer n (Figure 20). As an aid to sketching the graph of $g(x) = \sec x$, it is helpful to sketch the graph of $y = \cos x$ and then take the reciprocals of the ordinates to obtain points on the secant graph (Figure 20). The graph of g displays the fact that the range of the secant function is $\{y \mid y \leq -1 \text{ or } y \geq 1\}$. Notice the manner in which the graph of g increases or decreases by observing the behavior of the graph of $y = \cos x$ as it increases or decreases.

Figure 20

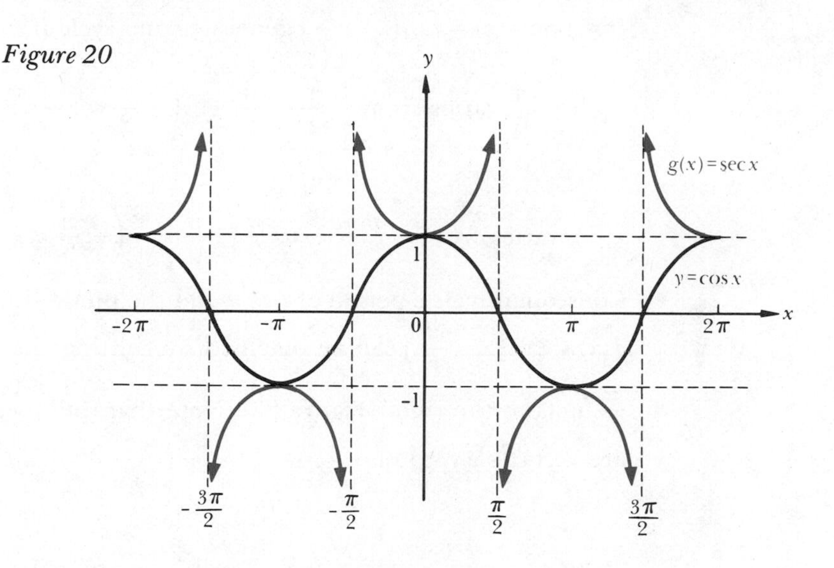

The graph of the cosecant function $h(x) = \csc x$ may be obtained in a similar fashion by using $\csc x = 1/\sin x$. The vertical asymptotes occur when $\sin x = 0$, or when $x = n\pi$, for any integer n (Figure 21). The graph reveals that the range of the cosecant function is $\{y \mid y \leq -1 \text{ or } y \geq 1\}$.

Figure 21

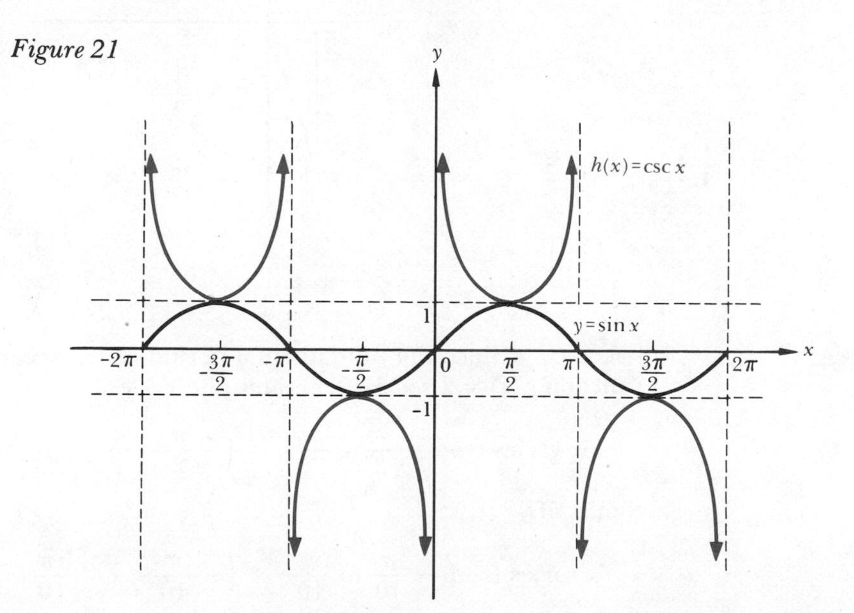

EXAMPLES Sketch the graph of each function over one period.

1 $f(x) = \tan\left(x - \frac{\pi}{6}\right)$

SOLUTION Since the fundamental period of the tangent function is π, the function $f(x) = \tan\left(x - \frac{\pi}{6}\right)$ generates one cycle if

$$x - \frac{\pi}{6} \text{ varies from } -\frac{\pi}{2} \text{ to } \frac{\pi}{2} \qquad \left(-\frac{\pi}{2} < x - \frac{\pi}{6} < \frac{\pi}{2}\right)$$

that is, if

$$x \text{ varies from } -\frac{\pi}{2} + \frac{\pi}{6} \text{ to } \frac{\pi}{2} + \frac{\pi}{6} \qquad \left(-\frac{\pi}{3} < x < \frac{2\pi}{3}\right)$$

Consequently, the period of f is π and the phase shift is $\pi/6$. The graph of $f(x) = \tan\left(x - \frac{\pi}{6}\right)$ can be obtained by shifting the graph of $y = \tan x$ by $\frac{\pi}{6}$ units to the right (Figure 22). Note that the lines $x = -\frac{\pi}{3}$ and $x = \frac{2\pi}{3}$ are vertical asymptotes.

Figure 22

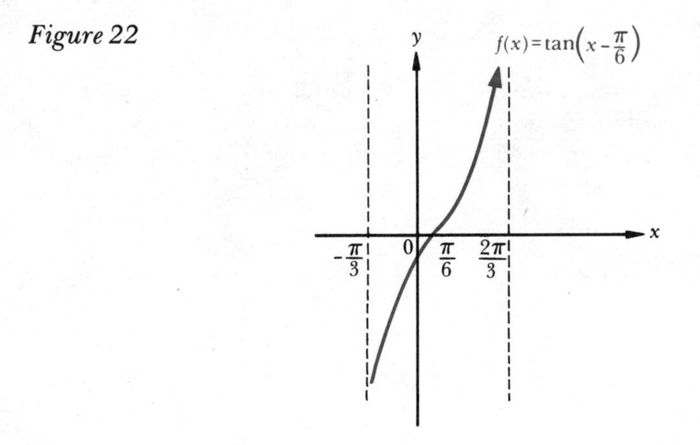

2 $g(x) = 2 \sec 5x$

SOLUTION Since the fundamental period of the secant function is 2π, the function $g(x) = 2 \sec 5x$ generates one cycle if

$$5x \text{ varies from } -\frac{\pi}{2} \text{ to } \frac{3\pi}{2} \qquad \left(-\frac{\pi}{2} < 5x < \frac{3\pi}{2}\right)$$

that is, if

$$x \text{ varies from } -\frac{\pi}{10} \text{ to } \frac{3\pi}{10} \qquad \left(-\frac{\pi}{10} < x < \frac{3\pi}{10}\right)$$

So the period of g is

$$\frac{3\pi}{10} - \left(-\frac{\pi}{10}\right) = \frac{2\pi}{5}$$

Since sec $5x$ is undefined for

$$5x = -\frac{\pi}{2} \quad \text{or} \quad 5x = \frac{\pi}{2} \quad \text{or} \quad 5x = \frac{3\pi}{2}$$

then the lines $x = -\frac{\pi}{10}$, $x = \frac{\pi}{10}$, and $x = \frac{3\pi}{10}$ are vertical asymptotes. The graph of $y = \sec 5x$ is shown in Figure 23. One cycle of the graph of $g(x) = 2 \sec 5x$ is obtained from the graph of $y = \sec 5x$ by multiplying the ordinate of each point by 2. The graph suggests that the range of g is $\{y \mid y \geq 2 \text{ or } y \leq -2\}$.

Figure 23

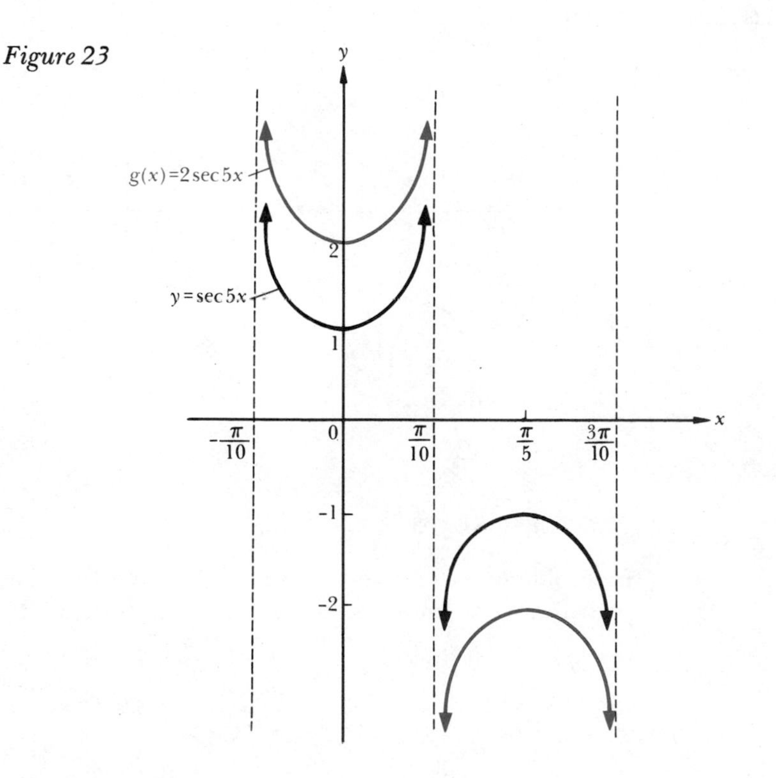

3 $h(x) = \csc\left(\frac{x}{4} - 5\right)$

SOLUTION Since $0 \leq \frac{x}{4} - 5 < 2\pi$ if $20 \leq x < 8\pi + 20$, the period of h is 8π. The graph of $y = \csc \frac{x}{4}$ in the interval $[0, 8\pi)$ is similar to the graph of

$f(x) = \csc x$ in the interval $[0, 2\pi)$ that appears in Figure 21; the vertical asymptotes of

$$y = \csc \frac{x}{4} = \frac{1}{\sin \frac{x}{4}}$$

occur when $\sin \frac{x}{4} = 0$, that is, when $x = 0, x = 4\pi$, and $x = 8\pi$ (Figure 24a). The graph of h is obtained by horizontally shifting the graph of $y = \csc \frac{x}{4}$ 20 units to the right (Figure 24b). The range of h is $\{y \mid y \geq 1 \text{ or } y \leq -1\}$.

Figure 24

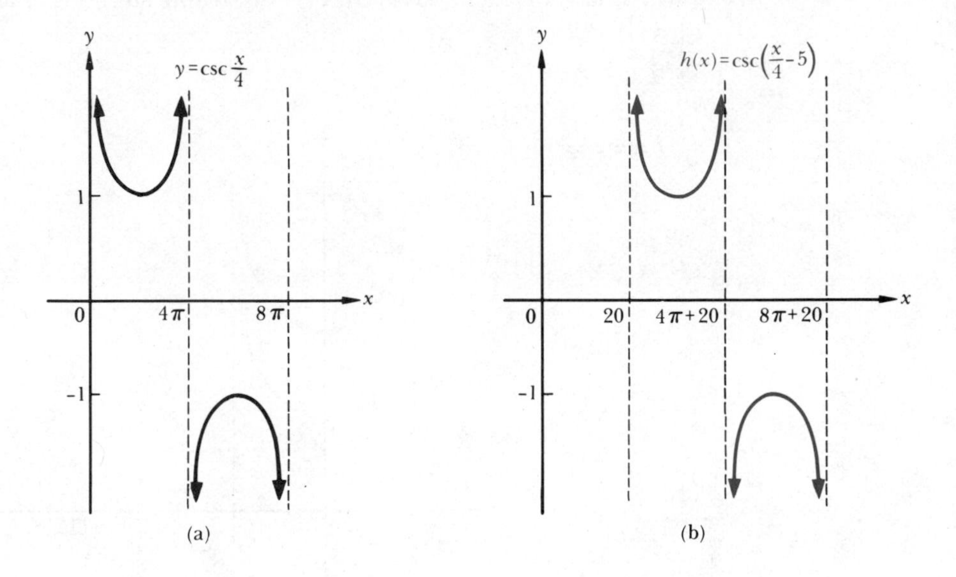

PROBLEM SET 3

In Problems 1–22, use the sine and cosine functions to graph a cycle of each function. Indicate the period, amplitude, and phase shift.

1 $f(x) = 10 \sin x$

2 $F(x) = \frac{2}{3} \sin x$

3 $g(x) = 2 \cos x$

4 $G(x) = 3 \cos(-x)$

5 $h(x) = \sin \frac{\pi x}{3}$

6 $H(x) = 2 \sin 2\pi x$

7 $F(x) = 5 \sin 10x$

8 $G(x) = 2 \cos \frac{\pi x}{2}$

9 $G(x) = 3 \cos 3\pi x$

10 $F(x) = \frac{1}{2} \sin \frac{\pi x}{4}$

11 $G(x) = \frac{5}{3} \sin(-6x)$

12 $H(x) = -3 \cos 5x$

13 $f(x) = -\pi \cos(-5x)$

14 $f(x) = -5 \sin \frac{2\pi x}{3}$

15 $H(x) = -\frac{1}{2} \sin \frac{3x}{2}$

16 $F(x) = |\cos x|$

17 $f(x) = \frac{1}{3} \sin\left(x - \frac{\pi}{6}\right)$

18 $G(x) = -5 \cos\left(x + \frac{\pi}{2}\right)$

19 $H(x) = -3 \cos\left(x + \frac{\pi}{4}\right)$

20 $f(x) = -4 \sin\left(\frac{x}{3} + \frac{\pi}{5}\right)$

21 $h(x) = 2 \sin(5 - 3x)$

22 $F(x) = -2 \cos(3 - 2x)$

23 How does the graph of $f(x) = \cos x$ display the fact that the cosine is an even function?

24 Assume that $f(x) = a \sin(kx + b)$, where a, b, and k are constants, and a and k are not equal to zero.
(a) Show that the amplitude is $|a|$.
(b) Show that the period is $\frac{2\pi}{|k|}$.
(c) Show that the phase shift is $-\frac{b}{k}$.

25 Discuss the behavior of the graph of $f(x) = \cot x = \cos x/\sin x$ in the interval $[0, 2\pi)$ by examining the graphs of the functions $g(x) = \cos x$ and $h(x) = \sin x$ in the interval $[0, 2\pi)$.
(a) As x increases from 0 to $\pi/2$
(b) As x increases from $\pi/2$ to π
(c) As x increases from π to $3\pi/2$
(d) As x increases from $3\pi/2$ to 2π

26 Use the graphs of the trigonometric functions to complete Table 3.

Table 3

	x increasing from			
	0 *to* $\frac{\pi}{2}$	$\frac{\pi}{2}$ *to* π	π *to* $\frac{3\pi}{2}$	$\frac{3\pi}{2}$ *to* 2π
$\tan x$	increasing		increasing	
$\cot x$		decreasing		
$\sec x$				decreasing
$\csc x$				

In Problems 27–32, sketch one period of the graph of each trigonometric function.

27 $f(x) = \tan\left(x - \frac{\pi}{3}\right)$

28 $g(x) = 2 \tan\left(\frac{x}{3} + 5\right)$

29 $F(x) = 8 \sec 2x$

30 $G(x) = \cot(2x - 1)$

31 $h(x) = 2 \csc(\pi x - \frac{1}{2})$

32 $F(x) = 3 \sec(4 - x)$

In Problems 33–36, make use of the facts developed in Section 3 to sketch, on the same coordinate system, the graphs of the four given functions over one period.

33 $y = \sin x$, $y = \sin 2x$, $y = \sin\left(2x - \frac{\pi}{2}\right)$, $y = 2 \sin\left(2x - \frac{\pi}{2}\right)$

34 $y = \cos x$, $y = \cos \frac{x}{2}$, $y = 3 \cos \frac{x}{2}$, $y = 1 + 3 \cos \frac{x}{2}$

35 $y = \cos x$, $y = \cos \frac{2\pi x}{3}$, $y = \frac{1}{2} \cos \frac{2\pi x}{3}$, $y = -\frac{1}{2} \cos \frac{2\pi x}{3}$

36 $y = \sin x$, $y = \sin 3\pi x$, $y = -2 \sin 3\pi x$, $y = 3 - 2 \sin 3\pi x$

37 In the condenser problem in the example on page 215, if $L = 0.9$ henry and $C = 10^{-5}$ farad, find
(a) the frequency of the circuit
(b) the time t_1 when $Q = 0$ for the first time
(c) the time t_2 when $Q = 0.5Q_0$ for the second time

38 A pendulum swings with a simple harmonic motion according to the equation

$$y = A \sin \sqrt{\frac{g}{L}}\, t$$

where A is the amplitude in feet, L is the length of the pendulum in feet, t is the time in seconds, and g is the acceleration of gravity (g is approximately equal to 32 feet per second per second). If the pendulum is 8 feet long and the amplitude is 1 foot, find the period and the frequency of the harmonic motion.

39 A simple harmonic motion is described by the equation

$$d = \sin \frac{5\pi t}{6}$$

Determine the period and the frequency of the harmonic motion, where t is in seconds.

40 The pressure of a traveling sound wave is given by the function

$$P(t) = 1.5 \sin \pi(x_0 - 330t)$$

where x_0 is a constant (in meters), t is elapsed time in seconds, and P is in newtons per square meter. Find the amplitude of the pressure and the frequency of the sound wave.

4 Trigonometric Functions of Angles

It is often necessary to consider trigonometric functions whose domains consist of angles rather than real numbers. This viewpoint is required in certain applications of mathematics to areas such as surveying and navigation. In this section, we define the trigonometric functions of angles, and then relate these functions to trigonometric (circular) functions defined on real numbers. We shall discover that the two types of trigonometric functions are essentially the same. We begin with a review of a few fundamental notions about angles.

4.1 Angle Measures

From plane geometry, an *angle* is determined by rotating a *ray* about its end point (the *vertex* of the angle) from some initial position (the *initial side* of the angle) to a terminal position (the *terminal side* of the angle) (Figure 1). If the angle is formed by a counterclockwise rotation, the angle is said to be an angle of positive sense, whereas if the angle is formed by a clockwise rotation, the angle is of negative sense. In Figure 1 the angle determined by Q, P, and R, $\angle QPR$, is a positive angle, whereas $\angle CAB$ is a negative angle. In this angle notation, the middle letter represents the vertex.

Figure 1

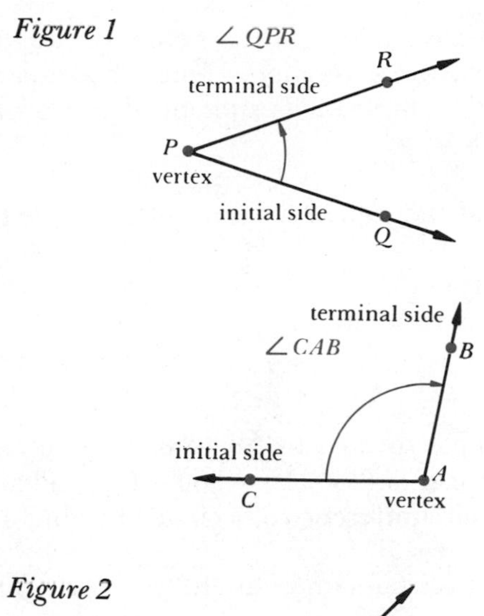

For convenience, Greek letters are often used to label angles. In Figure 2, α is the positive angle, $\angle DBC$, whereas θ is the negative angle, $\angle ABC$.

Figure 2

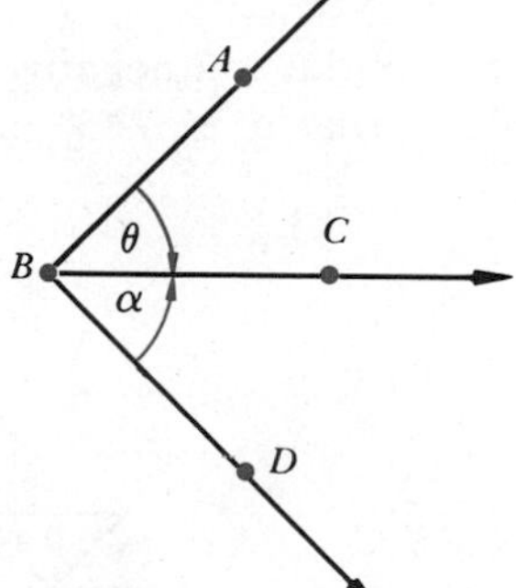

It is important to remember that an angle is determined by the *initial side*, the *terminal side*, and the *rotation* used to form it. For example, the angles α and β in Figure 3 have the same initial and terminal sides, yet they are different angles.

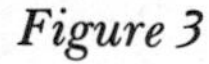
Figure 3

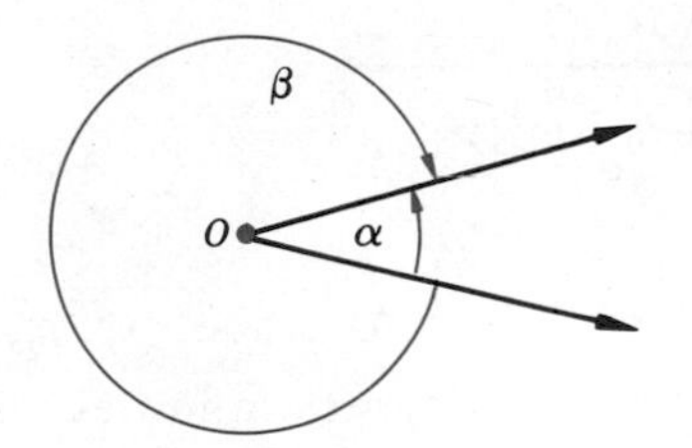

Angles are measured by using either degrees or radians. One *degree* (1°) is the measure of a positive angle that is formed by $\frac{1}{360}$ of one complete revolution (Figure 4a). Furthermore, a degree can be divided into 60 equal parts called *minutes* ($'$); a minute can be divided into 60 equal parts called *seconds* ($''$). On the other hand, one *radian* is the measure of a positive angle that intercepts an arc of length 1 on a circle of radius 1 (Figure 4b).

Figure 4

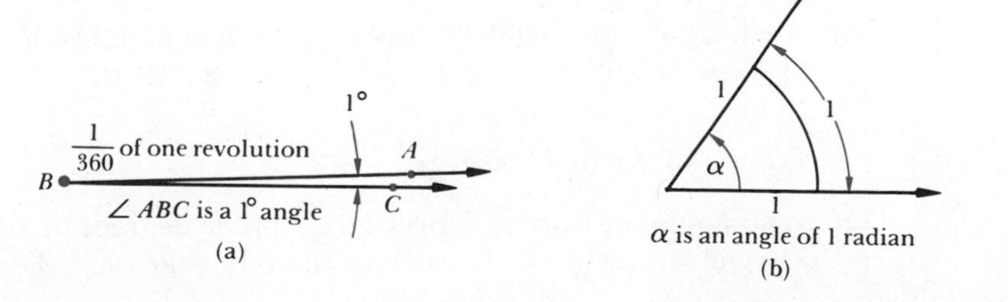

The angle measure is positive or negative according to whether the angle is formed by a counterclockwise or clockwise rotation. Hence the radian measure of an angle is the "directed" length of its subtended arc on a circle of radius 1.

EXAMPLES **1** Indicate both the degree measure and the radian measure of an angle α formed by

(a) One complete counterclockwise rotation
(b) One-quarter of a counterclockwise rotation
(c) One-eighth of a clockwise rotation

SOLUTION

(a) Since $1°$ is the measure of an angle formed by $\frac{1}{360}$ of one complete revolution, the degree measure of α is $(360)(1°) = 360°$. The radian measure of α is the length of the circumference of a circle of radius 1, namely, 2π radians (Figure 5a).
(b) The measure of angle α is expressed either as $\frac{1}{4}(360°) = 90°$ or $\frac{1}{4}(2\pi) = \pi/2$ radian (Figure 5b).
(c) Since $\frac{1}{8}(360°) = 45°$ or $\frac{1}{8}(2\pi) = \pi/4$, and α is a negative angle because of the clockwise rotation, it follows that α is $-45°$ or $-\pi/4$ radian (Figure 5c).

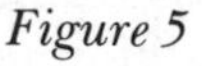
Figure 5

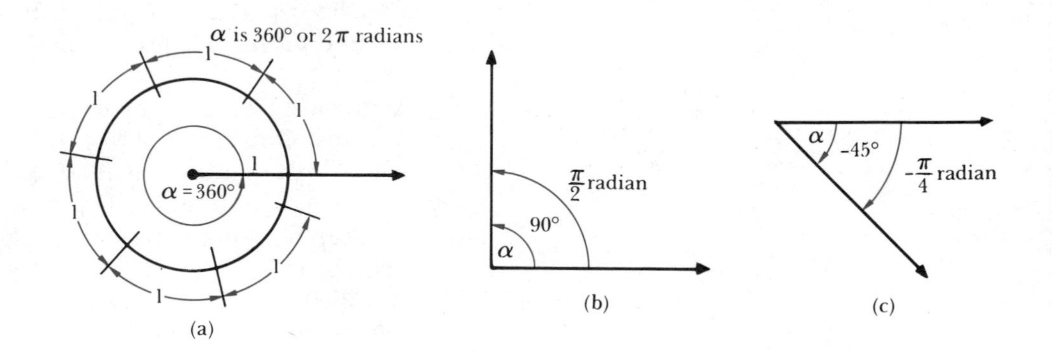

2 Express the following angles in terms of degrees, minutes, and seconds.

(a) $37.45°$ (b) $-84.32°$

SOLUTION

(a) $37.45° = 37° + 0.45°$; however, $0.45° = (0.45)(60') = 27'$, therefore, $37.45° = 37°27'0''$.
(b) $-84.32° = -(84° + 0.32°)$; however, $0.32° = (0.32)(60') = 19.20'$. Also, $0.20' = (0.20)(60'') = 12''$, so $-84.32° = -84°19'12''$.

4.2 Conversion of Angle Measures

Often an angle measure is expressed in either degrees or radians and we need to convert its given measure to the other measure. For example, an

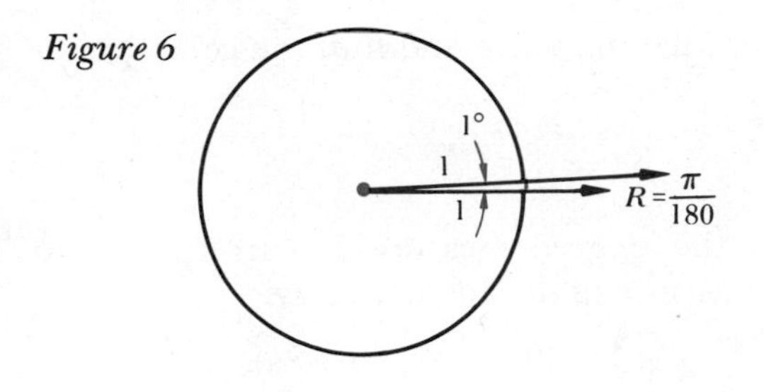

Figure 6

angle of 1° subtends an arc of length $\frac{1}{360}$ of the circumference of a circle with radius 1. In determining the radian measure R of this angle, we use the fact that R is the length of the subtended arc on a circle with radius 1, together with the fact that the circumference of this circle is 2π, to get

$$R = \frac{1}{360}(2\pi) = \frac{\pi}{180} \qquad \text{(Figure 6)}$$

Hence 1° corresponds to $\pi/180$ radians. This means that an angle of D degrees has a radian measure R that is given by

$$R = \frac{\pi}{180}D$$

This relationship between R and D can also be expressed by the formula

$$D = \frac{180}{\pi}R$$

which we get by solving the first equation for D in terms of R.

In precalculus courses, it is customary to omit the unit "radian." This practice will be followed in this textbook. Thus we write 60° to correspond to $\pi/3$, without using the word "radian" after $\pi/3$.

For example, an angle with radian measure 1 has a degree measure of $180°/\pi$ or approximately 57°19′29″.

EXAMPLES 1 Sketch the angle with the given degree measure and find its corresponding radian measure.

(a) 150° (b) 810° (c) −15°

SOLUTION Using the fact that the radian measure R corresponds to $(\pi/180)D$, where D is the degree measure of the angle, we have

(a) 150° corresponds to $\frac{\pi}{180}(150) = \frac{5\pi}{6}$ (Figure 7a)

(b) 810° corresponds to $\frac{\pi}{180}(810) = \frac{9\pi}{2}$ (Figure 7b)

(c) −15° corresponds to $\frac{\pi}{180}(-15) = -\frac{\pi}{12}$ (Figure 7c)

Figure 7

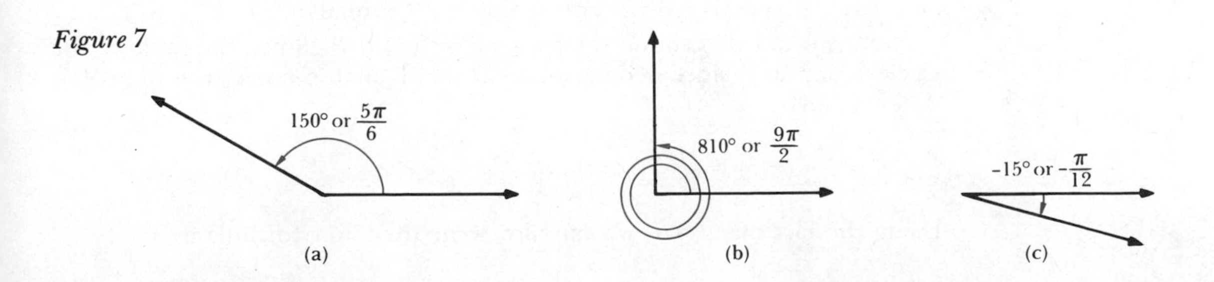

2 Sketch the angle with the given radian measure and find the corresponding degree measure.

(a) $\frac{\pi}{3}$ (b) $\frac{17\pi}{10}$ (c) $-\frac{5\pi}{12}$

SOLUTION Using the fact that the degree measure D corresponds to $(180/\pi)R$, where R is the radian measure of the angle, we have

(a) $\frac{\pi}{3}$ radians corresponds to $\frac{180}{\pi}\left(\frac{\pi}{3}\right) = 60°$ (Figure 8a)

(b) $\frac{17\pi}{10}$ radians corresponds to $\frac{180}{\pi}\left(\frac{17\pi}{10}\right) = 306°$ (Figure 8b)

(c) $-\frac{5\pi}{12}$ radians corresponds to $\frac{180}{\pi}\left(-\frac{5\pi}{12}\right) = -75°$ (Figure 8c)

Figure 8

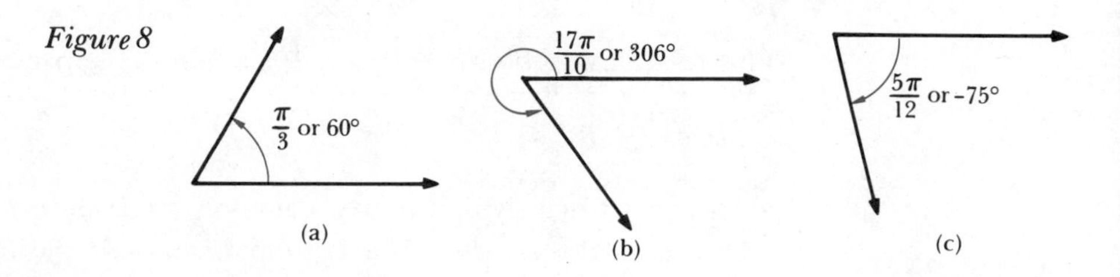

4.3 Arcs and Sectors of Circles

The radian measure of a central angle of a circle can be used to determine both the length of the arc the angle subtends and the area of the sector determined by the central angle. Suppose that θ is the central angle of a circle of radius r (Figure 9). Also assume that the angle θ has a radian measure t. (If θ is measured in degrees, we can always convert to radians.) Our task is to determine s, the length of $\widehat{AB}$. The length of the circumference of the circle is $2\pi r$; but, since θ determines an arc that is "$(t/2\pi)$th" of the circumference,

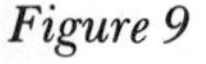
Figure 9

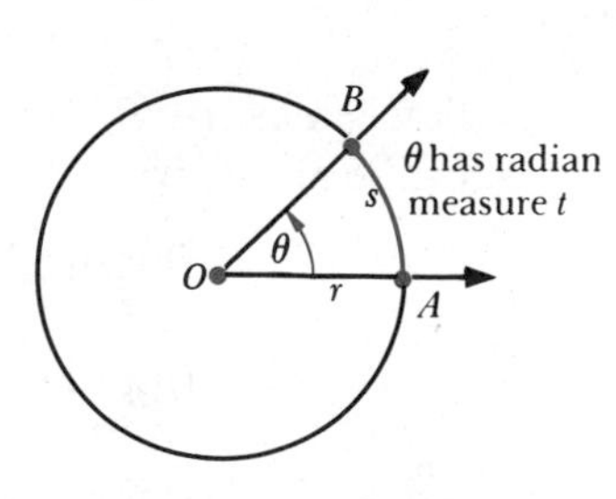

$$s = \frac{t}{2\pi}(2\pi r), \quad \text{that is,} \quad s = tr$$

Hence we have a formula for determining the length of an arc subtended on a circle of radius r by a central angle of t radians.

Next, we can determine the area of sector AOB. Since the area of the circle is πr^2 and since θ determines $(t/2\pi)$th of the circle, the area A is determined by

$$A = \frac{t}{2\pi}(\pi r^2) = \frac{tr^2}{2}, \quad \text{so that} \quad A = \frac{1}{2}r^2 t$$

Using the fact that $s = tr$, we can also write the latter formula as $A = \frac{1}{2}sr$.

EXAMPLES **1** Find the arc length and the area of the circular sector if the radius of the circle is 6 inches and the measure of the central angle is given as follows.

(a) $\frac{\pi}{6}$ (b) $70°$

SOLUTION

(a) $$s = tr = 6\left(\frac{\pi}{6}\right) = \pi \text{ inches}$$

and

$$A = \frac{1}{2}r^2t = \frac{1}{2}(36)\left(\frac{\pi}{6}\right) = 3\pi \text{ square inches} \qquad \text{(Figure 10a)}$$

(b) Since $70°$ corresponds to $70\pi/180 = 7\pi/18$ radians,

$$s = 6\left(\frac{7\pi}{18}\right) = \frac{7\pi}{3} \text{ inches}$$

and

$$A = \frac{1}{2}r^2t = \frac{1}{2}(36)\left(\frac{7\pi}{18}\right) = 7\pi \text{ square inches} \qquad \text{(Figure 10b)}$$

Figure 10

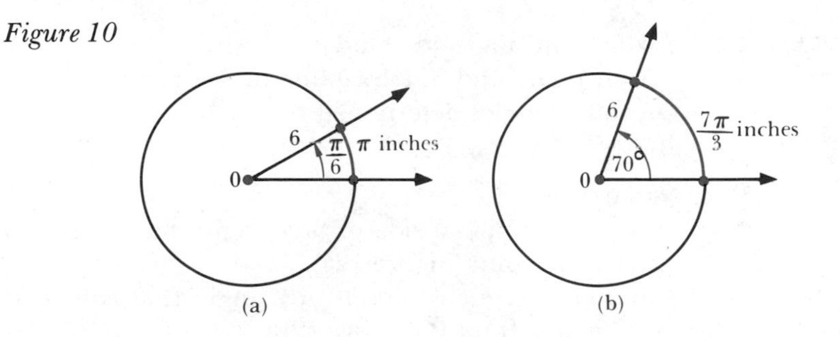

2 The tip of the minute hand of a clock travels $7\pi/10$ inches in 3 minutes. How long is the minute hand?

SOLUTION In 3 minutes the minute hand generates a central angle of

$$(3)\left(\frac{1}{60}\right)(360°) = 18° \qquad \text{(Why?)}$$

But an angle of $18°$ has a radian measure of

$$t = \left(\frac{\pi}{180}\right)(18) = \frac{\pi}{10}$$

so

$$\frac{7\pi}{10} = r\left(\frac{\pi}{10}\right)$$

Hence, $r = 7$, which is the length of the minute hand in inches.

4.4 Functions Defined on Angles

An angle is in *standard position* if it is placed on a cartesian coordinate system with its vertex at the origin and with the initial side coinciding with the positive x axis. For example, an 80° angle in standard position has its terminal side in quadrant I (Figure 11a), whereas an angle of radian measure $-3\pi/5$ in standard position has its terminal side in quadrant III (Figure 11b). [Notice that $-3\pi/5$ corresponds to $(180/\pi)(-3\pi/5) = -108°$.]

Figure 11

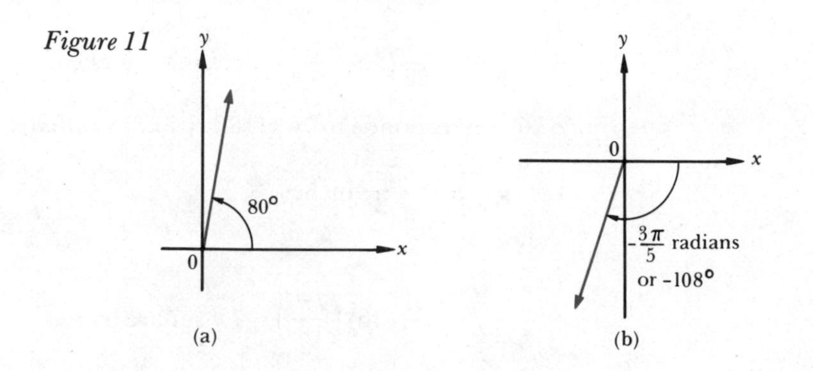

EXAMPLE Sketch an angle in standard position whose terminal side contains the given point. Indicate by θ one of the positive angles and by ϕ one of the negative angles determined by the terminal side of the angle sketched.
(a) $(3, 4)$ (b) $(-4, 3)$ (c) $(-2, -3)$ (d) $(4, -1)$

SOLUTION

(a) The terminal side of any angle that contains the point $(3, 4)$ lies in quadrant I (Figure 12a).
(b) The terminal side of any angle that contains the point $(-4, 3)$ lies in quadrant II (Figure 12b).
(c) The terminal side of any angle that contains the point $(-2, -3)$ lies in quadrant III (Figure 12c).
(d) The terminal side of any angle that contains the point $(4, -1)$ lies in quadrant IV (Figure 12d).

Figure 12

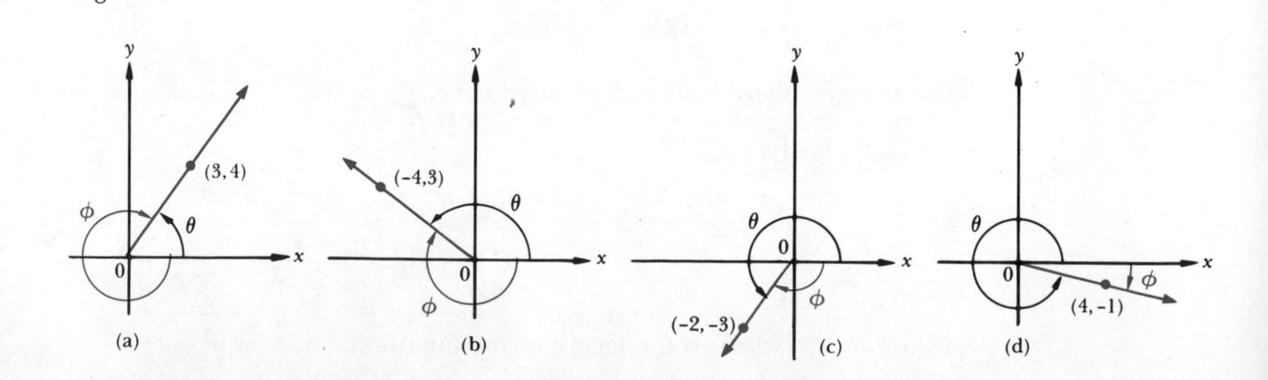

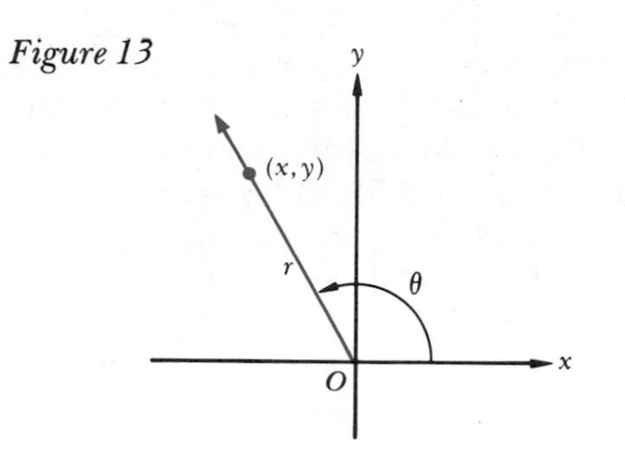

Now, suppose that θ is an angle in standard position and further suppose that (x, y) is any point other than $(0, 0)$ on the terminal side of θ (Figure 13). We can compute the distance r between (x, y) and $(0, 0)$ by using the distance formula:

$$r = \sqrt{(x-0)^2 + (y-0)^2} = \sqrt{x^2 + y^2}$$

What we want here are functions that relate θ, r, x, and y. These functions, the trigonometric functions of angles, are defined as follows.

DEFINITION 1 TRIGONOMETRIC FUNCTIONS OF ANGLES

Let θ be an angle in standard position and let (x, y) be any point other than $(0, 0)$ on the terminal side of θ, and let $r = \sqrt{x^2 + y^2}$ (Figure 13). Then the six trigonometric functions are defined on the angle θ as follows:

$$\sin\theta = \frac{y}{r} \qquad \csc\theta = \frac{r}{y}, \quad y \neq 0$$

$$\cos\theta = \frac{x}{r} \qquad \sec\theta = \frac{r}{x}, \quad x \neq 0$$

$$\tan\theta = \frac{y}{x}, \quad x \neq 0 \qquad \cot\theta = \frac{x}{y}, \quad y \neq 0$$

For example, suppose that θ is a $45°$ angle and that (x, y) is a point on the terminal side of θ other than $(0, 0)$ (Figure 14). Then the triangle determined by the three points $(0, 0)$, (x, y), and $(x, 0)$ is isosceles. (Why?) Thus, $x = y$ and $r = \sqrt{x^2 + x^2} = \sqrt{2x^2} = \sqrt{2}x$. By Definition 1 of the trigonometric functions of angles, we have

Figure 14

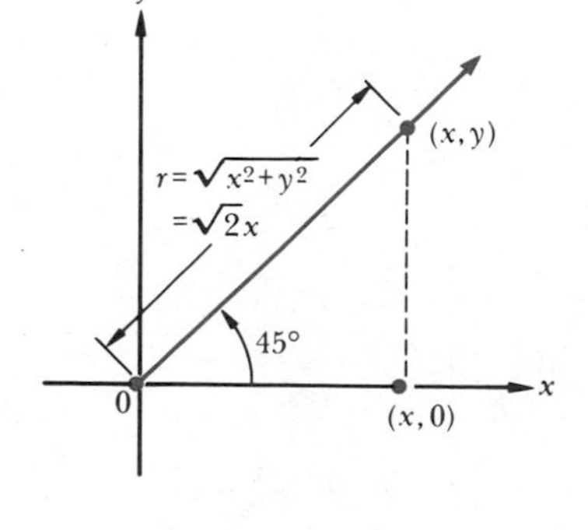

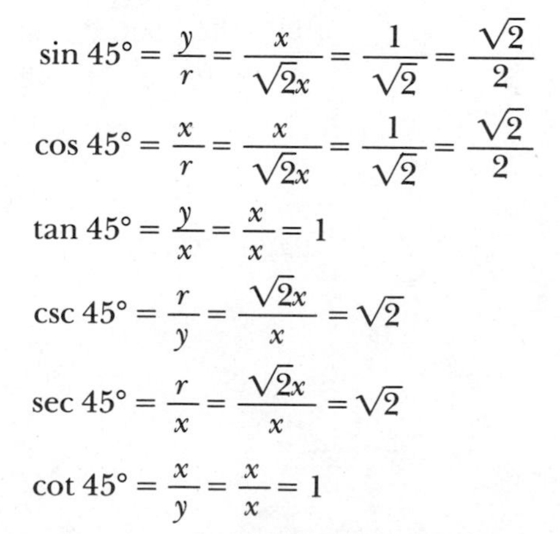

$$\sin 45° = \frac{y}{r} = \frac{x}{\sqrt{2}x} = \frac{1}{\sqrt{2}} = \frac{\sqrt{2}}{2}$$

$$\cos 45° = \frac{x}{r} = \frac{x}{\sqrt{2}x} = \frac{1}{\sqrt{2}} = \frac{\sqrt{2}}{2}$$

$$\tan 45° = \frac{y}{x} = \frac{x}{x} = 1$$

$$\csc 45° = \frac{r}{y} = \frac{\sqrt{2}x}{x} = \sqrt{2}$$

$$\sec 45° = \frac{r}{x} = \frac{\sqrt{2}x}{x} = \sqrt{2}$$

$$\cot 45° = \frac{x}{y} = \frac{x}{x} = 1$$

It is important to understand that the values of the six trigonometric functions for the $45°$ angle depend only on the *position* of the terminal side of the angle, in the sense that the results are the same no matter what particular point is selected on the terminal side to evaluate the six functions [other than (0, 0), of course]. For example, the point (1, 1), with $r = \sqrt{2}$, will yield the same result as the point (5, 5), with $r = 5\sqrt{2}$.

In general, if (x, y) and (x_1, y_1) are two *different* points in quadrant I, both of which lie on the terminal side of θ (Figure 15), so that $r = \sqrt{x^2 + y^2}$ and $r_1 = \sqrt{x_1^2 + y_1^2}$, because of the similar triangles, $\triangle OBP_1$ and $\triangle OAP$, we have these equal ratios:

Figure 15

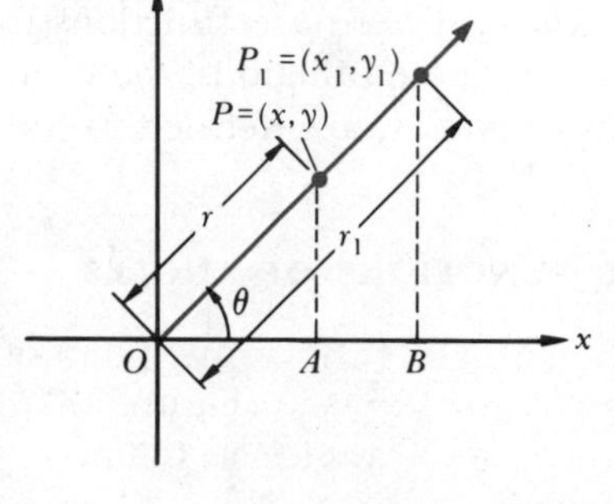

$$\frac{y_1}{r_1} = \frac{y}{r} \qquad \frac{x_1}{r_1} = \frac{x}{r} \qquad \frac{y_1}{x_1} = \frac{y}{x}$$

$$\frac{x_1}{y_1} = \frac{x}{y} \qquad \frac{r_1}{x_1} = \frac{r}{x} \qquad \frac{r_1}{y_1} = \frac{r}{y}$$

Hence the values of the trigonometric functions are the same no matter what two points are selected on the terminal side of the angle in quadrant I. Furthermore, by a similar construction, we can show that the above results are true for all angles regardless of the quadrant in which the terminal side lies (see Problem 40).

EXAMPLES 1 Evaluate each of the following.

(a) $\sin 180°$ and $\cos 180°$ (b) $\sin(-\pi/2)$ and $\cos(-\pi/2)$

SOLUTION First, select points on the terminal sides of the 180° angle and the $-\pi/2$ angle after the angles are placed in standard position (Figure 16).

(a) Here $(-3, 0)$ is selected on the terminal side of the 180° angle, so that $x = -3$, $y = 0$, and $r = 3$; thus

$$\sin 180° = 0/3 = 0 \quad \text{and} \quad \cos 180° = -3/3 = -1$$

(b) The point $(0, -2)$ is selected on the terminal side of the $-\pi/2$ angle, so that $x = 0$, $y = -2$, and $r = 2$; thus

$$\sin(-\pi/2) = -2/2 = -1 \quad \text{and} \quad \cos(-\pi/2) = 0/2 = 0$$

Figure 16

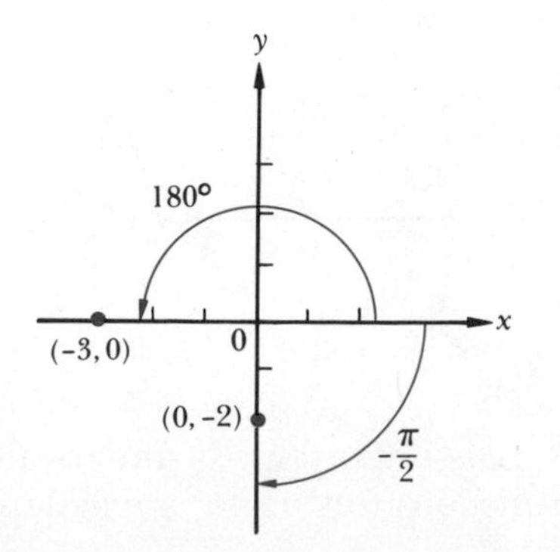

2 Evaluate the trigonometric functions of θ if θ has $(3, -5)$ on its terminal side. Is θ unique?

SOLUTION Figure 17 demonstrates that there are infinitely many different angles whose terminal side contains the point $(3, -5)$. Hence, θ is not unique.

Using the fact that $r = \sqrt{3^2 + (-5)^2} = \sqrt{34}$, we have

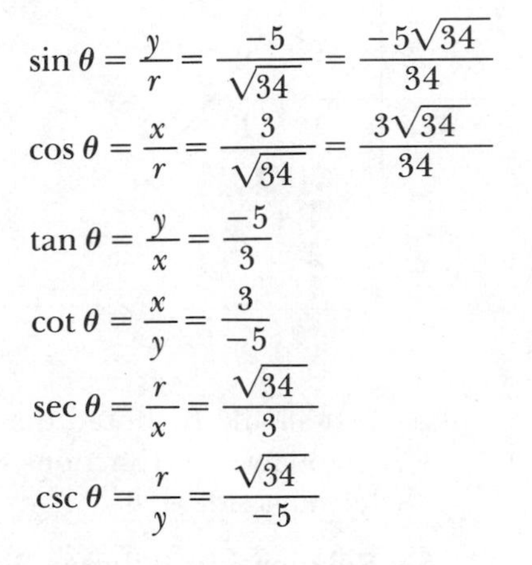

$$\sin\theta = \frac{y}{r} = \frac{-5}{\sqrt{34}} = \frac{-5\sqrt{34}}{34}$$
$$\cos\theta = \frac{x}{r} = \frac{3}{\sqrt{34}} = \frac{3\sqrt{34}}{34}$$
$$\tan\theta = \frac{y}{x} = \frac{-5}{3}$$
$$\cot\theta = \frac{x}{y} = \frac{3}{-5}$$
$$\sec\theta = \frac{r}{x} = \frac{\sqrt{34}}{3}$$
$$\csc\theta = \frac{r}{y} = \frac{\sqrt{34}}{-5}$$

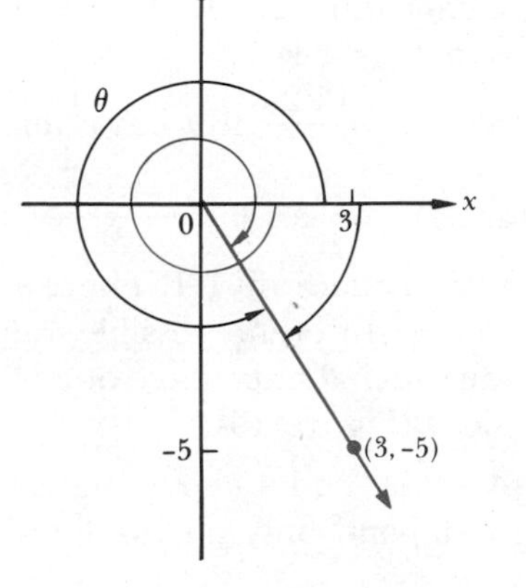

Figure 17

3 If $\tan\theta = -\frac{4}{3}$ and θ has its terminal side in quadrant II, find the values of the remaining trigonometric functions of θ. Is θ unique?

SOLUTION One point on the terminal side of θ is $(-3, 4)$. Figure 18 illustrates that there are infinitely many different angles whose terminal side contains the point $(-3, 4)$. Hence, θ is not unique.

Also,

$$r = \sqrt{(-3)^2 + 4^2} = 5$$

for this point, so that by Definition 1 of the trigonometric functions of angles, we have

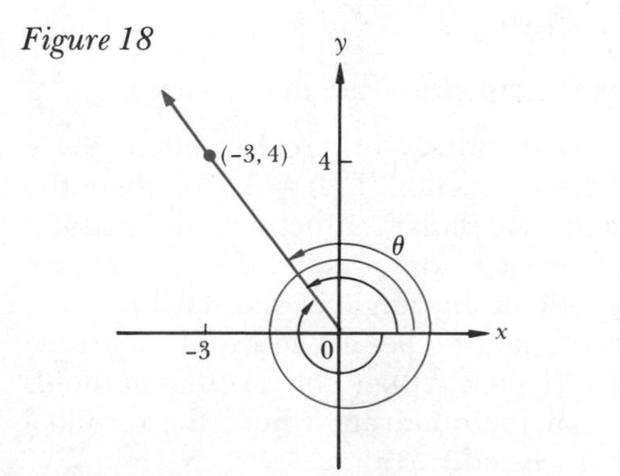

Figure 18

$$\sin\theta = \frac{y}{r} = \frac{4}{5}$$
$$\cos\theta = \frac{x}{r} = \frac{-3}{5}$$
$$\cot\theta = \frac{x}{y} = \frac{-3}{4}$$
$$\sec\theta = \frac{r}{x} = \frac{5}{-3}$$
$$\csc\theta = \frac{r}{y} = \frac{5}{4}$$

Figure 19

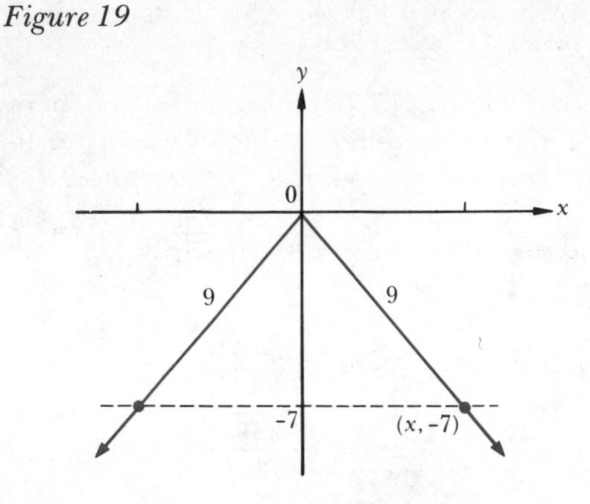

4 In which quadrant is the terminal side of θ found if $\sin\theta = -\frac{7}{9}$? Evaluate $\cos\theta$.

SOLUTION

$$\sin\theta = \frac{y}{r} = \frac{-7}{9}$$

from which we can deduce the two possibilities shown in Figure 19. If $9 = \sqrt{x^2 + 49}$, then $x^2 = 32$, so that $x = \pm 4\sqrt{2}$.

Hence, $\cos\theta = \frac{4\sqrt{2}}{9}$, if θ has a terminal side in quadrant IV, or $\cos\theta = \frac{-4\sqrt{2}}{9}$, if θ has a terminal side in quadrant III. There are actually infinitely many different possible values for θ; however, any such θ must have one of these two terminal sides (Figure 19).

It should be noted that in Examples 2, 3, and 4 above the values of the trigonometric functions of the angle depend only on the location of the terminal side of θ.

4.5 Relationship Between Trigonometric Functions of Angles and Trigonometric Functions of Real Numbers

The trigonometric functions of angles and the trigonometric (circular) functions of real numbers are so closely related that we use the term trigonometric function regardless of whether real numbers or angles are employed for the domain. The relationship between the two types of trigonometric functions is established in the following theorem.

THEOREM 1

If T is any one of the six trigonometric functions defined on angles, and C is the trigonometric (circular) function with the same name, but defined on real numbers, then

$$T(\theta) = C(t)$$

where θ is an angle with radian measure t.

Figure 20

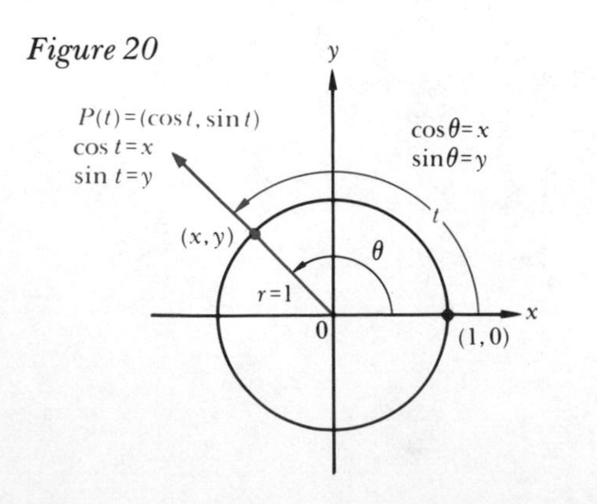

PROOF Suppose that t is a real number. Since the wrapping function $P(t) = (x, y)$, then the trigonometric (circular) functions defined on real numbers yield $\cos t = x$ and $\sin t = y$ (Figure 20). Now let θ be *any* angle in standard position with its terminal side being the ray through the point $P(t)$ (Figure 20). (The argument holds regardless of the quadrant where the terminal side of θ is located.)

By Definition 1, we have $\cos\theta = x/r$ and $\sin\theta = y/r$, where $(x, y) \neq (0, 0)$ is a point on the terminal side of θ and $r = \sqrt{x^2 + y^2}$. But with (x, y) selected as above, we have $r = 1$, so $\cos\theta = x/1 = x$ and $\sin\theta = y/1 = y$ (Figure 20). Thus

$$\cos\theta = \cos t \quad \text{and} \quad \sin\theta = \sin t$$

Notice that θ is the angle and, by the definition of radian measure, t equals the radian measure of θ. Here, $\cos\theta$ and $\sin\theta$ are the values of the trigonometric function on angles, whereas $\cos t$ and $\sin t$ are the values of the trigonometric (circular) functions on real numbers. Similarly, where t is the radian measure of θ,

$$\tan\theta = \frac{y}{x} = \frac{\sin t}{\cos t} = \tan t \qquad \cot\theta = \frac{x}{y} = \frac{\cos t}{\sin t} = \cot t$$

$$\sec\theta = \frac{1}{x} = \frac{1}{\cos t} = \sec t \qquad \csc\theta = \frac{1}{y} = \frac{1}{\sin t} = \csc t$$

Because of Theorem 1, it follows that *all* the properties established for the trigonometric functions on real numbers become the same properties for the trigonometric functions defined on angles [domain and range (see Problem 70), signs of values (see Problem 50), relationships between trigonometric functions (see Problem 48), and periodicity]. For example, the trigonometric functions of angles are periodic functions of period $360°$ or 2π. Thus the value of any trigonometric function of the angles $45°$, $405°$, $765°$, and $-315°$ is the same. For instance,

$$\cos 45° = \cos 405° = \cos 765° = \cos(-315°)$$

EXAMPLES Use Theorem 1 to find the values of the six trigonometric functions of the given angle.

1 (a) $\theta = 60°$ (b) $\theta = 420°$

SOLUTION

(a) Since $60°$ is the degree measure of an angle of $\pi/3$ radians (Figure 21), we have

$$\sin 60° = \sin\frac{\pi}{3} = \frac{\sqrt{3}}{2}$$

$$\cos 60° = \cos\frac{\pi}{3} = \frac{1}{2}$$

$$\tan 60° = \tan\frac{\pi}{3} = \sqrt{3}$$

$$\csc 60° = \csc\frac{\pi}{3} = \frac{2}{\sqrt{3}} = \frac{2\sqrt{3}}{3}$$

$$\sec 60° = \sec\frac{\pi}{3} = 2$$

$$\cot 60° = \cot\frac{\pi}{3} = \frac{1}{\sqrt{3}} = \frac{\sqrt{3}}{3}$$

Figure 21

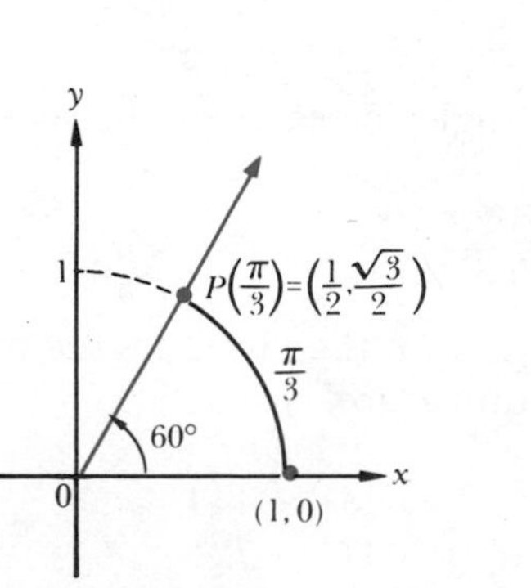

(b) The terminal side of 420° is the same as the terminal side of 60°, so

$$\sin 420° = \sin 60° = \frac{\sqrt{3}}{2}$$

$$\cos 420° = \cos 60° = \frac{1}{2}$$

The values of the remaining trigonometric functions of angle 420° are the same as those for angle 60° [see part (a) above].

2 (a) $\theta = 150°$ (b) $-210°$

SOLUTION

(a) Since 150° is the degree measure of an angle of measure $5\pi/6$ radians (Figure 22), it follows that

Figure 22

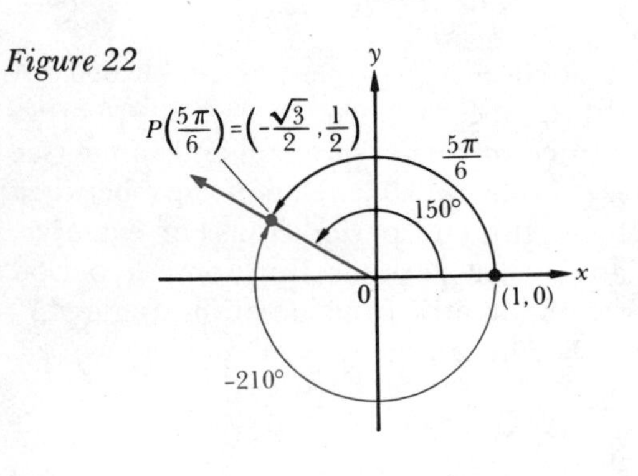

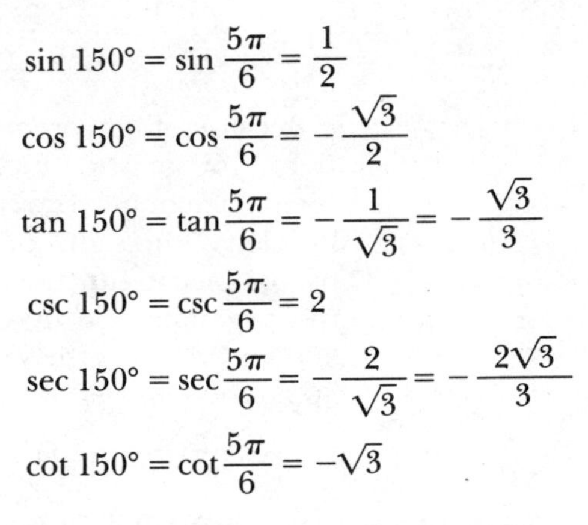

$$\sin 150° = \sin \frac{5\pi}{6} = \frac{1}{2}$$

$$\cos 150° = \cos \frac{5\pi}{6} = -\frac{\sqrt{3}}{2}$$

$$\tan 150° = \tan \frac{5\pi}{6} = -\frac{1}{\sqrt{3}} = -\frac{\sqrt{3}}{3}$$

$$\csc 150° = \csc \frac{5\pi}{6} = 2$$

$$\sec 150° = \sec \frac{5\pi}{6} = -\frac{2}{\sqrt{3}} = -\frac{2\sqrt{3}}{3}$$

$$\cot 150° = \cot \frac{5\pi}{6} = -\sqrt{3}$$

(b) The terminal side of $-210°$ is the same as the terminal side of 150°, so the trigonometric function values of the $-210°$ angle are the same as those for the 150° angle. For instance,

$$\sin(-210°) = \sin 150° = \frac{1}{2} \quad \text{and} \quad \tan(-210°) = \tan 150° = -\frac{\sqrt{3}}{3}$$

PROBLEM SET 4

In Problems 1–6, express each angle in terms of degrees, minutes, and seconds.

1 87.35° **2** $-62.45°$ **3** $-25.55°$
4 67.32° **5** 65.37° **6** $-81.41°$

In Problems 7–18, sketch the angle with the given degree measure in standard position and give its radian measure.

7 40° **8** 75° **9** 240° **10** 330°
11 $-95°$ **12** $-220°$ **13** $-444°$ **14** $-100°$
15 1080° **16** $-3050°$ **17** $-72°$ **18** $-420°$

In Problems 19–30, sketch the angle with the given radian measure in standard position and give its degree measure. (Use $\pi = 3.14$.)

19 $\frac{2\pi}{3}$ **20** $\frac{11\pi}{6}$ **21** $\frac{7\pi}{18}$ **22** $\frac{121\pi}{360}$

23 $-\frac{7\pi}{12}$ **24** $\frac{43\pi}{6}$ **25** $-\frac{4\pi}{9}$ **26** $\frac{7\pi}{9}$

27 -5π **28** $-\frac{13\pi}{6}$ **29** 4 **30** -2

In Problems 31–33, indicate both the degree measure and the radian measure of an angle α formed by

31 three-eighths of a clockwise rotation

32 two and one-sixth of a counterclockwise rotation

33 three-fourths of a counterclockwise rotation

34 Explain why an angle of 1 radian is larger than angle of 1°.

In Problems 35–39, find the arc length and the area of the circular sector if the radius r of the circle and the measure of the central angle θ are as given.

35 $r = 7$ inches, $\theta = \frac{3\pi}{14}$ **36** $r = 4$ inches, $\theta = \frac{2\pi}{5}$

37 $r = 6$ centimeters, $\theta = 50°$ **38** $r = 9$ centimeters, $\theta = 275°$

39 $r = 7$ centimeters, $\theta = \pi°$

40 With reference to Figure 23,

(a) Suppose that (x, y) and (x_1, y_1) are two different points in quadrant II on the terminal side of θ other than (0, 0). Show that

$$\frac{y_1}{r_1} = \frac{y}{r} \qquad \frac{x_1}{r_1} = \frac{x}{r} \qquad \frac{y_1}{x_1} = \frac{y}{x}$$

$$\frac{x_1}{y_1} = \frac{x}{y} \qquad \frac{r_1}{x_1} = \frac{r}{x} \qquad \frac{r_1}{y_1} = \frac{r}{y}$$

Figure 23

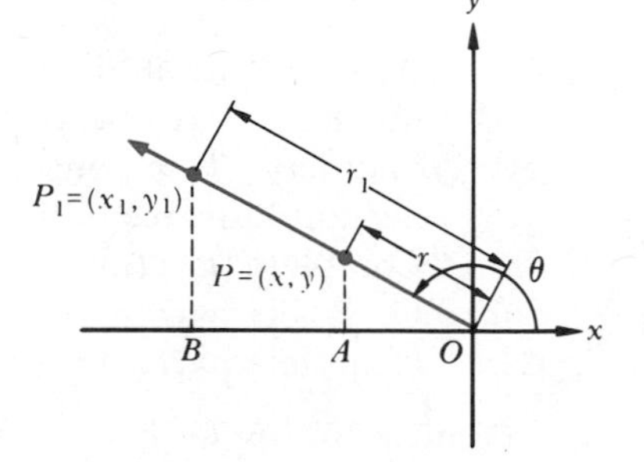

(b) Show that the ratios in part (a) also hold if the points are in quadrant III.

(c) Show that the ratios in part (a) also hold if the points are in quadrant IV.

In Problems 41–47, determine the values of the trigonometric functions of θ in each case. Sketch any two possible angles θ in standard position for each case.

41 θ has $(-5, 0)$ on its terminal side. What is the measure of θ?
42 θ has $(-3, -4)$ on its terminal side.
43 θ has $(7, -10)$ on its terminal side.
44 θ has $(0, 1)$ on its terminal side. What is the measure of θ?
45 θ has $(-1, \sqrt{3})$ on its terminal side. What is the measure of θ?
46 $(x, -4)$ is 11 units from $(0, 0)$ and is on the terminal side of θ.
47 $(3x, x)$ is on the terminal side of θ and $x \neq 0$.

48 Use Definition 1 on page 231 to prove each of the following relations.

(a) $\tan\theta = \dfrac{\sin\theta}{\cos\theta}$ (b) $\cot\theta = \dfrac{\cos\theta}{\sin\theta}$ (c) $\sec\theta = \dfrac{1}{\cos\theta}$

(d) $\csc\theta = \dfrac{1}{\sin\theta}$ (e) $\sin^2\theta + \cos^2\theta = 1$

49 Complete Table 1

Table 1

θ Measure							
Degrees	*Radians*	$\sin\theta$	$\cos\theta$	$\tan\theta$	$\cot\theta$	$\sec\theta$	$\csc\theta$
0°							
	$\pi/6$						
45°		$\sqrt{2}/2$					
60°							
	$\pi/2$						
120°							
135°							
150°							
	π						

50 (a) Assume that angle θ in standard position has a terminal side in quadrant I. Indicate the sign of the value of each of the trigonometric functions of θ, then compare your results to those in Section 2 on page 200.
(b) Complete part (a) for θ in quadrant II.
(c) Complete part (a) for θ in quadrant III.
(d) Complete part (a) for θ in quadrant IV.

51 Draw a 30° angle in standard position. Give three other angles with the same terminal side as the 30° angle. Determine the values of the trigonometric functions of 30°. What can you say about the values of the trigonometric functions of the other three angles? Explain.

52 Explain why the sine, tangent, cotangent, and cosecant functions of angles are odd and the cosine and secant functions of angles are even.

In Problems 53–58, construct each angle in standard position, with its terminal side in the indicated quadrant, then evaluate the other five trigonometric functions of θ.

53 $\sin\theta = \frac{15}{17}$, quadrant II

54 $\cos\theta = \frac{\sqrt{3}}{2}$, quadrant IV

55 $\tan\theta = -1$, quadrant IV

56 $\cot\theta = \sqrt{3}$, quadrant III

57 $\sec\theta = -\frac{13}{5}$, quadrant II

58 $\csc\theta = \sqrt{2}$, quadrant I

In Problems 59–69, use Theorem 1 to determine the values of the six trigonometric functions of the given angle θ.

59 $\theta = 330°$

60 $\theta = 315°$

61 $\theta = 135°$

62 $\theta = -45°$

63 $\theta = -240°$

64 $\theta = -120°$

65 $\theta = 780°$

66 $\theta = -750°$

67 $\theta = 585°$

68 $\theta = -405°$

69 $\theta = 3720°$

70 (a) Use Definition 1 on page 231 and Problem 48 to find the domain of each of the trigonometric functions. Describe the domains in terms of the radian measures of the angles and in terms of the degree measures of the angles. How do the domains described in these terms compare to the domains of the corresponding circular functions?

(b) Determine the ranges of the trigonometric functions, and then compare your results to the ranges of the circular functions.

71 The length of each chain supporting the seat of a baby swing is 8 feet. When the swing is at its highest forward and backward points, the radian measure of the angle determined by one of these chains is $5\pi/6$. How far does the baby travel in one trip between these high points; that is, what is the length of the arc generated by one swing between the two highest points?

72 A windshield wiper blade is 36 centimeters long, and rotates through an angle of $3\pi/4$. The tip of the wiper blade is 46 centimeters from the base of the arm that moves it back and forth. Determine the area of the surface that is swept by the blade.

73 Find the length of a pendulum if the tip of the pendulum traces an arc of $11\pi/6$ units and the measure of the angle that this arc subtends is $\pi/5$ radians.

5 Evaluation of the Trigonometric Functions

Since the trigonometric functions are periodic functions of period 2π (or 360°), two angles whose radian measures (or degree measures) differ by an integral multiple of 2π (or 360°) determine angles with the same termi-

Figure 1

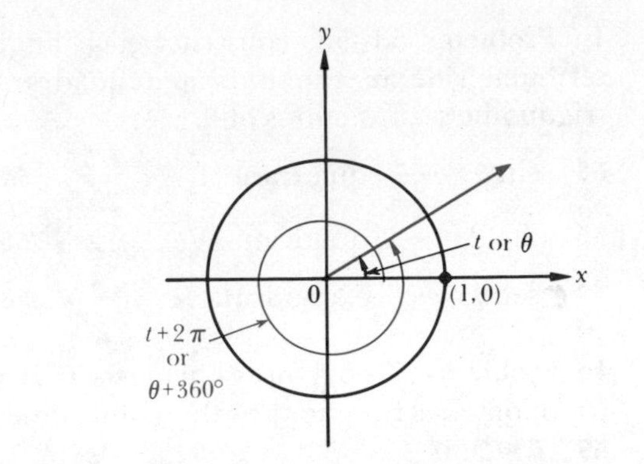

nal side (Figure 1). Thus, to evaluate the trigonometric functions of an angle whose radian measure is less than 0 or greater than 2π (or whose degree measure is less than 0° or greater than 360°), we may add or subtract an appropriate multiple of 2π (or 360°) in order to reduce the evaluation of the functions to an angle between 0 and 2π (or 0° and 360°).

For example,

$$\sin\frac{46\pi}{5} = \sin\left(\frac{6\pi}{5} + 4\cdot 2\pi\right) = \sin\frac{6\pi}{5}$$
$$\cos 860° = \cos(140° + 2\cdot 360°) = \cos 140°$$
$$\tan(-240°) = \tan(-240° + 360°) = \tan 120°$$

EXAMPLE Reduce the evaluation of the given trigonometric function of the given angle to an evaluation of the same function of an angle between 0 and 2π or 0° and 360°. Sketch each angle in standard position. (Use $\pi = 3.14$.)

(a) $\cos 7$ (b) $\sin 15$ (c) $\tan 552°$ (d) $\sec(-80°)$

SOLUTION Since the trigonometric functions are periodic functions of period 2π (or 6.28) or 360°, it follows that

(a) $\cos 7 = \cos(0.72 + 6.28) = \cos 0.72$ (Figure 2a)
(b) $\sin 15 = \sin[2.44 + 2(6.28)] = \sin 2.44$ (Figure 2b)
(c) $\tan 552° = \tan(192° + 360°) = \tan 192°$ (Figure 2c)
(d) $\sec(-80°) = \sec(-80° + 360°) = \sec 280°$ (Figure 2d)

Figure 2

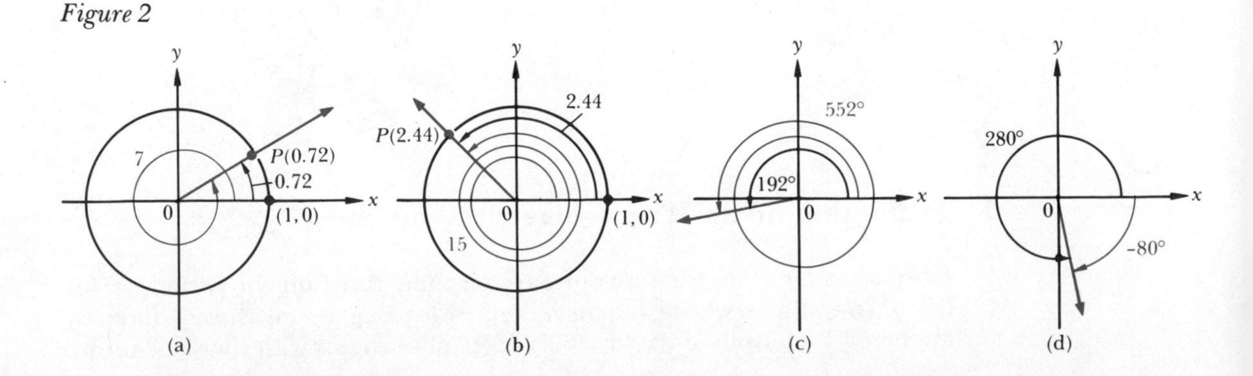

5.1 Reference Angles

We next show that the evaluation of a trigonometric function of any angle can be reduced to an evaluation of the function at 0, $\pi/2$, π, or $3\pi/2$ (0°, 90°, 180°, or 270°) or at an appropriate acute angle (with a possible adjustment in sign). This practice is very important, since trigonometric tables for radian and degree measures normally list only the trigonometric functions for acute angles.

Figure 3 shows the various situations that can occur if the terminal side of a given angle α lies in any one of the quadrants. If α has radian measure t (or degree measure θ), then the *reference angle* t_R (or θ_R) associated with α is the *positive acute angle* formed by the terminal side of α and the x axis, as specified in Figure 3.

Figure 3

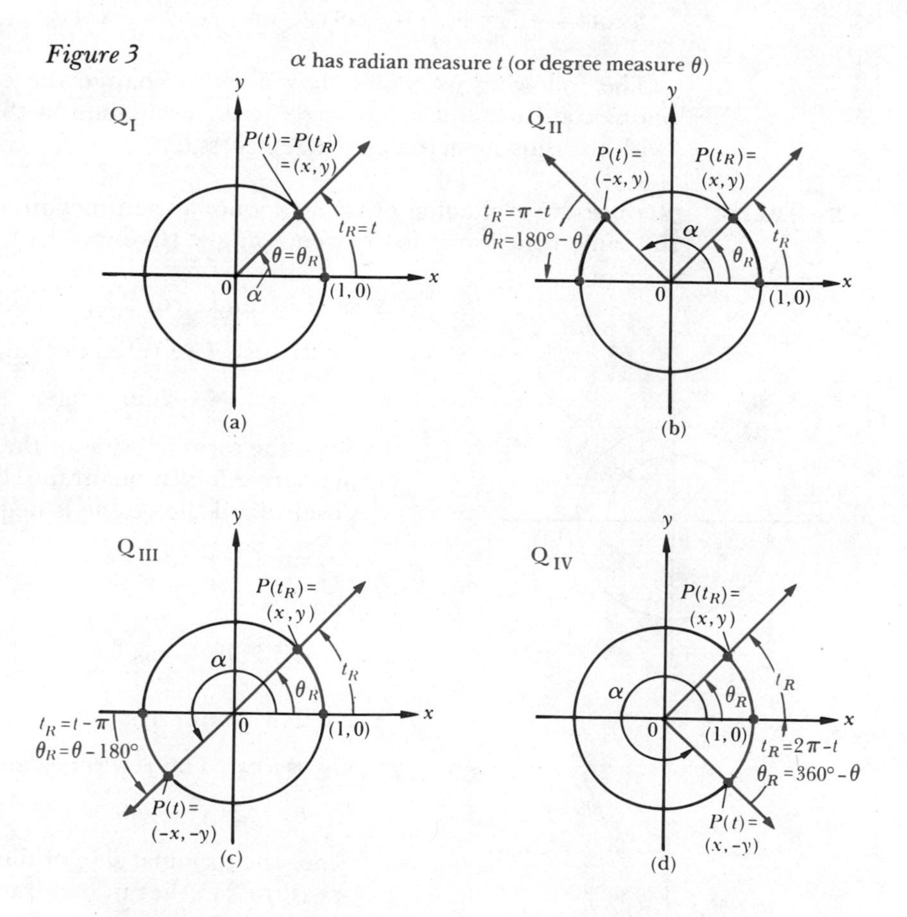

It is always true that the trigonometric function values of α and the trigonometric function values of the reference angle of α are the same in absolute value, but they may differ in sign (according to the function being evaluated and the quadrant where the terminal side of α lies). (See Section 2.2 on page 200).

We verify this assertion for the situation where the terminal side of α lies in quadrant II (Figure 3b) as follows (see Problem 24 for the situations where the terminal side lies in quadrants III and IV). By the definitions of the trigonometric functions given on pages 186 and 231, we have

$$\cos t = -x = -\cos t_R \quad \text{or} \quad \cos\theta = -\cos\theta_R$$
$$\sin t = y = \sin t_R \quad \text{or} \quad \sin\theta = \sin\theta_R$$
$$\tan t = \frac{y}{-x} = -\frac{y}{x} = -\tan t_R \quad \text{or} \quad \tan\theta = -\tan\theta_R$$
$$\sec t = \frac{1}{-x} = -\frac{1}{x} = -\sec t_R \quad \text{or} \quad \sec\theta = -\sec\theta_R$$
$$\csc t = \frac{1}{y} = \csc t_R \quad \text{or} \quad \csc\theta = \csc\theta_R$$
$$\cot t = \frac{-x}{y} = -\frac{x}{y} = -\cot t_R \quad \text{or} \quad \cot\theta = -\cot\theta_R$$

The following examples show how to change the evaluation of trigonometric functions of an angle to an evaluation at the reference angle (with an adjustment in sign where necessary).

EXAMPLES Reduce the evaluation of each trigonometric function to an evaluation of the same function at its reference angle. (Use $\pi = 3.14$.)

Figure 4

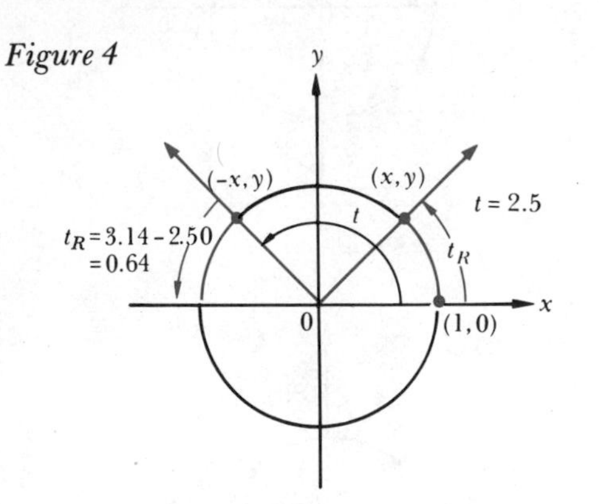

1 sin 2.5 and cos 2.5

SOLUTION The reference angle is

$$t_R = 3.14 - 2.50 = 0.64 \qquad \text{(Figure 4)}$$

Since the terminal side of the angle with radian measure 2.5 is in quadrant II, where the sine is positive and the cosine is negative, we have

$$\sin 2.5 = \sin 0.64$$

and

$$\cos 2.5 = -\cos 0.64$$

Figure 5

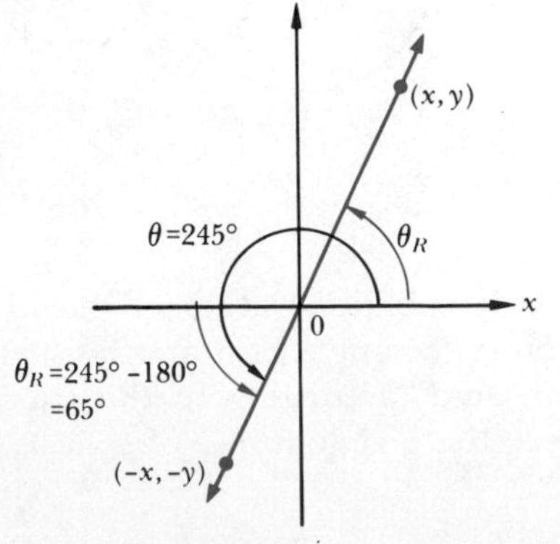

2 tan 245° and cot 245°

SOLUTION The reference angle is

$$\theta_R = 245° - 180° = 65° \qquad \text{(Figure 5)}$$

Since the terminal side of the angle with degree measure 245° lies in quadrant III, and the tangent and cotangent functions are both positive there, we have

$$\tan 245° = \tan 65°$$

and

$$\cot 245° = \cot 65°$$

3 sec 24.84 and csc 24.84

SOLUTION Using the periodicity of the secant and cosecant functions, we have

$$\sec 24.84 = \sec[6 + 3(6.28)] = \sec 6$$

and

$$\csc 24.84 = \csc[6 + 3(6.28)] = \csc 6$$

The reference angle t_R for 6 is

$$t_R = 6.28 - 6 = 0.28$$

The terminal side of the angle with radian measure 6 lies in quadrant IV (Figure 6), where the secant is positive and the cosecant is negative, so

$$\sec 6 = \sec 0.28$$

and

$$\csc 6 = -\csc 0.28$$

so that

$$\sec 24.84 = \sec 6 = \sec 0.28$$

and

$$\csc 24.84 = \csc 6 = -\csc 0.28$$

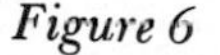
Figure 6

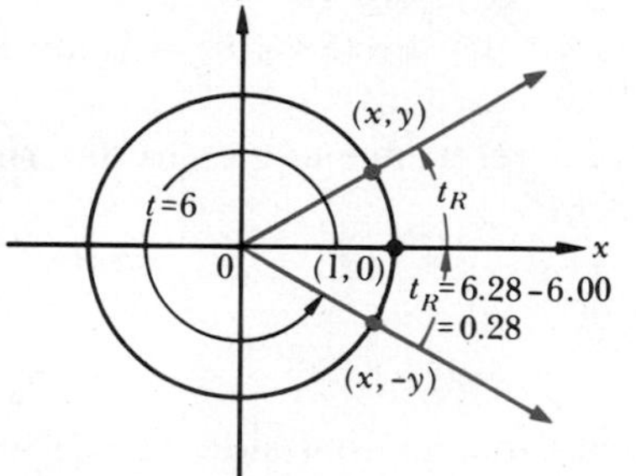

4 sin(−220°) and cos(−220°)

SOLUTION The terminal side of the angle with measure −220° coincides with the terminal side of the angle with measure 140°, which lies in quadrant II. Thus, the reference angle θ_R for −220° is the same as the reference angle for 140°, so

$$\theta_R = 180° - 140° = 40° \qquad \text{(Figure 7)}$$

Also, the sine is positive and the cosine is negative in quadrant II, so

$$\sin(-220°) = \sin 140° = \sin 40°$$

and

$$\cos(-220°) = \cos 140° = -\cos 40°$$

Figure 7

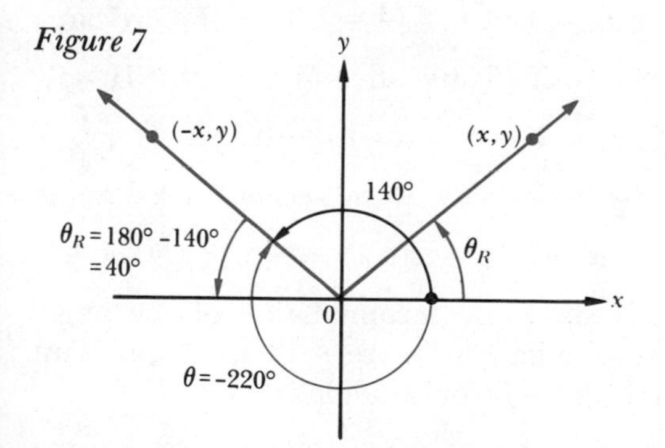

5.2 Values of Trigonometric Functions—Tables

The evaluation of a trigonometric function can always be reduced to an evaluation at a reference angle t_R, where $0 < t_R < \pi/2$ (or θ_R, where $0° < \theta_R < 90°$), with the proper adjustment in sign, or at 0, $\pi/2$, π, or $3\pi/2$ (0°, 90°, 180°, or 270°). The only problem that remains is to determine the values of the trigonometric functions for the reference angle.

For certain special values, such as $\pi/6$, $\pi/4$, and $\pi/3$ (or 30°, 45°, and 60°), the problem has been solved (see Problem Set 1, Problem 68). For other values, there are tables of values (for radian measures use Appendix Table III, and for degree measures use Appendix Table IV), which have been constructed by using advanced mathematical techniques. As with the logarithm tables, these tables give *approximations* of the values. Whenever Appendix Table III or IV is used, the equal sign is understood to mean "approximately equal to."

EXAMPLES 1 Use Appendix Table III for radian measures and Appendix Table IV for degree measures to determine each of the following values.

(a) $\cos 0.65$ (b) $\sin 0.84$ (c) $\tan 0.76$
(d) $\cot 43°$ (e) $\sec(75°10')$ (f) $\csc(83°40')$

SOLUTION

(a) $\cos 0.65 = 0.7961$ (b) $\sin 0.84 = 0.7446$
(c) $\tan 0.76 = 0.9505$ (d) $\cot 43° = 1.072$
(e) $\sec(75°10') = 3.906$ (f) $\csc(83°40') = 1.006$

2 Use Appendix Tables III and IV and reference angles to determine each of the following values. (Use $\pi = 3.14$.)

(a) $\cos 2.14$ (b) $\sec 41.8$ (c) $\cot(-4.1)$
(d) $\sin 339°$ (e) $\csc 965°$

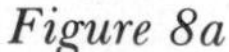
Figure 8a

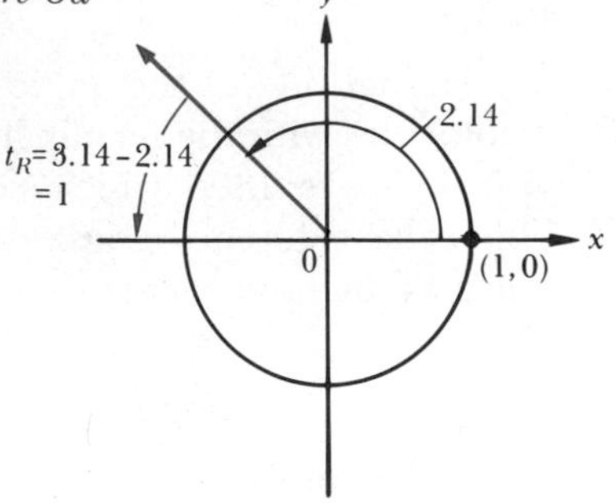

Figure 8b

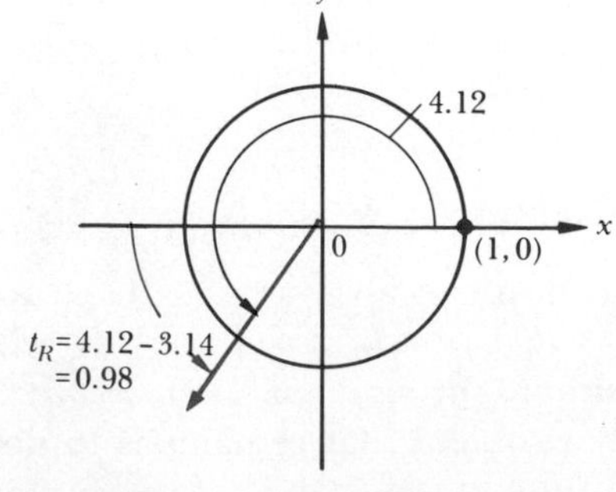

SOLUTION

(a) Since the terminal side of the angle with radian measure 2.14 lies in quadrant II, the reference angle is

$$t_R = 3.14 - 2.14 = 1.0 \qquad \text{(Figure 8a)}$$

The cosine is negative in quadrant II, so

$$\cos 2.14 = -\cos 1 = -0.5403$$

(b) The periodicity of the secant is used to get

$$\sec 41.8 = \sec[4.12 + 6(6.28)] = \sec 4.12$$

But since the terminal side of the angle with radian measure 4.12 is in quadrant III, the reference angle is

$$t_R = 4.12 - 3.14 = 0.98 \qquad \text{(Figure 8b)}$$

The secant is negative in quadrant III, so

$$\sec 41.8 = \sec 4.12$$
$$= -\sec 0.98 = -1.795$$

(c) The terminal side of the angle with radian measure -4.1 is in quadrant II and coincides with the terminal side of the angle

Figure 8c

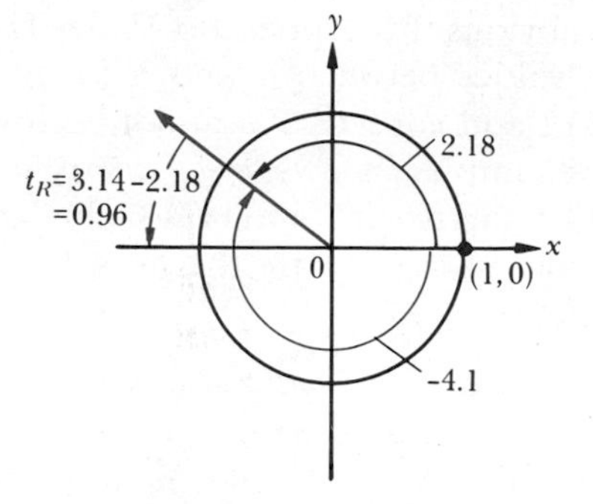

with radian measure 2.18 (Figure 8c). The reference angle is

$$t_R = 3.14 - 2.18 = 0.96$$

so

$$\cot(-4.1) = \cot 2.18 = -\cot 0.96 = -0.7001$$

since the cotangent is negative in quadrant II.

(d) Since the terminal side of an angle with degree measure 339° is in quadrant IV, the reference angle is

$$\theta_R = 360° - 339° = 21° \qquad \text{(Figure 8d)}$$

The sine is negative in quadrant IV, thus

$$\sin 339° = -\sin 21° = -0.3584$$

(e) The periodicity of the cosecant is used to get

$$\csc 965° = \csc[245° + 2(360°)] = \csc 245°$$

Since the terminal side of 245° is in quadrant III (Figure 8e), the reference angle is

$$\theta_R = 245° - 180° = 65°$$

The cosecant is negative in quadrant III, so

$$\csc 245° = -\csc 65° = -1.103$$

Hence $\csc 965° = -1.103$.

Figure 8d *Figure 8e*

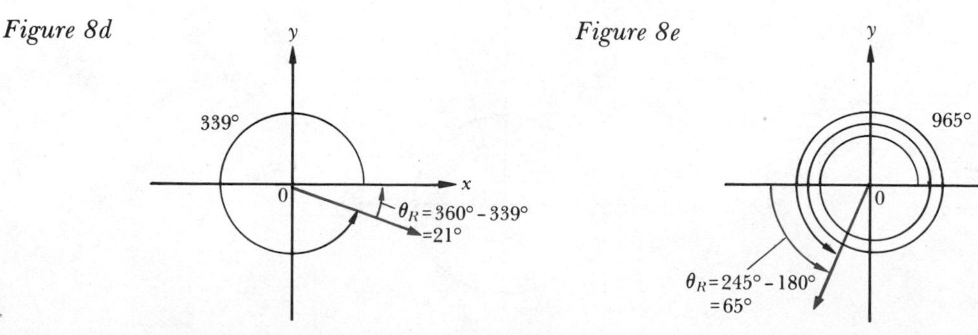

It is obvious that Appendix Tables III and IV do not contain entries for *all* the values between 0 and $\pi/2$, or between 0 and 90°. For example, sin 1.514 and cos(31°24′) cannot be found in these tables. As we shall see in the examples below, *linear interpolation* (see Chapter 4, Section 4) can be used to arrive at some reasonably accurate approximations.

First, let us summarize the steps for finding the trigonometric values of an angle.

1. Locate the quadrant where the terminal side of the given angle lies.
2. Determine the reference angle, an acute angle formed by the terminal side of the given angle and the x axis.
3. Use Appendix Table III if radians are being used, or Appendix Table IV if degrees are being used, and linear interpolation (if necessary) to find the trigonometric function value of the reference angle.
4. Attach the appropriate sign (+ or −) to the number obtained in step 3, depending on whether the desired trigonometric function value of the given angle is positive or negative in the quadrant in which its terminal side lies.

EXAMPLES Use Appendix Table III or IV and linear interpolation to approximate each value. (Use $\pi = 3.142$.)

1 sin 1.514

SOLUTION The terminal side of an angle of radian measure 1.514 is in quadrant I, so the reference angle $t_R = 1.514$ (Figure 9).

Appendix Table III, together with linear interpolation, can be used to find sin 1.514 as follows:

$$
\begin{array}{ccc}
 & t & \sin t \\
 & 1.51 & 0.9982 \\
0.01 \left[0.006 \left[\begin{array}{c} 1.514 \\ 1.52 \end{array} \right. \right. & & \left. \left. \begin{array}{c} ? \\ 0.9987 \end{array} \right] d \right] 0.0005
\end{array}
\qquad \frac{d}{0.0005} = \frac{0.006}{0.01}
$$

Hence, $d = 0.0003$ and

$$\sin 1.514 = 0.9987 - 0.0003 = 0.9984$$

Figure 9

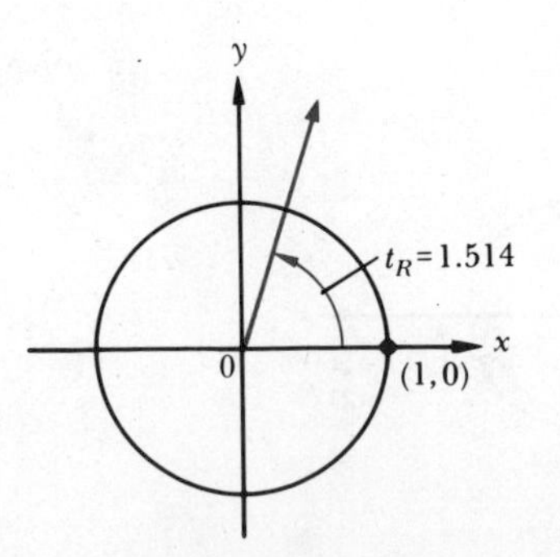

2 $\cos(31°24')$

SOLUTION The reference angle $\theta_R = 31°24'$ (Figure 10). Since $31°24'$ is between $31°20'$ and $31°30'$, by using Appendix Table IV and linear interpolation, we get

$$10'\left[4'\begin{bmatrix}31°20' \\ 31°24'\end{bmatrix} \atop 31°30'\right] \quad \begin{array}{c}\theta \\ 31°20' \\ 31°24' \\ 31°30'\end{array} \qquad \begin{array}{c}\cos\theta \\ 0.8542 \\ ? \\ 0.8526\end{array}\quad d,\ 0.0016 \qquad \frac{d}{0.0016} = \frac{4}{10}$$

and d is approximately 0.0006, so

$$\cos(31°24') = 0.8542 - 0.0006 = 0.8536$$

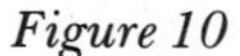

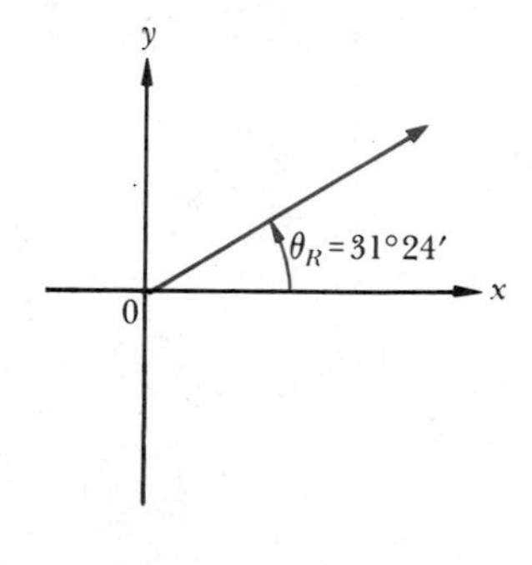

Figure 10

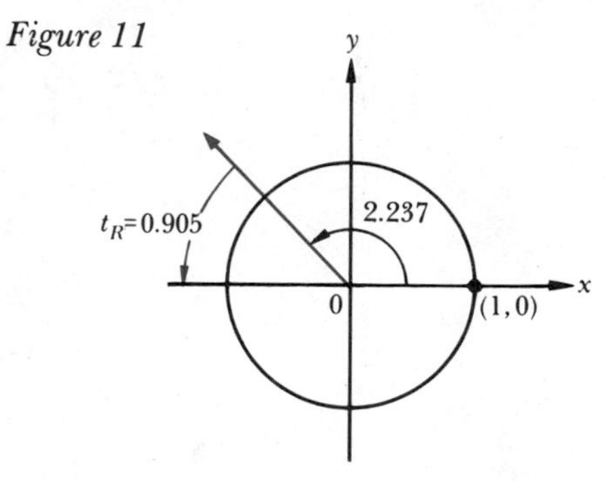

Figure 11

3 cot 8.521

SOLUTION

$$\cot 8.521 = \cot(2.237 + 6.284) = \cot 2.237$$

However, $\cot 2.237 = -\cot 0.905$, since the reference angle is

$$t_R = 3.142 - 2.237 = 0.905$$

and 2.237 is in quadrant II, where the cotangent is negative (Figure 11). Using linear interpolation and Appendix Table III, we have

$$0.01\left[0.005 \quad \begin{array}{c}t \\ 0.90 \\ 0.905 \\ 0.91\end{array} \qquad \begin{array}{c}\cot t \\ 0.7936 \\ ? \\ 0.7774\end{array}\right]\ d,\ 0.0162 \qquad \frac{d}{0.0162} = \frac{0.005}{0.01}$$

so $d = 0.0081$ and

$$\cot 0.905 = 0.7774 + 0.0081 = 0.7855$$

Hence

$$\cot 8.521 = \cot 2.237 = -\cot 0.905 = -0.7855$$

4 $\sec(-118°42')$

SOLUTION Since $-118°42'$ has the same terminal side as $241°18'$ and is located in quadrant III, $\sec(-118°42')$ is negative and the reference angle is $\theta_R = 241°18' - 180° = 61°18'$, and $\sec(-118°42') = -\sec(61°18')$ (Figure 12). Using Appendix Table IV and linear interpolation, we have

		θ	$\sec\theta$		
$10'$	$8'$	$61°10'$	2.074	d	0.011
		$61°18'$	$?$		
		$61°20'$	2.085		

$$\frac{d}{0.011} = \frac{8}{10}$$

and d is approximately 0.009, so

$$\sec(-118°42') = -\sec 61°18' = -(2.074 + 0.009) = -2.083$$

Figure 12

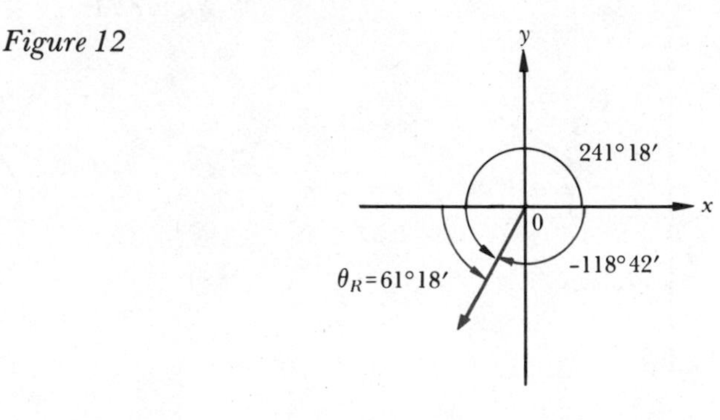

5 $\tan(-920°25')$

SOLUTION Since the tangent is an odd function,

$$\tan(-920°25') = -\tan(920°25')$$

Using periodicity, we have

$$-\tan(920°25') = -\tan[200°25' + 2(360°)] = -\tan(200°25')$$

An angle of $200°25'$ in standard position has its terminal side in quadrant III. The reference angle $\theta_R = 200°25' - 180° = 20°25'$ (Figure 13). Since the tangent is positive in quadrant III, we have $-\tan(200°25') = -\tan(20°25')$, and $\tan(20°25')$ can be approximated by linear interpolation as follows:

		θ	$\tan\theta$		
$10'$	$5'$	$20°20'$	0.3706	d	0.0033
		$20°25'$	$?$		
		$20°30'$	0.3739		

$$\frac{d}{0.0033} = \frac{5}{10}$$

so $d = 0.0017$. Hence, $\tan(20°25') = 0.3706 + 0.0017 = 0.3723$ and

$$\tan(-920°25') = -\tan(200°25') = -\tan(20°25') = -0.3723$$

Figure 13

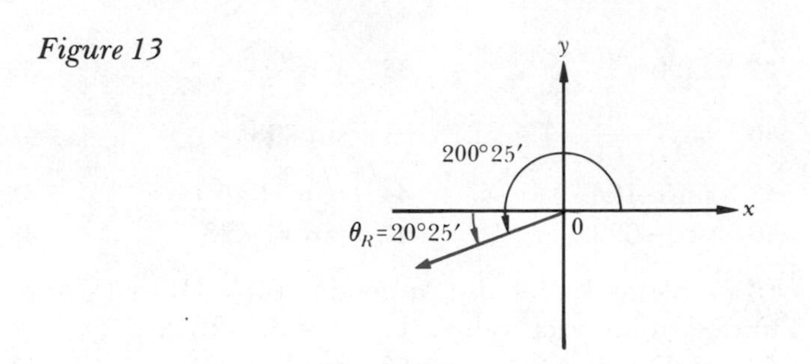

As with logarithms, the computation of the values of the trigonometric functions is easier with the use of a calculator than with the use of the tables and linear interpolation. At this level, it is strongly recommended that the student perform the calculations *both* ways, if a suitable calculator is available. The use of the tables will reinforce the *concepts* associated with evaluating the trigonometric functions and the repeated use of a calculator will enable the student to acquire a useful skill.

PROBLEM SET 5

In Problems 1–12, reduce the evaluation of the given function of the given angle to an evaluation of the same function at an angle between 0 and 2π for radian measures or between 0° and 360° for degree measures. (Use $\pi = 3.14$.)

1 $\sin 8.82$ **2** $\cos 7.76$ **3** $\tan 6.92$
4 $\cot 910°$ **5** $\sec(-390°)$ **6** $\csc(800°17')$
7 $\cos(-32.1)$ **8** $\sin(-10.53)$ **9** $\cot 12.15$
10 $\sec(-85°12')$ **11** $\cos(700°52')$ **12** $\tan(-200°10')$

In Problems 13–23, use Appendix Table III or IV, depending on whether the angle is measured in radians or degrees, to determine each of the following values.

13 $\sin 0.18$ **14** $\cos 0.64$ **15** $\tan(-1.31)$
16 $\sin 59°$ **17** $\cos 53°$ **18** $\cot(85°50')$
19 $\cot(-0.44)$ **20** $\sec 1.51$ **21** $\sin(-70°30')$
22 $\sec(51°20')$ **23** $\csc(-19°40')$

24 Use Figures 3c and 3d to verify the assertion on page 241 for quadrants III and IV, respectively.

In Problems 25–48, reduce the evaluation of each trigonometric function to an evaluation of the same function at its reference angle. Then use Appendix Table III or IV to determine each of the following values. (Use $\pi = 3.14$.)

25 $\cos 3$ **26** $\sin 5.8$ **27** $\tan 2$
28 $\cot 405°$ **29** $\sec(-770°)$ **30** $\csc(-750°)$
31 $\tan 510°$ **32** $\cot(-15)$ **33** $\sin(-14.23)$
34 $\cos(-1237°)$ **35** $\sin 850°$ **36** $\cot 1375°$

37 sin 60 **38** $\cot \frac{36\pi}{5}$ **39** sec(−495°)

40 $\cos\left(-\frac{49\pi}{11}\right)$ **41** sin 798° **42** tan 64

43 sin(−1045°) **44** cos(−625°) **45** cos 15.42

46 sin(−623°) **47** tan 119.23 **48** sec(−495°)

In Problems 49–59, use Appendix Table III or IV and linear interpolation to determine each value. (Use $\pi = 3.142$.)

49 cos 1.588 **50** csc(167°45′) **51** sin(−4.233)

52 sec(121°27′) **53** tan(326°15′) **54** cot 5.151

55 cot(760°38′) **56** sin(−12.468) **57** sec(−326°18′)

58 tan(103°25′) **59** cos 12.407

60 If available, use a calculator to recalculate the values in Problems 49–59. Compare the answers determined with the calculator to the answers determined by using Appendix Table III or IV.

REVIEW PROBLEM SET

In Problems 1–12, display the location of each point on the unit circle and determine the quadrant, if any, in which each point lies. (Use $\pi = 3.14$.)

1 $P\left(-\frac{3\pi}{2}\right)$ **2** $P\left(\frac{2\pi}{5}\right)$ **3** $P\left(-\frac{7\pi}{5}\right)$

4 $P(3.51)$ **5** $P(-4.32)$ **6** $P(-6.19)$

7 $P\left(\frac{3\pi}{7} + 2\pi\right)$ **8** $P(0.5 + 6.28)$ **9** $P(10.2)$

10 $P(-11)$ **11** $P(-77)$ **12** $P(127)$

In Problems 13–24, use the properties of the wrapping function of Sections 1 and 2 to determine each value (refer to Figure 14 on page 192).

13 $P(3\pi)$ **14** $P\left(\frac{9\pi}{2}\right)$ **15** $P\left(-\frac{9\pi}{2}\right)$

16 $P\left(-\frac{\pi}{4} + 8\pi\right)$ **17** $P\left(-\frac{\pi}{3} + 2\pi\right)$ **18** $P\left(-\frac{5\pi}{6} - 4\pi\right)$

19 $P\left(-\frac{7\pi}{3}\right)$ **20** $P\left(\frac{79\pi}{3}\right)$ **21** $P\left(\frac{29\pi}{6}\right)$

22 $P(105\pi)$ **23** $P\left(-\frac{97\pi}{4}\right)$ **24** $P\left(-\frac{121\pi}{3}\right)$

In Problems 25–33, refer to Figure 14 on page 192, if necessary, to find each value.

25 $\sec\left(-\frac{5\pi}{3}\right)$ **26** $\cot \frac{37\pi}{6}$ **27** $\cos\left(-\frac{47\pi}{6}\right)$

28 $\sin \frac{89\pi}{2}$ **29** $\tan \frac{14\pi}{3}$ **30** $\csc\left(-\frac{25\pi}{4}\right)$

31 $\cos\left(-\frac{125\pi}{6}\right)$ **32** $\sec \frac{325\pi}{3}$ **33** $\tan \frac{721\pi}{4}$

34 State the quadrant in which $P(t)$ lies if
(a) $\sin t < 0$ and $\cos t < 0$
(b) $\sec t > 0$ and $\cot t < 0$

In Problems 35–38, find the values of the indicated trigonometric functions.

35 $\csc t$ and $\cot t$ if $\sin t = 2/3$ and $P(t)$ is in quadrant II
36 $\sec t$ and $\tan t$ if $\cos t = 3/8$ and $P(t)$ is in quadrant IV
37 $\sin t$ and $\cos t$ if $\csc t = 5$ and $P(t)$ is in quadrant I
38 $\cot t$ and $\cos t$ if $\tan t = 5/12$ and $P(t)$ is in quadrant III

In Problems 39–47, use the known graphs of the trigonometric functions to graph one cycle of each function. Indicate the amplitude, the period, and the phase shift.

39 $f(x) = \frac{1}{3}\cos x$
40 $g(x) = \sin \frac{\pi}{3}x$
41 $h(x) = 2\sin(-2x)$
42 $F(x) = \frac{1}{4}\sin\left(x - \frac{\pi}{6}\right)$
43 $G(x) = \frac{1}{5}\cos\left(2x + \frac{\pi}{8}\right)$
44 $H(x) = -2\tan\left(\frac{x}{2} - 1\right)$
45 $f(x) = \frac{3}{2}\cot x$
46 $g(x) = 4\csc(x - 1)$
47 $h(x) = 3\sec 4x$

48 An iron ball of mass 10 kilograms is attached to a spring that oscillates freely in a simple harmonic motion defined by the equation

$$y = 2\sin\sqrt{\frac{k}{m}}\, t$$

Find the period and frequency of oscillation if $k = 4$ and m represents the mass.

In Problems 49–54, sketch the angle in standard position and give its radian measure.

49 $290°$ **50** $-105°$ **51** $820°$
52 $-70°$ **53** $-12°$ **54** $190°$

In Problems 55–60, sketch the angle in standard position and give its degree measure. (Use $\pi = 3.14$.)

55 $\frac{5\pi}{3}$ **56** $\frac{17\pi}{5}$ **57** $-\frac{13\pi}{4}$
58 $-\frac{18\pi}{7}$ **59** -4.7 **60** 5.82

In Problems 61–64, find the length of the circular arc and the area of the circular sector that are determined by a central angle θ in a circle of radius r.

61 $r = 7$ centimeters and $\theta = 75°$
62 $r = 2$ inches and $\theta = \frac{8\pi}{9}$
63 $r = 10$ inches and $\theta = \frac{\pi}{5}$
64 $r = 4$ centimeters and $\theta = 340°$

65 A circle of radius 4 inches contains a central angle θ that intercepts an arc of 10 inches. Express θ in radian measure and in degree measure. Find the area of the sector whose central angle is θ.

66 A bicycle with wheels of radius 13 inches travels 52 feet. What is the radian measure of the angle through which a wheel of the bicycle has turned?

In Problems 67–70, find the values of the trigonometric functions of the angle θ.

67 θ has $(-6, -8)$ located on its terminal side.
68 θ has $(-7, 24)$ located on its terminal side.
69 θ has $(5, -6)$ located on its terminal side.
70 θ has $(-8, -17)$ located on its terminal side.

In Problems 71–74, construct each angle in standard position, with its terminal side in the indicated quadrant, and evaluate the other trigonometric functions of θ.

71 $\tan \theta = \frac{6}{5}$, quadrant III
72 $\sec \theta = \frac{5}{4}$, quadrant IV
73 $\cot \theta = -\frac{25}{7}$, quadrant II
74 $\sin \theta = \frac{12}{13}$, quadrant I

In Problems 75–78, determine the values of the six trigonometric functions of the given angle θ without using a table.

75 $\theta = -330°$ **76** $\theta = 135°$ **77** $\theta = 210°$ **78** $\theta = -\frac{17\pi}{2}$

In Problems 79–95, use the steps outlined on page 246 to determine the following values. (Use $\pi = 3.14$.)

79 $\sin 3$
80 $\sin(-208°)$
81 $\cos 4.2$
82 $\cos(-3.92)$
83 $\tan(-209°40')$
84 $\tan 8.2$
85 $\cot(604°20')$
86 $\cot(-26.37)$
87 $\sec 18.71$
88 $\sec(-407°30')$
89 $\csc(-26.37)$
90 $\csc(248°5')$
91 $\sin 5.115$
92 $\cot(658°23')$
93 $\cos(743°56')$
94 $\sec(-19.213)$
95 $\tan(-241°28')$

96 Use a calculator to perform each evaluation in Problems 79–95. Compare the answers determined by the calculator to the answers determined above.

CHAPTER 6

Analytic Trigonometry

The basic properties and the graphs of the trigonometric functions discussed in Chapter 5 are used here to investigate various aspects of trigonometry. The topics covered include the fundamental identities, trigonometric formulas and equations, right triangle trigonometry, polar coordinates, and the laws of sines and cosines for triangles. Inverse trigonometric functions and an introduction to vectors and their applications are also presented.

1 Fundamental Identities

An equation that is true for all values of the variable for which both sides of the equation are defined is called an *identity*. In Chapter 5, we derived some of the basic *trigonometric identities,* called the *fundamental identities,* that relate the trigonometric functions to one another. These identities, listed below, hold for all values of t in the domains of the functions involved.

1 $\tan t = \dfrac{\sin t}{\cos t}$

2 $\cot t = \dfrac{\cos t}{\sin t}$

3 $\sec t = \dfrac{1}{\cos t}$

4 $\csc t = \dfrac{1}{\sin t}$

5 $\cot t = \dfrac{1}{\tan t}$

6 $\sin^2 t + \cos^2 t = 1$

7 $\cos(-t) = \cos t$

8 $\sin(-t) = -\sin t$

9 $\tan(-t) = -\tan t$

10 $\cot(-t) = -\cot t$

11 $\sec(-t) = \sec t$

12 $\csc(-t) = -\csc t$

13 $\tan^2 t + 1 = \sec^2 t$

14 $\cot^2 t + 1 = \csc^2 t$

The following examples illustrate some uses of the fundamental identities.

EXAMPLES 1 Express each of the following in terms of x if $\sec\theta = x/2$, for $x \neq 0$, and θ is an angle in standard position with its terminal side in quadrant I.

(a) $\cos\theta$ (b) $\sin\theta$ (c) $\tan\theta$

SOLUTION

(a) By Identity 3,

$$\cos\theta = \frac{1}{\sec\theta} = \frac{1}{x/2} = \frac{2}{x}$$

(b) Since the terminal side of θ is in quadrant I, Identity 6 yields

$$\sin\theta = \sqrt{1-\cos^2\theta} = \sqrt{1-\left(\frac{2}{x}\right)^2} = \sqrt{1-\frac{4}{x^2}}$$

$$= \sqrt{\frac{x^2-4}{x^2}} = \frac{\sqrt{x^2-4}}{x}$$

(c) By Identity 1,

$$\tan\theta = \frac{\sin\theta}{\cos\theta} = \frac{(\sqrt{x^2-4})/x}{2/x} = \frac{\sqrt{x^2-4}}{2}$$

2 Simplify each of the following expressions.

(a) $\sin^2 315° + \cos^2 315°$

(b) $\csc^2\left(\frac{\theta}{3} + 120°\right) - \cot^2\left(\frac{\theta}{3} + 120°\right)$

SOLUTION

(a) Since $\sin^2\theta + \cos^2\theta = 1$, then $\sin^2 315° + \cos^2 315° = 1$.

(b) Since $\cot^2\theta + 1 = \csc^2\theta$, then $\csc^2\theta - \cot^2\theta = 1$, so that

$$\csc^2\left(\frac{\theta}{3} + 120°\right) - \cot^2\left(\frac{\theta}{3} + 120°\right) = 1$$

3 Write each of the following expressions in terms of $\sin\theta$.

(a) $\dfrac{\sec\theta}{\tan\theta + \cot\theta}$ (b) $(\tan\theta + \cot\theta)\cot\theta$

SOLUTION

(a)
$$\frac{\sec\theta}{\tan\theta+\cot\theta} = \frac{\frac{1}{\cos\theta}}{\frac{\sin\theta}{\cos\theta}+\frac{\cos\theta}{\sin\theta}} = \frac{\frac{1}{\cos\theta}}{\frac{\sin^2\theta+\cos^2\theta}{\sin\theta\cos\theta}} = \frac{\frac{1}{\cos\theta}}{\frac{1}{\sin\theta\cos\theta}}$$

$$= \frac{\sin\theta\cos\theta}{\cos\theta} = \sin\theta$$

(b)
$$(\tan\theta+\cot\theta)\cot\theta = \left(\frac{1}{\cot\theta}+\cot\theta\right)\cot\theta = \left(\frac{1+\cot^2\theta}{\cot\theta}\right)\cot\theta$$

$$= 1+\cot^2\theta = \csc^2\theta = \frac{1}{\sin^2\theta}$$

1.1 Verifying Trigonometric Identities

Identities can often be used to change an expression involving trigonometric functions into a different but equivalent form that is more suitable for the purpose at hand.

The following suggestions are useful in proving that equations are identities.

1. Simplify one side until it is the same as the other side.
2. Simplify each side separately until both sides are the same.
3. To simplify a side of an identity:
   (a) If the side contains a sum or difference of fractions, it may be helpful to combine them into a single fraction.
   (b) If the numerator of a fraction has more than one term, it may be helpful to split the fraction into a sum of fractions.
4. If in doubt, write the expressions in terms of sines and cosines and then simplify.

EXAMPLES Prove that each equation is an identity.

1 $\sec^2\theta\csc^2\theta = \csc^2\theta + \sec^2\theta$

PROOF

$$\begin{aligned}\sec^2\theta\csc^2\theta &= (1+\tan^2\theta)(1+\cot^2\theta)\\ &= 1+\cot^2\theta+\tan^2\theta+\tan^2\theta\cot^2\theta\\ &= 1+\cot^2\theta+\tan^2\theta+\tan^2\theta\,\frac{1}{\tan^2\theta}\\ &= 1+\cot^2\theta+\tan^2\theta+1 = \csc^2\theta+\sec^2\theta\end{aligned}$$

2 $1-\tan^4\theta = 2\sec^2\theta - \sec^4\theta$

PROOF

$$\begin{aligned}2\sec^2\theta-\sec^4\theta &= \sec^2\theta(2-\sec^2\theta)\\ &= \sec^2\theta[2-(\tan^2\theta+1)]\\ &= (1+\tan^2\theta)(1-\tan^2\theta)\\ &= 1-\tan^4\theta\end{aligned}$$

3 $\dfrac{\sin\theta}{\csc\theta-\cot\theta} = 1+\cos\theta$

PROOF

$$\begin{aligned}\frac{\sin\theta}{\csc\theta-\cot\theta} &= \frac{\sin\theta}{\dfrac{1}{\sin\theta}-\dfrac{\cos\theta}{\sin\theta}} = \frac{\sin^2\theta}{1-\cos\theta} = \frac{1-\cos^2\theta}{1-\cos\theta}\\ &= \frac{(1-\cos\theta)(1+\cos\theta)}{1-\cos\theta} = 1+\cos\theta\end{aligned}$$

4 $\dfrac{\cos(-t)}{1 - \sin t} = \dfrac{1 - \sin(-t)}{\cos t}$

PROOF Since the cosine function is even and the sine is odd, we have $\cos(-t) = \cos t$ and $\sin(-t) = -\sin t$. Hence it is enough to prove that

$$\frac{\cos t}{1 - \sin t} = \frac{1 + \sin t}{\cos t}$$

Thus

$$\begin{aligned}\frac{\cos t}{1 - \sin t} &= \frac{\cos t(1 + \sin t)}{(1 - \sin t)(1 + \sin t)} \\ &= \frac{\cos t(1 + \sin t)}{1 - \sin^2 t} \\ &= \frac{\cos t(1 + \sin t)}{\cos^2 t} \\ &= \frac{1 + \sin t}{\cos t}\end{aligned}$$

PROBLEM SET 1

In Problems 1–2, assume that θ is an angle in standard position with its terminal side in quadrant I. Express the other five trigonometric functions of θ in terms of x.

1 $\sin \theta = x$

2 $\tan \theta = x$

In Problems 3–16, use the fundamental identities to simplify each expression.

3 $\sin \dfrac{25\pi}{8} \cot \dfrac{25\pi}{8} - \cos \dfrac{25\pi}{8}$

4 $(\tan 370° + \cot 370°) - \sec 370° \csc 370°$

5 $\sin^4 15 + 2 \sin^2 15 \cos^2 15 + \cos^4 15$

6 $(1 - \cos \theta)(1 + \cos \theta)$

7 $\cos^2 100t + \sin^2 100t$

8 $\dfrac{\sin(-t)}{\cos(-t)}$

9 $\dfrac{\cos t + \sin^2 t \sec t}{\sec t}$

10 $\dfrac{1}{1 - \cos \theta} + \dfrac{1}{1 + \cos \theta}$

11 $\dfrac{1 + \tan^2 \theta}{\csc^2 \theta}$

12 $\dfrac{\sec t \csc t}{\tan t + \cot t}$

13 $\dfrac{1 - \cos^2 t}{\sin t}$

14 $\dfrac{\cos \theta \csc \theta}{\csc^2 \theta - 1}$

15 $\dfrac{\csc \theta}{\tan \theta + \cot \theta}$

16 $\sec^2 t - \tan^2 t$

In Problems 17–20, write each expression in terms of sines and/or cosines.

17 $\cos\theta + \tan\theta\sin\theta$

18 $\dfrac{1}{(\cot\theta + \tan\theta)\cos\theta}$

19 $(\sec t + \csc t)^2 \cot t$

20 $\dfrac{\tan t}{\sin t(1 + \tan^2 t)}$

In Problems 21–40, prove that each of the given equations is an identity.

21 $\tan t \csc t = \sec t$

22 $\sin t \cot t = \cos t$

23 $\sec^2\theta(1 - \sin^2\theta) = 1$

24 $\cot^2\theta(\sec^2\theta - 1) = 1$

25 $\sin t(\csc t - \sin t) = \cos^2 t$

26 $\sin t \cos t \cot t = 1 - \sin^2 t$

27 $\cos^2\theta - \sin^2\theta = 2\cos^2\theta - 1$

28 $\cos^4\theta - \sin^4\theta = 1 - 2\sin^2\theta$

29 $(\cos t - \sin t)^2 + (\cos t + \sin t)^2 = 2$

30 $(a\cos t + b\sin t)^2 + (-a\sin t + b\cos t)^2 = a^2 + b^2$

31 $\dfrac{\sin(-t)}{1 - \cos(-t)} = \dfrac{1 + \cos(-t)}{\sin(-t)}$

32 $\dfrac{\sin t}{1 - \cos t} = \csc t + \cot t$

33 $(\tan\theta + \cot\theta)^2 = \sec^2\theta\csc^2\theta$

34 $\dfrac{\sin\theta}{\tan\theta} + \dfrac{\cos\theta}{\cot\theta} = \sin\theta + \cos\theta$

35 $\csc t + \cot t = \dfrac{1}{\csc t - \cot t}$

36 $\dfrac{1 + \tan t}{\sin t} - \sec t = \csc t$

37 $\tan^2\theta - \sin^2\theta = \dfrac{\sin^4\theta}{\cos^2\theta}$

38 $-\sin^2\theta(1 - \csc^2\theta) = \cos^2\theta$

39 $\dfrac{\sin t}{1 + \cos(-t)} + \dfrac{\sin t}{1 - \cos(-t)} = -2\csc(-t)$

40 $\dfrac{1}{\cos^2\theta} + 1 + \dfrac{\sin^2\theta}{\cos^2\theta} = 2\sec^2\theta$

2 Trigonometric Formulas

In this section we develop some important identities involving differences or sums of two angles. For example, suppose that we wish to compute the value of $\cos(t + s)$. One *might* say that $\cos(t + s) = \cos t + \cos s$, but this is *not* the general case. For example, $\cos(\pi/2 + \pi/2) = \cos\pi = -1$, whereas $\cos(\pi/2) + \cos(\pi/2) = 0$. Trigonometric formulas for double and half angles are also developed in this section.

Although the proofs of the following theorems are given in terms of the real numbers s and t, the results are also true if s and t are interpreted as angles.

2.1 Difference and Sum Formulas

THEOREM 1 COSINE OF A DIFFERENCE OR SUM

(i) $\cos(t - s) = \cos t \cos s + \sin t \sin s$

(ii) $\cos(t + s) = \cos t \cos s - \sin t \sin s$

PROOF

(i) For convenience, we illustrate the case where $0 < s < \pi/2 < t < \pi$ (Figure 1). However, the results are true for all possible values of s and t.

Figure 1

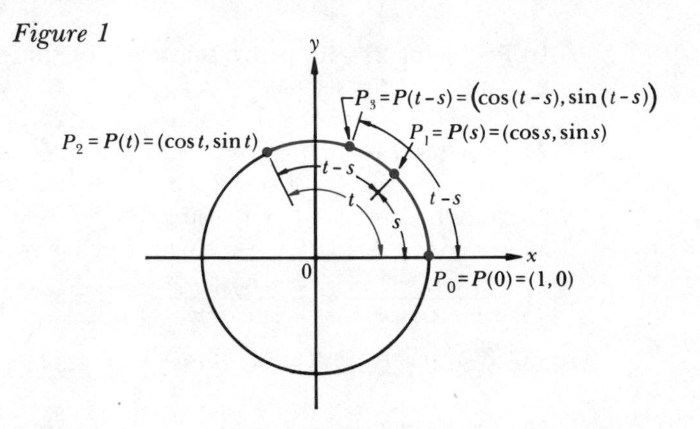

By construction, since the length of the arc $\widehat{P_0P_3}$ is equal to the length of the arc $\widehat{P_1P_2}$, it follows from geometry that the chord $\overline{P_0P_3}$ has the same length as the chord $\overline{P_1P_2}$. Using the distance formula, we have

$$\begin{aligned} |\overline{P_0P_3}|^2 &= [\cos(t-s) - 1]^2 + [\sin(t-s) - 0]^2 \\ &= \cos^2(t-s) - 2\cos(t-s) + 1 + \sin^2(t-s) \end{aligned}$$

and

$$\begin{aligned} |\overline{P_1P_2}|^2 &= (\cos t - \cos s)^2 + (\sin t - \sin s)^2 \\ &= \cos^2 t - 2\cos t \cos s + \cos^2 s + \sin^2 t - 2\sin t \sin s + \sin^2 s \end{aligned}$$

Using the facts that $\cos^2(t-s) + \sin^2(t-s) = 1$, $\cos^2 s + \sin^2 s = 1$, and $\cos^2 t + \sin^2 t = 1$, and equating $|\overline{P_1P_2}|^2$ with $|\overline{P_0P_3}|^2$, we get

$$2 - 2\cos(t-s) = 2 - 2\cos t \cos s - 2\sin t \sin s$$

so

$$-2\cos(t-s) = -2\cos t \cos s - 2\sin t \sin s$$

or

$$\cos(t-s) = \cos t \cos s + \sin t \sin s$$

(ii) $$\begin{aligned} \cos(t+s) &= \cos[t - (-s)] \\ &= \cos t \cos(-s) + \sin t \sin(-s) \qquad \text{(Theorem 1, i)} \end{aligned}$$

Since

$$\cos(-s) = \cos s \quad \text{and} \quad \sin(-s) = -\sin s$$

we have

$$\cos(t+s) = \cos t \cos s - \sin t \sin s$$

THEOREM 2

(i) $\cos\left(\frac{\pi}{2} - t\right) = \sin t$

(ii) $\sin\left(\frac{\pi}{2} - t\right) = \cos t$

PROOF Apply Theorem 1(i) to get

(i) $\cos\left(\frac{\pi}{2} - t\right) = \cos\frac{\pi}{2}\cos t + \sin\frac{\pi}{2}\sin t = 0 + \sin t = \sin t$

(ii) $\sin\left(\frac{\pi}{2} - t\right) = \cos\left[\frac{\pi}{2} - \left(\frac{\pi}{2} - t\right)\right]$ (Theorem 2, i)

$= \cos\left(\frac{\pi}{2} - \frac{\pi}{2} + t\right) = \cos t$

Notice that if Theorem 2 is applied to angles measured in degrees, then the identities are written as

$$\cos(90° - \theta) = \sin\theta \quad \text{and} \quad \sin(90° - \theta) = \cos\theta$$

THEOREM 3 SINE OF A SUM OR DIFFERENCE

(i) $\sin(t + s) = \sin t \cos s + \cos t \sin s$
(ii) $\sin(t - s) = \sin t \cos s - \cos t \sin s$

PROOF

(i) $\sin(t + s) = \cos\left[\frac{\pi}{2} - (t + s)\right]$ (Theorem 2, i)

$= \cos\left[\left(\frac{\pi}{2} - t\right) - s\right]$

$= \cos\left(\frac{\pi}{2} - t\right)\cos s + \sin\left(\frac{\pi}{2} - t\right)\sin s$ (Theorem 1, i)

$= \sin t \cos s + \cos t \sin s$

(ii) $\sin(t - s) = \sin[t + (-s)]$

$= \sin t \cos(-s) + \cos t \sin(-s)$ (Theorem 3, i)

$= \sin t \cos s - \cos t \sin s$ (Why?)

EXAMPLES 1 Simplify each expression.

(a) $\cos\left(\frac{\pi}{2} + t\right)$

(b) $\sin(90° + \theta)$

(c) $\sin 27° \cos 18° + \cos 27° \sin 18°$

(d) $\cos 25° \cos 35° - \sin 25° \sin 35°$

SOLUTION

(a) Using Theorem 1(ii), we have

$$\cos\left(\frac{\pi}{2} + t\right) = \cos\frac{\pi}{2}\cos t - \sin\frac{\pi}{2}\sin t = 0 - \sin t = -\sin t$$

(b) Using Theorem 3(i), we have

$$\sin(90° + \theta) = \sin 90° \cos\theta + \cos 90° \sin\theta = \cos\theta + 0 = \cos\theta$$

(c) Using Theorem 3(i), we have

$$\sin 27° \cos 18° + \cos 27° \sin 18° = \sin(27° + 18°) = \sin 45° = \frac{\sqrt{2}}{2}$$

(d) Using Theorem 1(ii), we have

$$\cos 25° \cos 35° - \sin 25° \sin 35° = \cos(25° + 35°) = \cos 60° = 1/2$$

2 Use the fact that $7\pi/12 = (\pi/3) + (\pi/4)$ to evaluate $\cos(7\pi/12)$.

SOLUTION

$$\cos \frac{7\pi}{12} = \cos\left(\frac{\pi}{3} + \frac{\pi}{4}\right) = \cos \frac{\pi}{3} \cos \frac{\pi}{4} - \sin \frac{\pi}{3} \sin \frac{\pi}{4}$$
$$= \left(\frac{1}{2}\right)\left(\frac{\sqrt{2}}{2}\right) - \left(\frac{\sqrt{3}}{2}\right)\left(\frac{\sqrt{2}}{2}\right) = \frac{\sqrt{2}}{4} - \frac{\sqrt{6}}{4} = \frac{\sqrt{2} - \sqrt{6}}{4}$$

3 Use the fact that $105° = 150° - 45°$ to evaluate $\sin 105°$.

SOLUTION

$$\sin 105° = \sin(150° - 45°)$$
$$= \sin 150° \cos 45° - \cos 150° \sin 45°$$
$$= \left(\frac{1}{2}\right)\left(\frac{\sqrt{2}}{2}\right) - \left(-\frac{\sqrt{3}}{2}\right)\left(\frac{\sqrt{2}}{2}\right) = \frac{\sqrt{2}}{4} + \frac{\sqrt{6}}{4} = \frac{\sqrt{2} + \sqrt{6}}{4}$$

4 If $\sin t = 3/5$, where $\pi/2 < t < \pi$, and $\cos s = 4/5$, where $0 < s < \pi/2$, determine each of the following values.
(a) $\sin s$ (b) $\cos t$ (c) $\sin(t + s)$ (d) $\cos(t + s)$

SOLUTION

(a) An angle s with radian measure between 0 and $\pi/2$ has its terminal side in quadrant I, so

$$\sin s = \sqrt{1 - \cos^2 s} = \sqrt{1 - \frac{16}{25}} = \frac{3}{5}$$

(b) Since an angle t with radian measure between $\pi/2$ and π has its terminal side in quadrant II,

$$\cos t = -\sqrt{1 - \sin^2 t} = -\sqrt{1 - \frac{9}{25}} = -\frac{4}{5}$$

(c) $\sin(t + s) = \sin t \cos s + \cos t \sin s = \left(\frac{3}{5}\right)\left(\frac{4}{5}\right) + \left(-\frac{4}{5}\right)\left(\frac{3}{5}\right) = 0$

(d) $\cos(t + s) = \cos t \cos s - \sin t \sin s$

$$= \left(-\frac{4}{5}\right)\left(\frac{4}{5}\right) - \left(\frac{3}{5}\right)\left(\frac{3}{5}\right) = -\frac{16}{25} - \frac{9}{25} = -\frac{25}{25} = -1$$

5 Verify that each of the following is an identity.
(a) $\sin(t + \pi) = -\sin t$ (b) $\cos(\theta - 180°) = -\cos\theta$

SOLUTION

(a) $\sin(t + \pi) = \sin t \cos \pi + \cos t \sin \pi = -\sin t + 0 = -\sin t$

(b) $\cos(\theta - 180°) = \cos \theta \cos 180° + \sin \theta \sin 180° = -\cos \theta + 0 = -\cos \theta$

6 Prove that

(a) $\tan(t + s) = \dfrac{\tan t + \tan s}{1 - \tan t \tan s}$ (b) $\tan(t - s) = \dfrac{\tan t - \tan s}{1 + \tan t \tan s}$

PROOF

(a) $\tan(t + s) = \dfrac{\sin(t + s)}{\cos(t + s)} = \dfrac{\sin t \cos s + \sin s \cos t}{\cos t \cos s - \sin t \sin s}$

Dividing both the numerator and the denominator of the right-hand side of the above equation by $\cos t \cos s$, we get

$$\tan(t + s) = \frac{\dfrac{\sin t \cos s}{\cos t \cos s} + \dfrac{\sin s \cos t}{\cos s \cos t}}{\dfrac{\cos t \cos s}{\cos t \cos s} - \dfrac{\sin t \sin s}{\cos t \cos s}}$$

Simplifying the above expression, we have

$$\tan(t + s) = \frac{\tan t + \tan s}{1 - \tan t \tan s}$$

(b) $\tan(t - s) = \tan[t + (-s)] = \dfrac{\tan t + \tan(-s)}{1 - \tan t \tan(-s)}$

Using $\tan(-s) = -\tan s$, we have

$$\tan(t - s) = \frac{\tan t - \tan s}{1 + \tan t \tan s}$$

7 Prove that $f(x) = \tan x$ is a periodic function with period π.

PROOF

$$\tan(x + \pi) = \frac{\tan x + \tan \pi}{1 - \tan x \tan \pi} = \frac{\tan x + 0}{1 - 0} = \tan x$$

It can be shown that π is the *smallest* positive period of the tangent function, hence π is the *fundamental period* of $f(x) = \tan x$.

2.2 Double and Half Angle Formulas

Special cases of the formulas derived above are used to establish two important classes of trigonometric formulas known as *double angle formulas* and *half angle formulas.*

THEOREM 4 DOUBLE ANGLE FORMULAS

(i) $\cos 2t = \cos^2 t - \sin^2 t = 2\cos^2 t - 1 = 1 - 2\sin^2 t$

(ii) $\sin 2t = 2 \sin t \cos t$

PROOF

(i) $\cos 2t = \cos(t + t)$
$= \cos t \cos t - \sin t \sin t$ (Theorem 1, ii)
$= \cos^2 t - \sin^2 t$

However, since $\cos^2 t + \sin^2 t = 1$, we also have

$$\cos 2t = \cos^2 t - (1 - \cos^2 t) = 2\cos^2 t - 1$$

or

$$\cos 2t = (1 - \sin^2 t) - \sin^2 t = 1 - 2\sin^2 t$$

(ii) $\sin 2t = \sin(t + t)$
$= \sin t \cos t + \cos t \sin t$
$= 2 \sin t \cos t$

THEOREM 5 HALF ANGLE FORMULAS

(i) $\cos^2 \frac{s}{2} = \frac{1 + \cos s}{2}$

(ii) $\sin^2 \frac{s}{2} = \frac{1 - \cos s}{2}$

PROOF

(i) Upon substituting s for $2t$ in Theorem 4(i), we have

$$\cos s = 2\cos^2 \frac{s}{2} - 1 \quad \text{so} \quad \cos^2 \frac{s}{2} = \frac{1 + \cos s}{2}$$

(ii) Upon substituting s for $2t$ in Theorem 4(i), we have

$$\cos s = 1 - 2\sin^2 \frac{s}{2} \quad \text{so} \quad \sin^2 \frac{s}{2} = \frac{1 - \cos s}{2}$$

Notice that by taking the square root of both sides of the above identities, we have

$$\cos \frac{s}{2} = \pm\sqrt{\frac{1 + \cos s}{2}} \quad \text{and} \quad \sin \frac{s}{2} = \pm\sqrt{\frac{1 - \cos s}{2}}$$

where the sign is determined by the quadrant in which the terminal side of an angle in standard position with measure $s/2$ lies.

Alternative forms of the half angle formulas are

$$\cos^2 t = \frac{1 + \cos 2t}{2} \quad \text{and} \quad \sin^2 t = \frac{1 - \cos 2t}{2}$$

EXAMPLES 1 Write each of the following expressions as a single term involving an angle twice as large as the given angle.

(a) $2 \sin 5\theta \cos 5\theta$
(b) $\cos^2 3t - \sin^2 3t$
(c) $2\cos^2 4t - 1$
(d) $1 - 2\sin^2 6\theta$

SOLUTION

(a) $2 \sin 5\theta \cos 5\theta = \sin[2(5\theta)] = \sin 10\theta$ (Theorem 4, ii)

(b) $\cos^2 3t - \sin^2 3t = \cos[2(3t)] = \cos 6t$ (Theorem 4, i)

(c) $2 \cos^2 4t - 1 = \cos[2(4t)] = \cos 8t$ (Theorem 4, i)

(d) $1 - 2 \sin^2 6\theta = \cos[2(6\theta)] = \cos 12\theta$ (Theorem 4, i)

2 Use Theorem 5 to evaluate each of the following.

(a) $\sin 15°$ (b) $\cos \frac{\pi}{12}$

SOLUTION

(a) Since $\sin 15° > 0$,

$$\sin 15° = \sin\left(\frac{30°}{2}\right) = \sqrt{\frac{1 - \cos 30°}{2}} = \sqrt{\frac{1 - (\sqrt{3}/2)}{2}}$$
$$= \sqrt{\frac{2 - \sqrt{3}}{4}} = \frac{\sqrt{2 - \sqrt{3}}}{2}$$

(b) Since $\cos \frac{\pi}{12} > 0$,

$$\cos \frac{\pi}{12} = \cos\left[\frac{1}{2}\left(\frac{\pi}{6}\right)\right] = \sqrt{\frac{1 + \cos(\pi/6)}{2}} = \sqrt{\frac{1 + (\sqrt{3}/2)}{2}}$$
$$= \sqrt{\frac{2 + \sqrt{3}}{4}} = \frac{\sqrt{2 + \sqrt{3}}}{2}$$

3 Simplify $\left(\sin \frac{\theta}{2} + \cos \frac{\theta}{2}\right)^2 - \sin \theta$.

SOLUTION

$$\left(\sin \frac{\theta}{2} + \cos \frac{\theta}{2}\right)^2 - \sin \theta$$
$$= \sin^2 \frac{\theta}{2} + 2 \sin \frac{\theta}{2} \cos \frac{\theta}{2} + \cos^2 \frac{\theta}{2} - \sin \theta$$
$$= 1 + 2 \sin \frac{\theta}{2} \cos \frac{\theta}{2} - \sin \theta = 1 + \sin\left[2\left(\frac{\theta}{2}\right)\right] - \sin \theta$$
$$= 1 + \sin \theta - \sin \theta = 1$$

4 Prove that each of the following equations is an identity.

(a) $\cos 2\theta + 2 \sin^2 \theta = 1$ (b) $\frac{2 \cos 2t}{\sin 2t - 2 \sin^2 t} = \cot t + 1$

PROOF

(a) $\cos 2\theta + 2 \sin^2 \theta = \cos^2 \theta - \sin^2 \theta + 2 \sin^2 \theta = \cos^2 \theta + \sin^2 \theta = 1$

(b)
$$\frac{2 \cos 2t}{\sin 2t - 2 \sin^2 t} = \frac{2(\cos^2 t - \sin^2 t)}{2 \sin t \cos t - 2 \sin^2 t}$$
$$= \frac{2(\cos t - \sin t)(\cos t + \sin t)}{2 \sin t(\cos t - \sin t)}$$
$$= \frac{\cos t + \sin t}{\sin t} = \frac{\cos t}{\sin t} + 1 = \cot t + 1$$

5 Prove that

$$\tan 2t = \frac{2 \tan t}{1 - \tan^2 t}$$

PROOF Using Example 6(a) on page 261, we have

$$\begin{aligned}\tan 2t &= \tan(t + t)\\ &= \frac{\tan t + \tan t}{1 - \tan t \tan t}\\ &= \frac{2 \tan t}{1 - \tan^2 t}\end{aligned}$$

PROBLEM SET 2

In Problems 1–6, write each of the following expressions as a single function evaluated at a single angle.

1 $\sin 33° \cos 27° + \cos 33° \sin 27°$
2 $\sin 35° \cos 25° + \cos 35° \sin 25°$
3 $\cos 55° \cos 25° - \sin 55° \sin 25°$
4 $\sin 2\theta \cos \theta + \cos 2\theta \sin \theta$
5 $\sin(t - s) \cos s + \cos(t - s) \sin s$
6 $\dfrac{\tan 32° + \tan 43°}{1 - \tan 32° \tan 43°}$

In Problems 7–16, write each given expression as a single trigonometric function of θ.

7 $\sin\left(\frac{5\pi}{2} - \theta\right)$ **8** $\cos\left(\frac{5\pi}{2} - \theta\right)$ **9** $\tan(\pi + \theta)$

10 $\cot(90° - \theta)$ **11** $\cos(270° + \theta)$ **12** $\sec\left(\frac{\pi}{2} + \theta\right)$

13 $\csc(90° - \theta)$ **14** $\sin(270° - \theta)$ **15** $\cot(10\pi - \theta)$

16 $\tan(3\pi - \theta)$

In Problems 17–20, simplify each expression.

17 $\sin t \cos s - \sin\left(t + \frac{\pi}{2}\right) \sin(-s)$

18 $\cos\left(t - \frac{\pi}{2}\right) \tan s + \sin t \cot s$

19 $-\sin(\pi - t) + \cot t \sin\left(t - \frac{\pi}{2}\right)$

20 $\cos(\pi - t) - \tan t \cos\left(\frac{\pi}{2} - t\right)$

21 Use the fact that $5\pi/12 = (\pi/4) + (\pi/6)$ to evaluate the following expressions.

(a) $\sin \frac{5\pi}{12}$ (b) $\cos \frac{5\pi}{12}$ (c) $\tan \frac{5\pi}{12}$

22 Prove that $\sin(t+s)\sin(t-s) = \sin^2 t - \sin^2 s$.

23 If $\sin t = \frac{12}{13}$ and $\cos s = -\frac{4}{5}$, evaluate each of the following expressions, where $\pi/2 < t < \pi$ and $\pi/2 < s < \pi$.

(a) $\sin(t+s)$ (b) $\cos(t+s)$ (c) $\sin(t-s)$
(d) $\cos(t-s)$ (e) $\tan(t+s)$ (f) $\tan(t-s)$

24 (a) Prove that each of the following equations is an identity.

(i) $\sin(u+v) + \sin(u-v) = 2 \sin u \cos v$
(ii) $\sin(u+v) - \sin(u-v) = 2 \cos u \sin v$
(iii) $\cos(u-v) + \cos(u+v) = 2 \cos u \cos v$
(iv) $\cos(u-v) - \cos(u+v) = 2 \sin u \sin v$

(b) Use the results of part (a) to prove that each of the following equations is an identity. [*Hint:* Using part (a), let $s = u + v$ and $t = u - v$; then, $u = \dfrac{s+t}{2}$ and $v = \dfrac{s-t}{2}$. Substitute for u and v in terms of s and t.]

(i) $\sin s + \sin t = 2 \sin \dfrac{s+t}{2} \cos \dfrac{s-t}{2}$

(ii) $\sin s - \sin t = 2 \cos \dfrac{s+t}{2} \sin \dfrac{s-t}{2}$

(iii) $\cos s + \cos t = 2 \cos \dfrac{s+t}{2} \cos \dfrac{s-t}{2}$

(iv) $\cos t - \cos s = 2 \sin \dfrac{s+t}{2} \sin \dfrac{s-t}{2}$

In Problems 25–28, prove that each equation is an identity.

25 $\sin(30° + t) = \dfrac{1}{2} \cos t + \dfrac{\sqrt{3}}{2} \sin t$

26 $\tan(45° + t) = \dfrac{1 + \tan t}{1 - \tan t}$

27 $\sin(60° + t) - \cos(30° + t) = \sin t$

28 $\cos\left(\dfrac{\pi}{4} - t\right) = \dfrac{1}{\sqrt{2}}(\cos t + \sin t)$

In Problems 29–32, use the double angle and half angle formulas.

29 Express $\sin 80°$ in terms of functions of $40°$.

30 Express $\cos 8t$ in terms of functions of $4t$.

31 Express $\sin 10°$ in terms of a function of $20°$.

32 Express $\cos \dfrac{3\pi}{14}$ in terms of a function of $\dfrac{3\pi}{7}$.

In Problems 33–38, use the half angle formulas to find the exact value of each expression.

33 $\cos \dfrac{\pi}{8}$ **34** $\cos \dfrac{5\pi}{12}$ **35** $\sin\left(-\dfrac{7\pi}{12}\right)$

36 $\cos \dfrac{3\pi}{8}$ **37** $\sin 75°$ **38** $\cos 105°$

In Problems 39–44, write each expression as a single term involving an angle twice as large as the given angle.

39 $2 \sin 3t \cos 3t$

40 $1 - 2 \sin^2 37°$

41 $1 - 2 \sin^2 7t$

42 $2 \cos^2 8 - 1$

43 $2 \cos^2 \frac{\theta}{2} - 1$

44 $10 \sin 20\theta \cos 20\theta$

45 If $\cos t = -\frac{7}{25}$ and $\pi < t < \frac{3\pi}{2}$, find the exact value of each of the following expressions.

(a) $\sin 2t$
(b) $\cos 2t$
(c) $\tan 2t$
(d) $\sin(t/2)$
(e) $\cos(t/2)$

46 Given angles s and t, with measures such that $s + t = \pi/2$, prove that
(a) $\sin(s - t) = \cos 2t = -\cos 2s$
(b) $\cos(s - t) = \sin 2s = \sin 2t$

47 (a) Use $15° = 45° - 30°$ and Theorem 3 to compute $\sin 15°$.
(b) Compare the answer obtained in part (a) to the result of Example 2(a) on page 263.

48 (a) Use $\cot(t + s) = \frac{1}{\tan(t + s)}$ to derive a formula for $\cot(t + s)$ in terms of $\cot s$ and $\cot t$.
(b) Use the result of part (a) to determine a formula for $\cot 2t$.
(c) Use part (a) to prove that $f(x) = \cot x$ is a periodic function with period π.

In Problems 49–66, prove that each equation is an identity.

49 $\cot \theta - \tan \theta = 2 \cot 2\theta$

50 $\cos^2 \theta(1 - \tan^2 \theta) = \cos 2\theta$

51 $\frac{1 + \cos 2t}{\sin 2t} = \cot t$

52 $\csc t \sec t = 2 \csc 2t$

53 $\frac{\sin 3\theta}{\sin \theta} - \frac{\cos 3\theta}{\cos \theta} = 2$

54 $4 \sin^2 \theta \cos^2 \theta + \cos^2 2\theta = 1$

55 $\frac{\sin 4t}{\sin 2t} = 2 \cos 2t$

56 $\frac{\sin 5t}{\sin t} - \frac{\cos 5t}{\cos t} = 4 \cos 2t$

57 $\frac{1}{2} \sin \theta \tan \frac{\theta}{2} \csc^2 \frac{\theta}{2} = 1$

58 $\frac{\sin s - \sin t}{\cos s + \cos t} = \tan\left[\frac{1}{2}(s - t)\right]$

59 $2 \cos \frac{\theta}{2} = (1 + \cos \theta) \sec \frac{\theta}{2}$

60 $\tan \frac{\theta}{2} = \pm \sqrt{\frac{1 - \cos \theta}{1 + \cos \theta}}$

61 $\cos(t + s) \cos(t - s) = \cos^2 t - \sin^2 s$

62 $\frac{\sin(t - s)}{\sin t \sin s} = \cot s - \cot t$

63 $\frac{\cos^3 t - \sin^3 t}{\cos t - \sin t} = \frac{2 + \sin 2t}{2}$

64 $2 \sin \theta \cos^3 \theta + 2 \sin^3 \theta \cos \theta = \sin 2\theta$

65 $\sin^2 \theta \cos^2 \theta = \frac{1}{8}(1 - \cos 4\theta)$

66 $\cos^4 t = \frac{3}{8} + \frac{1}{2} \cos 2t + \frac{1}{8} \cos 4t$

3 Inverse Trigonometric Functions

In this section, we study inverse trigonometric functions, which aid in the solution of certain *simple* trigonometric equations, like $\sin t = \frac{3}{5}$, $\tan \theta = -1$, and $\cos t = \frac{1}{2}$. Somewhat more complicated equations, such as $\sin^2 t = \frac{3}{4}$, $\cos 2\theta - \sin \theta = 0$, and $\tan^2 t + 3 \tan t - 4 = 0$, which arise in many applications of trigonometric functions, will be solved later (in Section 4) by using trigonometric identities and algebraic techniques.

The repeating pattern of each of the graphs of the trigonometric functions means that a horizontal line drawn through the graph of any one of these functions intersects the graph "infinitely often." For example, if the function $y = \cos x$ and the line $y = \frac{1}{2}$ are both graphed on the same coordinate system, it can be seen that this horizontal line intersects the graph of the cosine function at more than one point (Figure 1). Thus the equation $\cos x = \frac{1}{2}$ has infinitely many roots.

Because of their periodic nature, *none* of the trigonometric functions is invertible. However, if we restrict the domain of each of these functions, we can construct invertible trigonometric functions.

Figure 1

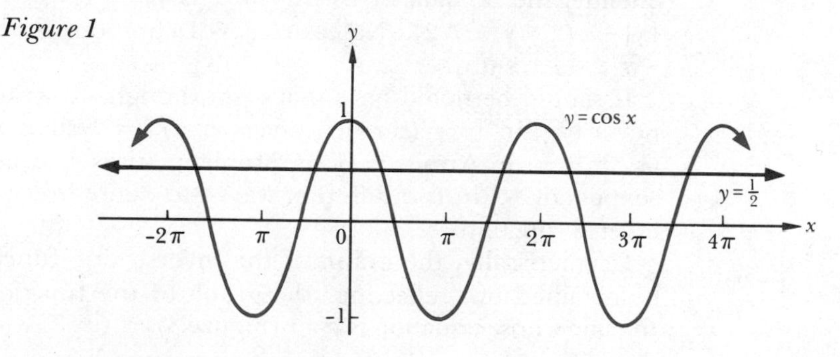

3.1 The Inverse Sine and Cosine Functions

We begin by constructing the function $f(x) = \text{Sin } x$ (notice that a capital letter S is used here to distinguish this function from the sine function). This is done by restricting the domain of $y = \sin x$ to the interval $[-\pi/2, \pi/2]$. The range is $\{y | -1 \leq y \leq 1\}$ (Figure 2). Since $f(x) = \text{Sin } x$ is increasing, it is invertible. The inverse is called the *Arcsine function.*

Figure 2

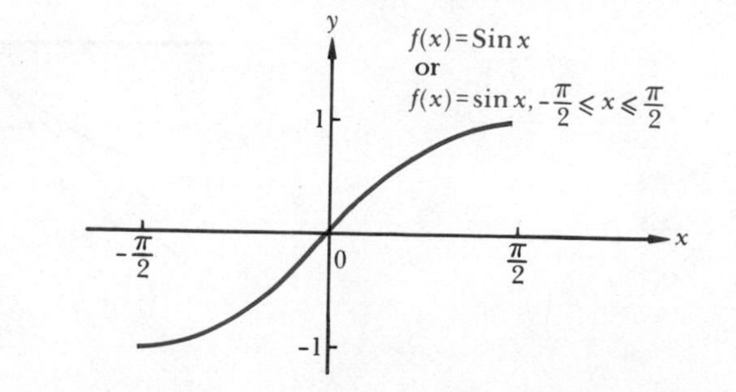

DEFINITION 1 ARCSINE FUNCTION

The *Arcsine function,* denoted by $y = \text{Arcsin}\, x$ or by $y = \sin^{-1} x$, is the inverse of the function $f(x) = \text{Sin}\, x$. It is defined by

$$y = \text{Arcsin}\, x \quad \text{if and only if} \quad x = \sin y \qquad \text{where } -\frac{\pi}{2} \leqslant y \leqslant \frac{\pi}{2}$$

For example,

$$\frac{\pi}{3} = \text{Arcsin}\,\frac{\sqrt{3}}{2} \qquad \text{is equivalent to} \qquad \frac{\sqrt{3}}{2} = \sin\frac{\pi}{3}$$

$$-\frac{\pi}{4} = \text{Arcsin}\left(-\frac{\sqrt{2}}{2}\right) \qquad \text{is equivalent to} \qquad -\frac{\sqrt{2}}{2} = \sin\left(-\frac{\pi}{4}\right)$$

$$0 = \sin^{-1} 0 \qquad \text{is equivalent to} \qquad 0 = \sin 0$$

Since the inverse of a function is formed by interchanging the numbers in the ordered pairs, the domain and the range of $f(x) = \text{Sin}\, x$ become, respectively, the range and the domain of the inverse function. Consequently, the domain of $y = \text{Arcsin}\, x$ is $\{x \mid -1 \leqslant x \leqslant 1\}$, and the range is $\{y \mid -\pi/2 \leqslant y \leqslant \pi/2\}$. Note that by Definition 1 above, it follows that $-\pi/2 \leqslant \text{Arcsin}\, x \leqslant \pi/2$.

It should be noted here that even though we write $\sin^2 x = (\sin x)^2$, we never use $\sin^{-1} x = (\sin x)^{-1}$, since $(\sin x)^{-1} = 1/\sin x$, which is *not* the same as $\sin^{-1} x$ or $\text{Arcsin}\, x$. This problem arises because we are using the "exponent" $^{-1}$ in two different ways—to denote reciprocals and to denote inverse functions.

Geometrically, the graph of the inverse sine function $y = \text{Arcsin}\, x$ can be obtained by "reflecting" the graph of the function $f(x) = \text{Sin}\, x$ across the line whose equation is $y = x$ (Figure 3).

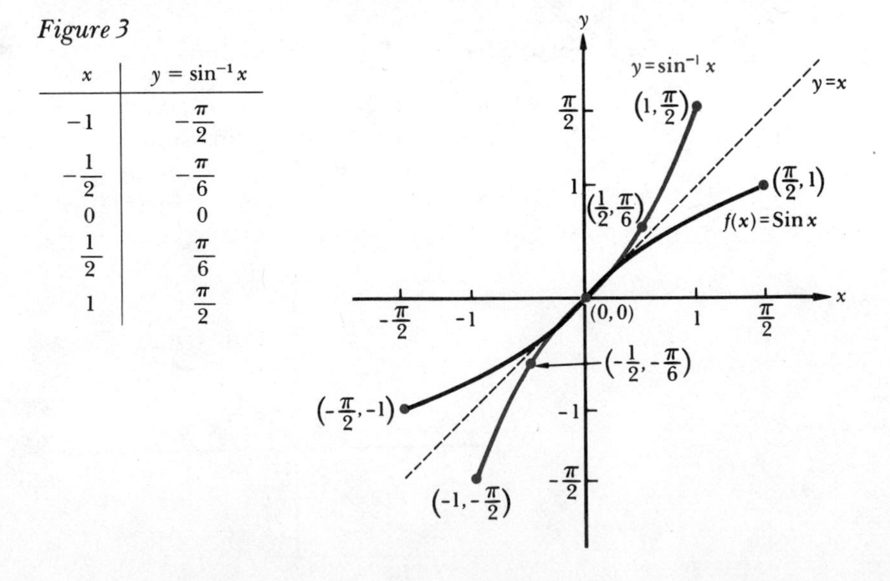

Figure 3

x	$y = \sin^{-1} x$
-1	$-\frac{\pi}{2}$
$-\frac{1}{2}$	$-\frac{\pi}{6}$
0	0
$\frac{1}{2}$	$\frac{\pi}{6}$
1	$\frac{\pi}{2}$

Figure 4

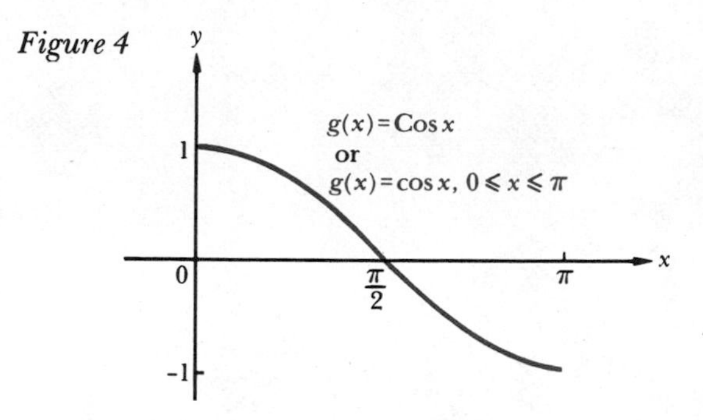

To define the inverse cosine function, we begin by constructing the function $g(x) = \text{Cos } x$ (notice the use of capital C) from $y = \cos x$ by restricting the domain to the interval $[0, \pi]$. The range is $\{y \mid -1 \leq y \leq 1\}$ (Figure 4). Since $g(x) = \text{Cos } x$ is a decreasing function, it is invertible. The inverse is called the *Arccosine function.*

DEFINITION 2 ARCCOSINE FUNCTION

The *Arccosine function,* denoted by $y = \text{Arccos } x$ or by $y = \cos^{-1} x$, is the inverse of the function $g(x) = \text{Cos } x$. It is defined by

$$y = \text{Arccos } x \quad \text{if and only if} \quad x = \cos y \qquad \text{where } 0 \leq y \leq \pi$$

For example,

$$\pi = \text{Arccos}(-1) \quad \text{is equivalent to} \quad -1 = \cos \pi$$

$$\frac{2\pi}{3} = \cos^{-1}\left(-\frac{1}{2}\right) \quad \text{is equivalent to} \quad -\frac{1}{2} = \cos\frac{2\pi}{3}$$

$$\frac{\pi}{2} = \cos^{-1} 0 \quad \text{is equivalent to} \quad 0 = \cos\frac{\pi}{2}$$

The domain of $y = \text{Arccos } x$ is the set $\{x \mid -1 \leq x \leq 1\}$, and the range is the set $\{y \mid 0 \leq y \leq \pi\}$. By Definition 2 above, we have $0 \leq \text{Arccos } x \leq \pi$.

The graph of $y = \cos^{-1} x$ can be obtained from the graph of $g(x) = \text{Cos } x$ by reflecting the latter graph across the line $y = x$ as illustrated in Figure 5.

Figure 5

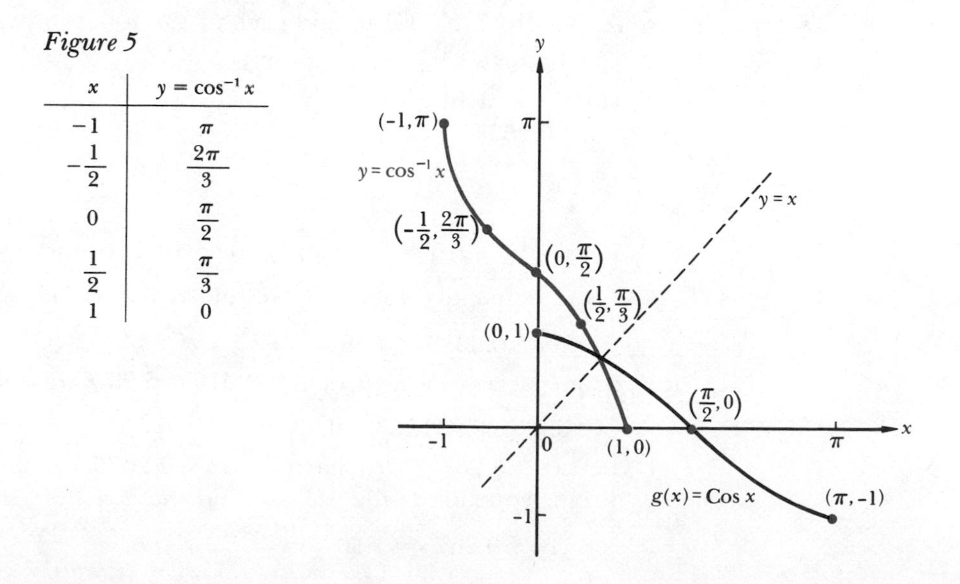

x	$y = \cos^{-1} x$
-1	π
$-\frac{1}{2}$	$\frac{2\pi}{3}$
0	$\frac{\pi}{2}$
$\frac{1}{2}$	$\frac{\pi}{3}$
1	0

EXAMPLES 1 Find each of the following values.

(a) $\text{Arccos}\frac{\sqrt{2}}{2}$

(b) $\sin^{-1}\left(-\frac{1}{2}\right)$

(c) $\tan\left[\cos^{-1}\left(-\frac{\sqrt{3}}{2}\right)\right]$

SOLUTION

(a) Let $\text{Arccos}\,\frac{\sqrt{2}}{2} = t.$ Then $\cos t = \frac{\sqrt{2}}{2},$ where $0 \leq t \leq \pi,$ so $t = \frac{\pi}{4}.$

(b) Let $\sin^{-1}\left(-\frac{1}{2}\right) = t.$ Then $\sin t = -\frac{1}{2},$ where $-\frac{\pi}{2} \leq t \leq \frac{\pi}{2},$ so $t = -\frac{\pi}{6}.$

Note that $t \neq \frac{11\pi}{6}$, because $\frac{11\pi}{6}$ is not between $-\frac{\pi}{2}$ and $\frac{\pi}{2}$.

(c) Let $\cos^{-1}\left(-\frac{\sqrt{3}}{2}\right) = t.$ Then $\cos t = -\frac{\sqrt{3}}{2},$ where $0 \leq t \leq \pi,$ so $t = \frac{5\pi}{6}.$

Thus, $\tan\left[\cos^{-1}\left(-\frac{\sqrt{3}}{2}\right)\right] = \tan\frac{5\pi}{6} = -\frac{\sqrt{3}}{3}.$

2 Use Appendix Table III to find each of the following values.

(a) $\sin^{-1} 0.8016$

(b) $\sin^{-1}(-0.8016)$

(c) $\cos^{-1} 0.2675$

SOLUTION

(a) Let $\sin^{-1} 0.8016 = t.$ Then $\sin t = 0.8016,$ where $-\frac{\pi}{2} \leq t \leq \frac{\pi}{2}.$ From Appendix Table III, we find that $t = 0.93$, so that

$$\sin^{-1} 0.8016 = 0.93$$

(b) From part (a), we have $\sin^{-1} 0.8016 = 0.93$. Consequently,

$$\sin^{-1}(-0.8016) = -0.93$$

(c) Let $\cos^{-1} 0.2675 = t.$ Then $\cos t = 0.2675,$ where $0 \leq t \leq \pi.$ From Appendix Table III, we find that $t = 1.30$, so that

$$\cos^{-1} 0.2675 = 1.30$$

3 Show that $\sin^{-1} x = \cos^{-1} \sqrt{1 - x^2}$ for $0 \leq x \leq 1$.

PROOF Let $t = \sin^{-1} x$. Then $\sin t = x$, where $0 \leq t \leq \dfrac{\pi}{2}$.
Since $\cos t = \sqrt{1 - \sin^2 t}$ for $0 \leq t \leq \pi/2$, we have

$$\cos t = \sqrt{1 - x^2} \quad \text{or} \quad t = \cos^{-1}\sqrt{1 - x^2}$$

and, by substitution,

$$\sin^{-1} x = \cos^{-1}\sqrt{1 - x^2}$$

4 If $\cos^{-1}(-\frac{3}{4}) = w$, find each of the following values.
(a) $\sin w$ (b) $\tan w$

SOLUTION

(a) Since $\cos^{-1}(-\frac{3}{4}) = w$, then $\cos w = -\frac{3}{4}$, where $\pi/2 \leq w \leq \pi$.
Using the identity $\sin^2 w + \cos^2 w = 1$, we have

$$\sin^2 w = 1 - \cos^2 w = 1 - (-\tfrac{3}{4})^2 = 1 - \tfrac{9}{16} = \tfrac{7}{16}$$

Since $\sin w$ is positive for $\pi/2 \leq w \leq \pi$,

$$\sin w = \frac{\sqrt{7}}{4}$$

(b) $$\tan w = \frac{\sin w}{\cos w} = \frac{\dfrac{\sqrt{7}}{4}}{-\dfrac{3}{4}} = -\frac{\sqrt{7}}{3}$$

5 If $y = 5 + 3\cos^{-1} 2x$, express x in terms of y.

SOLUTION

$$y = 5 + 3\cos^{-1} 2x \quad \text{can be written as} \quad \frac{y - 5}{3} = \cos^{-1} 2x$$

The latter equation is equivalent to

$$2x = \cos\left(\frac{y - 5}{3}\right) \qquad \text{where } 0 \leq \frac{y - 5}{3} \leq \pi$$

Thus

$$x = \frac{1}{2}\cos\left(\frac{y - 5}{3}\right)$$

6 Find the exact value of $\sin[2\cos^{-1}(-\frac{3}{5})]$.

SOLUTION

$$\begin{aligned}\sin[2\cos^{-1}(-\tfrac{3}{5})] &= 2\sin[\cos^{-1}(-\tfrac{3}{5})]\cos[\cos^{-1}(-\tfrac{3}{5})] \qquad \text{(Theorem 4, ii)}\\ &= 2\sqrt{1 - (-\tfrac{3}{5})^2}\cdot(-\tfrac{3}{5})\\ &= 2\sqrt{\tfrac{16}{25}}\cdot(-\tfrac{3}{5}) = 2(\tfrac{4}{5})(-\tfrac{3}{5}) = -\tfrac{24}{25}\end{aligned}$$

3.2 Other Inverse Trigonometric Functions

In order to obtain the inverse tangent function, construct $h(x) = \text{Tan}\, x$ (notice the use of capital T) by restricting the domain of $y = \tan x$ to the interval $(-\pi/2, \pi/2)$. The range is the set of real numbers (Figure 6). Since the function h is an increasing function, it is invertible. We call the inverse the *Arctangent function.*

Figure 6

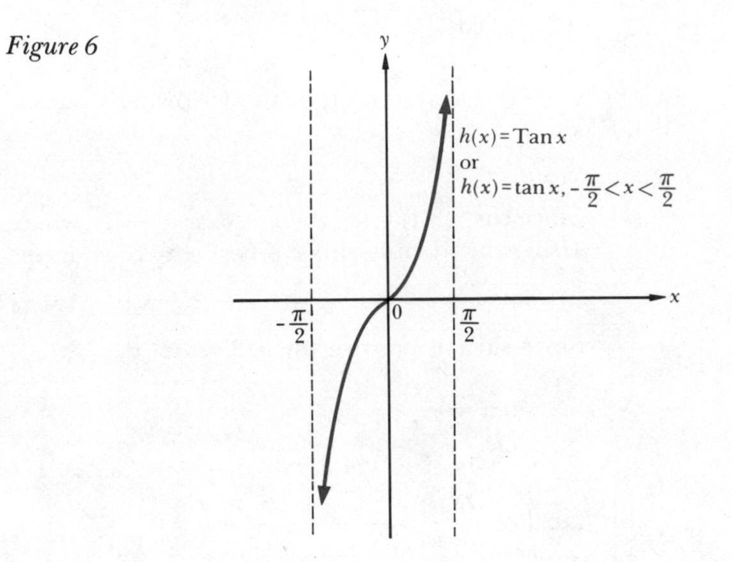

DEFINITION 3 ARCTANGENT FUNCTION

The *Arctangent function,* denoted by $y = \text{Arctan}\, x$ or by $y = \tan^{-1} x$, is the inverse function of $h(x) = \text{Tan}\, x$. It is defined by

$$y = \tan^{-1} x \quad \text{if and only if} \quad x = \tan y \qquad \text{where } -\frac{\pi}{2} < y < \frac{\pi}{2}$$

For example,

$$\frac{\pi}{4} = \tan^{-1} 1 \qquad \text{is equivalent to} \qquad 1 = \tan \frac{\pi}{4}$$

$$0 = \text{Arctan}\, 0 \qquad \text{is equivalent to} \qquad 0 = \tan 0$$

$$-\frac{\pi}{4} = \text{Arctan}(-1) \quad \text{is equivalent to} \quad -1 = \tan\left(-\frac{\pi}{4}\right)$$

The domain of the Arctangent function is the set of all real numbers, and the range is $\{y \mid -\pi/2 < y < \pi/2\}$. Also, by Definition 3, we have $-\pi/2 < \text{Arctan}\, x < \pi/2$. The graph of $y = \text{Arctan}\, x$ is displayed in Figure 7 (see Problem 26).

The definitions and the graphs of the inverses for the cotangent, secant, and cosecant functions are considered in Problems 34, 35, and 36.

Figure 7

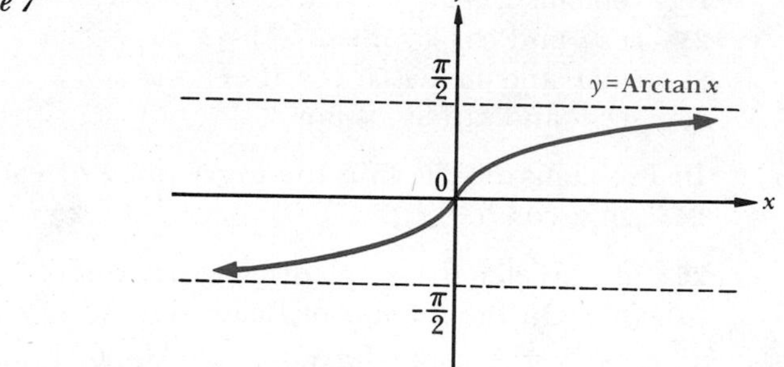

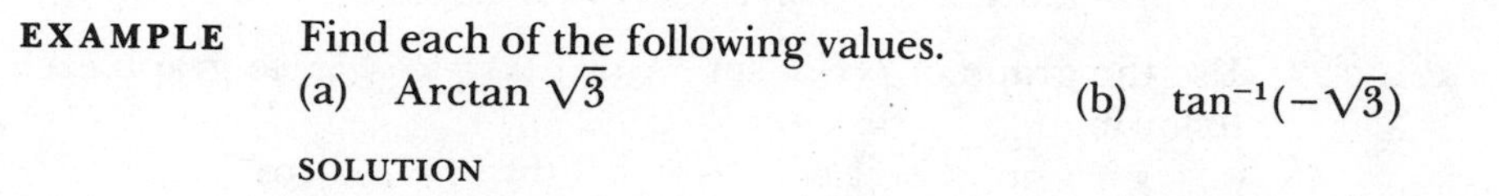

EXAMPLE Find each of the following values.

(a) Arctan $\sqrt{3}$ (b) $\tan^{-1}(-\sqrt{3})$

SOLUTION

(a) Let Arctan $\sqrt{3} = t$. Then $\tan t = \sqrt{3}$, where $-\frac{\pi}{2} < t < \frac{\pi}{2}$, so $t = \frac{\pi}{3}$.

(b) Let $\tan^{-1}(-\sqrt{3}) = w$. Then $\tan w = -\sqrt{3}$, where $-\frac{\pi}{2} < w < \frac{\pi}{2}$, so $w = -\frac{\pi}{3}$.

PROBLEM SET 3

In Problems 1–10, determine each value *without* the use of Appendix Table III.

1 Arcsin $\frac{1}{2}$

2 Arccos 1

3 $\cos^{-1}\left(-\frac{\sqrt{2}}{2}\right)$

4 $\sin^{-1}\left(-\frac{\sqrt{3}}{2}\right)$

5 $\sin^{-1}(-1)$

6 Arctan $\frac{\sqrt{3}}{3}$

7 $\tan^{-1}\left(-\frac{\sqrt{3}}{3}\right)$

8 $\sin^{-1}\left[\sin\left(-\frac{3\pi}{2}\right)\right]$

9 Arccos $\left(\sin\frac{\pi}{6}\right)$

10 $\tan^{-1}\left(\cos\frac{\pi}{2}\right)$

In Problems 11–20, use Appendix Table III to determine each value. (Use $\pi = 3.14$ where necessary.)

11 $\sin^{-1} 0.2182$

12 $\cos^{-1} 0.8628$

13 Arccos 0.2092

14 Arctan 1.072

15 Arcsin(−0.7771)

16 $\tan^{-1}(-0.5334)$

17 $\cos^{-1}(-0.8473)$

18 $\sin^{-1} 0.05$

19 $\tan^{-1} 3.01$

20 Arcsin(−0.9425)

In Problems 21–23, evaluate each expression.

21 $\cos t$ and $\cos 2t$, if $\sin^{-1}(\frac{4}{5}) = t$

22 $\sin w$ and $\sin 2w$, if $\cos^{-1}(-\frac{3}{5}) = w$

23 $\sec v$ and $\tan 2v$, if $\tan^{-1}(-\frac{1}{2}) = v$

In Problems 24–25, find the exact value of each expression.

24 $\sin[2 \cos^{-1}(-\frac{24}{25})]$

25 $\tan[2 \tan^{-1}(\frac{4}{3})]$

26 (a) Sketch the graph of $y = \tan x$, where $-\pi/2 < x < \pi/2$.
(b) On the same coordinate system, plot the points of the graph of $y = \text{Arctan}\, x$ for $x = -1, -\sqrt{3}, 0, 1$, and $\sqrt{3}/3$.
(c) Complete the graph of $y = \text{Arctan}\, x$ by reflecting the graph of $y = \tan x$, where $-\pi/2 < x < \pi/2$, across the line $y = x$.

27 Use the graph of $f(x) = \sin^{-1} x$ or $g(x) = \cos^{-1} x$ to graph each of the following.
(a) $y = \frac{1}{2} \sin^{-1} x$
(b) $y = -3 \cos^{-1} x$
(c) $y = 1 + \sin^{-1} x$
(d) $y = 2 - \frac{1}{3} \cos^{-1} x$

28 Given $f(x) = \sin^{-1} x$ and $g(x) = \sin x$.
(a) Find $h(x) = f[g(x)]$. What is the domain of h?
(b) Find $w(x) = g[f(x)]$. What is the domain of w?

In Problems 29–33, express x in terms of y.

29 $y = \cos^{-1} 2x$

30 $y = 2 \sin^{-1} 3x$

31 $y = 4 - 2 \sin^{-1} 4x$

32 $y = \tan^{-1}(2x + 1)$

33 $3y = 1 + 4 \cos^{-1}(x/2)$

34 *The Inverse Cotangent Function*
(a) Let $g(x) = \text{Cot}\, x$ be the function obtained by restricting the domain of $y = \cot x$ to the interval $(0, \pi)$. Graph g. Explain why g is invertible.
(b) Indicate the domain and range of g^{-1}.
(c) Graph g^{-1}.
(d) State a definition of the inverse cotangent function.

35 *The Inverse Secant Function*
(a) Let $h(x) = \text{Sec}\, x$ be the function obtained by restricting the domain of $y = \sec x$ to the intervals $[0, \pi/2)$ and $(\pi/2, \pi]$. Graph h. Explain why h is invertible.
(b) Indicate the domain and range of h^{-1}.
(c) Graph h^{-1}.
(d) State a definition of the inverse secant function.

36 *The Inverse Cosecant Function*
(a) Let $f(x) = \text{Csc}\, x$ be the function obtained by restricting the domain of $y = \csc x$ to the intervals $[-\pi/2, 0)$ and $(0, \pi/2]$. Graph f. Explain why f is invertible.
(b) Indicate the domain and range of f^{-1}.
(c) Graph f^{-1}.
(d) State a definition of the inverse cosecant function.

37 The magnitude M of a certain ray of polarized light is given by the equation $M = k \sin 2\theta$, where k is a constant and θ is the acute angle between the ray of light and the plane of the light source. Solve for θ in terms of M and k.

38 Consider the circuit problem on page 215, where

$$Q = Q_0 \sin\left(\frac{t}{\sqrt{LC}} - \frac{\pi}{2}\right)$$

Solve for time t in terms of Q, Q_0, L, and C, where $0 \leq t \leq \pi\sqrt{LC}$.

4 Trigonometric Equations

In Sections 1 and 2 we discussed trigonometric identities and formulas. These identities and formulas are equations that are true for *all* real numbers or angles at which the functions involved are defined. In this section we use the properties of the trigonometric functions, together with the identities, to solve conditional trigonometric equations that can yield *more than one solution.*

For example, the equation $\sin t = 1$ is true for $t = \pi/2$, since $\sin \pi/2 = 1$. But it is also true that $\sin t = \sin(t + 2\pi n)$, where n is any integer, because the sine function is a periodic function of period 2π. Therefore, $\sin t = 1$ is true for $t = \pi/2, 5\pi/2, 9\pi/2, \ldots$. The solution set of this equation is given by

$$\left\{t \mid t = \frac{\pi}{2} + 2\pi n,\ n \in I\right\}$$

EXAMPLES Find the solution set of each trigonometric equation.

1 $\sin t = \dfrac{1}{2}$, for $0 \leq t < 2\pi$

SOLUTION We know that

$$\sin \frac{\pi}{6} = \frac{1}{2}$$

Here $\pi/6$ is the reference angle, and since the sine is also positive for $\pi/2 < t < \pi$, we have

$$\sin \frac{5\pi}{6} = \frac{1}{2}$$

Thus the solution set of $\sin t = \dfrac{1}{2}$, for $0 \leq t < 2\pi$, is $\left\{\dfrac{\pi}{6}, \dfrac{5\pi}{6}\right\}$.

2 $\tan \theta = \sqrt{3}$, for $0 \leq \theta < 360°$

SOLUTION One number in the solution set is $60°$, since $\tan 60° = \sqrt{3}$.

Since the tangent is also positive for $180° < \theta < 270°$, we use the reference angle of $60°$ to get $\tan 240° = \sqrt{3}$.

Thus the solution set is $\{60°, 240°\}$.

3 $\sin t = -0.8, \quad \text{for } 0 \leq t < 2\pi \quad (\text{Use } \pi = 3.142.)$

SOLUTION We can find the reference angle t_R by using Appendix Table III and linear interpolation to solve $\sin t_R = 0.8$ as follows:

$$0.01\left[\begin{array}{ll} & t \qquad\qquad \sin t \\ & 0.92 \qquad 0.7956 \\ d\left[\begin{array}{l} t_R \\ 0.93 \end{array}\right. & \left.\begin{array}{l} 0.8000 \\ 0.8016 \end{array}\right]0.0016 \end{array}\right]0.0060 \qquad \frac{d}{0.01} = \frac{0.0016}{0.0060}$$

Hence, $d = 0.003$ (approximately) and $t_R = 0.927$. Since $\sin t$ is negative for $\pi < t \leq 3\pi/2$ and for $3\pi/2 \leq t < 2\pi$, we use the reference angle 0.927 to get $\sin 4.069 = -0.8$ and $\sin 5.357 = -0.8$.

Thus the solution set is $\{4.069, 5.357\}$.

4 $\sin 3\theta = 1$

SOLUTION Since $\sin 3\theta = 1$,

$$3\theta = 90°, 450°, 810°, \ldots \quad \text{or} \quad 3\theta = -270°, -630°, -990°, \ldots$$

so

$$\theta = 30°, 150°, 270°, \ldots \quad \text{or} \quad \theta = -90°, -210°, -330°, \ldots$$

Hence the solution set is

$$\{\theta \mid \theta = 30° + 120°k \text{ or } \theta = -90° - 120°k, \quad \text{for } k \text{ any positive integer}\}$$

5 $4\cos^2 t - 3 = 0, \quad \text{for } 0 \leq t < 2\pi$

SOLUTION Since $4\cos^2 t - 3 = 0$ is equivalent to $\cos^2 t = \frac{3}{4}$, then

$$\cos t = \frac{\sqrt{3}}{2} \quad \text{or} \quad \cos t = -\frac{\sqrt{3}}{2}$$

But the cosine is positive for t in quadrants I and IV, so the solution set of

$$\cos t = \frac{\sqrt{3}}{2} \quad \text{is} \quad \left\{\frac{\pi}{6}, \frac{11\pi}{6}\right\}$$

Since the cosine is negative for t in quadrants II and III, the solution set of

$$\cos t = -\frac{\sqrt{3}}{2} \quad \text{is} \quad \left\{\frac{5\pi}{6}, \frac{7\pi}{6}\right\}$$

The solutions of both $\cos t = \sqrt{3}/2$ and $\cos t = -\sqrt{3}/2$ are the solutions of $\cos^2 t = 3/4$.

Hence the solution set of $\cos^2 t = 3/4$, for $0 \leq t < 2\pi$, is

$$\left\{\frac{\pi}{6}, \frac{5\pi}{6}, \frac{7\pi}{6}, \frac{11\pi}{6}\right\}$$

6 $\sec^2\theta + \sec\theta - 2 = 0$, for $0° \leq \theta < 360°$

SOLUTION This equation can be factored as

$$(\sec\theta + 2)(\sec\theta - 1) = 0$$

Thus, $\sec^2\theta + \sec\theta - 2 = 0$ if and only if

$$\sec\theta + 2 = 0 \quad \text{or} \quad \sec\theta - 1 = 0$$

that is, if

$$\sec\theta = -2 \quad \text{or} \quad \sec\theta = 1$$

The solution set of $\sec\theta = -2$, for $0° \leq \theta < 360°$, is $\{120°, 240°\}$, and the solution set of $\sec\theta = 1$, for $0° \leq \theta < 360°$, is $\{0°\}$.

Therefore, the solution set of $\sec^2\theta + \sec\theta - 2 = 0$, for $0° \leq \theta < 360°$, is $\{0°, 120°, 240°\}$.

7 $\sin 2t - 2\sin t = 0$, for $0 \leq t < 2\pi$

SOLUTION First we use the identity for $\sin 2t$ to rewrite the equation as

$$2\sin t\cos t - 2\sin t = 0 \quad \text{or} \quad 2\sin t(\cos t - 1) = 0$$

so that

$$2\sin t = 0 \quad \text{or} \quad \cos t - 1 = 0$$

that is,

$$\sin t = 0 \quad \text{or} \quad \cos t = 1$$

The solution set of $\sin t = 0$, for $0 \leq t < 2\pi$, is $\{0, \pi\}$, and the solution set of $\cos t = 1$, for $0 \leq t < 2\pi$, is $\{0\}$.

Thus the solution set of $\sin 2t - 2\sin t = 0$, for $0 \leq t < 2\pi$, is $\{0, \pi\}$.

8 $2\sin^2 t - 11\sin t - 6 = 0$

SOLUTION This equation can be factored as

$$(2\sin t + 1)(\sin t - 6) = 0$$

Thus, $2\sin^2 t - 11\sin t - 6 = 0$ if and only if

$$2\sin t + 1 = 0 \quad \text{or} \quad \sin t - 6 = 0$$

that is, if $\sin t = -\frac{1}{2}$ or $\sin t = 6$. Since 6 is not in the range of the sine function, $\sin t = 6$ has no solution. The solution set of $\sin t = -\frac{1}{2}$ is

$$\left\{t \,\middle|\, t = \frac{11\pi}{6} + 2\pi k \text{ or } t = \frac{7\pi}{6} + 2\pi k, \quad \text{for } k \in I\right\}$$

Hence the solution set of $2\sin^2 t - 11\sin t - 6 = 0$ is

$$\left\{t \,\middle|\, t = \frac{11\pi}{6} + 2\pi k \text{ or } t = \frac{7\pi}{6} + 2\pi k, \quad \text{for } k \in I\right\}$$

9 Recall the circuit problem on page 215, where the equation was given by

$$Q = Q_0 \sin\left(\frac{t}{\sqrt{LC}} - \frac{\pi}{2}\right)$$

Find the *first* time when $Q = \frac{1}{2}Q_0$, if $L = 0.16$ henry, $C = 10^{-4}$ farad, and t represents the number of elapsed seconds. (Use $\pi = 3.142$.)

SOLUTION Using the given conditions, we get

$$\frac{1}{2}Q_0 = Q_0 \sin\left(\frac{t}{\sqrt{0.16 \cdot 10^{-4}}} - \frac{\pi}{2}\right)$$

which is equivalent to

$$\sin\left(\frac{t}{0.004} - \frac{\pi}{2}\right) = \frac{1}{2}$$

We know that

$$\sin\frac{\pi}{6} = \frac{1}{2}$$

Thus

$$\frac{t}{0.004} - \frac{\pi}{2} = \frac{t}{0.004} - 1.571 = 0.524$$

so that $t = 0.008$ seconds (approximately).

PROBLEM SET 4

In Problems 1–18, find the solution set of each trigonometric equation for $0 \leq t < 2\pi$.

1 $\sin t = \frac{\sqrt{3}}{2}$ **2** $\sin t = -\frac{1}{2}$ **3** $\tan t = -1$

4 $\cot t = -\sqrt{3}$ **5** $\sec t = \sqrt{2}$ **6** $\csc t = -\frac{2\sqrt{3}}{3}$

7 $\cos t = -0.4176$ **8** $\sin t = 0.8134$ **9** $\tan t = 0.6696$

10 $\cot t = 6.617$ **11** $2 \sin t - 1 = 0$ **12** $4 \sin^2 t - 3 = 0$

13 $\sin 2t = \sin t$ **14** $\sin 2t = \sqrt{2} \cos t$ **15** $\cos 2t = 2 \sin^2 t$

16 $\sqrt{3} \sec t = 2 \tan t$ **17** $\cos 2t = 1 - \sin t$ **18** $\sin 2t = \sqrt{2} \cos t$

In Problems 19–24, find the solution set of each trigonometric equation for $0 \leq \theta < 360°$.

19 $\sin \theta = -\frac{\sqrt{3}}{2}$ **20** $\tan 2\theta = 1$ **21** $\cos 3\theta = -\frac{\sqrt{2}}{2}$

22 $\cos 2\theta - \cos \theta = 0$ **23** $\csc 2\theta = \frac{2\sqrt{3}}{3}$ **24** $\cos \theta = -0.8880$

In Problems 25–32, find the solution set of each trigonometric equation for $0 \leq t < 2\pi$.

25 $5 \sin^2 t + 2 \sin t - 3 = 0$

26 $\sin^2 t - 2 \sin t + 1 = 0$

27 $2 - \sin t = 2 \cos^2 t$

28 $4 \cos^2 t - 5 \sin t \cot t - 6 = 0$

29 $2 \sin^2 t - \sin t - 1 = 0$

30 $4 \csc t - 8 = 0$

31 $2 \sec t + 4 = 0$

32 $3 \cot^2 t - 1 = 0$

In Problems 33–40, find the solution set of each trigonometric equation for $0 \leq \theta < 360°$.

33 $2 \cos^2 \theta - \cos \theta = 0$

34 $\tan^2 \theta + 3 \sec \theta - 3 = 0$

35 $\tan^2 \theta - 2 \tan \theta + 1 = 0$

36 $\cot^2 \theta - 5 \cot \theta + 4 = 0$

37 $2 \cos^2 \theta + \cos \theta = 0$

38 $2 \cos^2 \theta - \sin \theta = 1$

39 $\tan \theta - 3 \cot \theta = 0$

40 $\cos 2\theta + \sin 2\theta = 0$

In Problems 41–46, find the solution set of each trigonometric equation, where t is a real number and θ represents the degree measure.

41 $\sec^2 t - \tan t - 1 = 0$

42 $\sqrt{3} \csc^2 \theta + 2 \csc \theta = 0$

43 $2 \sin^2 \theta - \sqrt{3} \sin \theta = 0$

44 $\sec^2 \theta + 1 - 3 \tan \theta = 0$

45 $9(\cos^2 t + \sin t) = 11$

46 $3 \sin^2 t - 4 \sin t + 1 = 0$

In Problems 47–50, use the following information to solve for the unknown angle to the nearest 10 minutes (10′).

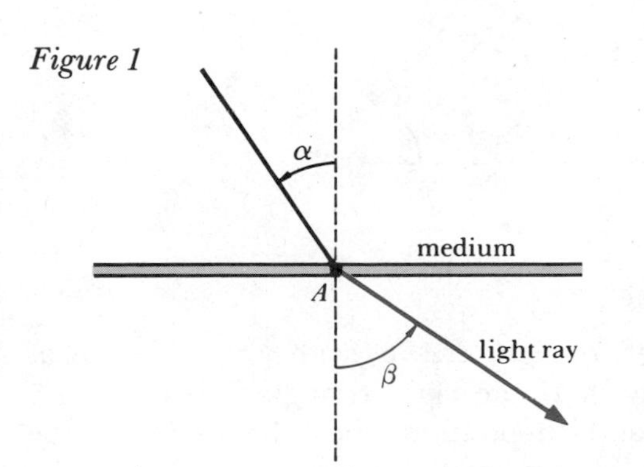

Figure 1

In physics there is a law, called *Snell's law,* which deals with the change in direction of a ray of light as it passes through a medium (Figure 1). In Figure 1 the dashed line is perpendicular to the surface of the medium. The angle α is called the *angle of incidence* and the angle β is called the *angle of refraction.* Snell's law states that, for a given medium,

$$\frac{\sin \alpha}{\sin \beta} = C, \quad \text{where } C \text{ is a constant}$$

The constant C is called the *index of refraction* of the medium.

47 Determine the angle of refraction of a light ray traveling through a diamond if the angle of incidence is 30° and the index of refraction is 2.42.

48 Find the angle of refraction of a light ray traveling through ice if the angle of incidence is 45° and the index of refraction is 1.31.

49 Find the angle of incidence of a light ray striking a rock salt crystal if the angle of refraction of the ray is 35°. The index of refraction of rock salt is 1.54.

50 Find the angle of incidence of a light ray striking the surface of turpentine in a container if the angle of refraction of the ray is 10°40′. The index of refraction of turpentine is 1.47.

5 Right Triangle Trigonometry and Polar Coordinates

In addition to being used to solve trigonometric equations, the trigonometric functions are used to establish relationships between the acute angles and sides of any right angle. These relationships enable us to find the measures of unknown sides or angles of right triangles and have applications in such fields as surveying and engineering.

After introducing a new coordinate system called the *polar system,* we shall see how the trigonometric functions also provide a way of relating the cartesian coordinates of a point in a plane and the polar coordinates of that point.

In what follows we make use of some standard notation to describe the sides and angles of any triangle. The triangle determined by the points A, B, and C is denoted by $\triangle ABC$. The angle at vertex A is denoted by the Greek letter alpha, α; the angle at vertex B is denoted by beta, β; the angle at vertex C is denoted by gamma, γ. The side opposite α is denoted by a; the side opposite β by b; and the side opposite γ by c (Figure 1). Also, recall from geometry that the sum of the degree measures of the angles in any triangle is 180°. Thus, in Figure 1, $\alpha + \beta + \gamma = 180°$.

Figure 1

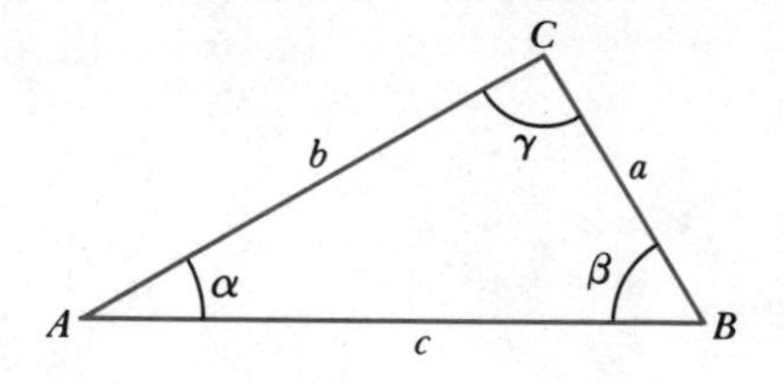

5.1 Right Triangle Trigonometry

Suppose that we are given a *right triangle,* $\triangle ABC$, with $\gamma = 90°$ and α as one of the acute angles (Figure 2). If the right triangle is placed on a coordinate system with α in standard position, then the values of the trigonometric functions of α may be expressed in terms of the sides of the right triangle.

Figure 2

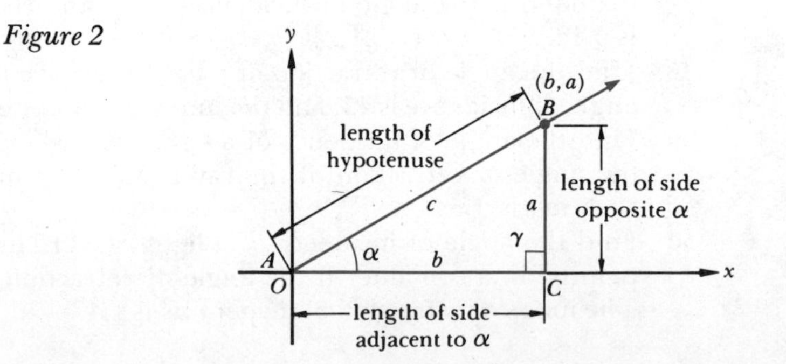

Using Definition 1 of the trigonometric functions of angles (Chapter 5, page 231), with $\theta = \alpha$, $r = c$, $x = b$, and $y = a$, we express the trigonometric functions of α in terms of the sides of right triangle $\triangle ABC$ as follows:

$$\sin\alpha = \frac{a}{c} = \frac{\text{length of side opposite } \alpha}{\text{length of hypotenuse}}$$

$$\cos\alpha = \frac{b}{c} = \frac{\text{length of side adjacent to } \alpha}{\text{length of hypotenuse}}$$

$$\tan\alpha = \frac{a}{b} = \frac{\text{length of side opposite } \alpha}{\text{length of side adjacent to } \alpha}$$

$$\cot\alpha = \frac{b}{a} = \frac{\text{length of side adjacent to } \alpha}{\text{length of side opposite } \alpha}$$

$$\sec\alpha = \frac{c}{b} = \frac{\text{length of hypotenuse}}{\text{length of side adjacent to } \alpha}$$

$$\csc\alpha = \frac{c}{a} = \frac{\text{length of hypotenuse}}{\text{length of side opposite } \alpha}$$

EXAMPLES 1 Determine the lengths of the unknown sides (to the nearest hundredth) of the right triangle, $\triangle ABC$, if $b = 4$ and $\alpha = 40°$.

Figure 3

SOLUTION From Figure 3 it is clear that we are attempting to find the lengths of a and c. Using Appendix Table IV, we find that

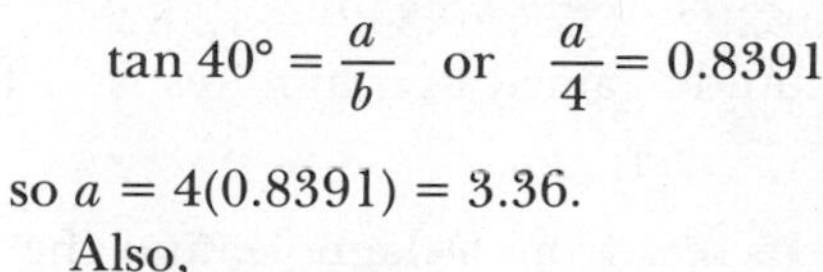

$$\tan 40° = \frac{a}{b} \quad \text{or} \quad \frac{a}{4} = 0.8391$$

so $a = 4(0.8391) = 3.36$.
Also,

$$\cos 40° = \frac{b}{c} \quad \text{or} \quad \frac{4}{c} = 0.7660$$

so $c = \dfrac{4}{0.7660} = 5.22$.

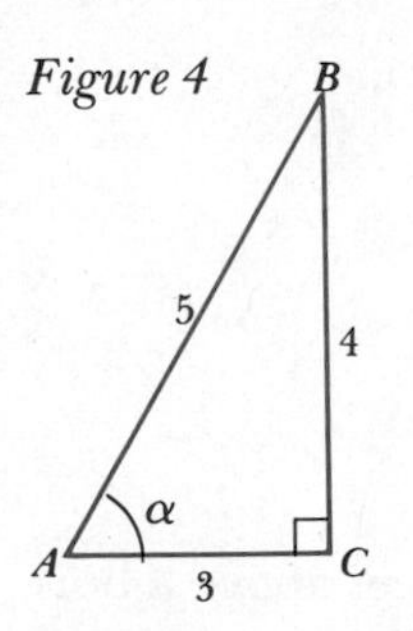

Figure 4

2 Determine the measure of angle α of the right triangle shown in Figure 4, if $a = 4$, $b = 3$, and $c = 5$.

SOLUTION From Figure 4, we see that $\sin\alpha = 4/5 = 0.8$, so that, by using linear interpolation, we have $\alpha = 53°8'$.

3 Use a right triangle to evaluate $\sin\left(\tan^{-1}\dfrac{3}{4}\right)$.

SOLUTION Let $t = \tan^{-1}\dfrac{3}{4}$. Using the definition of the inverse tangent,

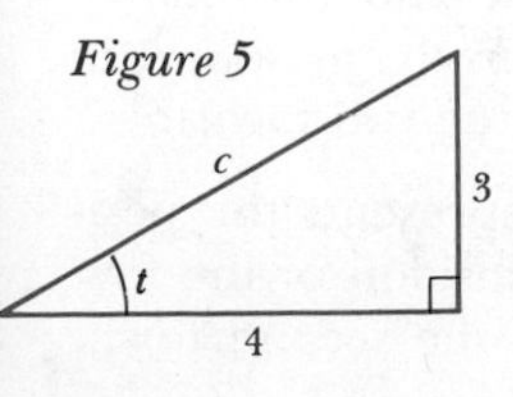

Figure 5

we have $\tan t = \dfrac{3}{4}$ (Figure 5). By the Pythagorean theorem,

$$c = \sqrt{3^2 + 4^2} = 5$$

Hence $\sin\left(\tan^{-1}\dfrac{3}{4}\right) = \sin t = \dfrac{3}{5}$.

4 A roadway rises 10 feet for every 200 feet along the horizon. Find the angle of inclination of the roadway.

Figure 6

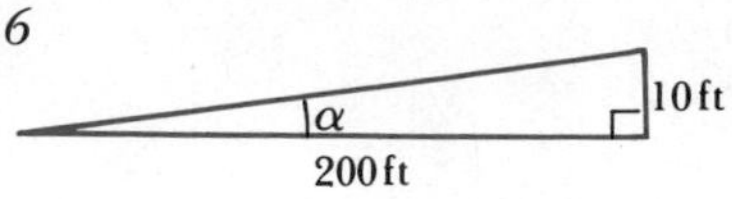

SOLUTION From Figure 6, it is clear that α represents the angle of inclination. Thus

$$\tan \alpha = \frac{10}{200} = 0.0500$$

so $\alpha = 2°52'$.

5 A jet fighter plane is climbing at an angle of 20° at a speed of 450 miles per hour. What is its gain in altitude in 5 minutes?

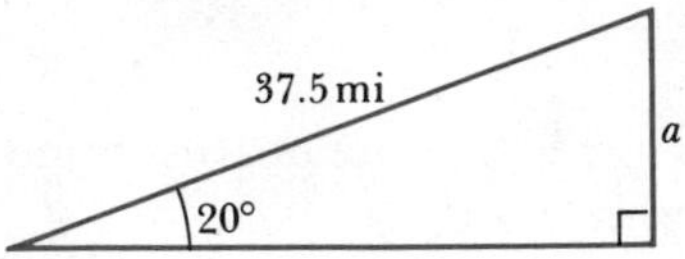

SOLUTION After 5 minutes the jet has traveled

$$450 \cdot 5 \cdot \frac{1}{60} = 37.5 \text{ miles}$$

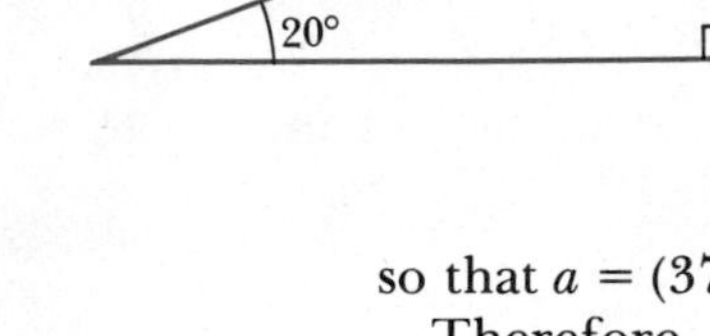

Figure 7

The altitude gain is represented by a in Figure 7. Thus

$$\sin 20° = \frac{a}{37.5}$$

so that $a = (37.5) \sin 20° = (37.5)(0.342) = 12.825$.

Therefore, the altitude gained in 5 minutes is 12.825 miles.

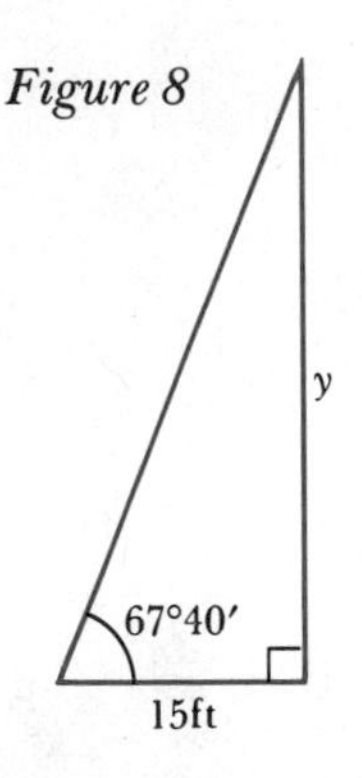

Figure 8

6 A man is standing 15 feet from a flagpole, and the foot of the pole is at eye level. The *angle of elevation* (the acute angle formed by the line of sight and a horizontal line passing through the position of the sighting) to the top of the flagpole is 67°40′. Find the height of the pole.

SOLUTION Let y be the height of the flagpole (Figure 8). Then

$$\tan 67°40' = \frac{y}{15} = 2.434$$

so $y = 15(2.434) = 36.51$.

Therefore, the height of the flagpole is 36.51 feet.

Figure 9

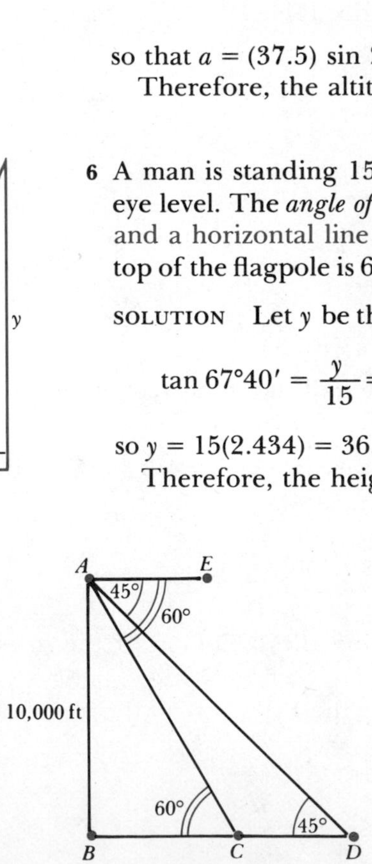

7 From a mountain top 10,000 feet above a horizontal plane, two towns were observed with *angles of depression* (the acute angle determined by the plane of the observer and the line of sight) of 60° and 45°, respectively, as shown in Figure 9. How far apart are the two towns?

SOLUTION In Figure 9, A represents the position of the observer, C the position of the first town, and D the position of the second town.

The angles of depression are $\angle DAE$ and $\angle CAE$. Using the fact that alternate interior angles formed by the transversal of two parallel lines are equal, we have

$$\angle BCA = \angle CAE = 60° \quad \text{and} \quad \angle BDA = \angle DAE = 45°$$

Hence

$$|\overline{BC}| = 10{,}000 \cot 60° = \frac{10{,}000}{\sqrt{3}} \text{ feet}$$

and

$$|\overline{BD}| = 10{,}000 \cot 45° = 10{,}000 \text{ feet}$$

so the distance between C and D is

$$|\overline{BD}| - |\overline{BC}| = 10{,}000 - \frac{10{,}000}{\sqrt{3}} = 4230 \text{ feet (approximately)}$$

5.2 Polar Coordinates

We have seen that points in the plane can be associated with pairs of real numbers by using a cartesian coordinate system, which is also called a *rectangular coordinate system.* Another way of associating pairs of numbers with points in the plane is based on a "grid" composed of concentric circles and rays emanating from the common center of the circles (Figure 10). Such a system is called a *polar coordinate system.* The frame of reference for this coordinate system consists of a fixed point O, called the *pole,* and a fixed ray called the *polar axis* (Figure 11). The position of a point P is uniquely determined by r and θ, where θ is any angle in standard position having the ray $\overrightarrow{OA}$ as its initial side and a ray on line OP as its terminal side, and r is the directed distance along the terminal side of θ from the pole O to P. The pair (r, θ) is called the *polar coordinates* of P.

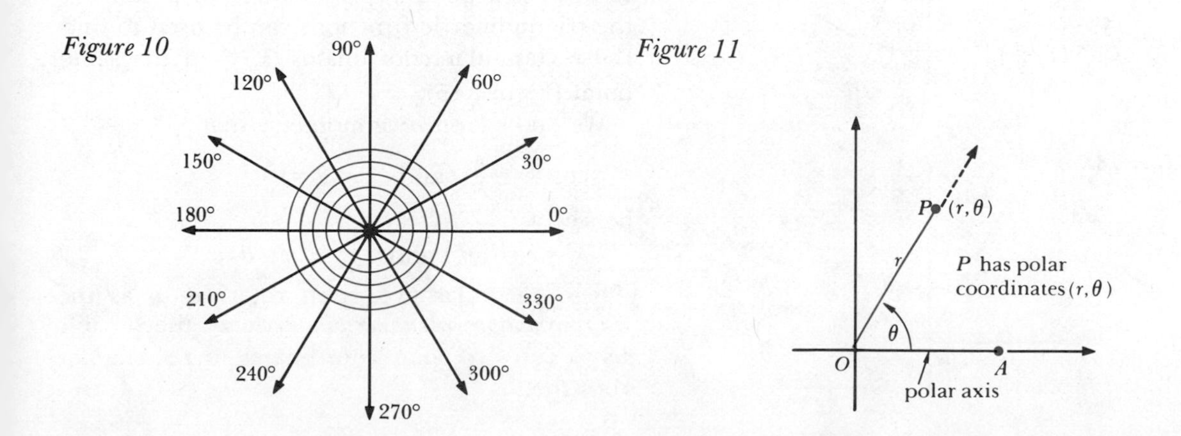

Figure 10

Figure 11

Each point in a polar coordinate system can be represented by infinitely many ordered pairs of numbers. For example, (2, 30°) and (2, 390°) and (2, −330°) each represent the same point (Figure 12).

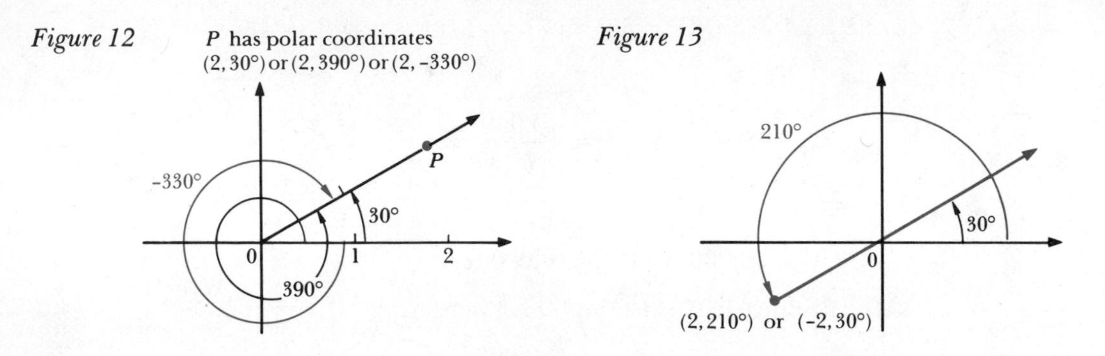

Figure 12 *Figure 13*

Also, r need not be positive. If $r < 0$, the point (r, θ) is determined by plotting $(|r|, \theta + 180°)$ or $(|r|, \theta + \pi)$, depending on whether θ is measured in degrees or radians.

For example, in Figure 13, (−2, 30°) is the same as (2, 210°).

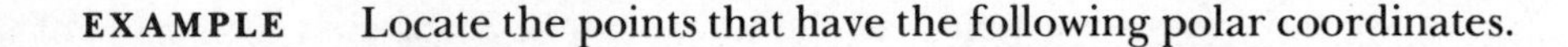

EXAMPLE Locate the points that have the following polar coordinates.

(a) (3, 70°) (b) (0, 0)

(c) $\left(7, \frac{7\pi}{5}\right)$ (d) (3, 100°)

(e) (π, π) (f) (3, 5)

(g) (5,0) (h) (−5, 15°)

SOLUTION The points are shown in Figure 14.

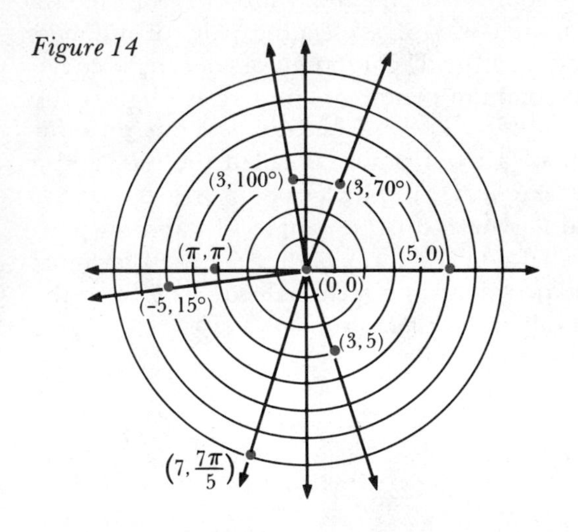

Figure 14

CONVERSION OF POLAR COORDINATES TO RECTANGULAR COORDINATES

If (r, θ) is a polar representation of a point P, the trigonometric functions can be used to find the rectangular coordinates (x, y) of the same point (Figure 15).

We know from trigonometry that

$$\cos\theta = \frac{x}{r} \quad \text{and} \quad \sin\theta = \frac{y}{r}$$

hence

$$x = r\cos\theta \quad \text{and} \quad y = r\sin\theta$$

These formulas are often referred to as the *transformation* or *conversion formulas;* they enable us to convert polar coordinates to rectangular coordinates.

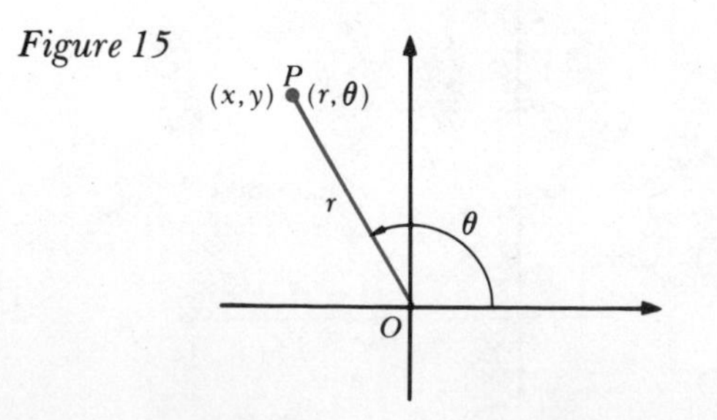

Figure 15

EXAMPLE Convert the given polar coordinates to rectangular coordinates.

(a) $(3, 60°)$ (b) $(-2, 180°)$ (c) $(4, -150°)$ (d) (π, π)

SOLUTION (See Figure 16.)

(a) $$x = r \cos \theta = 3 \cos 60° = \frac{3}{2}$$

and

$$y = r \sin \theta = 3 \sin 60° = \frac{3\sqrt{3}}{2}$$

Hence the rectangular coordinates are $\left(\frac{3}{2}, \frac{3\sqrt{3}}{2}\right)$.

(b) $$x = r \cos \theta = -2 \cos 180° = 2$$

and

$$y = r \sin \theta = -2 \sin 180° = 0$$

Hence the rectangular coordinates are $(2, 0)$.

Figure 16

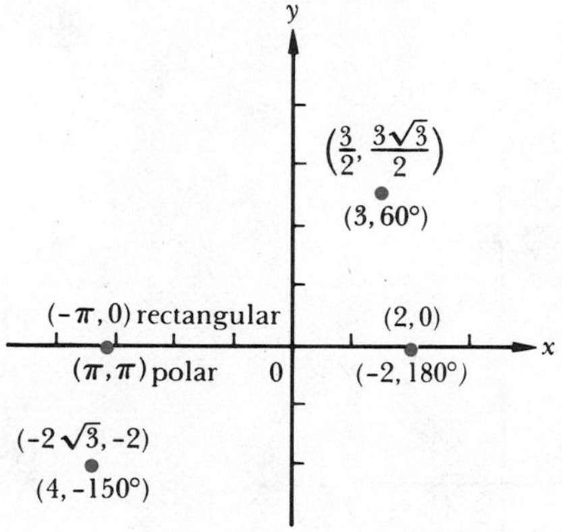

y
$\left(\frac{3}{2}, \frac{3\sqrt{3}}{2}\right)$
(3, 60°)
(−π, 0) rectangular
(π, π) polar
0
(2, 0)
(−2, 180°)
x
$(-2\sqrt{3}, -2)$
(4, −150°)

(c) $$x = r \cos \theta = 4 \cos(-150°) = 4\left(-\frac{\sqrt{3}}{2}\right) = -2\sqrt{3}$$

and

$$y = r \sin \theta = 4 \sin(-150°) = 4\left(-\frac{1}{2}\right) = -2$$

Hence the rectangular coordinates are $(-2\sqrt{3}, -2)$.

(d) $$x = r \cos \theta = \pi \cos \pi = -\pi$$

and

$$y = r \sin \theta = \pi \sin \pi = 0$$

Hence the rectangular coordinates are $(-\pi, 0)$.

CONVERSION OF RECTANGULAR COORDINATES TO POLAR COORDINATES

Assume that the rectangular coordinates of a point P are given by (x, y) and that any of the polar coordinates of the same point are given by (r, θ).

Figure 17

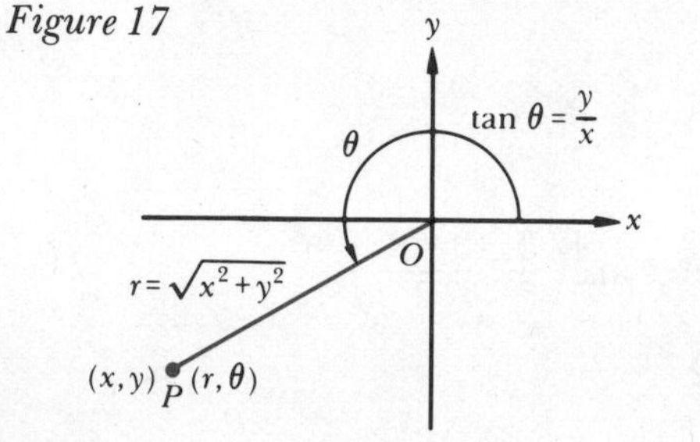

By the Pythagorean theorem,

$$r = \sqrt{x^2 + y^2}$$

From Definition 1 of the tangent function on page 231,

$$\tan\theta = \frac{y}{x} \qquad \text{(Figure 17)}$$

These two conversion formulas can be used to transform rectangular coordinates to polar coordinates.

EXAMPLE Convert the given rectangular coordinates to polar coordinates.
(a) $(-1, 1)$ (b $(3, -\sqrt{3})$ (c) $(-3, 4)$

Figure 18

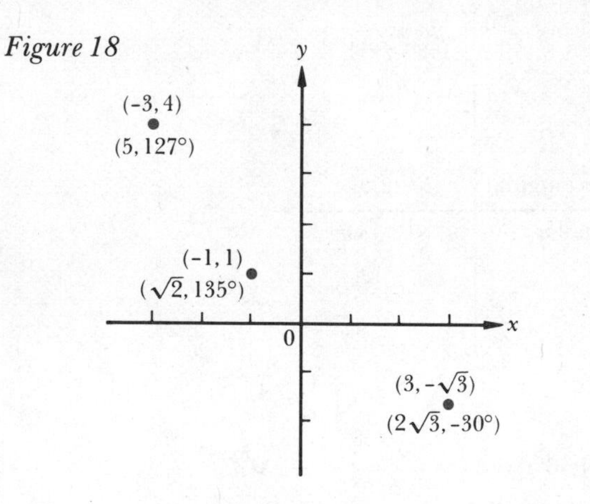

SOLUTION (See Figure 18.)

(a) $$r = \sqrt{x^2 + y^2}$$
$$= \sqrt{(-1)^2 + 1} = \sqrt{2}$$

and

$$\tan\theta = -1$$

The point is in quadrant II, so that $\tan\theta = -1$ implies that $\theta = 135°$. Thus, one pair of polar coordinates is $(\sqrt{2}, 135°)$, while another is $(\sqrt{2}, -225°)$.

(b) $$r = \sqrt{3^2 + (-\sqrt{3})^2} = \sqrt{9 + 3}$$
$$= \sqrt{12} = 2\sqrt{3}$$

and

$$\tan\theta = -\frac{\sqrt{3}}{3}$$

so $\theta = -30°$, since the point is in quadrant IV. Hence one possible pair of polar coordinates is given by $(2\sqrt{3}, -30°)$.

(c) $$r = \sqrt{(-3)^2 + 4^2} = 5$$

and

$$\tan\theta = -\frac{4}{3} = -1.3333$$

We find from Appendix Table IV that θ is approximately equal to 127°, so (5, 127°) is one pair of polar coordinates.

PROBLEM SET 5

In Problems 1–4, sketch $\triangle ABC$ with the specified parts.

1 $\alpha = 50°, \gamma = 80°, c = 1$ inch

2 $a = 5$ centimeters, $b = 3$ centimeters, $c = 4$ centimeters

3 $\alpha = 110°, \beta = 30°, c = 2$ centimeters

4 $a = 2$ centimeters, $b = 1$ centimeter, $\gamma = 150°$

In Problems 5–18, $\triangle ABC$ is a right triangle with $\gamma = 90°$. For each problem, sketch $\triangle ABC$ and then find all the missing parts. (Angles should be determined to the nearest 10′ and sides to the nearest hundredth.)

5 $a = 5, b = 3$

6 $a = 10, \alpha = 30°$

7 $c = 10, \beta = 50°$

8 $a = 7, c = 12$

9 $a = 17, \beta = 23°$

10 $c = 100, \alpha = 14°40'$

11 $a = b = 5$

12 $a = 5, \sin\alpha = \frac{1}{5}$

13 $c = 50, \tan\alpha = 3$

14 $c = 7, \sin\alpha = \frac{4}{7}$

15 $\alpha = \tan^{-1}\frac{8}{15}, c = 17$

16 $b = 10, \cos\alpha = \frac{3}{5}$

17 $c = 10\sqrt{2}, b = 10$

18 $\sec\alpha = 2, c = 6$

In Problems 19–22, use a right triangle to evaluate the expression.

19 $\sin\left(\cos^{-1}\frac{3}{5}\right)$

20 $\cos\left(\sin^{-1}\frac{2}{3}\right)$

21 $\cos\left(\tan^{-1}\frac{9}{2}\right)$

22 $\tan\left(\cos^{-1}\frac{\sqrt{5}}{3}\right)$

In Problems 23–32, locate each point with the given rectangular coordinates, then convert the given representation to polar coordinates. (Express the angle to the nearest 10′.)

23 $(-1, \sqrt{3})$ **24** $(-3, 0)$ **25** $(4, 3)$ **26** $(-6, 6\sqrt{3})$

27 $(5, 5)$ **28** $(0, -2)$ **29** $(-3, 3\sqrt{3})$ **30** $(2\sqrt{3}, -2)$

31 $(-5, -5)$ **32** $(-2, 5)$

In Problems 33–42, locate each point with the given polar coordinates, then convert the given representation to rectangular coordinates, where $r > 0$ and $0 \le \theta < 360°$.

33 $(6, 30°)$ **34** $(10, \pi/3)$ **35** $(7, 120°)$ **36** $(4, -\pi/6)$

37 $(-4, 90°)$ **38** $(-8, 45°)$ **39** $(2\pi, -\pi/6)$ **40** $(5, \pi/2)$

41 $(4, 210°)$ **42** $(4, -330°)$

43 If the angle of elevation of the top of a tower from a distance 200 feet away on the ground level is 60°, find the height of the tower.

44 What is the angle of elevation of the sun when a woman 2 meters tall casts a 3.1-meter shadow?

45 A railroad bridge is 25 feet above ground level. Find the angle of elevation of the bridge from a point on the ground 189.83 feet away from the bridge.

46 From the top of an observation post, the angle of depression of a ship is 30°. If the distance between the top of the observation post and the ship is 4500 feet, find the observation post's height above sea level.

47 From the top of a 300-foot tower, the angles of depression of the top and the bottom of a flagpole standing on a plane level with the base of the tower are observed to be 45° and 60°, respectively. Find the height of the flagpole.

48 From the top of a mountain, an observer viewed two boats sailing on a lake. The boats and the base of the mountain are on a straight line and the observer's eye is 2000 feet above the lake level. The angles of depression of the two boats are 18° and 14°, respectively. Find the distance between the boats.

49 From an altitude of 15,000 feet, a balloonist observes the angle of depression of the base of a building to be 20°. How far is the building from a point on the ground directly beneath the balloon?

50 A ladder 14 feet long is leaning against a building. The angle formed by the ladder and the ground is 65°. How far from the building is the foot of the ladder?

51 Assume that the opposite shores of a lake are relatively straight and parallel. Points A and B are located 3 miles apart and directly opposite each other across the lake. Point C is on the same shoreline as B, down the lake from B. A telephone company wishes to lay an underwater cable from A to C. If $\angle ACB$ is 48°, what is the length of the cable $\overline{AC}$?

52 An island is at point A, 6 miles offshore from the nearest point B on a straight beach. A store is at point C down the beach from B on the same shoreline. If $\angle ACB$ is 35°, find the distance from the store to the island.

53 A television antenna stands on top of a house that is located on level ground. From a point 14 feet from the base of the house, the angles of elevation of the top and bottom of the antenna measure 70° and 66°, respectively. How high is the house? How tall is the television antenna?

54 The angle of elevation of the top of a hill is 59° from a given point. After moving a distance of 170 feet in a horizontal line away from the point, the angle of elevation was found to be 52°. How high is the hill?

6 Law of Sines and Law of Cosines

In the preceding section, we saw how it is possible to use the trigonometric functions to relate the sides and angles of *right* triangles. Now, we use the trigonometric functions to derive two formulas—the *law of sines* and the *law of cosines*—that relate the sides and angles of triangles that are not necessarily right triangles.

6.1 Law of Sines

The law of sines is applicable to finding unknown parts of a triangle either (1) if we are given any two angles and any side, or (2) if we are given two sides and an angle opposite one of them. The first situation *always* yields a unique triangle, whereas the second situation, called the *ambiguous case,* may result in no triangle, one triangle, or two triangles with the given sides and angle.

THEOREM 1 LAW OF SINES

In any $\triangle ABC$,

$$\frac{\sin\alpha}{a} = \frac{\sin\beta}{b} = \frac{\sin\gamma}{c}$$

PROOF (Scc Figure 1.) The area of a triangle is equal to the product of one-half the base and the altitude. Draw altitudes h_1 and h_2. The area of $\triangle ABC = \frac{1}{2}ch_2 = \frac{1}{2}ah_1$, but

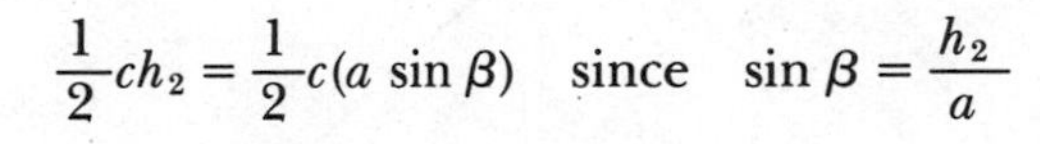

$$\frac{1}{2}ch_2 = \frac{1}{2}c(a\sin\beta) \quad \text{since} \quad \sin\beta = \frac{h_2}{a}$$

Figure 1

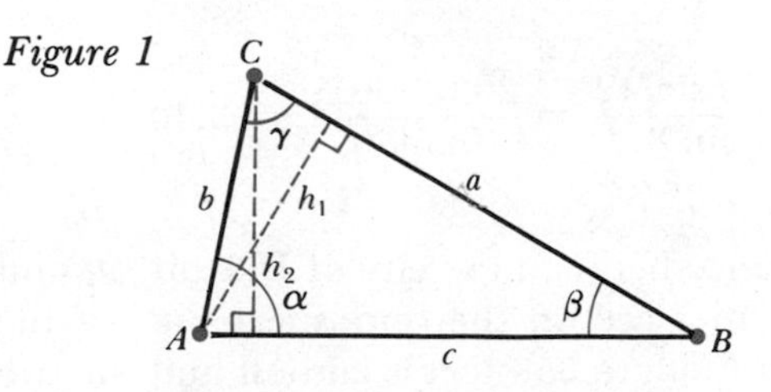

and

$$\frac{1}{2}ch_2 = \frac{1}{2}c(b\sin\alpha) \quad \text{since} \quad \sin\alpha = \frac{h_2}{b}$$

Also,

$$\frac{1}{2}ah_1 = \frac{1}{2}a(b\sin\gamma) \quad \text{since} \quad \sin\gamma = \frac{h_1}{b}$$

Hence

$$\frac{1}{2}ca\sin\beta = \frac{1}{2}cb\sin\alpha = \frac{1}{2}ab\sin\gamma$$

so

$$ca\sin\beta = cb\sin\alpha = ab\sin\gamma$$

After dividing the latter equations by abc, we get

$$\frac{\sin\alpha}{a} = \frac{\sin\beta}{b} = \frac{\sin\gamma}{c}$$

It is important to note that the law of sines can also be written as

$$\frac{a}{\sin\alpha} = \frac{b}{\sin\beta} = \frac{c}{\sin\gamma}$$

If any two angles and any side of a triangle are known, the law of sines can be used to determine either of the two remaining sides.

EXAMPLES Use the law of sines to determine the indicated parts of each triangle.

1 In $\triangle ABC$, $a = 12$, $\alpha = 45°$, and $\beta = 105°$. Find c.

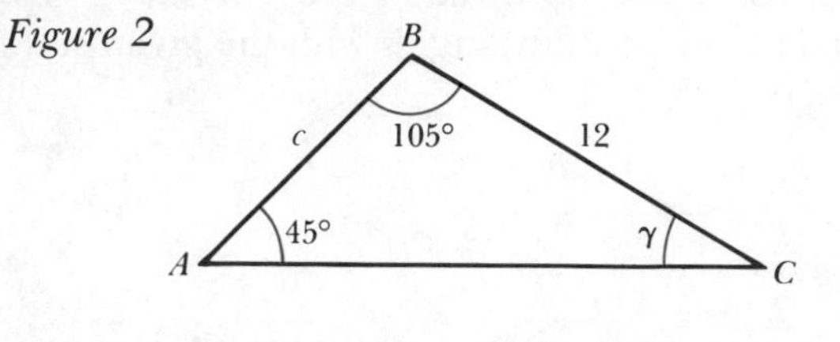

Figure 2

SOLUTION Since $\alpha + \beta + \gamma = 180°$, then $\gamma = 30°$ (Figure 2). Using the law of sines, we have

$$\frac{c}{\sin 30°} = \frac{12}{\sin 45°}$$

so

$$c = \frac{12 \sin 30°}{\sin 45°} = \frac{12(1/2)}{\sqrt{2}/2} = \frac{12}{\sqrt{2}} = 6\sqrt{2}$$

2 In $\triangle ABC$, if $a = 20$, $\gamma = 51°$, and $\beta = 42°$, find b to the nearest hundredth.

Figure 3

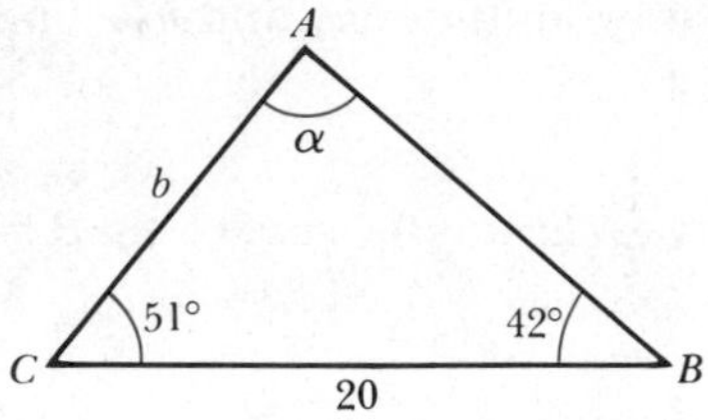

SOLUTION Since $\alpha + \beta + \gamma = 180°$, then $\alpha = 87°$ (Figure 3). Using the law of sines, we have

$$\frac{b}{\sin 42°} = \frac{20}{\sin 87°}$$

so

$$b = \frac{20 \sin 42°}{\sin 87°} = \frac{20(0.6691)}{0.9986} = 13.40$$

3 A surveyor wishes to find the distance between the city of Detroit, D, and the city of Windsor, W (the two cities are on the opposite banks of the Detroit River). From D a line, $\overline{DC}$, of length 550 feet is laid off and the two angles, $\angle CDW = 125°40'$ and $\angle DCW = 48°50'$, are measured. Find the distance between the two cities, that is, the length of $\overline{DW}$.

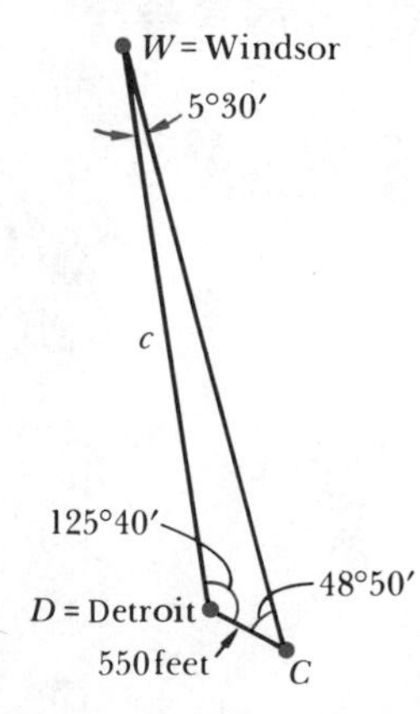

Figure 4

SOLUTION Let $|\overline{DW}| = c$ feet. In $\triangle WDC$ (Figure 4), the measure of $\angle DWC$ is $180° - (125°40' + 48°50') = 5°30'$. Using the law of sines, we have

$$\frac{c}{\sin 48°50'} = \frac{550}{\sin 5°30'}$$

so

$$c = \frac{550(\sin 48°50')}{\sin 5°30'} = \frac{550(0.7528)}{0.0958} = 4321.92$$

Therefore, the distance is approximately 4322 feet.

4 A tower 125 feet high is on a hill on the bank of a river. It is observed that the angle of depression from the top of the tower to a point on the opposite shore is $28°40'$ and the angle of depression from the base of the tower to the same point on the shore is $18°20'$. How wide is the river and how high is the hill (to the nearest foot)?

SOLUTION In Figure 5, $|\overline{BC}|$ represents the height of the tower, $|\overline{BD}|$ represents the height of the hill, and A is the point on the opposite shore. $|\overline{BC}| = 125$ feet. In $\triangle ABC$,

$$\angle BCA = 90° - 28°40' = 61°20'$$
$$\angle ABC = 90° + 18°20' = 108°20'$$
$$\angle CAB = 28°40' - 18°20' = 10°20'$$

Figure 5

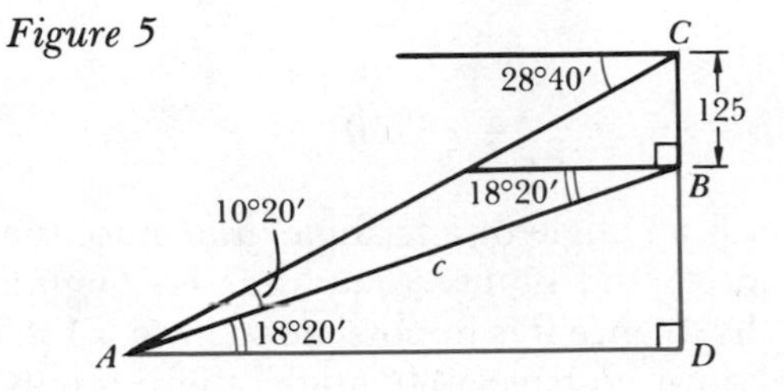

By the law of sines, we have

$$\frac{c}{\sin 61°20'} = \frac{125}{\sin 10°20'}$$

or

$$c = \frac{125 \sin 61°20'}{\sin 10°20'} = \frac{125(0.8774)}{0.1794} = 611$$

so

$$|\overline{BD}| = c \sin 18°20' = 611(0.3145) = 192.16$$

and

$$|\overline{AD}| = c \cos 18°20' = 611(0.9492) = 579.96$$

Thus, the river is approximately 580 feet wide and the hill 192 feet high.

A calculator is of great help in doing calculations such as those called for in the above examples.

THE AMBIGUOUS CASE

Let us examine the various cases that may occur if we are given two sides and an angle opposite one of them. In Figure 6, consider a, b, and α in each situation. It may be that there is no possible triangle with these parts (Figure 6a); there may be two different triangles with these parts (Figure 6b); or there may be only one triangle with these parts (Figure 6c).

Figure 6

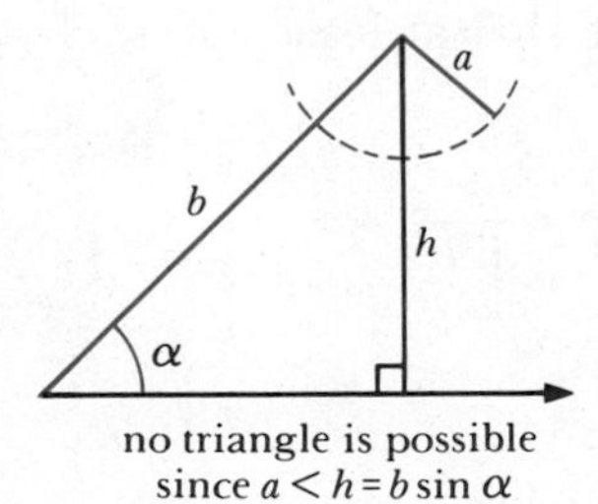

no triangle is possible since $a < h = b \sin \alpha$

(a)

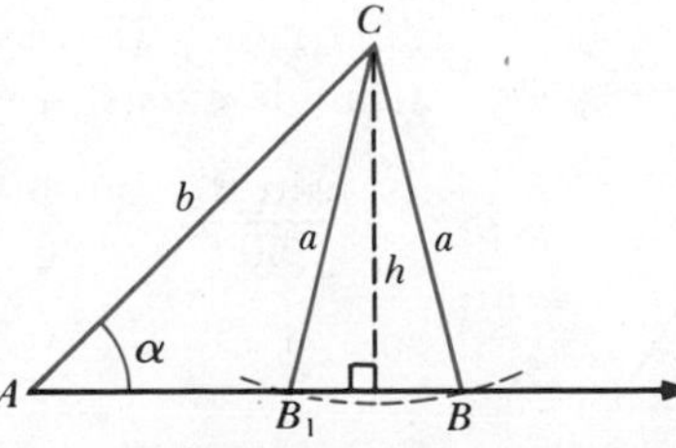

both $\triangle ABC$ and $\triangle AB_1C$ are possible since $h = b \sin \alpha < a < b$

(b)

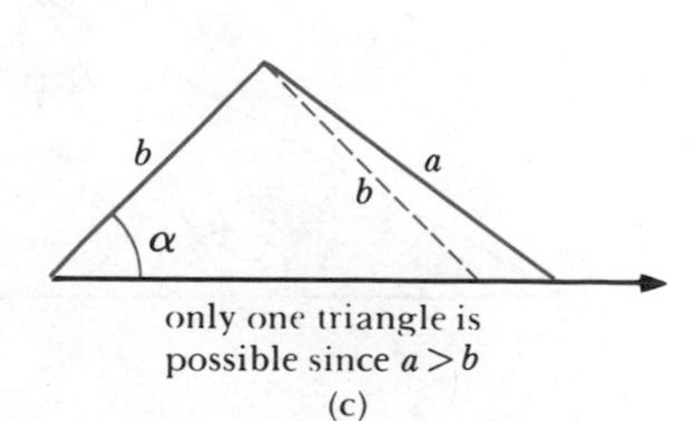

only one triangle is possible since $a > b$

(c)

Because there are three different possibilities here, we sometimes refer to this situation as the *ambiguous case.* The next examples show how the law of sines is used to solve situations occurring in the ambiguous case.

EXAMPLES In each of the following situations, find the indicated unknown parts of all possible triangles.

1 Find β to the nearest degree if $a = 20, b = 15$, and $\alpha = 30°$.

SOLUTION We shall attempt to find β, keeping in mind that there may be one, two, or no triangles. If we assume that at least one triangle exists, then, by the law of sines, we have

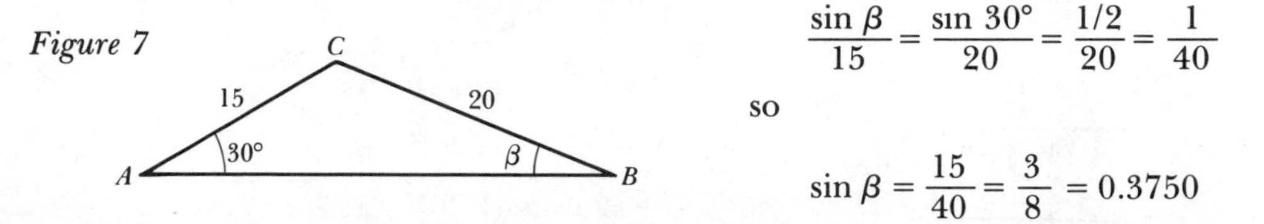

Figure 7

$$\frac{\sin\beta}{15} = \frac{\sin 30°}{20} = \frac{1/2}{20} = \frac{1}{40}$$

so

$$\sin\beta = \frac{15}{40} = \frac{3}{8} = 0.3750$$

Since β must be less than 180° (if it is an angle of a triangle) and since the sine is positive for acute (quadrant I) and obtuse (quadrant II) angles, β is approximately 22° or 158°. (Why?) Since it is impossible to have a 158° angle in a triangle that is already known to have a 30° angle, there is only one triangle possible, and $\beta = 22°$ (Figure 7).

2 If $a = 5, b = 20$, and $\alpha = 30°$, find β to the nearest degree.

SOLUTION Here again we assume that at least one such triangle exists, so

$$\frac{\sin\beta}{20} = \frac{\sin 30°}{5} = \frac{1/2}{5} = \frac{1}{10}$$

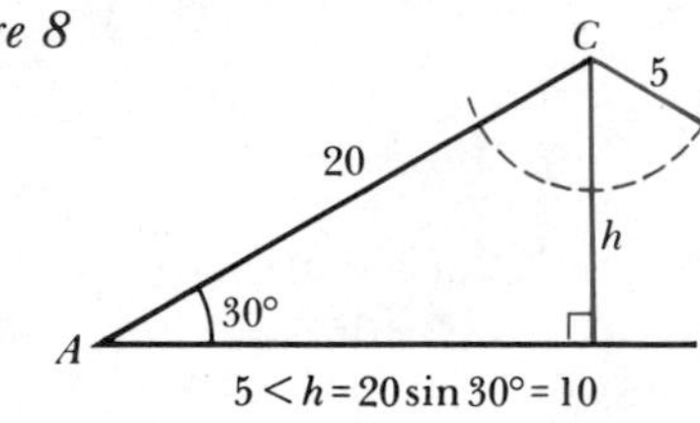

Figure 8

Hence

$$\sin\beta = 2$$

But, since $|\sin\theta| \leq 1$ for any angle, no such β exists; therefore, no triangle exists with the given parts (Figure 8).

3 Find β and γ to the nearest minute if $a = 30$, $b = 50$, and $\alpha = 30°$.

SOLUTION If we assume that at least one such triangle exists, we have

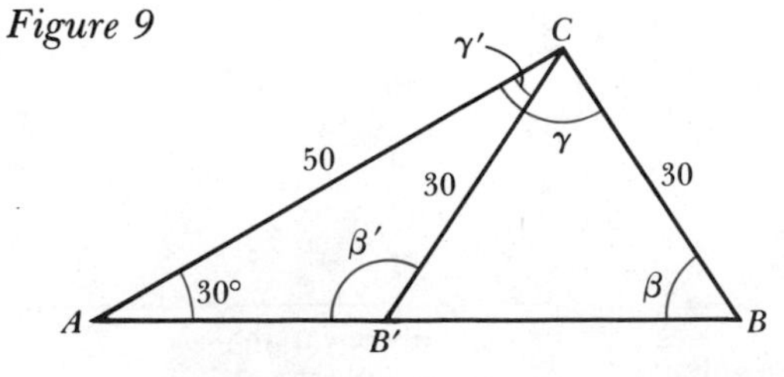

Figure 9

$$\frac{\sin\beta}{50} = \frac{\sin 30°}{30} = \frac{1/2}{30} = \frac{1}{60}$$

so

$$\sin\beta = \frac{5}{6} = 0.8333$$

Hence β is approximately 56°26′ or 123°34′. Since either value of β satisfies $\alpha + \beta < 180°$, we have two possible triangles, $\triangle ABC$ and $\triangle AB'C$ (Figure 9). One of the two possible triangles has angles 30°, 56°26′, and 93°34′, whereas the other triangle has angles 30°, 123°34′, and 26°26′.

6.2 Law of Cosines

The law of cosines is applicable to a triangle either (1) if we are given two sides and the included angle of a triangle, or (2) if we are given the three sides of a triangle.

THEOREM 2 LAW OF COSINES

In any $\triangle ABC$,

$$\text{(i)} \quad c^2 = a^2 + b^2 - 2ab \cos \gamma$$
$$\text{(ii)} \quad b^2 = a^2 + c^2 - 2ac \cos \beta$$
$$\text{(iii)} \quad a^2 = b^2 + c^2 - 2bc \cos \alpha$$

Only one of the above formulas need be memorized; the other two can be obtained by changing the letters. In general form, the theorem is stated as: The square of the length of any side of a triangle is equal to the sum of the squares of the lengths of the other two sides minus twice their product times the cosine of the angle included between these other two sides.

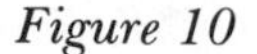

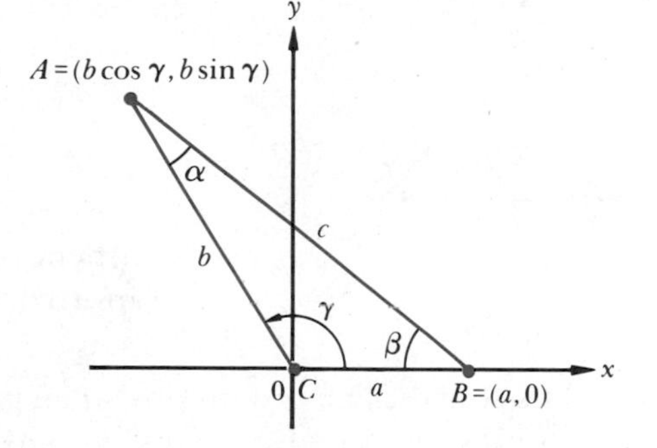

PROOF First, consider $\triangle ABC$ on a cartesian coordinate system with γ in standard position (Figure 10). A has coordinates $(b \cos \gamma, b \sin \gamma)$ (why?) and B has coordinates $(a, 0)$ (why?), so that by the distance formula, we have

$$\begin{aligned} c^2 &= (b \cos \gamma - a)^2 + (b \sin \gamma - 0)^2 \\ &= b^2 \cos^2 \gamma - 2ab \cos \gamma + a^2 + b^2 \sin^2 \gamma \\ &= b^2 \cos^2 \gamma + b^2 \sin^2 \gamma + a^2 - 2ab \cos \gamma \\ &= a^2 + b^2(\cos^2 \gamma + \sin^2 \gamma) - 2ab \cos \gamma \\ &= a^2 + b^2 - 2ab \cos \gamma \end{aligned}$$

In view of the fact that the location of the coordinate axes in the plane is merely a matter of convenience, the other two formulas in Theorem 2 are obtained in a similar manner, that is, by placing the other two angles in standard position and repeating the steps in the proof above.

If two sides and an included angle of a triangle are known, the law of cosines can be used to determine the third side, or if the three sides of a triangle are known, any of the angles can be found by using the law of cosines.

EXAMPLES Use the law of cosines to find the indicated part of each triangle.

1 In $\triangle ABC$, $a = 8$, $b = 6$, and $\gamma = 60°$; find c.

Figure 11

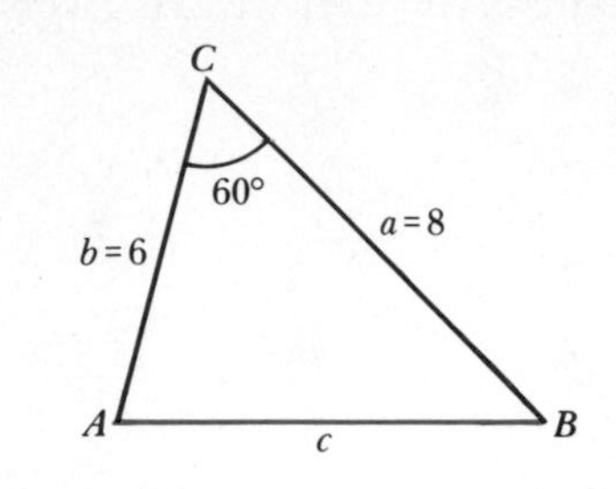

SOLUTION (See Figure 11.) Using the law of cosines, we have

$$\begin{aligned} c^2 &= a^2 + b^2 - 2ab \cos \gamma \\ &= 64 + 36 - 2(8)(6)(\tfrac{1}{2}) \\ &= 100 - 48 = 52 \end{aligned}$$

so $c = \sqrt{52} = 2\sqrt{13} = 7.21$.

2 In $\triangle ABC$, $a = 7$, $b = 4$, and $c = 5$; find α to the nearest degree.

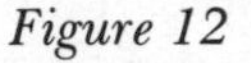

Figure 12

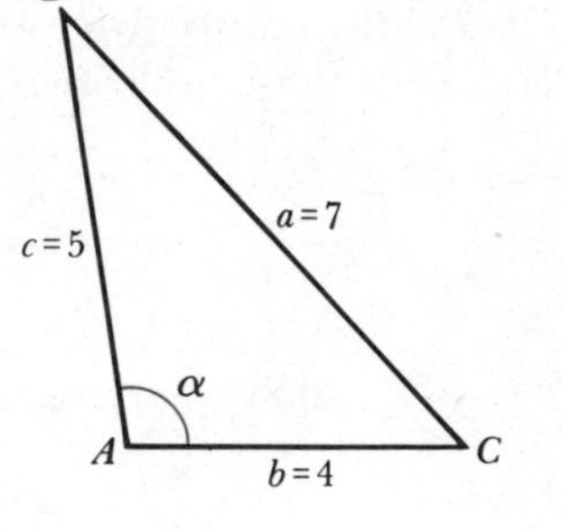

SOLUTION (See Figure 12.) Using the law of cosines, we have

$$\begin{aligned} a^2 &= b^2 + c^2 - 2bc \cos \alpha \\ 7^2 &= 4^2 + 5^2 - 2(4)(5) \cos \alpha \\ 49 &= 16 + 25 - 40 \cos \alpha \\ 49 &= 41 - 40 \cos \alpha \end{aligned}$$

so $\cos \alpha = -\frac{8}{40} = -0.2$.

Since $\cos \alpha$ is negative, $90° < \alpha < 180°$. From Appendix Table IV, we find that $\cos 78° = 0.2$ (approximately), so that $\alpha = 180° - 78° = 102°$.

3 The straight-line distance between two cities, A and B, cannot be measured directly because of a swamp located between them. A surveyor at city C is able to measure $|\overline{CA}| = 5.73$ miles, $|\overline{CB}| = 8.19$ miles, and $\angle ACB = 113°$. What is the distance from city A to city B (to the nearest hundredth)?

SOLUTION (See Figure 13.) Using the law of cosines, we have

$$\begin{aligned} |\overline{AB}|^2 &= |\overline{CB}|^2 + |\overline{CA}|^2 - 2|\overline{CB}||\overline{CA}| \cos \gamma \\ &= (8.19)^2 + (5.73)^2 - 2(8.19)(5.73) \cos 113° \\ &= 67.0761 + 32.8329 - (93.8574)(-0.3907) \\ &= 136.58 \end{aligned}$$

so $|\overline{AB}| = 11.69$. Therefore, the distance from city A to city B is approximately 11.69 miles.

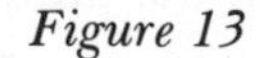

Figure 13

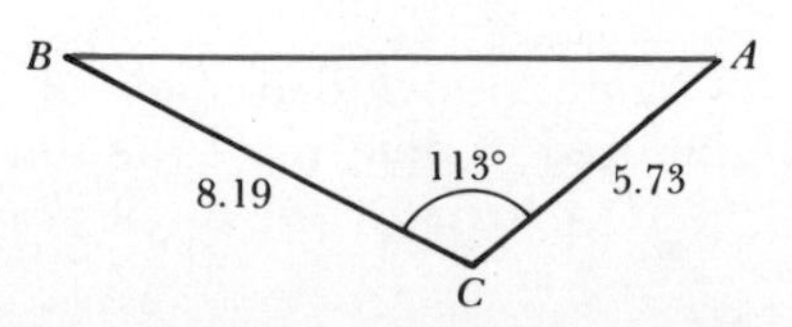

PROBLEM SET 6

In Problems 1–6, use the law of sines to find the specified side of $\triangle ABC$ to the nearest hundredth under the given conditions.

1 a if $\alpha = 60°$, $\beta = 45°$, and $b = 10$.
2 b if $\gamma = 75°$, $\beta = 25°$, and $c = 8$.
3 c if $\alpha = 50°$, $\beta = 100°$, and $a = 5$.
4 b if $\alpha = 60°$, $\beta = 30°$, and $c = 11$.
5 a if $\alpha = 100°20'$, $\beta = 5°30'$, and $b = 3.71$.
6 a if $\beta = 70°$ and $b = c = 17.13$.

In Problems 7–12, use the law of sines to find all the unknown parts of all possible triangles. Find angle measures to the nearest 10′ and sides to the nearest hundredth.

7 $b = 3, c = 5, \gamma = 30°$
8 $b = 17.5, a = 15.2, \alpha = 45°$
9 $\alpha = 50°, b = 17, a = 12$
10 $a = 182.5, b = 82.5, \alpha = 72°$
11 $a = 12, b = 8, \beta = 25°$
12 $a = 10, b = 50, \alpha = 22°$

In Problems 13–18, use the law of cosines to find the specified unknown parts of the given triangle. Find angle measures to the nearest 10′ and sides to the nearest hundredth.

13 a if $\alpha = 60°$, $b = 20$, and $c = 6$.
14 α if $a = 3$, $b = 4$, and $c = 6$.
15 β if $a = 144$, $b = 180$, and $c = 108$.
16 α, β, and γ if $a = 10$, $b = 15$, and $c = 20$.
17 a if $c = 39$, $b = 98$, and $\alpha = 17°$.
18 β if $a = 5$ and $b = c = 10$.

In Problems 19–24, find the specified unknown part, if possible, under the given conditions. Angles should be determined to the nearest 10′ and sides to the nearest hundredth.

19 a *if* $\alpha = 55°$, $\beta = 45°$, and $c = 7$.
20 γ if $a = 10$, $c = 12$, and $\alpha = 70°30'$.
21 c if $a = 26$, $b = 15$, and $\gamma = 105°$.
22 β if $c = 500$, $b = 513$, and $\gamma = 80°30'$.
23 The largest angle if $a = 7$, $b = 18$, and $c = 12$.
24 α if $a = 8$, $c = 19$, and $b = 9$.

25 An airplane is sighted simultaneously from two towns that are 3 miles apart. The angle of elevation from one of the towns is 40°50′; from the other town it is 75°. If the airplane is directly over a line from one town to the other, how far is the airplane from each town?

26 A parallelogram has sides of 15 and 7 inches, and one angle of 80°. Find the other angle and the length of the longest diagonal of the parallelogram.

27 Two railroad tracks meet at an angle of 15°30′. If two trains depart from the railroad station (where the tracks meet) at the same time, find how far apart the two trains are after $2\frac{1}{2}$ hours if the average speeds of the trains are 55 and 70 miles per hour, respectively.

28 Assume that one diagonal of a parallelogram is 80 inches long and that at one end it forms angles of 35° and 27°, respectively, with the two sides. Find the lengths of the sides of the parallelogram.

29 Find the area of a triangular plot of land whose sides are 125 feet, 200 feet, and 175 feet.

30 An airplane is sighted simultaneously from two towns that are 7 miles apart. How high above the ground is the airplane if the angles of sighting are 35° and 50°, respectively, and the airplane is directly over a line that runs between the two towns?

31 A sign is to be mounted on a building by using two pairs of steel brackets. Figure 14 displays a side view of one pair of brackets, $\overline{AB}$ and $\overline{BC}$, and the angles they make with the building. How long are $\overline{AB}$ and $\overline{BC}$ if the points where they are attached to the building are 7 feet apart?

Figure 14

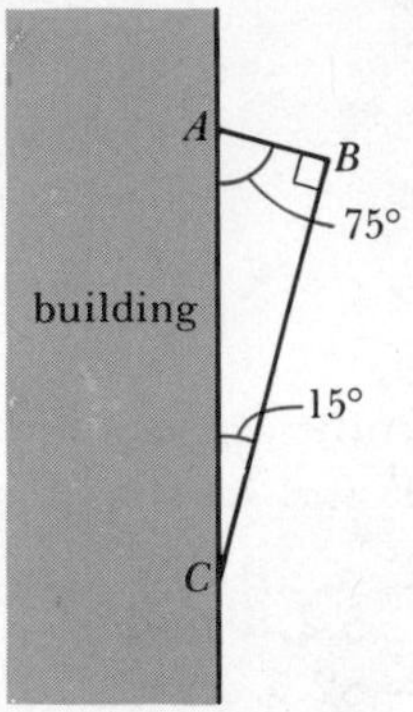

A, B, 75°, building, 15°, C

side view of one pair of brackets

7 Vectors and Trigonometry

Until now, we have dealt exclusively with quantities such as area, volume, angle, time, temperature, and speed that can be measured or represented by *single* real numbers. Since real numbers can be represented by points on a number scale, these quantities are often called *scalars.* In this section, we deal with quantities that have *both* magnitude and direction and thus *cannot* be described or represented by a *single* real number. Such quantities, which are called *vectors,* may, for example, represent force, velocity, acceleration, or displacement. We consider a few applications of vectors that utilize the tools of trigonometry.

7.1 Geometry of Vectors in the Plane

A *vector* in the plane is a line segment with a direction usually denoted by an arrowhead at the end of the segment (Figure 1). The length of the line segment is called the *magnitude* of the vector, and the orientation of the line segment is called the *direction* of the vector. The end point of the vector containing the arrowhead is called the *terminal point* of the vector, and the other end point is called the *initial point* of the vector. If a vector is determined by a directed line segment with the initial point A and the terminal point B, then it is denoted by **AB**; its magnitude is denoted by $|\mathbf{AB}|$ (Figure 1).

The *zero vector,* denoted by **0**, is a vector whose initial and terminal points are the same. Note that $|\mathbf{0}| = 0$.

Figure 1

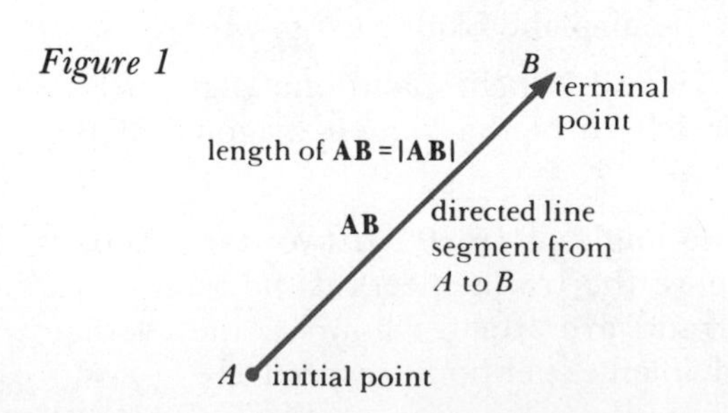

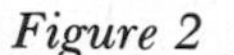

Figure 2

Two vectors are considered to be equal if they agree both in magnitude and in direction. Note that **AB** and **BA** (Figure 2) are *not* equal (even though their lengths are the same) because they are opposite in direction; that is, $|\mathbf{AB}| = |\mathbf{BA}|$, but $\mathbf{AB} \neq \mathbf{BA}$.

Lowercase boldface letters will be used to denote vectors. Hence we speak of "vector u" and write it as **u**.

ADDITION OF VECTORS

Suppose that we are given two vectors, **u** and **v** (Figure 3a). First, we "shift" **v** so that its initial point coincides with the terminal point of **u**. The vector having the same initial point as **u** and the same terminal point as **v** is defined to be the *sum* (*resultant vector*) $\mathbf{u} + \mathbf{v}$ of **u** and **v** (Figure 3b).

Figure 3

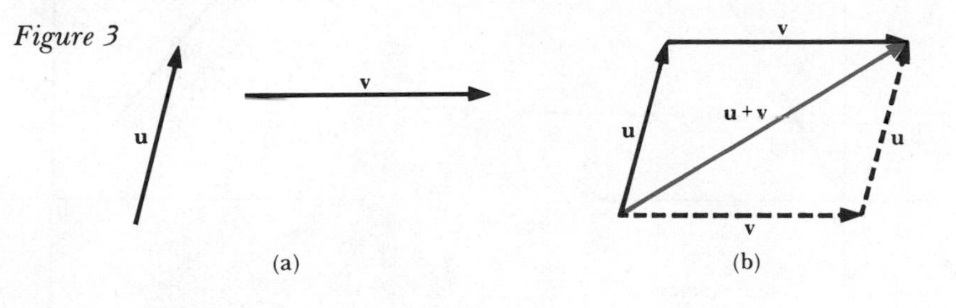

Figure 3b indicates that the resultant vector is, in a manner of speaking, a *diagonal vector* of the parallelogram "determined" by vectors **u** and **v**. Figure 3b also displays the fact that $\mathbf{u} + \mathbf{v} = \mathbf{v} + \mathbf{u}$; that is, vector addition is commutative.

SCALAR MULTIPLICATION AND SUBTRACTION OF VECTORS

Suppose that we are given a real number a and a vector **u**. The real number a in this context is a scalar; $a\mathbf{u}$ is defined to be the vector with magnitude $|a||\mathbf{u}|$, with the *same* direction as **u** if $a > 0$ and with the *opposite* direction if $a < 0$. This is illustrated in Figure 4, where $2\mathbf{u}$ is a vector in the same direction as **u**, with magnitude two times that of **u**, and $-\frac{1}{2}\mathbf{u}$ is a vector opposite in direction to **u**, with a length of $\frac{1}{2}$ that of **u**.

Now we can use vector addition and scalar multiplication to define *vector subtraction* as $\mathbf{u} - \mathbf{v} = \mathbf{u} + (-\mathbf{v})$ (Figure 5).

Figure 4 *Figure 5*

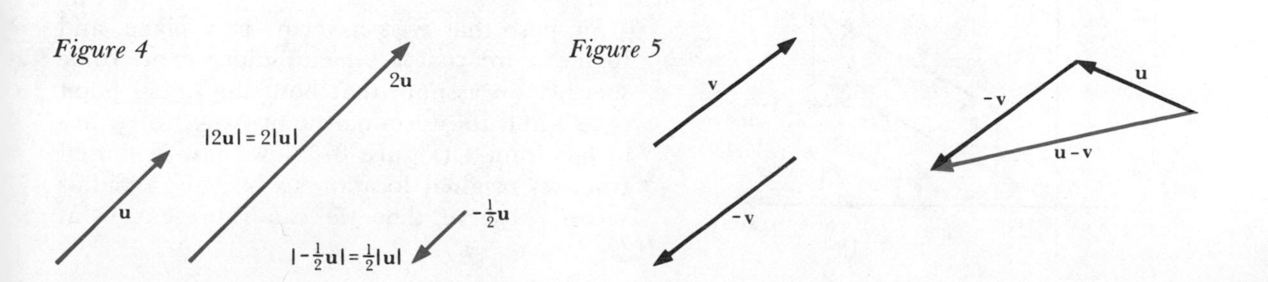

7.2 Analytic Representation of Vectors in the Plane

Suppose that vector **u** is positioned in a plane with a cartesian coordinate system so that the initial point of **u** is the origin and the terminal point is at point (a, b) (Figure 6). Then **u** is called a *radius vector* or *position vector;* we identify such a vector as $\mathbf{u} = \langle a, b \rangle$, and a and b are called the *x component* and the *y component* of **u**, respectively. Notice that

$$|\mathbf{u}| = \sqrt{a^2 + b^2}$$

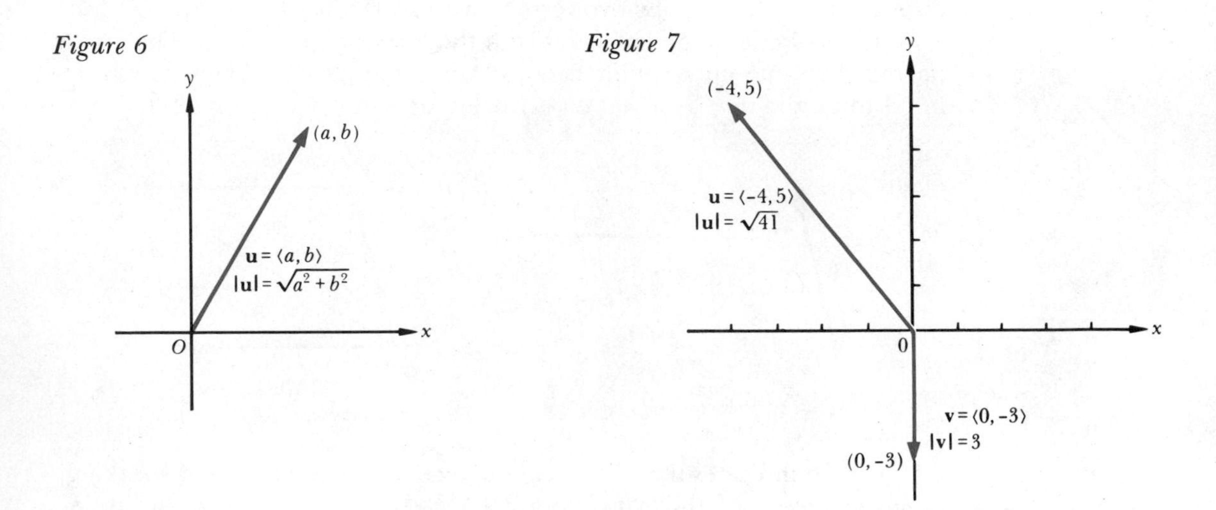

For example, $\mathbf{0} = \langle 0, 0 \rangle$. Also, $\mathbf{u} = \langle -4, 5 \rangle$ has the x component -4 and the y component 5, so

$$|\mathbf{u}| = \sqrt{(-4)^2 + 5^2} = \sqrt{41} \qquad \text{(Figure 7)}$$

All vectors with the x component 0 have terminal points on the y axis. (Why?) For example, $\mathbf{v} = \langle 0, -3 \rangle$ has its terminal point on the y axis, so

$$|\mathbf{v}| = \sqrt{0^2 + (-3)^2} = \sqrt{9} = 3 \qquad \text{(Figure 7)}$$

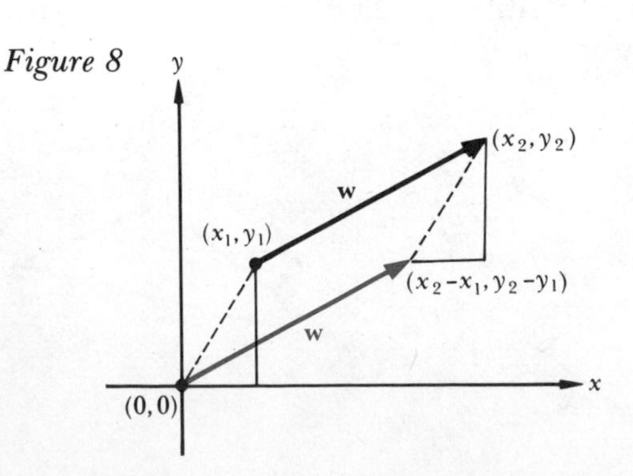

Figure 8

If $\mathbf{u}_1$ and $\mathbf{u}_2$ are *equal* position vectors, they have the same initial point (0, 0), so the components must also be equal; that is, if $\mathbf{u}_1 = \langle a_1, b_1 \rangle$ and $\mathbf{u}_2 = \langle a_2, b_2 \rangle$, then $\mathbf{u}_1 = \mathbf{u}_2$ whenever $a_1 = a_2$ and $b_1 = b_2$.

Suppose that **w** is a vector in a plane, and furthermore, that it is *not* a radius vector. To be specific, we assume that both the initial point (x_1, y_1) and the terminal point (x_2, y_2) of **w** are in quadrant I (Figure 8). Now, if **w** is shifted from its original location to become a radius vector, observe that we can represent **w** as $\langle x_2 - x_1, y_2 - y_1 \rangle$.

In general, if $\mathbf{w}$ is a vector with initial point (x_1, y_1) and terminal point (x_2, y_2), then $\mathbf{w} = \langle x_2 - x_1, y_2 - y_1 \rangle$.

If θ represents an angle formed by $\mathbf{u}$ and the positive x axis, θ is sometimes called a *direction angle* for $\mathbf{u}$.

EXAMPLES 1 Find the x and y components of the radius vector $\mathbf{u} = \langle x, y \rangle$ for each of the following situations. Assume that θ represents a direction angle.

(a) $|\mathbf{u}| = 5$, $\theta = 30°$

(b) $|\mathbf{u}| = 1$, $\theta = \frac{5\pi}{4}$

(c) $|\mathbf{u}| = 2$, $\theta = -50°$

Figure 9

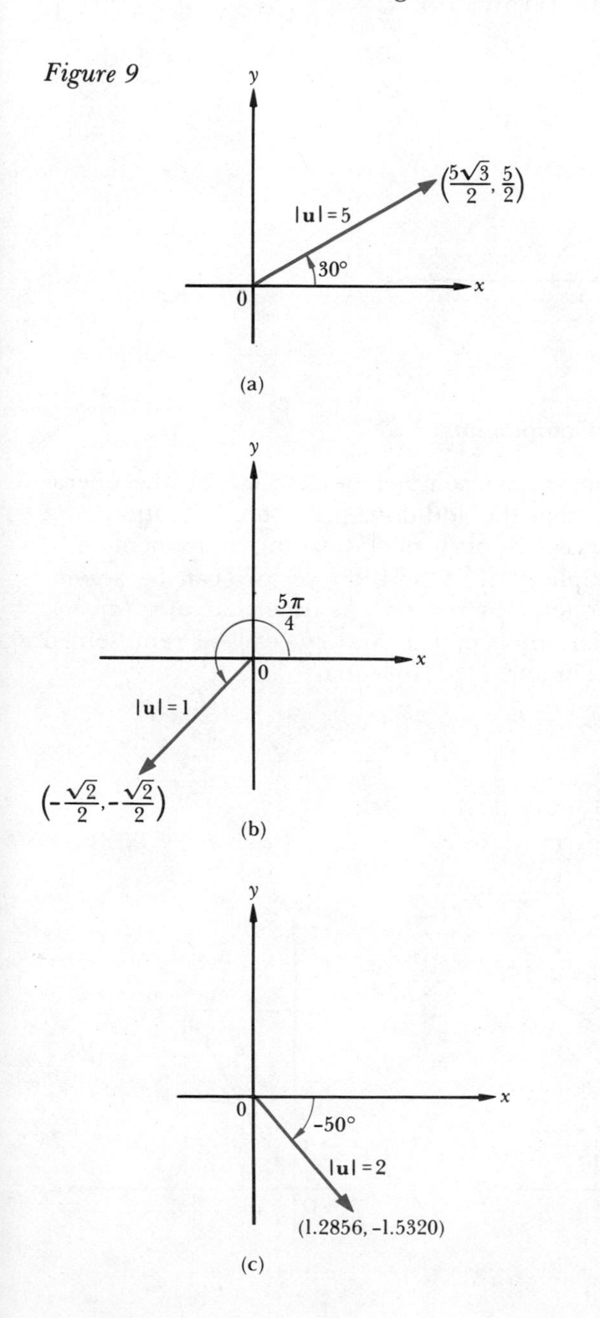

SOLUTION Figure 9 shows the locations of $\mathbf{u}$ for each situation. From trigonometry, we know that

$$\cos\theta = \frac{x}{|\mathbf{u}|} \quad \text{and} \quad \sin\theta = \frac{y}{|\mathbf{u}|}$$

so the x and y components of $\mathbf{u}$ are given by

$$x = |\mathbf{u}|\cos\theta \quad \text{and} \quad y = |\mathbf{u}|\sin\theta$$

Thus we have the following results:

(a) $$x = 5\cos 30° = 5\left(\frac{\sqrt{3}}{2}\right) = \frac{5\sqrt{3}}{2}$$

and

$$y = 5\sin 30° = 5\left(\frac{1}{2}\right) = \frac{5}{2} \quad \text{(Figure 9a)}$$

(b) $$x = 1\cos\frac{5\pi}{4} = -\frac{\sqrt{2}}{2}$$

and

$$y = 1\sin\frac{5\pi}{4} = -\frac{\sqrt{2}}{2} \quad \text{(Figure 9b)}$$

(c) Using Appendix Table IV, we have

$$\begin{aligned} x &= 2\cos(-50°) \\ &= 2\cos 50° \\ &= 2(0.6428) \\ &= 1.2856 \end{aligned}$$

and

$$\begin{aligned} y &= 2\sin(-50°) \\ &= -2\sin 50° \\ &= -2(0.7660) \\ &= -1.5320 \end{aligned}$$

(See Figure 9c.)

2 Assume that **u** is a vector whose initial point is $(-2, 3)$ and whose terminal point is $(5, 10)$. Represent **u** as a radius vector and then find $|\mathbf{u}|$ and θ, a direction angle for **u**.

SOLUTION $\mathbf{u} = \langle 5 - (-2), 10 - 3\rangle = \langle 7, 7\rangle$ so

$$|\mathbf{u}| = \sqrt{7^2 + 7^2} = \sqrt{98} = 7\sqrt{2}$$

and $\tan\theta = 7/7 = 1$, so $\theta = 45°$ (Figure 10).

Figure 10

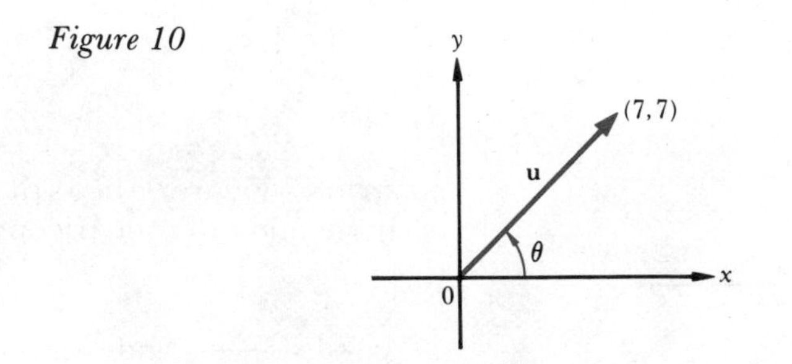

7.3 Vector Algebra in Terms of Components

The components of position vectors completely characterize the operations of vectors in the sense that the addition, scalar multiplication, and subtraction of position vectors can be performed by using components.

Addition and scalar multiplication of position vectors can be accomplished by using the components as follows. Assume that $\mathbf{u}_1 = \langle a_1, b_1\rangle$, $\mathbf{u}_2 = \langle a_2, b_2\rangle$, and c is a scalar; then $\mathbf{u}_1 + \mathbf{u}_2$ and $c\mathbf{u}_1$ can be represented geometrically as in Figures 11a and 11b. Note that

$$\mathbf{u}_1 + \mathbf{u}_2 = \langle a_1, b_1\rangle + \langle a_2, b_2\rangle = \langle a_1 + a_2, b_1 + b_2\rangle$$

and

$$c\mathbf{u}_1 = c\langle a_1, b_1\rangle = \langle ca_1, cb_1\rangle$$

Figure 11

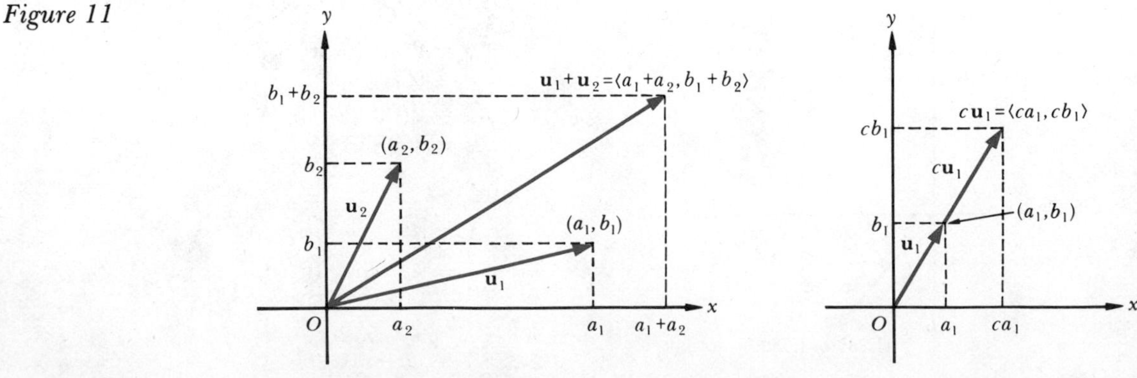

As before, $\mathbf{u} - \mathbf{v} = \mathbf{u} + (-\mathbf{v})$, so that if $\mathbf{u} = \langle a_1, b_1 \rangle$ and $\mathbf{v} = \langle a_2, b_2 \rangle$,

$$\begin{aligned} \mathbf{u} - \mathbf{v} &= \mathbf{u} + (-\mathbf{v}) \\ &= \langle a_1, b_1 \rangle + (-\langle a_2, b_2 \rangle) \\ &= \langle a_1, b_1 \rangle + \langle -a_2, -b_2 \rangle \\ &= \langle a_1 - a_2, b_1 - b_2 \rangle \end{aligned}$$

Thus,

$$\mathbf{u} - \mathbf{v} = \langle a_1 - a_2, b_1 - b_2 \rangle \qquad \text{(Figure 12)}$$

Figure 12

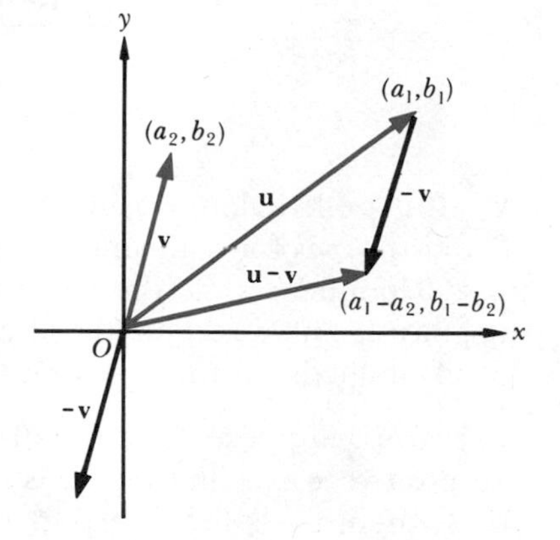

EXAMPLE Let $\mathbf{u} = \langle 3, 4 \rangle$ and $\mathbf{v} = \langle -5, 6 \rangle$. Find
(a) $\mathbf{u} + \mathbf{v}$ (b) $\mathbf{u} - \mathbf{v}$ (c) $4\mathbf{u} - 3\mathbf{v}$ (d) $|4\mathbf{u} - 3\mathbf{v}|$

SOLUTION

(a) $\mathbf{u} + \mathbf{v} = \langle 3, 4 \rangle + \langle -5, 6 \rangle = \langle -2, 10 \rangle$

(b) $\mathbf{u} - \mathbf{v} = \langle 3, 4 \rangle - \langle -5, 6 \rangle = \langle 8, -2 \rangle$

(c) $$\begin{aligned} 4\mathbf{u} - 3\mathbf{v} &= 4\langle 3, 4 \rangle - 3\langle -5, 6 \rangle \\ &= \langle 12, 16 \rangle + \langle 15, -18 \rangle = \langle 27, -2 \rangle \end{aligned}$$

(d) $$\begin{aligned} |4\mathbf{u} - 3\mathbf{v}| &= \sqrt{(27)^2 + (-2)^2} \\ &= \sqrt{729 + 4} = \sqrt{733} \end{aligned}$$

7.4 Applications of Vectors

The following examples illustrate how vectors, together with the tools of trigonometry, are used to solve velocity and force problems.

EXAMPLES 1 An airplane heading east at a still-air speed of 710 miles per hour is pushed off course by a wind blowing from the south at 65 miles per hour. Find the velocity vector of the flight of the airplane. Also, find the speed (the magnitude of the velocity vector) and a direction angle θ of the airplane's path to the nearest $10'$.

SOLUTION The velocity vector, denoted by **v**, is defined by $\mathbf{v} = \langle 710, 65 \rangle$ (Figure 13). The speed of the airplane is given by $|\mathbf{v}| = \sqrt{710^2 + 65^2} = 713$ miles per hour (approximately). The direction angle θ satisfies the equation $\tan \theta = 65/710 = 0.0915$ (approximately), so $\theta = 5°10'$.

Figure 13

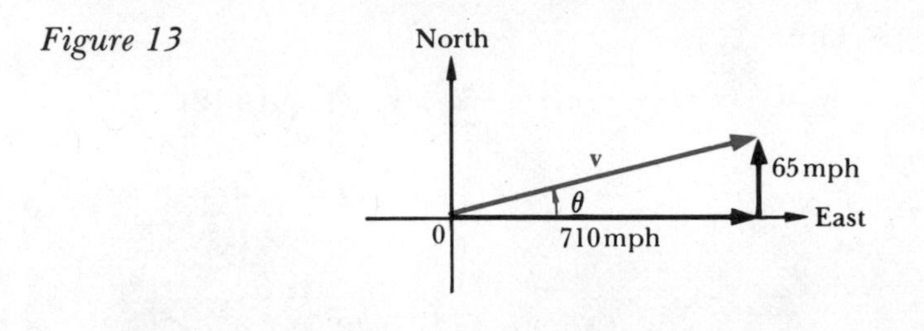

2 A raft is pulled along a canal by two ropes on opposite sides of the canal. The two ropes form an angle of 60° with each other; one pulls with a force of 200 pounds, while the force of the other is 350 pounds. What is the magnitude and direction (the angle that the resultant vector makes with the horizontal axis) of the resultant force?

SOLUTION Suppose that the 350-pound force is exerted in the direction of the positive x axis and that it is represented by vector **v**. The 200-pound force, then, has a 60° direction angle and thus is represented by vector **u** (Figure 14).

Figure 14

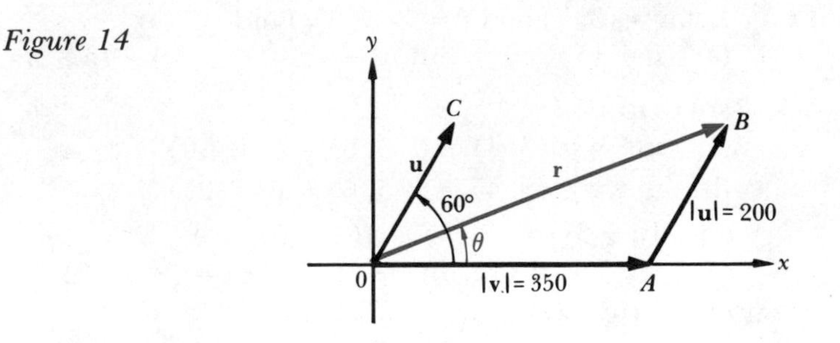

Therefore,

$$\begin{aligned}\mathbf{u} &= \langle 200 \cos 60°, 200 \sin 60° \rangle \\ &= \langle 100, 100\sqrt{3} \rangle\end{aligned}$$

and

$$\mathbf{v} = \langle 350, 0 \rangle$$

so the resultant vector is

$$\mathbf{r} = \mathbf{u} + \mathbf{v} = \langle 450, 100\sqrt{3} \rangle$$

and

$$|\mathbf{r}| = \sqrt{202{,}500 + 30{,}000} = \sqrt{232{,}500} = 482.18 \text{ pounds}$$

The angle θ between $\mathbf{v}$ and $\mathbf{r}$ is given by

$$\tan \theta = \frac{100\sqrt{3}}{450} = 0.3849$$

so $\theta = 21°$.

ALTERNATIVE SOLUTION This problem can also be solved without the use of vectors (see Figure 14). In $\triangle OAB$, $\angle OAB = 120°$, $|\overline{OA}| = 350$, and $|\overline{AB}| = 200$, so, by the law of cosines,

$$\begin{aligned} |\mathbf{r}|^2 = |\overline{OB}|^2 &= 350^2 + 200^2 - 2(350)(200) \cos 120° \\ &= 350^2 + 200^2 - 2(350)(200)(-\tfrac{1}{2}) \\ &= 122{,}500 + 40{,}000 + 70{,}000 \\ &= 232{,}500 \end{aligned}$$

that is,

$$|\mathbf{r}| = \sqrt{232{,}500} = 482.18 \text{ pounds}$$

By the law of sines,

$$\frac{\sin \theta}{200} = \frac{\sin 120°}{482.18} \quad \text{or} \quad \sin \theta = \frac{200 \sin 120°}{482.18} = 0.3592$$

so $\theta = 21°$.

3 Two forces, one of 10 pounds and the other of 15 pounds, are applied to the same object, yielding a resultant force of 18 pounds. To the nearest 10′, what angle does the resultant vector make with the vector representative of 15 pounds of force?

SOLUTION Assume that the force vectors are $\mathbf{u}$, $\mathbf{v}$, and $\mathbf{r}$, as shown in Figure 15. Then, by the law of cosines,

$$\begin{aligned} 10^2 &= 18^2 + 15^2 - 2(18)(15) \cos \theta \\ 100 &= 324 + 225 - 540 \cos \theta \end{aligned}$$

so

$$\cos \theta = \frac{449}{540} = 0.8315$$

Thus, $\theta = 33°40'$.

Figure 15

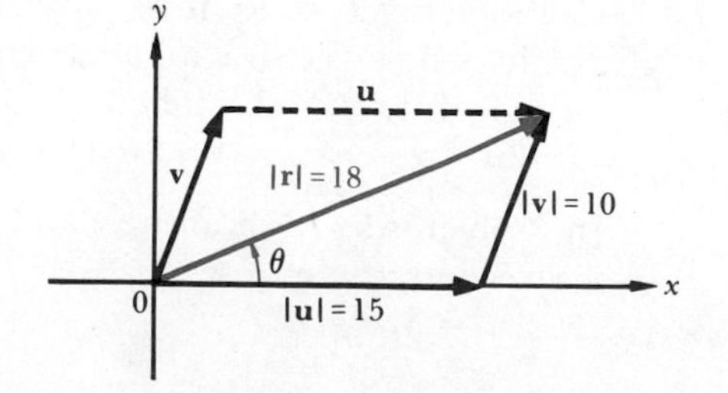

4 A boat heading 40° S of E at a still-water speed of 25 miles per hour is pushed off course by a water current of 20 miles per hour flowing in the direction of 50° S of W. Find the velocity vector of the boat's path. What is the speed of the boat? In what direction is the boat traveling? (Give the angle measure to the nearest 10′.)

SOLUTION The boat's velocity vector is the resultant vector of the boat's velocity **OS** in still water and the water current's velocity **OW** (Figure 16).

$$\begin{aligned}\mathbf{OS} &= \langle 25\cos(-40^\circ), 25\sin(-40^\circ)\rangle \\ &= \langle 25(0.7660), -25(0.6428)\rangle \\ &= \langle 19.15, -16.07\rangle\end{aligned}$$

Figure 16

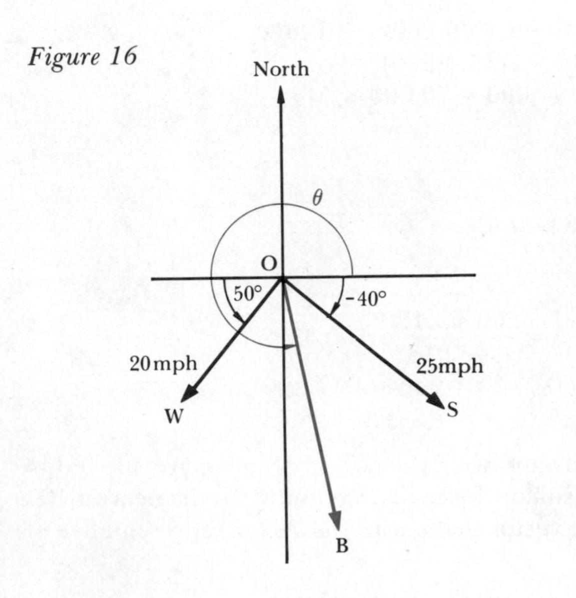

and

$$\begin{aligned}\mathbf{OW} &= \langle 20\cos 230^\circ, 20\sin 230^\circ\rangle \\ &= \langle -20\cos 50^\circ, -20\sin 50^\circ\rangle \\ &= \langle -20(0.6428), -20(0.7660)\rangle \\ &= \langle -12.86, -15.32\rangle\end{aligned}$$

so the velocity of the boat's path is given by

$$\begin{aligned}\mathbf{OB} &= \mathbf{OS} + \mathbf{OW} \\ &= \langle 19.15, -16.07\rangle + \langle -12.86, -15.32\rangle \\ &= \langle 6.29, -31.39\rangle\end{aligned}$$

The speed is given by

$$\begin{aligned}|\mathbf{OB}| &= \sqrt{39.56 + 985.33} \\ &= \sqrt{1024.89} \\ &= 32.01 \text{ miles per hour}\end{aligned}$$

The angle θ satisfies

$$\tan\theta = -\frac{31.39}{6.29} = -4.987$$

so θ is approximately $360^\circ - 78^\circ 40' = 281^\circ 20'$, and the velocity vector has a direction $78^\circ 40'$ S of E.

PROBLEM SET 7

In Problems 1–4, let **u** be a vector in the positive direction of the x axis and let **v** be a vector with a direction angle of 30°. In addition, assume that $|\mathbf{u}| = 3$ and $|\mathbf{v}| = 2$. Make a sketch of each of the following vectors.

1 $2\mathbf{u} + \mathbf{v}$ **2** $2\mathbf{u} + \frac{1}{2}\mathbf{v}$ **3** $\frac{1}{2}\mathbf{u} - 2\mathbf{v}$ **4** $-\mathbf{u} - 3\mathbf{v}$

In Problems 5–12, find the x and y components of $\mathbf{u} = \langle x, y\rangle$ if θ represents a direction angle of **u**.

5 $|\mathbf{u}| = 3, \theta = 0^\circ$ **6** $|\mathbf{u}| = 2, \theta = 180^\circ$

7 $|\mathbf{u}| = 5, \theta = 45°$
8 $|\mathbf{u}| = 1, \theta = 300°$
9 $|\mathbf{u}| = 1, \theta = -150°$
10 $|\mathbf{u}| = 4, \theta = -75°$
11 $|\mathbf{u}| = \frac{1}{2}, \theta = 100°10'$
12 $|\mathbf{u}| = 7, \theta = 253°20'$

In Problems 13–20, find the magnitude and a direction angle (to the nearest 10′) of each given vector.

13 $\langle -2, 2 \rangle$
14 $\langle \sqrt{3}, -1 \rangle$
15 $\langle -5, 0 \rangle$
16 $\langle 1, 3 \rangle$
17 $\langle -8, -6 \rangle$
18 $\langle 3, 4 \rangle$
19 $\langle 6, -5 \rangle$
20 $\langle -5, -1 \rangle$

In Problems 21–26, if **u** is a vector whose initial point is the first point given and whose terminal point is the second point given, represent **u** as a radius vector. Also find $|\mathbf{u}|$.

21 $(8, 6), (3, 4)$
22 $(-7, -6), (2, 3)$
23 $(-2, 6), (3, -5)$
24 $(-3, 2), (1, -3)$
25 $(3, 7), (-3, 1)$
26 $(1, -3), (5, -1)$

In Problems 27–36, use $\mathbf{u}_1 = \langle 3, 4 \rangle$, $\mathbf{u}_2 = \langle -2, 4 \rangle$, and $\mathbf{u}_3 = \langle 7, 8 \rangle$ to find the value of each expression.

27 $2\mathbf{u}_1 + 3\mathbf{u}_2$
28 $3\mathbf{u}_2 - \mathbf{u}_1$
29 $|\mathbf{u}_1| + |\mathbf{u}_2|$
30 $|\mathbf{u}_1 + \mathbf{u}_2|$
31 $\mathbf{u}_1 + (\mathbf{u}_2 + \mathbf{u}_3)$
32 $(\mathbf{u}_1 + \mathbf{u}_2) + \mathbf{u}_3$
33 $2\mathbf{u}_1 - 3\mathbf{u}_2$
34 $\mathbf{u}_2 - \mathbf{u}_1 - \mathbf{u}_3$
35 $3|\mathbf{u}_1 - \mathbf{u}_2|$
36 $3|\mathbf{u}_1| - 3|\mathbf{u}_2|$

37 Let $P_1 = (-6, -10)$ and $P_2 = (8, 8)$ be two points in a plane.
(a) Use vectors to find the point that is two-thirds of the way from P_1 to P_2.
(b) Use vectors to find the point that is $\sqrt{130}$ units from P_1 along the directed line from P_1 to P_2.

38 A *unit* vector is a vector of magnitude 1. Find k so that $k\mathbf{u}$ is a unit vector for each of the following vectors.
(a) $\mathbf{u} = \langle 3, 4 \rangle$
(b) $\mathbf{u} = \langle -2, 5 \rangle$

39 The *inner product* or *dot product* of **u** and **v**, denoted by $\mathbf{u} \cdot \mathbf{v}$, is defined as $\mathbf{u} \cdot \mathbf{v} = |\mathbf{u}||\mathbf{v}| \cos \theta$, where θ is the angle between **u** and **v** and $0° \leq \theta \leq 180°$. Find $\mathbf{u} \cdot \mathbf{v}$ for each of the following situations.
(a) If $|\mathbf{u}| = 3$, $|\mathbf{v}| = 2$, and $\theta = 60°$.
(b) If $|\mathbf{u}| = 5$, $|\mathbf{v}| = 1$, and $\theta = 150°$.

40 Use the definition in Problem 39 to prove that two nonzero vectors **u** and **v** are perpendicular (orthogonal) if and only if $\mathbf{u} \cdot \mathbf{v} = 0$.

41 Assume that $\mathbf{u} = \langle a, b \rangle$ and $\mathbf{v} = \langle c, d \rangle$. It can be proved that

$$\mathbf{u} \cdot \mathbf{v} = ac + bd$$

(see Problem 39). Use this result to determine $\mathbf{u} \cdot \mathbf{v}$ for each of the following situations.
(a) $\mathbf{u} = \langle 1, 1 \rangle$ and $\mathbf{v} = \langle 1, -1 \rangle$
(b) $\mathbf{u} = \langle 2, -3 \rangle$ and $\mathbf{v} = \langle -4, -1 \rangle$
(c) $\mathbf{u} = \langle 4, 2 \rangle$ and $\mathbf{v} = \langle -3, 5 \rangle$
(d) $\mathbf{u} = \langle 0, 3 \rangle$ and $\mathbf{v} = \langle 5, 0 \rangle$

42 Use the definition in Problem 39 and the formula in Problem 41 to find the angle (to the nearest $10'$) between the following pairs of vectors.

(a) $\langle 1, 1 \rangle$ and $\langle 1, -1 \rangle$ (b) $\langle \sqrt{3}, 1 \rangle$ and $\langle 0, 5 \rangle$

(c) $\langle 1, 2 \rangle$ and $\langle -1, 1 \rangle$ (d) $\langle -3, -4 \rangle$ and $\langle 2, -3 \rangle$

43 Suppose that a motor boat, which has a speed in still water of 12 miles per hour, heads west across a river that is flowing south at a speed of 3 miles per hour. Find the velocity vector of the boat. What is the speed of the boat? What is the angle (to the nearest $10'$) that the boat's path makes with the vector representing the current?

44 An airplane with a still-air speed of 150 miles per hour heads north and is pushed off course by a wind blowing from the west at a speed of 45 miles per hour. Find the velocity vector of the airplane. What is the speed of the airplane? In what direction is the airplane traveling? (Give the angle measure to the nearest $10'$.)

45 Two forces, one of 25 pounds and one of 8 pounds, act on an object at right angles to each other. Find the resultant force vector. What is the force and the angle (to the nearest $10'$) that the resultant vector forms with the 8-pound force?

46 A boat, with a still-water speed of 10 miles per hour, heading northeast, is pushed off course by a current of 3 miles per hour moving in a direction 60° S of W. What is the resulting speed of the boat? In what direction is it traveling? (Give the angle measure to the nearest $10'$.)

47 Two forces acting on the same object produce a resultant force of 17 pounds with an angle of 20° relative to a horizontal axis. Assume that one of the two forces is 20 pounds and has an angle of 50° relative to the horizontal axis. Find the other force and its direction angle (to the nearest $10'$) relative to the horizontal axis.

48 An airplane heading 30° E of N with a still-air speed of 550 miles per hour is affected by a wind of 50 miles per hour blowing from 25° S of W. Find the speed of the airplane and the direction the airplane is traveling. (Give the angle measure to the nearest $10'$.)

49 The still-air speed of an airplane is 350 miles per hour. The pilot is to travel in the direction of 40° N of E through a wind blowing at 45 miles per hour from 25° S of W. What direction should the pilot take, and what will be the actual speed in the desired direction?

REVIEW PROBLEM SET

In Problems 1–2, assume that θ is an angle in standard position with its terminal side in quadrant I. Express the other five trigonometric functions of θ in terms of x.

1 $\tan \theta = \frac{x}{2}$ **2** $\cos \theta = \sqrt{1 - x^2}$

In Problems 3–4, find the values of the other five trigonometric functions under the given conditions. Assume that θ is an angle in standard position.

3 $\cot\theta = -\frac{15}{8}$, terminal side of θ in quadrant II

4 $\sec\theta = 8$, terminal side of θ in quadrant IV

In Problems 5–14, write each expression as a single term involving only one angle.

5 $\sin 15^\circ \cos 10^\circ - \cos 15^\circ \sin 10^\circ$

6 $\cos\left(\frac{\pi}{2} - t\right)\tan\left(\frac{\pi}{2} - t\right)$

7 $\sec^2 58^\circ - 1$

8 $\csc(90^\circ - \theta)\cos\theta - \cot\theta$

9 $\dfrac{\tan 5 + \tan 1}{1 - \tan 5 \tan 1}$

10 $\sin(3\pi - t)$

11 $2\sin^2 1.3 - 1$

12 $\tan(270^\circ + \theta)$

13 $\sin 17^\circ \cos 17^\circ$

14 $2\cos^2\frac{\pi}{7} - 1$

In Problems 15–22, let $\sin t = \frac{3}{5}$, for $0 < t < \pi/2$, and $\cos s = -\frac{8}{17}$, for $\pi/2 < s < \pi$. Evaluate each given expression.

15 $\sin 2s$

16 $\tan 2t$

17 $\cos(t + s)$

18 $\cos\left(\frac{s}{2}\right)$

19 $\sin(t - s)$

20 $\cos 2t$

21 $\sin\left(\frac{t}{2}\right)$

22 $\sin^2 s - \cos^2 s$

In Problems 23–28, use the given information, together with an appropriate trigonometric formula, to evaluate each expression.

23 $\sin 22.5^\circ$; $22.5^\circ = \frac{1}{2}(45^\circ)$

24 $\cos 75^\circ$; $75^\circ = 45^\circ + 30^\circ$

25 $\cos\frac{11\pi}{12}$; $\frac{11\pi}{12} = \frac{1}{2}\left(\frac{11\pi}{6}\right)$

26 $\cos 75^\circ$; $75^\circ = \frac{1}{2}(150^\circ)$

27 $\tan 255^\circ$; $255^\circ = 210^\circ + 45^\circ$

28 $\sin\frac{\pi}{12}$; $\frac{\pi}{12} = \frac{\pi}{3} - \frac{\pi}{4}$

In Problems 29–40, prove that the given equation is an identity.

29 $\cos t \sin t \sec t \csc t = 1$

30 $\dfrac{\sin^4 t - \cos^4 t}{\sin^2 t - \cos^2 t} = 1$

31 $\sec\theta - \cos\theta = \sin\theta\tan\theta$

32 $\cot\theta\cos\theta + \sin\theta = \csc\theta$

33 $\cos^2 t - \sin^2 t = \dfrac{1 - \tan^2 t}{1 + \tan^2 t}$

34 $\dfrac{\sin^2 t \cos t + \cos^3 t}{\cot t} = \sin t$

35 $\dfrac{1 - \cos 2\theta}{\sin 2\theta} = \tan\theta$

36 $\sin\frac{\theta}{2}\cos\frac{\theta}{2} = \frac{\sin\theta}{2}$

37 $\dfrac{\sin t}{1 + \cos t} = \csc t - \cot t$

38 $\csc 2t + \cot 2t = \cot t$

39 $2\csc 2\theta \cot\theta = 1 + \cot^2\theta$

40 $\dfrac{\cos\theta}{\sec\theta + \tan\theta} = 1 - \sin\theta$

In Problems 41–50, evaluate the given expression.

41 $\tan^{-1} 1$

42 $\cos^{-1}\frac{1}{2}$

43 $\operatorname{Arcsin}(-\frac{1}{2})$

44 $\sin\left(\tan^{-1}\frac{\sqrt{3}}{3}\right)$

45 $\tan^{-1}\left(\sin\frac{\pi}{2}\right)$

46 $\cos[\operatorname{Arctan}(-1)]$

47 $\sec(\sin^{-1} \frac{2}{3})$

48 $\sin^{-1} 0.4969$

49 $\cos^{-1}(-0.6294)$

50 $\text{Arcsin}(-0.9959)$

In Problems 51–54, solve for x in terms of y.

51 $y = \sin^{-1}\left(\frac{x}{2}\right)$

52 $y = 2 \text{ Arctan}(x + 3)$

53 $y = 3 + \sec^{-1} 4x$

54 $y = \pi - \cos^{-1}(2x - 1)$

In Problems 55–60, find the solution set of each equation, if $0° \leq \theta < 360°$.

55 $\sin \theta = -1$

56 $\cos \theta = \frac{\sqrt{3}}{2}$

57 $\tan \theta = \sqrt{3}$

58 $\cos \theta = -\frac{\sqrt{2}}{2}$

59 $\csc \theta = -2$

60 $\sin \theta = \cos 2\theta$

In Problems 61–66, find the solution set of each equation, if $0 \leq t < 2\pi$.

61 $\sin t = \frac{\sqrt{2}}{2}$

62 $4 \sin^2 t - 3 = 0$

63 $2 \cos^2 t - \cos t - 1 = 0$

64 $2 \sin^2 t + \sqrt{2} \sin t = 0$

65 $3 \tan^2 t - \sqrt{3} \tan t = 0$

66 $2 \cos^2 t + 5 \cos t + 2 = 0$

In Problems 67–74, solve for the unknown parts of $\triangle ABC$ if $\gamma = 90°$. (Angles should be determined to the nearest 10′ and sides to the nearest hundredth.)

67 $a = 5$ and $b = 12$

68 $c = 15$ and $\alpha = 37°$

69 $b = 25$ and $\beta = 65°$

70 $a = 5$ and $c = 14$

71 $a = 17$ and $\beta = 51°$

72 $\sin \alpha = \frac{4}{5}$ and $c = 25$

73 $\tan \alpha = \frac{3}{4}$ and $a = 12$

74 $a = 6$ and $b = 8$

In Problems 75–80, locate the point with the given polar coordinates, then find the corresponding rectangular coordinates of that point, where $r > 0$ and $0° \leq \theta < 360°$.

75 $\left(5, \frac{\pi}{4}\right)$

76 $\left(2, -\frac{\pi}{6}\right)$

77 $(\sqrt{2}, -135°)$

78 $(3, \pi)$

79 $(3, 270°)$

80 $\left(-2, \frac{4\pi}{3}\right)$

In Problems 81–86, locate the point with the given rectangular coordinates, then find polar coordinates that correspond to that point.

81 $(-3, 0)$

82 $(-2, 2\sqrt{3})$

83 $(-5\sqrt{3}, -5)$

84 $(10, 10)$

85 $(0, -14)$

86 $(-15, 0)$

In Problems 87–96, find the unknown parts of the triangle with the given parts. (Angles should be measured to the nearest 10′ and sides to the nearest hundredth.)

87 $a = 5, b = 7$, and $\gamma = 30°$

88 $\alpha = 120°, c = 8$, and $b = 3$

89 $c = 10, \alpha = 45°$, and $\beta = 75°$

90 $a = 162, b = 215$, and $\beta = 110°$

91 $b = 4, c = 6$, and $\beta = 30°$

92 $a = 13.6, b = 7.82$, and $\alpha = 60°$

93 $a = 4.8, c = 4.3$, and $\alpha = 115°$

94 $b = 66.2, c = 42.3$, and $\alpha = 30°$

95 $a = 10, \beta = 42°$, and $\gamma = 51°$

96 $a = 4, c = 10$, and $\beta = 150°$

97 Two men standing 600 feet apart on level ground observe a balloon in the sky between them. The balloon is in the same vertical plane with the two men. The angles of elevation of the balloon are observed by the men to measure 75° and 48°, respectively. Find how high the balloon is above the ground.

98 Two points A and B are 50 feet apart on one side of a river. A point C across the river is located so that $\angle CAB$ is 70° and angle $\angle ABC$ is 80°. How wide is the river?

99 A guy wire attached to the top of a pole is 40 feet long and forms a 50° angle with the ground. How tall is the pole if it is tilted 15° out of line away from the guy line?

100 A diagonal of a parallelogram is 16 inches long and forms angles of 43° and 15°, respectively, with the two sides. How long are the sides of the parallelogram?

In Problems 101–104, let $\mathbf{u} = \langle 3, 4 \rangle$ and $\mathbf{v} = \langle 4, 3 \rangle$. Find each of the following vectors and represent them graphically.

101 $3\mathbf{u}$ **102** $\mathbf{v} - \mathbf{u}$ **103** $-2\mathbf{u} + \mathbf{v}$ **104** $3\mathbf{u} + 2\mathbf{v}$

In Problems 105–107, determine the components of **u** if θ is a direction angle of **u**.

105 $|\mathbf{u}| = 5$ and $\theta = 30°$ **106** $|\mathbf{u}| = 6$ and $\theta = 45°$

107 $|\mathbf{u}| = 8$ and $\theta = 150°$

108 Let $\mathbf{u} = \langle 2, 3 \rangle$, $\mathbf{v} = \langle -1, 4 \rangle$, and $\mathbf{w} = \langle 4, 5 \rangle$. Determine **z** such that $\mathbf{u} + \mathbf{v} = \mathbf{w} + \mathbf{z}$.

109 If **u** is a vector whose initial point is the first point given and whose terminal point is the second point given, write **u** in the form $\mathbf{u} = \langle a, b \rangle$ and find $|\mathbf{u}|$.
(a) $(-6, 8), (4, 3)$ (b) $(-7, 6), (3, -1)$

110 Let $\mathbf{u} = \langle -1, 4 \rangle$ and $\mathbf{v} = \langle 3, 5 \rangle$. Find each of the following.
(a) $\mathbf{u} \cdot \mathbf{v}$
(b) $\mathbf{v} \cdot \mathbf{u}$
(c) The angle between **u** and **v** (to the nearest 10′)

111 Two forces, one of 63 pounds and one of 45 pounds, yield a resultant force of 75 pounds. Find the angle between the two forces.

112 An airplane, which has a still-air speed of 550 miles per hour, heading 30° W of N, is affected by a wind blowing at 45 miles per hour from 45° S of W. Find the velocity vector of the airplane's path. What is the speed of the airplane and in what direction is it traveling?

CHAPTER 7

Analytic Geometry

In this chapter, we discuss certain types of relations that play a special role in geometry and a number of other applications. The graphs or curves that we consider are *circles, ellipses, hyperbolas,* and *parabolas.* These curves are called *conic sections,* because they are obtained by the sectioning or cutting of circular cones by planes. We introduce these curves by establishing standard equation forms of the graphs. In addition, we study the geometric properties of the conics.

1 Circle

Geometrically, a *circle* in a plane can be defined as the set of all points that are at a fixed distance r, called the *radius,* from a fixed point C, called the *center.* The length of a diameter of a circle is equal to $2r$. In Figure 1, P is a point of the circle, C is the center, and r is the length of the radius. $|\overline{AB}| = 2r$ is the length of a diameter. The distance formula can be used to write an equation whose graph is a circle.

Figure 1

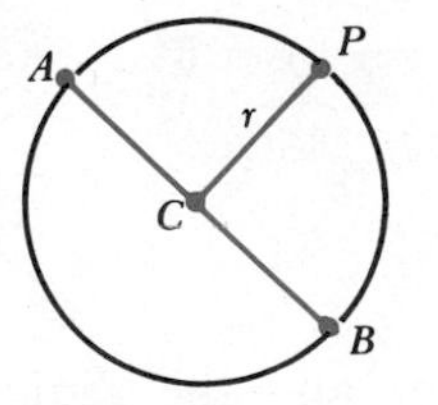

THEOREM 1 CIRCLE EQUATION

Let (h, k) be the center of a circle whose radius is r; then an equation of the circle is

$$(x - h)^2 + (y - k)^2 = r^2$$

PROOF Let $P = (x, y)$ represent *any* point on the circle whose center C is (h, k) (Figure 2). Then, by the distance formula,

Figure 2

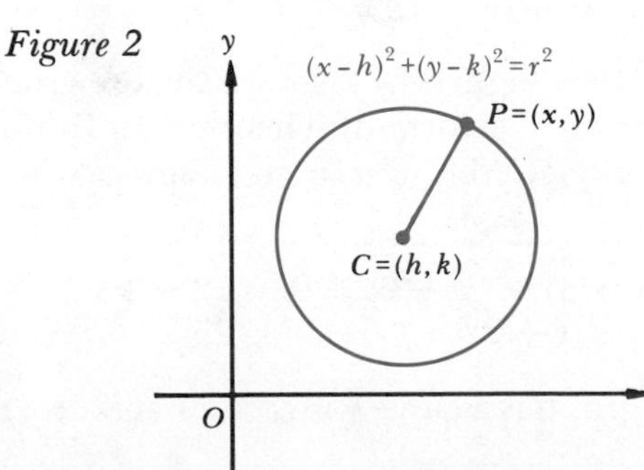

$$r = [(x - h)^2 + (y - k)^2]^{1/2}$$

After squaring both sides of this equation, we get

$$r^2 = (x - h)^2 + (y - k)^2$$

Thus any point on the circle has coordinates that satisfy this equation.

Conversely, if $P = (x, y)$ is a point that satisfies the above equation, we have

$$\sqrt{(x-h)^2 + (y-k)^2} = \sqrt{r^2} = r$$

so that P is r units from the point (h, k); thus P is a point on the circle.

Notice that this equation of the circle gives us an algebraic characterization that depends only on the center and the radius of the circle, namely, $(x-h)^2 + (y-k)^2 = r^2$, which defines a relation whose graph is a circle with center (h, k) and radius r.

EXAMPLES 1 Find an equation of a circle where $P_1 = (3, 7)$ and $P_2 = (-3, -1)$ are the end points of a diameter.

SOLUTION (See Figure 3.) The center of the circle, the midpoint of $\overline{P_1P_2}$, can be determined by the formula

$$\left(\frac{x_1 + x_2}{2}, \frac{y_1 + y_2}{2}\right)$$

Figure 3

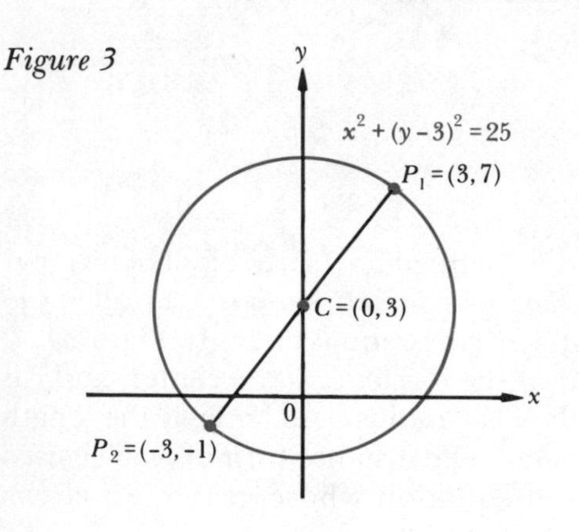

(see Chapter 2, Problem Set 1, Problem 28). Hence we have the center $C = (0, 3)$. The radius is

$$\begin{aligned} r &= \tfrac{1}{2}|\overline{P_1P_2}| \\ &= \tfrac{1}{2}\sqrt{(3+3)^2 + (7+1)^2} \\ &= \tfrac{1}{2}\sqrt{36 + 64} = 5 \end{aligned}$$

Therefore, an equation of the circle is given by

$$(x-0)^2 + (y-3)^2 = 5^2$$

or

$$x^2 + (y-3)^2 = 25$$

2 Find the center and the radius of the circle whose equation is

$$x^2 + y^2 - 4x + 6y - 12 = 0$$

Figure 4

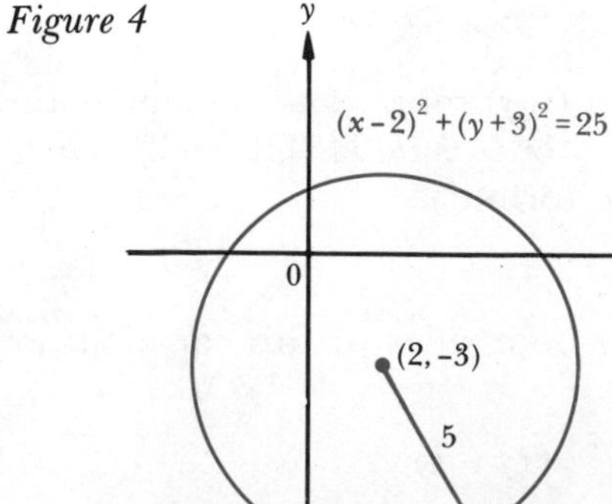

SOLUTION (See Figure 4.) First, we shall rewrite the equation in the form of Theorem 1. To do this, we "complete the square" as follows:

$$\begin{aligned} (x^2 - 4x + \quad) + (y^2 + 6y + \quad) &= 12 \\ (x^2 - 4x + 4) + (y^2 + 6y + 9) &= 12 + 4 + 9 \\ (x-2)^2 + (y+3)^2 &= 25 \end{aligned}$$

so that the graph is a circle with center at $(2, -3)$ and radius 5.

3 Find an equation of the circle that is tangent to the x axis and whose center is the point $(-3, 10)$.

SOLUTION An equation of the circle is $(x - h)^2 + (y - k) = r^2$, with $h = -3$ and $k = 10$. We need to determine r. Since the circle is tangent to the x axis, then, from plane geometry, its radius is perpendicular to the x axis (Figure 5). It follows that the circle contains the point $P = (-3, 0)$. (Why?) Therefore, $r = 10$ and an equation of the circle is

$$(x + 3)^2 + (y - 10)^2 = 100$$

Figure 5

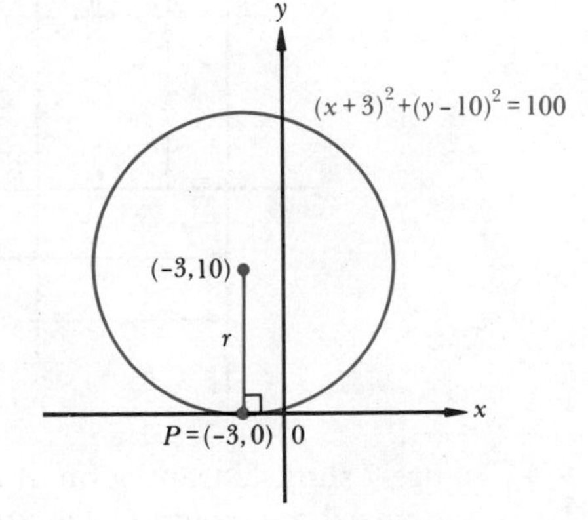

1.1 Translation of Axes

An equation of a curve can often be simplified by changing to a more appropriate coordinate system. In practice, this is usually accomplished by choosing one or both of the new coordinate axes to coincide with an axis of symmetry of the curve. For instance, an equation of the curve C in Figure 6 would probably be simplified by switching from the xy coordinate system to the $\bar{x}\bar{y}$ coordinate system as shown, since C is symmetric with respect to the $\bar{y}$ axis.

Figure 6

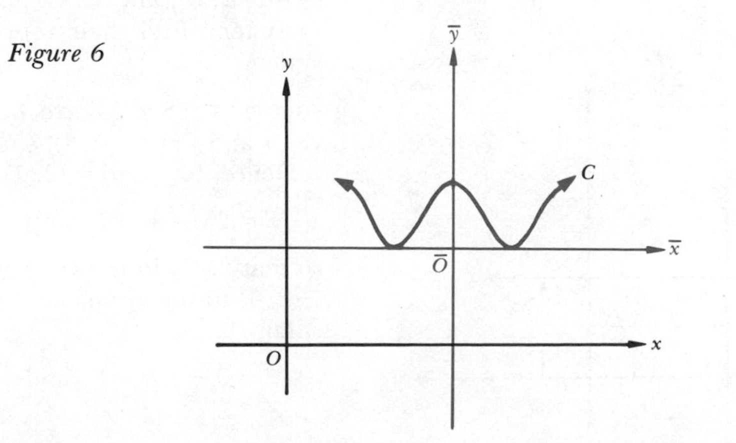

If two cartesian coordinate systems have corresponding axes that are parallel and have the same positive directions, then we say that these systems are obtained from one another by *translation*.

Figure 7

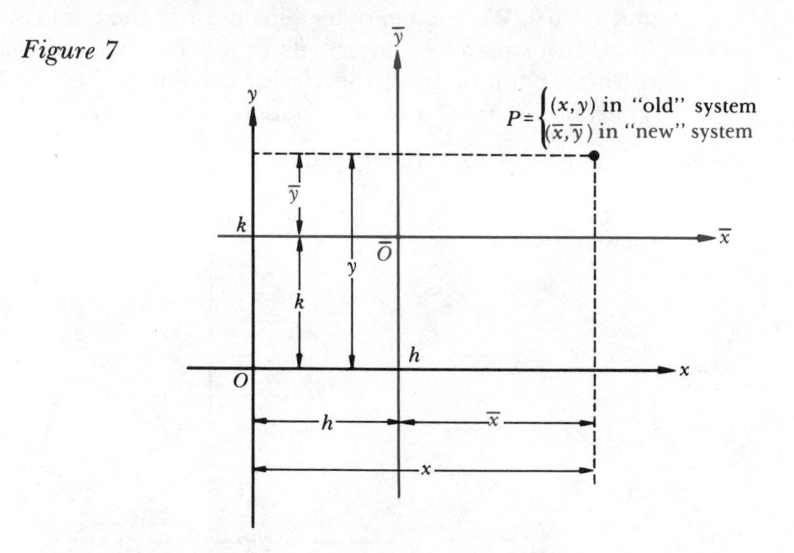

Figure 7 shows a translation of the "old" xy coordinate system to a "new" $\bar{x}\bar{y}$ system whose origin $\bar{O}$ has the "old" coordinates (h, k). Consider the point P having "old" coordinates (x, y), but having "new" coordinates $(\bar{x}, \bar{y})$. We have the following *rule for translation of cartesian coordinates:*

$$\begin{cases} x = \bar{x} + h \\ y = \bar{y} + k \end{cases} \quad \text{or} \quad \begin{cases} \bar{x} = x - h \\ \bar{y} = y - k \end{cases}$$

EXAMPLE Translate the xy axes to form the $\bar{x}\bar{y}$ axes, so that the origin $\bar{0}$ in the $\bar{x}\bar{y}$ system corresponds to $(4, -3)$ in the xy system. If $P_1 = (2, 1)$, $P_2 = (0, 1)$, $P_3 = (-2, 3)$, and $P_4 = (-3, 5)$ are given in the xy system, find their representations in the $\bar{x}\bar{y}$ system.

Figure 8

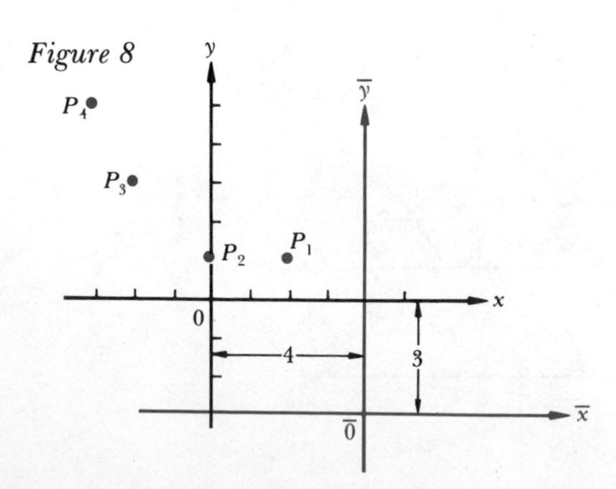

SOLUTION (See Figure 8.) Here we have $\bar{x} = x - 4$ and $\bar{y} = y - (-3)$, since $h = 4$ and $k = -3$. Hence, for $(x, y) = (2, 1)$,

$$\bar{x} = 2 - 4 = -2 \quad \text{and} \quad \bar{y} = 1 - (-3) = 4$$

so that $(2, 1)$ in the xy system is represented by $(-2, 4)$ in the $\bar{x}\bar{y}$ system.

For $(0, 1)$,

$$\bar{x} = 0 - 4 = -4 \quad \text{and} \quad \bar{y} = 1 - (-3) = 4$$

so that (0, 1) in the xy system is represented by $(-4, 4)$ in the $\bar{x}\bar{y}$ system.
For $(-2, 3)$,

$$\bar{x} = -2 - 4 = -6 \quad \text{and} \quad \bar{y} = 3 - (-3) = 6$$

so that $(-2, 3)$ in the xy system is represented by $(-6, 6)$ in the $\bar{x}\bar{y}$ system.
For $(-3, 5)$,

$$\bar{x} = -3 - 4 = -7 \quad \text{and} \quad \bar{y} = 5 - (-3) = 8$$

so that $(-3, 5)$ in the xy system is represented by $(-7, 8)$ in the $\bar{x}\bar{y}$ system.

Figure 9

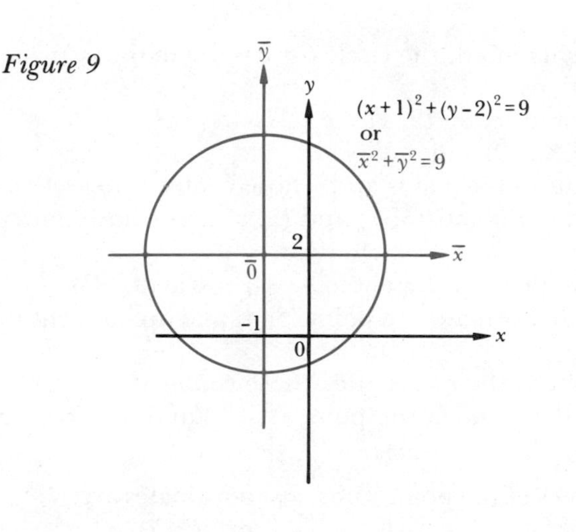

Notice that an equation of a curve in the plane depends not only on the set of points comprising this curve, but also on our choice of the coordinate system. For instance, the circle of radius 3 with center at $\bar{0}$ in Figure 9 has the equation $(x + 1)^2 + (y - 2)^2 = 9$ with respect to the xy coordinate system; however, the very same circle has the equation $\bar{x}^2 + \bar{y}^2 = 9$ with respect to the $\bar{x}\bar{y}$ coordinate system. This can be seen either by noticing that the center of the circle is the origin $\bar{0}$ in the $\bar{x}\bar{y}$ system, or by substituting $\bar{x} = x - h = x - (-1) = x + 1$ and $\bar{y} = y - k = y - 2$ into $\bar{x}^2 + \bar{y}^2 = 9$ to get the equation $(x + 1)^2 + (y - 2)^2 = 9$. Notice that a translation of the coordinate axes does not change the position of a curve in the plane or the shape of the curve—it only changes the *equation* of the curve.

EXAMPLE Find a translation of axes that transforms the equation of a circle given by $x^2 + y^2 - 4x + 2y - 4 = 0$ into the form $\bar{x}^2 + \bar{y}^2 = r^2$.

SOLUTION (See Figure 10.) Completing the square in both x and y in

$$x^2 + y^2 - 4x + 2y - 4 = 0$$

we get the equation

$$(x - 2)^2 + (y + 1)^2 = 9$$

So, if we let

$$\bar{x} = x - 2 \quad \text{and} \quad \bar{y} = y + 1$$

we get an equation of the circle in the form

$$\bar{x}^2 + \bar{y}^2 = 9$$

Figure 10

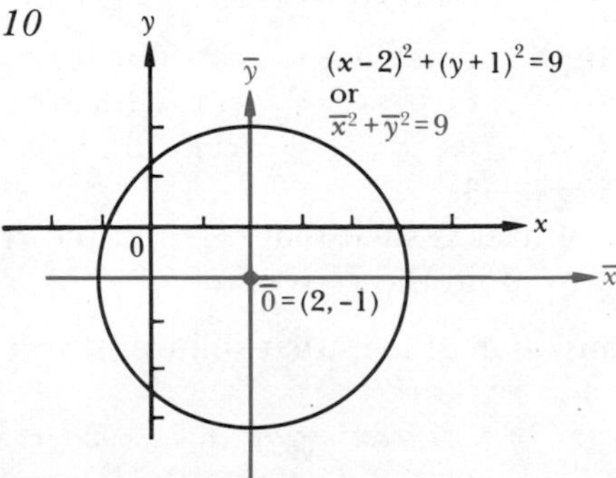

PROBLEM SET 1

In Problems 1–7, find the center and radius of each of the given circles. Also, sketch the graph of each circle.

1 $(x-3)^2+(y-1)^2=4$ **2** $(x-1)^2+(y+2)^2=16$

3 $x^2+y^2-3x+4y+4=0$ **4** $x^2+y^2+4x-6y=5$

5 $3x^2+3y^2-6x+9y=27$ **6** $4x^2+4y^2+4x-4y+1=0$

7 $4x^2+4y^2+12x+20y+25=0$

8 Prove that any circle with its center at the origin has an equation of the form $x^2+y^2=r^2$.

In Problems 9–16, find an equation of the circle (or circles) in the xy plane that satisfies the given conditions.

9 The circle with radius 3 and center (2, 1)

10 The circle with center (−3, −2) and radius 3

11 The circle that is tangent to the x axis and whose center is (5, −4)

12 The circle that contains the points (3, 7) and (5, 5) and whose center lies on the line $x-4y=1$

13 The circles of radius $\sqrt{10}$ that are tangent to $3x+y=6$ at (3, −3)

14 The circles of radius 2 that contain the point (3, 4) and are tangent to the circle $x^2+y^2=25$

15 The circle that is tangent to the x axis and whose center is (1, −7)

16 The circles of radius 5 that contain the point (3, −2) and with center on the line $2x-y+2=0$

17 Indicate the xy coordinates of the point whose $\bar{x}\bar{y}$ coordinates are

(a) (3, −4) (b) (2, 2) (c) (5, 7)

(d) (3, −5) (e) (0, 0) (f) (1, −3)

(i) if $\bar{x}=x+1$ and $\bar{y}=y-3$

(ii) if the translation of axes is such that (5, 6) in the xy system is represented by (−1, 3) in the $\bar{x}\bar{y}$ system

18 We say that a quantity is *invariant* under an operation if the quantity remains the same after the operation has been carried out. Prove that each of the following is invariant under a *translation.*

(a) The distance between $P_1=(x_1,y_1)$ and $P_2=(x_2,y_2)$

(b) The area of a triangle whose base is b and whose height is h

(c) The slope of a line joining $P_1=(x_1,y_1)$ and $P_2=(x_2,y_2)$

19 Indicate the $\bar{x}\bar{y}$ coordinates of the point whose xy coordinates are

(a) (2, −7) (b) (−5, 3) (c) (−3, 4)

(d) $(\frac{7}{4}, -\frac{13}{4})$ (e) (−7, −5) (f) (3, −7)

(i) if $\bar{x}=x+1$ and $\bar{y}=y-3$

(ii) if the translation of axes is such that (3, 4) in the xy system is represented by (0, 0) in the $\bar{x}\bar{y}$ system

In Problems 20–23, find the translation of axes that will transform each of the given equations into the form $\bar{x}^2+\bar{y}^2=r^2$.

20 $2x^2+2y^2+16x-7y=0$ **21** $3x^2+3y^2+7x-5y+3=0$

22 $x^2+y^2-8x-10y+40=0$ **23** $3x^2+3y^2-9x-7y-36=0$

24 We say that a relation R is symmetric with respect to the x axis if, whenever $(x, y) \in R$, then $(x, -y) \in R$ for all x in the domain of R. Discuss the symmetry of the circle $x^2 + y^2 = r^2$ with respect to the x axis, with respect to the y axis, and with respect to the origin.

2 Ellipse

With the tool of translations available, let us now study the *ellipse*. The importance of the ellipse is exemplified by the planets, which follow elliptical orbits about the sun, and satellites, which follow very nearly elliptical orbits about the planets.

Geometrically, an *ellipse* is defined to be the set of all points P in a plane such that the sum of the distances from P to two fixed points, called *foci*, is a constant. In Figure 1, F_1 and F_2 are the foci, P_1, P_2, and P_3 are points on the curve, and $d_1 + c_1 = d_2 + c_2 = d_3 + c_3 = k$, where k is the constant of the ellipse.

Figure 1

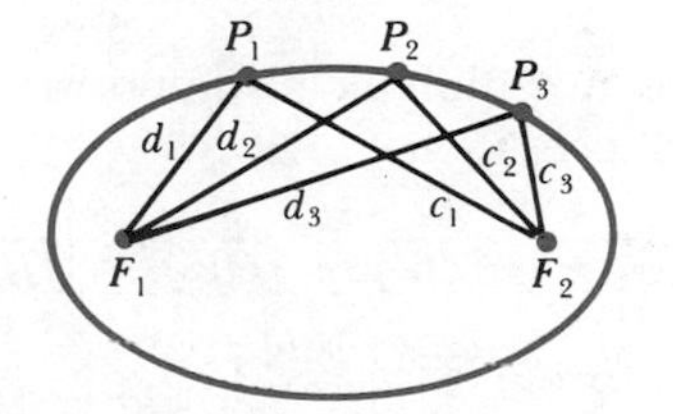

The midpoint of line segment $\overline{F_1F_2}$ is called the *center of the ellipse.* The ellipse is symmetric with respect to each of two perpendicular lines that intersect at its center (Figure 2).

The four points of intersection of the lines of symmetry and the ellipse are called *vertices* of the ellipse; the longer line segment determined by the vertices is called the *major axis,* whereas the shorter line segment determined by the vertices is called the *minor axis.* In Figure 3, V_1, V_2, V_3, and V_4 are the vertices, $\overline{V_1V_2}$ is the major axis, $\overline{V_3V_4}$ is the minor axis, and O is the center.

Figure 2

Figure 3

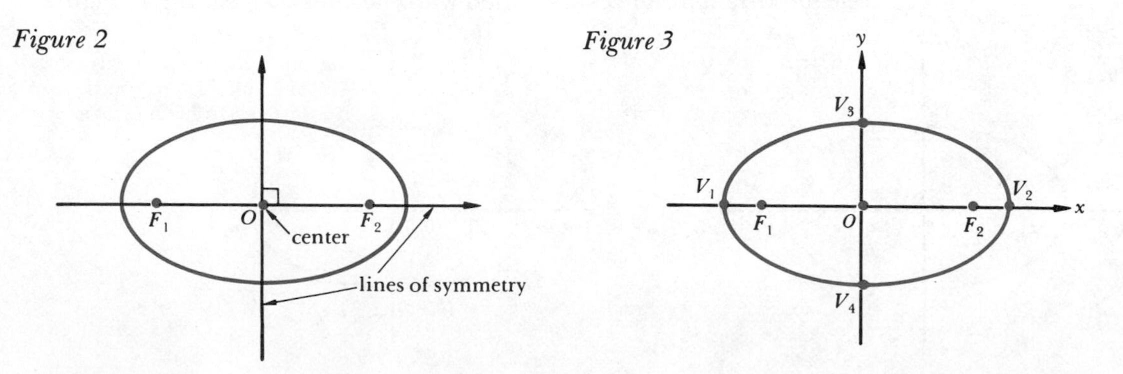

In order to express the geometric description of an ellipse in analytic terms, let us choose the coordinate system so that the foci are located at $(-c, 0)$ and $(c, 0)$, where $c > 0$. Also, we shall assume that the constant k, which is equal to the sum of the distances between a point on the ellipse and the foci, is $2a$; that is, $r_1 + r_2 = 2a$ (Figure 4). (This constant is written in the form $2a$ so that the equation of the ellipse will have a simpler form.) Notice that the major axis lies on the x axis; the minor axis lies on the y axis; the center of the ellipse is O.

Figure 4

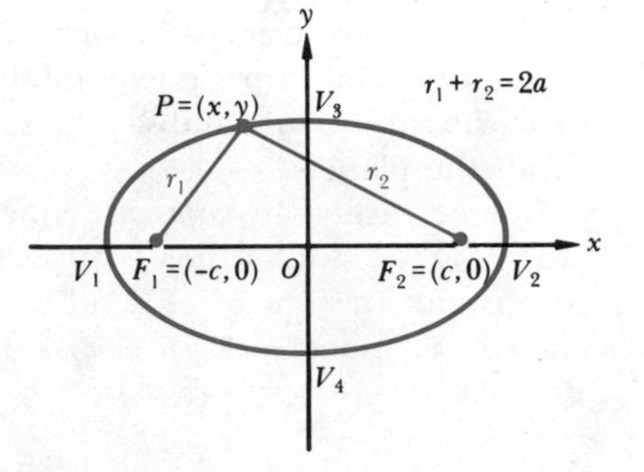

It is not difficult to demonstrate that $|\overline{V_1V_2}| = 2a$. Since V_2 is a point on the ellipse, then

$$|\overline{F_1O}| + |\overline{OV_2}| + |\overline{OV_2}| - |\overline{OF_2}| = 2a$$

Therefore,

$$c + |\overline{OV_2}| + |\overline{OV_2}| - c = 2|\overline{OV_2}| = 2a$$

so that $|\overline{OV_2}| = a$. By symmetry, $|\overline{OV_1}| = a$; hence $|\overline{V_1V_2}| = 2a$.

We next establish the relationship between the length of the major axis and the length of the minor axis. Let the distance $|\overline{V_3V_4}| = 2b$ (Figure 5). Since V_3 is a point on the ellipse, $|\overline{V_3F_1}| + |\overline{V_3F_2}| = 2a$. By symmetry, $|\overline{V_3F_1}| = |\overline{V_3F_2}|$, so $|\overline{V_3F_2}| = a$. From the right triangle, $\triangle OF_2V_3$, we obtain $a^2 = b^2 + c^2$, from which we conclude that $b < a$, and that the major axis is longer than the minor axis.

The important distances associated with the ellipse are shown in Figure 6.

Figure 5 *Figure 6*

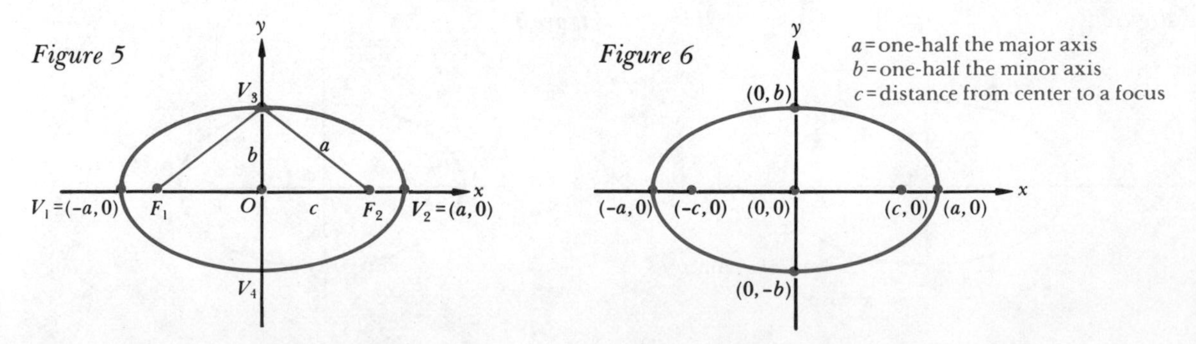

THEOREM 1 ELLIPSE EQUATION

An equation for the ellipse with foci at $F_1 = (-c, 0)$ and $F_2 = (c, 0)$ is

$$\frac{x^2}{a^2} + \frac{y^2}{b^2} = 1$$

where $a > b$ and $b^2 = a^2 - c^2$. Moreover, $2a$ is the length of the major axis and $2b$ is the length of the minor axis.

PROOF Assume that $P = (x, y)$ is any point on the ellipse (Figure 7). Then $|\overline{PF_1}| + |\overline{PF_2}| = 2a$, so that, by the distance formula,

$$\sqrt{(x + c)^2 + y^2} + \sqrt{(x - c)^2 + y^2} = 2a$$

That is,

$$\sqrt{(x + c)^2 + y^2} = 2a - \sqrt{(x - c)^2 + y^2}$$

Squaring both sides of the latter equation, we get

$$x^2 + 2xc + c^2 + y^2 = 4a^2 - 4a\sqrt{(x - c)^2 + y^2} + x^2 - 2cx + c^2 + y^2$$

so that

$$4cx - 4a^2 = -4a\sqrt{(x - c)^2 + y^2}$$

Figure 7

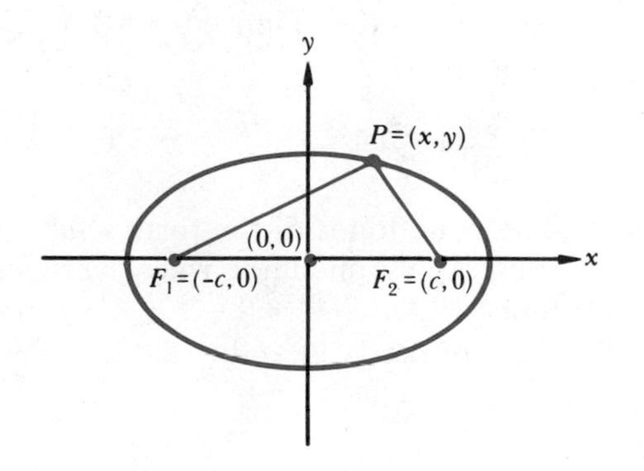

That is,

$$cx - a^2 = -a\sqrt{(x - c)^2 + y^2}$$

Squaring both sides of the equation again, we get

$$c^2x^2 - 2a^2cx + a^4 = a^2(x^2 - 2cx + c^2 + y^2)$$

so that

$$a^4 - a^2c^2 = (a^2 - c^2)x^2 + a^2y^2$$

and

$$(a^2 - c^2)x^2 + a^2y^2 = a^2(a^2 - c^2)$$

Dividing both sides of the equation by $a^2(a^2 - c^2)$, we get

$$\frac{x^2}{a^2} + \frac{y^2}{a^2 - c^2} = 1$$

Since $a^2 - c^2 = b^2$, we have $\dfrac{x^2}{a^2} + \dfrac{y^2}{b^2} = 1$. The graph of the ellipse is shown in Figure 8.

Figure 8

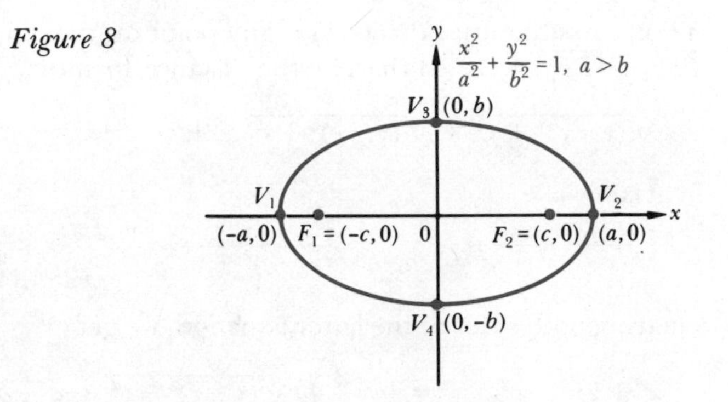

EXAMPLES 1 Given $x^2 + 9y^2 = 9$ as an equation of an ellipse, find the vertices and foci, and sketch the graph.

SOLUTION (See Figure 9.) After dividing both sides of the equation by 9, we get

$$\frac{x^2}{9} + \frac{y^2}{1} = 1$$

The equation is of the form $x^2/a^2 + y^2/b^2 = 1$, with $a^2 = 9$ and $b^2 = 1$. Thus the graph is an ellipse whose vertices are $(3, 0)$, $(-3, 0)$, $(0, 1)$, and $(0, -1)$. Since $c^2 = a^2 - b^2 = 9 - 1 = 8$, we have $c = 2\sqrt{2}$, so the coordinates of the foci are $(2\sqrt{2}, 0)$ and $(-2\sqrt{2}, 0)$.

Figure 9

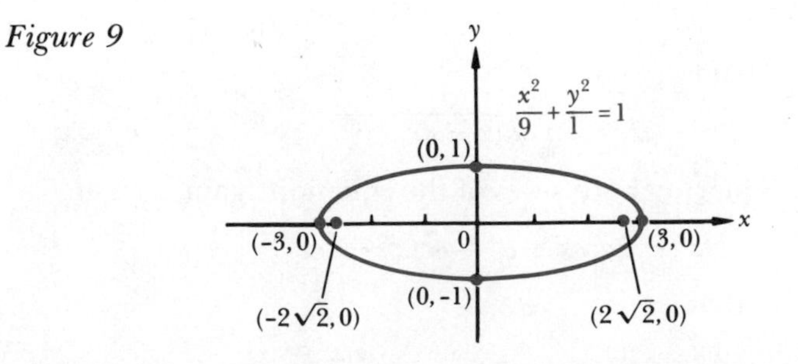

2 Find an equation of the ellipse with foci $(2, 0)$ and $(-2, 0)$ and vertices $(3, 0)$ and $(-3, 0)$, and sketch the graph of the ellipse.

SOLUTION (See Figure 10.) The ellipse has its center at the origin and its foci on the x axis; therefore, its equation is of the form $x^2/a^2 + y^2/b^2 = 1$, with $b^2 = a^2 - c^2 = 9 - 4 = 5$, so $b = \sqrt{5}$; thus the ellipse has an equation of the form $\frac{x^2}{9} + \frac{y^2}{5} = 1$.

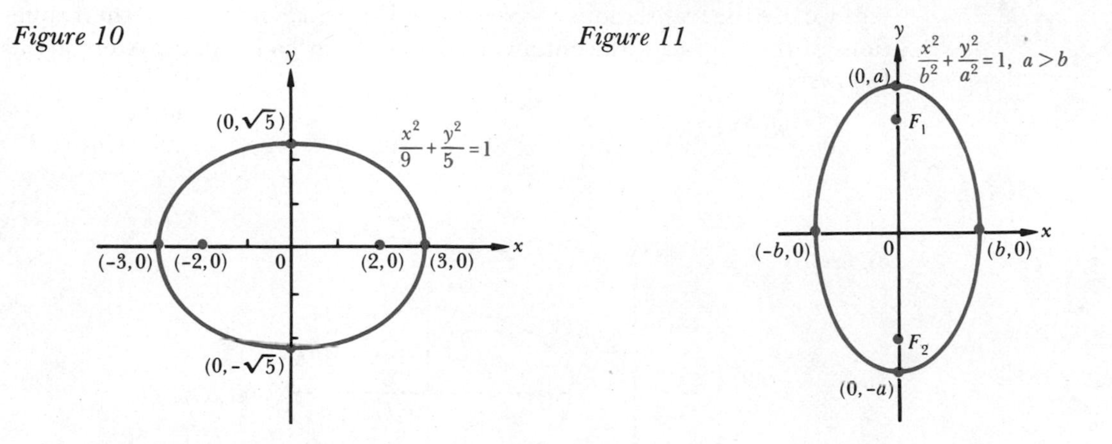

If the foci of an ellipse centered at the origin are on the y axis and $2a$ is the sum of the distances to the foci from each point on the ellipse, then, by symmetry, its equation will become

$$\frac{x^2}{b^2} + \frac{y^2}{a^2} = 1$$

and, since $a > b$, we still have $c^2 = a^2 - b^2$. The graph is shown in Figure 11.

EXAMPLE Given $4x^2 + y^2 = 4$ as the equation of an ellipse, find the vertices and foci, and sketch the graph.

SOLUTION (See Figure 12.) After dividing both sides of the equation by 4, we get $\frac{x^2}{1} + \frac{y^2}{4} = 1$.

Figure 12

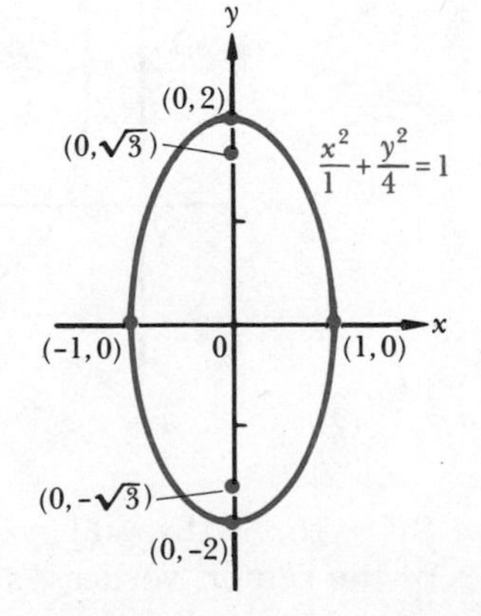

This equation is of the form $x^2/b^2 + y^2/a^2 = 1$, with $b^2 = 1$ and $a^2 = 4$. Thus the graph is an ellipse whose vertices are $(0, 2)$, $(0, -2)$, $(1, 0)$, and $(-1, 0)$. Since $c^2 = a^2 - b^2 = 4 - 1 = 3$, $c = \sqrt{3}$, and the coordinates of the foci are $(0, -\sqrt{3})$ and $(0, \sqrt{3})$.

If we use the translations $\bar{x} = x - h$ and $\bar{y} = y - k$, and if $a > b$, then equations of the ellipses with centers at (h, k) in the xy coordinate system are as follows:

(i) $\dfrac{(x-h)^2}{a^2} + \dfrac{(y-k)^2}{b^2} = 1$ if the major axis is horizontal **(Figure 13)**

Figure 13

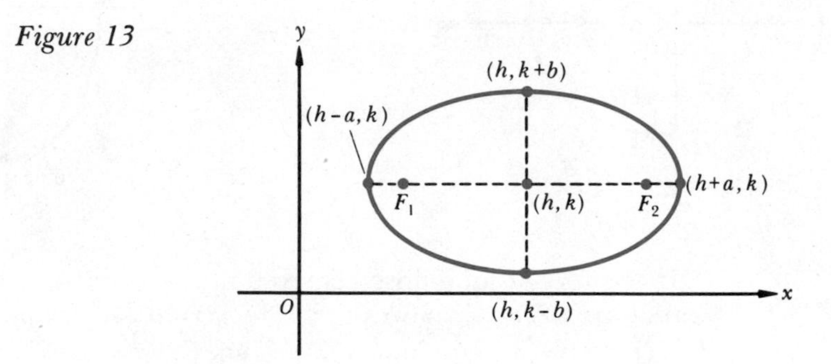

(ii) $\dfrac{(x-h)^2}{b^2} + \dfrac{(y-k)^2}{a^2} = 1$ if the major axis is vertical **(Figure 14)**

Figure 14

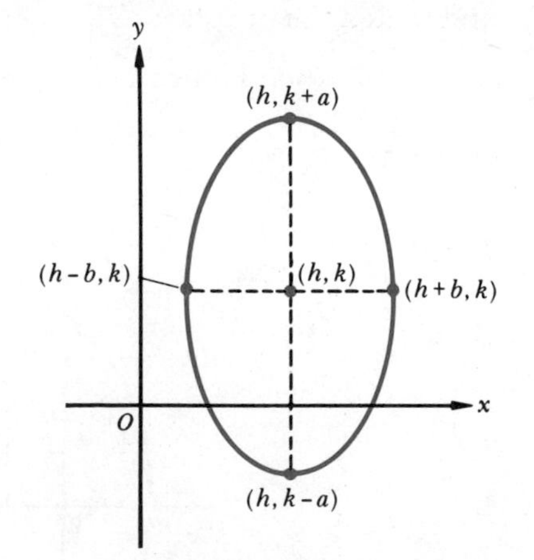

EXAMPLES 1 Given $4x^2 + 9y^2 + 16x - 18y - 11 = 0$ as an equation of an ellipse, find the coordinates of the center, vertices, and foci. Also, sketch the graph.

Figure 15

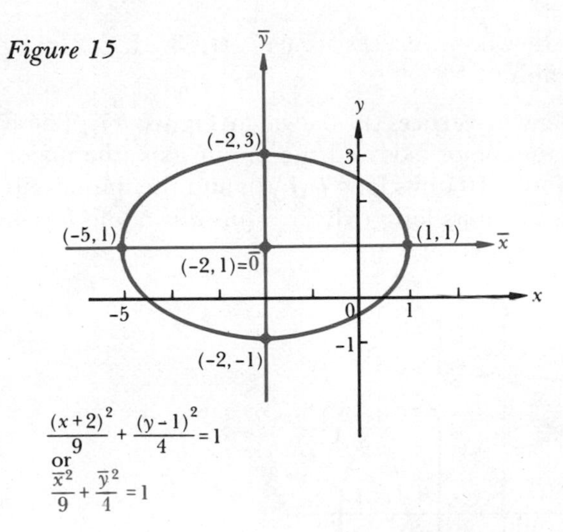

SOLUTION (See Figure 15.) First, complete the square to get

$$4(x^2 + 4x + 4) + 9(y^2 - 2y + 1) = 11 + 16 + 9$$

so that

$$4(x + 2)^2 + 9(y - 1)^2 = 36$$

or, equivalently,

$$\frac{(x + 2)^2}{9} + \frac{(y - 1)^2}{4} = 1$$

If we let $\bar{x} = x + 2$ and $\bar{y} = y - 1$, then

$$\frac{\bar{x}^2}{9} + \frac{\bar{y}^2}{4} = 1$$

The center is $(-2, 1)$ in the xy system.

Notice that $a^2 = 9$ and $b^2 = 4$, so $a = 3$ and $b = 2$, and the coordinates of the vertices are $(1, 1)$, $(-5, 1)$, $(-2, -1)$, and $(-2, 3)$, since the major axis is parallel to the x axis. Therefore, $c^2 = a^2 - b^2 = 9 - 4 = 5$, so that $c = \sqrt{5}$, and the coordinates of the foci are $(-2 + \sqrt{5}, 1)$ and $(-2 - \sqrt{5}, 1)$.

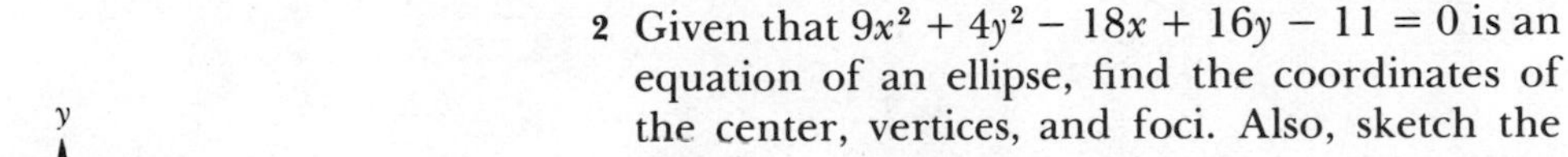

2 Given that $9x^2 + 4y^2 - 18x + 16y - 11 = 0$ is an equation of an ellipse, find the coordinates of the center, vertices, and foci. Also, sketch the graph.

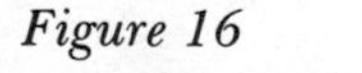

Figure 16

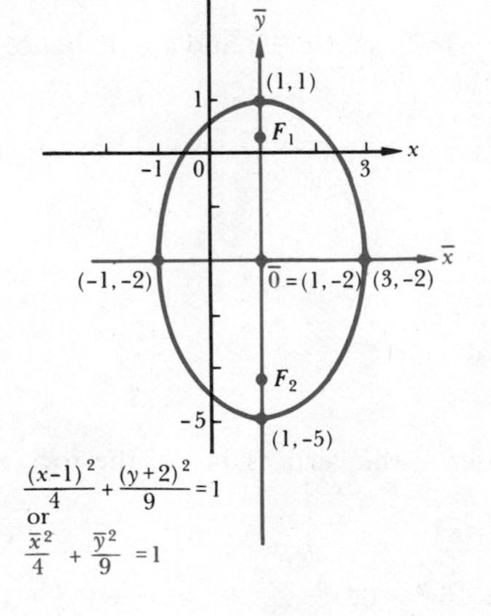

SOLUTION (See Figure 16.) First, complete the square to get

$$9(x^2 - 2x + 1) + 4(y^2 + 4y + 4) = 11 + 9 + 16$$

Hence

$$9(x - 1)^2 + 4(y + 2)^2 = 36$$

That is,

$$\frac{(x - 1)^2}{4} + \frac{(y + 2)^2}{9} = 1$$

If we let $\bar{x} = x - 1$ and $\bar{y} = y + 2$, we get

$$\frac{\bar{x}^2}{4} + \frac{\bar{y}^2}{9} = 1$$

The center is $(1, -2)$ in the xy system.

Here, $a^2 = 9$ and $b^2 = 4$, so that $a = 3$ and $b = 2$, and the coordinates of the vertices are $(1, 1)$, $(1, -5)$, $(-1, -2)$, and $(3, -2)$, since the major axis is parallel to the y axis. From the equation $c^2 = a^2 - b^2 = 9 - 4 = 5$, we have $c = \sqrt{5}$ and the coordinates of the foci are $(1, -2 + \sqrt{5})$ and $(1, -2 - \sqrt{5})$.

3 Write an equation of the ellipse whose vertices are $(-7, 1)$, $(3, 1)$, $(-2, -3)$, and $(-2, 5)$, and sketch its graph.

SOLUTION First, locate the given vertices as shown in Figure 17. These four points are the ends of the major axis and the minor axis; the major axis is parallel to the x axis and is 10 units long (why?), and the minor axis is parallel to the y axis and is 8 units long (why?). Thus $a = 5$ and $b = 4$.

Figure 17

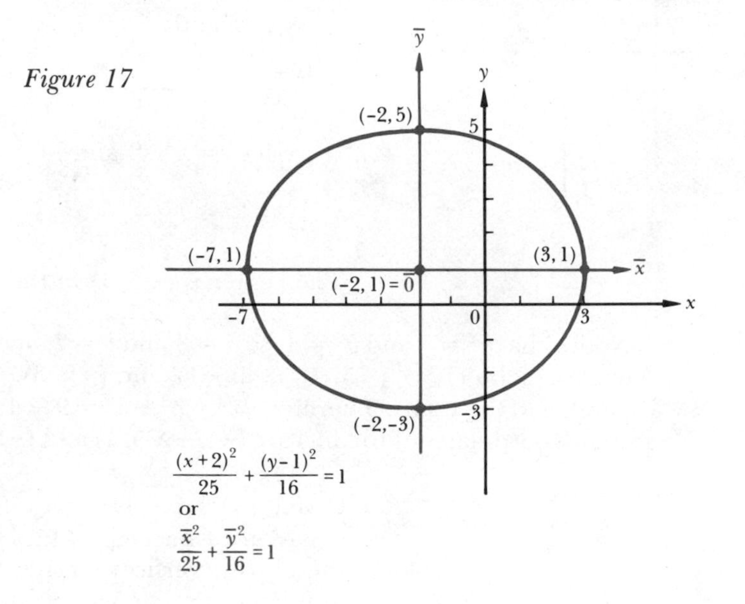

In this case the center is at $(-2, 1)$ (why?), so $h = -2$ and $k = 1$; hence an equation in the $\bar{x}\bar{y}$ system is

$$\frac{\bar{x}^2}{25} + \frac{\bar{y}^2}{16} = 1 \qquad \text{where} \quad \bar{x} = x + 2 \quad \text{and} \quad \bar{y} = y - 1$$

or

$$\frac{(x+2)^2}{25} + \frac{(y-1)^2}{16} = 1 \qquad \text{in the } xy \text{ system}$$

PROBLEM SET 2

In Problems 1–6, find the coordinates of the vertices and of the foci, and sketch the graph of each ellipse.

1 $\frac{x^2}{16} + \frac{y^2}{9} = 1$ **2** $\frac{y^2}{25} + \frac{x^2}{16} = 1$ **3** $\frac{y^2}{16} + \frac{x^2}{4} = 1$

4 $4x^2 + 9y^2 = 36$ **5** $4x^2 + 16y^2 = 64$ **6** $25x^2 + 9y^2 = 1$

In Problems 7–14, find the coordinates of the center, of the vertices, and of the foci, then sketch the graph of each ellipse.

7 $3(x-1)^2 + 4(y+2)^2 = 192$ **8** $x^2 + 4y^2 - 2x - 16y + 13 = 0$

9 $x^2 + 4y^2 + 2x - 8y + 1 = 0$ **10** $25(x-3)^2 + 4(y-1)^2 = 100$

11 $9x^2 + 4y^2 + 18x - 16y - 11 = 0$ **12** $9x^2 + y^2 - 18x + 2y + 9 = 0$
13 $16(x + 2)^2 + 25(y - 1)^2 = 400$ **14** $4x^2 + 9y^2 - 24x + 36y + 36 = 0$

In Problems 15–21, find an equation of the ellipse for each case.

15 Vertices at $(1, -2)$, $(5, -2)$, $(3, -7)$, and $(3, 3)$

16 Vertices at $(0, -1)$, $(12, -1)$, $(6, -4)$, and $(6, 2)$

17 Vertices at $(1, 1)$, $(5, 1)$, $(3, 6)$, and $(3, -4)$

18 Vertices at $(-5, 0)$ and $(5, 0)$, and containing the point $(4, 12/5)$

19 Foci at $(1, 4)$ and $(3, 4)$, and major axis 4 units long

20 Center at $(0, 0)$, axes parallel to coordinate axes, and containing the points $(3\sqrt{3}/2, 1)$ and $(2, 2\sqrt{5}/3)$

21 Center at $(-3, 1)$, major axis parallel to the y axis and 10 units long, and minor axis 2 units long

22 Discuss the symmetry of an ellipse of the form $\dfrac{x^2}{a^2} + \dfrac{y^2}{b^2} = 1$.

23 The segment cut by an ellipse from a line that contains a focus and is perpendicular to the major axis is called a *latus rectum* (or *focal chord*) of the ellipse.

(a) Show that $2b^2/a$ is the length of a latus rectum of an ellipse whose equation is $b^2x^2 + a^2y^2 = a^2b^2$.

(b) Find the length of a latus rectum of an ellipse whose equation is $9x^2 + 16y^2 = 144$.

24 An arch in the shape of the upper half of an ellipse with a horizontal major axis is to support a bridge 100 feet long over a river. The center of the arch is to be 25 feet above the surface of the river. Find an equation of the ellipse.

25 Consider the equation $\dfrac{x^2}{a^2} + \dfrac{y^2}{b^2} = 1$ of an ellipse. If $a = b$, describe the graph of the ellipse.

26 A mathematician has accepted a position at a university situated 6 miles from the straight shoreline of a large lake (Figure 18). The professor wishes to build a home that is half as far from the university as it is from the shore of the lake. The possible homesites satisfying this condition lie along a curve. Describe this curve and find its equation with respect to a coordinate system having the shoreline as the x axis and the university at the point $(0, 6)$ on the y axis.

Figure 18

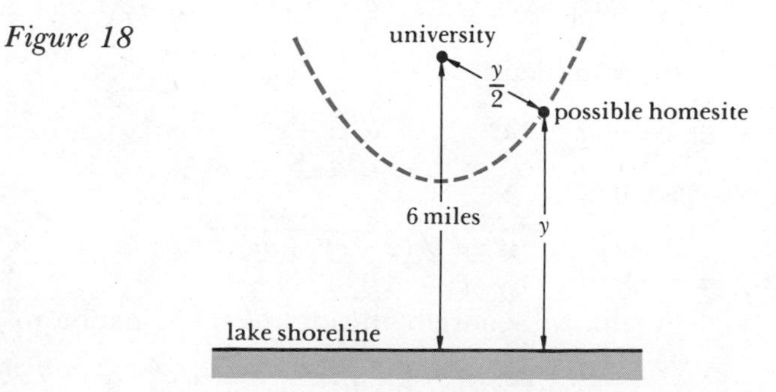

3 Hyperbola

The *hyperbola* will be developed in a manner similar to that used to develop the ellipse. Geometrically, a hyperbola is defined to be the set of points P in the plane such that the absolute value of the difference of distances from P to two distinct fixed points, called *foci,* is equal to a constant.

Figure 1

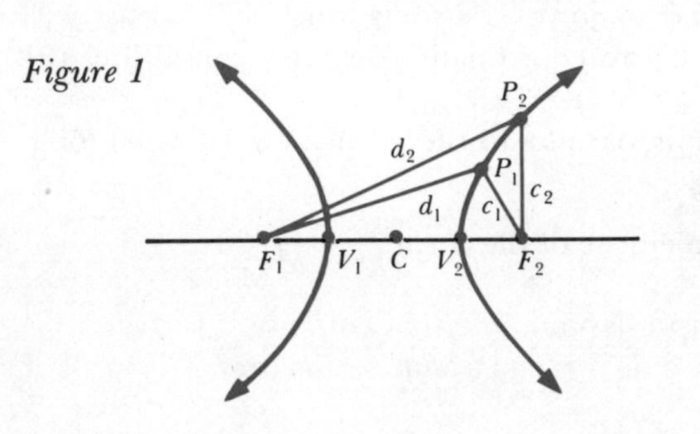

The line determined by the foci is a line of symmetry, and the midpoint of the line segment determined by the foci is the *center* of the hyperbola. The two points of intersection of the hyperbola with the line of symmetry are called the *vertices* of the hyperbola, and the line segment determined by the vertices is called the *transverse axis.* In Figure 1, F_1 and F_2 are the foci, C is the center, V_1 and V_2 are the vertices, the line determined by V_1 and V_2 is the transverse axis, and $|d_2 - c_2| = |d_1 - c_1| = k$, where k is the constant of the hyperbola.

Figure 2

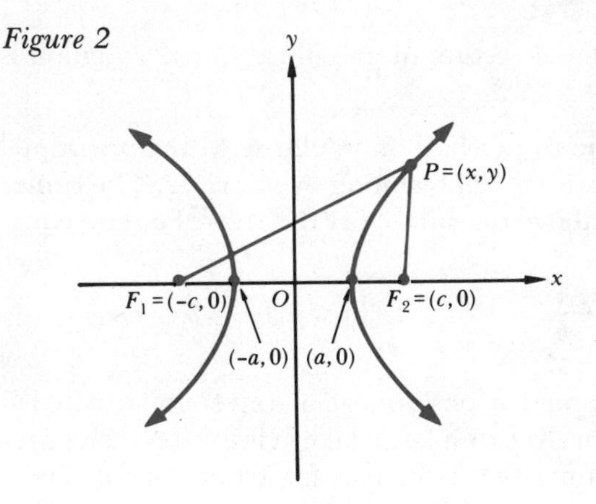

We determine an equation of the hyperbola by choosing the coordinate system so that the foci are $(-c, 0)$ and $(c, 0)$, with $c > 0$, and we assume the constant difference k to be $2a$. Thus, if $P = (x, y)$ is a point on the hyperbola, we have

$$\left| |\overline{PF_1}| - |\overline{PF_2}| \right| = 2a \qquad \text{(Figure 2)}$$

By using the distance formula, we get

$$\left| \sqrt{(x+c)^2 + y^2} - \sqrt{(x-c)^2 + y^2} \right| = 2a$$

Hence

$$\sqrt{(x+c)^2 + y^2} - \sqrt{(x-c)^2 + y^2} = \pm 2a$$

so that

$$\sqrt{(x+c)^2 + y^2} = \pm 2a + \sqrt{(x-c)^2 + y^2}$$

After squaring both sides of the equation, we get

$$x^2 + 2cx + c^2 + y^2 = 4a^2 \pm 4a\sqrt{(x-c)^2 + y^2} + x^2 - 2cx + c^2 + y^2$$

or, equivalently,

$$4cx - 4a^2 = \pm 4a\sqrt{(x-c)^2 + y^2}$$

so that

$$cx - a^2 = \pm a\sqrt{(x-c)^2 + y^2}$$

Again, we square both sides of the equation to get

$$c^2x^2 - 2a^2cx + a^4 = a^2(x^2 - 2cx + c^2 + y^2)$$

That is,

$$(c^2 - a^2)x^2 - a^2y^2 = a^2c^2 - a^4 = a^2(c^2 - a^2)$$

so that, after dividing both sides by $a^2(c^2 - a^2)$, we get

$$\frac{x^2}{a^2} - \frac{y^2}{c^2 - a^2} = 1$$

If we let $b^2 = c^2 - a^2$, the equation becomes

$$\frac{x^2}{a^2} - \frac{y^2}{b^2} = 1$$

Notice that by letting $y = 0$, we get $(a, 0)$ and $(-a, 0)$, the vertices of the hyperbola. Also notice that $a^2 + b^2 = c^2$, since we defined $b^2 = c^2 - a^2$.

Thus we have the following theorem.

THEOREM 1 HYPERBOLA EQUATION

An equation of the hyperbola with foci $F_1 = (-c, 0)$ and $F_2 = (c, 0)$ and the constant difference $k = 2a$ is

$$\frac{x^2}{a^2} - \frac{y^2}{b^2} = 1$$

where $b^2 = c^2 - a^2$.

3.1 Properties of the Hyperbola

The graph of the hyperbola with the equation $\frac{x^2}{a^2} - \frac{y^2}{b^2} = 1$ is symmetric with respect to both the x axis and the y axis. (Why?)

The x intercepts of the hyperbola $\frac{x^2}{a^2} - \frac{y^2}{b^2} = 1$ are found by letting $y = 0$ and solving for x to get $(-a, 0)$ and $(a, 0)$. If $x = 0$, then we have $y^2 = -b^2$. But such a value for y is not a real number (why?); hence we conclude that the hyperbola does *not* intersect the y axis.

If we write

$$\frac{x^2}{a^2} - \frac{y^2}{b^2} = 1 \quad \text{as} \quad y^2 = \frac{b^2}{a^2}(x^2 - a^2)$$

then

$$y = \pm \frac{b}{a} x \sqrt{1 - \frac{a^2}{x^2}}$$

Now, $1 - (a^2/x^2)$ approaches 1 as $|x|$ gets very large, so the larger x is in absolute value, the closer the graph of the hyperbola is to the lines whose equations are given by

$$y = \frac{b}{a}x \quad \text{and} \quad y = -\frac{b}{a}x$$

These lines are *asymptotes* of the hyperbola. The asymptotes are not part of the hyperbola; they are easy to draw if we construct the rectangle of dimensions $2a$ and $2b$, as in Figure 3.

Figure 3

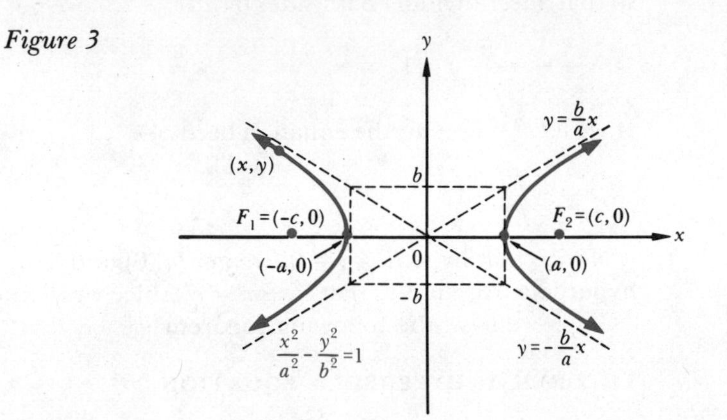

EXAMPLES 1 Given $\frac{x^2}{16} - \frac{y^2}{9} = 1$ as an equation of a hyperbola, find the coordinates of the foci and vertices, and equations of the asymptotes; sketch the graph.

SOLUTION Since $a = 4$ and $b = 3$, from Theorem 1, $c = \sqrt{4^2 + 3^2} = 5$; so the coordinates of the foci are $(5, 0)$ and $(-5, 0)$ and those of the vertices are $(4, 0)$ and $(-4, 0)$. Equations of the asymptotes are given by

$$y = \pm \frac{b}{a}x = \pm \frac{3}{4}x \qquad \text{(Figure 4)}$$

Figure 4

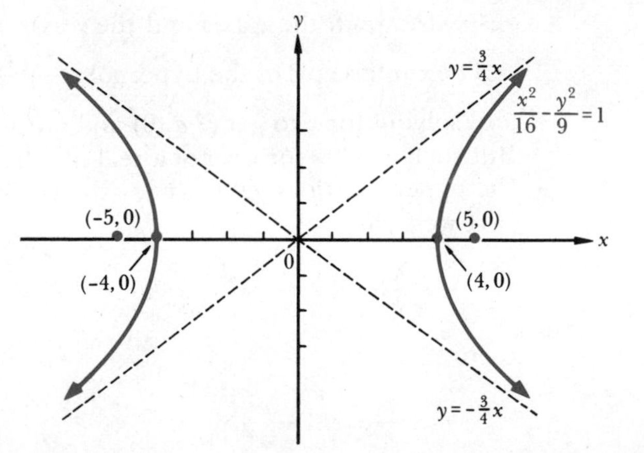

2 Given the hyperbola $25x^2 - 16y^2 = 400$, find the coordinates of the vertices, of the foci, and equations of the asymptotes. Also sketch the graph.

SOLUTION Divide both sides of the equation by 400 to get $\frac{x^2}{16} - \frac{y^2}{25} = 1$.

Figure 5

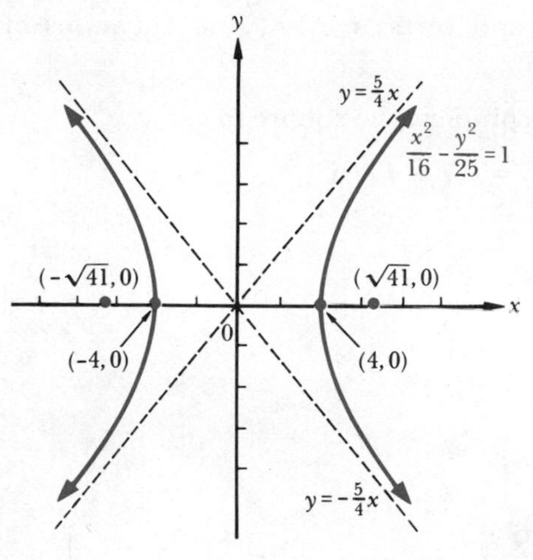

The vertices are (4, 0) and (−4, 0). Since $c^2 = a^2 + b^2 = 16 + 25 = 41, c = \sqrt{41}$, so that the foci are $(-\sqrt{41}, 0)$ and $(\sqrt{41}, 0)$. Equations of the asymptotes are given by

$$y = \pm \frac{b}{a}x = \pm \frac{5}{4}x \qquad \text{(Figure 5)}$$

Equations of hyperbolas with center (0, 0) are:

(i) $\frac{x^2}{a^2} - \frac{y^2}{b^2} = 1$ (transverse axis on x axis)

(ii) $\frac{y^2}{a^2} - \frac{x^2}{b^2} = 1$ (transverse axis on y axis)

Equations of their respective asymptotes are:

$$y = \pm \frac{b}{a}x \quad \text{and} \quad y = \pm \frac{a}{b}x$$

As with the ellipse, if the center of a hyperbola is translated so that the point (h, k) is the center, equations of the hyperbola are:

(i) $\frac{(x-h)^2}{a^2} - \frac{(y-k)^2}{b^2} = 1$ with its transverse axis parallel to the x axis and asymptotes $y - k = \pm \frac{b}{a}(x - h)$ (Figure 6)

(ii) $\frac{(y-k)^2}{a^2} - \frac{(x-h)^2}{b^2} = 1$ with its transverse axis parallel to the y axis and asymptotes $y - k = \pm \frac{a}{b}(x - h)$ (Figure 7)

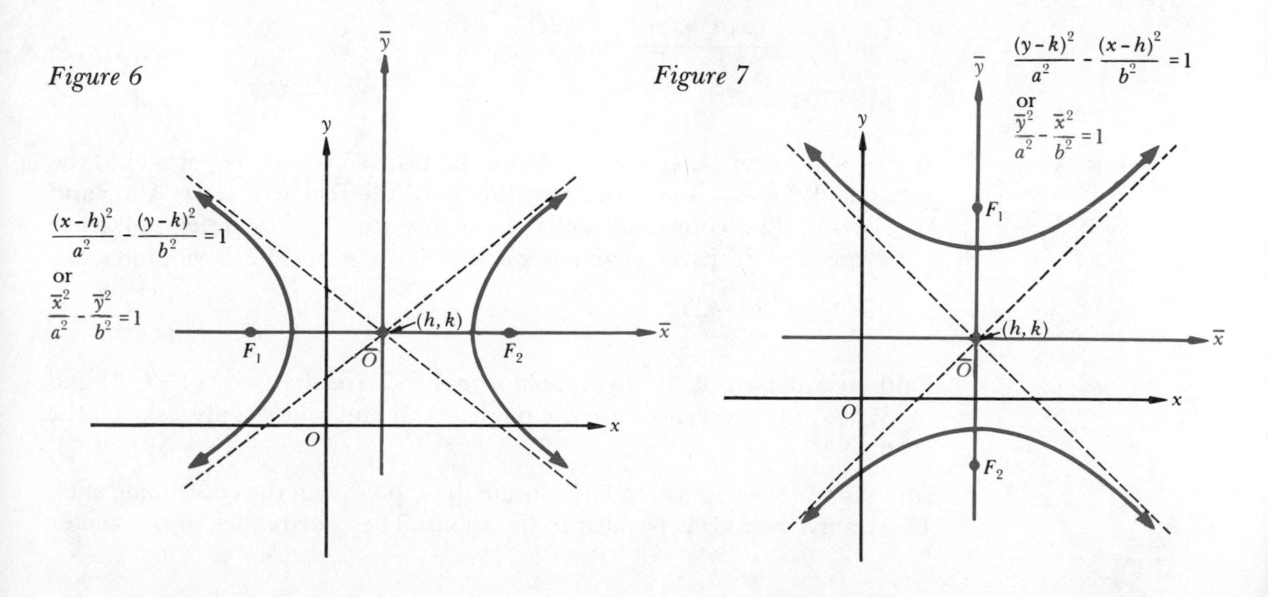

Figure 6

Figure 7

EXAMPLES 1 Given $4x^2 - y^2 - 8x + 2y + 7 = 0$ as an equation of a hyperbola, find the coordinates of the center, foci, and vertices. Also, find equations of the asymptotes and sketch the graph.

SOLUTION (See Figure 8.) First, complete the square to get

$$4(x^2 - 2x + 1) - (y^2 - 2y + 1) = -7 + 4 - 1$$

that is,

$$4(x - 1)^2 - (y - 1)^2 = -4$$

so that

$$\frac{(y - 1)^2}{4} - \frac{(x - 1)^2}{1} = 1$$

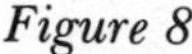

Figure 8

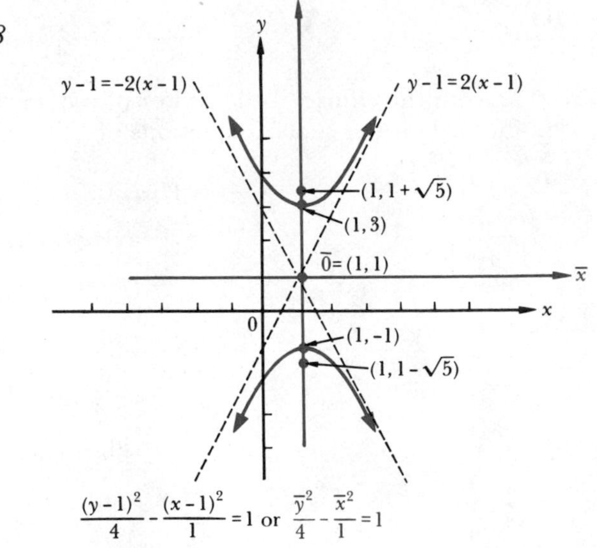

Hence the center $(h, k) = (1, 1)$. Since the transverse axis is parallel to the y axis, and $c^2 = 4 + 1 = 5$, the coordinates of the foci are $(1, 1 + \sqrt{5})$ and $(1, 1 - \sqrt{5})$. The coordinates of the vertices are $(1, -1)$ and $(1, 3)$, since $a = 2$ and the transverse axis is on the line $x = 1$. The asymptotes are

$$y - 1 = \pm 2(x - 1)$$

2 Find an equation of the hyperbola whose foci are the points $(-\frac{9}{2}, 3)$ and $(\frac{1}{2}, 3)$, and whose vertices are the points $(0, 3)$ and $(-4, 3)$; also, sketch the graph.

SOLUTION (See Figure 9.) First, locate these points on the coordinate axes. The transverse axis is parallel to the x axis. The coordinates of the center are given by $((-\frac{9}{2}+\frac{1}{2})/2, (3+3)/2) = (-2, 3)$, so that $a = 2$ and $c = \frac{5}{2}$. Since $b^2 = c^2 - a^2 = \frac{25}{4} - 4 = \frac{9}{4}$, then $b = \frac{3}{2}$. Hence an equation in the $\bar{x}\bar{y}$ system is

$$\frac{\bar{x}^2}{4} - \frac{\bar{y}^2}{9/4} = 1 \qquad \text{with } \bar{x} = x + 2 \text{ and } \bar{y} = y - 3$$

or, equivalently, an equation in the xy system is

$$\frac{(x+2)^2}{4} - \frac{(y-3)^2}{9/4} = 1$$

Figure 9

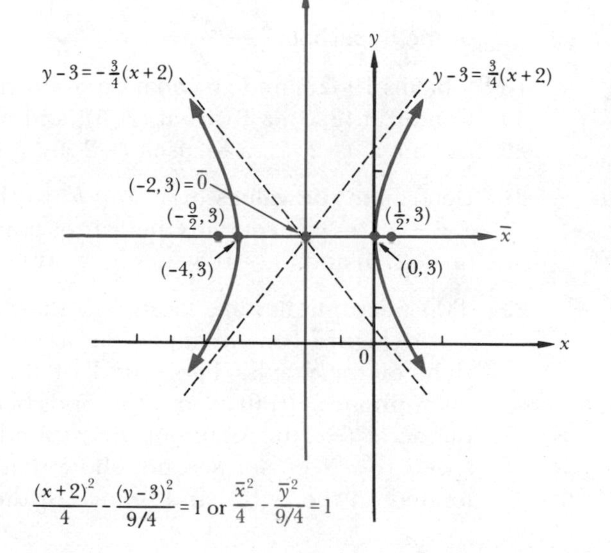

PROBLEM SET 3

In Problems 1–6, find the coordinates of the vertices and of the foci; find equations of the asymptotes, and sketch the graph of each hyperbola.

1 $36x^2 - 49y^2 = 1764$ **2** $16y^2 - 4x^2 = 48$ **3** $y^2 - x^2 = 9$

4 $36x^2 - 9y^2 = 1$ **5** $x^2 - 9y^2 = 9$ **6** $9x^2 - y^2 = 9$

7 Find an equation of the hyperbola with vertices at $(-16, 0)$ and $(16, 0)$ and asymptotes $y = \pm\frac{5}{4}x$.

8 Discuss the symmetry of the graph of a hyperbola.

In Problems 9–16, find the coordinates of the center, of the vertices, and of the foci; find equations of the asymptotes, and sketch the graph of each hyperbola.

9 $\frac{(x-5)^2}{25} - \frac{(y-4)^2}{4} = 1$ **10** $\frac{(y-2)^2}{16} - \frac{(x+3)^2}{25} = 1$

11 $\frac{(x+2)^2}{16} - \frac{(y-3)^2}{9} = 1$ **12** $4x^2 - y^2 + 8x - 2y + 6 = 0$

13 $4x^2 - 3y^2 - 32x + 6y + 73 = 0$
14 $4x^2 - 9y^2 - 32x + 36y + 27 = 0$
15 $9x^2 - 25y^2 + 72x - 100y + 269 = 0$
16 $9x^2 - 16y^2 - 90x - 256y = 223$

17 Find an equation of the hyperbola with vertices at $(-1, 4)$ and $(-1, 6)$ and foci at $(-1, 3)$ and $(-1, 7)$.

18 The line segment cut by a hyperbola from a line that contains a focus and is perpendicular to the transverse axis is called a *focal chord* (or *latus rectum*) of the hyperbola. Show that the length of the focal chord of the hyperbola $\frac{x^2}{a^2} - \frac{y^2}{b^2} = 1$ is $\frac{2b^2}{a}$.

In Problems 19–20, find an equation of the hyperbola in each case.

19 Center at $(2, 3)$, a focus at $(2, 5)$, and a focal chord of 6 units
20 Center at $(-2, 1)$, a focus at $(-2, 6)$, and a focal chord of 32/3 units

21 Determine the values of a^2 and b^2 so that the graph of the equation $b^2x^2 - a^2y^2 = a^2b^2$ contains the pair of points
(a) $(2, 5)$ and $(3, -10)$ (b) $(4, 3)$ and $(-7, 6)$

22 Two microphones are located at the points $(-c, 0)$ and $(c, 0)$ on the x axis (Figure 10). An explosion occurs at an unknown point P to the right of the y axis. The sound of the explosion is detected by the microphone at $(c, 0)$ exactly t seconds before it is detected by the microphone at $(-c, 0)$. Assuming that sound travels in air at the constant speed of v feet per second, show that the point P must have been located on the right-hand branch of the hyperbola whose equation is

$$\frac{x^2}{a^2} - \frac{y^2}{b^2} = 1$$

where

$$a = \frac{vt}{2} \quad \text{and} \quad b = \frac{\sqrt{4c^2 - v^2t^2}}{2}$$

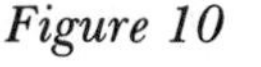
Figure 10

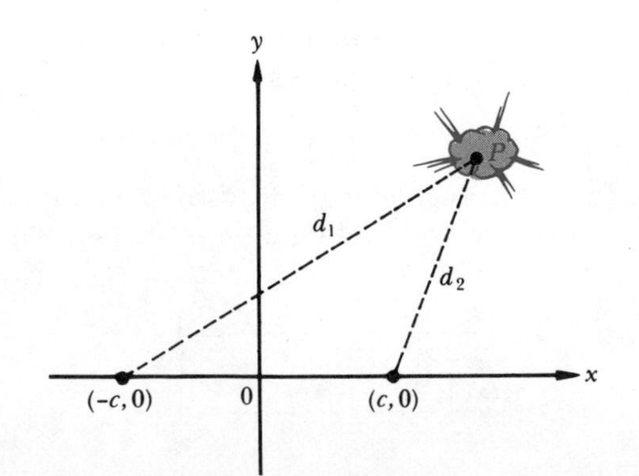

4 Parabola

Geometrically, a *parabola* is defined to be the set of all points P such that the distance from P to a fixed point, called the *focus,* is equal to the distance from P to a fixed line, called the *directrix.* In Figure 1, the points P_1, P_2, and P_3 are on the parabola; hence, $d_1 = c_1$, $d_2 = c_2$, and $d_3 = c_3$. It is not difficult to derive an equation of a parabola from this geometric description.

Figure 1

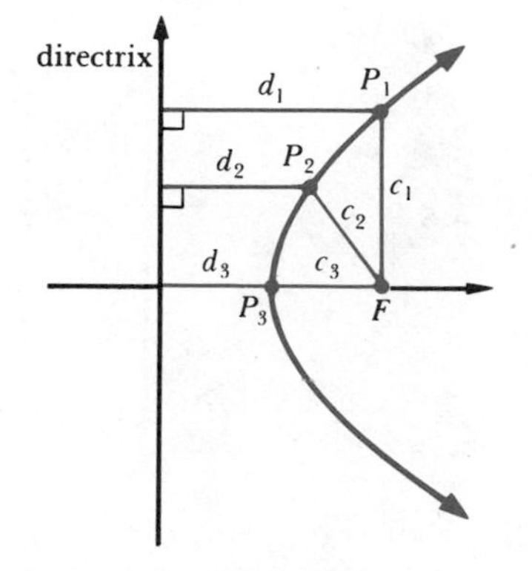

THEOREM 1 PARABOLA EQUATION

Let $P = (x, y)$ be any point on a parabola with focus $(c, 0)$ and directrix $x = -c$, where $c > 0$; then its equation is $y^2 = 4cx$ (Figure 2).

PROOF (See Figure 2.) Since $P = (x, y)$ is equidistant from the focus and directrix, we have $|\overline{PF}| = |\overline{PD}|$, so

$$\sqrt{(x - c)^2 + y^2} = |x + c|$$

After squaring both sides of the equation, we get

$$(x - c)^2 + y^2 = (x + c)^2 \quad \text{or} \quad x^2 - 2cx + c^2 + y^2 = x^2 + 2cx + c^2$$

so that, $y^2 = 4cx$.

Figure 2

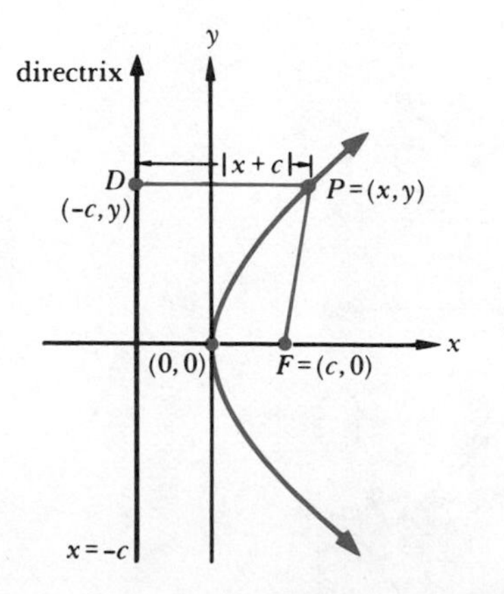

The line passing through the focus of the parabola perpendicular to the directrix is called the *axis of symmetry;* the point on the axis of symmetry midway between the directrix and focus is called the *vertex* of the parabola; the line segment with end points on the parabola and that is perpendicular to the axis at the focus is called a *focal chord* (or *latus rectum*) of the parabola. In Figure 3, V is the vertex, F is the focus, and $\overline{AB}$ is the focal chord.

Figure 3

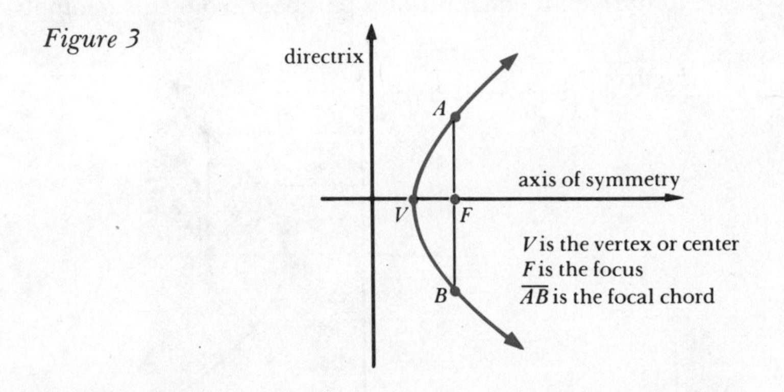

Notice that the graph of $y^2 = 4cx$ has the x axis as its axis of symmetry. Also, no point of the graph lies to the left of the y axis, for, if $x < 0$, we have $y^2 < 0$ (why?) and thus y is not real. The directrix is parallel to the y axis, and the focus is a point to the right of the vertex. This case, which might be described by saying that the parabola "opens to the right," is one of four possible cases that could have been developed. These cases are as follows (assume that $c > 0$ in each case):

CASE 1 $y^2 = 4cx$

The focus is at $(c, 0)$ on the x axis, the directrix is parallel to the y axis, and the vertex is $(0, 0)$. The parabola opens to the right (Figure 4).

CASE 2 $y^2 = -4cx$

The focus is at $(-c, 0)$ on the x axis, the directrix is parallel to the y axis, and the vertex is $(0, 0)$. The parabola opens to the left (Figure 5).

Figure 4 *Figure 5*

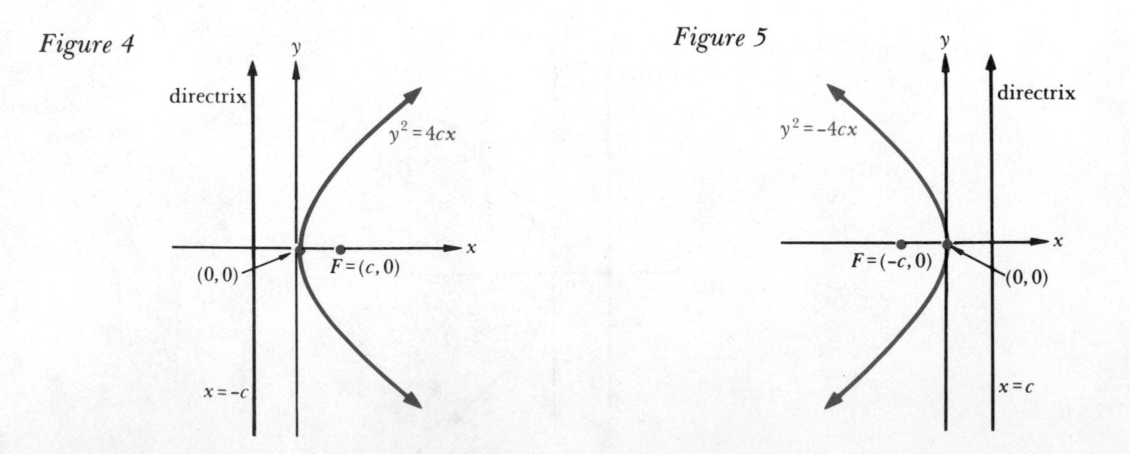

CASE 3 $x^2 = 4cy$

The focus is at $(0, c)$ on the y axis, the directrix is parallel to the x axis, and the vertex is $(0, 0)$. The parabola opens upward (Figure 6).

CASE 4 $x^2 = -4cy$

The focus is at $(0, -c)$ on the y axis, the directrix is parallel to the x axis, and the vertex is $(0, 0)$. The parabola opens downward (Figure 7).

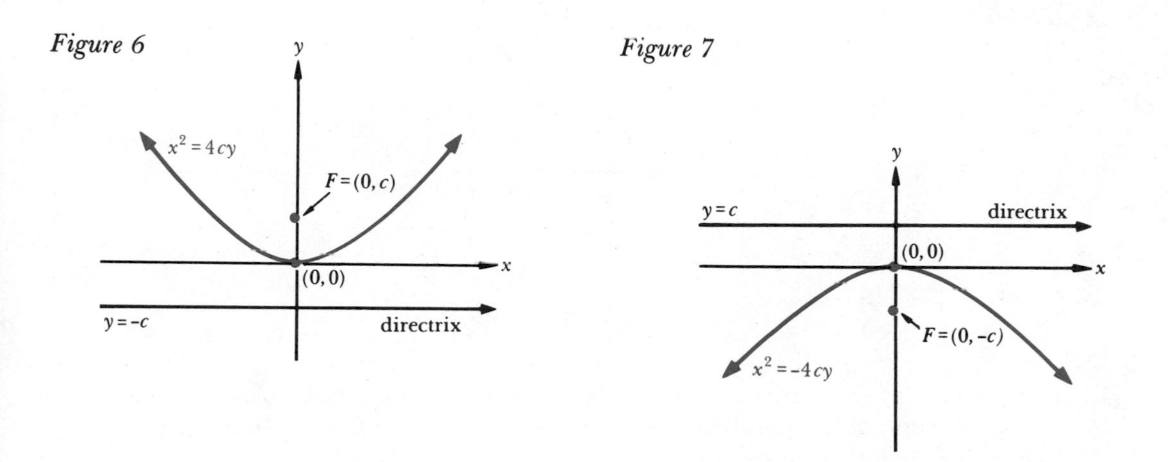

Figure 6

Figure 7

EXAMPLES 1 Find the focus and directrix of the parabola $x^2 = 12y$ and sketch its graph.

SOLUTION This is an example of Case 3. The graph is symmetric with respect to the y axis because $(-x, y)$ lies on the graph whenever (x, y) lies on the graph. Since x^2 is always nonnegative, then $y \geqslant 0$, and the parabola opens upward. Since $4c = 12$, then $c = 3$; therefore, the focus is $(0, 3)$, and the equation of the directrix is $y = -3$ (Figure 8).

Figure 8

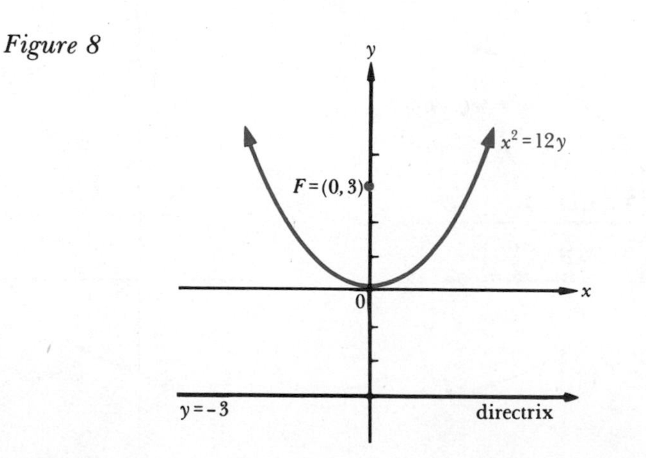

2 A parabola with its vertex at the origin has its focus at (6, 0). Find an equation of the parabola and sketch its graph.

SOLUTION Since the focus is (6, 0), the directrix is the line $x = -6$. The axis of symmetry in this case is the x axis; hence an equation of the parabola is $y^2 = 4cx$. Since $c = 6$, the equation is $y^2 = 24x$. This is an example of Case 1. The graph is shown in Figure 9.

Figure 9

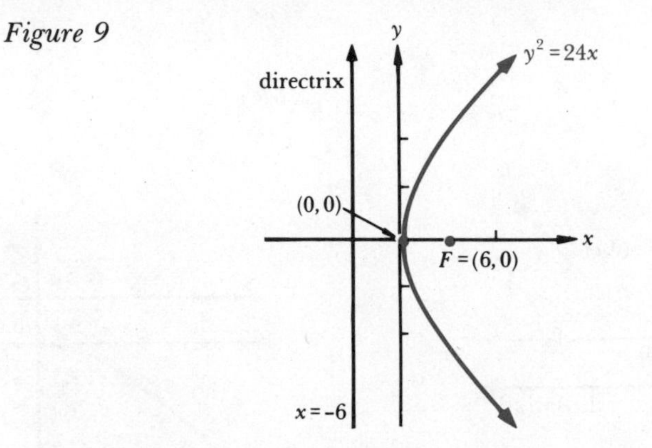

If the vertex of a parabola is translated to the point (h, k), then an equation of the parabola can take any one of the following four forms (assume that $c > 0$ in each case):

CASE 1 $(y - k)^2 = 4c(x - h)$
The vertex of the parabola is (h, k), the focus is $(h + c, k)$, and the directrix is the line $x = h - c$ (Figure 10).

CASE 2 $(y - k)^2 = -4c(x - h)$
The vertex of the parabola is (h, k), the focus is $(h - c, k)$, and the directrix is the line $x = h + c$ (Figure 11).

Figure 10 *Figure 11*

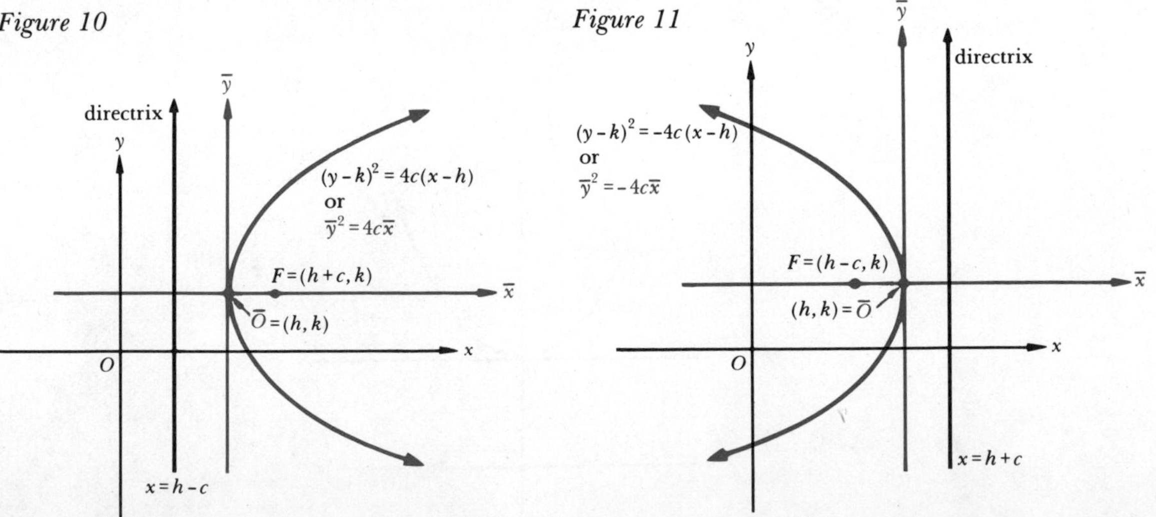

CASE 3 $(x - h)^2 = 4c(y - k)$

The vertex of the parabola is (h, k), the focus is $(h, k + c)$, and the directrix is the line $y = k - c$ (Figure 12).

CASE 4 $(x - h)^2 = -4c(y - k)$

The vertex of the parabola is (h, k), the focus is $(h, k - c)$, and the directrix is the line $y = k + c$ (Figure 13).

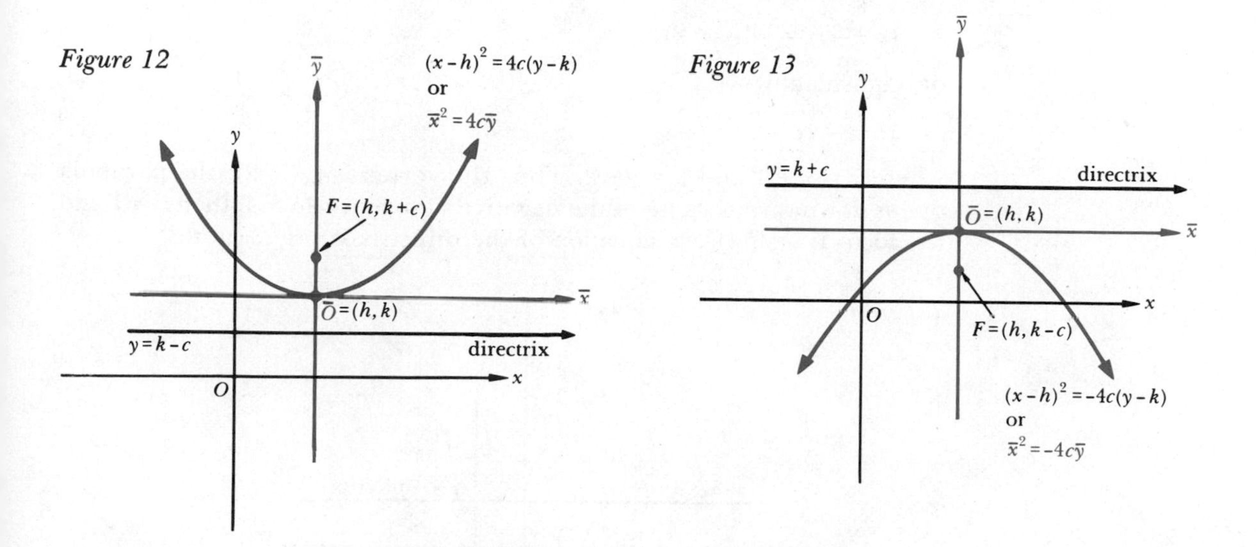

EXAMPLES 1 For the parabola $y^2 + 2y - 8x - 3 = 0$, find the vertex, the focus, an equation of the directrix, and the length of the focal chord. Also, sketch the graph.

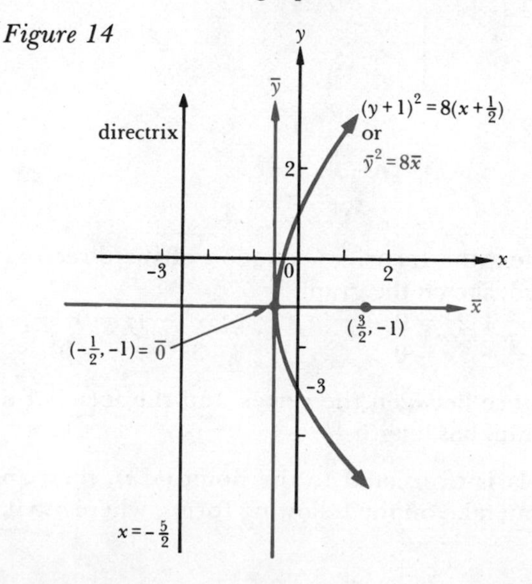

SOLUTION (See Figure 14.) Complete the square in $y^2 + 2y - 8x - 3 = 0$ to get

$$\begin{aligned} y^2 + 2y + 1 &= 8x + 3 + 1 \\ &= 8x + 4 \end{aligned}$$

Hence,

$$(y + 1)^2 = 8(x + \tfrac{1}{2})$$

or, equivalently,

$$\bar{y}^2 = 8\bar{x}$$

where $\bar{y} = y + 1$ and $\bar{x} = x + \frac{1}{2}$; therefore, the vertex is $(-\frac{1}{2}, -1)$. Since $4c = 8$, it follows that $c = 2$; hence, we have the focus at $(-\frac{1}{2} + 2, -1) = (\frac{3}{2}, -1)$. An equation of the directrix is $x = -\frac{1}{2} - 2 = -\frac{5}{2}$. The length of the focal chord (see Problem 7) is $4c = 8$

2 Consider the equation of a parabola $x^2 + 2x + 4y - 7 = 0$. Find the vertex, the focus, and an equation of the directrix. Also, sketch the graph.

SOLUTION (See Figure 15.) After completing the square, we have

$$\begin{aligned} x^2 + 2x + 1 &= -4y + 7 + 1 \\ &= -4y + 8 \end{aligned}$$

Hence,

$$(x + 1)^2 = -4(y - 2)$$

or, equivalently,

$$\bar{x}^2 = -4\bar{y}$$

where $\bar{x} = x + 1$ and $\bar{y} = y - 2$. Thus the vertex is $(-1, 2)$; the parabola opens downward because of the negative sign. Since $4c = 4$, then $c = 1$ and the focus is $(-1, 1)$. An equation of the directrix is $y = 2 + 1 = 3$.

Figure 15

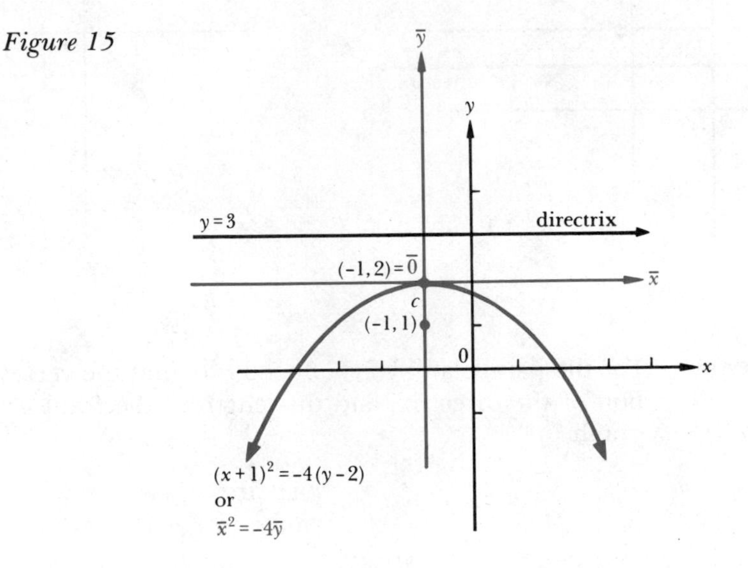

PROBLEM SET 4

In Problems 1–6, find the vertex, the focus, an equation of the directrix, and the length of the focal chord; sketch the graph.

1 $y^2 = 8x$	2 $x^2 + 2y = 0$	3 $x^2 - 4y = 0$
4 $y^2 + 4x = 0$	5 $y^2 + 5x = 0$	6 $3x^2 - 2y = 0$

7 Show that, if c is the distance between the vertex and the focus of a parabola, then its latus rectum has length $4c$.

8 If the vertex of a parabola is translated to the point (h, k), then an equation of the parabola can take on the following forms, where $c > 0$.

(a) $(y - k)^2 = 4c(x - h)$
(b) $(y - k)^2 = -4c(x - h)$
(c) $(x - h)^2 = 4c(y - k)$
(d) $(x - h)^2 = -4c(y - k)$
Sketch the graph of each parabola when $(h, k) = (3, 1)$ and $c = 2$.

In Problems 9–14, find the vertex, the focus, an equation of the directrix, and the length of the focal chord; sketch the graph.

9 $(x - 2)^2 = -6(y - 1)$
10 $(y + 3)^2 = 4(x + 1)$
11 $y^2 + 2x + 2y + 7 = 0$
12 $2y^2 + 2x + 6y + 9 = 0$
13 $3x^2 + 12x - 6y + 13 = 0$
14 $x^2 + 4x + 5y - 11 = 0$

In Problems 15–18, find an equation of the parabola in each case.

15 Vertex at (1, 2) and focus at (3, 2)
16 Vertex at (3, 4) and directrix $x = 1$
17 Focus at (0, 4) and directrix $x = 2$
18 Vertex at (2, −5) and directrix $y = 3$

19 Find an equation of the parabola with vertex at (1, 3), containing the point (5, 7), and whose axis is parallel to the y axis.

20 Find an equation of the parabola with focus (−3, −5) and directrix $x = 5$.

21 Find an equation of the line through the points on the parabola $y^2 = 3x$ whose ordinates are 2 and 3.

22 A bridge 400 meters long is held up by a parabolic main cable (Figure 16). The main cable is 100 meters above the roadway at the ends and 4 meters above the roadway at the center. Vertical supporting cables run at 50-meter intervals along the roadway. Find the lengths of these vertical cables. (*Hint:* Set up an xy coordinate system with vertical y axis and having the vertex of the parabola 4 units above the origin.)

Figure 16

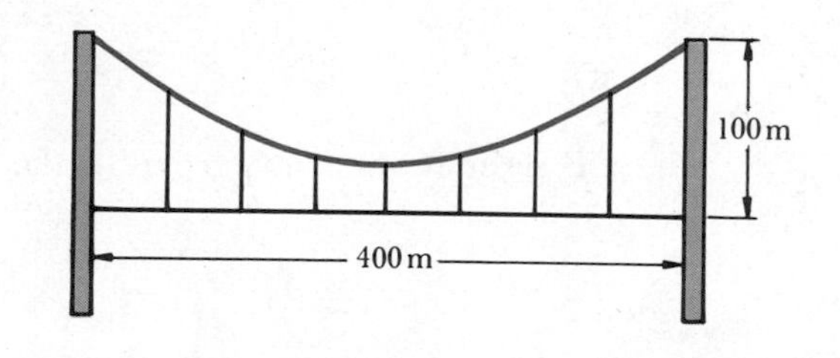

23 The surface of a roadway over a stone bridge follows a parabolic curve with the vertex in the middle of the bridge. The span of the bridge is 60 feet and the road surface is 1 foot higher in the middle than at the ends. How much higher than the ends is a point on the roadway 15 feet from an end?

5 Conics

We have developed *analytically* equations of the circle, the ellipse, the hyperbola, and the parabola, and we have also discussed the properties of their graphs. In our approach, we derived the equations using geometric properties. In this section we shall give a general approach that simultaneously applies to the ellipse, the hyperbola, and the parabola. These three graphs are called *conics* because they are determined by the intersections of planes with cones as illustrated in Figure 1.

Figure 1

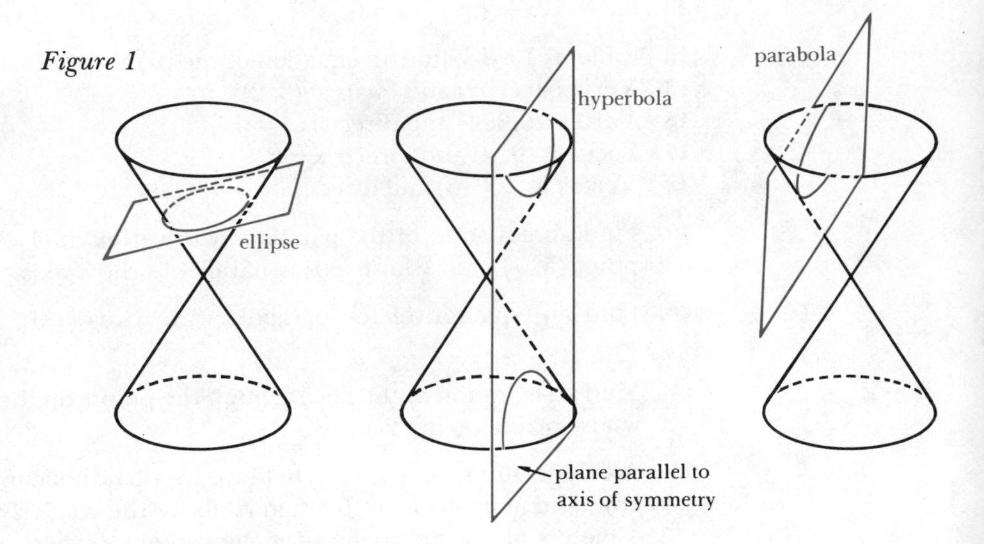

DEFINITION 1 CONIC

A *conic* is determined by a given point F (a *focus*), a given line d not containing F (the *directrix* associated with F), and a positive number e (the *eccentricity*). The conic contains a point P if and only if

$$\frac{|\overline{FP}|}{|\overline{PD}|} = e$$

where D is the foot of the perpendicular from P to D (Figure 2).

Figure 2

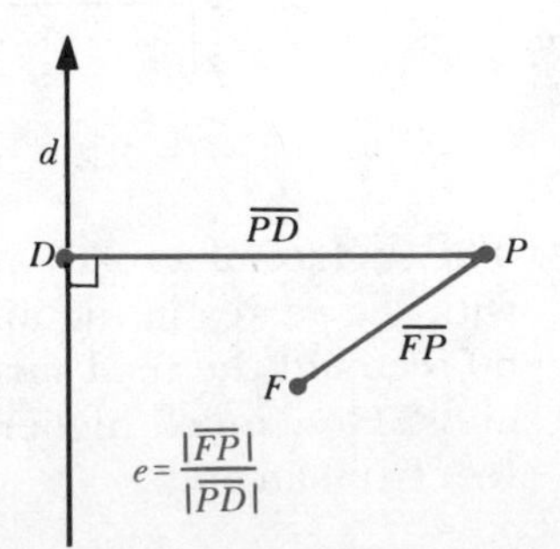

If $|\overline{FP}| = r_1$ and $|\overline{PD}| = r_2$, then

$$\frac{|\overline{FP}|}{|\overline{PD}|} = \frac{r_1}{r_2} = e$$

EXAMPLES 1 Assume that eccentricity $e = 1$, the focus is $(-2, 0)$, and the directrix is $x = 3$. Find an equation of the conic and sketch the conic.

SOLUTION (See Figure 3.) If $P = (x, y)$ is a point of the conic, then $\frac{r_1}{r_2} = 1$ (why?); that is, $r_1 = r_2$ or, equivalently, $r_1^2 = r_2^2$. By the distance formula,

$$(x + 2)^2 + y^2 = (x - 3)^2$$

so that, the conic is the parabola

$$y^2 = -10(x - \tfrac{1}{2})$$

Figure 3

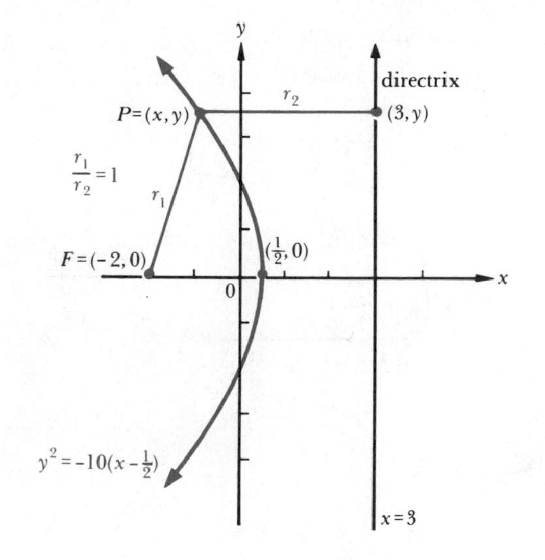

Figure 4

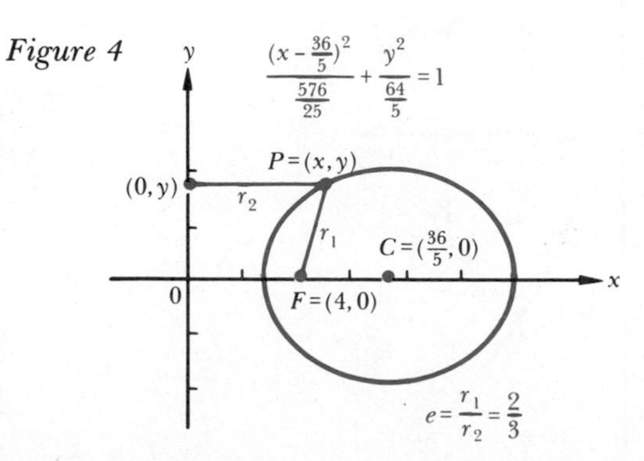

2 Find an equation of the conic that has eccentricity $e = 2/3$, the point $(4, 0)$ as a focus, and the y axis as its directrix.

SOLUTION (See Figure 4.) By the definition of a conic, we have $e = \frac{r_1}{r_2} = \frac{2}{3}$ or, equivalently, $3r_1 = 2r_2$. Since $r_1^2 = (x - 4)^2 + y^2$ and $r_2^2 = x^2$, then the expression $9r_1^2 = 4r_2^2$ implies that

$$9(x^2 - 8x + 16 + y^2) = 4x^2$$

or

$$5x^2 - 72x + 9y^2 + 144 = 0$$

However, upon completing the square, we have

$$5\left[x^2 - \frac{72}{5}x + \left(\frac{36}{5}\right)^2\right] + 9y^2 = -144 + \frac{(36)^2}{5}$$

That is,

$$5\left(x - \frac{36}{5}\right)^2 + 9y^2 = \frac{576}{5}$$

Hence the conic is the ellipse

$$\frac{\left(x - \frac{36}{5}\right)^2}{\frac{576}{25}} + \frac{y^2}{\frac{64}{5}} = 1$$

3 Find an equation of the conic with focus (2, 0), directrix $x = 0$, and with eccentricity $e = 2$.

SOLUTION (See Figure 5.) Since $e = r_1/r_2 = 2$, $r_1 = 2r_2$; that is, $r_1{}^2 = 4r_2{}^2$, so that $(x - 2)^2 + y^2 = 4x^2$ (why?) or, equivalently, $3x^2 + 4x - y^2 = 4$. After completing the square, we get

$$3\left(x + \frac{2}{3}\right)^2 - y^2 = \frac{16}{3}$$

or, equivalently, the hyperbola

$$\frac{\left(x + \frac{2}{3}\right)^2}{\frac{16}{9}} - \frac{y^2}{\frac{16}{3}} = 3$$

Figure 5

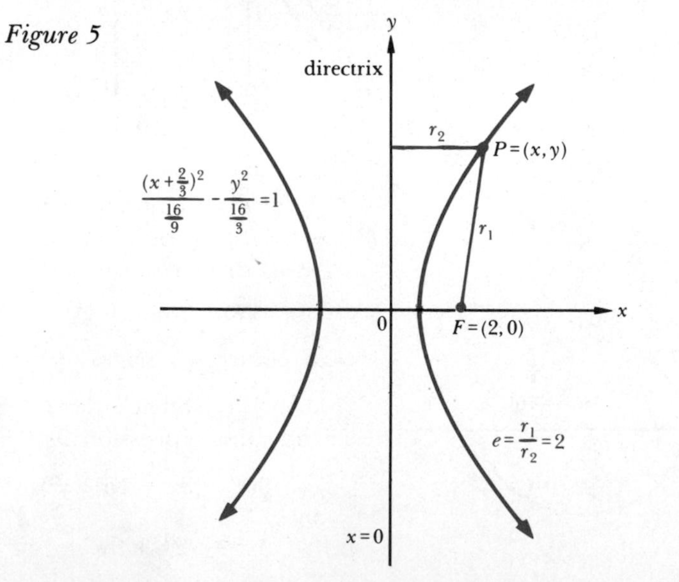

From the geometric definition of a *parabola* (see Section 4), it can be seen that the eccentricity $e = 1$ (Example 1 above). Also, it can be shown that if $e < 1$, the conic is an ellipse (Example 2), and if $e > 1$, the conic is a hyperbola (Example 3).

Finally, we state the following theorem without proof.

THEOREM 1

Consider ellipses and hyperbolas of the form

$$\frac{x^2}{a^2} \pm \frac{y^2}{b^2} = 1$$

where $c^2 = a^2 - b^2$ for the ellipse and $c^2 = a^2 + b^2$ for the hyperbola. The eccentricity $e = \frac{c}{a}$ and equations of the directrices are given by

$$x = -\frac{a^2}{c} \quad \text{and} \quad x = \frac{a^2}{c}$$

EXAMPLES 1 Find the eccentricity and the directrices of the hyperbola whose equation is $x^2 - 2y^2 = 8$.

SOLUTION (See Figure 6.) The equation $x^2 - 2y^2 = 8$ in standard form is $\frac{x^2}{8} - \frac{y^2}{4} = 1$. This is an equation of a hyperbola where $a = 2\sqrt{2}$ and $b = 2$; thus, $c = \sqrt{a^2 + b^2} = \sqrt{8 + 4} = 2\sqrt{3}$. Thus, by Theorem 1, the eccentricity

$$e = \frac{c}{a} = \frac{2\sqrt{3}}{2\sqrt{2}} = \frac{\sqrt{6}}{2}$$

Equations of the directrices are

$$x = -\frac{a^2}{c} = -\frac{8}{2\sqrt{3}} = -\frac{4\sqrt{3}}{3} \quad \text{and} \quad x = \frac{a^2}{c} = \frac{8}{2\sqrt{3}} = \frac{4\sqrt{3}}{3}$$

Figure 6

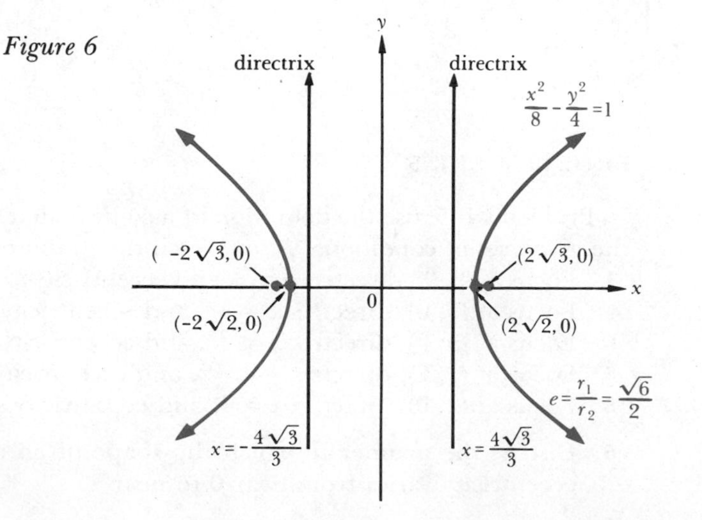

2 Find the eccentricity of the ellipse whose equation is $4x^2 + 9y^2 = 36$.

SOLUTION (See Figure 7.) The equation $4x^2 + 9y^2 = 36$ in standard form is $\frac{x^2}{9} + \frac{y^2}{4} = 1$. This is an equation of an ellipse where $a = 3$ and $b = 2$, so that we have $c = \sqrt{a^2 - b^2} = \sqrt{9 - 4} = \sqrt{5}$. Thus, by Theorem 1, the eccentricity

Figure 7

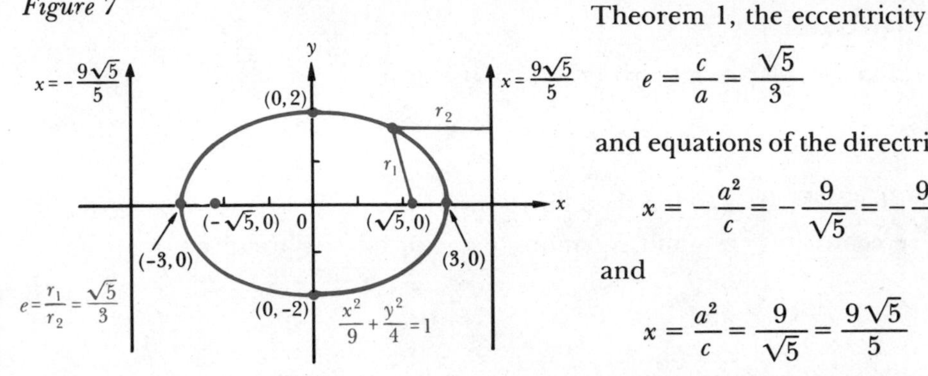

$$e = \frac{c}{a} = \frac{\sqrt{5}}{3}$$

and equations of the directrices are

$$x = -\frac{a^2}{c} = -\frac{9}{\sqrt{5}} = -\frac{9\sqrt{5}}{5}$$

and

$$x = \frac{a^2}{c} = \frac{9}{\sqrt{5}} = \frac{9\sqrt{5}}{5}$$

3 Find an equation of the parabola that has focus (1, 3) and directrix $y = -2$, and then graph the parabola.

Figure 8

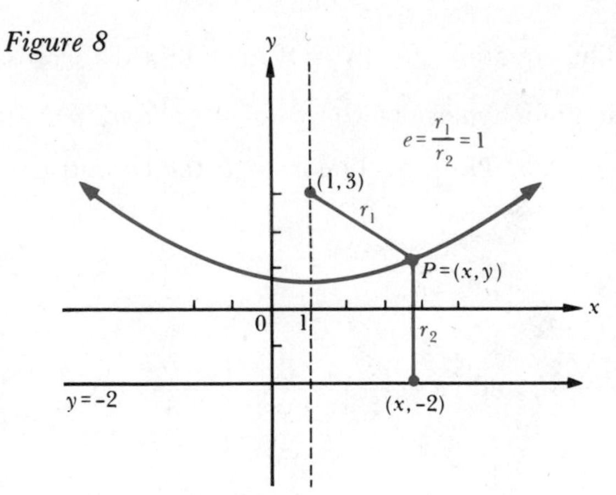

SOLUTION (See Figure 8.) The conic is a parabola, so the eccentricity $e = 1$, so that $r_1^2 = r_2^2$ (why?); that is,

$$(x - 1)^2 + (y - 3)^2 = (y + 2)^2$$

Hence

$$(x - 1)^2 = 4(\tfrac{5}{2})(y - \tfrac{1}{2})$$

Therefore, an equation of the parabola is

$$(x - 1)^2 = 10(y - \tfrac{1}{2})$$

PROBLEM SET 5

In Problems 1–5, use the definition of a conic to determine an equation for the given set of conditions. Also, sketch the graph of the conic.

1 Focus at (3, 0), directrix $y = \frac{5}{3}$, and eccentricity $e = 3/\sqrt{5}$

2 Focus at (5, 0), directrix $x = \frac{16}{5}$, and eccentricity $e = \frac{5}{4}$

3 Focus at (3, 0), directrix $x = \frac{25}{3}$, and eccentricity $e = \frac{3}{5}$

4 Focus at (1, 2), directrix $y = -2$, and eccentricity $e = \frac{1}{2}$

5 Focus at (2, 0), directrix $x = 4$, and eccentricity $e = 1$

6 Discuss the manner in which the shape of an ellipse changes as the eccentricity varies from near 0 to near 1.

In Problems 7–14, find the eccentricity and equations of the directrices for the given conic.

7 $25x^2 + 16y^2 = 400$

8 $x^2 + 2y^2 = 1$

9 $x^2 + 3y^2 = 4$

10 $y^2 - 8y + 3x + 5 = 0$

11 $9x^2 - 16y^2 + 144 = 0$

12 $4x^2 - 4y^2 + 1 = 0$

13 $3x^2 - 5x + y^2 + 22y = 1$

14 $x^2 + 16x - y + 7 = 0$

15 Given the equation $25x^2 - 4y^2 + 50x - 12y + 116 = 0$, find the center, the foci, the vertices, equations of the asymptotes, and the eccentricity. Also, sketch the graph.

16 We know that the eccentricity $e = \dfrac{c}{a} < 1$ for the ellipse $\dfrac{x^2}{a^2} + \dfrac{y^2}{b^2} = 1$, where $c^2 = a^2 - b^2$. Now, as e becomes closer and closer to zero in value, $c/a = e$ approaches zero; hence c^2 approaches zero or a approaches b in value (why?). What is the shape of an ellipse in which a and b are close in value? What if $a = b$? Does this give an indication of how to define a conic with eccentricity $e = 0$? Use sketches and examples to answer these questions.

17 Suppose that the foci of a particular ellipse lie midway between the center and the vertices.

(a) Find the eccentricity.

(b) If the length of the major axis is $2a$, find the length of the minor axis and the distance from the center to the directrices.

(c) Sketch the graph.

18 Except for minor perturbations, the orbit of the earth is an ellipse with the sun at one focus. The least and greatest distances from the earth to the sun have a ratio of $\frac{29}{30}$. Find the eccentricity of this elliptical orbit.

REVIEW PROBLEM SET

In Problems 1–3, find an equation of the circle in each of the given cases and sketch the graph.

1 Center at $(4, -3)$ and radius 5

2 Tangent to the y axis and center at $(4, -3)$

3 Center at $(1, 2)$ and containing the point $(3, 1)$

4 If $\bar{x} = x - 2$ and $\bar{y} = y + 4$, find the xy coordinates of the point whose $\bar{x}\bar{y}$ coordinates are

(a) $(-5, 2)$ (b) $(5, -3)$ (c) $(-7, -1)$

5 If $\bar{x} = x + 3$, and $\bar{y} = y - 1$, find the $\bar{x}\bar{y}$ coordinates of the point whose xy coordinates are

(a) $(6, 1)$ (b) $(-2, 9)$ (c) $(-1, -4)$

6 The end points of the base of a triangle are $(-2, 0)$ and $(2, 0)$, and the sum of the lengths of the other two sides is 6. Find an equation of the set of all points that are possible vertices.

In Problems 7–14, find an equation of the ellipse satisfying each given set of conditions and sketch the graph.

7 Vertices at $(-5, 0)$ and $(5, 0)$ and foci at $(-3, 0)$ and $(3, 0)$
8 Foci at $(-1, 0)$ and $(1, 0)$ and length of minor axis $2\sqrt{2}$
9 Foci at $(0, -6)$ and $(0, 6)$ and eccentricity $e = \frac{1}{2}$
10 Vertices at $(-5, 0)$ and $(5, 0)$ and eccentricity $e = \frac{3}{5}$
11 Focus at $(2, 0)$, eccentricity $e = \frac{2}{3}$, and directrix $2x - 9 = 0$
12 Focus at $(0, -4)$, eccentricity $e = \frac{2}{3}$, and directrix $y + 9 = 0$
13 Vertex at $(10, 0)$, center at $(0, 0)$, and eccentricity $e = \frac{3}{5}$
14 Center at $(0, 0)$, vertex at $(0, 3)$, and eccentricity $e = \frac{2}{5}$

In Problems 15–19, find the coordinates of the center, the vertices, and the foci of the given ellipse. Also find the eccentricity and equations of the directrices.

15 $\frac{x^2}{8} + \frac{y^2}{12} = 1$
16 $144x^2 + 169y^2 = 24{,}336$
17 $9x^2 + 25y^2 + 18x - 50y - 191 = 0$
18 $3x^2 + 4y^2 - 28x - 16y + 48 = 0$
19 $9x^2 + 4y^2 + 72x - 48y + 144 = 0$

20 Show that the eccentricity e of the ellipse whose equation is given by $b^2x^2 + a^2y^2 = a^2b^2$, for $a > b$, satisfies $b^2 = a^2(1 - e^2)$.

In Problems 21–24, find an equation of the hyperbola in each case and sketch the graph.

21 Vertices at $(-2, 0)$ and $(2, 0)$ and foci at $(-3, 0)$ and $(3, 0)$
22 Foci at $(-10, 0)$ and $(10, 0)$ and eccentricity $e = \frac{5}{4}$
23 Vertices at $(-15, 0)$ and $(15, 0)$ and asymptotes $5y = \pm 4x$
24 Containing the point $(5, 9)$ and with asymptotes $y = \pm x$

In Problems 25–28, find the coordinates of the center, the vertices, and the foci of the given hyperbola. Also find the eccentricity and equations of the asymptotes. Sketch the graph.

25 $x^2 - 9y^2 = 72$
26 $y^2 - 9x^2 = 54$
27 $x^2 - 4y^2 + 4x + 24y - 48 = 0$
28 $4y^2 - x^2 - 24y + 2x + 34 = 0$

In Problems 29–32, find an equation of the parabola in each case.

29 Focus at $(-4, 0)$ and directrix $y = 3$
30 Focus at $(0, 3)$ and directrix $x = \frac{5}{2}$
31 Focus at $(1, -\frac{5}{2})$ and directrix $y = -4$
32 Focus at $(3, 0)$, containing point $(2, 2\sqrt{2})$, the x axis as the axis of symmetry, and opening to the right

33 Find an equation of the line containing the points with y coordinates of 2 and 8, respectively, that lie on the parabola $y^2 = 8x$.

34 Consider the equation $(b^2 - k)x^2 + (a^2 - k)y^2 = 1$, where $a \geq b$.
(a) For what choices of k is the graph an ellipse?
(b) For what choices of k is the graph a hyperbola?
(c) For what choices of k is the graph a circle?

35 Show that the hyperbolas whose equations are

$$\frac{x^2}{4} - \frac{y^2}{9} = 1 \quad \text{and} \quad \frac{y^2}{9} - \frac{x^2}{4} = 1$$

have the same set of asymptotes, and sketch them on the same coordinate system.

36 A square with sides parallel to the coordinate axes is inscribed in the ellipse $9x^2 + 16y^2 = 100$. Find the coordinates of the vertices and the area of the square.

37 Show that if a point P is equidistant from the y axis and the point $(6, 0)$, then its coordinates (x, y) must satisfy the equation $y^2 = 12(x - 3)$. Is the converse true?

38 Let d_1 be the distance from point $P = (x, y)$ to point $A = (-10, 0)$ and d_2 be the distance from point $P = (x, y)$ to point $B = (10, 0)$. Show that if $d_1 - d_2 = 12$, then the coordinates of P must satisfy the equation $\frac{x^2}{36} - \frac{y^2}{64} = 1$. Is it true that every point P whose coordinates satisfy this equation has $d_1 - d_2 = 12$?

39 Let d_1 be the distance from the point $P = (x, y)$ to the point $A = (1, 0)$, and d_2 be the distance from the point $P = (x, y)$ to the point $B = (9, 0)$. Derive an equation that the coordinates of point P must satisfy if $d_1 + d_2 = 10$.

40 Write an equation of the ellipse that contains the point $(3, 1)$ and whose center is at the origin, whose axes of symmetry are the coordinate axes, and whose major axis is three times the length of the minor axis.

41 The earth moves in an elliptical orbit with eccentricity 0.017, major axis 185.8 million miles, and the sun at one focus. When the earth is positioned on the major axis, how close is it to the sun?

42 Except for minor perturbations, a satellite orbiting the earth moves in an elliptical orbit with the center of the earth at one focus. Suppose that a satellite at perigee (the point of its orbit nearest to the center of the earth) is 400 kilometers from the surface of the earth and at apogee (the point in its orbit farthest from the center of the earth) is 600 kilometers from the surface of the earth. Assume that the earth is a sphere of radius 6371 kilometers. Find the semi-minor axis b of the elliptical orbit (one-half the minor axis).

43 Two points are 2000 feet apart. At one of these points the report of a cannon is heard 1 second later than at the other. By means of the definition of a hyperbola, show that the cannon is somewhere on a particular hyperbola, then after making a suitable choice of axes, write its equation. (Consider the velocity of sound to be 1100 feet per second.)

CHAPTER 8

Systems of Equations and Linear Programming

In Chapter 3, Section 1, we established the fact that two nonparallel lines intersect if they have different slopes. To find the point of intersection, it is necessary to solve a system of linear equations. In this chapter we begin with two methods for solving these systems—the *substitution method* and the *elimination method.* Next, we see how it is possible to solve linear systems by using matrices. In addition, determinants are introduced and then applied to solving linear systems by using *Cramer's rule.* A survey of linear programming and the solution of systems involving second-degree equations is also presented.

1 Systems of Linear Equations

The graphs of linear functions can be used to investigate the common solutions of linear equations. A set of two or more linear equations is called a *system of linear equations.* The *solution set* of a linear system containing two equations with two variables is the set of all ordered pairs of numbers that satisfy simultaneously each of the equations in the system. (To *solve* a system of linear equations means to find the solution set of the system.)

Geometrically, this set of ordered pairs of numbers corresponds to the points of intersection of the graphs of the two linear equations, provided that such points exist. There are three possible situations.

1 If the graphs of the two equations intersect at one point, we say that the system is *independent.*

For example, let us consider the linear system

$$\begin{cases} 3x + 4y = 12 \\ 3x - 8y = 0 \end{cases}$$

The graphs of the two corresponding linear functions, $y = -\frac{3}{4}x + 3$ and $y = \frac{3}{8}x$, are shown in Figure 1. The solution set of the system is $\{(\frac{8}{3}, 1)\}$, and so the point of intersection of the two linear equations is $(\frac{8}{3}, 1)$.

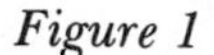
Figure 1

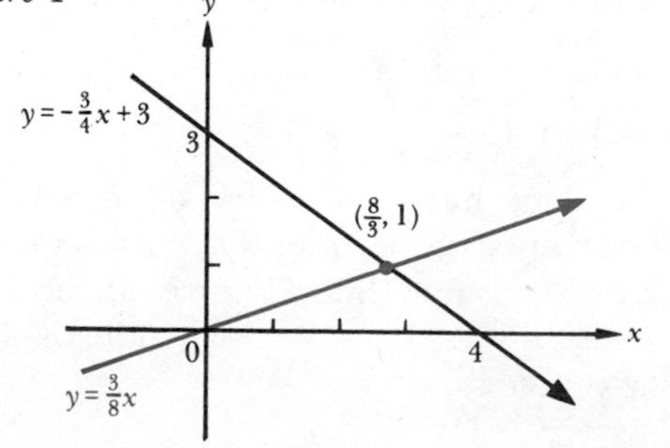

2 If the graphs of linear equations are parallel lines, the intersection is the null (empty) set and we say the system is *inconsistent*.

For example, consider the linear system

$$\begin{cases} x + 2y = 4 \\ 3x + 6y = -3 \end{cases}$$

Note that if the two lines are expressed in the slope-intercept form, the system becomes

$$\begin{cases} y = -\frac{1}{2}x + 2 \\ y = -\frac{1}{2}x - \frac{1}{2} \end{cases}$$

The slopes are equal but the intercepts are different, so we have two parallel lines (Figure 2). Since no point of intersection exists, the solution set is empty and the system is inconsistent.

Figure 2

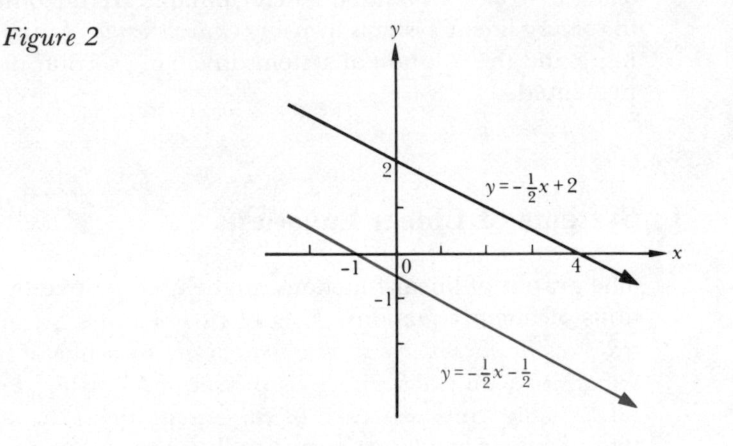

3 If the graphs of two linear equations coincide, the intersection is every point of each line. In this case the system is said to be *dependent*.

For example, the linear system

$$\begin{cases} 3x - y = -1 \\ 6x - 2y = -2 \end{cases}$$

can be written in the slope-intercept form as

$$\begin{cases} y = 3x + 1 \\ y = 3x + 1 \end{cases}$$

Here the lines have the same slope with the same intercepts; consequently, each equation represents the same line (Figure 3), and the solution set is $\{(x, y) \mid y = 3x + 1\}$. This is a dependent system.

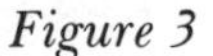
Figure 3

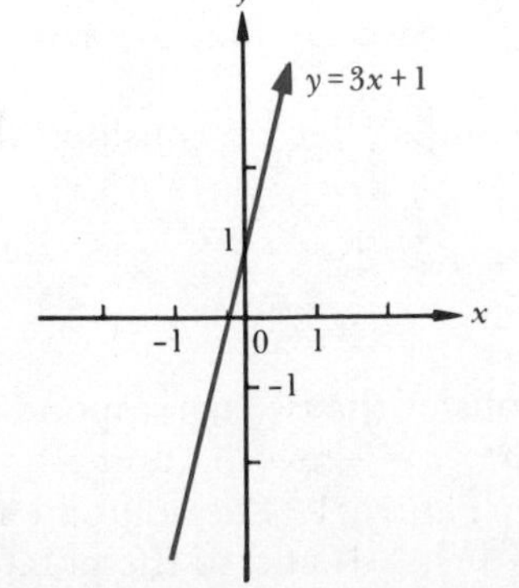

Although the *type* of solution set of a linear system of two equations in two variables can be determined by investigating their graphs, it is usually difficult to arrive at the solution by using the graphs. For this reason, we now introduce two *algebraic methods* for solving linear systems—the *substitution method* and the *elimination method.*

1.1 The Substitution Method

To solve a linear system of two equations in *two* variables by the substitution method, we use one of the equations to express one of the variables in terms of the other, and then substitute this expression into the remaining equation to obtain an equation in *one* variable.

To illustrate the substitution method, consider the system

$$\begin{cases} 3x + y = -1 \\ x + 2y = 8 \end{cases}$$

We begin by solving for y in terms of x in the first equation. This gives us $y = -3x - 1$. Next, we substitute this latter expression for y into the second equation to obtain

$$x + 2(-3x - 1) = 8$$

which simplifies to

$$x - 6x - 2 = 8 \quad \text{or} \quad x = -2$$

To find the corresponding value of y, we replace x by -2 in the equation $y = -3x - 1$ to get $y = -3(-2) - 1 = 5$.

Hence, the solution set is $\{(-2, 5)\}$.

EXAMPLES 1 Find the solution set of the following linear systems by using the substitution method.

$$\begin{cases} 4x - y = 2 \\ 3x + y = 5 \end{cases}$$

SOLUTION Solving the first equation for y in terms of x yields $y = 4x - 2$. Replacing y with $4x - 2$ in the second equation of the system gives us $3x + (4x - 2) = 5$, so $7x = 7$ or $x = 1$. The corresponding value of y is obtained by substituting 1 for x in the equation $y = 4x - 2$, so that

$$y = 4(1) - 2 = 2$$

CHECK To check the solution, replace x by 1 and y by 2 in the original system to get

$$\begin{cases} 4(1) - (2) = 2 \\ 3(1) + (2) = 5 \end{cases}$$

Thus, the solution set is $\{(1, 2)\}$.

2 A farmer sold one grade of wheat at $4.20 a bushel and another grade of wheat at $5.50 a bushel. How many bushels of each kind did he sell if he received $8251 for 1850 bushels?

SOLUTION Let x represent the number of bushels sold at $4.20 and y represent the number of bushels sold at $5.50. Then

$$\begin{cases} x + y = 1850 \\ 4.20x + 5.50y = 8251 \end{cases}$$

Solving the first equation for x, we have $x = 1850 - y$. Next, by substituting $1850 - y$ for x in the second equation, we obtain

$$4.20(1850 - y) + 5.50y = 8251$$

so that

$$y = 370$$

Thus the corresponding value of x is

$$x = 1850 - y = 1850 - 370 = 1480$$

Hence, there were 370 bushels sold at $5.50 and 1480 bushels sold at $4.20.

Equations such as $3x + 4y - 2z = 5$ and $5x_1 - 7x_2 + 9x_3 = 11$ are called *linear equations in three variables.* Systems of linear equations involving three variables can also be solved by the method of substitution. The following example illustrates this technique.

EXAMPLE Find the solution set of the system

$$\begin{cases} x + y + z = 3 \\ x - 2y + 4z = 6 \\ 2x - y + z = 11 \end{cases}$$

SOLUTION Solve the first equation for x in terms of y and z to get

$$x = 3 - y - z$$

Replacing x by $3 - y - z$ in the second and third equations yields the following linear system in two variables.

$$\begin{cases} (3 - y - z) - 2y + 4z = 6 \\ 2(3 - y - z) - y + z = 11 \end{cases}$$

or

$$\begin{cases} -y + z = 1 \\ -3y - z = 5 \end{cases}$$

Applying the substitution method to the latter linear system, we have $z = y + 1$, so $-3y - (y + 1) = 5$ or $y = -\frac{3}{2}$, and $z = (-\frac{3}{2}) + 1 = -\frac{1}{2}$. Thus, $x = 3 - (-\frac{3}{2}) - (-\frac{1}{2}) = 5$.

The solution set of the original linear system is $\{(5, -\frac{3}{2}, -\frac{1}{2})\}$.

1.2 The Elimination Method

In the *elimination method,* the addition of two equations results in an equation in which only one of the variables appears.

For example, consider the system

$$\begin{cases} 2x + y = 4 \\ x - y = 5 \end{cases}$$

Adding the corresponding sides of the two equations yields $3x = 9$ (note that y has been eliminated), so that $x = 3$. To find y, we substitute 3 for x in either equation to get $y = -2$. Thus, the solution set is $\{(3, -2)\}$.

It should be noted that in applying the elimination method it is sometimes necessary first to multiply one or both equations by nonzero constants in order to be able to eliminate a variable. This is illustrated in the following examples.

EXAMPLES Use the elimination method to find the solution set of each linear system.

1 $$\begin{cases} x - y = 1 \\ 3x + 2y = 8 \end{cases}$$

SOLUTION Multiply the first equation by -3 to get

$$\begin{cases} -3x + 3y = -3 \\ 3x + 2y = 8 \end{cases}$$

Then, add the corresponding sides of the first equation and the second equation to obtain $5y = 5$ or $y = 1$. Substituting $y = 1$ into the first equation in the original system yields $x = 2$.

Thus, the solution set is $\{(2, 1)\}$.

2 $$\begin{cases} 3x - 2y = 3 \\ 2x + 5y = -1 \end{cases}$$

SOLUTION In order to be able to eliminate the variable x, multiply the first equation by 2 and the second equation by -3 to get

$$\begin{cases} 6x - 4y = 6 \\ -6x - 15y = 3 \end{cases}$$

Adding the corresponding sides of the first and second equations, we have $-19y = 9$ or $y = -\frac{9}{19}$. To find x, we substitute $-\frac{9}{19}$ for y in $3x - 2y = 3$ to get $3x - 2(-\frac{9}{19}) = 3$ or $x = \frac{13}{19}$.

Here, the solution set is $\{(\frac{13}{19}, -\frac{9}{19})\}$.

In the two examples above, we solved the system of linear equations in two variables by first eliminating one of the variables. This method is basically a scheme for replacing one system of linear equations with an "equivalent system" that is easier to solve. Two linear systems are said to be *equivalent* if they have the same solution set. By performing one or more of the following operations on a system of equations, we obtain an equivalent system.

1 Interchange the positions of two equations in the system.
2 Replace an equation in the system by a nonzero multiple of itself. (A *multiple* of an equation is the equation that results when we multiply both sides of the equation by the same number.)
3 Replace an equation in the system by the sum of a nonzero multiple of that equation and a nonzero multiple of another equation of the system.

These operations are illustrated in the following example.

EXAMPLE Use the elimination method to find the solution set of the system. Display all equations in the equivalent system in each step of the process.

$$\text{(A)} \quad \begin{cases} 2x_1 + x_2 - 2x_3 = 10 \\ 3x_1 + 2x_2 + 2x_3 = 1 \\ 5x_1 + 4x_2 + 3x_3 = 4 \end{cases}$$

SOLUTION Let us refer to the given system as system (A). First, replace the second equation in system (A) by the sum of -3 times the first equation and 2 times the second equation to obtain system (B):

$$\text{(B)} \quad \begin{cases} 2x_1 + x_2 - 2x_3 = 10 \\ x_2 + 10x_3 = -28 \\ 5x_1 + 4x_2 + 3x_3 = 4 \end{cases}$$

System (B) is *equivalent* to system (A). In the next step, replace the third equation of (B) by the sum of -5 times the first equation and 2 times the third equation to obtain the equivalent system (C):

$$\text{(C)} \quad \begin{cases} 2x_1 + x_2 - 2x_3 = 10 \\ x_2 + 10x_3 = -28 \\ 3x_2 + 16x_3 = -42 \end{cases}$$

Now, replace the third equation of (C) by the sum of the third equation and -3 times the second equation to get the equivalent system (D):

$$\text{(D)} \quad \begin{cases} 2x_1 + x_2 - 2x_3 = 10 \\ x_2 + 10x_3 = -28 \\ -14x_3 = 42 \end{cases}$$

Next, multiply the third equation by $-\frac{1}{14}$, and add -1 times the second equation to the first equation, to obtain the equivalent system (E):

$$\text{(E)} \quad \begin{cases} 2x_1 - 12x_3 = 38 \\ x_2 + 10x_3 = -28 \\ x_3 = -3 \end{cases}$$

Finally, multiply the first equation by $\frac{1}{2}$ to obtain the equivalent system (F):

$$\text{(F)} \quad \begin{cases} x_1 \qquad - \ 6x_3 = \ \ 19 \\ \qquad x_2 + 10x_3 = -28 \\ \qquad\qquad\quad x_3 = \ -3 \end{cases}$$

From the third equation, $x_3 = -3$, so, after substituting into the second equation, we get $x_2 = 2$; then, after substituting into the first equation, we obtain $x_1 = 1$. Thus, the solution set is $\{(1, 2, -3)\}$.

PROBLEM SET 1

In Problems 1–6, sketch the graph of each equation in the given linear system on the same coordinate system. Use the graphs to determine whether the system is independent, inconsistent, or dependent.

1 $\begin{cases} 3x - 2y = 1 \\ x + y = 5 \end{cases}$ **2** $\begin{cases} y = 2x - 3 \\ 4x - 2y = 6 \end{cases}$

3 $\begin{cases} 2x + 6y = 7 \\ x + 3y = 1 \end{cases}$ **4** $\begin{cases} 2x = y + 3 \\ 4x - 2y = 5 \end{cases}$

5 $\begin{cases} x + 2y = 3 \\ 2x - 3y = 1 \end{cases}$ **6** $\begin{cases} 3x + 2y = 11 \\ -2x + y = 2 \end{cases}$

In Problems 7–16, use the substitution method to find the solution set of the given linear system.

7 $\begin{cases} 2x - y = 5 \\ x + 3y = 13 \end{cases}$ **8** $\begin{cases} 3x + y = 4 \\ 7x - y = 6 \end{cases}$ **9** $\begin{cases} y = x - 2 \\ 2y = x - 3 \end{cases}$

10 $\begin{cases} 3x + 5y = 1 \\ x - 4y = -6 \end{cases}$ **11** $\begin{cases} x + y = 1 \\ 5x - y = 13 \end{cases}$ **12** $\begin{cases} 3x - 4y = -5 \\ 2x + 2y = 3 \end{cases}$

13 $\begin{cases} x + y = 5 \\ x + z = 1 \\ y + z = 2 \end{cases}$ **14** $\begin{cases} x - 3y = -11 \\ 2y - 5z = 26 \\ 7x - 3z = -2 \end{cases}$

15 $\begin{cases} x + y + 2z = 11 \\ x - y + z = 3 \\ 2x + y + 3z = 17 \end{cases}$ **16** $\begin{cases} x + 3y - z = 4 \\ 3x - 2y + 4z = 11 \\ 2x + y + 3z = 13 \end{cases}$

In Problems 17–30, use the elimination method to find the solution set of the given linear system.

17 $\begin{cases} 2x - y = 1 \\ x + y = 2 \end{cases}$ **18** $\begin{cases} x + y = 0 \\ x - y = 1 \end{cases}$

19 $\begin{cases} x + 3y = 9 \\ x - y = 1 \end{cases}$ **20** $\begin{cases} x + y = 1 \\ y - 2x = 3 \end{cases}$

21 $\begin{cases} 2x + 4y = 3 \\ -x + y = 3 \end{cases}$

22 $\begin{cases} 3y + 2z = 5 \\ 2y = -3z + 1 \end{cases}$

23 $\begin{cases} 3x + 2y = 4 \\ 5x + 3y = 7 \end{cases}$

24 $\begin{cases} \frac{1}{2}x + \frac{1}{3}y = 13 \\ \frac{1}{5}x + \frac{1}{8}y = 5 \end{cases}$

25 $\begin{cases} x + y + 2z = 4 \\ x + y - 2z = 0 \\ x - y = 0 \end{cases}$

26 $\begin{cases} x + y + z = 2 \\ x + 2y - z = 4 \\ 2x - y + z = 0 \end{cases}$

27 $\begin{cases} x + y + z = 6 \\ x - y + 2z = 12 \\ 2x + y + z = 1 \end{cases}$

28 $\begin{cases} x + y + 2z = 4 \\ x - 5y + z = 5 \\ 3x - 4y + 7z = 24 \end{cases}$

29 $\begin{cases} 2x + y - 3z = 9 \\ x - 2y + 4z = 5 \\ 3x + y - 2z = 15 \end{cases}$

30 $\begin{cases} x + 3y - 2z = -21 \\ 7x - 5y + 4z = 31 \\ 2x + y + 3z = 17 \end{cases}$

In Problems 31–43, use a system of linear equations to solve each problem.

31 The sum of two numbers is 12. If one of the numbers is multiplied by 5 and the other is multiplied by 8, the sum of the products is 75. Find the numbers.

32 In a given two-digit number, the tens digit is 5 more than the units digit, and the number is 63 more than the sum of its digits. Find the number.

33 A farmer sold two grades of wheat, one grade at $4.25 per bushel and another grade at $5.50 per bushel. How many bushels of each kind did he sell if he received $15,078.75 for 3000 bushels?

34 The rowing team of a certain school practices on a river that flows at a relatively constant speed. They can row downstream at a rate of 7 meters per second and upstream at a rate of 3 meters per second. Find the team's rate in still water and the rate of the river's current.

35 A man invested a total of $4000 in securities. Part of the money was invested at 5 percent, the rest at 6 percent. His annual income from both investments was $230. Find the amount of money he invested at each rate.

36 A woman has 11 coins, consisting of nickels and dimes. She has three more dimes than nickels. How many of each type of coin does she have if the total value of the coins is 90 cents?

37 Two cars start from the same point, at the same time, and travel in opposite directions. One car travels at a rate that is 10 miles per hour faster than the other. After 3 hours, they are 330 miles apart. What is the speed of each car?

38 The perimeter of an isosceles triangle is 26 centimeters. The base is 2 centimeters longer than one of the equal sides of the triangle. Find the length of each side of the triangle.

39 A car that traveled at an average speed of 40 miles per hour went 35 miles farther than one that traveled at 55 miles per hour. The slower car traveled 2 hours longer than the faster one. What was the length of time each car traveled?

40 Three men bought a grocery store for $100,000. If the second man invested twice as much as the first, and the third invested $20,000 more than the second, how much did each invest?

41 A woman bought three different kinds of common stock for $20,000, one paying a 6 percent annual dividend, one paying a 7 percent annual dividend, and the other paying an 8 percent annual dividend. At the end of the first year, the sum of the dividends from the 6 percent and the 7 percent stocks was $940, and the sum of the dividends from the 6 percent and the 8 percent stocks was $720. How much did she invest in each kind of stock?

42 The perimeter of a triangle is 45 inches. One side is twice as long as the shortest side, and the third side is 5 inches longer than the shortest side. Find the lengths of the three sides of the triangle.

43 A chemist at work in his laboratory has containers of the same acid in two different strengths. Six parts of the first mixed with four parts of the second gives a mixture that is 86 percent acid, and four parts of the first mixed with six parts of the second gives a mixture that is 84 percent acid. What is the percentage of purity of each acid?

2 Matrices and Row Reduction

Notice that in the elimination method there is no reason to continue writing the variables. All we need do is maintain a record of the coefficients that belong to each variable in each equation. The standard device used for doing this is a *matrix*.

A *matrix* is a rectangular array of numbers. The *size* of a matrix is described by specifying the number of rows and columns. The numbers occurring in a matrix are called the *entries* of the matrix.

For example, the matrix

$$A = \begin{bmatrix} 1 & 2 \\ 3 & -1 \end{bmatrix}$$

has two rows and two columns; we say that A is a 2×2 matrix and that 1, 2, 3, and -1 are the entries of A. The matrix

$$B = \begin{bmatrix} 1 & 3 & 5 \\ -2 & 5 & 6 \end{bmatrix}$$

has two rows and three columns, so B is a 2×3 matrix and its entries are 1, 3, 5, -2, 5, and 6. The matrix

$$C = \begin{bmatrix} 3 \\ 1 \\ 2 \end{bmatrix}$$

is a 3×1 matrix with entries 3, 1, and 2. Notice that when the size of a matrix is specified, the number of rows is given first and the number of columns is given second.

An entry of a matrix is identified by using subscripts to indicate its row position and its column position, as in matrix A displayed below:

$$A = \begin{bmatrix} a_{11} & a_{12} & a_{13} & a_{14} \\ a_{21} & a_{22} & a_{23} & a_{24} \\ a_{31} & a_{32} & a_{33} & a_{34} \\ a_{41} & a_{42} & a_{43} & a_{44} \end{bmatrix}$$

The entry of the second row and the second column is denoted by a_{22}, while the entry of the first row and the fourth column is a_{14}.

For example, if

$$B = \begin{bmatrix} 3 & 2 & -1 \\ 4 & 1 & 5 \end{bmatrix}$$

then $b_{11} = 3$, $b_{12} = 2$, $b_{13} = -1$, $b_{21} = 4$, $b_{22} = 1$, and $b_{23} = 5$.

Now, let us examine how matrix notation, together with the elimination method, can be used to solve linear systems.

Suppose that we are to find the solution set of the linear system

$$\text{(A)} \quad \begin{cases} 3x - y = 1 \\ x + 2y = 0 \end{cases}$$

The matrix form of (A) is written as

$$A = \left[\begin{array}{cc|c} 3 & -1 & 1 \\ 1 & 2 & 0 \end{array}\right]$$

and is called an *augmented matrix.* An augmented matrix can be used to "maintain a record" of the coefficients of the linear equations when the elimination method is applied to solve the linear system (A). (A vertical arrow ↓ will be used to indicate that the systems are equivalent.)

System	*Matrix Form of System*
(A) $\begin{cases} 3x - y = 1 \\ x + 2y = 0 \end{cases}$	$A = \left[\begin{array}{cc\|c} 3 & -1 & 1 \\ 1 & 2 & 0 \end{array}\right]$
↓ First, multiply the second equation by -3.	↓ First, multiply the second row by -3.
(B) $\begin{cases} 3x - y = 1 \\ -3x - 6y = 0 \end{cases}$	$B = \left[\begin{array}{cc\|c} 3 & -1 & 1 \\ -3 & -6 & 0 \end{array}\right]$
↓ Next, replace equation two with the sum of equations one and two.	↓ Next, replace row two with the sum of rows one and two.

(C) $\begin{cases} 3x - y = 1 \\ \quad - 7y = 1 \end{cases}$ $\qquad C = \left[\begin{array}{cc|c} 3 & -1 & 1 \\ 0 & -7 & 1 \end{array}\right]$

↓ Multiply equation two by $-\frac{1}{7}$. ↓ Multiply row two by $-\frac{1}{7}$.

(D) $\begin{cases} 3x - y = 1 \\ \quad y = -\frac{1}{7} \end{cases}$ $\qquad D = \left[\begin{array}{cc|c} 3 & -1 & 1 \\ 0 & 1 & -\frac{1}{7} \end{array}\right]$

↓ Replace equation one with the sum of equations one and two. ↓ Replace row one with the sum of row one and row two.

(E) $\begin{cases} 3x = \frac{6}{7} \\ \quad y = -\frac{1}{7} \end{cases}$ $\qquad E = \left[\begin{array}{cc|c} 3 & 0 & \frac{6}{7} \\ 0 & 1 & -\frac{1}{7} \end{array}\right]$

↓ Multiply equation one by $\frac{1}{3}$. ↓ Multiply row one by $\frac{1}{3}$.

(F) $\begin{cases} x = \frac{2}{7} \\ \quad y = -\frac{1}{7} \end{cases}$ $\qquad F = \left[\begin{array}{cc|c} 1 & 0 & \frac{2}{7} \\ 0 & 1 & -\frac{1}{7} \end{array}\right]$

Hence, the solution set of the linear system (A) is $\{(\frac{2}{7}, -\frac{1}{7})\}$.

The numbers to the left of the dashed line represent the coefficients of x and y, so the solution set is $\{(\frac{2}{7}, -\frac{1}{7})\}$.

EXAMPLE Find the solution set of the system of linear equations

$$\text{(A)} \begin{cases} x_1 - 2x_2 + 3x_3 = -1 \\ 2x_1 - x_2 + 2x_3 = 2 \\ 3x_1 + x_2 + 2x_3 = 3 \end{cases}$$

by using the elimination method and displaying the corresponding matrix.

SOLUTION The matrix form of system (A) is written as

$$A = \left[\begin{array}{ccc|c} 1 & -2 & 3 & -1 \\ 2 & -1 & 2 & 2 \\ 3 & 1 & 2 & 3 \end{array}\right]$$

The elimination method proceeds as follows.

System / *Matrix Form of System*

(A) $\begin{cases} x_1 - 2x_2 + 3x_3 = -1 \\ 2x_1 - x_2 + 2x_3 = 2 \\ 3x_1 + x_2 + 2x_3 = 3 \end{cases}$ $\qquad A = \left[\begin{array}{ccc|c} 1 & -2 & 3 & -1 \\ 2 & -1 & 2 & 2 \\ 3 & 1 & 2 & 3 \end{array}\right]$

↓ First, replace the second equation by the sum of the second equation and -2 times the first equation. ↓ First, replace the second row by the sum of the second row and -2 times the first row.

(B) $\begin{cases} x_1 - 2x_2 + 3x_3 = -1 \\ 3x_2 - 4x_3 = 4 \\ 3x_1 + x_2 + 2x_3 = 3 \end{cases}$ $B = \left[\begin{array}{rrr|r} 1 & -2 & 3 & -1 \\ 0 & 3 & -4 & 4 \\ 3 & 1 & 2 & 3 \end{array}\right]$

Next, replace equation three with the sum of the third equation and −3 times equation one. | Next, replace row three with the sum of the third row and −3 times row one.

(C) $\begin{cases} x_1 - 2x_2 + 3x_3 = -1 \\ 3x_2 - 4x_3 = 4 \\ 7x_2 - 7x_3 = 6 \end{cases}$ $C = \left[\begin{array}{rrr|r} 1 & -2 & 3 & -1 \\ 0 & 3 & -4 & 4 \\ 0 & 7 & -7 & 6 \end{array}\right]$

In the next step, we multiply the second equation by $\frac{1}{3}$. | In the next step, we multiply the second row by $\frac{1}{3}$.

(D) $\begin{cases} x_1 - 2x_2 + 3x_3 = -1 \\ x_2 - \frac{4}{3}x_3 = \frac{4}{3} \\ 7x_2 - 7x_3 = 6 \end{cases}$ $D = \left[\begin{array}{rrr|r} 1 & -2 & 3 & -1 \\ 0 & 1 & -\frac{4}{3} & \frac{4}{3} \\ 0 & 7 & -7 & 6 \end{array}\right]$

Now, replace the third equation with the sum of equation three and − 7 times equation two. | Now, replace the third row with the sum of row three and − 7 times row two.

(E) $\begin{cases} x_1 - 2x_2 + 3x_3 = -1 \\ x_2 - \frac{4}{3}x_3 = \frac{4}{3} \\ \frac{7}{3}x_3 = -\frac{10}{3} \end{cases}$ $E = \left[\begin{array}{rrr|r} 1 & -2 & 3 & -1 \\ 0 & 1 & -\frac{4}{3} & \frac{4}{3} \\ 0 & 0 & \frac{7}{3} & -\frac{10}{3} \end{array}\right]$

Next, replace equation one with equation one plus 2 times equation two. | Next, replace row one with row one plus 2 times row two.

(F) $\begin{cases} x_1 + \frac{1}{3}x_3 = \frac{5}{3} \\ x_2 - \frac{4}{3}x_3 = \frac{4}{3} \\ \frac{7}{3}x_3 = -\frac{10}{3} \end{cases}$ $F = \left[\begin{array}{rrr|r} 1 & 0 & \frac{1}{3} & \frac{5}{3} \\ 0 & 1 & -\frac{4}{3} & \frac{4}{3} \\ 0 & 0 & \frac{7}{3} & -\frac{10}{3} \end{array}\right]$

In the next step, we multiply the third equation by $\frac{3}{7}$. | In the next step, we multiply the third row by $\frac{3}{7}$.

(G) $\begin{cases} x_1 + \frac{1}{3}x_3 = \frac{5}{3} \\ x_2 - \frac{4}{3}x_3 = \frac{4}{3} \\ x_3 = -\frac{10}{7} \end{cases}$ $G = \left[\begin{array}{rrr|r} 1 & 0 & \frac{1}{3} & \frac{5}{3} \\ 0 & 1 & -\frac{4}{3} & \frac{4}{3} \\ 0 & 0 & 1 & -\frac{10}{7} \end{array}\right]$

Replace equation one with the sum of equation one and $-\frac{1}{3}$ times equation three. | Replace row one with the sum of row one and $-\frac{1}{3}$ times row three.

$$\text{(H)} \begin{cases} x_1 \qquad\qquad = \frac{15}{7} \\ x_2 - \frac{4}{3}x_3 = \frac{4}{3} \\ \qquad\quad x_3 = -\frac{10}{7} \end{cases}$$

$$H = \left[\begin{array}{ccc|c} 1 & 0 & 0 & \frac{15}{7} \\ 0 & 1 & -\frac{4}{3} & \frac{4}{3} \\ 0 & 0 & 1 & -\frac{10}{7} \end{array}\right]$$

Finally, replace equation two with the sum of equation two and $\frac{4}{3}$ times equation three.

Finally, replace row two with the sum of row two and $\frac{4}{3}$ times row three.

$$\text{(I)} \begin{cases} x_1 \qquad = \frac{15}{7} \\ x_2 \qquad = -\frac{4}{7} \\ \qquad x_3 = -\frac{10}{7} \end{cases}$$

$$I = \left[\begin{array}{ccc|c} 1 & 0 & 0 & \frac{15}{7} \\ 0 & 1 & 0 & -\frac{4}{7} \\ 0 & 0 & 1 & -\frac{10}{7} \end{array}\right]$$

Hence, the solution set of system (A) is $\{(\frac{15}{7}, -\frac{4}{7}, -\frac{10}{7})\}$.

Hence, the solution set is read from the augmented matrix as $\{(\frac{15}{7}, -\frac{4}{7}, -\frac{10}{7})\}$.

Notice that in solving the system of linear equations (A), the system was replaced by an augmented matrix A whose elements are the coefficients and constants occurring in the equations. Notice also that we can work with the augmented matrix instead of the actual equations by performing the following operations on the augmented matrix [compare with equation operations (1) through (3) on page 354]:

1′ Interchange two rows of the matrix ($R_i \leftrightarrow R_j$).

2′ Multiply the elements in a row of the matrix by a nonzero number ($R_i \to kR_i$, where $k \neq 0$).

3′ Replace a row with the sum of a nonzero multiple of itself and a nonzero multiple of another row ($R_i \to kR_j + cR_i$, where $c \neq 0$ and $k \neq 0$).

Operations (1′), (2′), and (3′) are called the *elementary row operations.* The resulting matrix I in the above example is called a row-reduced echelon matrix. In general, a matrix is a *row-reduced echelon matrix* if all of the following conditions hold:

(i) The first nonzero entry in each row is 1; all other entries in that column are zeros. (That is, if a column contains 1 as a leading entry of some row, then the remaining entries in that column must be zero.)

(ii) Any rows that consist entirely of zeros are below all rows that do not consist entirely of zeros.

(iii) The first nonzero entry in each row is to the right of the first nonzero entry in the preceding row.

EXAMPLES Find the solution set of the given linear system by performing elementary row operations.

1 $\begin{cases} 4x + y = 2 \\ x + 4y = 1 \end{cases}$

SOLUTION The matrix form of the system is

$$\left[\begin{array}{cc|c} 4 & 1 & 2 \\ 1 & 4 & 1 \end{array}\right]$$

First, multiply row one by $\frac{1}{4}$, that is, $R_1 \to \frac{1}{4}R_1$, to get

$$\left[\begin{array}{cc|c} 1 & \frac{1}{4} & \frac{1}{2} \\ 1 & 4 & 1 \end{array}\right]$$

Next, replace row two with the sum of row two and -1 times row one $[R_2 \to R_2 + (-1)R_1]$ to get

$$\left[\begin{array}{cc|c} 1 & \frac{1}{4} & \frac{1}{2} \\ 0 & \frac{15}{4} & \frac{1}{2} \end{array}\right]$$

Multiply row two by $\frac{4}{15}$, that is, $R_2 \to \frac{4}{15}R_2$, to get

$$\left[\begin{array}{cc|c} 1 & \frac{1}{4} & \frac{1}{2} \\ 0 & 1 & \frac{2}{15} \end{array}\right]$$

Finally, replace row one with the sum of row one and $-\frac{1}{4}$ times row two $[R_1 \to R_1 + (-\frac{1}{4})R_2]$ to get

$$\left[\begin{array}{cc|c} 1 & 0 & \frac{7}{15} \\ 0 & 1 & \frac{2}{15} \end{array}\right]$$

Hence the solution set is $\{(\frac{7}{15}, \frac{2}{15})\}$.

2 $$\begin{cases} x_1 + 2x_2 - 3x_3 = 6 \\ 2x_1 - x_2 + 4x_3 = 2 \\ 4x_1 + 3x_2 - 2x_3 = 14 \end{cases}$$

SOLUTION The matrix form of the system is

$$\left[\begin{array}{ccc|c} 1 & 2 & -3 & 6 \\ 2 & -1 & 4 & 2 \\ 4 & 3 & -2 & 14 \end{array}\right]$$

First, replace the second row by the sum of -2 times the first row and the second row $[R_2 \to (-2)R_1 + R_2]$ to obtain the equivalent matrix

$$\left[\begin{array}{ccc|c} 1 & 2 & -3 & 6 \\ 0 & -5 & 10 & -10 \\ 4 & 3 & -2 & 14 \end{array}\right]$$

Next, replace the third row by the sum of -4 times the first row and the third row $[R_3 \to (-4)R_1 + R_3]$ to obtain

$$\left[\begin{array}{ccc|c} 1 & 2 & -3 & 6 \\ 0 & -5 & 10 & -10 \\ 0 & -5 & 10 & -10 \end{array}\right]$$

Since the second row R_2 and the third row R_3 are identical, the corresponding linear system will have identical equations for equation two and equation three, namely,

$$-5x_2 + 10x_3 = -10$$

In this case, the solution set contains an infinite number of ordered triples, and the system is dependent.

3 $$\begin{cases} 3x_1 - x_2 + x_3 = 1 \\ 7x_1 + x_2 - x_3 = 6 \\ 2x_1 + x_2 - x_3 = 2 \end{cases}$$

SOLUTION The augmented matrix that corresponds to this particular linear system is

$$\left[\begin{array}{rrr|r} 3 & -1 & 1 & 1 \\ 7 & 1 & -1 & 6 \\ 2 & 1 & -1 & 2 \end{array}\right]$$

which can be reduced by performing the elementary row operations that are specified below each arrow:

$$\left[\begin{array}{rrr|r} 3 & -1 & 1 & 1 \\ 7 & 1 & -1 & 6 \\ 2 & 1 & -1 & 2 \end{array}\right] \xrightarrow[R_3 \to R_1 + R_3]{} \left[\begin{array}{rrr|r} 3 & -1 & 1 & 1 \\ 7 & 1 & -1 & 6 \\ 5 & 0 & 0 & 3 \end{array}\right]$$

$$\xrightarrow[R_2 \to R_1 + R_2]{} \left[\begin{array}{rrr|r} 3 & -1 & 1 & 1 \\ 10 & 0 & 0 & 7 \\ 5 & 0 & 0 & 3 \end{array}\right]$$

$$\xrightarrow[R_2 \to R_2 + (-2)R_3]{} \left[\begin{array}{rrr|r} 3 & -1 & 1 & 1 \\ 0 & 0 & 0 & 1 \\ 5 & 0 & 0 & 3 \end{array}\right]$$

But R_2 implies that $0 \cdot x_1 + 0 \cdot x_2 + 0 \cdot x_3 = 1$, that is, $0 = 1$, which, of course, is impossible. Hence the solution set is empty, and the system is inconsistent.

In summary, the procedure for solving a linear system is to form an augmented matrix and proceed to reduce it to echelon form.

1 If the resulting augmented echelon matrix has no row with its first nonzero entry in the last column, then the system of linear equations has one solution (the system is unique), or it has infinitely many solutions (the system is dependent).

2 If the resulting augmented echelon matrix has a row with its first nonzero entry appearing in the last column, then the system of linear equations has no solution (the system is inconsistent).

PROBLEM SET 2

In Problems 1–11, solve the linear system by using the elimination method. Show the corresponding matrix form of the system in each step of the process.

1 $\begin{cases} 3x + y = 14 \\ 2x - y = 1 \end{cases}$

2 $\begin{cases} 4x + 3y = 15 \\ 3x + 5y = 14 \end{cases}$

3 $\begin{cases} -2x + 3y = 8 \\ 2x - y = 5 \end{cases}$

4 $\begin{cases} x - 2y = 5 \\ 3x - 6y = 4 \end{cases}$

5 $\begin{cases} \frac{1}{2}x + \frac{1}{6}y = \frac{2}{3} \\ 3x + y = 4 \end{cases}$

6 $\begin{cases} x - 2y - 4 = 0 \\ x + y + 3 = 0 \end{cases}$

7 $\begin{cases} x + y + z = 6 \\ 3x - y + 2z = 7 \\ 2x + 3y - z = 5 \end{cases}$

8 $\begin{cases} 2x + 3y + z = 6 \\ x - 2y + 3z = -3 \\ 3x + y - z = 8 \end{cases}$

9 $\begin{cases} x + y + 2z = 4 \\ x + y - 2z = 0 \\ x - y = 0 \end{cases}$

10 $\begin{cases} x + y + z = 4 \\ x - y + 2z = 8 \\ 2x + y - z = 3 \end{cases}$

11 $\begin{cases} 2x + y - z = 7 \\ y - x = 1 \\ z - y = 1 \end{cases}$

12 Suppose that

$$A = \begin{bmatrix} -4 & 0 & 1 \\ 2 & 3 & -1 \\ 5 & 2 & 8 \end{bmatrix}$$

(a) What is the size of matrix A?

(b) Use subscript notation to identify each of the entries of matrix A.

In Problems 13–20, each row-reduced echelon matrix is the augmented matrix form of a corresponding linear system of equations. Find the solution of each system.

13 $\left[\begin{array}{cc|c} 1 & 0 & 1 \\ 0 & 1 & 1 \end{array}\right]$

14 $\left[\begin{array}{ccc|c} 1 & 0 & 0 & 0 \\ 0 & 1 & 0 & 0 \\ 0 & 0 & 1 & 2 \end{array}\right]$

15 $\left[\begin{array}{ccc|c} 1 & 0 & 0 & 3 \\ 0 & 1 & 0 & 0 \\ 0 & 0 & 1 & 5 \end{array}\right]$

16 $\left[\begin{array}{cccc|c} 1 & 0 & -1 & 2 & 3 \\ 0 & 1 & -2 & 1 & 4 \end{array}\right]$

17 $\left[\begin{array}{ccc|c} 1 & 0 & 1 & 3 \\ 0 & 1 & 0 & 4 \\ 0 & 0 & 0 & 0 \end{array}\right]$

18 $\left[\begin{array}{cccc|c} 1 & 0 & 0 & 0 & 3 \\ 0 & 1 & 0 & 0 & -2 \\ 0 & 0 & 0 & 1 & -2 \\ 0 & 0 & 1 & 0 & 5 \end{array}\right]$

19 $\left[\begin{array}{ccccc|c} 1 & -1 & 0 & -2 & 0 & 1 \\ 0 & 0 & 1 & 3 & 0 & 1 \\ 0 & 0 & 0 & 0 & 1 & 1 \end{array}\right]$ **20** $\left[\begin{array}{ccc|c} 0 & 0 & 0 & 0 \\ 0 & 0 & 0 & 0 \\ 0 & 0 & 0 & 1 \end{array}\right]$

In Problems 21–30, find the augmented matrix, reduce it to a row-reduced echelon matrix, and determine the solution if it is unique. If the solution is not unique, indicate whether the system is dependent or inconsistent.

21 $\begin{cases} 4x - y = 5 \\ -3x + 2y = 0 \end{cases}$ **22** $\begin{cases} 4x + 3y = 240 \\ -x + y = 10 \end{cases}$

23 $\begin{cases} x + y = 14 \\ 3x - y = 6 \end{cases}$ **24** $\begin{cases} 2x - 8y = 7 \\ 3x + 2y = 9 \end{cases}$

25 $\begin{cases} x - 2y + z = -1 \\ 3x + y - 2z = 4 \\ y - z = 1 \end{cases}$ **26** $\begin{cases} x + y - 2z = 3 \\ 3x - y + z = 5 \\ 3x + 3y - 6z = 9 \end{cases}$

27 $\begin{cases} 2x + y + z = 1 \\ 4x + 2y + 3z = 1 \\ -2x - y + z = 2 \end{cases}$ **28** $\begin{cases} x + y + z = 0 \\ 2x - y - 4z = 15 \\ x - 2y - z = 7 \end{cases}$

29 $\begin{cases} 2x - 3y + z = 4 \\ x - 4y - z = 3 \\ x - 9y - 4z = 5 \end{cases}$ **30** $\begin{cases} 2x + 3y - z = -2 \\ x - y + 2z = 4 \\ 3x + y + z = 7 \end{cases}$

3 Determinants

Next, we study a function called the *determinant,* which can be used to solve linear systems that have unique solutions. The determinant is a function that associates each square matrix with a real number.

This section surveys some of the relevant properties of the determinant. A complete discussion of determinants is usually reserved for a course in linear algebra.

The *determinant* is a function that has the set of square matrices as its domain and the set of real numbers as its range. If A is a square matrix, the determinant of A is denoted by det A or by $|A|$.

The determinant of the 2×2 matrix

$$A = \begin{bmatrix} a_{11} & a_{12} \\ a_{21} & a_{22} \end{bmatrix}$$

is defined to be the number $a_{11}a_{22} - a_{21}a_{12}$, and we write

$$\det A = \det\begin{bmatrix} a_{11} & a_{12} \\ a_{21} & a_{22} \end{bmatrix} = a_{11}a_{22} - a_{21}a_{12}$$

or

$$|A| = \begin{vmatrix} a_{11} & a_{12} \\ a_{21} & a_{22} \end{vmatrix} = a_{11}a_{22} - a_{21}a_{12}$$

Thus,

$$\begin{vmatrix} 5 & 3 \\ -3 & -6 \end{vmatrix} = 5(-6) - (-3)(3) = -30 + 9 = -21$$

Next, we consider the determinant of the 3×3 matrix

$$A = \begin{bmatrix} a_{11} & a_{12} & a_{13} \\ a_{21} & a_{22} & a_{23} \\ a_{31} & a_{32} & a_{33} \end{bmatrix}$$

We define $\det A$ or $|A|$ by

$$\det A = |A| = \begin{vmatrix} a_{11} & a_{12} & a_{13} \\ a_{21} & a_{22} & a_{23} \\ a_{31} & a_{32} & a_{33} \end{vmatrix}$$

$$= a_{11} \begin{vmatrix} a_{22} & a_{23} \\ a_{32} & a_{33} \end{vmatrix} - a_{12} \begin{vmatrix} a_{21} & a_{23} \\ a_{31} & a_{33} \end{vmatrix} + a_{13} \begin{vmatrix} a_{21} & a_{22} \\ a_{31} & a_{32} \end{vmatrix}$$

For example,

$$\begin{vmatrix} 3 & 2 & 7 \\ -1 & 5 & 3 \\ 2 & -3 & -6 \end{vmatrix} = 3 \begin{vmatrix} 5 & 3 \\ -3 & -6 \end{vmatrix} - 2 \begin{vmatrix} -1 & 3 \\ 2 & -6 \end{vmatrix} + 7 \begin{vmatrix} -1 & 5 \\ 2 & -3 \end{vmatrix}$$

$$= 3(-30 + 9) - 2(6 - 6) + 7(3 - 10)$$
$$= 3(-21) - 2(0) + 7(-7)$$
$$= -63 - 49$$
$$= -112$$

3.1 Properties of Determinants

The determinant possesses properties that can be used to simplify the task of its evaluation. Although we shall be restricting the investigation of these properties to 2×2 and 3×3 matrices, it is important to realize that the properties hold for evaluating the determinant of *any* square matrix. These properties are stated in the following theorems.

THEOREM 1

A common factor that appears in all entries in a row of a matrix can be factored out of the row when evaluating the determinant of the matrix.

PROOF (for $n = 2$)

$$\begin{vmatrix} ka_{11} & ka_{12} \\ a_{21} & a_{22} \end{vmatrix} = ka_{11}a_{22} - ka_{21}a_{12}$$
$$= k(a_{11}a_{22} - a_{21}a_{12})$$
$$= k \begin{vmatrix} a_{11} & a_{12} \\ a_{21} & a_{22} \end{vmatrix}$$

For example,

$$\begin{vmatrix} 6 & 9 \\ 1 & 4 \end{vmatrix} = 3 \begin{vmatrix} 2 & 3 \\ 1 & 4 \end{vmatrix}$$

and

$$\begin{vmatrix} 15 & 45 & 60 \\ 1 & 2 & -1 \\ 2 & 4 & 8 \end{vmatrix} = 15 \begin{vmatrix} 1 & 3 & 4 \\ 1 & 2 & -1 \\ 2 & 4 & 8 \end{vmatrix}$$
$$= (15)(2) \begin{vmatrix} 1 & 3 & 4 \\ 1 & 2 & -1 \\ 1 & 2 & 4 \end{vmatrix}$$

THEOREM 2

If two (not necessarily adjacent) rows of a square matrix are interchanged, the values of the determinants of the two matrices differ only in the algebraic sign.

PROOF (for $n = 2$)

$$\begin{vmatrix} a_{11} & a_{12} \\ a_{21} & a_{22} \end{vmatrix} = a_{11}a_{22} - a_{21}a_{12}$$
$$= -(a_{21}a_{12} - a_{11}a_{22})$$
$$= - \begin{vmatrix} a_{21} & a_{22} \\ a_{11} & a_{12} \end{vmatrix}$$

For example,

$$\begin{vmatrix} 2 & 3 \\ 4 & 1 \end{vmatrix} = -10 \quad \text{whereas} \quad \begin{vmatrix} 4 & 1 \\ 2 & 3 \end{vmatrix} = 10$$

and

$$\begin{vmatrix} 1 & -1 & 0 \\ 3 & 0 & 4 \\ 2 & 1 & 5 \end{vmatrix} = 3 \quad \text{whereas} \quad \begin{vmatrix} 2 & 1 & 5 \\ 3 & 0 & 4 \\ 1 & -1 & 0 \end{vmatrix} = -3$$

THEOREM 3

If any nonzero multiple of one row is added to any other row of a square matrix, the value of the determinant is unaltered.

[For the proof of this theorem, see Problem 16(a).]

For example, consider

$$\begin{vmatrix} 1 & 0 & 2 \\ 4 & 6 & 1 \\ -1 & 0 & -1 \end{vmatrix}$$

If we multiply the third row by 2 and add the result to the first row, we get

$$\begin{vmatrix} -1 & 0 & 0 \\ 4 & 6 & 1 \\ -1 & 0 & -1 \end{vmatrix}$$

and we are assured by Theorem 3 that the latter determinant has the same value as the original. Note that this operation affects only the first row, whereas the other two rows remain the same. But we need not stop here; indeed, we can add the third row to the second row in the latter determinant to obtain

$$\begin{vmatrix} -1 & 0 & 0 \\ 3 & 6 & 0 \\ -1 & 0 & -1 \end{vmatrix}$$

Again, this does not change the value of the determinant. Determinants such as the last one, which contain many zero entries, are relatively easy to evaluate; hence, Theorem 3 simplifies the task of calculating determinants.

EXAMPLES Use the three theorems above to evaluate each of the following determinants.

1 $\begin{vmatrix} 1 & 0 & 2 \\ 4 & 6 & 1 \\ -1 & 0 & -1 \end{vmatrix}$

SOLUTION

$$\begin{vmatrix} 1 & 0 & 2 \\ 4 & 6 & 1 \\ -1 & 0 & -1 \end{vmatrix} = \begin{vmatrix} -1 & 0 & 0 \\ 4 & 6 & 1 \\ -1 & 0 & -1 \end{vmatrix} \qquad \begin{matrix}\text{(Theorem 3)} \\ (R_1 \to R_1 + 2R_3)\end{matrix}$$

$$= \begin{vmatrix} -1 & 0 & 0 \\ 3 & 6 & 0 \\ -1 & 0 & -1 \end{vmatrix} \qquad \begin{matrix}\text{(Theorem 3)} \\ (R_2 \to R_2 + R_3)\end{matrix}$$

$$= \begin{vmatrix} -1 & 0 & 0 \\ 3 & 6 & 0 \\ 0 & 0 & -1 \end{vmatrix} \qquad \text{(Theorem 3)} \; [R_3 \to R_3 + (-1)R_1]$$

$$= 3 \begin{vmatrix} -1 & 0 & 0 \\ 1 & 2 & 0 \\ 0 & 0 & -1 \end{vmatrix} \qquad \text{(Theorem 1)}$$

$$= (3)(-1) \begin{vmatrix} 1 & 0 & 0 \\ 1 & 2 & 0 \\ 0 & 0 & -1 \end{vmatrix} \qquad \text{(Theorem 1)}$$

$$= (3)(-1)(-1) \begin{vmatrix} 1 & 0 & 0 \\ 1 & 2 & 0 \\ 0 & 0 & 1 \end{vmatrix} \qquad \text{(Theorem 1)}$$

$$= 3 \begin{vmatrix} 1 & 0 & 0 \\ 0 & 2 & 0 \\ 0 & 0 & 1 \end{vmatrix} \qquad \text{(Theorem 3)} \; [R_2 \to R_2 + (-1)R_1]$$

$$= (3)(2) \begin{vmatrix} 1 & 0 & 0 \\ 0 & 1 & 0 \\ 0 & 0 & 1 \end{vmatrix} \qquad \text{(Theorem 1)}$$

$$= 6 \cdot 1 = 6 \qquad \text{(Why?)}$$

2 $\begin{vmatrix} 3 & 1 & -1 \\ 0 & 2 & 4 \\ -1 & 4 & 2 \end{vmatrix}$

SOLUTION

$$\begin{vmatrix} 3 & 1 & -1 \\ 0 & 2 & 4 \\ -1 & 4 & 2 \end{vmatrix} = 2 \begin{vmatrix} 3 & 1 & -1 \\ 0 & 1 & 2 \\ -1 & 4 & 2 \end{vmatrix} \qquad \text{(Theorem 1)}$$

$$= 2 \begin{vmatrix} 0 & 13 & 5 \\ 0 & 1 & 2 \\ -1 & 4 & 2 \end{vmatrix} \qquad \text{(Theorem 3)} \; (R_1 \to R_1 + 3R_3)$$

$$= 2 \begin{vmatrix} 0 & 13 & 5 \\ 0 & 1 & 2 \\ -1 & 3 & 0 \end{vmatrix} \qquad \text{(Theorem 3)} \; [R_3 \to R_3 + (-1)R_2]$$

$$= (2)(-1) \begin{vmatrix} 0 & 13 & 5 \\ 0 & 1 & 2 \\ 1 & -3 & 0 \end{vmatrix} \qquad \text{(Theorem 1)}$$

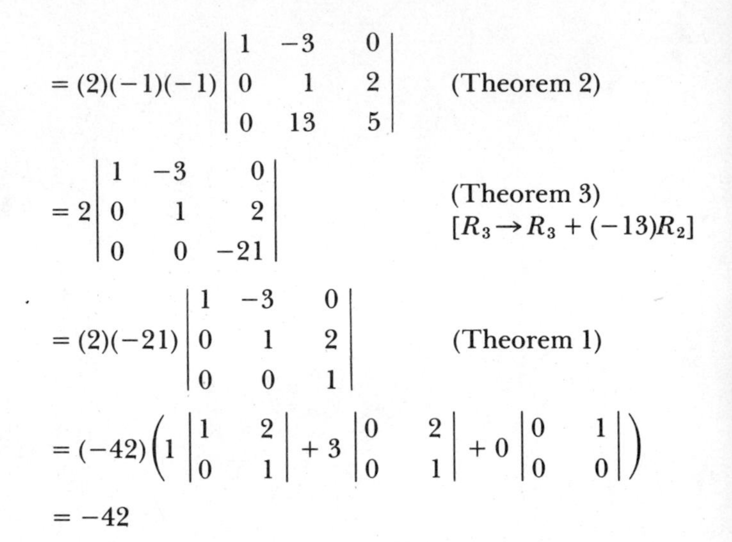

$$= (2)(-1)(-1)\begin{vmatrix} 1 & -3 & 0 \\ 0 & 1 & 2 \\ 0 & 13 & 5 \end{vmatrix} \qquad \text{(Theorem 2)}$$

$$= 2\begin{vmatrix} 1 & -3 & 0 \\ 0 & 1 & 2 \\ 0 & 0 & -21 \end{vmatrix} \qquad \begin{array}{l}\text{(Theorem 3)} \\ [R_3 \to R_3 + (-13)R_2]\end{array}$$

$$= (2)(-21)\begin{vmatrix} 1 & -3 & 0 \\ 0 & 1 & 2 \\ 0 & 0 & 1 \end{vmatrix} \qquad \text{(Theorem 1)}$$

$$= (-42)\left(1\begin{vmatrix} 1 & 2 \\ 0 & 1 \end{vmatrix} + 3\begin{vmatrix} 0 & 2 \\ 0 & 1 \end{vmatrix} + 0\begin{vmatrix} 0 & 1 \\ 0 & 0 \end{vmatrix}\right)$$

$$= -42$$

3.2 Cramer's Rule

We have investigated methods for computing determinants of 2×2 and 3×3 matrices. *Cramer's rule* provides us with a technique for using determinants to solve systems of linear equations that have unique solutions. Although Cramer's rule is not always the most practical way to solve linear systems, we shall use the rule as an example of an application of determinants.

Before stating Cramer's rule, let us establish some useful notation. Suppose that we are given a linear system (S) containing the same number of equations as unknowns.

$$\text{(S)} \quad \begin{cases} a_{11}x_1 + a_{12}x_2 + \cdots + a_{1n}x_n = c_1 \\ a_{21}x_1 + a_{22}x_2 + \cdots + a_{2n}x_n = c_2 \\ \cdots\cdots\cdots\cdots\cdots\cdots\cdots\cdots \\ a_{n1}x_1 + a_{n2}x_2 + \cdots + a_{nn}x_n = c_n \end{cases}$$

The determinant of the matrix of coefficients occurring in the system is called the determinant of the coefficient matrix and is denoted by D. Hence

$$D = \begin{vmatrix} a_{11} & a_{12} & \cdots & a_{1n} \\ a_{21} & a_{22} & \cdots & a_{2n} \\ \cdots & \cdots & \cdots & \cdots \\ a_{n1} & a_{n2} & \cdots & a_{nn} \end{vmatrix}$$

D_j will be used to denote the determinant of the matrix obtained by replacing the jth column in D by the column of constant terms in the system, the column on the right of the linear system (S), so that

$$D_j = \begin{vmatrix} a_{11} & a_{12} & \cdots & c_1 & \cdots & a_{1n} \\ a_{21} & a_{22} & \cdots & c_2 & \cdots & a_{2n} \\ \cdots & \cdots & \cdots & \cdots & \cdots & \cdots \\ a_{n1} & a_{n2} & \cdots & c_n & \cdots & a_{nn} \end{vmatrix}$$

(the boxed column of c's is the jth column)

For example, for the linear system

$$\begin{cases} 2x + 3y = 1 \\ x - y = 2 \end{cases}$$

$$D = \begin{vmatrix} 2 & 3 \\ 1 & -1 \end{vmatrix} = -5$$

$$D_1 = \begin{vmatrix} 1 & 3 \\ 2 & -1 \end{vmatrix} = -7 \quad \text{and} \quad D_2 = \begin{vmatrix} 2 & 1 \\ 1 & 2 \end{vmatrix} = 3$$

For the system

$$\begin{cases} 3x - y + 3z = 1 \\ x + y = 4 \\ -5x + 7y - 2z = -2 \end{cases}$$

$$D = \begin{vmatrix} 3 & -1 & 3 \\ 1 & 1 & 0 \\ -5 & 7 & -2 \end{vmatrix} = 28$$

$$D_1 = \begin{vmatrix} 1 & -1 & 3 \\ 4 & 1 & 0 \\ -2 & 7 & -2 \end{vmatrix} = 80 \qquad D_2 = \begin{vmatrix} 3 & 1 & 3 \\ 1 & 4 & 0 \\ -5 & -2 & -2 \end{vmatrix} = 32$$

and

$$D_3 = \begin{vmatrix} 3 & -1 & 1 \\ 1 & 1 & 4 \\ -5 & 7 & -2 \end{vmatrix} = -60$$

THEOREM 4 CRAMER'S RULE

Let (S) be the system of n linear equations in n unknowns described on page 370. Let D be the determinant of the coefficient matrix of (S). If $D \neq 0$, then the system (S) has exactly one solution, $(x_1, x_2, \ldots, x_n)$, where

$$x_j = \frac{D_j}{D} \qquad \text{for } j = 1, 2, \ldots, n$$

and D_j is the determinant defined above.

The proof of Cramer's rule is beyond the scope of this textbook, but we shall consider its application to cases in which $n = 2$ and $n = 3$.

For $n = 2$, Cramer's rule indicates that the solution of the system

$$\text{(A)} \quad \begin{cases} a_{11}x_1 + a_{12}x_2 = c_1 \\ a_{21}x_1 + a_{22}x_2 = c_2 \end{cases}$$

is given by (x_1, x_2), where

$$x_1 = \frac{\begin{vmatrix} c_1 & a_{12} \\ c_2 & a_{22} \end{vmatrix}}{\begin{vmatrix} a_{11} & a_{12} \\ a_{21} & a_{22} \end{vmatrix}} \quad \text{and} \quad x_2 = \frac{\begin{vmatrix} a_{11} & c_1 \\ a_{21} & c_2 \end{vmatrix}}{\begin{vmatrix} a_{11} & a_{12} \\ a_{21} & a_{22} \end{vmatrix}} \quad \text{if} \quad \begin{vmatrix} a_{11} & a_{12} \\ a_{21} & a_{22} \end{vmatrix} \neq 0$$

For $n = 3$, Cramer's rule can be applied to the system

$$\text{(B)} \quad \begin{cases} a_{11}x_1 + a_{12}x_2 + a_{13}x_3 = c_1 \\ a_{21}x_1 + a_{22}x_2 + a_{23}x_3 = c_2 \\ a_{31}x_1 + a_{32}x_2 + a_{33}x_3 = c_3 \end{cases}$$

First, we determine

$$D = \begin{vmatrix} a_{11} & a_{12} & a_{13} \\ a_{21} & a_{22} & a_{23} \\ a_{31} & a_{32} & a_{33} \end{vmatrix}$$

$$D_1 = \begin{vmatrix} c_1 & a_{12} & a_{13} \\ c_2 & a_{22} & a_{23} \\ c_3 & a_{32} & a_{33} \end{vmatrix} \qquad D_2 = \begin{vmatrix} a_{11} & c_1 & a_{13} \\ a_{21} & c_2 & a_{23} \\ a_{31} & c_3 & a_{33} \end{vmatrix}$$

and

$$D_3 = \begin{vmatrix} a_{11} & a_{12} & c_1 \\ a_{21} & a_{22} & c_2 \\ a_{31} & a_{32} & c_3 \end{vmatrix}$$

If $D \neq 0$, then $x_1 = \dfrac{D_1}{D}$, $x_2 = \dfrac{D_2}{D}$, and $x_3 = \dfrac{D_3}{D}$, and (x_1, x_2, x_3) is the solution of system (B).

Notice that Cramer's rule has nothing to say about the existence of solutions in the case in which $D = 0$. Actually, if $D = 0$, either the system has no solution (inconsistent) or the system has an infinite number of different solutions (dependent).

EXAMPLES Use Cramer's rule to solve each of the following linear systems, if possible.

1 $\begin{cases} 3x + 4y = 12 \\ 3x - 8y = 0 \end{cases}$

SOLUTION Here

$$D = \begin{vmatrix} 3 & 4 \\ 3 & -8 \end{vmatrix} = -24 - 12 = -36$$

$$D_1 = \begin{vmatrix} 12 & 4 \\ 0 & -8 \end{vmatrix} = -96 \quad \text{and} \quad D_2 = \begin{vmatrix} 3 & 12 \\ 3 & 0 \end{vmatrix} = -36$$

Since $D \neq 0$, then

$$x = \frac{D_1}{D} = \frac{-96}{-36} = \frac{8}{3} \quad \text{and} \quad y = \frac{D_2}{D} = \frac{-36}{-36} = 1$$

The solution set is $\{(\frac{8}{3}, 1)\}$.

2 $\begin{cases} x + 2y = 4 \\ 3x + 6y = -3 \end{cases}$

SOLUTION Here

$$D = \begin{vmatrix} 1 & 2 \\ 3 & 6 \end{vmatrix} = 0$$

Since $D = 0$, Cramer's rule is not applicable. Notice that if we graph the lines represented by these equations, we find that they are parallel.

3 Suppose that 4 pounds of a mixture of three different types of candy is made. One of the three types costs \$1.40 per pound, the second costs \$1.30 per pound, and the third costs \$1.50 per pound. Assume that the total cost of the 4-pound mixture is \$5.50 and that the mixture contains twice as much of the \$1.30 candy as of the \$1.40 candy. How many pounds of each type is in the mixture?

SOLUTION Let x be the number of pounds of the \$1.40 candy, let y be the number of pounds of \$1.30 candy, and let z be the number of pounds of \$1.50 candy. The values of x, y, and z are found by solving the system

$$\begin{cases} x + y + z = 4 \\ 1.40x + 1.30y + 1.50z = 5.50 \\ y = 2x \end{cases}$$

which is equivalent to the system

$$\begin{cases} x + y + z = 4 \\ 14x + 13y + 15z = 55 \\ 2x - y = 0 \end{cases}$$

Using Cramer's rule, we have

$$D = \begin{vmatrix} 1 & 1 & 1 \\ 14 & 13 & 15 \\ 2 & -1 & 0 \end{vmatrix} = 5$$

$$D_1 = \begin{vmatrix} 4 & 1 & 1 \\ 55 & 13 & 15 \\ 0 & -1 & 0 \end{vmatrix} = 5 \qquad D_2 = \begin{vmatrix} 1 & 4 & 1 \\ 14 & 55 & 15 \\ 2 & 0 & 0 \end{vmatrix} = 10$$

$$D_3 = \begin{vmatrix} 1 & 1 & 4 \\ 14 & 13 & 55 \\ 2 & -1 & 0 \end{vmatrix} = 5$$

so that

$$x = \frac{D_1}{D} = \frac{5}{5} = 1 \qquad y = \frac{D_2}{D} = \frac{10}{5} = 2 \qquad z = \frac{D_3}{D} = \frac{5}{5} = 1$$

Hence, the mixture contains 1 pound of \$1.40 candy, 2 pounds of \$1.30 candy, and 1 pound of \$1.50 candy.

PROBLEM SET 3

In Problems 1–6, evaluate each determinant.

1 $\begin{vmatrix} -1 & 3 \\ -7 & 4 \end{vmatrix}$

2 $\begin{vmatrix} 2 & 3 \\ 9 & 4 \end{vmatrix}$

3 $\begin{vmatrix} 2 & -1 & 3 \\ 9 & -7 & 4 \\ 11 & -6 & 2 \end{vmatrix}$

4 $\begin{vmatrix} 3 & -1 & 2 \\ 0 & 1 & -5 \\ 6 & 7 & 4 \end{vmatrix}$

5 $\begin{vmatrix} 2 & 2 & 2 \\ 3 & 3 & 3 \\ 4 & 4 & 4 \end{vmatrix}$

6 $\begin{vmatrix} \frac{1}{2} & 4 & 7 \\ 1 & -1 & 2 \\ 3 & 2 & 5 \end{vmatrix}$

In Problems 7–10, solve for x.

7 $\begin{vmatrix} x & -x \\ 5 & 3 \end{vmatrix} = 2$

8 $\begin{vmatrix} x & 4 & 5 \\ 0 & 1 & x \\ 5 & 2 & 1 \end{vmatrix} = 7$

9 $\begin{vmatrix} x & 0 & 0 \\ 3 & 1 & 2 \\ 0 & 4 & 1 \end{vmatrix} = 5$

10 $\begin{vmatrix} 5x & 0 & 1 \\ 2x & 1 & 2 \\ 3x & 2 & 3 \end{vmatrix} = 0$

In Problems 11–15, show why each statement is true, not by evaluating each side, but by citing which theorems from Section 3 have been used.

11 $\begin{vmatrix} 4 & 5 \\ 3 & -2 \end{vmatrix} = -\begin{vmatrix} 3 & -2 \\ 4 & 5 \end{vmatrix}$

12 $\begin{vmatrix} 3 & 0 & 1 \\ 1 & 1 & 2 \\ 3 & 0 & 1 \end{vmatrix} = \begin{vmatrix} 0 & 0 & 0 \\ 1 & 1 & 2 \\ 3 & 0 & 1 \end{vmatrix}$

13 $\begin{vmatrix} 3 & -6 & 2 \\ 5 & -3 & 0 \\ 0 & 9 & 18 \end{vmatrix} = 9 \begin{vmatrix} 3 & -6 & 2 \\ 5 & -3 & 0 \\ 0 & 1 & 2 \end{vmatrix}$

14 $\begin{vmatrix} 2 & 4 & 12 \\ -1 & 0 & 3 \\ 1 & 0 & 6 \end{vmatrix} = 18 \begin{vmatrix} 1 & 2 & 6 \\ -1 & 0 & 3 \\ 0 & 0 & 1 \end{vmatrix}$

15 $\begin{vmatrix} 1 & 1 & 1 \\ 3 & 3 & 3 \\ 2 & 2 & 2 \end{vmatrix} = 6 \begin{vmatrix} 0 & 0 & 0 \\ 0 & 0 & 0 \\ 1 & 1 & 1 \end{vmatrix}$

16 (a) Prove Theorem 3 for $n = 2$. *Hint:* Compare

$$\begin{vmatrix} a_{11} & a_{12} \\ a_{21} & a_{22} \end{vmatrix} \quad \text{and} \quad \begin{vmatrix} a_{11} + ca_{21} & a_{12} + ca_{22} \\ a_{21} & a_{22} \end{vmatrix}$$

(b) Prove, for $n = 2$, that if two rows of a matrix are the same, the determinant is zero.

(c) Prove, for $n = 2$, that if all the entries in one row of a matrix are zeros, the determinant is zero.

In Problems 17–21, use Theorems 1, 2, and 3 to evaluate each of the determinants.

17 $\begin{vmatrix} -1 & 0 & 2 \\ 0 & 0 & 0 \\ -1 & 5 & 1 \end{vmatrix}$

18 $\begin{vmatrix} 3 & 1 & 1 \\ -1 & 0 & 3 \\ 2 & 1 & 1 \end{vmatrix}$

19 $\begin{vmatrix} 2 & 1 & 3 \\ 1 & 2 & 1 \\ 4 & 0 & 0 \end{vmatrix}$

20 $\begin{vmatrix} 20 & 12 & 8 \\ 5 & 3 & 2 \\ 5 & 7 & 2 \end{vmatrix}$

21 $\begin{vmatrix} 1 & 0 & 2 \\ 1 & -3 & 0 \\ 0 & 3 & 1 \end{vmatrix}$

22 (a) *Principle of duality.* Consider Theorems 1, 2, and 3. If we replace the word "row" with the word "column," then the theorems are still true. Rewrite the three theorems with this substitution.

(b) Use the theorems of part (a) to evaluate the determinants in Problems 17–21.

In Problems 23–32, use Cramer's rule to find the solution set of each system, if possible.

23 $\begin{cases} 2x - y = 0 \\ x + y = 1 \end{cases}$

24 $\begin{cases} -3x + y = 3 \\ -2x - y = -5 \end{cases}$

25 $\begin{cases} x + y = 0 \\ x - y = 0 \end{cases}$

26 $\begin{cases} 3x + y = 1 \\ 9x + 3y = -4 \end{cases}$

27 $\begin{cases} 2x_1 - x_2 + x_3 = 3 \\ -x_1 + 2x_2 - x_3 = 1 \\ 3x_1 + x_2 + 2x_3 = -1 \end{cases}$

28 $\begin{cases} 3x + 2z = 8 - 2y \\ x - 5y + 6z = 8 \\ 6x - 8z = 4 \end{cases}$

29 $\begin{cases} x + y + 2z = 4 \\ x + y - 2z = 0 \\ x - y = 0 \end{cases}$

30 $\begin{cases} 2x_1 - 3x_2 = 4 \\ x_1 + x_2 - 2x_3 = 1 \\ x_1 - x_2 - x_3 = 5 \end{cases}$

31 $\begin{cases} x + y + z = 4 \\ x - y + 2z = 8 \\ 2x + y - z = 3 \end{cases}$

32 $\begin{cases} 2x + 3y + z = 6 \\ x - 2y + 3z = -3 \\ 3x + y - z = 8 \end{cases}$

In Problems 33–37, set up a linear system that will serve as a model of the situation. Then, use Cramer's rule to solve the system.

33 Twice the sum of two numbers is 30, and three times the smaller equals twice the larger. Determine the numbers.

34 The total cost of a watch, chain, and ring together was $225. The watch cost $50 more than the chain, and the ring cost $25 more than the watch and chain together. What was the cost of each item of jewelry?

35 How much cream containing 25 percent butterfat and milk containing 10 percent butterfat must be mixed to make 30 gallons of a mixture containing 20 percent butterfat?

36 A rectangular field is 40 meters longer than it is wide, and the length of the fence around it is 560 meters. What are the dimensions of the field?

37 The sum of three numbers is 5. The first number is twice the second number, and the third number equals the sum of the other two numbers. Find the three numbers.

4 Systems of Linear Inequalities and Linear Programming

Linear programming is the name given to a field of mathematics that deals, in general, with the problem of finding the maximum or minimum value of a given linear expression, where the variables are subject to certain linear constraints. In this section, we restrict our attention to linear expressions with *two variables,* although it is possible to use linear programming for expressions containing more than two variables.

Much of the initial development of the concepts and techniques of linear programming is credited to George B. Dantzig, who was working on a number of applied problems for the U.S. Air Force in 1947. Since that time the techniques have been used to solve problems in the fields of health, transportation, economics, engineering, and agriculture.

In order to understand the basic notions of linear programming, it is necessary to investigate systems of linear inequalities by extending the idea of linear systems covered in the preceding sections.

4.1 Systems of Linear Inequalities

It is generally true that the graph of a linear equation in two variables divides the plane into two disjoint *half planes.* For example, the graph of $y + 2x = 3$ divides the plane into the two regions R_1 and R_2 shown in Figure 1. *All* points that satisfy the inequalities $y + 2x > 3$ and $y + 2x < 3$ lie either in R_1 or R_2. To determine such points and the region in which they lie, all we need do is *test one point* in either region. For example, the point (4, 0) lies in R_1 and, since the coordinates satisfy $y + 2x > 3$, we conclude that R_1 contains *all* points that satisfy $y + 2x > 3$. Similarly, R_2 contains all points whose coordinates satisfy $y + 2x < 3$.

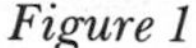
Figure 1

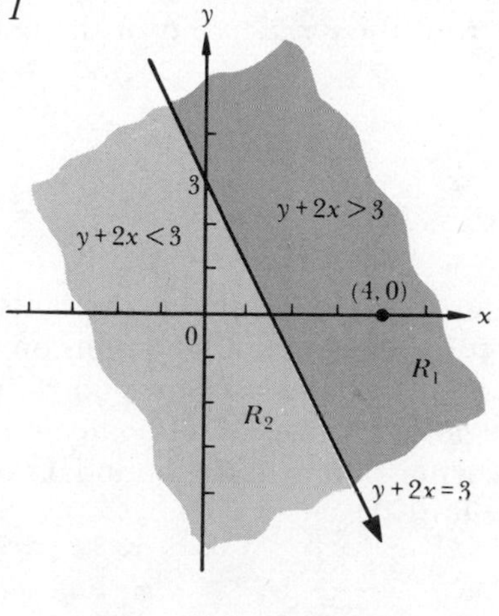

In summary, we have the following:

Condition	*Location of Points*
$y + 2x = 3$	on the line
$y + 2x > 3$	R_1
$y + 2x < 3$	R_2

EXAMPLE Sketch the region containing all the points whose coordinates satisfy the inequality $-3x + y > -2$.

SOLUTION First, we sketch the graph of the associated linear equation $-3x + y = -2$ (Figure 2). Next, we select point (0, 0) to test for the region. Since $x = 0$ and $y = 0$ satisfy $-3x + y > -2$, we conclude that *all* the points above the line satisfy $-3x + y > -2$. The region in question is shaded in Figure 2. Notice that a dashed line is used to represent the graph of the equation $-3x + y = -2$; this emphasizes that the line is *not* included in the region satisfying the inequality.

Figure 2

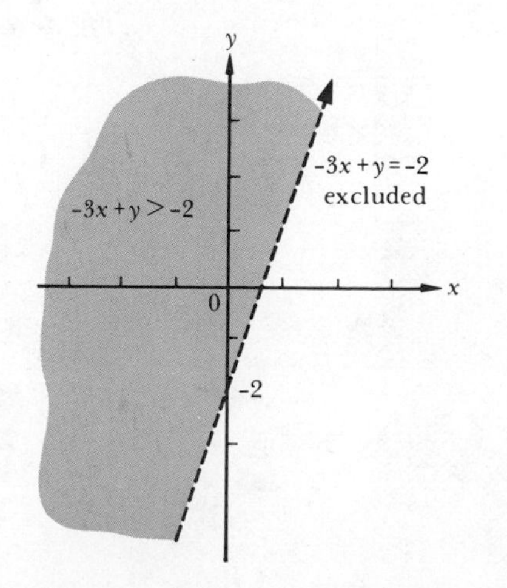

The region consisting of points whose coordinates satisfy a *system of linear inequalities* can be found by finding the intersection of the regions that are the graphs of the respective inequalities in the system. This technique is illustrated in the following examples.

EXAMPLES Sketch the region containing *all* points whose coordinates satisfy the given system of linear inequalities. Also, find the *corner points* of the boundaries of the region.

1 $\begin{cases} x \geqslant 0 \\ y \geqslant 0 \\ x + y \leqslant 1 \end{cases}$

SOLUTION (See Figure 3.) The region for $x \geqslant 0$ contains all points on or to the right of the y axis; the region for $y \geqslant 0$ contains all points on or above the x axis; and the region for $x + y \leqslant 1$ contains all points on or below the line $x + y = 1$. Thus the desired region is the set of all points in the intersection of the three regions. The corner points of the boundary of the triangular region are (0, 0), (1, 0), and (0, 1).

2 $\begin{cases} y < x + 2 \\ x + y \leqslant 4 \end{cases}$

SOLUTION (See Figure 4.) The graph of $y < x + 2$ contains all points below the line $y = x + 2$. The graph of $x + y \leqslant 4$ contains all points on or below the line $x + y = 4$. Thus the solution of the given system contains all points in the intersection of the two regions. The corner point of the boundary can be found by solving the linear system

$$\begin{cases} y = x + 2 \\ x + y = 4 \end{cases}$$

to get (1, 3).

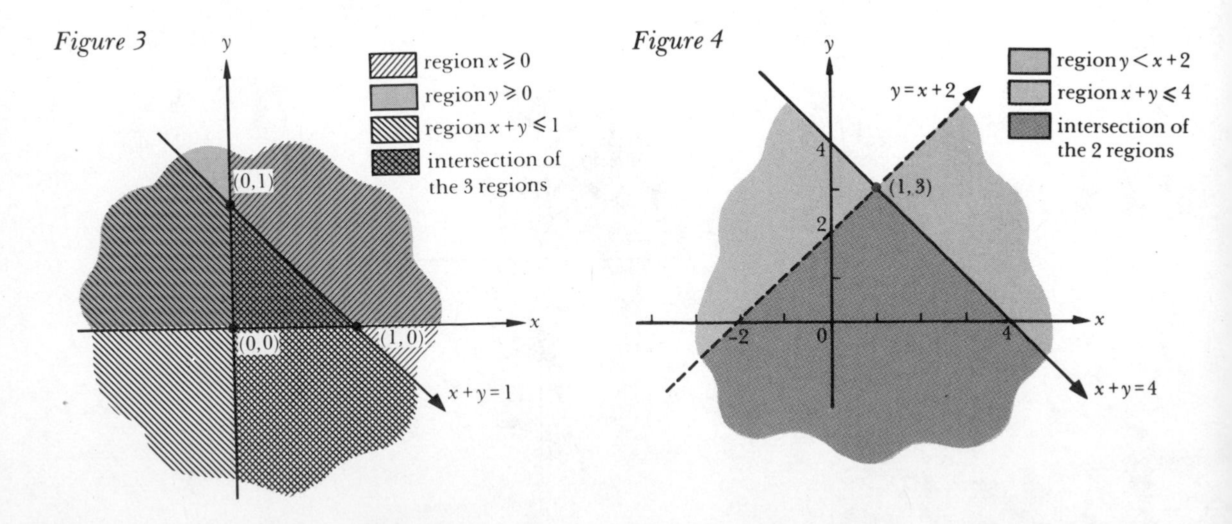

4.2 Linear Programming

Suppose that we are given the problem of finding the maximum value of the linear expression $3x + 2y$, where x and y are subject to the conditions of the following system of linear inequalities:

$$\begin{cases} x \geq 0 \\ y \geq 0 \\ x + y \leq 6 \\ x + 3y \leq 10 \end{cases}$$

These four inequalities are referred to as the *constraints* or *constraint set* of the problem.

The constraint set is graphed and the corner points are labeled as shown in Figure 5. This region represents *all* possible values for (x, y) that satisfy the system of inequalities (the constraint set).

Figure 5

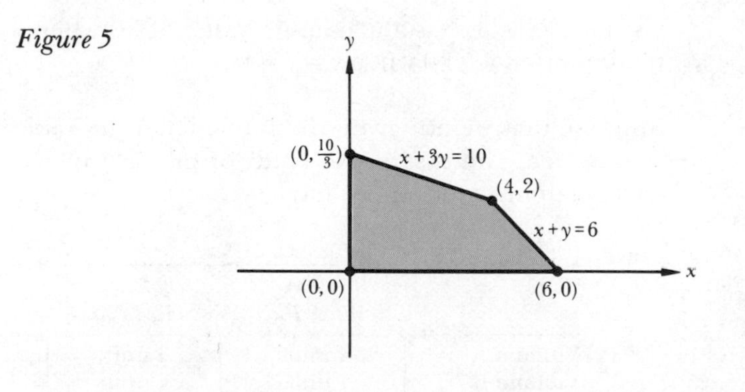

The theory of linear programming establishes the fact that the *maximum* or *minimum* value of a linear expression occurs at a corner point of the boundary of the constraint set.

Table 1 indicates the value of $3x + 2y$ for each corner point.

Table 1

Corner Point	*Value of* $3x + 2y$
(0, 0)	0
(6, 0)	18
(4, 2)	16
$(0, \frac{10}{3})$	$\frac{20}{3}$

Thus, by using linear programming, the maximum value of the expression $3x + 2y$ under the given constraints is 18; this occurs when $x = 6$ and $y = 0$.

The next few examples illustrate the use of linear programming techniques and their practical application.

EXAMPLES 1 Find the maximum and minimum values of $2x + 3y$ under the constraints

$$\begin{cases} x \geqslant 0 \\ y \geqslant 0 \\ 2y + x \leqslant 16 \\ x - y \leqslant 10 \end{cases}$$

Figure 6

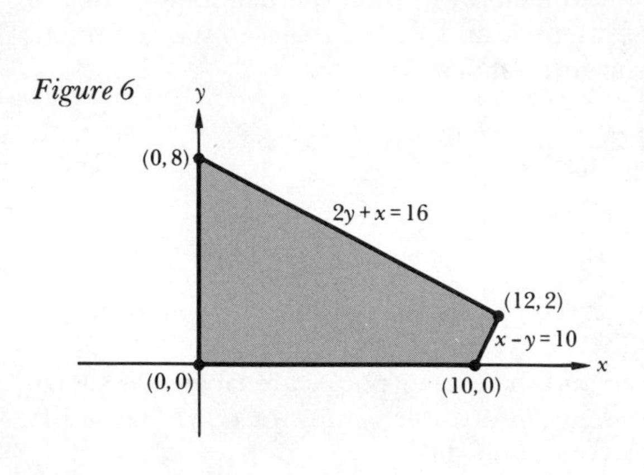

SOLUTION The constraint set is graphed and the corner points are labeled in Figure 6. Table 2 gives the values of $2x + 3y$ at the corner points.

Table 2

Corner Point	*Value of* $2x + 3y$
(0, 0)	0
(10, 0)	20
(12, 2)	30
(0, 8)	24

Thus, $2x + 3y$ has a maximum value of 30 when $x = 12$ and $y = 2$ and a minimum value of 0 when $x = y = 0$.

2 Suppose that we are given the information in Table 3 regarding the vitamin content and cost of two kinds of pills as well as the minimum units of each vitamin required per day.

Table 3

	White Pill	*Red Pill*	*Minimum Units Required per Day*
Vitamin A	4 units	1 unit	8
Vitamin B	1 unit	1 unit	5
Vitamin C	2 units	7 units	20
Cost per pill	7¢	14¢	

How many of each kind of pill should be taken in order to fulfill the minimum daily requirements and at the same time minimize the cost?

SOLUTION If we let x represent the number of white pills and y represent the number of red pills, then we must minimize $7x + 14y$ under the following constraints:

$$\begin{cases} x \geqslant 0 \\ y \geqslant 0 \\ 4x + y \geqslant 8 \\ x + y \geqslant 5 \\ 2x + 7y \geqslant 20 \end{cases}$$

Figure 7 shows the graph of the constraint set, together with the corner points found by solving systems of linear equations. The corner points yield the values for $7x + 14y$ given in Table 4.

Figure 7

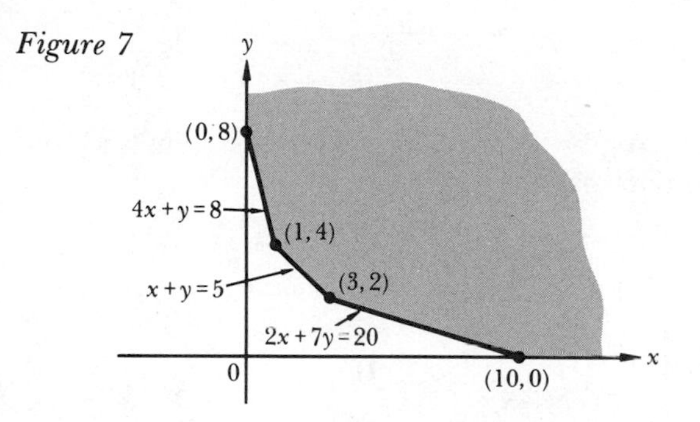

Table 4

Corner Point	*Value of* $7x + 14y$
(0, 8)	\$1.12
(1, 4)	\$0.63
(3, 2)	\$0.49
(10, 0)	\$0.70

Thus, three white and two red pills should be taken to fulfill the minimum daily requirement at a minimum cost.

3 A company produces automobile panels on two different production lines. The total work force provides 500 hours of production per week and the production budget is restricted to \$3000 per week. The newer production line takes $\frac{1}{2}$ hour to produce a single part at \$5 per part, whereas the older line requires 2 hours at \$4 per part. How many panels should be produced on each line in order to maximize productivity each week?

SOLUTION Let x represent the number of panels produced by the new line and y the number of panels produced by the older line. Thus we need to maximize $x + y$ under the constraints

$$\begin{cases} x \geq 0 \\ y \geq 0 \\ \frac{1}{2}x + 2y \leq 500 \\ 5x + 4y \leq 3000 \end{cases}$$

Figure 8

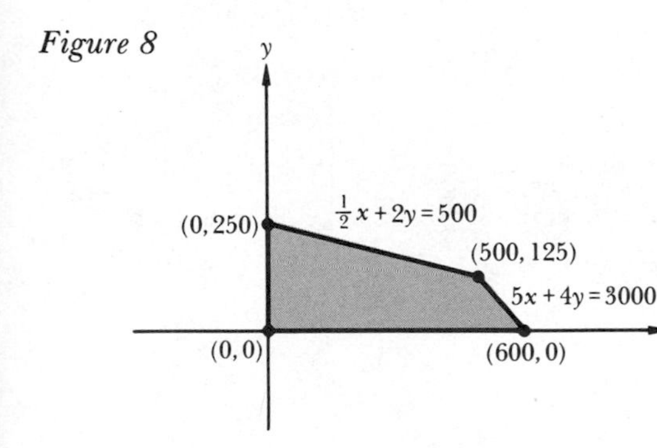

The constraints are graphed in Figure 8 and the corner points are labeled. Table 5 gives the values of $x + y$ at the corner points.

Table 5

Corner Point	*Value of* $x + y$
(0, 0)	0
(600, 0)	600
(500, 125)	625
(0, 250)	250

Thus production is maximized at 625 panels per week if the newer line produces 500 parts and the older line produces 125 parts per week.

PROBLEM SET 4

In Problems 1–10, sketch the region in the plane determined by the given inequality.

1 $x \geq 2$ **2** $y < -1$ **3** $y \geq -\frac{1}{2}x$ **4** $3y \leq 2x + 1$

5 $2x - 3y < 0$ **6** $y > 2 - x$ **7** $x - 2y \leq 1$

8 $y + 2x \geq 5$ **9** $2x + 3y < 4$ **10** $-3x - 4y > 12$

In Problems 11–20, sketch the region defined by the given system and find the corner points of the boundary of the region.

11 $\begin{cases} x < 3 \\ y \geq 2 \end{cases}$ **12** $\begin{cases} 1 \leq x < 2 \\ -1 < y \leq 2 \end{cases}$ **13** $\begin{cases} y \geq x \\ y \leq 2 \end{cases}$

14 $\begin{cases} y \leq x \\ y \geq -1 \end{cases}$ **15** $\begin{cases} x + y \leq 2 \\ y - 1 > 2x \end{cases}$ **16** $\begin{cases} y - x \leq -1 \\ 3y - x > 4 \end{cases}$

17 $\begin{cases} x + y \leq 3 \\ x - y \geq 3 \end{cases}$ **18** $\begin{cases} x + 2y > 5 \\ 5x - y < 9 \end{cases}$

19 $\begin{cases} 2x - 3y \leq -3 \\ 5x - 2y \geq 9 \end{cases}$ **20** $\begin{cases} y - 2x \leq 4 \\ y + 2x \geq 3 \end{cases}$

In Problems 21–30, graph the region defined by each constraint system, label each corner point, and then find the maximum and minimum values of the given linear expression over the region.

21 $3x + 5y$ $\begin{cases} x \geq 0 \\ y \geq 0 \\ 2x + y \leq 6 \end{cases}$

22 $2x - y$ $\begin{cases} 0 \leq x \leq 4 \\ 0 \leq y \leq 3 \end{cases}$

23 $2x + y$ $\begin{cases} x \geq 0 \\ y \geq 0 \\ 4x + y \leq 36 \\ 4x + 3y \leq 60 \end{cases}$

24 $15x + 25y$ $\begin{cases} x \geq 0 \\ y \geq 0 \\ x + y \leq 50 \\ 2x - y \leq 40 \\ -3x + y \leq 10 \end{cases}$

25 $7x - 3y$ $\begin{cases} x \geq 0 \\ y \leq 4 \\ x + y \geq 1 \\ x - y \leq 1 \end{cases}$

26 $4x + 2y$ $\begin{cases} x \geq 0 \\ y \geq 0 \\ x + 3y \leq 15 \\ 2x + y \leq 10 \end{cases}$

27 $5x + 4y$ $\begin{cases} x \geq 0 \\ y \geq 0 \\ x + 2y \geq 3 \\ 2y \leq 5 - x \end{cases}$

28 $x + 2y$ $\begin{cases} y \geq 0 \\ x + y \geq 1 \\ x + 2y \leq 2 \end{cases}$

29 $3x + 4y$ $\begin{cases} 2x + y \geq 2 \\ x + 2y \geq 2 \\ x + y \leq 2 \end{cases}$

30 $x + 5y$ $\begin{cases} x \geq 0 \\ y \geq 0 \\ 5x + 4y \leq 2000 \\ 5x + 12y \leq 3000 \end{cases}$

In Problems 31–35, use linear programming to solve each problem.

31 Suppose that because of limited storage capacity, a restaurant owner can order no more than 200 pounds of ground beef per week for making hamburgers and tacos. Each hamburger contains $\frac{1}{3}$ pound of ground beef, while each taco contains $\frac{1}{4}$ pound. His profit is 15 cents on each hamburger and 20 cents on each taco. Labor cost is 8 cents for each hamburger and 12 cents for each taco. If he is willing to pay at most $60 for labor costs, how many tacos and hamburgers should he sell to maximize his profits?

32 A refinery produces a combined maximum of 25,000 barrels of gasoline and diesel oil per day. The refinery must produce at least 5000 barrels of diesel oil each day. If the profit is $6.50 per barrel of gasoline and $4.00 per barrel of diesel oil, find the maximum profit and how many barrels of each product must be made to yield this maximum.

33 A boat builder has available 100 units of wood, 160 units of plastic, and 400 units of fiberglass. Each regular boat produced requires 1 unit of wood, 2 units of plastic, and 2 units of fiberglass and yields a profit of $1000. By comparison, each deluxe model produced requires 1 unit of wood, 1 unit of plastic, and 5 units of fiberglass and yields a $1500 profit. How many boats of each type should be produced in order to maximize the profit?

34 Suppose that each serving of a special food contains 2 units of vitamin B and 5 units of iron and that each glass of a special drink contains 4 units of vitamin B and 2 units of iron. A minimum of 80 units of vitamin B and 60 units of iron must be provided each day. How much of the food and drink need be consumed in order to meet the daily requirements and at the same time minimize costs if each serving of the food is $1.00 and each drink is $0.80?

35 A manufacturer makes two types of fertilizer. One type is 80 percent nitrogen and 20 percent potassium. Another type is 60 percent nitrogen and 40 percent potassium. He needs at least 30 tons of the first type and at least 50 tons of the second type. In attempting to make as much fertilizer as possible, how many tons of each type should he produce if he has 100 tons of nitrogen and 50 tons of potassium?

5 Systems with Quadratic Equations

The methods that were introduced in Section 1 can be used to solve systems containing second-degree equations. If A is the set of ordered pairs of numbers that satisfy one equation in a system containing two equations, and B is the set of ordered pairs of numbers satisfying the second equation in the system, then the *solution set* of the system is $A \cap B$—namely, the set of all ordered pairs common to A and B.

For example, to solve the system of equations

$$\begin{cases} x + y = 7 \\ x^2 + y^2 = 25 \end{cases}$$

we use the substitution method. Solve the first equation for y in terms of x to obtain $y = 7 - x$. Then, substitute the result into the second equation, $x^2 + y^2 = 25$. Thus, we have $x^2 + (7 - x)^2 = 25$ or $x^2 - 7x + 12 = 0$, so that $x = 3$ or $x = 4$. Hence, when $x = 3$, then $y = 7 - 3 = 4$, and when $x = 4$, then $y = 7 - 4 = 3$, and the solution set is $\{(3, 4), (4, 3)\}$.

If we had used the equation $x^2 + y^2 = 25$ to find values of y corresponding to $x = 3$ and $x = 4$, we would have obtained the *four* ordered pairs (3, 4), (3, −4), (4, 3), and (4, −3). However, only *two* ordered pairs, (3, 4) and (4, 3), satisfy *both* equations of the system.

Hence, in solving systems involving one first-degree equation and one second-degree equation, the final substitution should be made in *both* equations.

The two ordered pairs in the solution set of the above system of equations can be illustrated graphically. The graph of $x^2 + y^2 = 25$ is a circle and that of $x + y = 7$ is a straight line (Figure 1). Graphing both equations on the same coordinate axes, we find that they intersect at exactly two points.

Figure 1

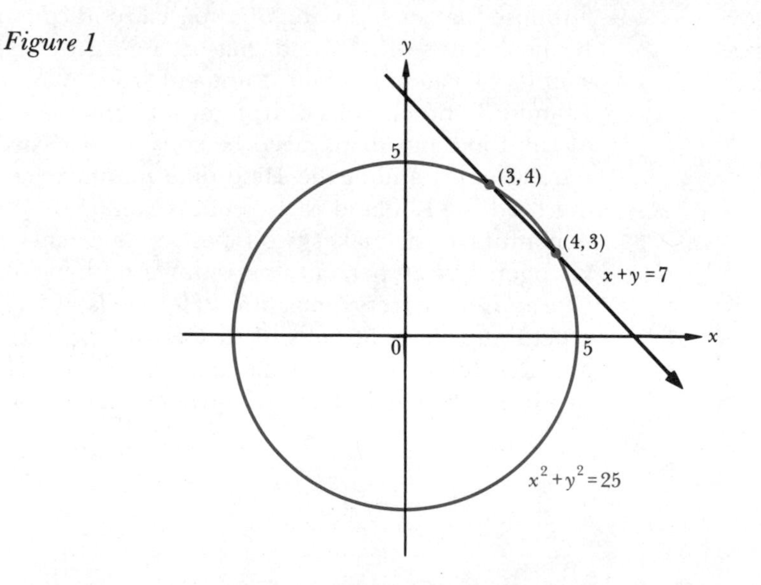

Sketching the graph of each equation on the same coordinate system is helpful in determining the *number* of ordered pairs to expect when solving the given system. This is illustrated in the following examples.

EXAMPLES For each of the following systems, use the graphs of the equations to determine the number of ordered pairs in the solution set; then find the solution set of the system of equations.

1 $\begin{cases} 2x + 3y = 8 \\ 2x^2 - 3y^2 = -10 \end{cases}$

SOLUTION From the graphs of the two given equations we see that there are two points of intersection, one in the first quadrant and one in the second quadrant (Figure 2). These points of intersection of the two graphs are in the solution set of the system of equations. Solving the linear equation for x, we obtain

$$x = \frac{8 - 3y}{2}$$

Figure 2

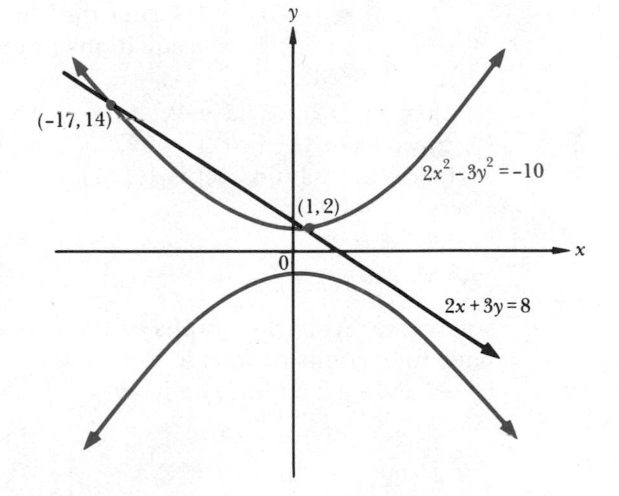

Substituting $\frac{8 - 3y}{2}$ for x into the second-degree equation, we obtain

$$2\left(\frac{8 - 3y}{2}\right)^2 - 3y^2 = -10$$

so that

$$3y^2 - 48y + 84 = 0 \quad \text{or} \quad y^2 - 16y + 28 = 0$$

By factoring, we obtain $(y - 2)(y - 14) = 0$, so that $y = 2$ or $y = 14$.

When $y = 2$,

$$x = \frac{8 - 3(2)}{2} = 1$$

and when $y = 14$,

$$x = \frac{8 - 3(14)}{2} = -17$$

Hence, the solution set of the system is $\{(1, 2), (-17, 14)\}$.

Figure 3

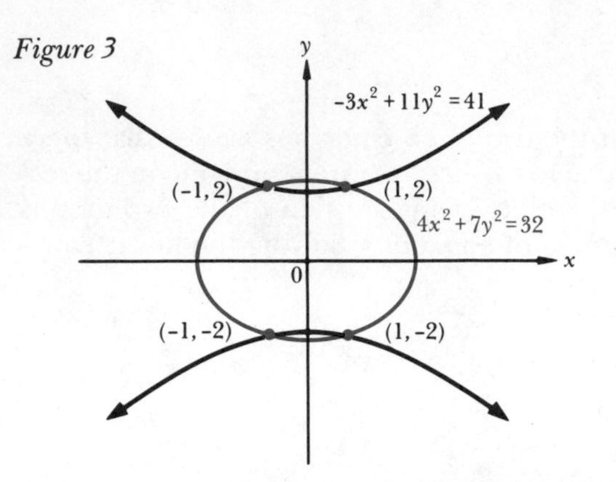

2 $\begin{cases} 4x^2 + 7y^2 = 32 \\ -3x^2 + 11y^2 = 41 \end{cases}$

SOLUTION From the graphs of the two given equations, it is evident that we have four points of intersection (Figure 3). Thus, we conclude that there are four ordered pairs that satisfy the system. Multiplying the first equation by 3 and the second by 4 results in

$$\begin{cases} 12x^2 + 21y^2 = 96 \\ -12x^2 + 44y^2 = 164 \end{cases}$$

Using the elimination method, we add the two equations to get $65y^2 = 260$, so that $y^2 = 4$; that is, $y = -2$ or $y = 2$.

When $y = -2$, we have $4x^2 + 7(-2)^2 = 32$, so $x = -1$ or $x = 1$, and when $y = 2$, we have $4x^2 + 7(2)^2 = 32$, so $x = -1$ or $x = 1$.

Hence, the solution set is $\{(1, 2), (-1, 2), (1, -2), (-1, -2)\}$.

3 $\begin{cases} x^2 + 2y^2 = 22 \\ 2x^2 + y^2 = 17 \end{cases}$

SOLUTION From the graphs of the two given equations it is clear that we have four points of intersection (Figure 4). Multiplying the first equation by -2 and then adding, we have

$$\begin{cases} -2x^2 - 4y^2 = -44 \\ 2x^2 + y^2 = 17 \end{cases}$$

Using the elimination method, we add the two equations to get $-3y^2 = -27$, so that $y^2 = 9$, so $y = 3$ or $y = -3$.

When $y = 3$, we have $x^2 + 2(3)^2 = 22$, so that $x = 2$ or $x = -2$, and when $y = -3$, we have $x^2 + 2(-3)^2 = 22$, so $x = 2$ or $x = -2$.

Hence, the solution set is $\{(2, 3), (-2, 3), (2, -3), (-2, -3)\}$.

Figure 4

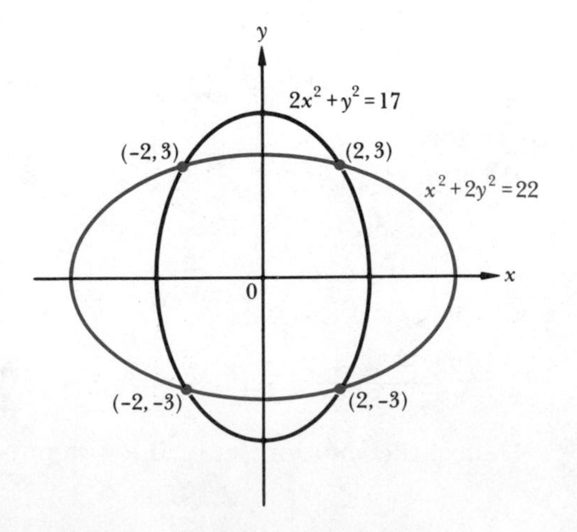

PROBLEM SET 5

In Problems 1–30, use the graphs of the equations to determine the number of ordered pairs in the solution set; then find the solution set of the system of equations.

1 $\begin{cases} x - y = 1 \\ x^2 + y^2 = 5 \end{cases}$

2 $\begin{cases} x - 2y = 3 \\ x^2 - y^2 = 24 \end{cases}$

3 $\begin{cases} 3x - y = 2 \\ x^2 + y^2 = 20 \end{cases}$

4 $\begin{cases} x + y = 3 \\ 3x^2 - y^2 = \frac{9}{2} \end{cases}$

5 $\begin{cases} 3x + 2y = 1 \\ 3x^2 - y^2 = -4 \end{cases}$

6 $\begin{cases} x + y = 6 \\ x^2 + y^2 = 20 \end{cases}$

7 $\begin{cases} 5x - 3y = 10 \\ x^2 - y^2 = 6 \end{cases}$

8 $\begin{cases} 2x + y = 10 \\ xy = 12 \end{cases}$

9 $\begin{cases} 2x + 3y = 7 \\ x^2 + y^2 + 4y + 4 = 0 \end{cases}$

10 $\begin{cases} x - y + 4 = 0 \\ x^2 + 3y^2 = 12 \end{cases}$

11 $\begin{cases} 5x - y = 21 \\ y = x^2 - 5x + 4 \end{cases}$

12 $\begin{cases} x^2 - 25y^2 = 20 \\ 2x^2 + 25y^2 = 88 \end{cases}$

13 $\begin{cases} x - y^2 = 0 \\ x^2 + 2y^2 = 24 \end{cases}$

14 $\begin{cases} 3x^2 - 8y^2 = 40 \\ 5x^2 + y^2 = 81 \end{cases}$

15 $\begin{cases} 2x^2 - 3y^2 = 6 \\ 3x^2 + 2y^2 = 35 \end{cases}$

16 $\begin{cases} x^2 - y^2 = 7 \\ x^2 + y^2 = 25 \end{cases}$

17 $\begin{cases} x^2 + 9y^2 = 33 \\ x^2 + y^2 = 25 \end{cases}$

18 $\begin{cases} x^2 + 5y^2 = 70 \\ 3x^2 - 5y^2 = 30 \end{cases}$

19 $\begin{cases} 4x^2 - y^2 = 4 \\ 4x^2 + \frac{5}{3}y^2 = 36 \end{cases}$

20 $\begin{cases} x^2 - 2y^2 = 17 \\ 2x^2 + y^2 = 54 \end{cases}$

21 $\begin{cases} 2x^2 - 3y^2 = 20 \\ x^2 + 2y = 20 \end{cases}$

22 $\begin{cases} 4x^2 + 3y^2 = 43 \\ 3x^2 - y^2 = 3 \end{cases}$

23 $\begin{cases} x^2 - 2y^2 = 1 \\ x^2 + 4y^2 = 25 \end{cases}$

24 $\begin{cases} 2x^2 - 5y^2 + 8 = 0 \\ x^2 - 7y^2 + 4 = 0 \end{cases}$

25 $\begin{cases} x^2 + 4y = 8 \\ x^2 + y^2 = 5 \end{cases}$

26 $\begin{cases} 3x - 2y = 9 \\ 9x = y^2 \end{cases}$

27 $\begin{cases} x^2 + y^2 = 16 \\ x^2 - y^2 = -34 \end{cases}$

28 $\begin{cases} x^2 - 4y^2 = -15 \\ -x^2 + 3y^2 = 11 \end{cases}$

29 $\begin{cases} x^2 + y^2 = 25 \\ (x - 5)^2 + y^2 = 9 \end{cases}$

30 $\begin{cases} x^2 - y = 0 \\ x^2 + (y - 6)^2 = 36 \end{cases}$

REVIEW PROBLEM SET

In Problems 1–6, use the graphs of the equations in each system to determine whether the system is dependent, inconsistent, or independent. Use the substitution method to find the solution set of each independent system.

1 $\begin{cases} y = -2x + 2 \\ y = x - 4 \end{cases}$

2 $\begin{cases} y = 5x + 2 \\ 10x - 2y + 4 = 0 \end{cases}$

3 $\begin{cases} 3x + 2y = 1 \\ 3x - 2y = 2 \end{cases}$

4 $\begin{cases} x + y = 4 \\ 2x - y = 8 \end{cases}$

5 $\begin{cases} \frac{1}{2}x + \frac{1}{3}y = \frac{1}{5} \\ x - \frac{1}{4}y = 7 \end{cases}$

6 $\begin{cases} 0.3x - 0.7y = 1 \\ 2.3x + y = -2 \end{cases}$

In Problems 7–8, use the substitution method to find the solution set of each linear system.

7 $\begin{cases} x + y = 1 \\ 3x - y + z = 16 \\ 2y + 3z = 11 \end{cases}$

8 $\begin{cases} x + y + z = 1 \\ x - y + z = 1 \\ x - y - z = 1 \end{cases}$

In Problems 9–12, use the elimination method to find the solution set of each linear system.

9 $\begin{cases} x - y = 3 \\ 2x + y = 3 \end{cases}$

10 $\begin{cases} 5x + 2y = 3 \\ 2x - 3y = 5 \end{cases}$

11 $\begin{cases} x - y + 2z = 0 \\ 3x + y + z = 2 \\ 2x - y + 5z = 5 \end{cases}$

12 $\begin{cases} 3x + 2y - z = -4 \\ x - y + 2z = 13 \\ 5x + 3y - 4z = -15 \end{cases}$

In Problems 13–16, use row reduction to find the solution set of each linear system.

13 $\begin{cases} 4x - y = -4 \\ x + 2y = 6 \end{cases}$

14 $\begin{cases} 2x + 3y = 9 \\ 5x - 2y = 5 \end{cases}$

15 $\begin{cases} 3x - y + 2z = 1 \\ x - 2y + 4z = 2 \\ x - y + z = 0 \end{cases}$

16 $\begin{cases} x + 2y + 3z = 4 \\ -x - 4y + z = 1 \\ x + y + z = 0 \end{cases}$

In Problems 17–20, evaluate each determinant.

17 $\begin{vmatrix} 1 & -1 \\ 3 & 2 \end{vmatrix}$

18 $\begin{vmatrix} \sin 25^\circ & \cos 25^\circ \\ -\cos 25^\circ & \sin 25^\circ \end{vmatrix}$

19 $\begin{vmatrix} 4 & 2 & 1 \\ 5 & 7 & 1 \\ 6 & 2 & 3 \end{vmatrix}$

20 $\begin{vmatrix} 1 & 2 & 1 \\ 1 & 3 & 4 \\ 1 & 4 & 9 \end{vmatrix}$

21 Use Cramer's rule to find the solution set of each system of linear equations in Problems 13–16.

22 Solve for x.

(a) $\begin{vmatrix} x & 0 & 0 \\ 3 & x & 2 \\ 0 & 4 & 1 \end{vmatrix} = 5$ (b) $\begin{vmatrix} -2 & 1 & x \\ 1 & x+1 & -2 \\ x-6 & 3 & -1 \end{vmatrix} = 16$

In Problems 23–26, graph the region defined by each constraint system, locate each corner point, and then find the maximum and minimum values of the given linear expression over the region specified.

23 $7x + 3y$

$$\begin{cases} x \geqslant 0 \\ y \geqslant 0 \\ 7x + 2y \leqslant 14 \end{cases}$$

24 $2x + y$

$$\begin{cases} x \geqslant 1 \\ y \geqslant 2 \\ x + 2y \leqslant 10 \end{cases}$$

25 $x + 5y$

$$\begin{cases} x \geqslant 0 \\ y \geqslant 0 \\ x + y \leqslant 4 \\ 4x + y \leqslant 7 \end{cases}$$

26 $x + y$

$$\begin{cases} x \geqslant 0 \\ y \geqslant 0 \\ 3y - 2x \leqslant 6 \\ 3y + 4x \leqslant 24 \end{cases}$$

In Problems 27–30, use the graphs of the equations to determine the number of ordered pairs in the solution set; then find the solution set of the system of equations.

27 $\begin{cases} 3x - 4y = 25 \\ x^2 + y^2 = 25 \end{cases}$

28 $\begin{cases} 2x - y = 2 \\ x^2 + 2y^2 = 12 \end{cases}$

29 $\begin{cases} x + y^2 = 6 \\ x^2 + y^2 = 36 \end{cases}$

30 $\begin{cases} 3x^2 - 2y^2 = 35 \\ 7x^2 + 5y^2 = 43 \end{cases}$

In Problems 31–34, use systems of linear equations to solve each problem.

31 A textbook publisher mailed out 160 letters, some requiring 13 cents postage and the rest requiring 18 cents postage. If the total bill was \$25.80, find the number sent at each rate.

32 The specific gravity of an object is defined to be its weight in air divided by its loss of weight when submerged in water. An object made partly of gold (specific gravity 16) and partly of silver (specific gravity 10.8) weighs 8 grams in air and 7.3 grams when submerged in water. How many grams of gold and how many grams of silver are in the object?

33 A coin collection containing nickels, dimes, and quarters consists of 35 coins altogether. If there are twice as many nickels as quarters, and one-fourth as many dimes as nickels, find the number of each type of coin in the collection.

34 Find the values of a, b, and c in the equation $y = ax^2 + bx + c$ if its graph contains the points $(1, 3)$, $(3, 5)$, and $(-1, 9)$.

In Problems 35–36, use linear programming to solve each problem.

35 A cabinetmaker produces a deluxe model as well as a regular model of a cabinet to house a stereo. There is enough lumber in stock to produce a total of 300 deluxe models, at a profit of \$45 per cabinet, or 400 regular models, at a profit of \$25 per cabinet. If he can produce no more than 600 cabinets altogether, how many of each type should he make in order to maximize his profit?

36 A farmer ships 800 crates of produce by truck to market each week. He must deliver at least 200 crates of potatoes, at a profit of \$2 per box, at least 100 crates of beans, at a profit of \$1 a box, and at most 200 crates of tomatoes, at a profit of \$3 a crate. How should the truck be loaded to maximize profit?

CHAPTER 9

Complex Numbers and Discrete Algebra

In Chapter 3, we discussed the real solutions of quadratic equations, that is, equations of the form $ax^2 + bx + c = 0$, where a, b, and c are real numbers and $a \neq 0$. At times, we encounter quadratic equations such as $x^2 + 1 = 0$. However, there is no real number x that satisfies this equation, since there is no real number whose square is -1. In this chapter, we discuss *complex numbers,* which enable us to solve equations like $x^2 + 1 = 0$. Complex numbers occur in many practical applications of mathematics in such fields as engineering, physics, and chemistry. Other topics covered in the present chapter include the theory of equations, mathematical induction, the binomial theorem, and finite sums and series.

1 Complex Numbers

A *complex number* is a number of the form $a + bi$, where a and b are real numbers. The symbol i denotes a number whose square is -1; that is, i is defined by the equation $i^2 = -1$. Therefore, i is *not* a real number. The algebra of complex numbers is defined in terms of the algebra of real numbers as follows.

Let $z_1 = a_1 + b_1 i$ and $z_2 = a_2 + b_2 i$ be complex numbers.

Equality: $z_1 = z_2$ if and only if $a_1 = a_2$ and $b_1 = b_2$.

Addition: $z_1 + z_2 = (a_1 + a_2) + (b_1 + b_2)i$.

Multiplication: $z_1 \cdot z_2 = (a_1 a_2 - b_1 b_2) + (a_1 b_2 + a_2 b_1)i$.

For $z = a + bi$, a is called the *real part* of z and b is called the *imaginary part* of z. It would perhaps be better to identify a as the "non-i part" and b as the "i part" of the complex number, but the choice of words "real part" and "imaginary part" is accepted today for historical reasons. Hence two complex numbers are equal if and only if the real parts are equal and the imaginary parts are equal; the real part of the sum of two complex numbers is the sum of the real parts and the imaginary part of the sum of two complex numbers is the sum of the imaginary parts.

The set of complex numbers will be denoted by C, so in set notation we have $C = \{a + bi \mid a, b \in R, i^2 = -1\}$.

If the imaginary part of a complex number is 0, we shall consider the number to be real; hence $a + 0i$ will be considered to be the real number a. In this sense, R can be considered to be a proper subset of C.

EXAMPLES 1 Find the sum and product of each of the following pairs of complex numbers and identify the real and imaginary parts of each result.

(a) $3i, 3i$ (b) $5 + 6i, 9 + 3i$ (c) $4 - 2i, -3 + i$

SOLUTION

(a) $$3i + 3i = (0 + 0) + (3 + 3)i = 6i$$

The real part of $6i$ is 0 and the imaginary part is 6.

$$3i \cdot 3i = (0 + 3i)(0 + 3i) = (0 \cdot 0 - 3 \cdot 3) + (0 \cdot 3 + 3 \cdot 0)i = -9$$

The real part of -9 is -9 and the imaginary part is 0.

(b) $$(5 + 6i) + (9 + 3i) = 14 + 9i$$

Here, 14 is the real part and 9 is the imaginary part of $14 + 9i$.

$$(5 + 6i)(9 + 3i) = (45 - 18) + (54 + 15)i = 27 + 69i$$

Here, 27 is the real part and 69 is the imaginary part of $27 + 69i$.

(c) $$(4 - 2i) + (-3 + i) = 1 - i$$

Here, 1 is the real part and -1 is the imaginary part of $1 - i$.

$$(4 - 2i)(-3 + i) = (-12 + 2) + (6 + 4)i = -10 + 10i$$

The real part of $-10 + 10i$ is -10 and the imaginary part is 10.

2 Write each of the following complex numbers in the form of $a + bi$. Identify the real and imaginary parts of each result.

(a) $(1 + 2i)^2$ (b) $i^4 + 3i^3$

SOLUTION

(a) $(1 + 2i)^2 = (1 + 2i)(1 + 2i) = (1 - 4) + (2 + 2)i = -3 + 4i$. The real part of $-3 + 4i$ is -3 and the imaginary part is 4.

(b) $i^4 + 3i^3 = i^2 \cdot i^2 + 3i^2 i = (-1)(-1) + 3(-1)i = 1 - 3i$. The real part of $1 - 3i$ is 1 and the imaginary part is -3.

In Example 1(a) above, we have seen one of the properties that does not hold in R but holds in C: If $x \in R$, $x^2 \geqslant 0$, whereas it is possible to have $z \in C$ such that $z^2 < 0$; recall that $i^2 = -1$.

If the order properties of R did hold in C, then, by the Trichotomy Principle,

$$i = 0 \quad \text{or} \quad i < 0 \quad \text{or} \quad i > 0$$

But, if $i = 0$, then $i \cdot i = 0$ implies that $-1 = 0$; hence $i \neq 0$. On the other hand, if $i > 0$, then $i^2 > 0$ implies that $-1 > 0$ since $i^2 = -1$; hence $i \not> 0$. Finally, if $i < 0$, then $i^2 > 0$, or $-1 > 0$ since $i^2 = -1$; hence $i \not< 0$. Consequently, the Trichotomy Principle does *not* hold in C; that is, the order relation that exists in R does not exist in C.

The domain of functions can be extended to the set of complex numbers. For example, consider the function f whose domain is the set of complex numbers and whose rule of formation is $f(z) = 3z + 1$; then, we have $f(i) = 3i + 1$ and $f(1 - i) = 3(1 - i) + 1 = 4 - 3i$.

1.1 Properties of Addition and Multiplication

Since we defined the operations of addition and multiplication on C in terms of the corresponding operations on R, it is not surprising that addition and multiplication on C have those properties of addition and multiplication on R listed in Chapter 1, Section 1.

Assume that $z_1, z_2, z_3 \in C$; then the following properties hold.

1 CLOSURE OF ADDITION AND MULTIPLICATION

(i) $z_1 + z_2 \in C$
(ii) $z_1 \cdot z_2 \in C$

2 COMMUTATIVITY OF ADDITION AND MULTIPLICATION

(i) $z_1 + z_2 = z_2 + z_1$
(ii) $z_1 z_2 = z_2 z_1$

3 ASSOCIATIVITY OF ADDITION AND MULTIPLICATION

(i) $(z_1 + z_2) + z_3 = z_1 + (z_2 + z_3)$
(ii) $(z_1 \cdot z_2) \cdot z_3 = z_1 \cdot (z_2 \cdot z_3)$

4 DISTRIBUTIVE PROPERTIES

(i) $z_1 \cdot (z_2 + z_3) = z_1 \cdot z_2 + z_1 \cdot z_3$
(ii) $(z_1 + z_2) \cdot z_3 = z_1 \cdot z_3 + z_2 \cdot z_3$

5 IDENTITIES

(i) There exists $0 \in C$ such that $z + 0 = 0 + z = z$ for every $z \in C$.
(ii) There exists $1 \in C$ such that $z \cdot 1 = 1 \cdot z = z$ for every $z \in C$.

6 INVERSES

(i) If $z \in C$, then there exists $-z \in C$ such that

$$z + (-z) = (-z) + z = 0$$

(ii) If $z \in C$ and $z \neq 0$, then there exists $z^{-1} \in C$ such that

$$z \cdot z^{-1} = z^{-1} \cdot z = 1$$

PROOF OF PROPERTY 6(i): Suppose that $z = a + bi$. By letting $-z = -a - bi$, we have $(a + bi) + (-a - bi) = 0 + 0i = 0$.

For the proofs of the other properties, see Problems 41–45.

1.2 Subtraction of Complex Numbers

If $z_1, z_2 \in C$, then $z_1 - z_2$, that is, the *difference* of z_1 and z_2, is defined as $z_1 - z_2 = z_1 + (-z_2)$.

For example, if $z_1 = 7 + 2i$ and $z_2 = 1 - 5i$, then

$$\begin{aligned} z_1 - z_2 &= (7 + 2i) - (1 - 5i) = (7 + 2i) + (-1 + 5i) \\ &= (7 - 1) + (2 + 5)i = 6 + 7i \end{aligned}$$

EXAMPLES Find $z_1 - z_2$ for each of the following pairs of complex numbers.

1 $z_1 = 7 + 4i$ and $z_2 = 3 + 5i$

SOLUTION

$$z_1 - z_2 = (7 + 4i) - (3 + 5i) = (7 + 4i) + (-3 - 5i) = 4 - i$$

2 $z_1 = 4 - 5i$ and $z_2 = -5 + 7i$

SOLUTION

$$z_1 - z_2 = (4 - 5i) - (-5 + 7i) = (4 - 5i) + (5 - 7i) = 9 - 12i$$

1.3 Division of Complex Numbers

The *conjugate* of a complex number $z = a + bi$, written $\bar{z}$ (read "z bar"), is defined as $\bar{z} = a - bi$.

For example, the conjugate of $3 + 3i$ is $3 - 3i$; the conjugate of the number -4 or $-4 + 0i$ is $-4 - 0i = -4$; and the conjugate of the number $5i = 0 + 5i$ is $0 - 5i = -5i$.

EXAMPLE Show that if z is a complex number, then $z\bar{z}$ is a nonnegative real number.

SOLUTION Let $z = a + bi$; then $\bar{z} = a - bi$, so that

$$\begin{aligned} z\bar{z} &= (a + bi)(a - bi) \\ &= (a^2 + b^2) + (ab - ab)i \\ &= (a^2 + b^2) + (0 \cdot i) \\ &= (a^2 + b^2) \in R \qquad \text{since } a \text{ and } b \text{ are real numbers} \end{aligned}$$

Since $a^2 \geq 0$ and $b^2 \geq 0$ for any $a, b \in R$, we have $a^2 + b^2 \geq 0$, so $z\bar{z} \geq 0$.

If $z_1 = a_1 + b_1 i$, for $z_1 \neq 0$, and $z_2 = a_2 + b_2 i$, then the *quotient* $z_2 \div z_1$ is given by

$$\frac{z_2}{z_1} = \frac{z_2 \cdot \bar{z}_1}{z_1 \cdot \bar{z}_1}$$

or, equivalently, by

$$\frac{a_2 + b_2 i}{a_1 + b_1 i} = \frac{(a_2 + b_2 i)(a_1 - b_1 i)}{(a_1 + b_1 i)(a_1 - b_1 i)} = \frac{(a_2 a_1 + b_2 b_1) + (a_1 b_2 - b_1 a_2)i}{a_1{}^2 + b_1{}^2}$$

Notice that the result is a complex number of the form $a + bi$, with

$$a = \frac{a_2 a_1 + b_2 b_1}{a_1^2 + b_1^2} \quad \text{and} \quad b = \frac{a_1 b_2 - b_1 a_2}{a_1^2 + b_1^2}$$

and that the denominator is a real number.

For example, if $z_2 = 2 + 3i$, then

$$\frac{1}{z_2} = \frac{1}{2+3i} = \frac{1}{2+3i} \cdot \frac{2-3i}{2-3i} = \frac{2-3i}{4+9} = \frac{2}{13} - \frac{3}{13}i$$

If $z_1 = -2 + 3i$ and $z_2 = 1 - i$, then

$$\frac{z_1}{z_2} = \frac{-2+3i}{1-i} \cdot \frac{1+i}{1+i} = \frac{-5+i}{1+1} = -\frac{5}{2} + \frac{1}{2}i$$

EXAMPLES 1 Show that if z is a complex number, then $z + \bar{z}$ is a real number.

SOLUTION Let $z = a + bi$; then $\bar{z} = a - bi$, so that

$$z + \bar{z} = (a + a) + (b - b)i = 2a$$

which is a real number because a is a real number.

2 Let $z = 3 + 5i$. Find $\bar{z}$, $\bar{z} + z$, and $(z - \bar{z})/i$.

SOLUTION If $z = 3 + 5i$, then $\bar{z} = 3 - 5i$, so that

$$\bar{z} + z = (3 - 5i) + (3 + 5i) = 6$$

and

$$\frac{z - \bar{z}}{i} = \frac{(3+5i) - (3-5i)}{i} = \frac{10i}{i} = 10$$

3 Write each of the following quotients in the form $a + bi$.

(a) $\dfrac{1}{3-2i}$ (b) $\dfrac{1+i}{1-i}$ (c) $\dfrac{6+9i}{1-2i}$

SOLUTION

(a) $\dfrac{1}{3-2i} = \dfrac{1}{3-2i} \cdot \dfrac{3+2i}{3+2i} = \dfrac{3+2i}{9+4} = \dfrac{3}{13} + \dfrac{2}{13}i$

(b) $\dfrac{1+i}{1-i} = \dfrac{1+i}{1-i} \cdot \dfrac{1+i}{1+i} = \dfrac{(1+i)^2}{2} = \dfrac{2i}{2} = i = 0 + 1i$

(c) $\dfrac{6+9i}{1-2i} = \dfrac{6+9i}{1-2i} \cdot \dfrac{1+2i}{1+2i} = \dfrac{-12+21i}{5} = -\dfrac{12}{5} + \dfrac{21}{5}i$

4 If $z \neq 0$, show that $1/z$ is the multiplicative inverse of z; that is, show that $z \cdot z^{-1} = 1$ if $z^{-1} = 1/z$.

SOLUTION If $z = a + bi \neq 0$, then

$$\frac{1}{z} = \frac{1}{a+bi} \cdot \frac{a-bi}{a-bi} = \frac{a-bi}{a^2+b^2}$$

so that

$$z \cdot \frac{1}{z} = (a + bi) \cdot \frac{a-bi}{a^2+b^2} = \frac{a^2+b^2}{a^2+b^2} = 1$$

PROBLEM SET 1

In Problems 1–26, perform the indicated operations and write the answer in the form $a + bi$.

1 $(2 + 3i) + (4 + 5i)$

2 $(-1 + 2i) + (3 + 5i)$

3 $(2 + i) - (4 + 3i)$

4 $(-4 + 2i) - (3 + 2i)$

5 $(6 + i)(5 - 3i)$

6 $(3 - 4i)(7 - 3i)$

7 $(-7 + 2i)(-7 - 2i)$

8 $(-2 + 3i)(-2 - 3i)$

9 $i^{27} + i^5 - i^9$

10 $i^{18} - 3i^7$

11 $i^{14} - 3i^5$

12 $i^{102} + 5i^{51}$

13 $(3 - i)^2$

14 $(1 + 3i)^2$

15 $(2 - 3i)^4$

16 $(1 - \sqrt{2}i)^3$

17 $\dfrac{6}{7i}$

18 $-\dfrac{3}{5i^3}$

19 $4i^{-13}$

20 $\dfrac{5 + 8i}{3i}$

21 $\dfrac{7 - 3i}{5i}$

22 $\dfrac{3 - i}{2 + 5i}$

23 $\dfrac{3 + 5i}{4 - 3i}$

24 $\dfrac{7 + 4i}{3 + 5i}$

25 $\dfrac{2i}{(1 + i)^4}$

26 $\dfrac{2i^4}{(6 - i)^2}$

In Problems 27–28, find the real numbers x and y that satisfy each of the equations.

27 $x - 3 + 2yi = 8i$

28 $3x - y + ix - 2iy = 6 - 3i$

In Problems 29–32, find $\bar{z}$, the real part of z, the imaginary part of z, and $\dfrac{1}{z}$ for each of the numbers.

29 $z = 2 + \sqrt{3}i$

30 $z = 1 - \dfrac{1}{2}i$

31 $z = (3 - 2i)^2$

32 $z = 2i$

In Problems 33–38, prove each statement.

33 If $\bar{z} = z$, then z is a real number.

34 $\overline{z_1 z_2} = \bar{z}_1 \bar{z}_2$

35 $z + \bar{z} = 0$ if and only if the real part of z is 0.

36 $\overline{\left(\dfrac{z_1}{z_2}\right)} = \dfrac{\bar{z}_1}{\bar{z}_2}$, for $z_2 \neq 0$

37 $\overline{z_1 + z_2} = \bar{z}_1 + \bar{z}_2$

38 $\bar{\bar{z}} = z$

39 Let f be a function whose domain is the set of complex numbers and whose rule is given by $f(z) = z^2 + 5z + i$. Find

(a) $f(2 - i)$

(b) $f(2 + i)$

40 Let z_1 and z_2 be complex numbers, and let "Re z" denote the *real part* of z. Show that

$$\text{Re}\left(\frac{z_1}{z_1 + z_2}\right) + \text{Re}\left(\frac{z_2}{z_1 + z_2}\right) = 1$$

In Problems 41–45, use the properties of addition and multiplication on R that are listed in Chapter 1, Section 1, to prove each property.

41 C is closed under addition and multiplication
42 The associativity of addition and multiplication on C
43 The commutativity of addition and multiplication on C
44 The distributive properties on C
45 The identity properties of addition and multiplication on C

2 Geometric Representation of Complex Numbers

Each ordered pair of real numbers (a, b) can be associated with the complex number $z = a + bi$, and each complex number $z = a + bi$ can be associated with the ordered pair of real numbers (a, b). Because of this one-to-one correspondence between the set of complex numbers and the set of ordered pairs of real numbers, we use the points in the plane associated with the ordered pairs of real numbers to represent the complex numbers. For example, the ordered pairs $(2, -3)$, $(5, 2)$, and (e, π) are used to represent complex numbers $z_1 = 2 - 3i$, $z_2 = 5 + 2i$, and $z_3 = e + \pi i$, respectively, as points in the plane (Figure 1). The plane on which the complex numbers are represented is called the *complex plane*; the horizontal axis (x axis) is called the *real axis*, and the vertical axis (y axis) is called the *imaginary axis.*

Thus, complex numbers of the form $z = bi$ are represented by points of the form $(0, b)$, that is, as points on the imaginary axis; whereas complex numbers of the form $z = a$ are represented by points of the form $(a, 0)$, that is, as points on the real axis.

Figure 1

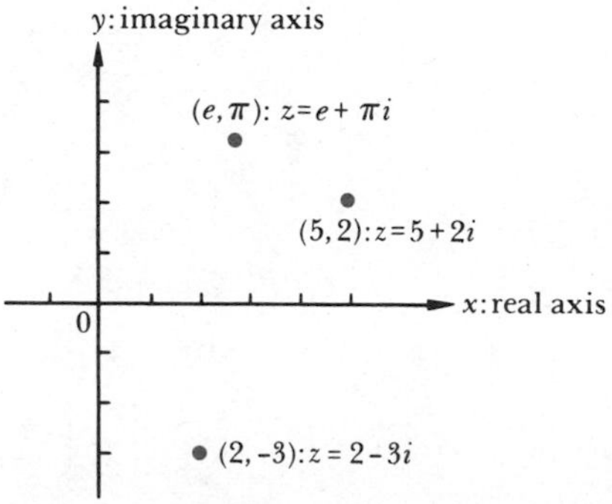

y: imaginary axis
(e, π): $z = e + \pi i$
$(5, 2)$: $z = 5 + 2i$
x: real axis
0
$(2, -3)$: $z = 2 - 3i$

2.1 The Modulus of a Complex Number

If $z = a + bi$, then the *absolute value* or *length* or *modulus* of z, written $|z|$, is defined by

$$|z| = \sqrt{a^2 + b^2}$$

The modulus of $z = a + bi$ is the distance between the origin and the point (a, b) (Figure 2). Notice that $|z| = \sqrt{z \cdot \bar{z}}$ (see Problem 20).

Figure 2

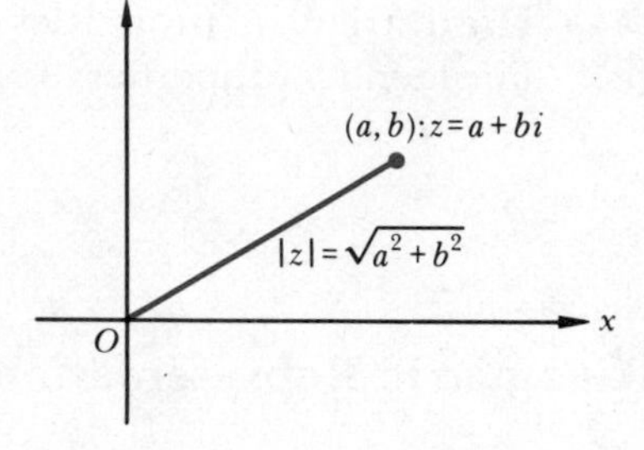

EXAMPLES 1 Let $z = 1 + \sqrt{3}i$. Find $|z|$ and show that $z\bar{z} = |z|^2$.

SOLUTION (See Figure 3.)

$$|z| = \sqrt{1^2 + (\sqrt{3})^2} = \sqrt{1 + 3} = \sqrt{4} = 2$$

and

$$z\bar{z} = (1 + \sqrt{3}i)(1 - \sqrt{3}i) = 1^2 + (\sqrt{3})^2 = 4 = |z|^2$$

Figure 3

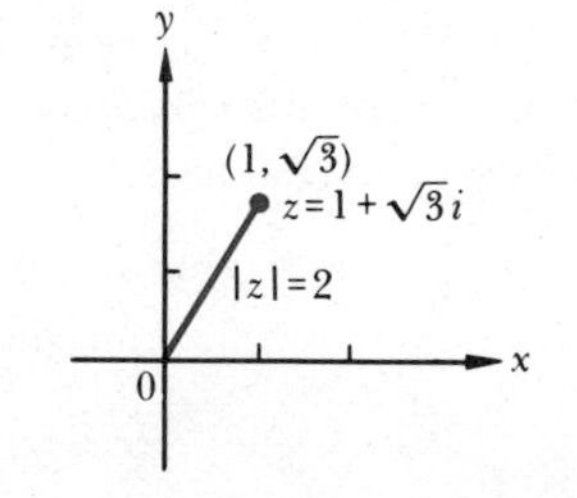

2 Let $z_1 = 4 + 3i$ and $z_2 = \sqrt{3} - i$. Find

(a) $|z_1|$ (b) $|z_2|$

(c) $|z_1 z_2|$ (d) $\left|\dfrac{z_1}{z_2}\right|$

SOLUTION

(a) $|z_1| = \sqrt{16 + 9} = \sqrt{25} = 5$

(b) $|z_2| = \sqrt{(\sqrt{3})^2 + (-1)^2} = \sqrt{4} = 2$

(c)
$$\begin{aligned} |z_1 z_2| &= |(4 + 3i)(\sqrt{3} - i)| \\ &= |(4\sqrt{3} + 3) + (3\sqrt{3} - 4)i| \\ &= \sqrt{(4\sqrt{3} + 3)^2 + (3\sqrt{3} - 4)^2} = 10 \end{aligned}$$

(d)
$$\begin{aligned} \left|\frac{z_1}{z_2}\right| &= \left|\frac{4 + 3i}{\sqrt{3} - i}\right| = \left|\frac{(4 + 3i)(\sqrt{3} + i)}{3 + 1}\right| = \left|\frac{(4\sqrt{3} - 3) + (3\sqrt{3} + 4)i}{4}\right| \\ &= \sqrt{\left(\frac{4\sqrt{3} - 3}{4}\right)^2 + \left(\frac{3\sqrt{3} + 4}{4}\right)^2} = \frac{5}{2} \end{aligned}$$

3 Let z_1 and z_2 be complex numbers; then $|z_1 z_2| = |z_1|\,|z_2|$.

PROOF (This is a generalization of the results of the first three parts of Example 2 above.)

$$\begin{aligned} |z_1 z_2| &= \sqrt{(z_1 z_2)(\overline{z_1 z_2})} = \sqrt{(z_1 z_2)(\bar{z}_1 \bar{z}_2)} = \sqrt{(z_1 \bar{z}_1)(z_2 \bar{z}_2)} \\ &= \sqrt{|z_1|^2 |z_2|^2} = |z_1|\,|z_2| \end{aligned}$$

4 Let $z = x + iy$. Geometrically describe the set of all complex numbers z such that $|z - 1| = 1$.

SOLUTION

$$\begin{aligned} |z - 1| &= |(x + iy) - 1| = |(x - 1) + iy| \\ &= \sqrt{(x - 1)^2 + y^2} \end{aligned}$$

so that $|z - 1| = 1$ is equivalent to

$$(x - 1)^2 + y^2 = 1$$

In other words, the distance between any point (x, y) and the point $(1, 0)$ is always 1. Thus, the points (x, y) lie on a circle with center $(1, 0)$ and radius 1 (Figure 4), and the graph of $|z - 1| = 1$ is the circle with center at $(1, 0)$ and radius 1.

Figure 4

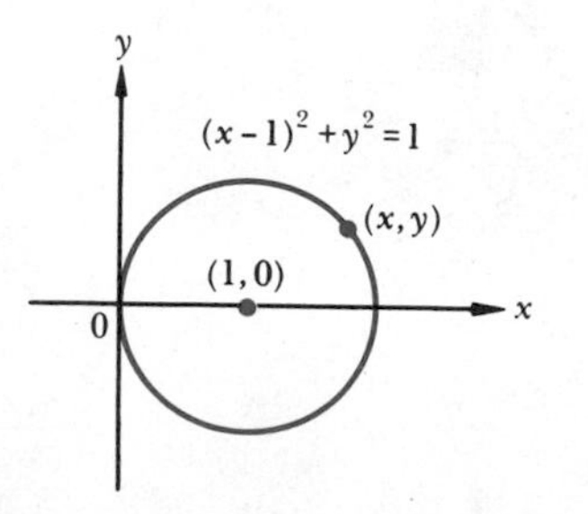

2.2 Polar Form of Complex Numbers

A complex number $z = x + iy$ can be written in the form

$$\begin{aligned} z &= r \cos \theta + ir \sin \theta \\ &= r(\cos \theta + i \sin \theta) \end{aligned}$$

where $x = r \cos \theta$, $y = r \sin \theta$, and r is the modulus of z (**Figure 5**). The expression $r(\cos \theta + i \sin \theta)$ is called the *polar form* or *trigonometric form* of the complex number z. The number θ in this representation is called an *argument* of the complex number z. ("Argument" means a polar angle associated with z and has nothing to do with its meaning in English grammar.) Notice that θ is not unique, since $r(\cos \theta + i \sin \theta) = r(\cos \theta_1 + i \sin \theta_1)$ holds whenever $\theta - \theta_1$ is an integer multiple of 2π.

Hence *two complex numbers are equal if and only if their moduli are equal and their arguments differ by a multiple of* 2π. We can write the complex number z in the form

Figure 5

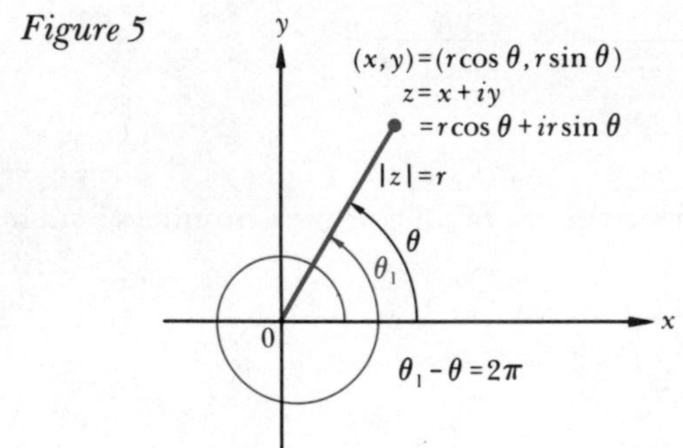

$$\begin{aligned} z &= x + iy \\ &= r[\cos(\theta + 2\pi k) + i \sin(\theta + 2\pi k)] \qquad k \in I \end{aligned}$$

For example, the complex number $z = 1 + i$ can be represented in polar form as

$$z = \sqrt{2}\left(\cos \frac{\pi}{4} + i \sin \frac{\pi}{4}\right)$$

or as

$$z = \sqrt{2}\left(\cos \frac{9\pi}{4} + i \sin \frac{9\pi}{4}\right)$$

since $r = \sqrt{1^2 + 1^2} = \sqrt{2}$ and since both $\pi/4$ and $9\pi/4$ have the same terminal sides and both satisfy $1 = \sqrt{2} \cos \theta$ and $1 = \sqrt{2} \sin \theta$.

The complex number $z = 1 + i$ can also be represented in polar form as

$$\begin{aligned} z &= \sqrt{2}\left[\cos\left(-\frac{7\pi}{4}\right) + i \sin\left(-\frac{7\pi}{4}\right)\right] \\ &= \sqrt{2}\left(\cos \frac{7\pi}{4} - i \sin \frac{7\pi}{4}\right) \qquad \text{(Figure 6)} \end{aligned}$$

Figure 6

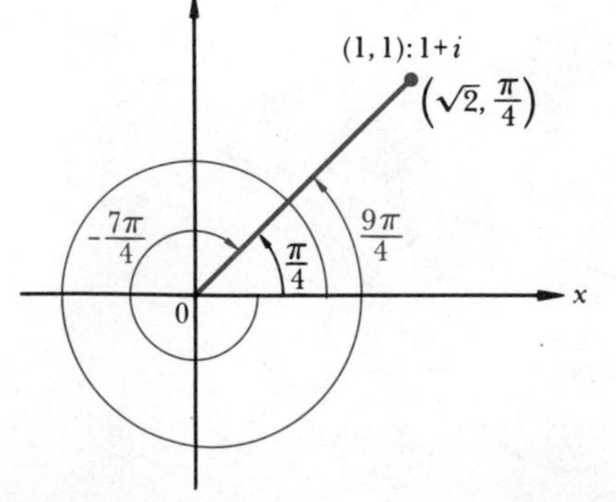

Notice that

$$\left(\sqrt{2}, \frac{\pi}{4}\right), \left(\sqrt{2}, \frac{9\pi}{4}\right), \text{ and } \left(\sqrt{2}, -\frac{7\pi}{4}\right)$$

are possible polar coordinates of the point with rectangular coordinates (1, 1).

EXAMPLES 1 Change each of the following complex numbers from polar form to rectangular form.

(a) $z = 2[\cos(\pi/3) + i\sin(\pi/3)]$
(b) $z = 4[\cos(-\pi/6) + i\sin(-\pi/6)]$
(c) $z = 8[\cos(\pi/2) + i\sin(\pi/2)]$

SOLUTION

(a) $$z = x + iy = 2\cos\frac{\pi}{3} + i\cdot 2\sin\frac{\pi}{3}$$
$$= 2\cdot\frac{1}{2} + i\cdot 2\,\frac{\sqrt{3}}{2} = 1 + \sqrt{3}i$$

(b) $$z = x + iy = 4\cos\left(-\frac{\pi}{6}\right) + i\cdot 4\sin\left(-\frac{\pi}{6}\right)$$
$$= 4\,\frac{\sqrt{3}}{2} + i\cdot 4\left(-\frac{1}{2}\right) = 2\sqrt{3} - 2i$$

(c) $$z = x + iy = 8\cos\frac{\pi}{2} + i\cdot 8\sin\frac{\pi}{2}$$
$$= 8\cdot 0 + i\cdot 8(1) = 8i$$

2 Change each of the following complex numbers from rectangular form to a polar form.

(a) $z = 2$ (b) $z = 2i$ (c) $z = -\sqrt{3} - i$

SOLUTION (See Figure 7.)

(a) $z = 2 = 2 + 0i = r(\cos\theta + i\sin\theta)$, where $r = \sqrt{2^2 + 0^2} = 2$ and an argument is $\theta = 0$, so that $z = 2(\cos 0 + i\sin 0)$.

(b) $z = 2i = 0 + 2i = r(\cos\theta + i\sin\theta)$, where $r = \sqrt{0 + 2^2} = 2$. One measure of θ is $\frac{\pi}{2}$. Hence $z = 2\left(\cos\frac{\pi}{2} + i\sin\frac{\pi}{2}\right)$.

(c) $z = r(\cos\theta + i\sin\theta)$, where $r = \sqrt{(-\sqrt{3})^2 + (-1)^2} = 2$, and θ satisfies $\tan\theta = \frac{1}{\sqrt{3}}$ with the terminal side of θ in quadrant III, so that one value of θ is $\frac{7\pi}{6}$. Hence $z = 2\left(\cos\frac{7\pi}{6} + i\sin\frac{7\pi}{6}\right)$.

Figure 7

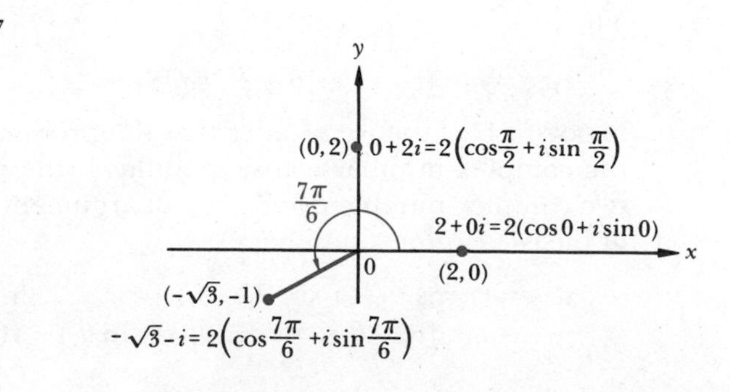

2.3 Multiplication and Division of Complex Numbers in Polar Form

Consider the complex numbers $z_1 = 1 + i$ and $z_2 = 1 + \sqrt{3}i$ (Figure 8). The polar representations of z_1 and z_2 can be given by

$$z_1 = \sqrt{2}\left(\cos\frac{\pi}{4} + i \sin\frac{\pi}{4}\right) \quad \text{and} \quad z_2 = 2\left(\cos\frac{\pi}{3} + i \sin\frac{\pi}{3}\right)$$

Figure 8

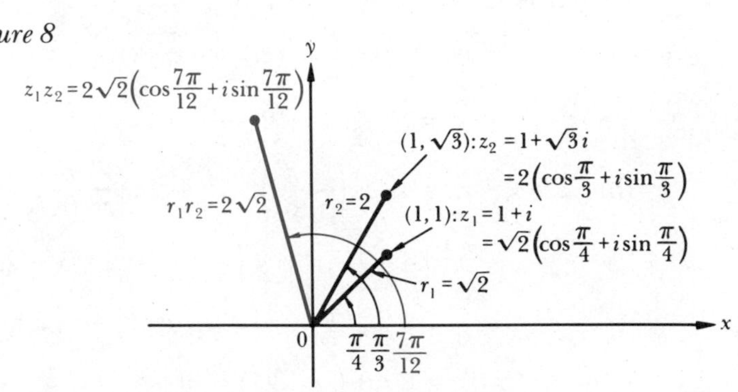

The modulus of z_1 is $r_1 = \sqrt{2}$ and an argument is $\pi/4$; the modulus of z_2 is $r_2 = 2$ and an argument is $\pi/3$. Multiplying, we find that z_1z_2 equals

$$2\sqrt{2}\left[\left(\cos\frac{\pi}{4}\cos\frac{\pi}{3} - \sin\frac{\pi}{4}\sin\frac{\pi}{3}\right) + i\left(\sin\frac{\pi}{4}\cos\frac{\pi}{3} + \sin\frac{\pi}{3}\cos\frac{\pi}{4}\right)\right]$$

This result can be simplified, by using the trigonometric identities, to

$$2\sqrt{2}\left[\cos\left(\frac{\pi}{4} + \frac{\pi}{3}\right) + i\sin\left(\frac{\pi}{4} + \frac{\pi}{3}\right)\right] = 2\sqrt{2}\left(\cos\frac{7\pi}{12} + i\sin\frac{7\pi}{12}\right)$$

Notice in this example that the modulus of the product z_1z_2 satisfies the equation $|z_1z_2| = |z_1||z_2| = r_1r_2 = 2\sqrt{2}$ and an argument of z_1z_2 is $\pi/4 + \pi/3 = 7\pi/12$. The following theorem generalizes the results of this example.

THEOREM 1

Suppose that z_1 and z_2 are complex numbers in polar form such that

$$z_1 = r_1(\cos\theta_1 + i\sin\theta_1) \quad \text{and} \quad z_2 = r_2(\cos\theta_2 + i\sin\theta_2)$$

Then

$$z_1z_2 = r_1r_2[\cos(\theta_1 + \theta_2) + i\sin(\theta_1 + \theta_2)]$$

PROOF (This theorem states that the product of two complex numbers is the complex number whose modulus is the product of the moduli of the two complex numbers, and with an argument that is the sum of arguments of the two complex numbers.)

$$\begin{aligned} z_1z_2 &= r_1(\cos\theta_1 + i\sin\theta_1) \cdot r_2(\cos\theta_2 + i\sin\theta_2) \\ &= r_1r_2[(\cos\theta_1\cos\theta_2 - \sin\theta_1\sin\theta_2) + i(\cos\theta_1\sin\theta_2 + \cos\theta_2\sin\theta_1)] \end{aligned}$$

Using the trigonometric identities for $\cos(\theta_1 + \theta_2)$ and $\sin(\theta_1 + \theta_2)$, we can write this latter result as

$$z_1 z_2 = r_1 r_2[\cos(\theta_1 + \theta_2) + i \sin(\theta_1 + \theta_2)]$$

EXAMPLES 1 Let $z_1 = 7(\cos 25° + i \sin 25°)$ and $z_2 = 3(\cos 35° + i \sin 35°)$. Find
(a) $z_1 z_2$ in rectangular form (b) z_1^2 in polar form

SOLUTION

(a) The modulus of $z_1 z_2$ is $r_1 r_2 = 7(3) = 21$ and an argument of $z_1 z_2$ is $\theta_1 + \theta_2 = 25° + 35° = 60°$. Hence,

$$\begin{aligned} z_1 z_2 &= 21(\cos 60° + i \sin 60°) \\ &= 21\left(\frac{1}{2} + i\frac{\sqrt{3}}{2}\right) \\ &= \frac{21}{2} + i\,\frac{21\sqrt{3}}{2} \end{aligned}$$

(b) The modulus of z_1^2 is $r_1^2 = 49$ and an argument is $2\theta_1 = 2(25°) = 50°$. Hence, $z_1^2 = 49(\cos 50° + i \sin 50°)$.

2 Convert $z_1 = 1 + i$ and $z_2 = 2 - 2\sqrt{3}i$ to polar form, and then compute $z_1 z_2$ in polar form and in rectangular form.

SOLUTION (See Figure 9.) The modulus of z_1 is $r_1 = \sqrt{1 + 1} = \sqrt{2}$, and the modulus of z_2 is $r_2 = \sqrt{4 + 12} = 4$. An argument of z_1 is $\theta_1 = \pi/4$, and an argument of z_2 is $\theta_2 = -\pi/3$. Hence,

$$\begin{aligned} z_1 z_2 &= 4\sqrt{2}\left\{\cos\left[\left(\frac{\pi}{4}\right) + \left(-\frac{\pi}{3}\right)\right] + i \sin\left[\left(\frac{\pi}{4}\right) + \left(-\frac{\pi}{3}\right)\right]\right\} \\ &= 4\sqrt{2}\left[\cos\left(-\frac{\pi}{12}\right) + i \sin\left(-\frac{\pi}{12}\right)\right] \end{aligned}$$

so that the rectangular form is

$$z_1 z_2 = 4\sqrt{2}[0.9659 - i(0.2588)] = 3.8636\sqrt{2} - 1.0352\sqrt{2}i$$

Figure 9

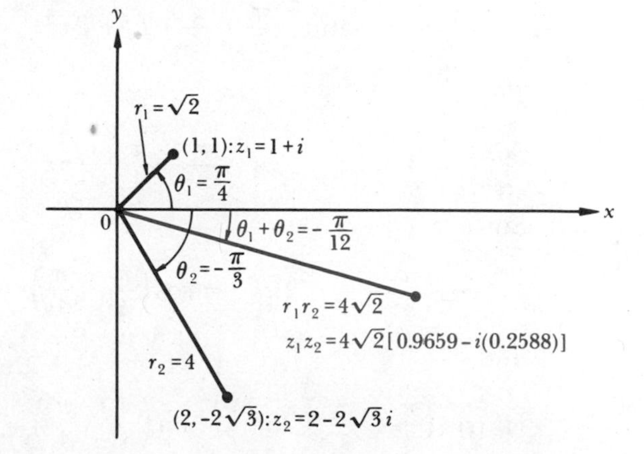

To find the quotient z_1/z_2 of two complex numbers

$$z_1 = 6 + 6\sqrt{3}i \quad \text{and} \quad z_2 = 2\sqrt{2} + 2\sqrt{2}i$$

we can first locate z_1 and z_2 as shown in Figure 10. Polar representations of z_1 and z_2 are

$$z_1 = 12\left(\cos\frac{\pi}{3} + i\sin\frac{\pi}{3}\right) \quad \text{and} \quad z_2 = 4\left(\cos\frac{\pi}{4} + i\sin\frac{\pi}{4}\right)$$

We are now looking for a complex number $z_3 = r_3(\cos\theta_3 + i\sin\theta_3)$ such that $z_1 = z_2z_3$. Hence, if we use Theorem 1, we see that r_3 and θ_3 can be chosen so that

$$r_1(\cos\theta_1 + i\sin\theta_1) = r_2r_3[\cos(\theta_2 + \theta_3) + i\sin(\theta_2 + \theta_3)]$$

Figure 10

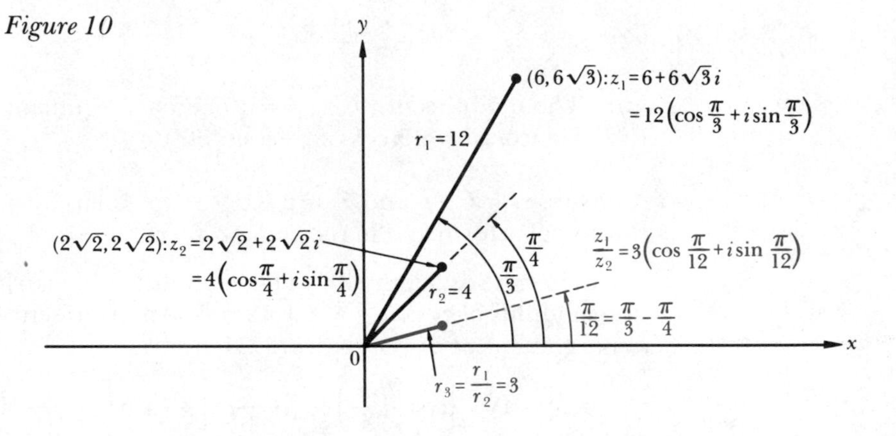

That is,

$$12\left(\cos\frac{\pi}{3} + i\sin\frac{\pi}{3}\right) = 4r_3\left[\cos\left(\frac{\pi}{4} + \theta_3\right) + i\sin\left(\frac{\pi}{4} + \theta_3\right)\right]$$

But two complex numbers are equal if and only if their moduli are equal and their arguments differ by a multiple of 2π. Thus

$$12 = 4r_3 \quad \text{and} \quad \frac{\pi}{3} = \frac{\pi}{4} + \theta_3 + 2\pi k \qquad k \in I$$

That is,

$$r_3 = 3 \quad \text{and} \quad \theta_3 = \frac{\pi}{3} - \frac{\pi}{4} - 2\pi k \qquad k \in I$$

Since

$$\cos\left(\frac{\pi}{3} - \frac{\pi}{4} - 2\pi k\right) = \cos\left(\frac{\pi}{3} - \frac{\pi}{4}\right)$$

and

$$\sin\left(\frac{\pi}{3} - \frac{\pi}{4} - 2\pi k\right) = \sin\left(\frac{\pi}{3} - \frac{\pi}{4}\right)$$

the polar form of the quotient z_1/z_2 can be written as

$$\frac{z_1}{z_2} = \frac{12}{4}\left[\cos\left(\frac{\pi}{3} - \frac{\pi}{4}\right) + i \sin\left(\frac{\pi}{3} - \frac{\pi}{4}\right)\right]$$
$$= 3\left(\cos\frac{\pi}{12} + i \sin\frac{\pi}{12}\right)$$

This example is generalized as follows.

THEOREM 2

Let $z_1 = r_1(\cos \theta_1 + i \sin \theta_1)$ and $z_2 = r_2(\cos \theta_2 + i \sin \theta_2)$; then, if $z_2 \neq 0$

$$\frac{z_1}{z_2} = \frac{r_1}{r_2}[\cos(\theta_1 - \theta_2) + i \sin(\theta_1 - \theta_2)]$$

The proof of this theorem is left as an exercise for the student (see Problem 28).

EXAMPLES 1 Let $z_1 = 4(\cos 80° + i \sin 80°)$ and $z_2 = 2(\cos 50° + i \sin 50°)$. Find z_1/z_2 and express the result in rectangular form.

SOLUTION Since $z_2 \neq 0$, we use Theorem 2 to get

$$\frac{z_1}{z_2} = \frac{4}{2}[\cos(80° - 50°) + i \sin(80° - 50°)]$$
$$= 2(\cos 30° + i \sin 30°)$$
$$= 2\left(\frac{\sqrt{3}}{2} + i\frac{1}{2}\right)$$
$$= \sqrt{3} + i$$

2 Express $z_1 = -1 + i$ and $z_2 = -4i$ in polar form, and then compute z_1/z_2 in polar form. Give the answer in rectangular form also.

SOLUTION (See Figure 11.) The modulus of z_1 is $r_1 = \sqrt{2}$ and an argument is $3\pi/4$. (Why?) The modulus of z_2 is $r_2 = 4$, and an argument is $3\pi/2$. Using Theorem 2, we have

Figure 11

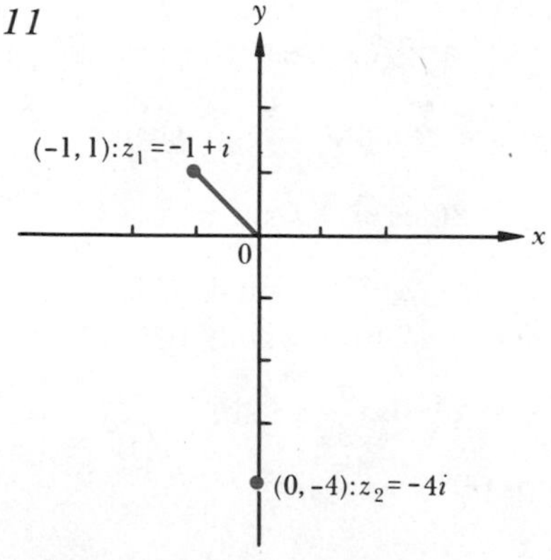

$$\frac{z_1}{z_2} = \frac{\sqrt{2}}{4}\left[\cos\left(\frac{3\pi}{4} - \frac{3\pi}{2}\right) + i \sin\left(\frac{3\pi}{4} - \frac{3\pi}{2}\right)\right]$$
$$= \frac{\sqrt{2}}{4}\left[\cos\left(-\frac{3\pi}{4}\right) + i \sin\left(-\frac{3\pi}{4}\right)\right]$$
$$= \frac{\sqrt{2}}{4}\left(\cos\frac{3\pi}{4} - i \sin\frac{3\pi}{4}\right)$$
$$= \frac{\sqrt{2}}{4}\left(-\frac{1}{\sqrt{2}} - i\frac{1}{\sqrt{2}}\right)$$
$$= -\frac{1}{4} - \frac{1}{4}i$$

PROBLEM SET 2

In Problems 1–8, let $z_1 = 2 + 7i$ and $z_2 = 5 - 2i$; find the value of each expression.

1 $|z_1|$ **2** $|z_2|$ **3** $|z_1 z_2|$ **4** $|z_1||z_2|$

5 $\left|\dfrac{z_1}{z_2}\right|$ **6** $\dfrac{|z_1|}{|z_2|}$ **7** $|z_1 \bar{z}_1|$ **8** $|3z_1 - 2z_2|$

In Problems 9–12, let $z = x + yi$; geometrically describe the set of all complex numbers that satisfy the given condition.

9 $|z| = 1$ **10** $|z| = 2$ **11** $|z - i| = 1$

12 $|z + 1| = |z - 2|$

In Problems 13–19, represent each of the complex numbers graphically, and then express the number in the rectangular form $a + bi$.

13 $z = 2(\cos 10° + i \sin 10°)$ **14** $z = 10\left(\cos \dfrac{3\pi}{4} + i \sin \dfrac{3\pi}{4}\right)$

15 $z = 3[\cos(-75°) + i \sin(-75°)]$ **16** $z = 2\left(\cos \dfrac{\pi}{2} + i \sin \dfrac{\pi}{2}\right)$

17 $z = 4(\cos 0° + i \sin 0°)$

18 $z = 7\left[\cos\left(-\dfrac{3\pi}{2}\right) + i \sin\left(-\dfrac{3\pi}{2}\right)\right]$

19 $z = 2\left(\cos \dfrac{\pi}{4} + i \sin \dfrac{\pi}{4}\right)$

20 Prove that $|z| = \sqrt{z\bar{z}}$.

In Problems 21–27, express each complex number in polar form, and then represent each of them graphically.

21 $-1 - i$ **22** 7 **23** $-2i$ **24** $\dfrac{\sqrt{3}}{2} + \dfrac{1}{2}i$

25 $-\sqrt{3} - i$ **26** $3 + 4i$ **27** $-\dfrac{1}{2} + \dfrac{\sqrt{3}}{2}i$

28 Prove Theorem 2 by generalizing the example preceding its statement on page 405.

In Problems 29–34, find $z_1 z_2$ and z_1/z_2 and express the answers in both polar and rectangular form.

29 $z_1 = 5(\cos 170° + i \sin 170°)$ and $z_2 = \cos 55° + i \sin 55°$

30 $z_1 = 4\left(\cos \dfrac{3\pi}{4} + i \sin \dfrac{3\pi}{4}\right)$ and $z_2 = 2(\cos \pi + i \sin \pi)$

31 $z_1 = 2(\cos 50° + i \sin 50°)$ and $z_2 = 3(\cos 40° + i \sin 40°)$

32 $z_1 = 5(\cos 30° + i \sin 30°)$ and $z_2 = 6(\cos 240° + i \sin 240°)$

33 $z_1 = 4\left(\cos \dfrac{5\pi}{6} + i \sin \dfrac{5\pi}{6}\right)$ and $z_2 = 2\left(\cos \dfrac{\pi}{3} + i \sin \dfrac{\pi}{3}\right)$

34 $z_1 = \cos 30° + i \sin 30°$ and $z_2 = \cos 60° + i \sin 60°$

In Problems 35–38, convert $z_1 = -1 - i$ and $z_2 = -4 + 4\sqrt{3}i$ to polar form, and then compute each of the following values in polar form. Convert the answers to rectangular form.

35 z_1z_2 **36** $(z_1z_2)^2$ **37** $\dfrac{z_1}{z_2}$ **38** $\left(\dfrac{z_1}{z_2}\right)^2$

3 Powers and Roots of Complex Numbers

If $z = r(\cos\theta + i\sin\theta)$, then

$$z^2 = r \cdot r[\cos(\theta + \theta) + i\sin(\theta + \theta)] = r^2(\cos 2\theta + i\sin 2\theta)$$

But $z^3 = z^2 \cdot z$, so that

$$z^3 = r^2 \cdot r[\cos(2\theta + \theta) + i\sin(2\theta + \theta)] = r^3(\cos 3\theta + i\sin 3\theta)$$

If we repeat the process onc more time, we get

$$z^4 = z^3 \cdot z = r^4(\cos 4\theta + i\sin 4\theta)$$

This scheme for repeated multiplication of a complex number in polar form is generalized in the following theorem, which is given without proof.

THEOREM 1 DEMOIVRE'S THEOREM

Let $z = r(\cos\theta + i\sin\theta)$. Then,

$$z^n = r^n(\cos n\theta + i\sin n\theta), \quad \text{for } n \text{ a positive integer}$$

EXAMPLES 1 Use DeMoivre's theorem to determine each of the following values.

(a) $[3(\cos 60° + i\sin 60°)]^4$ in rectangular form
(b) $(1 + i)^{20}$ in rectangular form

SOLUTION

(a) By DeMoivre's theorem,

$$[3(\cos 60° + i\sin 60°)]^4 = 3^4(\cos 240° + i\sin 240°)$$
$$= 81\left(-\frac{1}{2} - \frac{i\sqrt{3}}{2}\right) = -\frac{81}{2} - \frac{81\sqrt{3}}{2}i$$

(b) The complex number $1 + i$ can be expressed in polar form as

$$\sqrt{2}\left(\cos\frac{\pi}{4} + i\sin\frac{\pi}{4}\right)$$

where the modulus is $\sqrt{2}$ and an argument is $\pi/4$. According to DeMoivre's theorem,

$$(1 + i)^{20} = \left[\sqrt{2}\left(\cos\frac{\pi}{4} + i\sin\frac{\pi}{4}\right)\right]^{20} = 2^{10}(\cos 5\pi + i\sin 5\pi)$$
$$= 1024(-1 + 0i) = -1024$$

2 Use DeMoivre's theorem to express $\cos 2\theta$ and $\sin 2\theta$ in terms of $\sin\theta$ and $\cos\theta$.

SOLUTION By DeMoivre's theorem for $n = 2$, we have

$$\begin{aligned}\cos 2\theta + i\sin 2\theta &= (\cos\theta + i\sin\theta)^2 \\ &= \cos^2\theta + i\cdot 2\sin\theta\cos\theta - \sin^2\theta \\ &= (\cos^2\theta - \sin^2\theta) + i\cdot 2\sin\theta\cos\theta\end{aligned}$$

Since the two complex numbers are equal, the real parts are equal; that is,

$$\cos 2\theta = \cos^2\theta - \sin^2\theta$$

and the imaginary parts are also equal; that is,

$$\sin 2\theta = 2\sin\theta\cos\theta$$

3.1 Roots

DeMoivre's theorem is useful in finding the nth roots of a complex number. If $w = R(\cos\phi + i\sin\phi)$, and $z = r(\cos\theta + i\sin\theta)$ is any root of $z^n = w$, where n is a positive integer, then, by DeMoivre's theorem, it follows that

$$[r(\cos\theta + i\sin\theta)]^n = r^n(\cos n\theta + i\sin n\theta) = R(\cos\phi + i\sin\phi)$$

so that

$$r^n = R \quad\text{and}\quad n\theta = \phi + 2k\pi \qquad (n\theta = \phi + 360°k \text{ if degrees are used})$$

Hence

$$z = r(\cos\theta + i\sin\theta)$$

is a root of $z^n = w$ whenever

$$r = \sqrt[n]{R} \qquad \text{(note that } R \geqslant 0)$$

and

$$\theta = \frac{\phi}{n} + \frac{2k\pi}{n} \qquad \left(\theta = \frac{\phi}{n} + \frac{360°k}{n} \text{ if degrees are used}\right)$$

where $k = 0, \pm 1, \pm 2, \pm 3, \ldots$.

Thus it would appear that there are an infinite number of roots for the equation $z^n = w$. However, for n a positive integer, there are only n distinct roots. These n roots can be determined by letting k take on the values $0, 1, 2, 3, 4, \ldots, n-1$. This is true because if we let $k = n$, the argument θ will be

$$\frac{\phi}{n} + \frac{2n\pi}{n} = \frac{\phi}{n} + 2\pi$$

and this angle has the same terminal side as ϕ/n, the value of θ where $n = 0$;

similarly, the value of θ obtained by letting $k = n + 1$ gives an angle with the same terminal side as $(\phi/n) + (2\pi/n)$, the value of θ when $k = 1$; and so on. In summary, the n distinct roots of the equation $z^n = w$, where $w = R(\cos \phi + i \sin \phi)$, are given by

$$z_k = \sqrt[n]{R}\left[\cos\left(\frac{\phi}{n} + \frac{2\pi k}{n}\right) + i \sin\left(\frac{\phi}{n} + \frac{2\pi k}{n}\right)\right] \qquad k = 0, 1, 2, \ldots, n - 1$$

or, if degrees are used for the angles,

$$z_k = \sqrt[n]{R}\left[\cos\left(\frac{\phi}{n} + \frac{360°k}{n}\right) + i \sin\left(\frac{\phi}{n} + \frac{360°k}{n}\right)\right] \qquad k = 0, 1, 2, \ldots, n - 1$$

The points representing these n roots are equally spaced around the circle with center at (0, 0) and with radius $\sqrt[n]{R}$ (since each root has modulus $\sqrt[n]{R}$).

EXAMPLES 1 Determine the square roots of $-i$; that is, solve $z^2 = -i$, and represent the two roots geometrically.

SOLUTION First, we determine $R(\cos \phi + i \sin \phi)$, a polar representation of $-i$. Here $R = \sqrt{0^2 + (-1)^2} = 1$ and $\phi = 270°$ (Figure 1). Thus we obtain the roots $z = r(\cos \theta + i \sin \theta)$ by using the formula $r = \sqrt{1} = 1$ as well as $\theta = (270°/2) + (360°k/2) = 135° + 180°k$, for $k = 0$ and 1. The roots are

$$z_0 = 1(\cos 135° + i \sin 135°) = -\frac{\sqrt{2}}{2} + i\frac{\sqrt{2}}{2} \qquad \text{for} \quad k = 0$$

$$z_1 = 1(\cos 315° + i \sin 315°) = \frac{\sqrt{2}}{2} - i\frac{\sqrt{2}}{2} \qquad \text{for} \quad k = 1$$

Figure 1

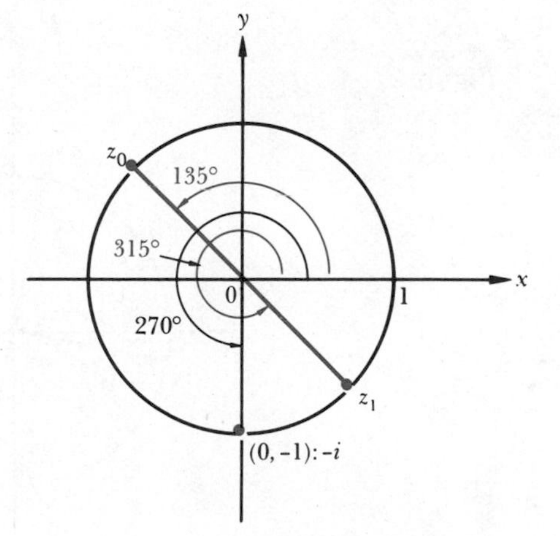

The square roots of $-i$ are equally spaced on the circumference of a circle of radius 1 and the arguments of the roots differ by an angle of π radians or 180° (Figure 1).

2 Find the fourth roots of $1 + i$, that is, solve $z^4 = 1 + i$. Express the roots in polar form and represent them geometrically.

SOLUTION As our first step, we determine $R(\cos \phi + i \sin \phi)$, one polar representation of $1 + i$. Here $R = \sqrt{2}$ and $\phi = \pi/4$. Finally, we obtain the roots $z = r(\cos \theta + i \sin \theta)$ by using the formulas for r and θ.

$$r = \sqrt[4]{\sqrt{2}} = \sqrt[8]{2} \quad \text{and} \quad \theta = \frac{\pi}{4 \cdot 4} + \frac{2\pi k}{4} = \frac{\pi}{16} + \frac{\pi k}{2} \qquad k = 0, 1, 2, 3$$

The roots are

$$z_0 = \sqrt[8]{2}\left(\cos\frac{\pi}{16} + i \sin\frac{\pi}{16}\right) \qquad \text{for } k = 0$$

$$z_1 = \sqrt[8]{2}\left(\cos\frac{9\pi}{16} + i \sin\frac{9\pi}{16}\right) \qquad \text{for } k = 1$$

$$z_2 = \sqrt[8]{2}\left(\cos\frac{17\pi}{16} + i \sin\frac{17\pi}{16}\right) \qquad \text{for } k = 2$$

$$z_3 = \sqrt[8]{2}\left(\cos\frac{25\pi}{16} + i \sin\frac{25\pi}{16}\right) \qquad \text{for } k = 3$$

The fourth roots of $1 + i$ are equally spaced on the circumference of a circle of radius $\sqrt[8]{2}$ and their arguments differ by angles of $\pi/2$ radians (Figure 2). Notice that if we were to set $k = 4$, we would get

$$z_4 = \sqrt[8]{2}\left(\cos\frac{33\pi}{16} + i \sin\frac{33\pi}{16}\right)$$

which is the same number as z_0. In general, if $k = 4, 5, 6, \ldots,$ we would get a repetition of the roots that we have already found.

Figure 2

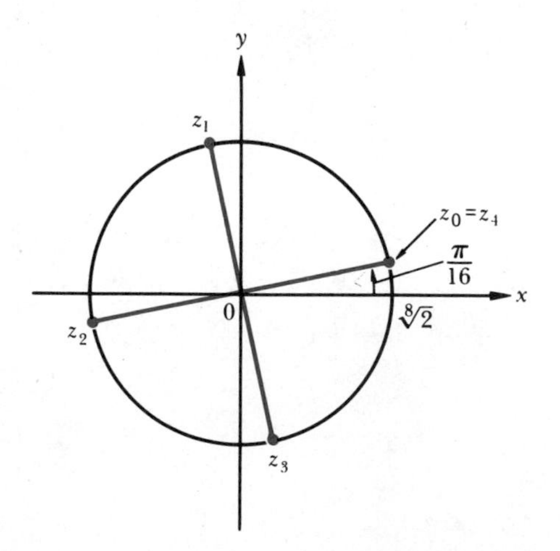

3 The solutions to $z^n = 1$ are called the nth *roots of unity*. Find the fifth roots of unity and express them in polar form. Represent the roots geometrically.

SOLUTION The number 1 can be written as $1 = 1(\cos 0 + i \sin 0)$ in polar form. We obtain the five roots $z = r(\cos \theta + i \sin \theta)$ by using the formulas

$$r = \sqrt[5]{1} = 1 \quad \text{and} \quad \theta = \frac{0}{5} + \frac{2\pi k}{5} \qquad k = 0, 1, 2, 3, 4$$

The roots are

$$z_0 = 1(\cos 0 + i \sin 0) \qquad \text{for } k = 0$$

$$z_1 = 1\left(\cos \frac{2\pi}{5} + i \sin \frac{2\pi}{5}\right) \qquad \text{for } k = 1$$

$$z_2 = 1\left(\cos \frac{4\pi}{5} + i \sin \frac{4\pi}{5}\right) \qquad \text{for } k = 2$$

$$z_3 = 1\left(\cos \frac{6\pi}{5} + i \sin \frac{6\pi}{5}\right) \qquad \text{for } k = 3$$

$$z_4 = 1\left(\cos \frac{8\pi}{5} + i \sin \frac{8\pi}{5}\right) \qquad \text{for } k = 4$$

Observe that all the fifth roots of unity are on a unit circle and their arguments differ by multiples of $2\pi/5$ radians (Figure 3). Again, $k = 5, 6, 7, \ldots$ would give us a repetition of the roots we have already found.

Figure 3

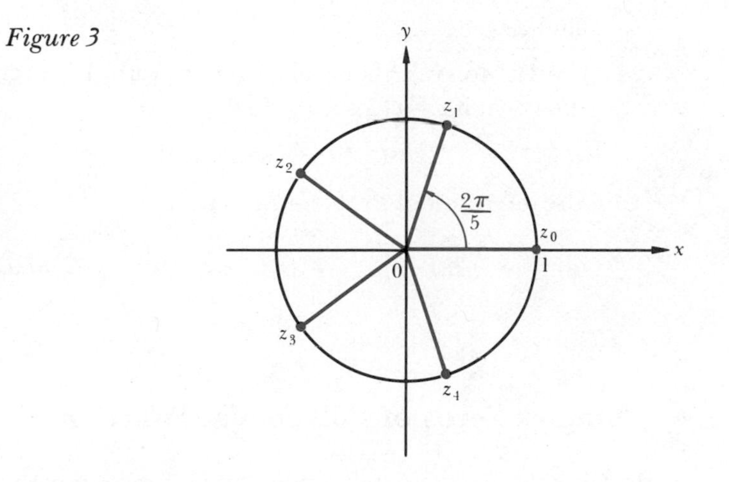

PROBLEM SET 3

In Problems 1–6, use DeMoivre's theorem to compute each of the powers. Express the answer in the rectangular form $a + bi$.

1 $(\cos 30° + i \sin 30°)^7$

2 $(\cos 15° + i \sin 15°)^8$

3 $\left[2\left(\cos \frac{\pi}{6} + i \sin \frac{\pi}{6}\right)\right]^{10}$

4 $\left[3\left(\cos \frac{\pi}{18} + i \sin \frac{\pi}{18}\right)\right]^6$

5 $\left[2\left(\cos \frac{5\pi}{4} + i \sin \frac{5\pi}{4}\right)\right]^8$

6 $[4(\cos 36° + i \sin 36°)]^5$

In Problems 7–14, express each of the complex numbers in polar form, then use DeMoivre's theorem to calculate the indicated powers. Express the result in the rectangular form $a + bi$.

7 $(5 + 5i)^6$ **8** $(1 + i\sqrt{3})^5$ **9** $(\sqrt{3} - i)^4$

10 $\left(-\frac{1}{2} - i\frac{\sqrt{3}}{2}\right)^8$ **11** $(\sqrt{3} + i)^{30}$ **12** $(1 + i)^{50}$

13 $\left(\frac{1}{\sqrt{2}} + i\frac{1}{\sqrt{2}}\right)^{100}$ $\left(\frac{1}{2} + i\frac{\sqrt{3}}{2}\right)^{30}$ **14** $\dfrac{(\sqrt{3} + i)^3}{(1 - \sqrt{3}i)^3}$

In Problems 15–20, find all the roots in each equation and represent them geometrically.

15 $z^2 = i$ **16** $z^2 = 3 - 3i$ **17** $z^3 = 8$

18 $z^3 = i$ **19** $z^4 = -16$ **20** $z^4 = -8 - 8\sqrt{3}i$

21 Find the fifth roots of -32 and represent them geometrically.

22 Use DeMoivre's theorem to derive formulas for $\cos 3\theta$ and $\sin 3\theta$. [*Hint:* $\cos 3\theta + i \sin 3\theta = (\cos\theta + i\sin\theta)^3$.]

23 Let

$$z_1 = \left[2\left(\cos\frac{\pi}{8} + i\sin\frac{\pi}{8}\right)\right]^4 \quad \text{and} \quad z_2 = \left[4\left(\cos\frac{\pi}{12} + i\sin\frac{\pi}{12}\right)\right]^4$$

Find $z_1 z_2$.

24 Use DeMoivre's theorem, together with Theorem 2 on page 405, to show that if $z = r(\cos\theta + i\sin\theta)$, then

$$z^{-n} = r^{-n}[\cos(-n\theta) + i\sin(-n\theta)]$$

25 Use the result of Problem 24 to find

(a) $\left(\frac{\sqrt{3}}{2} + i\frac{1}{2}\right)^{-5}$ (b) $(-2 + 2i)^{-3}$

4 Complex Zeros of Polynomial Functions

In Chapter 3 we saw that a polynomial function with real coefficients does not always have real number zeros. In particular, a quadratic polynomial function $f(x) = ax^2 + bx + c$, where a, b, and c are real numbers and $a \neq 0$, has real zeros if and only if the discriminant, $b^2 - 4ac$, is nonnegative. These real zeros can be found by using the quadratic formula,

$$x = \frac{-b \pm \sqrt{b^2 - 4ac}}{2a}$$

If the zeros of the quadratic polynomial function are not real numbers, that is, if $b^2 - 4ac < 0$, the zeros are complex numbers and the quadratic formula still can be used. For example, the zeros of $f(x) = 2x^2 + x + 1$ can be found by using the quadratic formula as follows:

$$x = \frac{-1 \pm \sqrt{1-8}}{4} = \frac{-1 \pm \sqrt{-7}}{4} = \frac{-1 \pm \sqrt{7}i}{4}$$

to get

$$-\frac{1}{4} + \frac{\sqrt{7}}{4}i \quad \text{and} \quad -\frac{1}{4} - \frac{\sqrt{7}}{4}i$$

The polynomial function $f(x) = x^3 - 6x^2 + 13x - 10$ can be factored as $f(x) = (x-2)(x^2 - 4x + 5)$, so that, after using the quadratic formula, we find that the zeros of f are 2, $2 - i$, and $2 + i$, and

$$f(x) = (x-2)[x - (2-i)][x - (2+i)]$$

In general, all polynomial functions are factorable as the product of linear factors in the complex domain, and the factorization is based on the zeros of the polynomial functions.

Notice, for example, that

$$2x^2 + x + 1 = 2\left(x - \frac{-1+\sqrt{7}i}{4}\right)\left(x - \frac{-1-\sqrt{7}i}{4}\right)$$

This result follows as a corollary of the *fundamental theorem of algebra,* whose proof depends upon methods generally considered beyond the scope of this textbook. We shall state the theorem without proof.

THEOREM 1 FUNDAMENTAL THEOREM OF ALGEBRA

If $f(x)$ is a polynomial of degree $n \geqslant 1$ with complex coefficients, then there is a complex number c such that $f(c) = 0$.

We shall assume this fundamental theorem in proving the following theorems.

THEOREM 2 FACTORIZATION THEOREM

If $f(x) = a_n x^n + a_{n-1}x^{n-1} + \cdots + a_1 x + a_0$, where $a_n \neq 0$ and n is a positive integer, then

$$f(x) = a_n(x - c_1)(x - c_2) \cdots (x - c_n)$$

where the numbers c_j are complex numbers.

PROOF According to the fundamental theorem of algebra, $f(x) = 0$ has a root c_1, so that, by the factor theorem (see Chapter 3, Section 4),

$$f(x) = (x - c_1)Q_1(x)$$

where $Q_1(x)$ is a polynomial of degree $n - 1$, so it has a zero c_2 if $n - 1 \geqslant 1$, and, as above,

$$Q_1(x) = (x - c_2)Q_2(x)$$

so that

$$f(x) = (x - c_1)(x - c_2)Q_2(x)$$

where $Q_2(x)$ has degree $n - 2$. Continuing the process, we get

$$f(x) = (x - c_1)(x - c_2) \cdots (x - c_n)Q_n(x)$$

where $Q_n(x)$ has degree 0; that is, $Q_n(x)$ is a constant, which we denote by Q_n. Multiplying out this expression for $f(x)$, it is seen that the coefficient of x^n is Q_n; hence $Q_n = a_n$ and the theorem is proved.

THEOREM 3

If $f(x)$ is a polynomial of degree n, where $n \neq 0$, then $f(x)$ has, at most, n zeros.

PROOF By the factorization theorem,

$$f(x) = a_n(x - c_1)(x - c_2) \cdots (x - c_n)$$

Clearly, the numbers $c_1, c_2, \ldots, c_n$ are zeros of f. Also, if $f(c) = 0$, for $c \neq c_i$ and $i = 1, \ldots, n$, then

$$f(x) = a_n(x - c_1)(x - c_2) \cdots (x - c_n)(x - c)$$

so the degree of $f(x)$ is $n + 1$ (why?), which contradicts our assumption that $f(x)$ is a polynomial of degree n and so there can be no more than n zeros.

Notice that the zeros need not be distinct. For example, the function $f(x) = x^2 - 4x + 4$ has only one zero, 2, and we say that the function has 2 as a *double* zero. In general, if

$$f(x) = (x - c)^s Q(x) \qquad \text{and} \qquad Q(c) \neq 0$$

we say that c is a zero of *multiplicity s*. For example, $f(x) = (x - 1)^2(x - 2)$ has 1 as a zero of multiplicity 2 and 2 as a zero of multiplicity 1.

THEOREM 4 CONJUGATE ROOT THEOREM

If $f(z)$ is a polynomial of degree n, with $n \neq 0$ and with *real* coefficients, and if $f(z_0) = 0$, where $z_0 = a + bi$, then $f(\bar{z}_0) = 0$.

PROOF Let $f(z) = a_n z^n + a_{n-1}z^{n-1} + \cdots + a_1 z + a_0$ be a polynomial with real coefficients. Since $z_0 = a + bi$ is a root of $f(z) = 0$, then

$$f(z_0) = a_n z_0^n + a_{n-1}z_0^{n-1} + \cdots + a_1 z_0 + a_0 = 0$$

so that

$$\begin{aligned} \overline{f(z_0)} &= \overline{a_n z_0^n + a_{n-1}z_0^{n-1} + \cdots + a_1 z_0 + a_0} \\ &= \overline{a_n z_0^n} + \overline{a_{n-1}z_0^{n-1}} + \cdots + \overline{a_1 z_0} + \overline{a_0} = \overline{0} = 0 \end{aligned}$$

Recall that the conjugate of the sum of complex numbers is the same as the sum of the conjugates of the complex numbers (see Problem Set 1, Problem 37). Also,

$$\overline{a_n z_0^n} = \overline{a_n}\,\overline{z_0^n}, \qquad \overline{a_{n-1} z_0^{n-1}} = \overline{a_{n-1}}\,\overline{z_0^{n-1}}, \ldots, \overline{a_1 z_0} = \overline{a}_1 \overline{z}_0$$

and

$$\overline{a_n}\,\overline{z_0^n} = \overline{a}_n \overline{z}_0^n, \qquad \overline{a_{n-1}}\,\overline{z_0^{n-1}} = \overline{a_{n-1}}\,\overline{z}_0^{n-1}, \ldots, \overline{a}_1\overline{z_0} = \overline{a}_1 \overline{z}_0$$

(See Problem Set 1, Problem 34.) Since the conjugate of a real number is the real number itself, we have $\bar{a}_0 = a_0$, $\bar{a}_1 = a_1, \ldots, \bar{a}_n = a_n$. Hence

$$f(\bar{z}_0) = a_n \bar{z}_0^n + a_{n-1}\bar{z}_0^{n-1} + \cdots + a_1\bar{z}_0 + a_0 = \overline{f(z_0)} = 0$$

so that $\bar{z}_0$ is also a root of $f(z) = 0$.

For example, the polynomial function $f(x) = x^2 - 4x + 5$ has two zeros, one of which is the complex number $2 + i$. By the conjugate root theorem, $2 - i$ is also a zero of $f(x) = x^2 - 4x + 5$, as shown by the following multiplication:

$$\begin{aligned}[x - (2 + i)][x - (2 - i)] &= x^2 - (2 + i)x - (2 - i)x + (2 + i)(2 - i) \\ &= x^2 - (2 + i + 2 - i)x + 2^2 + 1 \\ &= x^2 - 4x + 5\end{aligned}$$

EXAMPLES 1 Form $f(x)$, a polynomial with real coefficients, that has the following numbers as its zeros: $-\frac{1}{2}$, $1 + i$, and 1 as a double zero.

SOLUTION Since $1 + i$ is a root of $f(x) = 0$, it follows from the conjugate root theorem that $1 - i$ is also a root; therefore,

$$f(x) = (x + \tfrac{1}{2})(x - 1)^2[x - (1 + i)][x - (1 - i)]$$

has the given zeros. After multiplying, we get

$$f(x) = x^5 - \tfrac{7}{2}x^4 + 5x^3 - \tfrac{5}{2}x^2 - x + 1$$

2 Determine the multiplicity of the zeros of the polynomial

$$f(x) = x^4 - 4x^3 + 5x^2 - 4x + 4$$

SOLUTION With the help of synthetic division, the polynomial can be factored as

$$x^4 - 4x^3 + 5x^2 - 4x + 4 = (x - 2)^2(x + i)(x - i)$$

so that 2 is a double zero and i and $-i$ are each zeros of multiplicity 1.

PROBLEM SET 4

In Problems 1–2, use the zeros to write each polynomial function in factored form

1 $f(x) = 2x^2 - x - 2$ **2** $f(x) = x^2 - 3x - 3$

In Problems 3–7, find a polynomial function with real coefficients that has the given numbers as its zeros.

3 $2, 3, i$

4 $2, 2, 1+i, 1-i$

5 $1-3i, 1-3i, 1+3i, 1+3i$

6 $i, i, 0, 1, 2i$

7 $2, 2, \frac{1}{2}(-1+i\sqrt{3})$

8 Show that the following equation has no root.

$$\frac{1}{x-3} + \frac{1}{x-2} - \frac{x-2}{x-3} = 0$$

Why does this not contradict the fundamental theorem of algebra?

9 Given that $-1+i$ is a zero of $f(x) = x^4 + 2x^3 - 4x - 4$, find all the other zeros of f.

10 Given that i is a double zero of $f(x) = 2x^6 + x^5 + 2x^3 - 6x^2 + x - 4$, find all the other zeros of f.

In Problems 11–13, determine whether the given number is a zero of the given polynomial function. If it is, find its multiplicity.

11 $f(x) = x^4 - x^3 - 18x^2 + 52x - 40, x = 2$

12 $f(x) = 4x^6 + 4x^5 + 9x^4 + 8x^3 + 6x^2 + 4x + 1, x = i$

13 $f(x) = 9x^4 - 12x^3 + 13x^2 - 12x + 4, x = \frac{2}{3}$

5 Mathematical Induction

Suppose we were given the task to prove the following assertion:

$$1 + 2 + 3 + \cdots + n = \frac{n(n+1)}{2} \qquad \text{for any positive integer } n$$

Observe that if $n = 1$, then $1 = 1(1+1)/2$, which is true; if $n = 2$, then $1 + 2 = 2(2+1)/2$, which is also true; if $n = 3$, then $1 + 2 + 3 = 3(3+1)/2$, which is again true.

These tested values of n can only give us an *impression* that the statement is true. Simply testing these values is not adequate to establish a formal proof of the statement for all possible values of n.

Let us consider another example. Suppose that we were to prove or disprove the following statement:

$$1 + 4 + 7 + \cdots + (3n-2) = \frac{n(3n-1)}{2} \qquad \text{for any positive integer } n$$

We can begin to test the equation for specific values of n as we did in the first example. If $n = 1$, then the statement becomes $1 = 1(3 \cdot 1 - 1)/2$, which is true; if $n = 2$, then we have $1 + 4 = 2(3 \cdot 2 - 1)/2$, which is also true; if $n = 3$, then we have $1 + 4 + 7 = 3(3 \cdot 3 - 1)/2$, which is true. Again, however, our testing process is not enough to give us a generalized proof.

In order to establish a formal proof of each of the above statements, we need a method of proof that makes use of the *principle of mathematical induction.*

5.1 Principle of Mathematical Induction

Suppose that $S_1, S_2, S_3, \ldots$ is a sequence of assertions; that is, suppose that for each positive integer n we have a corresponding assertion S_n. Assume that the following two conditions hold:

(i) S_1 is true.

(ii) For each fixed positive integer k, the truth of S_k implies the truth of S_{k+1}.

Then it follows that every assertion $S_1, S_2, S_3, \ldots$ is true; that is, S_n is true for all positive integers.

Let us see how this principle can be applied to prove the statement where S_n is the assertion:

$$1 + 2 + 3 + \cdots + n = \tfrac{1}{2}n(n + 1)$$

So far as we know, S_n may be true for certain values of n and false for other values of n. In order to show that S_n is, in fact, true for all values of n, we need to verify the following conditions:

(i) S_1 is true.

(ii) If S_k is true, then S_{k+1} is also true, where k is a fixed positive integer.

Condition (i) can be verified by direct computation, for S_1 is the assertion that $1 = (\tfrac{1}{2})(1)(1 + 1)$, which is clearly true.

To prove condition (ii), we must show that S_k implies S_{k+1}; that is, we must show that if S_k is assumed to be true, then S_{k+1} must be true. To this end, assume that S_k is true; that is, assume that the assertion

$$1 + 2 + 3 + \cdots + k = (\tfrac{1}{2})k(k + 1)$$

is true. Since S_k is a true assertion, we can add $(k + 1)$ to both sides of this equation, to get

$$\begin{aligned} 1 + 2 + 3 + \cdots + k + (k + 1) &= (\tfrac{1}{2})k(k + 1) + (k + 1) \\ &= (k + 1)(\tfrac{1}{2}k + 1) \\ &= (k + 1)\left(\frac{k + 2}{2}\right) \\ &= (\tfrac{1}{2})(k + 1)(k + 2) \end{aligned}$$

But the latter assertion is precisely S_{k+1}. Hence, we have proved condition (ii). Thus, by the principle of mathematical induction, we conclude that S_n is true for any positive integer n; therefore,

$$1 + 2 + 3 + \cdots + n = (\tfrac{1}{2})n(n + 1) \qquad \text{for any positive integer } n$$

Let us investigate other examples to see how the principle of mathematical induction is especially valuable in proving certain statements in algebra that hold for all positive integers.

EXAMPLES Use mathematical induction to prove each of the following assertions for any positive integer n.

1 $1 + 4 + 7 + \cdots + (3n - 2) = \dfrac{n(3n - 1)}{2}$

PROOF Let S_n represent the above assertion. Using the principle of mathematical induction, we have

(i) S_1 becomes $1 = 1(3 \cdot 1 - 1)/2$, which is true.

(ii) Assume that S_k is true; that is, assume that, for any fixed positive integer k,

$$1 + 4 + 7 + \cdots + (3k - 2) = \frac{k(3k - 1)}{2}$$

We are to prove that S_{k+1} is true; that is, we must prove that, for each positive integer k,

$$1 + 4 + 7 + \cdots + (3k - 2) + (3k + 1) = \frac{(k + 1)(3k + 2)}{2}$$

Since S_k is true, we can add $(3k + 1)$ to each side of the first equation to get

$$\begin{aligned} 1 + 4 + 7 + \cdots + (3k - 2) + (3k + 1) &= \frac{k(3k - 1)}{2} + (3k + 1) \\ &= \frac{3k^2 - k + 2(3k + 1)}{2} \\ &= \frac{3k^2 + 5k + 2}{2} \\ &= \frac{(k + 1)(3k + 2)}{2} \end{aligned}$$

Hence S_{k+1} is true and we have proved condition (ii). We may now conclude that S_n is true; that is, for any positive integer,

$$1 + 4 + 7 + \cdots + (3n - 2) = \frac{n(3n - 1)}{2}$$

is true.

2 $1^2 + 2^2 + 3^2 + \cdots + n^2 = (\frac{1}{6})n(n + 1)(2n + 1)$

PROOF Let S_n represent the above assertion. Using the principle of mathematical induction,

(i) S_1 becomes $1^2 = (\frac{1}{6})(1)(1 + 1)(2 \cdot 1 + 1) = (\frac{1}{6})(1)(2)(3) = 1$, which is true.

(ii) Assume that S_k is true; that is, assume that, for any fixed positive integer k,

$$1^2 + 2^2 + 3^2 + \cdots + k^2 = (\tfrac{1}{6})k(k + 1)(2k + 1)$$

We must now prove that S_{k+1} is true where S_{k+1} is the assertion

$$1^2 + 2^2 + 3^2 + \cdots + k^2 + (k + 1)^2 = (\tfrac{1}{6})(k + 1)(k + 2)(2k + 3)$$

Adding $(k + 1)^2$ to both sides of S_k we have

$$\begin{aligned} 1^2 + 2^2 + 3^2 + \cdots + k^2 + (k + 1)^2 &= (\tfrac{1}{6})k(k + 1)(2k + 1) + (k + 1)^2 \\ &= (k + 1)[(\tfrac{1}{6})k(2k + 1) + (k + 1)] \\ &= (k + 1)\left(\frac{2k^2 + k + 6k + 6}{6}\right) \\ &= (k + 1)\left(\frac{2k^2 + 7k + 6}{6}\right) \\ &= (\tfrac{1}{6})(k + 1)(k + 2)(2k + 3) \end{aligned}$$

Hence, S_{k+1} is true, and we have proved condition (ii). Thus, by the principle of mathematical induction, we conclude that S_n is true for any positive integer n. That is, for any positive integer n,

$$1^2 + 2^2 + 3^2 + \cdots + n^2 = (\tfrac{1}{6})n(n + 1)(2n + 1)$$

is true.

5.2 Binomial Expansions

The principle of mathematical induction is used to prove a theorem that provides a technique (other than repeated multiplication) for expanding positive integer powers of binomial expressions, such as $(x + 3)^{10}$ and $(2x - y)^8$. This theorem is called the *binomial theorem.*

Before investigating this theorem, let us consider some examples of expanding the binomial $a + b$ to positive integer powers.

$$\begin{aligned} (a + b)^1 &= a + b \\ (a + b)^2 &= a^2 + 2ab + b^2 \\ (a + b)^3 &= a^3 + 3a^2b + 3ab^2 + b^3 \\ (a + b)^4 &= a^4 + 4a^3b + 6a^2b^2 + 4ab^3 + b^4 \\ (a + b)^5 &= a^5 + 5a^4b + 10a^3b^2 + 10a^2b^3 + 5ab^4 + b^5 \end{aligned}$$

Notice that the following pattern holds for the terms in the expansion of $(a + b)^n$, where n is a positive integer.

(i) There are $n + 1$ terms; the "first term" is a^n; the "last term" is b^n.

(ii) The power of a decreases by 1 for each term, and the power of b increases by 1 for each term. In any case, the sum of the exponents of a and b is n for each term.

The pattern for the coefficients is easier to detect if the following notation is used. We define $0!$ (zero factorial) as $0! = 1$. If n is a positive integer, $n!$ (n factorial) is defined as $n! = n(n-1)(n-2)\cdots 3\cdot 2\cdot 1$. For example,

$$5! = 5\cdot 4\cdot 3\cdot 2\cdot 1 = 120 \quad \text{and} \quad 7! = 7\cdot 6\cdot 5! = (42)(120) = 5040$$

Now, if $0 \leq k \leq n$, where k and n are integers, then the binomial coefficient,

$$\binom{n}{k} \text{ is given by } \binom{n}{k} = \frac{n!}{k!(n-k)!}$$

For example,

$$\binom{5}{3} = \frac{5!}{3!(5-3)!} = \frac{5!}{3!2!} = 10$$

$$\binom{17}{14} = \frac{17!}{14!(17-14)!} = \frac{17!}{14!3!} = 680$$

$$\binom{5}{2} = \frac{5!}{2!(5-2)!} = \frac{5!}{2!3!} = 10$$

The binomial theorem formalizes the patterns that occur in binomial power expansions.

THEOREM 1 BINOMIAL THEOREM

Let a and b be real numbers and let n be a positive integer; then

$$(a+b)^n = \binom{n}{0}a^n + \binom{n}{1}a^{n-1}b + \cdots + \binom{n}{k}a^{n-k}b^k + \cdots + \binom{n}{n}b^n$$

PROOF Use the principle of mathematical induction.

(i) S_1 is certainly true since $(a+b)^1 = a^1 + b^1$.

(ii) We must show that if S_n is true, then S_{n+1} is also true. [Notice that we are using n instead of k for (ii).]

To this end, assume that S_n is true; that is, assume that

$$(a+b)^n = \binom{n}{0}a^n + \binom{n}{1}a^{n-1}b + \cdots + \binom{n}{k}a^{n-k}b^k + \cdots + \binom{n}{n}b^n$$

for n a positive integer. After multiplying both sides by $(a+b)$, we obtain

$$\begin{aligned}(a+b)^n(a+b) &= (a+b)\left[\binom{n}{0}a^n + \binom{n}{1}a^{n-1}b + \cdots + \binom{n}{k}a^{n-k}b^k + \cdots + \binom{n}{n}b^n\right] \\ &= \binom{n}{0}(a^{n+1} + a^n b) + \binom{n}{1}(a^n b + a^{n-1}b^2) \\ &\quad + \cdots + \binom{n}{k}(a^{n-k+1}b^k + a^{n-k}b^{k+1}) + \cdots + \binom{n}{n}(ab^n + b^{n+1}) \\ &= \binom{n}{0}a^{n+1} + \left[\binom{n}{0} + \binom{n}{1}\right]a^n b + \left[\binom{n}{1} + \binom{n}{2}\right]a^{n-1}b^2 \\ &\quad + \cdots + \left[\binom{n}{k-1} + \binom{n}{k}\right]a^{n+1-k}b^k + \cdots + \binom{n}{n}b^{n+1}\end{aligned}$$

But

$$\binom{n}{k-1} + \binom{n}{k} = \frac{n!}{(k-1)!(n-k+1)!} + \frac{n!}{k!(n-k)!}$$
$$= \frac{n!k + n!(n-k+1)}{k!(n-k+1)!} = \frac{n!(n+1)}{k!(n+1-k)!}$$
$$= \frac{(n+1)!}{k!(n+1-k)!} = \binom{n+1}{k}$$

so that

$$(a+b)^{n+1} = \binom{n+1}{0}a^{n+1} + \binom{n+1}{1}a^n b$$
$$+ \cdots + \binom{n+1}{k}a^{n+1-k}b^k + \cdots + \binom{n+1}{n+1}b^{n+1}$$

But the latter assertion is precisely S_{n+1}, and the proof is complete.

EXAMPLES Use the binomial theorem to expand the following binomials.

1 $(x+y)^5$

SOLUTION

$$(x+y)^5 = \binom{5}{0}x^5 + \binom{5}{1}x^4y + \binom{5}{2}x^3y^2 + \binom{5}{3}x^2y^3 + \binom{5}{4}xy^4 + \binom{5}{5}y^5$$
$$= x^5 + \frac{5!}{1!4!}x^4y + \frac{5!}{2!3!}x^3y^2 + \frac{5!}{3!2!}x^2y^3 + \frac{5!}{4!1!}xy^4 + y^5$$
$$= x^5 + 5x^4y + 10x^3y^2 + 10x^2y^3 + 5xy^4 + y^5$$

2 $(x-3)^4$

SOLUTION

$$(x-3)^4 = [x+(-3)]^4$$
$$= \binom{4}{0}x^4 + \binom{4}{1}x^3(-3) + \binom{4}{2}x^2(-3)^2 + \binom{4}{3}x(-3)^3 + \binom{4}{4}(-3)^4$$
$$= x^4 + 4x^3(-3) + 6x^2(-3)^2 + 4x(-3)^3 + (-3)^4$$
$$= x^4 - 12x^3 + 54x^2 - 108x + 81$$

3 $(3x^2 - \frac{1}{2}\sqrt{y})^4$

SOLUTION By the binomial theorem,

$$\left(3x^2 - \frac{1}{2}\sqrt{y}\right)^4 = \binom{4}{0}(3x^2)^4 + \binom{4}{1}(3x^2)^3\left(-\frac{1}{2}\sqrt{y}\right) + \binom{4}{2}(3x^2)^2\left(-\frac{1}{2}\sqrt{y}\right)^2$$
$$+ \binom{4}{3}(3x^2)\left(-\frac{1}{2}\sqrt{y}\right)^3 + \binom{4}{4}\left(-\frac{1}{2}\sqrt{y}\right)^4$$
$$= (3x^2)^4 - 4(3x^2)^3(\tfrac{1}{2}\sqrt{y}) + 6(3x^2)^2(\tfrac{1}{2}\sqrt{y})^2$$
$$- 4(3x^2)(\tfrac{1}{2}\sqrt{y})^3 + (\tfrac{1}{2}\sqrt{y})^4$$
$$= 81x^8 - 54x^6\sqrt{y} + \tfrac{27}{2}x^4y - \tfrac{3}{2}x^2y^{3/2} + \tfrac{1}{16}y^2$$

Notice the "symmetry" of the values of the coefficients of the binomial expansions.

$$(a+b)^2 = 1a^2 + 2ab + 1b^2$$

$$(a+b)^3 = 1a^3 + 3a^2b + 3ab^2 + 1b^3$$

and

$$(a+b)^4 = 1a^4 + 4a^3b + 6a^2b^2 + 4ab^3 + 1b^4$$

In general,

$$(a+b)^n = \binom{n}{0}a^n + \binom{n}{1}a^{n-1}b + \binom{n}{2}a^{n-2}b^2 + \cdots + \binom{n}{n-2}a^2b^{n-2} + \binom{n}{n-1}ab^{n-1} + \binom{n}{n}b^n$$

where

$$\binom{n}{k} = \binom{n}{n-k}$$

since

$$\binom{n}{k} = \frac{n!}{k!(n-k)!}$$

and

$$\binom{n}{n-k} = \frac{n!}{(n-k)![n-(n-k)]!} = \frac{n!}{(n-k)!k!}$$

4 Find the sixth term of the expansion of $(2x - y^2)^8$.

SOLUTION The sixth term is

$$\binom{8}{5}(2x)^3(-y^2)^5 = -\binom{8}{5}8x^3y^{10}$$
$$= -\frac{8\cdot 7\cdot 6}{3\cdot 2\cdot 1}8x^3y^{10}$$
$$= -448x^3y^{10}$$

PROBLEM SET 5

In Problems 1–8, use mathematical induction to prove the given assertion for all positive integers n.

1 $1 + 3 + 5 + \cdots + (2n-1) = n^2$

2 $1^3 + 2^3 + 3^3 + \cdots + n^3 = \frac{1}{4}n^2(n+1)^2$

3 $2 + 4 + 6 + \cdots + 2n = n^2 + n$

4 $1^2 + 3^2 + 5^2 + \cdots + (2n-1)^2 = \frac{1}{3}n(2n-1)(2n+1)$

5 $1 \cdot 2 + 2 \cdot 3 + 3 \cdot 4 + \cdots + n(n+1) = \frac{1}{3}n(n+1)(n+2)$

6 $\dfrac{1}{1 \cdot 2} + \dfrac{1}{2 \cdot 3} + \dfrac{1}{3 \cdot 4} + \cdots + \dfrac{1}{n(n+1)} = \dfrac{n}{n+1}$

7 $4 + 4^2 + 4^3 + \cdots + 4^n = \frac{4}{3}(4^n - 1)$

8 $x^0 + x^1 + x^2 + \cdots + x^n = \dfrac{1 - x^{n+1}}{1 - x}$, for $x \neq 1$

In Problems 9–17, evaluate the given expressions.

9 $\binom{15}{10}$ **10** $\binom{n}{n}$ **11** $\binom{6}{2}$

12 $\binom{15}{5}$ **13** $\binom{n}{0}$ **14** $\binom{6}{5}$

15 $\binom{15}{3}$ **16** $\binom{n}{1}$ **17** $\binom{6}{4}$

18 (a) Show that

$$\binom{n}{r} + \binom{n}{r+1} = \binom{n+1}{r+1}$$

(b) Let $a = b = 1$ in the expansion of $(a + b)^n$, and find the sum

$$\binom{n}{0} + \binom{n}{1} + \binom{n}{2} + \cdots + \binom{n}{n}$$

In Problems 19–26, use the binomial theorem to expand each expression as specified.

19 $(x + 3)^5$, all terms **20** $(2z + x)^4$, all terms

21 $(x - 2)^4$, all terms **22** $\left(\dfrac{1}{a} + \dfrac{x}{2}\right)^3$, all terms

23 $(x + y)^{12}$, first four terms **24** $(x - 3y)^7$, first four terms

25 $(a^{3/2} - 2x^2)^8$, first four terms **26** $(x + \frac{1}{2})^{10}$, first four terms

In Problems 27–32, use the binomial theorem to find the indicated term.

27 $\left(\dfrac{x^2}{2} + a\right)^{15}$, 4th term **28** $(y^2 - 2z)^{10}$, 6th term

29 $\left(2x^2 - \dfrac{a^2}{3}\right)^9$, 7th term **30** $(x + \sqrt{a})^{12}$, middle term

31 $\left(a + \dfrac{x^2}{3}\right)^9$, term containing x^{12} **32** $\left(2\sqrt{y} - \dfrac{x}{2}\right)^{10}$, term containing y^4

6 Finite Sums and Series

We discover in this section an abbreviated way, called *sigma notation,* to write finite sums of a recursive nature such as those sums produced by the binomial theorem. In addition, we survey geometric series that are, in a sense, special types of "infinite sums."

6.1 Finite Sums

The symbol $\sum_{k=1}^{n} a_k$ is called *sigma notation* and is an abbreviated way of writing the finite sum $a_1 + a_2 + \cdots + a_n$; that is, $\sum_{k=1}^{n} a_k = a_1 + a_2 + \cdots + a_n$. Here a_k represents a function where k, called the *index*, represents the domain variable. The domain is a subset of the set of integers.

For example,

$$\sum_{k=1}^{4} \frac{1}{k} = 1 + \frac{1}{2} + \frac{1}{3} + \frac{1}{4} = \frac{25}{12} \qquad \text{with } a_k = \frac{1}{k}$$

EXAMPLES In Examples 1–3, evaluate each finite sum.

1 $\sum_{k=1}^{3} (4k^2 - 3k)$

SOLUTION Here $a_k = 4k^2 - 3k$. To find the indicated sum, we substitute the integers 1, 2, and 3 for k, then add the resulting numbers. Thus

$$\begin{aligned}\sum_{k=1}^{3} (4k^2 - 3k) &= [4(1^2) - 3(1)] + [4(2^2) - 3(2)] + [4(3^2) - 3(3)] \\ &= 1 + 10 + 27 = 38\end{aligned}$$

2 $\sum_{k=3}^{6} k(k-2)$

SOLUTION Here $a_k = k(k-2)$. Notice here that the index begins with 3. To find the indicated sum, we substitute the integers 3, 4, 5, and 6 in succession, then add the resulting numbers. Thus

$$\begin{aligned}\sum_{k=3}^{6} k(k-2) &= [3(3-2)] + [4(4-2)] + [5(5-2)] + [6(6-2)] \\ &= 3 + 8 + 15 + 24 = 50\end{aligned}$$

3 $\sum_{k=2}^{5} \frac{k-1}{k+1}$

SOLUTION Here $a_k = \frac{k-1}{k+1}$, so

$$\begin{aligned}\sum_{k=2}^{5} \frac{k-1}{k+1} &= \left(\frac{2-1}{2+1}\right) + \left(\frac{3-1}{3+1}\right) + \left(\frac{4-1}{4+1}\right) + \left(\frac{5-1}{5+1}\right) \\ &= \frac{1}{3} + \frac{2}{4} + \frac{3}{5} + \frac{4}{6} = \frac{21}{10}\end{aligned}$$

4 Write each of the following finite sums in sigma notation.

(a) $1 + \frac{1}{2} + \frac{1}{4} + \frac{1}{8} + \frac{1}{16}$

(b) The binomial theorem formula:

$$(a+b)^n = \binom{n}{0}a^n + \binom{n}{1}a^{n-1}b + \cdots + \binom{n}{k}a^{n-k}b^k + \cdots + \binom{n}{n}b^n$$

SOLUTION

(a) $1 + \frac{1}{2} + \frac{1}{4} + \frac{1}{8} + \frac{1}{16} = (\frac{1}{2})^0 + (\frac{1}{2})^1 + (\frac{1}{2})^2 + (\frac{1}{2})^3 + (\frac{1}{2})^4 = \sum_{k=0}^{4} (\frac{1}{2})^k$

or, equivalently, $1 + \frac{1}{2} + \frac{1}{4} + \frac{1}{8} + \frac{1}{16} = \sum_{k=1}^{5} (\frac{1}{2})^{k-1}$. (Why?)

(b) $$\binom{n}{0}a^n + \binom{n}{1}a^{n-1}b + \cdots + \binom{n}{k}a^{n-k}b^k + \cdots + \binom{n}{n}b^n$$
$$= \binom{n}{0}a^n b^0 + \binom{n}{1}a^{n-1}b^1 + \cdots + \binom{n}{k}a^{n-k}b^k + \cdots + \binom{n}{n}a^{n-n}b^n$$
$$= \sum_{k=0}^{n} \binom{n}{k}a^{n-k}b^k$$

so that $(a + b)^n = \sum_{k=0}^{n} \binom{n}{k}a^{n-k}b^k$.

6.2 Geometric Series

An indicated sum of the terms of a sequence, such as

$$a_1 + a_2 + a_3 + \cdots + a_n + \cdots$$

is called an *infinite series* or a *series*. Using the sigma notation, we can write this series more compactly as $\sum_{k=1}^{\infty} a_k$. The numbers a_1, a_2, a_3, and so forth, are called the terms of the series, and a_n is called the *nth term* of the series. Here, we shall be concerned with determining the "sum" of a series of the form

$$\sum_{k=1}^{\infty} ar^{k-1} = a + ar + ar^2 + \cdots + ar^{n-1} + \cdots$$

where a and r are constants and $|r| < 1$. A series of this form is called a *geometric series.*

For example,

$$\sum_{k=1}^{\infty} (\tfrac{1}{2})^{k-1} = 1 + \tfrac{1}{2} + \tfrac{1}{4} + \tfrac{1}{8} + \cdots + (\tfrac{1}{2})^k + \cdots$$

is a geometric series in which $a = 1$ and $r = \frac{1}{2}$.

The notion of an "infinite sum" is defined as follows.

Given that $\sum_{k=1}^{\infty} a_k$ is an infinite series. First, *partial sums* are formed:

$$s_1 = a_1$$
$$s_2 = a_1 + a_2$$
$$s_3 = a_1 + a_2 + a_3$$
$$\cdots\cdots\cdots\cdots\cdots\cdots$$
$$s_n = a_1 + a_2 + \cdots + a_n = \sum_{k=1}^{n} a_k$$

Then the sum S of $\sum_{k=1}^{\infty} a_k$, written $S = \sum_{k=1}^{\infty} a_k$, is defined to be the "limit value" that s_n approaches as "n approaches infinity," if there is a (finite) limit value.

Let us use this notion to determine the formula for the sum of a geometric series

$$\sum_{k=1}^{\infty} ar^{k-1} \qquad \text{where } a \text{ and } r \text{ are constants and } |r| < 1$$

Here, $a_k = ar^{k-1}$, so

$$s_1 = a$$

$$s_2 = a + ar$$

$$s_3 = a + ar + ar^2$$

$$\cdots\cdots\cdots\cdots\cdots\cdots\cdots\cdots$$

$$s_n = a + ar + ar^2 + ar^3 + \cdots + ar^{n-1}$$

Upon multiplying both sides of this last equation by r, we get

$$rs_n = ar + ar^2 + ar^3 + \cdots + ar^n$$

Next, subtract this new equation from the preceding one to get

$$s_n - rs_n = a - ar^n$$

so that

$$(1 - r)s_n = a - ar^n$$

Hence

$$s_n = \frac{a - ar^n}{1 - r} \qquad \text{where } |r| < 1$$

However, the latter equation can be written as

$$s_n = \frac{a}{1 - r} - \frac{ar^n}{1 - r}$$

Since $|r| < 1$, we see, intuitively, that as n becomes increasingly large (as n approaches infinity), r^n approaches 0. [Consider what happens, for example, to the values of $(\frac{1}{4})^n$ as n becomes larger and larger by examining the graph of $f(n) = (\frac{1}{4})^n$ (see Chapter 4, Section 1, page 149).]

Consequently, s_n approaches $a/(1 - r)$ as n approaches infinity, so that

$$\sum_{k=1}^{\infty} ar^{k-1} = \frac{a}{1 - r} \qquad \text{where } |r| < 1$$

For example,

$$\sum_{k=1}^{\infty} (\tfrac{1}{2})^{k-1} = \frac{1}{1 - \frac{1}{2}} = 2$$

The next example shows that geometric series have an interesting application in connection with the repeating decimals discussed in Chapter 1.

EXAMPLE Use geometric series to find the rational number that corresponds to each of the following decimals.

(a) $0.\overline{3}$ (b) $0.\overline{24}$

SOLUTION

(a) From the expression $0.\overline{3}$ we obtain the geometric series

$$0.\overline{3} = \frac{3}{10} + \frac{3}{100} + \frac{3}{1000} + \cdots + \frac{3}{10^n} + \cdots$$
$$= \frac{3}{10} + \frac{3}{10}\left(\frac{1}{10}\right) + \frac{3}{10}\left(\frac{1}{10}\right)^2 + \cdots + \frac{3}{10}\left(\frac{1}{10}\right)^{n-1} + \cdots$$
$$= \sum_{k=1}^{\infty} \frac{3}{10}\left(\frac{1}{10}\right)^{k-1}$$

Here $a = \frac{3}{10}$ and $r = \frac{1}{10}$, so the sum is given by

$$0.\overline{3} = \frac{\frac{3}{10}}{1 - \frac{1}{10}} = \frac{1}{3}$$

(b) $0.\overline{24}$ can be written as

$$0.\overline{24} = \frac{24}{100} + \frac{24}{10{,}000} + \frac{24}{1{,}000{,}000} + \cdots + \frac{24}{(100)^n} + \cdots$$
$$= \sum_{k=1}^{\infty} \left(\frac{24}{100}\right)\left(\frac{1}{100}\right)^{k-1}$$

We have a geometric series in which $a = \frac{24}{100}$ and $r = \frac{1}{100}$. Hence

$$0.\overline{24} = \frac{\frac{24}{100}}{1 - \frac{1}{100}} = \frac{24}{99} = \frac{8}{33}$$

PROBLEM SET 6

In Problems 1–10, find the numerical value of each finite sum.

1 $\sum_{k=1}^{5} k$

2 $\sum_{k=0}^{4} 3^{2k}$

3 $\sum_{k=1}^{3} (2k + 1)$

4 $\sum_{k=1}^{4} k^k$

5 $\sum_{k=1}^{5} \frac{1}{k(k + 1)}$

6 $\sum_{k=0}^{4} \frac{2^k}{(k + 1)}$

7 $\sum_{k=2}^{5} 2^{k-2}$

8 $\sum_{k=1}^{5} (3k^2 - 5k + 1)$

9 $\sum_{k=1}^{100} 5$

10 $\sum_{k=1}^{4} \frac{3}{k}$

In Problems 11–12, use mathematical induction to prove each assertion for all positive integers n.

11 $\sum_{k=1}^{n} k = \frac{n(n + 1)}{2}$

12 $\sum_{k=1}^{n} (2k - 1) = n^2$

In Problems 13–16, express each finite sum in sigma notation.

13 $1 + 4 + 7 + 10 + 13$

14 $\frac{1}{2} + \frac{1}{4} + \frac{1}{8} + \frac{1}{16} + \frac{1}{32}$

15 $\frac{3}{5} + \frac{9}{25} + \frac{27}{125} + \frac{81}{625}$

16 $\frac{1}{6} + \frac{2}{11} + \frac{3}{16} + \frac{4}{21}$

In Problems 17–22, determine whether the statement is true or false. Give the reason.

17 $\sum_{k=0}^{100} k^3 = \sum_{k=1}^{100} k^3$

18 $\sum_{k=0}^{100} 2 = 200$

19 $\sum_{k=0}^{100} (k + 2) = \left(\sum_{k=0}^{100} k\right) + 2$

20 $\sum_{k=1}^{20} (8k)^2 = 64 \sum_{k=2}^{21} (k - 1)^2$

21 $\sum_{k=0}^{99} (k + 1)^2 = \sum_{k=1}^{100} k^2$

22 $\sum_{k=0}^{100} k^2 = \left(\sum_{k=0}^{100} k\right)^2$

In Problems 23–26, find the sum of the given geometric series.

23 $\frac{1}{3} + \frac{1}{9} + \frac{1}{27} + \cdots + (\frac{1}{3})^n + \cdots$

24 $\frac{2}{3} + \frac{4}{9} + \frac{8}{27} + \cdots + (\frac{2}{3})^n + \cdots$

25 $\frac{1}{5} + \frac{1}{25} + \frac{1}{125} + \cdots + (\frac{1}{5})^n + \cdots$

26 $\frac{9}{8} + \frac{9}{64} + \frac{9}{512} + \cdots + 9(\frac{1}{8})^n + \cdots$

In Problems 27–32, use geometric series to find the rational number represented by the given decimal.

27 $0.\overline{32}$

28 $0.04\overline{9}$

29 $0.\overline{46}$

30 $0.\overline{072}$

31 $3.\overline{561}$

32 $32.\overline{4218}$

REVIEW PROBLEM SET

In Problems 1–8, perform the indicated operations and write the answer in the form $a + bi$.

1 $(3 - 2i) + (7 - 3i)$

2 $(3 - \sqrt{7}i)(3 + \sqrt{7}i)$

3 $(5 + 12i) + (-5 - 3i) - (1 - 3i)$

4 $(4 - i) - (7 - 3i) + (3 - i)$

5 $(5 + 3i)(3 - 5i)$

6 $\dfrac{3 + 7i}{2 - 3i}$

7 $\dfrac{4 - \sqrt{3}i}{2 + \sqrt{3}i}$

8 $\dfrac{3 - 5i}{4i}$

In Problems 9–14, solve each of the equations for x and y, where x and y are real numbers.

9 $5x + 15i = 15 - yi$

10 $(2x + 3) + (y - 3)i = 0$

11 $-3 + 17i = x + 3yi$

12 $3x - 2i + 7 + 3yi = 0$

13 $x + iy = \dfrac{3 - 2i}{2 - 3i}$

14 $3x + 5yi = \dfrac{1 + 3i}{2 + i}$

In Problems 15–22, find the modulus and an argument of each complex number.

15 $z = 5 + 5i$

16 $z = \sqrt{3} + i$

17 $z = 6\sqrt{3} + 6i$

18 $z = 8i$

19 $z = 2 + 2i$

20 $z = -1 + \sqrt{3}i$

21 $z = 3 + 4i$

22 $z = (2 + 2\sqrt{3}i)^{10}$

In Problems 23–26, find z_1z_2 and z_1/z_2 for each pair of complex numbers. Write the results in polar form and in rectangular form.

23 $z_1 = 2(\cos \pi + i \sin \pi)$ and $z_2 = 3[\cos(\pi/2) + i \sin(\pi/2)]$

24 $z_1 = 6(\cos 230° + i \sin 230°)$ and $z_2 = 3(\cos 75° + i \sin 75°)$

25 $z_1 = 6(\cos 110° + i \sin 110°)$ and $z_2 = 2(\cos 212° + i \sin 212°)$

26 $z_1 = 14(\cos 305° + i \sin 305°)$ and $z_2 = 7(\cos 65° + i \sin 65°)$

In Problems 27–31, use DeMoivre's theorem to write each expression in polar form and rectangular form.

27 $(\cos 60° + i \sin 60°)^5$

28 $(1 + i)^{40}$

29 $[\sqrt{2}/2 + i(\sqrt{2}/2)]^{100}$

30 $(\cos 0° + i \sin 0°)^{150}$

31 $(\sqrt{3} + i)^{30}$

32 Use DeMoivre's theorem to find expressions for $\cos 5\theta$ and $\sin 5\theta$. [*Hint:* $\cos 5\theta + i \sin 5\theta = (\cos \theta + i \sin \theta)^5$.]

33 Let $z = \dfrac{10(\cos 17° + i \sin 17°)^{10}}{(1 + i)^2}$. Express z in the form $a + bi$.

34 Let $z = \cos(2\pi/5) + i \sin(2\pi/5)$. Show that

$$\left| \frac{z^2 - z^3}{z^4 - z^5} \right| = 1$$

In Problems 35–38, find all the roots of each equation.

35 $z^3 = -64$

36 $z^4 = -8i$

37 $z^4 = 1 + i$

38 $z^4 = 8 - 8i$

In Problems 39–42, find a polynomial function with real coefficients that has the given numbers as its zeros.

39 $-2, -2, -3, 3$

40 $1, 1, 3, -2, 5$

41 $2, 2 - i$

42 $i, 1 + i$

In Problems 43–44, use mathematical induction to prove each formula.

43 $2 + 2^2 + 2^3 + \cdots + 2^n = 2(2^n - 1)$

44 $1 \cdot 3 + 2 \cdot 4 + 3 \cdot 5 + \cdots + n(n + 2) = \frac{1}{6}n(n + 1)(2n + 7)$

In Problems 45–48, use the binomial theorem to expand each expression.

45 $(3x + y)^4$

46 $(3x + \sqrt{x})^5$

47 $\left(2x + \dfrac{1}{y}\right)^3$

48 $\left(x - \dfrac{1}{x}\right)^8$

In Problems 49–54, find the numerical value of each finite sum.

49 $\sum_{k=1}^{5} k(2k - 1)$

50 $\sum_{k=1}^{4} 2k^2(k - 3)$

51 $\sum_{k=2}^{6} (k + 1)(k + 2)$

52 $\sum_{k=5}^{10} (2k - 1)^2$

53 $\sum_{k=1}^{6} 3^{k+1}$

54 $\sum_{k=4}^{7} \frac{1}{k(k-3)}$

In Problems 55–58, find the sum of each geometric series.

55 $\frac{1}{10} + \frac{1}{100} + \frac{1}{1000} + \cdots + (\frac{1}{10})^n + \cdots$

56 $\frac{3}{5} + \frac{9}{25} + \frac{27}{125} + \cdots + (\frac{3}{5})^n + \cdots$

57 $\frac{3}{4} + \frac{9}{16} + \frac{27}{64} + \cdots + (\frac{3}{4})^n + \cdots$

58 $(\frac{2}{3})^2 + (\frac{2}{3})^4 + (\frac{2}{3})^6 + \cdots + (\frac{2}{3})^{2n} + \cdots$

APPENDIX

APPENDIX

Tables

TABLE I COMMON LOGARITHMS

n	0.00	0.01	0.02	0.03	0.04	0.05	0.06	0.07	0.08	0.09
1.0	.0000	.0043	.0086	.0128	.0170	.0212	.0253	.0294	.0334	.0374
1.1	.0414	.0453	.0492	.0531	.0569	.0607	.0645	.0682	.0719	.0755
1.2	.0792	.0828	.0864	.0899	.0934	.0969	.1004	.1038	.1072	.1106
1.3	.1139	.1173	.1206	.1239	.1271	.1303	.1335	.1367	.1399	.1430
1.4	.1461	.1492	.1523	.1553	.1584	.1614	.1644	.1673	.1703	.1732
1.5	.1761	.1790	.1818	.1847	.1875	.1903	.1931	.1959	.1987	.2014
1.6	.2041	.2068	.2095	.2122	.2148	.2175	.2201	.2227	.2253	.2279
1.7	.2304	.2330	.2355	.2380	.2405	.2430	.2455	.2480	.2504	.2529
1.8	.2553	.2577	.2601	.2625	.2648	.2672	.2695	.2718	.2742	.2765
1.9	.2788	.2810	.2833	.2856	.2878	.2900	.2923	.2945	.2967	.2989
2.0	.3010	.3032	.3054	.3075	.3096	.3118	.3139	.3160	.3181	.3201
2.1	.3222	.3243	.3263	.3284	.3304	.3324	.3345	.3365	.3385	.3404
2.2	.3424	.3444	.3464	.3483	.3502	.3522	.3541	.3560	.3579	.3598
2.3	.3617	.3636	.3655	.3674	.3692	.3711	.3729	.3747	.3766	.3784
2.4	.3802	.3820	.3838	.3856	.3874	.3892	.3909	.3927	.3945	.3962
2.5	.3979	.3997	.4014	.4031	.4048	.4065	.4082	.4099	.4116	.4133
2.6	.4150	.4166	.4183	.4200	.4216	.4232	.4249	.4265	.4281	.4298
2.7	.4314	.4330	.4346	.4362	.4378	.4393	.4409	.4425	.4440	.4456
2.8	.4472	.4487	.4502	.4518	.4533	.4548	.4564	.4579	.4594	.4609
2.9	.4624	.4639	.4654	.4669	.4683	.4698	.4713	.4728	.4742	.4757
3.0	.4771	.4786	.4800	.4814	.4829	.4843	.4857	.4871	.4886	.4900
3.1	.4914	.4928	.4942	.4955	.4969	.4983	.4997	.5011	.5024	.5038
3.2	.5051	.5065	.5079	.5092	.5105	.5119	.5132	.5145	.5159	.5172
3.3	.5185	.5198	.5211	.5224	.5237	.5250	.5263	.5276	.5289	.5302
3.4	.5315	.5328	.5340	.5353	.5366	.5378	.5391	.5403	.5416	.5428
3.5	.5441	.5453	.5465	.5478	.5490	.5502	.5514	.5527	.5539	.5551
3.6	.5563	.5575	.5587	.5599	.5611	.5623	.5635	.5647	.5658	.5670
3.7	.5682	.5694	.5705	.5717	.5729	.5740	.5752	.5763	.5775	.5786
3.8	.5798	.5809	.5821	.5832	.5843	.5855	.5866	.5877	.5888	.5899
3.9	.5911	.5922	.5933	.5944	.5955	.5966	.5977	.5988	.5999	.6010
4.0	.6021	.6031	.6042	.6053	.6064	.6075	.6085	.6096	.6107	.6117
4.1	.6128	.6138	.6149	.6160	.6170	.6180	.6191	.6201	.6212	.6222
4.2	.6232	.6243	.6253	.6263	.6274	.6284	.6294	.6304	.6314	.6325
4.3	.6335	.6345	.6355	.6365	.6375	.6385	.6395	.6405	.6415	.6425
4.4	.6435	.6444	.6454	.6464	.6474	.6484	.6493	.6503	.6513	.6522
4.5	.6532	.6542	.6551	.6561	.6571	.6580	.6590	.6599	.6609	.6618
4.6	.6628	.6637	.6646	.6656	.6665	.6675	.6684	.6693	.6702	.6712
4.7	.6721	.6730	.6739	.6749	.6758	.6767	.6776	.6785	.6794	.6803
4.8	.6812	.6821	.6830	.6839	.6848	.6857	.6866	.6875	.6884	.6893
4.9	.6902	.6911	.6920	.6928	.6937	.6946	.6955	.6964	.6972	.6981

n	0.00	0.01	0.02	0.03	0.04	0.05	0.06	0.07	0.08	0.09
5.0	.6990	.6998	.7007	.7016	.7024	.7033	.7042	.7050	.7059	.7067
5.1	.7076	.7084	.7093	.7101	.7110	.7118	.7126	.7135	.7143	.7152
5.2	.7160	.7168	.7177	.7185	.7193	.7202	.7210	.7218	.7226	.7235
5.3	.7243	.7251	.7259	.7267	.7275	.7284	.7292	.7300	.7308	.7316
5.4	.7324	.7332	.7340	.7348	.7356	.7364	.7372	.7380	.7388	.7396
5.5	.7404	.7412	.7419	.7427	.7435	.7443	.7451	.7459	.7466	.7474
5.6	.7482	.7490	.7497	.7505	.7513	.7520	.7528	.7536	.7543	.7551
5.7	.7559	.7566	.7574	.7582	.7589	.7597	.7604	.7612	.7619	.7627
5.8	.7634	.7642	.7649	.7657	.7664	.7672	.7679	.7686	.7694	.7701
5.9	.7709	.7716	.7723	.7731	.7738	.7745	.7752	.7760	.7767	.7774
6.0	.7782	.7789	.7796	.7803	.7810	.7818	.7825	.7832	.7839	.7846
6.1	.7853	.7860	.7868	.7875	.7882	.7889	.7896	.7903	.7910	.7917
6.2	.7924	.7931	.7938	.7945	.7952	.7959	.7966	.7973	.7980	.7987
6.3	.7993	.8000	.8007	.8014	.8021	.8028	.8035	.8041	.8048	.8055
6.4	.8062	.8069	.8075	.8082	.8089	.8096	.8102	.8109	.8116	.8122
6.5	.8129	.8136	.8142	.8149	.8156	.8162	.8169	.8176	.8182	.8189
6.6	.8195	.8202	.8209	.8215	.8222	.8228	.8235	.8241	.8248	.8254
6.7	.8261	.8267	.8274	.8280	.8287	.8293	.8299	.8306	.8312	.8319
6.8	.8325	.8331	.8338	.8344	.8351	.8357	.8363	.8370	.8376	.8382
6.9	.8388	.8395	.8401	.8407	.8414	.8420	.8426	.8432	.8439	.8445
7.0	.8451	.8457	.8463	.8470	.8476	.8482	.8488	.8494	.8500	.8506
7.1	.8513	.8519	.8525	.8531	.8537	.8543	.8549	.8555	.8561	.8567
7.2	.8573	.8579	.8585	.8591	.8597	.8603	.8609	.8615	.8621	.8627
7.3	.8633	.8639	.8645	.8651	.8657	.8663	.8669	.8675	.8681	.8686
7.4	.8692	.8698	.8704	.8710	.8716	.8722	.8727	.8733	.8739	.8745
7.5	.8751	.8756	.8762	.8768	.8774	.8779	.8785	.8791	.8797	.8802
7.6	.8808	.8814	.8820	.8825	.8831	.8837	.8842	.8848	.8854	.8859
7.7	.8865	.8871	.8876	.8882	.8887	.8893	.8899	.8904	.8910	.8915
7.8	.8921	.8927	.8932	.8938	.8943	.8949	.8954	.8960	.8965	.8971
7.9	.8976	.8982	.8987	.8993	.8998	.9004	.9009	.9015	.9020	.9025
8.0	.9031	.9036	.9042	.9047	.9053	.9058	.9063	.9069	.9074	.9079
8.1	.9085	.9090	.9096	.9101	.9106	.9112	.9117	.9122	.9128	.9133
8.2	.9138	.9143	.9149	.9154	.9159	.9165	.9170	.9175	.9180	.9186
8.3	.9191	.9196	.9201	.9206	.9212	.9217	.9222	.9227	.9232	.9238
8.4	.9243	.9248	.9253	.9258	.9263	.9269	.9274	.9279	.9284	.9289
8.5	.9294	.9299	.9304	.9309	.9315	.9320	.9325	.9330	.9335	.9340
8.6	.9345	.9350	.9355	.9360	.9365	.9370	.9375	.9380	.9385	.9390
8.7	.9395	.9400	.9405	.9410	.9415	.9420	.9425	.9430	.9435	.9440
8.8	.9445	.9450	.9455	.9460	.9465	.9469	.9474	.9479	.9484	.9489
8.9	.9494	.9499	.9504	.9509	.9513	.9518	.9523	.9528	.9533	.9538
9.0	.9542	.9547	.9552	.9557	.9562	.9566	.9571	.9576	.9581	.9586
9.1	.9590	.9595	.9600	.9605	.9609	.9614	.9619	.9624	.9628	.9633
9.2	.9638	.9643	.9647	.9652	.9657	.9661	.9666	.9671	.9675	.9680
9.3	.9685	.9689	.9694	.9699	.9703	.9708	.9713	.9717	.9722	.9727
9.4	.9731	.9736	.9741	.9745	.9750	.9754	.9759	.9763	.9768	.9773
9.5	.9777	.9782	.9786	.9791	.9795	.9800	.9805	.9809	.9814	.9818
9.6	.9823	.9827	.9832	.9836	.9841	.9845	.9850	.9854	.9859	.9863
9.7	.9868	.9872	.9877	.9881	.9886	.9890	.9894	.9899	.9903	.9908
9.8	.9912	.9917	.9921	.9926	.9930	.9934	.9939	.9943	.9948	.9952
9.9	.9956	.9961	.9965	.9969	.9974	.9978	.9983	.9987	.9991	.9996

TABLE II NATURAL LOGARITHMS

t	0.00	0.01	0.02	0.03	0.04	0.05	0.06	0.07	0.08	0.09
1.0	0.0000	0.0100	0.0198	0.0296	0.0392	0.0488	0.0583	0.0677	0.0770	0.0862
1.1	0.0953	0.1044	0.1133	0.1222	0.1310	0.1398	0.1484	0.1570	0.1655	0.1740
1.2	0.1823	0.1906	0.1989	0.2070	0.2151	0.2231	0.2311	0.2390	0.2469	0.2546
1.3	0.2624	0.2700	0.2776	0.2852	0.2927	0.3001	0.3075	0.3148	0.3221	0.3293
1.4	0.3365	0.3436	0.3507	0.3577	0.3646	0.3716	0.3784	0.3853	0.3920	0.3988
1.5	0.4055	0.4121	0.4187	0.4253	0.4318	0.4383	0.4447	0.4511	0.4574	0.4637
1.6	0.4700	0.4762	0.4824	0.4886	0.4947	0.5008	0.5068	0.5128	0.5188	0.5247
1.7	0.5306	0.5365	0.5423	0.5481	0.5539	0.5596	0.5653	0.5710	0.5766	0.5822
1.8	0.5878	0.5933	0.5988	0.6043	0.6098	0.6152	0.6206	0.6259	0.6313	0.6366
1.9	0.6419	0.6471	0.6523	0.6575	0.6627	0.6678	0.6729	0.6780	0.6831	0.6881
2.0	0.6931	0.6981	0.7031	0.7080	0.7130	0.7178	0.7227	0.7275	0.7324	0.7372
2.1	0.7419	0.7467	0.7514	0.7561	0.7608	0.7655	0.7701	0.7747	0.7793	0.7839
2.2	0.7885	0.7930	0.7975	0.8020	0.8065	0.8109	0.8154	0.8198	0.8242	0.8286
2.3	0.8329	0.8372	0.8416	0.8459	0.8502	0.8544	0.8587	0.8629	0.8671	0.8713
2.4	0.8755	0.8796	0.8838	0.8879	0.8920	0.8961	0.9002	0.9042	0.9083	0.9123
2.5	0.9163	0.9203	0.9243	0.9282	0.9322	0.9361	0.9400	0.9439	0.9478	0.9517
2.6	0.9555	0.9594	0.9632	0.9670	0.9708	0.9746	0.9783	0.9821	0.9858	0.9895
2.7	0.9933	0.9969	1.0006	1.0043	1.0080	1.0116	1.0152	0.0188	1.0225	1.0260
2.8	1.0296	1.0332	1.0367	1.0403	1.0438	1.0473	1.0508	1.0543	1.0578	1.0613
2.9	1.0647	1.0682	1.0716	1.0750	1.0784	1.0818	1.0852	1.0886	1.0919	1.0953
3.0	1.0986	1.1019	1.1053	1.1086	1.1119	1.1151	1.1184	1.1217	1.1249	1.1282
3.1	1.1314	1.1346	1.1378	1.1410	1.1442	1.1474	1.1506	1.1537	1.1569	1.1600
3.2	1.1632	1.1663	1.1694	1.1725	1.1756	1.1787	1.1817	1.1848	1.1878	1.1909
3.3	1.1939	1.1970	1.2000	1.2030	1.2060	1.2090	1.2119	1.2149	1.2179	1.2208
3.4	1.2238	1.2267	1.2296	1.2326	1.2355	1.2384	1.2413	1.2442	1.2470	1.2499
3.5	1.2528	1.2556	1.2585	1.2613	1.2641	1.2669	1.2698	1.2726	1.2754	1.2782
3.6	1.2809	1.2837	1.2865	1.2892	1.2920	1.2947	1.2975	1.3002	1.3029	1.3056
3.7	1.3083	1.3110	1.3137	1.3164	1.3191	1.3218	1.3244	1.3271	1.3297	1.3324
3.8	1.3350	1.3376	1.3403	1.3429	1.3455	1.3481	1.3507	1.3533	1.3558	1.3584
3.9	1.3610	1.3635	1.3661	1.3686	1.3712	1.3737	1.3762	1.3788	1.3813	1.3838
4.0	1.3863	1.3888	1.3913	1.3938	1.3962	1.3987	1.4012	1.4036	1.4061	1.4085
4.1	1.4110	1.4134	1.4159	1.4183	1.4207	1.4231	1.4255	1.4279	1.4303	1.4327
4.2	1.4351	1.4375	1.4398	1.4422	1.4446	1.4469	1.4493	1.4516	1.4540	1.4563
4.3	1.4586	1.4609	1.4633	1.4656	1.4679	1.4702	1.4725	1.4748	1.4770	1.4793
4.4	1.4816	1.4839	1.4861	1.4884	1.4907	1.4929	1.4952	1.4974	1.4996	1.5019
4.5	1.5041	1.5063	1.5085	1.5107	1.5129	1.5151	1.5173	1.5195	1.5217	1.5239
4.6	1.5261	1.5282	1.5304	1.5326	1.5347	1.5369	1.5390	1.5412	1.5433	1.5454
4.7	1.5476	1.5497	1.5518	1.5539	1.5560	1.5581	1.5602	1.5623	1.5644	1.5665
4.8	1.5686	1.5707	1.5728	1.5748	1.5769	1.5790	1.5810	1.5831	1.5851	1.5872
4.9	1.5892	1.5913	1.5933	1.5953	1.5974	1.5994	1.6014	1.6034	1.6054	1.6074
5.0	1.6094	1.6114	1.6134	1.6154	1.6174	1.6194	1.6214	1.6233	1.6253	1.6273
5.1	1.6292	1.6312	1.6332	1.6351	1.6371	1.6390	1.6409	1.6429	1.6448	1.6467
5.2	1.6487	1.6506	1.6525	1.6544	1.6563	1.6582	1.6601	1.6620	1.6639	1.6658
5.3	1.6677	1.6696	1.6715	1.6734	1.6752	1.6771	1.6790	1.6808	1.6827	1.6845
5.4	1.6864	1.6882	1.6901	1.6919	1.6938	1.6956	1.6974	1.6993	1.7011	1.7029

t	0.00	0.01	0.02	0.03	0.04	0.05	0.06	0.07	0.08	0.09
5.5	1.7047	1.7066	1.7084	1.7102	1.7120	1.7138	1.7156	1.7174	1.7192	1.7210
5.6	1.7228	1.7246	1.7263	1.7281	1.7299	1.7317	1.7334	1.7352	1.7370	1.7387
5.7	1.7405	1.7422	1.7440	1.7457	1.7475	1.7492	1.7509	1.7527	1.7544	1.7561
5.8	1.7579	1.7596	1.7613	1.7630	1.7647	1.7664	1.7682	1.7699	1.7716	1.7733
5.9	1.7750	1.7766	1.7783	1.7800	1.7817	1.7834	1.7851	1.7867	1.7884	1.7901
6.0	1.7918	1.7934	1.7951	1.7967	1.7984	1.8001	1.8017	1.8034	1.8050	1.8066
6.1	1.8083	1.8099	1.8116	1.8132	1.8148	1.8165	1.8181	1.8197	1.8213	1.8229
6.2	1.8245	1.8262	1.8278	1.8294	1.8310	1.8326	1.8342	1.8358	1.8374	1.8390
6.3	1.8406	1.8421	1.8437	1.8453	1.8469	1.8485	1.8500	1.8516	1.8532	1.8547
6.4	1.8563	1.8579	1.8594	1.8610	1.8625	1.8641	1.8656	1.8672	1.8687	1.8703
6.5	1.8718	1.8733	1.8749	1.8764	1.8779	1.8795	1.8810	1.8825	1.8840	1.8856
6.6	1.8871	1.8886	1.8901	1.8916	1.8931	1.8946	1.8961	1.8976	1.8991	1.9006
6.7	1.9021	1.9036	1.9051	1.9066	1.9081	1.9095	1.9110	1.9125	1.9140	1.9155
6.8	1.9169	1.9184	1.9199	1.9213	1.9228	1.9242	1.9257	1.9272	1.9286	1.9301
6.9	1.9315	1.9330	1.9344	1.9359	1.9373	1.9387	1.9402	1.9416	1.9430	1.9445
7.0	1.9459	1.9473	1.9488	1.9502	1.9516	1.9530	1.9544	1.9559	1.9573	1.9587
7.1	1.9601	1.9615	1.9629	1.9643	1.9657	1.9671	1.9685	1.9699	1.9713	1.9727
7.2	1.9741	1.9755	1.9769	1.9782	1.9796	1.9810	1.9824	1.9838	1.9851	1.9865
7.3	1.9879	1.9892	1.9906	1.9920	1.9933	1.9947	1.9961	1.9974	1.9988	2.0001
7.4	2.0015	2.0028	2.0042	2.0055	2.0069	2.0082	2.0096	2.0109	2.0122	2.0136
7.5	2.0149	2.0162	2.0176	2.0189	2.0202	2.0215	2.0229	2.0242	2.0255	2.0268
7.6	2.0282	2.0295	2.0308	2.0321	2.0334	2.0347	2.0360	2.0373	2.0386	2.0399
7.7	2.0412	2.0425	2.0438	2.0451	2.0464	2.0477	2.0490	2.0503	2.0516	2.0528
7.8	2.0541	2.0554	2.0567	2.0580	2.0592	2.0605	2.0618	2.0631	2.0643	2.0665
7.9	2.0669	2.0681	2.0694	2.0707	2.0719	2.0732	2.0744	2.0757	2.0769	2.0782
8.0	2.0794	2.0807	2.0819	2.0832	2.0844	2.0857	2.0869	2.0882	2.0894	2.0906
8.1	2.0919	2.0931	2.0943	2.0956	2.0968	2.0980	2.0992	2.1005	2.1017	2.1029
8.2	2.1041	2.1054	2.1066	2.1078	2.1090	2.1102	2.1114	2.1126	2.1138	2.1150
8.3	2.1163	2.1175	2.1187	2.1199	2.1211	2.1223	2.1235	2.1247	2.1258	2.1270
8.4	2.1282	2.1294	2.1306	2.1318	2.1330	2.1342	2.1353	2.1365	2.1377	2.1389
8.5	2.1401	2.1412	2.1424	2.1436	2.1448	2.1459	2.1471	2.1483	2.1494	2.1506
8.6	2.1518	2.1529	2.1541	2.1552	2.1564	2.1576	2.1587	2.1599	2.1610	2.1622
8.7	2.1633	2.1645	2.1656	2.1668	2.1679	2.1691	2.1702	2.1713	2.1725	2.1736
8.8	2.1748	2.1759	2.1770	2.1782	2.1793	2.1804	2.1815	2.1827	2.1838	2.1849
8.9	2.1861	2.1872	2.1883	2.1894	2.1905	2.1917	2.1928	2.1939	2.1950	2.1961
9.0	2.1972	2.1983	2.1994	2.2006	2.2017	2.2028	2.2039	2.2050	2.2061	2.2072
9.1	2.2083	2.2094	2.2105	2.2116	2.2127	2.2138	2.2148	2.2159	2.2170	2.2181
9.2	2.2192	2.2203	2.2214	2.2225	2.2235	2.2246	2.2257	2.2268	2.2279	2.2289
9.3	2.2300	2.2311	2.2322	2.2332	2.2343	2.2354	2.2364	2.2375	2.2386	2.2396
9.4	2.2407	2.2418	2.2428	2.2439	2.2450	2.2460	2.2471	2.2481	2.2492	2.2502
9.5	2.2513	2.2523	2.2534	2.2544	2.2555	2.2565	2.2576	2.2586	2.2597	2.2607
9.6	2.2618	2.2628	2.2638	2.2649	2.2659	2.2670	2.2680	2.2690	2.2701	2.2711
9.7	2.2721	2.2732	2.2742	2.2752	2.2762	2.2773	2.2783	2.2793	2.2803	2.2814
9.8	2.2824	2.2834	2.2844	2.2854	2.2865	2.2875	2.2885	2.2895	2.2905	2.2915
9.9	2.2925	2.2935	2.2946	2.2956	2.2966	2.2976	2.2986	2.2996	2.3006	2.3016

TABLE III TRIGONOMETRIC FUNCTIONS—RADIAN MEASURE

t	sin *t*	cos *t*	tan *t*	cot *t*	sec *t*	csc *t*
.00	.0000	1.0000	.0000	—	1.000	—
.01	.0100	1.0000	.0100	99.997	1.000	100.00
.02	.0200	.9998	.0200	49.993	1.000	50.00
.03	.0300	.9996	.0300	33.323	1.000	33.34
.04	.0400	.9992	.0400	24.987	1.001	25.01
.05	.0500	.9988	.0500	19.983	1.001	20.01
.06	.0600	.9982	.0601	16.647	1.002	16.68
.07	.0699	.9976	.0701	14.262	1.002	14.30
.08	.0799	.9968	.0802	12.473	1.003	12.51
.09	.0899	.9960	.0902	11.081	1.004	11.13
.10	.0998	.9950	.1003	9.967	1.005	10.02
.11	.1098	.9940	.1104	9.054	1.006	9.109
.12	.1197	.9928	.1206	8.293	1.007	8.353
.13	.1296	.9916	.1307	7.649	1.009	7.714
.14	.1395	.9902	.1409	7.096	1.010	7.166
.15	.1494	.9888	.1511	6.617	1.011	6.692
.16	.1593	.9872	.1614	6.197	1.013	6.277
.17	.1692	.9856	.1717	5.826	1.015	5.911
.18	.1790	.9838	.1820	5.495	1.016	5.586
.19	.1889	.9820	.1923	5.200	1.018	5.295
.20	.1987	.9801	.2027	4.933	1.020	5.033
.21	.2085	.9780	.2131	4.692	1.022	4.797
.22	.2182	.9759	.2236	4.472	1.025	4.582
.23	.2280	.9737	.2341	4.271	1.027	4.386
.24	.2377	.9713	.2447	4.086	1.030	4.207
.25	.2474	.9689	.2553	3.916	1.032	4.042
.26	.2571	.9664	.2660	3.759	1.035	3.890
.27	.2667	.9638	.2768	3.613	1.038	3.749
.28	.2764	.9611	.2876	3.478	1.041	3.619
.29	.2860	.9582	.2984	3.351	1.044	3.497
.30	.2955	.9553	.3093	3.233	1.047	3.384
.31	.3051	.9523	.3203	3.122	1.050	3.278
.32	.3146	.9492	.3314	3.018	1.053	3.179
.33	.3240	.9460	.3425	2.920	1.057	3.086
.34	.3335	.9428	.3537	2.827	1.061	2.999
.35	.3429	.9394	.3650	2.740	1.065	2.916
.36	.3523	.9359	.3764	2.657	1.068	2.839
.37	.3616	.9323	.3879	2.578	1.073	2.765
.38	.3709	.9287	.3994	2.504	1.077	2.696
.39	.3802	.9249	.4111	2.433	1.081	2.630
.40	.3894	.9211	.4228	2.365	1.086	2.568
.41	.3986	.9171	.4346	2.301	1.090	2.509
.42	.4078	.9131	.4466	2.239	1.095	2.452
.43	.4169	.9090	.4586	2.180	1.100	2.399
.44	.4259	.9048	.4708	2.124	1.105	2.348

t	$\sin t$	$\cos t$	$\tan t$	$\cot t$	$\sec t$	$\csc t$
.45	.4350	.9004	.4831	2.070	1.111	2.299
.46	.4439	.8961	.4954	2.018	1.116	2.253
.47	.4529	.8916	.5080	1.969	1.122	2.208
.48	.4618	.8870	.5206	1.921	1.127	2.166
.49	.4706	.8823	.5334	1.875	1.133	2.125
.50	.4794	.8776	.5463	1.830	1.139	2.086
.51	.4882	.8727	.5594	1.788	1.146	2.048
.52	.4969	.8678	.5726	1.747	1.152	2.013
$\frac{\pi}{6}$	.5000	.8660	.5774	1.732	1.155	2.000
.53	.5055	.8628	.5859	1.707	1.159	1.978
.54	.5141	.8577	.5994	1.668	1.166	1.945
.55	.5227	.8525	.6131	1.631	1.173	1.913
.56	.5312	.8473	.6269	1.595	1.180	1.883
.57	.5396	.8419	.6410	1.560	1.188	1.853
.58	.5480	.8365	.6552	1.526	1.196	1.825
.59	.5564	.8309	.6696	1.494	1.203	1.797
.60	.5646	.8253	.6841	1.462	1.212	1.771
.61	.5729	.8196	.6989	1.431	1.220	1.746
.62	.5810	.8139	.7139	1.401	1.229	1.721
.63	.5891	.8080	.7291	1.372	1.238	1.697
.64	.5972	.8021	.7445	1.343	1.247	1.674
.65	.6052	.7961	.7602	1.315	1.256	1.652
.66	.6131	.7900	.7761	1.288	1.266	1.631
.67	.6210	.7838	.7923	1.262	1.276	1.610
.68	.6288	.7776	.8087	1.237	1.286	1.590
.69	.6365	.7712	.8253	1.212	1.297	1.571
.70	.6442	.7648	.8423	1.187	1.307	1.552
.71	.6518	.7584	.8595	1.163	1.319	1.534
.72	.6594	.7518	.8771	1.140	1.330	1.517
.73	.6669	.7452	.8949	1.117	1.342	1.500
.74	.6743	.7385	.9131	1.095	1.354	1.483
.75	.6816	.7317	.9316	1.073	1.367	1.467
.76	.6889	.7248	.9505	1.052	1.380	1.452
.77	.6961	.7179	.9697	1.031	1.393	1.437
.78	.7033	.7109	.9893	1.011	1.407	1.422
$\frac{\pi}{4}$	.7071	.7071	1.000	1.000	1.414	1.414
.79	.7104	.7038	1.009	.9908	1.421	1.408
.80	.7174	.6967	1.030	.9712	1.435	1.394
.81	.7243	.6895	1.050	.9520	1.450	1.381
.82	.7311	.6822	1.072	.9331	1.466	1.368
.83	.7379	.6749	1.093	.9146	1.482	1.355
.84	.7446	.6675	1.116	.8964	1.498	1.343

TABLE III TRIGONOMETRIC FUNCTIONS—RADIAN MEASURE

t	$\sin t$	$\cos t$	$\tan t$	$\cot t$	$\sec t$	$\csc t$
.85	.7513	.6600	1.138	.8785	1.515	1.331
.86	.7578	.6524	1.162	.8609	1.533	1.320
.87	.7643	.6448	1.185	.8437	1.551	1.308
.88	.7707	.6372	1.210	.8267	1.569	1.297
.89	.7771	.6294	1.235	.8100	1.589	1.287
.90	.7833	.6216	1.260	.7936	1.609	1.277
.91	.7895	.6137	1.286	.7774	1.629	1.267
.92	.7956	.6058	1.313	.7615	1.651	1.257
.93	.8016	.5978	1.341	.7458	1.673	1.247
.94	.8076	.5898	1.369	.7303	1.696	1.238
.95	.8134	.5817	1.398	.7151	1.719	1.229
.96	.8192	.5735	1.428	.7001	1.744	1.221
.97	.8249	.5653	1.459	.6853	1.769	1.212
.98	.8305	.5570	1.491	.6707	1.795	1.204
.99	.8360	.5487	1.524	.6563	1.823	1.196
1.00	.8415	.5403	1.557	.6421	1.851	1.188
1.01	.8468	.5319	1.592	.6281	1.880	1.181
1.02	.8521	.5234	1.628	.6142	1.911	1.174
1.03	.8573	.5148	1.665	.6005	1.942	1.166
1.04	.8624	.5062	1.704	.5870	1.975	1.160
$\frac{\pi}{3}$	.8660	.5000	1.732	.5774	2.000	1.155
1.05	.8674	.4976	1.743	.5736	2.010	1.153
1.06	.8724	.4889	1.784	.5604	2.046	1.146
1.07	.8772	.4801	1.827	.5473	2.083	1.140
1.08	.8820	.4713	1.871	.5344	2.122	1.134
1.09	.8866	.4625	1.917	.5216	2.162	1.128
1.10	.8912	.4536	1.965	.5090	2.205	1.122
1.11	.8957	.4447	2.014	.4964	2.249	1.116
1.12	.9001	.4357	2.066	.4840	2.295	1.111
1.13	.9044	.4267	2.120	.4718	2.344	1.106
1.14	.9086	.4176	2.176	.4596	2.395	1.101
1.15	.9128	.4085	2.234	.4475	2.448	1.096
1.16	.9168	.3993	2.296	.4356	2.504	1.091
1.17	.9208	.3902	2.360	.4237	2.563	1.086
1.18	.9246	.3809	2.427	.4120	2.625	1.082
1.19	.9284	.3717	2.498	.4003	2.691	1.077
1.20	.9320	.3624	2.572	.3888	2.760	1.073
1.21	.9356	.3530	2.650	.3773	2.833	1.069
1.22	.9391	.3436	2.733	.3659	2.910	1.065
1.23	.9425	.3342	2.820	.3546	2.992	1.061
1.24	.9458	.3248	2.912	.3434	3.079	1.057
1.25	.9490	.3153	3.010	.3323	3.171	1.054
1.26	.9521	.3058	3.113	.3212	3.270	1.050
1.27	.9551	.2963	3.224	.3102	3.375	1.047
1.28	.9580	.2867	3.341	.2993	3.488	1.044
1.29	.9608	.2771	3.467	.2884	3.609	1.041

t	$\sin t$	$\cos t$	$\tan t$	$\cot t$	$\sec t$	$\csc t$
1.30	.9636	.2675	3.602	.2776	3.738	1.038
1.31	.9662	.2579	3.747	.2669	3.878	1.035
1.32	.9687	.2482	3.903	.2562	4.029	1.032
1.33	.9711	.2385	4.072	.2456	4.193	1.030
1.34	.9735	.2288	4.256	.2350	4.372	1.027
1.35	.9757	.2190	4.455	.2245	4.566	1.025
1.36	.9779	.2092	4.673	.2140	4.779	1.023
1.37	.9799	.1994	4.913	.2035	5.014	1.021
1.38	.9819	.1896	5.177	.1931	5.273	1.018
1.39	.9837	.1798	5.471	.1828	5.561	1.017
1.40	.9854	.1700	5.798	.1725	5.883	1.015
1.41	.9871	.1601	6.165	.1622	6.246	1.013
1.42	.9887	.1502	6.581	.1519	6.657	1.011
1.43	.9901	.1403	7.055	.1417	7.126	1.010
1.44	.9915	.1304	7.602	.1315	7.667	1.009
1.45	.9927	.1205	8.238	.1214	8.299	1.007
1.46	.9939	.1106	8.989	.1113	9.044	1.006
1.47	.9949	.1006	9.887	.1011	9.938	1.005
1.48	.9959	.0907	10.983	.0910	11.029	1.004
1.49	.9967	.0807	12.350	.0810	12.390	1.003
1.50	.9975	.0707	14.101	.0709	14.137	1.003
1.51	.9982	.0608	16.428	.0609	16.458	1.002
1.52	.9987	.0508	19.670	.0508	19.695	1.001
1.53	.9992	.0408	24.498	.0408	24.519	1.001
1.54	.9995	.0308	32.461	.0308	32.476	1.000
1.55	.9998	.0208	48.078	.0208	48.089	1.000
1.56	.9999	.0108	92.620	.0108	92.626	1.000
1.57	1.0000	.0008	1255.8	.0008	1255.8	1.000
$\frac{\pi}{2}$	1.0000	.0000	—	.0000	—	1.000

TABLE IV TRIGONOMETRIC FUNCTIONS—DEGREE MEASURE

Degrees	Sin	Csc	Tan	Cot	Sec	Cos	
0° 0′	.0000	——	.0000	——	1.000	1.0000	90° 0′
10′	029	343.8	029	343.8	000	000	50′
20′	058	171.9	058	171.9	000	000	40′
30′	.0087	114.6	.0087	114.6	1.000	1.0000	30′
40′	116	85.95	116	85.94	000	0.9999	20′
50′	145	68.76	145	68.75	000	999	10′
1° 0′	.0175	57.30	.0175	57.29	1.000	.9998	89° 0′
10′	204	49.11	204	49.10	000	998	50′
20′	233	42.98	233	42.96	000	997	40′
30′	.0262	38.20	.0262	38.19	1.000	.9997	30′
40′	291	34.38	291	34.37	000	996	20′
50′	320	31.26	320	31.24	001	995	10′
2° 0′	.0349	28.65	.0349	28.64	1.001	.9994	88° 0′
10′	378	26.45	378	26.43	001	993	50′
20′	407	24.56	407	24.54	001	992	40′
30′	.0436	22.93	.0437	22.90	1.001	.9990	30′
40′	465	21.49	466	21.47	001	989	20′
50′	494	20.23	495	20.21	001	988	10′
3° 0′	.0523	19.11	.0524	19.08	1.001	.9986	87° 0′
10′	552	18.10	553	18.07	002	985	50′
20′	581	17.20	582	17.17	002	983	40′
30′	.0610	16.38	.0612	16.35	1.002	.9981	30′
40′	640	15.64	641	15.60	002	980	20′
50′	669	14.96	670	14.92	002	978	10′
4° 0′	.0698	14.34	.0699	14.30	1.002	.9976	86° 0′
10′	727	13.76	729	13.73	003	974	50′
20′	756	13.23	758	13.20	003	971	40′
30′	.0785	12.75	.0787	12.71	1.003	.9969	30′
40′	814	12.29	816	12.25	003	967	20′
50′	843	11.87	846	11.83	004	964	10′
5° 0′	.0872	11.47	.0875	11.43	1.004	.9962	85° 0′
10′	901	11.10	904	11.06	004	959	50′
20′	929	10.76	934	10.71	004	957	40′
30′	.0958	10.43	.0963	10.39	1.005	.9954	30′
40′	.0987	10.13	.0992	10.08	005	951	20′
50′	.1016	9.839	.1022	9.788	005	948	10′
6° 0′	.1045	9.567	.1051	9.514	1.006	.9945	84° 0′
10′	074	9.309	080	9.255	006	942	50′
20′	103	9.065	110	9.010	006	939	40′
30′	.1132	8.834	.1139	8.777	1.006	.9936	30′
40′	161	8.614	169	8.556	007	932	20′
50′	190	8.405	198	8.345	007	929	10′
7° 0′	.1219	8.206	.1228	8.144	1.008	.9925	83° 0′
10′	248	8.016	257	7.953	008	922	50′
20′	276	7.834	287	7.770	008	918	40′
30′	.1305	7.661	.1317	7.596	1.009	.9914	30′
40′	334	7.496	346	7.429	009	911	20′
50′	363	7.337	376	7.269	009	907	10′
8° 0′	.1392	7.185	.1405	7.115	1.010	.9903	82° 0′
	Cos	Sec	Cot	Tan	Csc	Sin	Degrees

Degrees	Sin	Csc	Tan	Cot	Sec	Cos	
8° 0′	.1392	7.185	.1405	7.115	1.010	.9903	82° 0′
10′	421	7.040	435	6.968	010	899	50′
20′	449	6.900	465	6.827	011	894	40′
30′	.1478	6.765	.1495	6.691	1.011	.8980	30′
40′	507	6.636	524	6.561	012	886	20′
50′	536	6.512	554	6.435	012	881	10′
9° 0′	.1564	6.392	.1584	6.314	1.012	.9877	81° 0′
10′	593	277	614	197	013	872	50′
20′	622	166	644	6.084	013	868	40′
30′	.1650	6.059	.1673	5.976	1.014	.9863	30′
40′	679	5.955	703	871	014	858	20′
50′	708	855	733	769	015	853	10′
10° 0′	.1736	5.759	.1763	5.671	1.015	.9848	80° 0′
10′	765	665	793	576	016	843	50′
20′	794	575	823	485	016	838	40′
30′	.1822	5.487	.1853	5.396	1.017	.9833	30′
40′	851	403	883	309	018	827	20′
50′	880	320	914	226	018	822	10′
11° 0′	.1908	5.241	.1944	5.145	1.019	.9816	79° 0′
10′	937	164	.1974	5.066	019	811	50′
20′	965	089	.2004	4.989	020	805	40′
30′	.1994	5.016	.2035	4.915	1.020	.9799	30′
40′	.2022	4.945	065	843	021	793	20′
50′	051	876	095	773	022	787	10′
12° 0′	.2079	4.810	.2126	4.705	1.022	.9781	78° 0′
10′	108	745	156	638	023	775	50′
20′	136	682	186	574	024	769	40′
30′	.2164	4.620	.2217	4.511	1.024	.9763	30′
40′	193	560	247	449	025	757	20′
50′	221	502	278	390	026	750	10′
13° 0′	.2250	4.445	.2309	4.331	1.026	.9744	77° 0′
10′	278	390	339	275	027	737	50′
20′	306	336	370	219	028	730	40′
30′	.2334	4.284	.2401	4.165	1.028	.9724	30′
40′	363	232	432	113	029	717	20′
50′	391	182	462	061	030	710	10′
14° 0′	.2419	4.134	.2493	4.011	1.031	.9703	76° 0′
10′	447	086	524	3.962	031	696	50′
20′	476	4.039	555	914	032	698	40′
30′	.2504	3.994	.2586	3.867	1.033	.9681	30′
40′	532	950	617	821	034	674	20′
50′	560	906	648	776	034	667	10′
15° 0′	.2588	3.864	.2679	3.732	1.035	.9659	75° 0′
10′	616	822	711	689	036	652	50′
20′	644	782	742	647	037	644	40′
30′	.2672	3.742	.2773	3.606	1.038	.9636	30′
40′	700	703	805	566	039	628	20′
50′	728	665	836	526	039	621	10′
16° 0′	.2756	3.628	.2867	3.487	1.040	.9613	74° 0′
	Cos	Sec	Cot	Tan	Csc	Sin	Degrees

TABLE IV TRIGONOMETRIC FUNCTIONS—DEGREE MEASURE

Degrees	Sin	Csc	Tan	Cot	Sec	Cos	
16° 0′	.2756	3.628	.2867	3.487	1.040	.9613	74° 0′
10′	784	592	899	450	041	605	50′
20′	812	556	931	412	042	596	40′
30′	.2840	3.521	.2962	3.376	1.043	.9588	30′
40′	868	487	.2994	340	044	580	20′
50′	896	453	3026	305	045	572	10′
17° 0′	.2924	3.420	.3057	3.271	1.046	.9563	73° 0′
10′	952	388	089	237	047	555	50′
20′	.2979	357	121	204	048	546	40′
30′	.3007	3.326	.3153	3.172	1.048	.9537	30′
40′	035	295	185	140	049	528	20′
50′	062	265	217	108	050	520	10′
18° 0′	.3090	3.236	.3249	3.078	1.051	.9511	72° 0′
10′	118	207	281	047	052	502	50′
20′	145	179	314	3.018	053	492	40′
30′	.3173	3.152	.3346	2.989	1.054	.9483	30′
40′	201	124	378	960	056	474	20′
50′	228	098	411	932	057	465	10′
19° 0′	.3256	3.072	.3443	2.904	1.058	.9455	71° 0′
10′	283	046	476	877	059	446	50′
20′	311	3.021	508	850	060	436	40′
30′	.3338	2.996	.3541	2.824	1.061	.9426	30′
40′	365	971	574	798	062	417	20′
50′	393	947	607	773	063	407	10′
20° 0′	.3420	2.924	.3640	2.747	1.064	.9397	70° 0′
10′	448	901	673	723	065	387	50′
20′	475	878	706	699	066	377	40′
30′	.3502	2.855	.3739	2.675	1.068	.9367	30′
40′	529	833	772	651	069	356	20′
50′	557	812	805	628	070	346	10′
21° 0′	.3584	2.790	.3839	2.605	1.071	.9336	69° 0′
10′	611	769	872	583	072	325	50′
20′	638	749	906	560	074	315	40′
30′	.3665	2.729	.3939	2.539	1.075	.9304	30′
40′	692	709	.3973	517	076	293	20′
50′	719	689	.4006	496	077	283	10′
22° 0′	.3746	2.669	.4040	2.475	1.079	.9272	68° 0′
10′	773	650	074	455	080	261	50′
20′	800	632	108	434	081	250	40′
30′	.3827	2.613	.4142	2.414	1.082	.9239	30′
40′	854	595	176	394	084	228	20′
50′	881	577	210	375	085	216	10′
23° 0′	.3907	2.559	.4245	2.356	1.086	.9205	67° 0′
10′	934	542	279	337	088	194	50′
20′	961	525	314	318	089	182	40′
30′	.3987	2.508	.4348	2.300	1.090	.9171	30′
40′	.4014	491	383	282	092	159	20′
50′	041	475	417	264	093	147	10′
24° 0′	.4067	2.459	.4452	2.246	1.095	.9135	66° 0′
	Cos	Sec	Cot	Tan	Csc	Sin	Degrees

Degrees	Sin	Csc	Tan	Cot	Sec	Cos	
24° 0′	.4067	2.459	.4452	2.246	1.095	.9135	66° 0′
10′	094	443	487	229	096	124	50′
20′	120	427	522	211	097	112	40′
30′	.4147	2.411	.4557	2.194	1.099	.9100	30′
40′	173	396	592	177	100	088	20′
50′	200	381	628	161	102	075	10′
25° 0′	.4226	2.366	.4663	2.145	1.103	.9063	65° 0′
10′	253	352	699	128	105	051	50′
20′	279	337	734	112	106	038	40′
30′	.4305	2.323	.4770	2.097	1.108	.9026	30′
40′	331	309	806	081	109	013	20′
50′	358	295	841	066	111	.9001	10′
26° 0′	.4384	2.281	.4877	2.050	1.113	.8988	64° 0′
10′	410	268	913	035	114	975	50′
20′	436	254	950	020	116	962	40′
30′	.4462	2.241	.4986	2.006	1.117	.8949	30′
40′	488	228	.5022	1.991	119	936	20′
50′	514	215	059	977	121	923	10′
27° 0′	.4540	2.203	.5095	1.963	1.122	.8910	63° 0′
10′	566	190	132	949	124	897	50′
20′	592	178	169	935	126	884	40′
30′	.4617	2.166	.5206	1.921	1.127	.8870	30′
40′	643	154	243	907	129	857	20′
50′	669	142	280	894	131	843	10′
28° 0′	.4695	2.130	.5317	1.881	1.133	.8829	62° 0′
10′	720	118	354	868	134	.816	50′
20′	746	107	392	855	136	802	40′
30′	.4772	2.096	.5430	1.842	1.138	.8788	30′
40′	797	085	467	829	140	774	20′
50′	823	074	505	816	142	760	10′
29° 0′	.4848	2.063	.5543	1.804	1.143	.8746	61° 0′
10′	874	052	581	792	145	732	50′
20′	899	041	619	780	147	718	40′
30′	.4924	2.031	.5658	1.767	1.149	.8704	30′
40′	950	020	696	756	151	689	20′
50′	.4975	010	735	744	153	675	10′
30° 0′	.5000	2.000	.5774	1.732	1.155	.8660	60° 0′
10′	025	1.990	812	720	157	646	50′
20′	050	980	851	709	159	631	40′
30′	.5075	1.970	.5890	1.698	1.161	.8616	30′
40′	100	961	930	686	163	601	20′
50′	125	951	.5969	675	165	587	10′
31° 0′	.5150	1.942	.6009	1.664	1.167	.8572	59° 0′
10′	175	932	048	653	169	557	50′
20′	200	923	088	643	171	542	40′
30′	.5225	1.914	.6128	1.632	1.173	.8526	30′
40′	250	905	168	621	175	511	20′
50′	275	896	208	611	177	496	10′
32° 0′	.5299	1.887	.6249	1.600	1.179	.8480	58° 0′
	Cos	Sec	Cot	Tan	Csc	Sin	Degrees

TABLE IV TRIGONOMETRIC FUNCTIONS—DEGREE MEASURE

Degrees	Sin	Csc	Tan	Cot	Sec	Cos	
32° 0′	.5299	1.887	.6249	1.600	1.179	.8480	58° 0′
10′	324	878	289	590	181	465	50′
20′	348	870	330	580	184	450	40′
30′	.5373	1.861	.6371	1.570	1.186	.8434	30′
40′	398	853	412	560	188	418	20′
50′	422	844	453	550	190	403	10′
33° 0′	.5446	1.836	.6494	1.540	1.192	.8387	57° 0′
10′	471	828	536	530	195	371	50′
20′	495	820	577	520	197	355	40′
30′	.5519	1.812	.6619	1.511	1.199	.8339	30′
40′	544	804	661	501	202	323	20′
50′	568	796	703	1.492	204	307	10′
34° 0′	.5592	1.788	.6745	1.483	1.206	.8290	56° 0′
10′	616	781	787	473	209	274	50′
20′	640	773	830	464	211	258	40′
30′	.5664	1.766	.6873	1.455	1.213	.8241	30′
40′	688	758	916	446	216	225	20′
50′	712	751	.6959	437	218	208	10′
35° 0′	.5736	1.743	.7002	1.428	1.221	.8192	55° 0′
10′	760	736	046	419	223	175	50′
20′	783	729	089	411	226	158	40′
30′	.5807	1.722	.7133	1.402	1.228	.8141	30′
40′	831	715	177	393	231	124	20′
50′	854	708	221	385	233	107	10′
36° 0′	.5878	1.701	.7265	1.376	1.236	.8090	54° 0′
10′	901	695	310	368	239	073	50′
20′	925	688	355	360	241	056	40′
30′	.5948	1.681	.7400	1.351	1.244	.8039	30′
40′	972	675	445	343	247	021	20′
50′	.5995	668	490	335	249	.8004	10′
37° 0′	.6018	1.662	.7536	1.327	1.252	.7986	53° 0′
10′	041	655	581	319	255	969	50′
20′	065	649	627	311	258	951	40′
30′	.6088	1.643	.7673	1.303	1.260	.7934	30′
40′	111	636	720	295	263	916	20′
50′	134	630	766	288	266	898	10′
38° 0′	.6157	1.624	.7813	1.280	1.269	.7880	52° 0′
10′	180	618	860	272	272	862	50′
20′	202	612	907	265	275	844	40′
30′	.6225	1.606	.7954	1.257	1.278	.7826	30′
40′	248	601	.8002	250	281	808	20′
50′	271	595	050	242	284	790	10′
39° 0′	.6293	1.589	.8098	1.235	1.287	.7771	51° 0′
10′	316	583	146	228	290	753	50′
20′	338	578	195	220	293	735	40′
30′	.6361	1.572	.8243	1.213	1.296	.7716	30′
40′	383	567	292	206	299	698	20′
50′	406	561	342	199	302	679	10′
40° 0′	.6428	1.556	.8391	1.192	1.305	.7660	50° 0′
	Cos	Sec	Cot	Tan	Csc	Sin	Degrees

Degrees	Sin	Csc	Tan	Cot	Sec	Cos	
40° 0′	.6428	1.556	.8391	1.192	1.305	.7660	50° 0′
10′	450	550	441	185	309	642	50′
20′	472	545	491	178	312	623	40′
30′	.6494	1.540	.8541	1.171	1.315	.7604	30′
40′	517	535	591	164	318	585	20′
50′	539	529	642	157	322	566	10′
41° 0′	.6561	1.524	.8693	1.150	1.325	.7547	49° 0′
10′	583	519	744	144	328	528	50′
20′	604	514	796	137	332	509	40′
30′	.6626	1.509	.8847	1.130	1.335	.7490	30′
40′	648	504	899	124	339	470	20′
50′	670	499	.8952	117	342	451	10′
42° 0′	.6691	1.494	.9004	1.111	1.346	.7431	48° 0′
10′	713	490	057	104	349	412	50′
20′	734	485	110	098	353	392	40′
30′	.6756	1.480	.9163	1.091	1.356	.7373	30′
40′	777	476	217	085	360	353	20′
50′	799	471	271	079	364	333	10′
43° 0′	.6820	1.466	.9325	1.072	1.367	.7314	47° 0′
10′	841	462	380	066	371	294	50′
20′	862	457	435	060	375	274	40′
30′	.6884	1.453	.9490	1.054	1.379	.7254	30′
40′	905	448	545	048	382	234	20′
50′	926	444	601	042	386	214	10′
44° 0′	.6947	1.440	.9657	1.036	1.390	.7193	46° 0′
10′	967	435	713	030	394	173	50′
20′	.6988	431	770	024	398	153	40′
30′	.7009	1.427	.9827	1.018	1.402	.7133	30′
40′	030	423	884	012	406	112	20′
50′	050	418	.9942	006	410	092	10′
45° 0′	.7071	1.414	1.000	1.000	1.414	.7071	45° 0′
	Cos	Sec	Cot	Tan	Csc	Sin	Degrees

TABLE V POWERS AND ROOTS

Number	Square	Square Root	Cube	Cube Root	Number	Square	Square Root	Cube	Cube Root
1	1	1.000	1	1.000	51	2,601	7.141	132,651	3.708
2	4	1.414	8	1.260	52	2,704	7.211	140,608	3.733
3	9	1.732	27	1.442	53	2,809	7.280	148,877	3.756
4	16	2.000	64	1.587	54	2,916	7.348	157,464	3.780
5	25	2.236	125	1.710	55	3,025	7.416	166,375	3.803
6	36	2.449	216	1.817	56	3,136	7.483	175,616	3.826
7	49	2.646	343	1.913	57	3,249	7.550	185,193	3.849
8	64	2.828	512	2.000	58	3,364	7.616	195,112	3.871
9	81	3.000	729	2.080	59	3,481	7.681	205,379	3.893
10	100	3.162	1,000	2.154	60	3,600	7.746	216,000	3.915
11	121	3.317	1,331	2.224	61	3,721	7.810	226,981	3.936
12	144	3.464	1,728	2.289	62	3,844	7.874	238,328	3.958
13	169	3.606	2,197	2.351	63	3,969	7.937	250,047	3.979
14	196	3.742	2,744	2.410	64	4,096	8.000	262,144	4.000
15	225	3.873	3,375	2.466	65	4,225	8.062	274,625	4.021
16	256	4.000	4,096	2.520	66	4,356	8.124	287,496	4.041
17	289	4.123	4,913	2.571	67	4,489	8.185	300,763	4.062
18	324	4.243	5,832	2.621	68	4,624	8.246	314,432	4.082
19	361	4.359	6,859	2.668	69	4,761	8.307	328,509	4.102
20	400	4.472	8,000	2.714	70	4,900	8.367	343,000	4.121
21	441	4.583	9,261	2.759	71	5,041	8.426	357,911	4.141
22	484	4.690	10,648	2.802	72	5,184	8.485	373,248	4.160
23	529	4.796	12,167	2.844	73	5,329	8.544	389,017	4.179
24	576	4.899	13,824	2.884	74	5,476	8.602	405,224	4.198
25	625	5.000	15,625	2.924	75	5,625	8.660	421,875	4.217
26	676	5.099	17,576	2.962	76	5,776	8.718	438,976	4.236
27	729	5.196	19,683	3.000	77	5,929	8.775	456,533	4.254
28	784	5.292	21,952	3.037	78	6,084	8.832	474,552	4.273
29	841	5.385	24,389	3.072	79	6,241	8.888	493,039	4.291
30	900	5.477	27,000	3.107	80	6,400	8.944	512,000	4.309
31	961	5.568	29,791	3.141	81	6,561	9.000	531,441	4.327
32	1,024	5.657	32,768	3.175	82	6,724	9.055	551,368	4.344
33	1,089	5.745	35,937	3.208	83	6,889	9.110	571,787	4.362
34	1,156	5.831	39,304	3.240	84	7,056	9.165	592,704	4.380
35	1,225	5.916	42,875	3.271	85	7,225	9.220	614,125	4.397
36	1,296	6.000	46,656	3.302	86	7,396	9.274	636,056	4.414
37	1,369	6.083	50,653	3.332	87	7,569	9.327	658,503	4.431
38	1,444	6.164	54,872	3.362	88	7,744	9.381	681,472	4.448
39	1,521	6.245	59,319	3.391	89	7,921	9.434	704,969	4.465
40	1,600	6.325	64,000	3.420	90	8,100	9.487	729,000	4.481
41	1,681	6.403	68,921	3.448	91	8,281	9.539	753,571	4.498
42	1,764	6.481	74,088	3.476	92	8,464	9.592	778,688	4.514
43	1,849	6.557	79,507	3.503	93	8,649	9.644	804,357	4.531
44	1,936	6.633	85,184	3.530	94	8,836	9.695	830,584	4.547
45	2,025	6.708	91,125	3.557	95	9,025	9.747	857,375	4.563
46	2,116	6.782	97,336	3.583	96	9,216	9.798	884,736	4.579
47	2,209	6.856	103,823	3.609	97	9,409	9.849	912,673	4.595
48	2,304	6.928	110,592	3.634	98	9,604	9.899	941,192	4.610
49	2,401	7.000	117,649	3.659	99	9,801	9.950	970,299	4.626
50	2,500	7.071	125,000	3.684	100	10,000	10.000	1,000,000	4.642

Answers to Selected Problems

Chapter 1

PROBLEM SET 1, page 7

1 T **3** T **5** T **7** T

9 $A = \{x | x \text{ is an even counting number}\} = \{2, 4, 6, 8, \ldots\}$

11 $A = \{x | x = 3k, k = 1, 2, 3, \ldots\} = \{3, 6, 9, 12, \ldots\}$

13 2 subsets; $\{3\}$; proper subset, $\varnothing$ **15** 8 subsets; $\{6, 8, 10\}$; proper subsets, $\varnothing$, $\{6\}$, $\{8\}$, $\{10\}$, $\{6, 8\}$, $\{8, 10\}$, $\{6, 10\}$ **17** 16 subsets; $\{5, 6, 7, 9\}$; proper subsets, $\varnothing$, $\{5\}$, $\{6\}$, $\{7\}$, $\{9\}$, $\{5, 6\}$, $\{5, 7\}$, $\{5, 9\}$, $\{6, 7\}$, $\{6, 9\}$, $\{7, 9\}$, $\{5, 6, 7\}$, $\{6, 7, 9\}$, $\{5, 7, 9\}$, $\{5, 6, 9\}$ **19** $\{1, 4, 5, 6, 7, 8\}$ **21** $\{1, 7\}$ **23** $\{1, 4, 5, 7\}$ **25** $\{1, 4, 7\}$ **27** $\{4, 5, 6, 7, 8\}$ **29** $\{0, 1, 2, 3, \ldots\}$ **31** 0.6 **33** $-0.8\overline{3}$ **35** $-2.\overline{3}$ **37** $\frac{27}{100}$ **39** $-\frac{1}{8}$ **41** $\frac{5}{3}$ **43** $-\frac{1216}{333}$ **45** Closure under addition **47** Multiplication is commutative **49** Addition is associative **51** Distributive **53** Distributive **55** Rule 3(iii) **57** Cancellation property of multiplication **59** Zero property

PROBLEM SET 2, page 13

1 6^6 **3** $2x^3y^2 - 3x^3$ **5** Monomial; degree 2; 4 **7** Binomial; degree 3; 5, 4 **9** $4xy^2$ **11** $4x^2 + 6x - 2$ **13** $2x^2 + 8x - 14$ **15** $6x^2 + 7x - 3$ **17** $45x^2 - xy - 28y^2$ **19** $2x^3 + 7x^2 + 17x + 7$ **21** $x^4 - x^3 - 5x^2 - 21x + 54$ **23** $4x^2 + 4xy + y^2$ **25** $64y^2 - 128yz + 64z^2$ **27** $x^2 - 9y^2$ **29** $25r^2 - 49s^2$ **31** $27 + y^3$ **33** $3x(3x + 1)$ **35** $12x^2y(x - 4y)$ **37** $(m + 1)(x + y)$ **39** $(1 + 3y)(1 - 3y)$ **41** $(4x - 5y)(4x + 5y)$ **43** $(x + 3y)(x - 3y)(x^2 + 9y^2)$ **45** $(x + y + a - b)(x + y - a + b)$ **47** $(4 + y)(16 - 4y + y^2)$ **49** $(y + 3)(y^2 + 3)$ **51** $(x - 9)(x - 7)$ **53** $(6 + x)(2 - x)$ **55** $(3x - 1)(x + 2)$ **57** $(4x - 1)(3x + 5)$ **59** $(4 + x)(3 - 2x)$ **61** $\{2\}$ **63** $\{0\}$ **65** $\{1\}$ **67** $\{5\}$ **69** $\{9\}$ **71** 84 children's tickets, 252 adult's tickets **73** 100 pounds **75** \$2100 at 6 percent, \$1800 at 7 percent

PROBLEM SET 3, page 20

1 $\frac{1}{5xy}$ **3** $\frac{2x + 3}{3x}$ **5** $\frac{1}{7x + 5y}$ **7** $\frac{x + 4}{x - 2}$ **9** $\frac{3}{5(x + 3)}$ **11** $7a + 2$ **13** $\frac{4x(x - 1)}{(3x - 5)(x + 2)}$ **15** $\frac{2(x + 2)}{3(x + 1)(x - 2)}$ **17** $\frac{2x - 5}{x^2 - 25}$ **19** $\frac{-4x + 3}{(x - 2)(x - 3)(x + 3)}$ **21** $\frac{2(x^2 + 2)}{x(x - 6)(x + 1)}$ **23** $\frac{-x^2 - 4x}{x^2 - 4}$ **25** x **27** $\{2\}$ **29** $\{2\}$ **31** $\varnothing$ **33** $\left\{-\frac{a}{5}\right\}$ **35** $\{3\}$ **37** 3 minutes

PROBLEM SET 4, page 26

1 $\frac{3}{32}$ **3** $\frac{1}{320}$ **5** $2^{12} = 4096$ **7** $\frac{64}{729}$ **9** $\frac{y^8}{x^2}$ **11** x **13** $\frac{b^2 - a^2}{ab}$ **15** $\frac{b + a}{b - a}$ **17** $x^{1/3}$ **19** $a^{101/24}$ **21** $\frac{x}{y^{1/2}}$ **23** $\frac{x^3y^4}{3}$ **25** 3 **27** 4 **29** $4x\sqrt{2x}$

31 $2x^2y^3\sqrt[4]{x^3y}$ **33** $\dfrac{1}{x}$ **35** $\dfrac{1}{4}$ **37** $\sqrt{2}$ **39** $5(\sqrt{3}+\sqrt{2})$ **41** $\dfrac{x(\sqrt{x}+\sqrt{y})}{x-y}$

43 $\dfrac{\sqrt{3}+1}{2}$ **45** $\{\frac{14}{3}\}$ **47** $\varnothing$ **49** $\{1\}$ **51** $\{\frac{2}{5}\}$ **53** $\{4\}$

PROBLEM SET 5, page 34

1

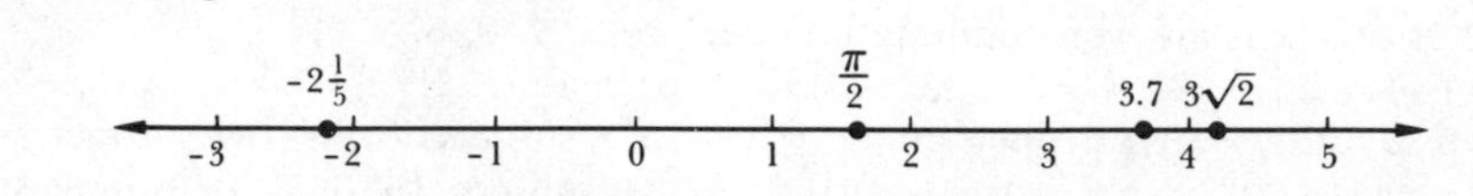

3 Addition property **5** Addition property **7** Transitive property

9 Multiplication property **11** Multiplication property **13** Multiplication property

15 Transitive property

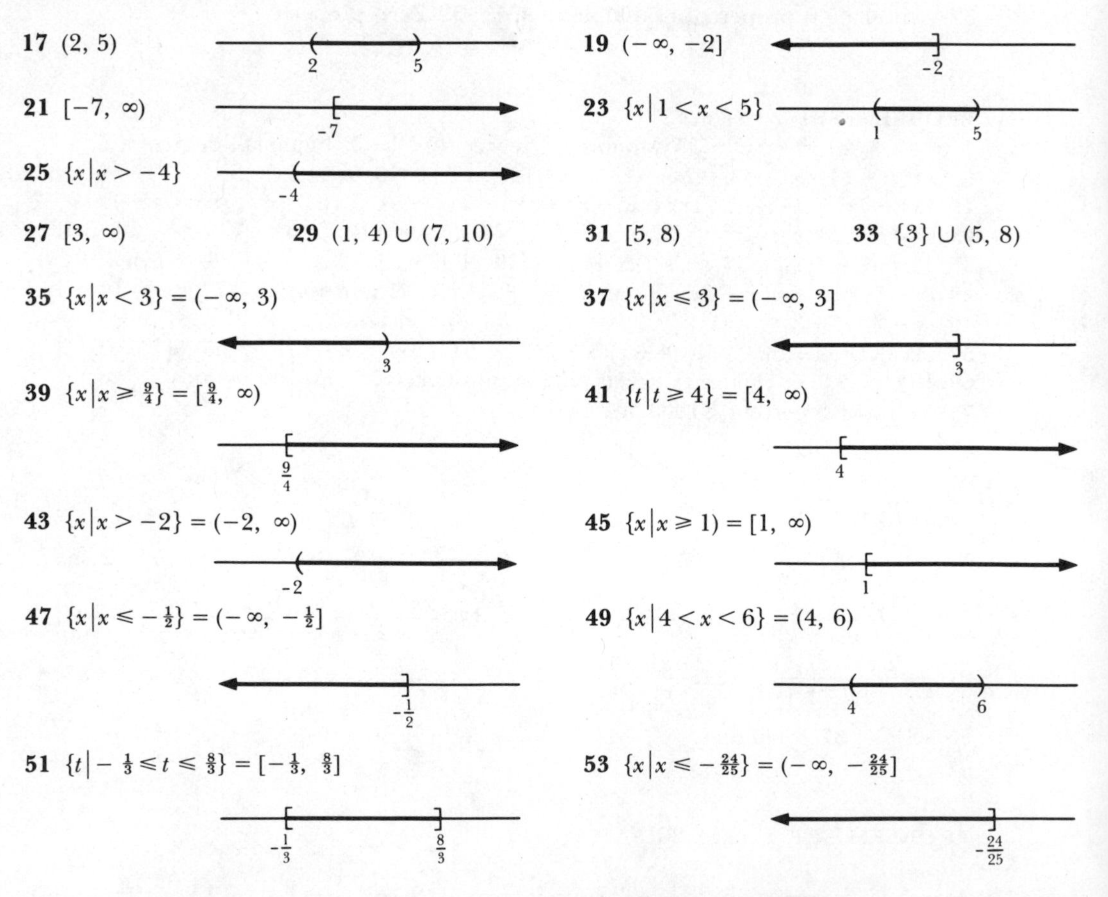

17 $(2, 5)$

19 $(-\infty, -2]$

21 $[-7, \infty)$

23 $\{x \mid 1 < x < 5\}$

25 $\{x \mid x > -4\}$

27 $[3, \infty)$ **29** $(1, 4) \cup (7, 10)$ **31** $[5, 8)$ **33** $\{3\} \cup (5, 8)$

35 $\{x \mid x < 3\} = (-\infty, 3)$

37 $\{x \mid x \leqslant 3\} = (-\infty, 3]$

39 $\{x \mid x \geqslant \frac{9}{4}\} = [\frac{9}{4}, \infty)$

41 $\{t \mid t \geqslant 4\} = [4, \infty)$

43 $\{x \mid x > -2\} = (-2, \infty)$

45 $\{x \mid x \geqslant 1) = [1, \infty)$

47 $\{x \mid x \leqslant -\frac{1}{2}\} = (-\infty, -\frac{1}{2}]$

49 $\{x \mid 4 < x < 6\} = (4, 6)$

51 $\{t \mid -\frac{1}{3} \leqslant t \leqslant \frac{8}{3}\} = [-\frac{1}{3}, \frac{8}{3}]$

53 $\{x \mid x \leqslant -\frac{24}{25}\} = (-\infty, -\frac{24}{25}]$

59 9 or more years **61** 39 or more months

63 84 or more tires per week

PROBLEM SET 6, page 40

1 7 **3** 7 **5** 12 **7** 25 **9** $\begin{cases} 1 \text{ if } x > -1 \\ -1 \text{ if } x < -1 \end{cases}$ **11** $\{-4, 4\}$ **13** $\{-2, 2\}$
15 $\{-5, 5\}$ **17** $\{0\}$ **19** $\{-7, 3\}$ **21** $\varnothing$ **23** $\{-\frac{7}{3}, 1\}$ **25** $\{10, -5\}$ **27** $\{-1\}$
29 $\{-\frac{1}{2}\}$

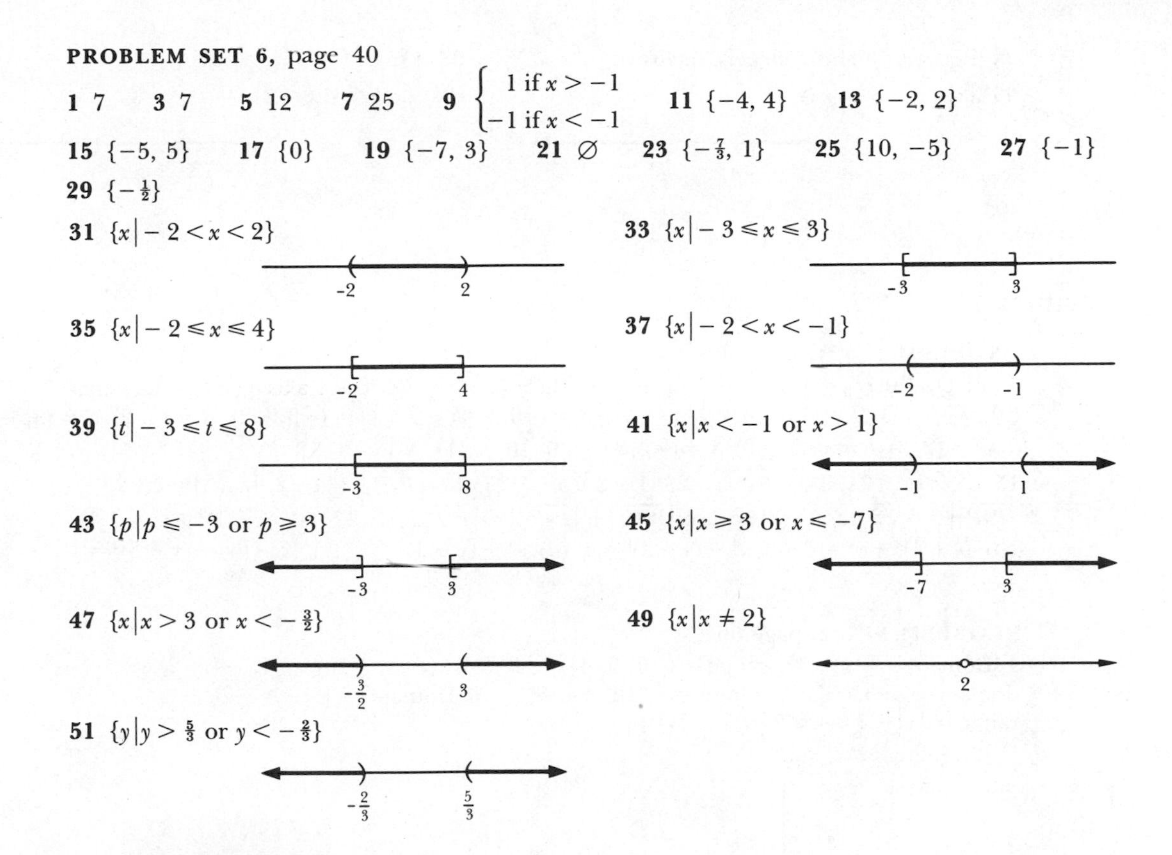

31 $\{x \mid -2 < x < 2\}$

33 $\{x \mid -3 \leq x \leq 3\}$

35 $\{x \mid -2 \leq x \leq 4\}$

37 $\{x \mid -2 < x < -1\}$

39 $\{t \mid -3 \leq t \leq 8\}$

41 $\{x \mid x < -1 \text{ or } x > 1\}$

43 $\{p \mid p \leq -3 \text{ or } p \geq 3\}$

45 $\{x \mid x \geq 3 \text{ or } x \leq -7\}$

47 $\{x \mid x > 3 \text{ or } x < -\frac{3}{2}\}$

49 $\{x \mid x \neq 2\}$

51 $\{y \mid y > \frac{5}{3} \text{ or } y < -\frac{2}{3}\}$

REVIEW PROBLEM SET, page 41

1 F **3** F **5** F **7** T **9** $\{5, 10, 11\}$; proper subsets $\varnothing$, $\{5\}$, $\{10\}$, $\{11\}$, $\{5, 10\}$, $\{10, 11\}$, $\{5, 11\}$ **11** $\{c\}$ **13** $\{a, c, d, e, f, g\}$ **15** $\{a, c\}$
17 $\{x \mid x = 15k, k \text{ is a positive integer}\}$ **19** 0.175 **21** -0.5 **23** $\frac{4}{25}$
25 $-\frac{31}{990}$ **27** Closed under addition **29** Multiplication is commutative
31 Multiplicative identity **33** Multiplication property **35** Property 3 iv **37** $3x^2 - 2x + 3$
39 $4x^2 + 6x - 4$ **41** $2x^4 - 5x^3 + 7x^2 - 5x + 2$ **43** $28 + xy - 15x^2y^2$ **45** $x^4 - 16$
47 $x^6 + x^3y + x^2y^2 - x^4y - xy^2 - y^3$ **49** $13x^2y^2(2x + 3x^3y^2 - 4y)$ **51** $(y + 11)(y - 11)$
53 $(5y - 9z)(5y + 9z)$ **55** $(x + 4)(x^2 - 4x + 16)$ **57** $(x - 8)(x + 7)$
59 (a) $\{\frac{11}{5}\}$ (b) $\{-\frac{2}{7}\}$ **61** $\frac{x - 4}{x}$ **63** $\frac{x + 1}{x - 3}$ **65** $\frac{5x - 9}{(x - 1)(x - 3)(x - 4)}$
67 $\frac{1}{x - 1}$ **69** (a) $\{25\}$ (b) $\{\frac{11}{3}\}$ **71** x^6y^9 **73** $\frac{1}{x^2y^{4/3}}$ **75** $4x^2y^5z^7\sqrt{2xz}$
77 $2x^2y^4\sqrt[6]{2xy}$ **79** $13 - 2\sqrt{42}$ **81** (a) $\{2\}$ (b) $\varnothing$ **83** Multiplication property
85 Multiplication property

87 (a) $[-3, 2]$ (b) $[2, \infty)$

-3 2 2

89 $\{x \mid x > 4\} = (4, \infty)$

4

91 $\{x \mid x \geq \frac{4}{3}\} = [\frac{4}{3}, \infty)$

$\frac{4}{3}$

93 Between 0 and 7 tickets, inclusive **95** $\{\frac{8}{3}, -\frac{16}{3}\}$ **97** $\{\frac{1}{3}\}$

99 $\{x \mid x < -3 \text{ or } x > \frac{7}{3}\}$ **101** $\{x \mid -7 \leq x \leq 2\}$

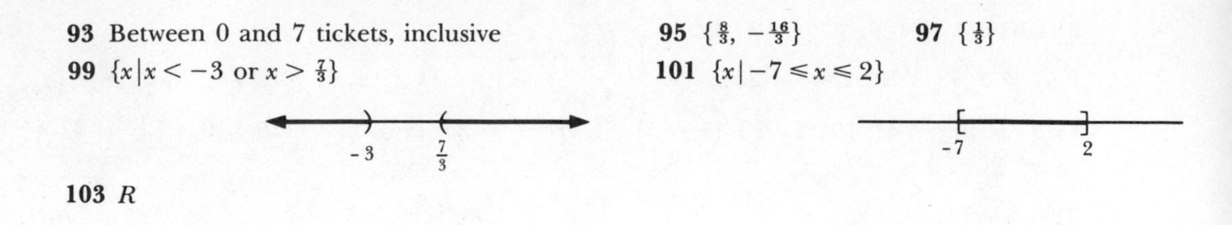

103 R

Chapter 2

PROBLEM SET 1, page 50

1 (a) Q_{I} (b) Q_{II} (c) Q_{IV} (d) Q_{IV} (e) On the y axis (f) On the y axis (g) On the x axis (h) Q_{III} **3** $Q = (1, -4), R = (-1, 4), S = (-1, -4)$ **5** $Q = (-3, -2), R = (3, 2), S = (3, -2)$ **7** $Q = (2, 3), R = (-2, -3), S = (-2, 3)$ **9** 10 **11** $\sqrt{17}$ **13** $3\sqrt{17}$ **15** $\frac{1}{2}\sqrt{41}$ **71** $2\sqrt{1 + t^2}$ **23** $1 + 3\sqrt{5}$ **25** (b) $|\overline{P_1P_2}| = 4\sqrt{2}$, $|\overline{P_2P_3}| = 2\sqrt{2}$, $|\overline{P_1P_3}| = 6\sqrt{2}$ (c) Yes, because $|\overline{P_1P_2}| + |\overline{P_2P_3}| = |\overline{P_1P_3}|$ **27** (a) $(x - 1)^2 + y^2 = 1$ (b) $(x + 1)^2 + y^2 = 1$ (c) $x^2 + (y - 1)^2 = 1$ (d) $x^2 + (y + 1)^2 = 1$ (e) $(x - 2)^2 + (y - 3)^2 = 1$

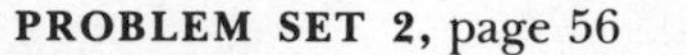

PROBLEM SET 2, page 56

1 Domain = $\{1, 2, 3\}$, range = $\{-1, 2, 4\}$ **3** $R_3 = \{(-3, -12), (-1, -4), (2, 8)\}$, domain = $\{-3, -1, 2\}$, range = $\{-12, -4, 8\}$ **5** Domain = $\{x \mid x \geq -2\}$, range = $\{y \mid -1 \leq y \leq 2\}$

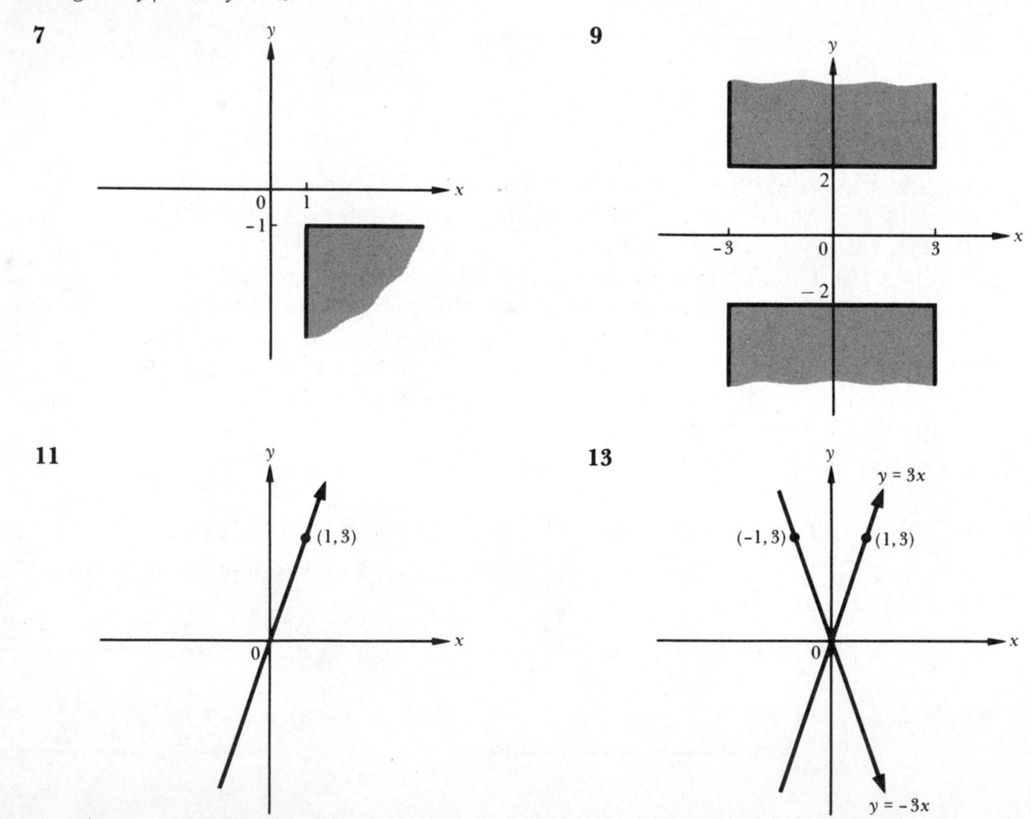

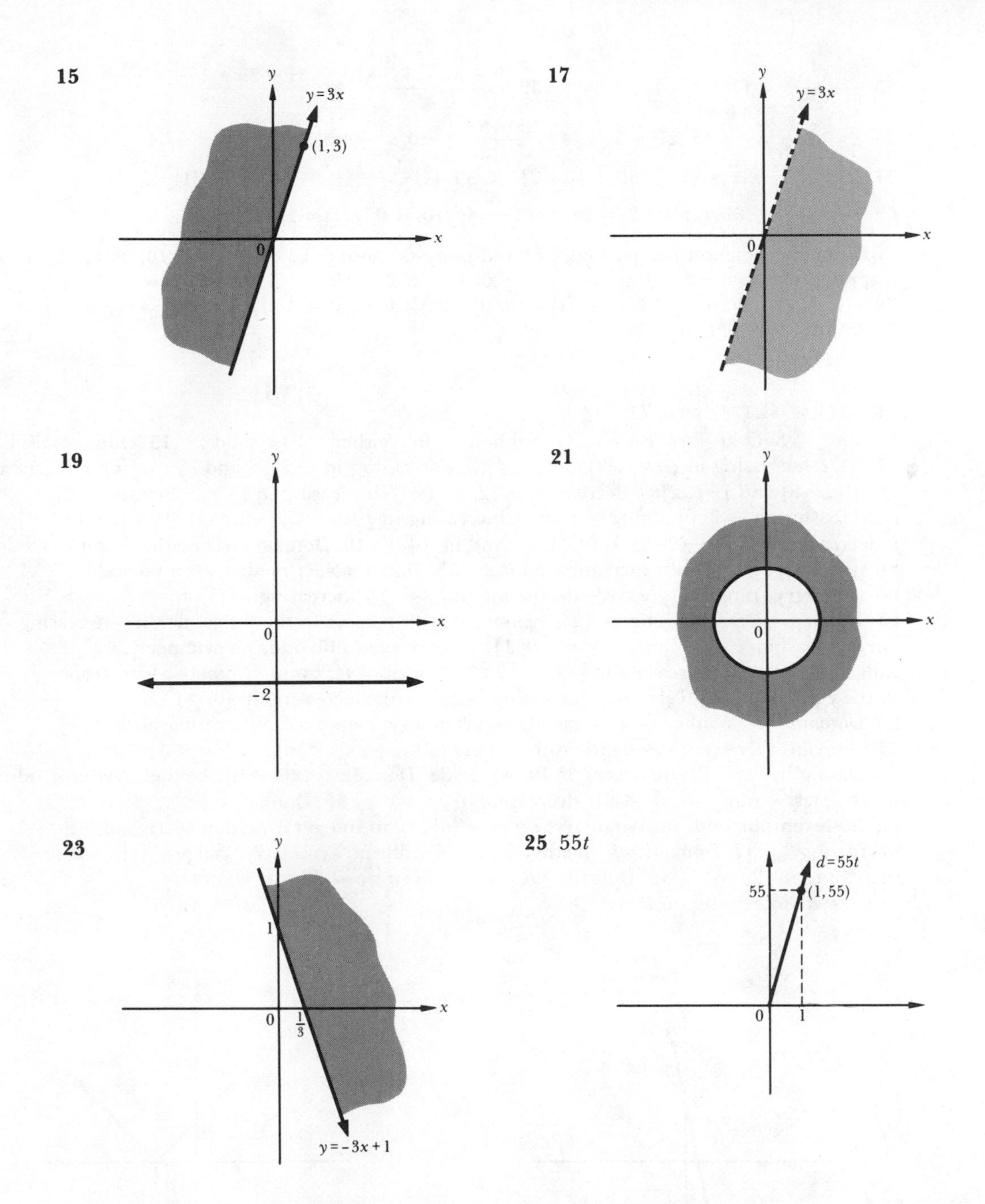

PROBLEM SET 3, page 62

1 $f(x) = 7 - x$ **3** $f(x) = 3x^2 - 2$ **5** $f(x) = \dfrac{4}{x+3}$ **7** $f(x) = \sqrt{9 - x^2}$ **9** 5

11 $-\dfrac{3}{2}$ **13** 11 **15** 1 **17** $\dfrac{3}{2}$ **19** 2 **21** $2a + 7$ **23** $\dfrac{3a-1}{a-2}$

25 $1 + \sqrt{a+2}$ **27** $2a + 14$ **29** $\dfrac{7a-9}{a-2}$ **31** $2ab + b^2 - 2b$ **33** $14x + 5$

35 $\dfrac{7x+3}{7x-2}$ **37** $7x^2 - 14x + 56$ **39** $\dfrac{2+5x+5a}{x+a}$ **41** $\dfrac{3x+3a+1}{1-2x-2a}$
43 $x^2 + 2x + 8$ **45** $2x^2 + 5$ **47** $\dfrac{x^2+3}{x^2-2}$ **49** $x^4 - 4x^3 + 20x^2 - 32x + 64$
51 R **53** $\{x \mid x \geq \frac{1}{2}\}$ **55** $\{x \mid x \neq 2\}$ **57** $\{x \mid x > -\frac{2}{3}\}$ **59** 0 **61** -5
63 $\dfrac{-1}{x(x+h)}$ **65** (a) $f(-2) = 14, f(-1) = 4, f(0) = 0, f(1) = 2, f(2) = 10$
(b) Function notation: see part (a); ordered-pair notation: $(-2, 14), (-1, 4), (0, 0), (1, 2), (2, 10)$; mapping notation: $-2 \to 14, -1 \to 4, 0 \to 0, 1 \to 2, 2 \to 10$ **67** (c) and (d)
69 $r = f(C) = C/2\pi$ **71** $V = f(x) = x(10 - 2x)(14 - 2x) = 140x - 48x^2 + 4x^3$
73 (a) 0°C (b) 30°C (c) $-\frac{160}{9}$°C (d) -25°C

PROBLEM SET 4, page 74

1 Even **3** Odd **5** Even **7** Neither **9** Neither **11** Odd **13** Odd **15** Even
17 (a) f increasing in $(-\infty, -2]$ and $[1, \frac{5}{2}]$; f decreasing in $[-2, 1]$ and $[\frac{5}{2}, \infty)$ (b) f increasing in $(-\infty, -3]$ and $[-1, 2]$; f decreasing in $[2, 3]$ (c) f increasing in $(-\infty, -2)$ and $[0, 2]$; f decreasing in $[-2, 0]$ and $(2, \infty)$ (d) f increasing in $[-6, -4]$, $[-2, -1]$, $[0, 1]$, and $[2, 4]$; f decreasing in $[-4, -2]$, $[-1, 0]$, $[1, 2]$, and $[4, 6]$ **19** Domain $= R$; neither even nor odd; no symmetry; range $= R$; increasing on R **21** Domain $= R$; neither even nor odd; no symmetry; range $= \{y \mid y \geq 0\}$; decreasing in $(-\infty, 1]$; increasing in $[1, \infty)$
23 Domain $= R$; even; range $= \{2\}$; symmetric with respect to the y axis; neither increasing nor decreasing **25** Domain $= \{x \mid x \geq 1\}$; neither even nor odd; no symmetry; range $= \{y \mid y \geq 0\}$; increasing in $[1, \infty)$ **27** Domain $= R$; even; symmetric with respect to the y axis; range $= \{y \mid y \leq 0\}$; increasing in $(-\infty, 0]$; decreasing in $[0, \infty)$
29 Domain $= R$; neither even nor odd; no symmetry; range $= R$; increasing on R
31 Domain $= R$; even; symmetric with respect to the y axis; range $= \{y \mid y \geq 4\}$; decreasing in $(-\infty, 0]$; increasing in $[0, \infty)$ **33** Domain $= \{x \mid x \geq 0\}$; neither even nor odd; no symmetry; range $= \{y \mid y \leq 0\}$; decreasing in $[0, \infty)$ **35** Domain $= R$; neither even nor odd; no symmetry; range $= \{y \mid y$ is an integer$\}$; neither increasing nor decreasing **37** Domain $= R$; neither even nor odd; no symmetry; range $= \{y \mid y \geq 2\}$; increasing in $[2, \infty)$ **39** Domain $= R$; neither even nor odd; no symmetry; range $= R$; increasing on R

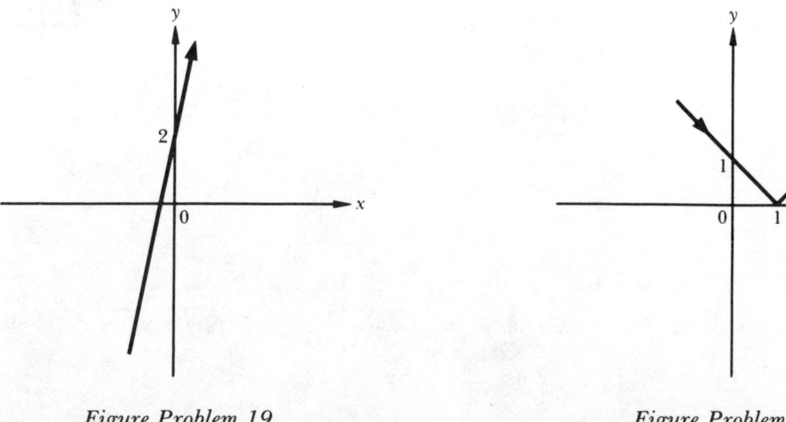

Figure Problem 19 *Figure Problem 21*

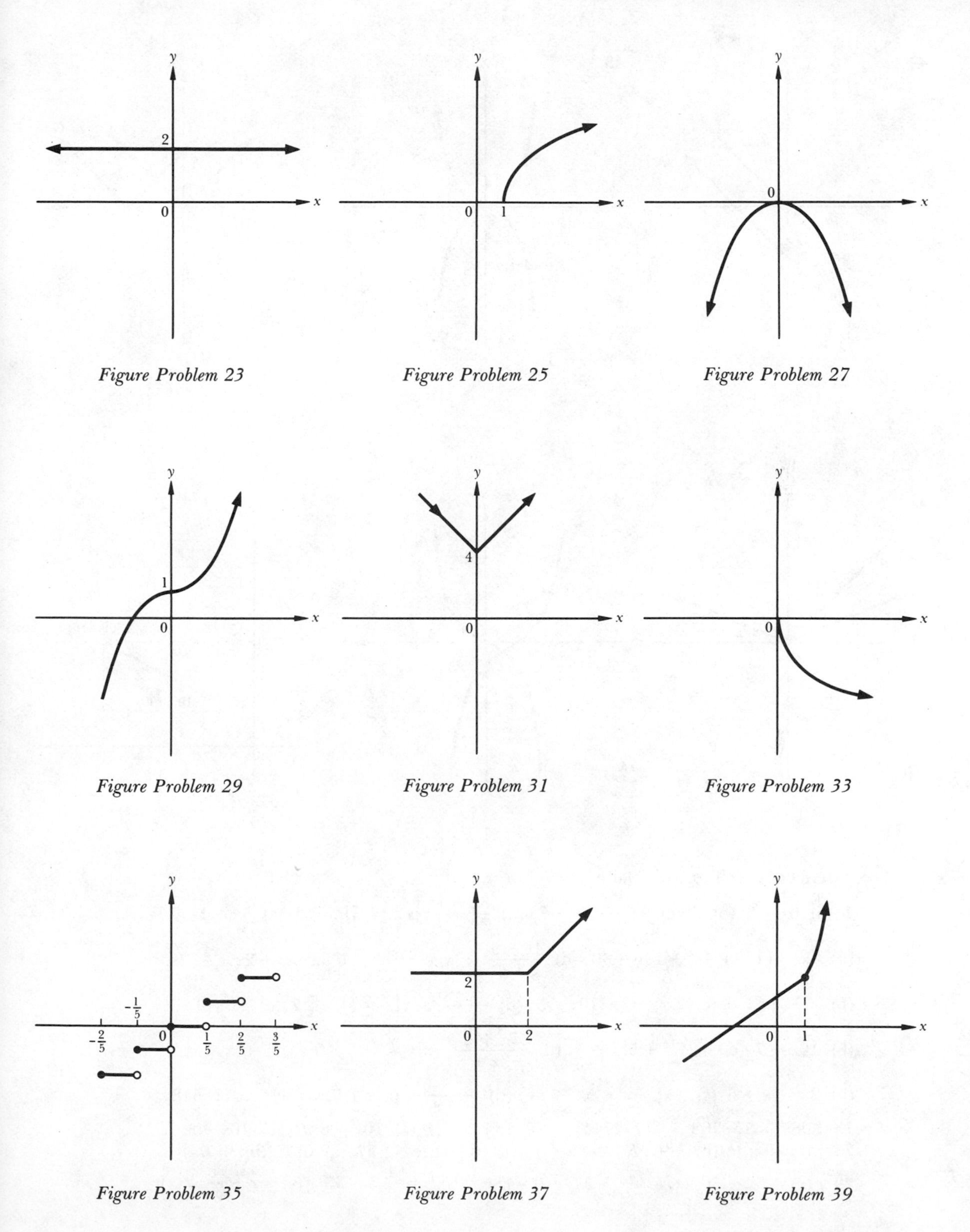

Figure Problem 23

Figure Problem 25

Figure Problem 27

Figure Problem 29

Figure Problem 31

Figure Problem 33

Figure Problem 35

Figure Problem 37

Figure Problem 39

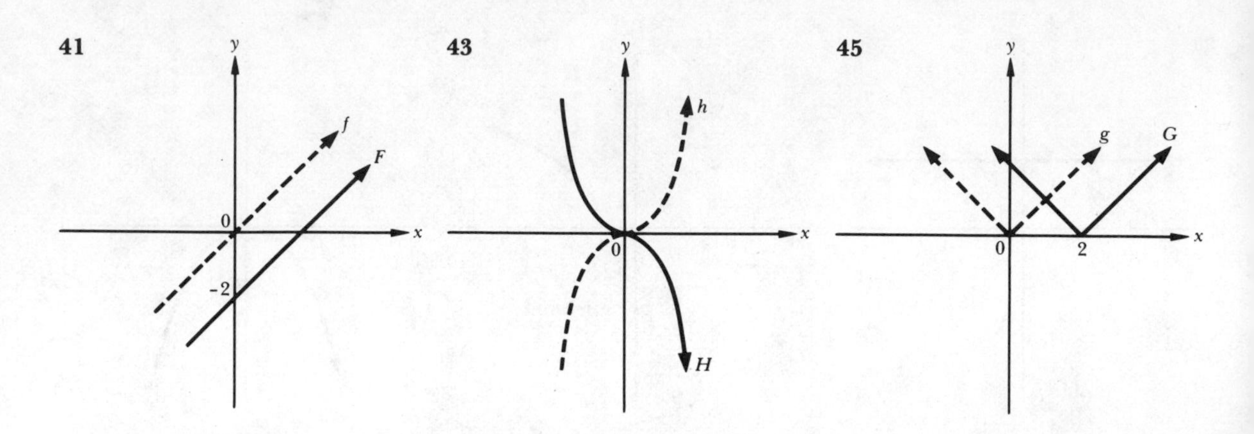

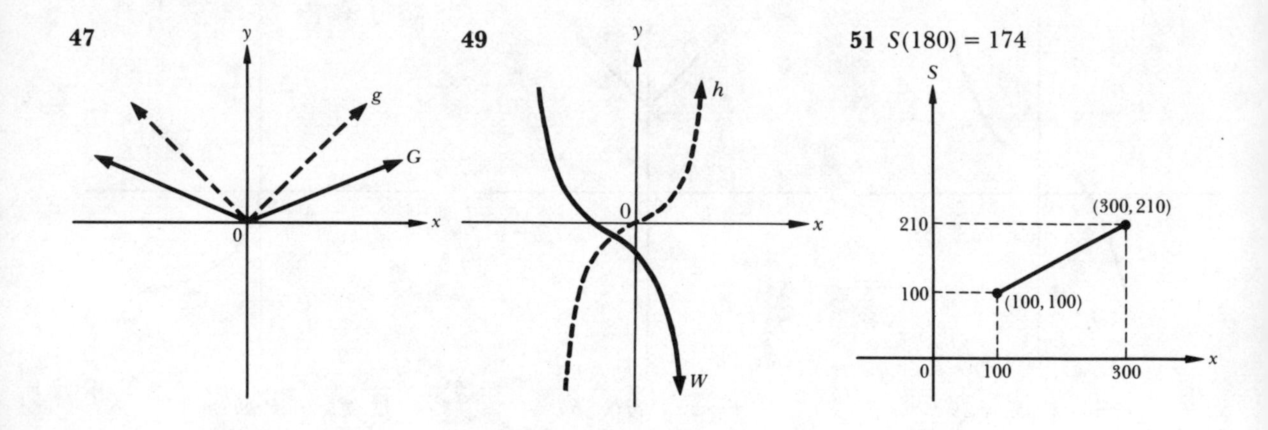

PROBLEM SET 5, page 80

1 (a) $6x - 6$ (b) 8 (c) $9x^2 - 18x - 7$ (d) $\dfrac{3x+1}{3x-7}$; $\{x \mid x \neq \frac{7}{3}\}$ **3** (a) $5x + 1$ (b) $3x - 11$ (c) $4x^2 + 19x - 30$ (d) $\dfrac{4x-5}{x+6}$; $\{x \mid x \neq -6\}$ **5** (a) $x^2 + 2x + 4$ (b) $x^2 - 2x + 6$ (c) $2x^3 - x^2 + 10x - 5$ (d) $\dfrac{x^2+5}{2x-1}$; $\{x \mid x \neq \frac{1}{2}\}$ **7** (a) $4x + 9$ (b) $10x - 7$ (c) $-21x^2 + 53x + 8$ (d) $\dfrac{7x+1}{-3x+8}$; $\{x \mid x \neq \frac{8}{3}\}$ **9** (a) $5 + x$ (b) $2x^2 - x + 5$ (c) $-x^4 + x^3 - 5x^2 + 5x$ (d) $\dfrac{x^2+5}{-x^2+x}$; $\{x \mid x \neq 0, x \neq 1\}$ **11** 518 **13** 398 **15** 268 **17** 2744 **19** 134 **21** (a) $10x - 6$; R (b) $10x - 3$; R **23** (a) $-9x$; R (b) $-9x$; R **25** (a) x; R (b) x; R **27** (a) 6; R (b) 9; R **29** $g(x) = 5x - 3$; $f(x) = x^3$ **31** $g(t) = t^2 - 2$, $f(t) = \dfrac{1}{t^2}$ **33** $g(x) = x + \dfrac{1}{x}$, $f(x) = x^5$

35 $-\frac{7}{2}$ **37** $x^4 + 4x^3 + 6x^2 + 4x + 2$; $x^4 + 4x^2 + 4$, $f \circ g \neq g \circ f$ **39** (a) $-15x + 7$ (b) $\frac{1}{3}$
41 (a) $x + 9$ (b) 4 **43** $a = -1$ **45** (a) T (b) T (c) T **47** (a) $V = 16\pi h$
(b) $h = 2t + 4$ (c) $V = 32\pi t + 64\pi$

49 $(S \circ C)(x) = 0.051\left(\frac{x^2 + 120x + 8000}{x}\right)$; \$57.53

PROBLEM SET 6, page 88

5 (a) Yes (b) No (c) No (d) Yes

7 $f^{-1}(x) = \frac{x + 7}{3}$ **9** $f^{-1}(x) = \frac{4}{3}(x - 5)$ **11** $f^{-1}(x) = -\frac{5}{x}$ **13** $f^{-1}(x) = \frac{\sqrt[3]{x}}{2}$

15 $f^{-1}(x) = -\sqrt{x}$ **19** (a) No inverse (b) No inverse **21** $f^{-1}(x) = \frac{x}{7} - \frac{5}{7}$

23 $f^{-1}(x) = -\frac{x}{3} + \frac{1}{3}$ **25** $f^{-1}(x) = \frac{3}{x}$ **27** No inverse

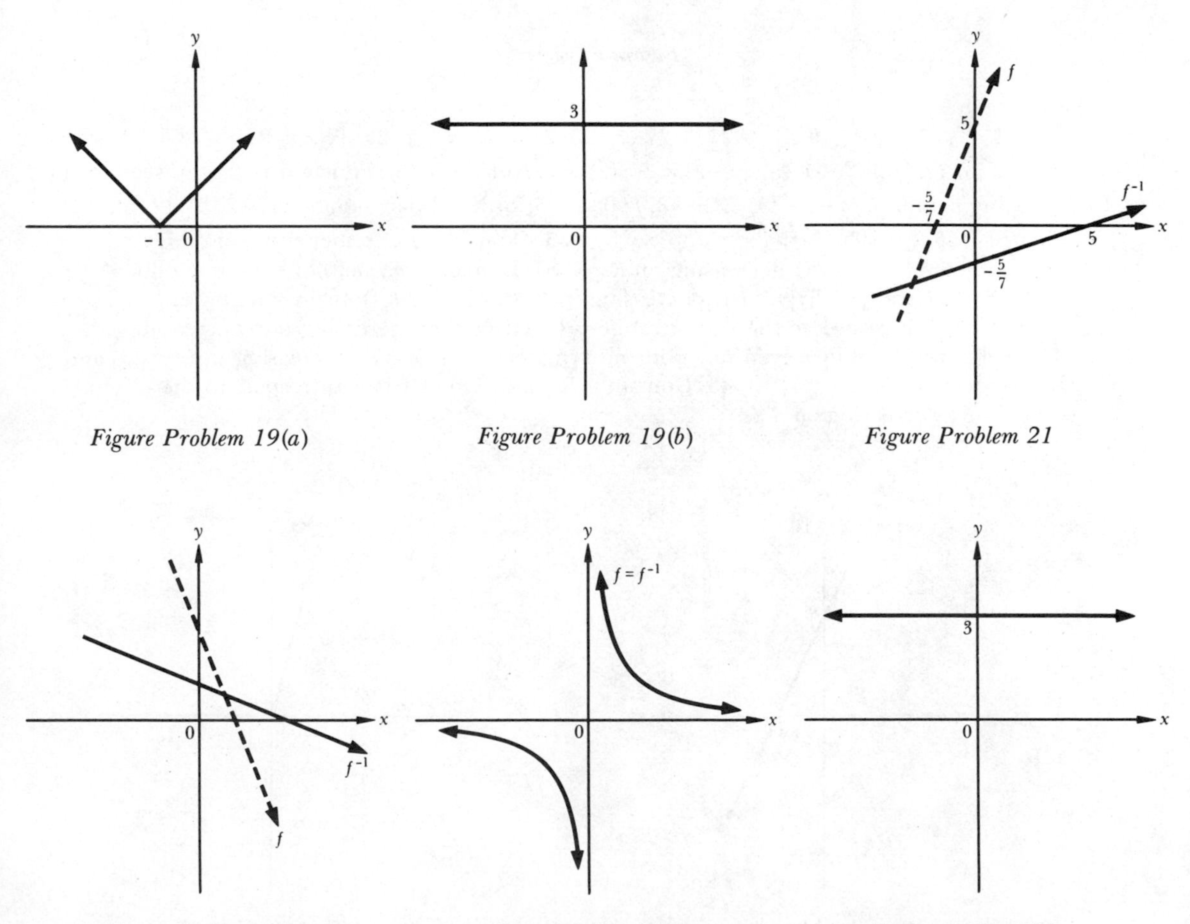

Figure Problem 19(a) *Figure Problem 19(b)* *Figure Problem 21*

Figure Problem 23 *Figure Problem 25* *Figure Problem 27*

REVIEW PROBLEM SET, page 89

1 (a) Q_{I} (b) Q_{II} (c) Q_{IV} (d) Q_{III} (e) Q_{IV} (f) on the x axis (g) on the x axis (h) on the y axis **3** $2\sqrt{10}$ **5** $4\sqrt{5}$ **9** $R = \{(-2, -12), (-1, -7), (0, -2)\ (1, 3), (2, 8)\}$; domain $= \{-2, -1, 0, 1, 2\}$; range $= \{-12, -7, -2, 3, 8\}$ **13** $g(x) = \dfrac{7}{5x - 2}$

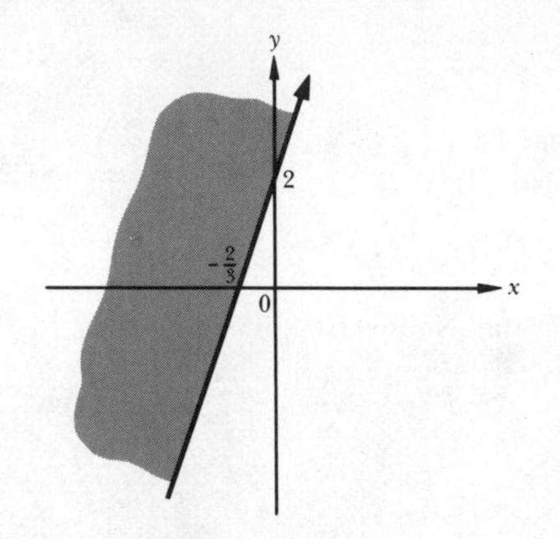

Figure Problem 11

15 -1 **17** 50 **19** 2 **21** 1/27 **23** 2 **25** x^3 **27** $\{x \mid x \neq 2\}$ **29** (a), (b), and (c) **31** (a) $f(-2) = 3, f(2) = 3, f(0) = \sqrt{5}$ (b) Function notation, see part (a); mapping notation, $f\colon -2 \to 3, f\colon 2 \to 3, f\colon 0 \to \sqrt{5}$; ordered-pair notation, $(-2, 3)$, $(2, 3)$, $(0, \sqrt{5})$ **33** $y = \sqrt{36 - x^2}$; $A = x\sqrt{36 - x^2}$ **35** Domain $= R$; neither even nor odd; no symmetry; range $= R$; decreasing on R **37** Domain $= R$; neither even nor odd; no symmetry; range $= \{y \mid y \geq 0\}$; decreasing in $(-\infty, 0]$ **39** Domain $= R$; even; symmetric with respect to the y axis; range $= \{-1\}$; neither increasing nor decreasing **41** Domain $= R$; neither even nor odd; no symmetry; range $= R$; increasing in $(-\infty, 0]$ and $[1, \infty)$, decreasing in $[0, 1]$ **43** Domain $= R$; odd; symmetric with respect to the origin; range $= R$; decreasing on R

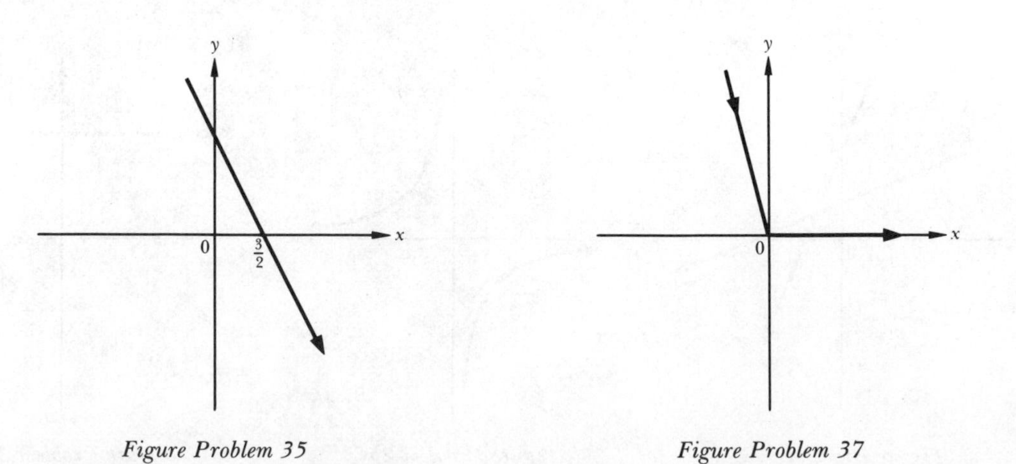

Figure Problem 35 *Figure Problem 37*

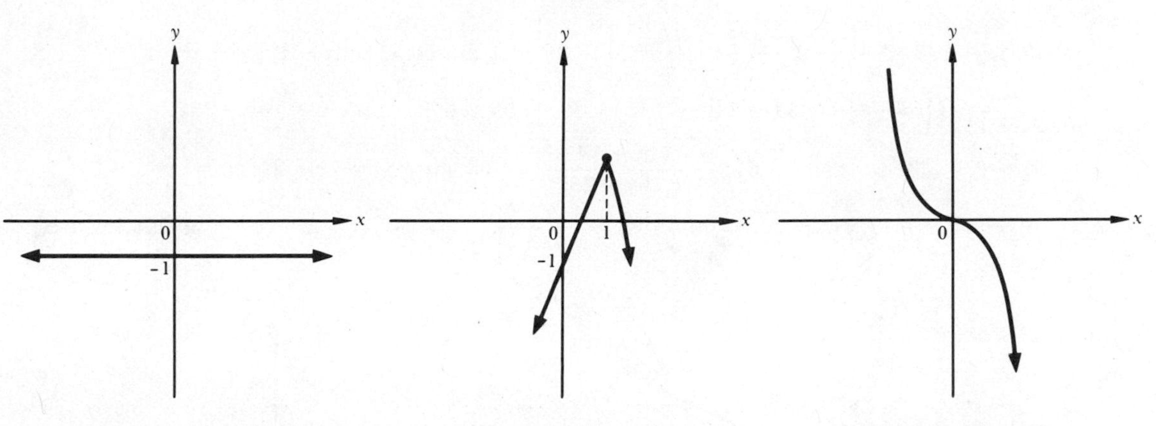

Figure Problem 39 *Figure Problem 41* *Figure Problem 43*

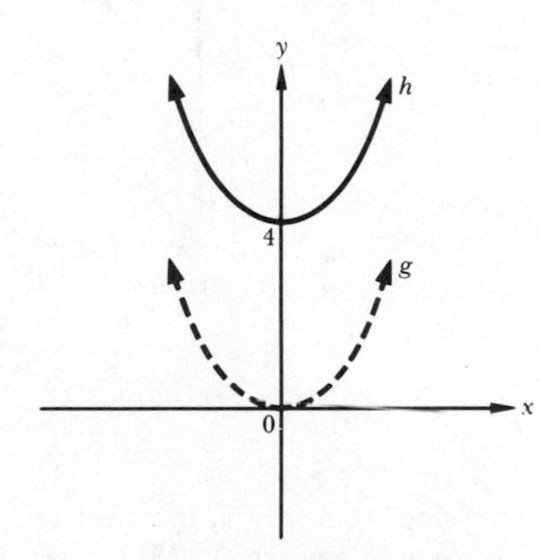

Figure Problem 45(a)

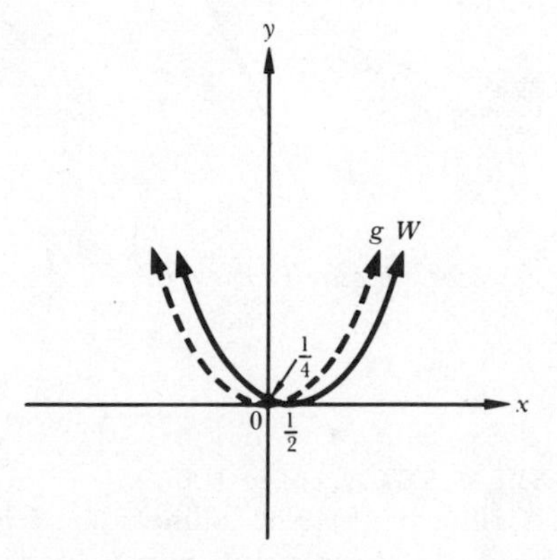

Figure Problem 45(b)

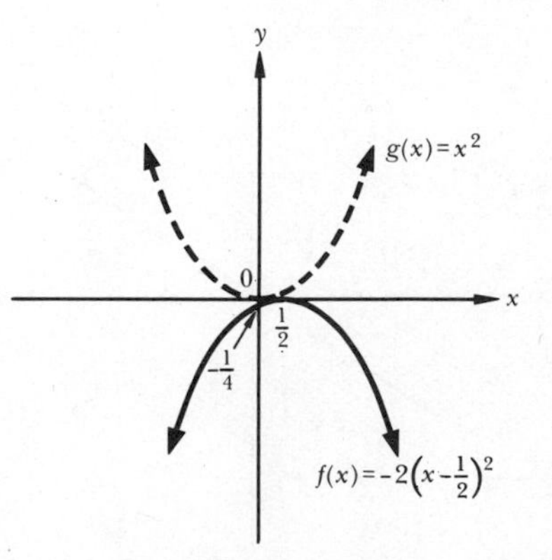

Figure Problem 45(c)

47 $2x + 1$, 3, $x^2 + x - 2$, $\frac{x+2}{x-1}$; $\{x \mid x \neq 1\}$ **49** $x^2 + 2x + 1$, $x^2 - 2x - 1$, $2x^3 + x^2$, $\frac{x^2}{2x+1}$; $\{x \mid x \neq -\frac{1}{2}\}$ **51** -11, -7, 18, $\frac{9}{2}$; R **53** -669 **55** $6 - 10x - 25x^2$ **57** -2 **59** $6 + 25x$ **61** $\frac{6 + 70x + 175x^2}{1 + 10x + 25x^2}$ **63** (a) $g(x) = 7x + 2, f(x) = x^5$ (b) $g(t) = t^2 + 17, f(t) = \sqrt{t}$ **65** $f^{-1}(x) = \frac{-x + 7}{13}$ **67** $f^{-1}(x) = 2x - 4$ **69** No inverse

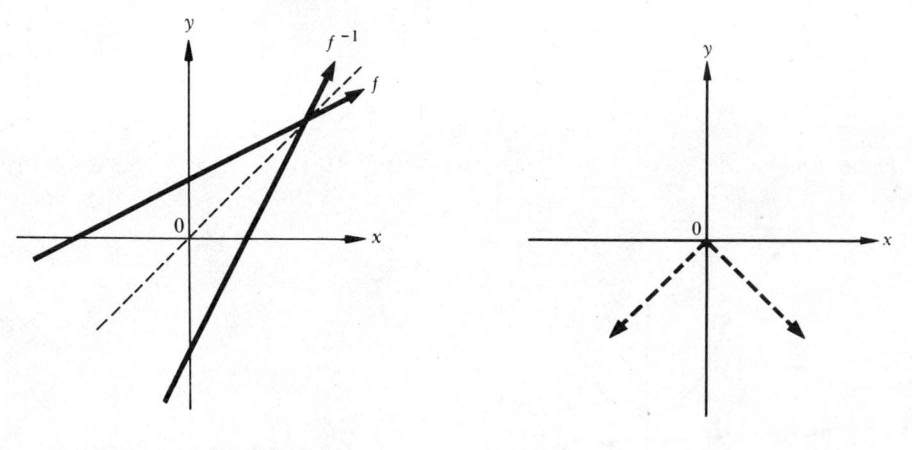

Figure Problem 67

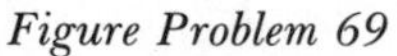

Figure Problem 69

Chapter 3

PROBLEM SET 1, page 105

1 (a) Collinear (b) Not collinear **3** $m = 5$, increasing **5** $m = \frac{3}{5}$, increasing **7** $m = -3$, decreasing **9** $m = 3$, increasing **11** $m = 0$, neither **13** $\frac{7}{3}$; rises to the right **15** -2; falls to the right **17** 0; neither **19** $(y - 1) = 3(x - 1)$ **21** $y = -2$ **23** $y - 3 = \frac{5}{2}(x - 3)$ **25** $y + 5 = \frac{2}{3}(x + 3)$ **27** Parallel **29** Intersect **33** \$166,250 **35** 2°C

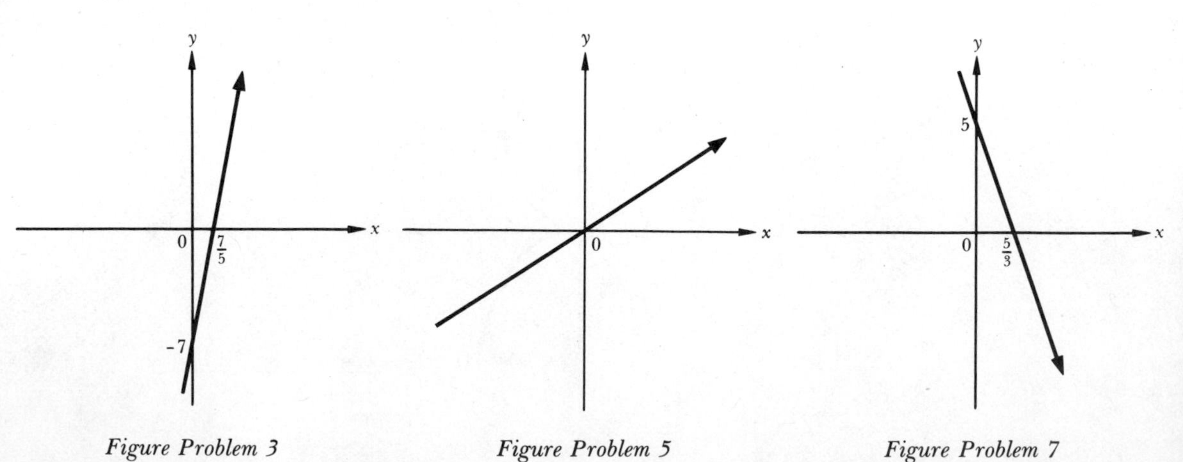

Figure Problem 3 *Figure Problem 5* *Figure Problem 7*

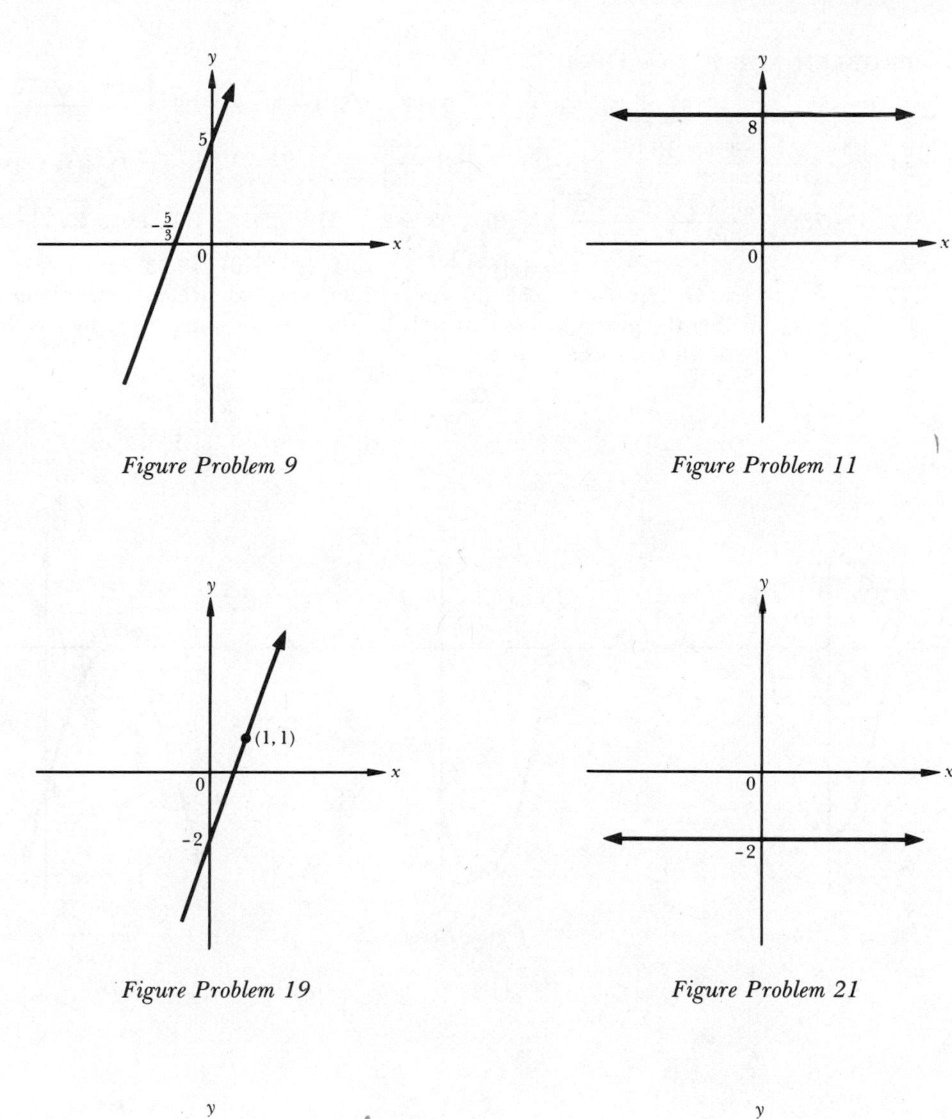

Figure Problem 9

Figure Problem 11

Figure Problem 19

Figure Problem 21

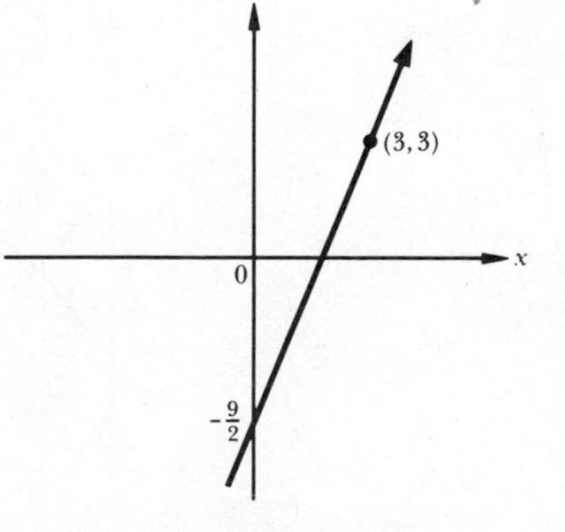

Figure Problem 23

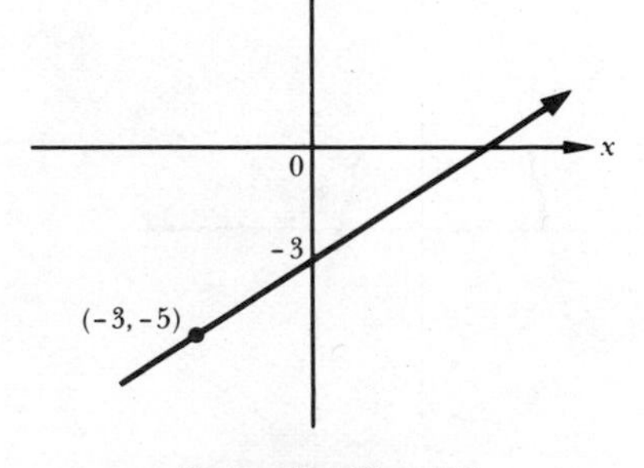

Figure Problem 25

PROBLEM SET 2, page 118

1 $\{0, -5\}$ **3** $\{2, 4\}$ **5** $\left\{\frac{1}{3}, \frac{1}{2}\right\}$ **7** $\{1 - \sqrt{3}, 1 + \sqrt{3}\}$ **9** $\left\{\frac{-3 - \sqrt{13}}{2}, \frac{-3 + \sqrt{13}}{2}\right\}$
11 $\left\{\frac{4 - \sqrt{10}}{3}, \frac{4 + \sqrt{10}}{3}\right\}$ **13** $D = 9$; $\left\{-1, \frac{1}{2}\right\}$ **15** $D = 128$; $\{6 - 4\sqrt{2}, 6 + 4\sqrt{2}\}$
17 $D = 73$; $\left\{\frac{5 - \sqrt{73}}{12}, \frac{5 + \sqrt{73}}{12}\right\}$ **19** $\{-3, -2, 2, 3\}$ **21** $\{-1, 0, 1\}$ **23** $\left\{-2, -1, \frac{1}{3}, \frac{2}{3}\right\}$
25 $\{-\frac{1}{4}\}$ **27** $\{y \mid y \geq -9\}$ **29** $\{y \mid y \geq -\frac{81}{4}\}$ **31** $\{y \mid y \leq 0\}$ **33** $\{y \mid y \leq \frac{25}{4}\}$
35 $\{y \mid y \geq -5\}$ **37** $\{y \mid y \leq \frac{1}{8}\}$ **39** (a) As $|a|$ increases, ax^2 stretches out along the y axis
(b) $k > 0$ means that the graph moves vertically up the y axis k units. $k < 0$ means that the graph moves vertically down the y axis k units.

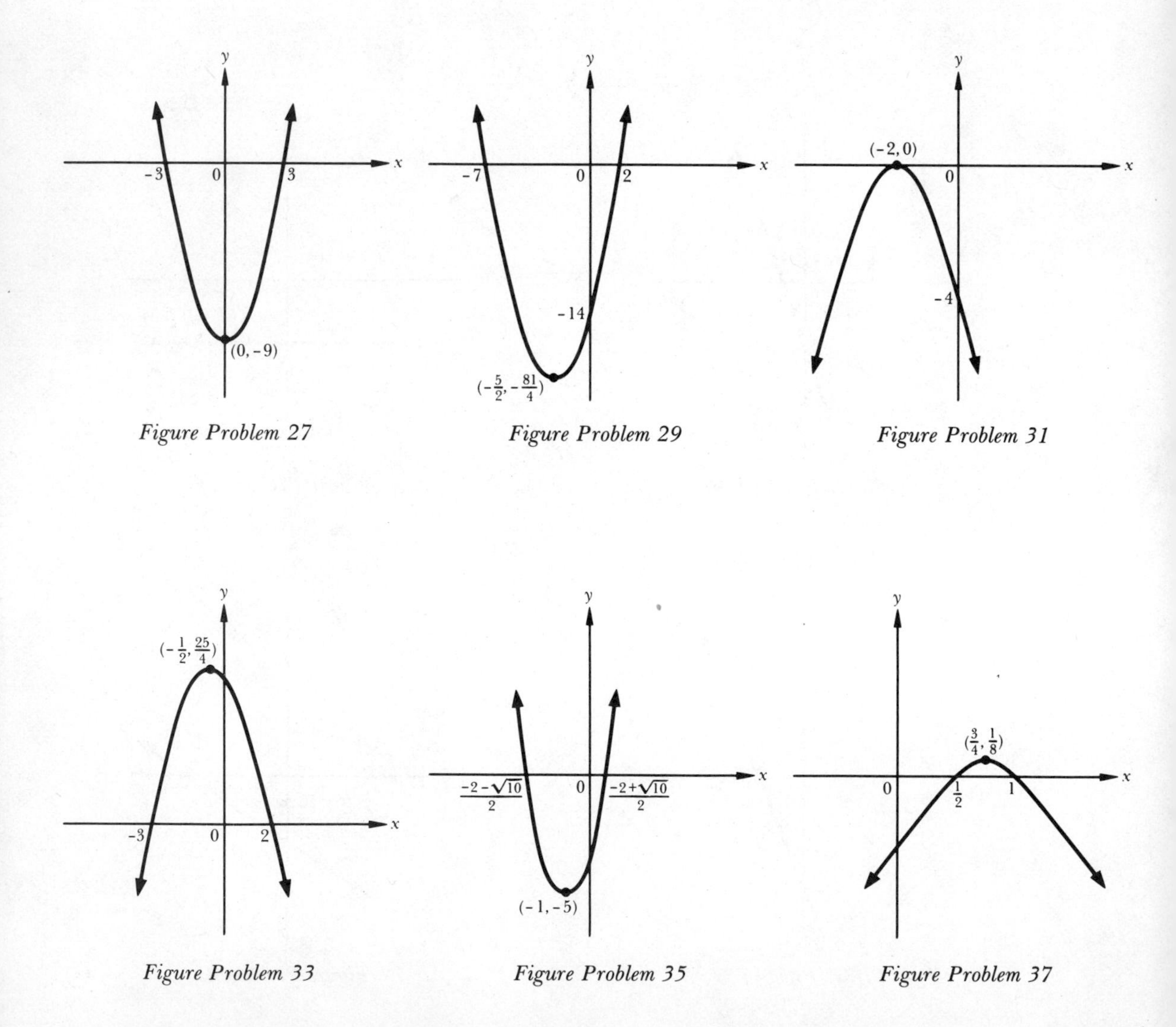

Figure Problem 27 *Figure Problem 29* *Figure Problem 31*

Figure Problem 33 *Figure Problem 35* *Figure Problem 37*

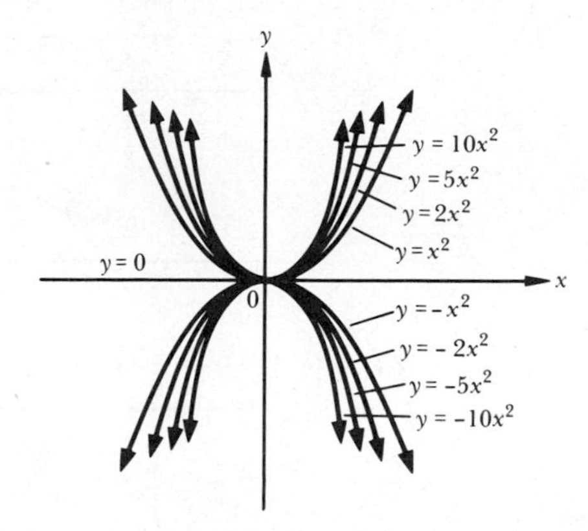

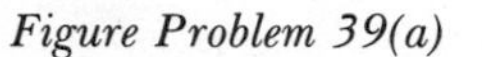
Figure Problem 39(a)

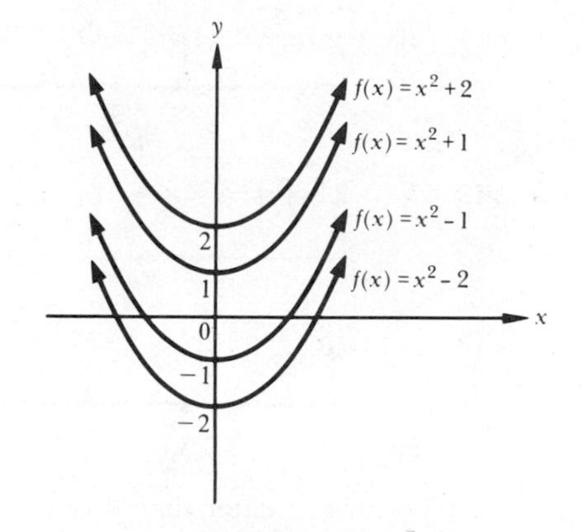

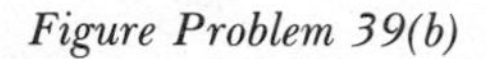
Figure Problem 39(b)

41 (a) 256 (b) 400 feet

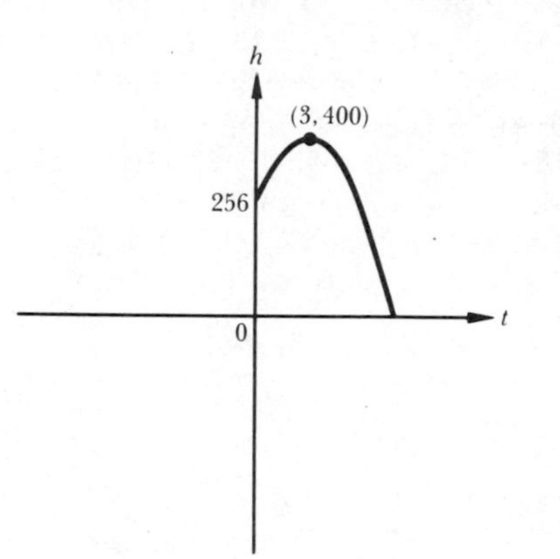

Figure Problem 41(c)

43 312.5 square yards **45** $A/2$

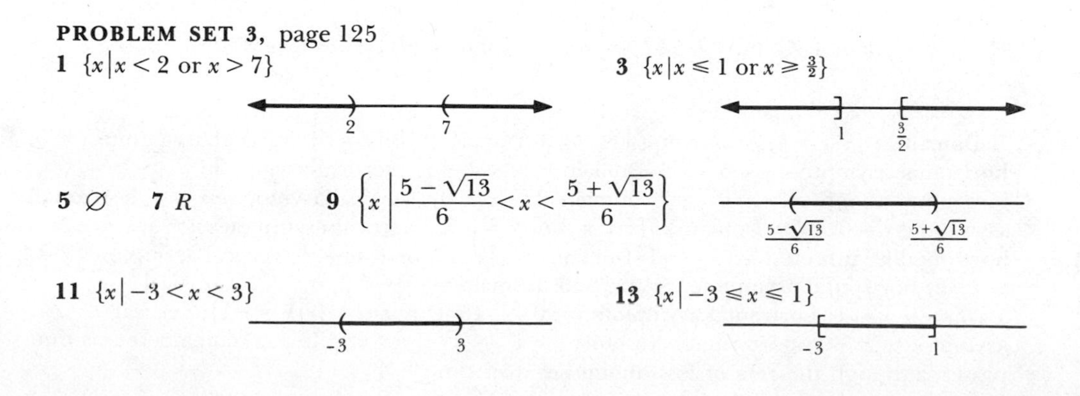

PROBLEM SET 3, page 125

1 $\{x \mid x < 2 \text{ or } x > 7\}$

3 $\{x \mid x \leq 1 \text{ or } x \geq \frac{3}{2}\}$

5 $\varnothing$ **7** R

9 $\left\{x \,\middle|\, \dfrac{5-\sqrt{13}}{6} < x < \dfrac{5+\sqrt{13}}{6}\right\}$

11 $\{x \mid -3 < x < 3\}$

13 $\{x \mid -3 \leq x \leq 1\}$

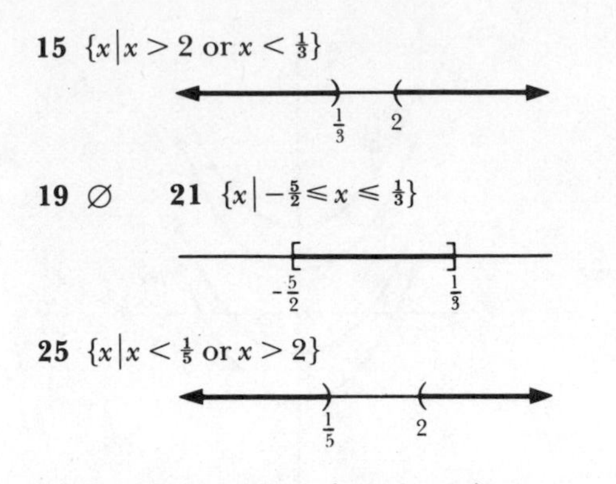

15 $\{x \mid x > 2 \text{ or } x < \frac{1}{3}\}$

19 $\varnothing$ **21** $\{x \mid -\frac{5}{2} \leq x \leq \frac{1}{3}\}$

25 $\{x \mid x < \frac{1}{5} \text{ or } x > 2\}$

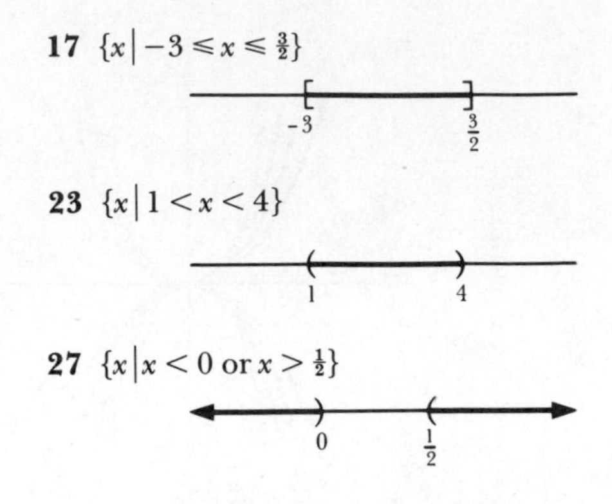

17 $\{x \mid -3 \leq x \leq \frac{3}{2}\}$

23 $\{x \mid 1 < x < 4\}$

27 $\{x \mid x < 0 \text{ or } x > \frac{1}{2}\}$

29 Width is greater than 4 centimeters

PROBLEM SET 4, page 136

1 Yes; degree 0; coefficient $\frac{1}{5}$ **3** Not a polynomial function **5** $\frac{2}{3}$ **7** 1, 2 **9** -2, 2 **11** -3, 0, 3 **13** -5, -1, 2, 5 **15** $f(x) = (x - 3)(5x^2 + 13x + 42) + 122$
17 $f(x) = (x - 2)(5x^4 + 7x^3 + 16x^2 + 33x + 59) + 121$
19 $f(x) = (x - 1)(-4x^5 - 4x^4 - 4x^3 - 9x^2 - 6x - 5) + 2$
21 $Q(x) = 3x^2 + 12x + 14, f(2) = 35$ **23** $Q(x) = 2x^2 - 4x + 3, f(\frac{1}{2}) = \frac{25}{2}$
25 $Q(x) = -3x^3 + 3x^2 - 3x + 8, f(-2) = -20$ **27** (a) -5 (b) -10 **29** -2, 1, 2
31 1 **33** $\frac{1}{3}$ **35** $-1, \frac{1}{2}, \frac{2}{5}$ **37** $-1, -\frac{1}{2}, 1, \frac{3}{2}$

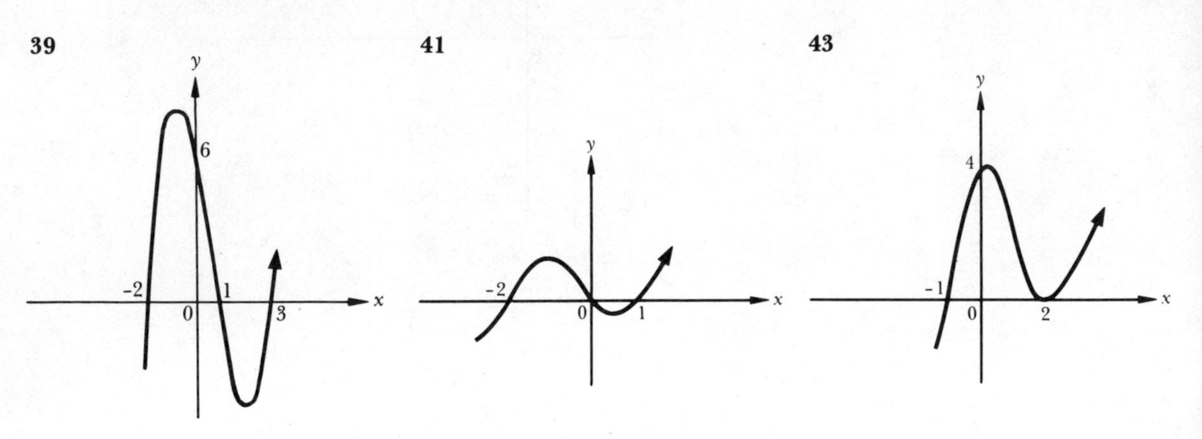

45 $\{x \mid x \leq -2 \text{ or } 1 \leq x \leq 3\}$ **47** $\{x \mid -2 < x < 0 \text{ or } x > 1\}$ **49** $\{x \mid x \leq -1\} \cup \{2\}$

PROBLEM SET 5, page 143

1 Domain = $\{x \mid x \neq \frac{2}{3}\}$; no asymptotes **3** Domain = $\{x \mid x \neq 0\}$; vertical asymptote $x = 0$; horizontal asymptote $y = 0$ **5** Domain = $\{x \mid x \neq -\frac{2}{3}\}$; vertical asymptote $x = -\frac{2}{3}$; horizontal asymptote $y = 0$ **7** Domain = $\{x \mid x \neq 0\}$; vertical asymptote $x = 0$; horizontal asymptote $y = 0$ **9** Domain = $\{x \mid x \neq 3 \text{ or } x \neq -3\}$; vertical asymptotes $x = 3$, $x = -3$; horizontal asymptote $y = 0$ **11** Domain = $\{x \mid x \neq 2 \text{ or } x \neq -2\}$; vertical asymptotes $x = 2$, $x = -2$; horizontal asymptote $y = 0$ **13** Domain = $\{x \mid x \neq 2 \text{ or } x \neq 1\}$; vertical asymptote $x = 1$; horizontal asymptote $y = 0$ **15** Domain = $\{x \mid x \neq -1\}$; vertical asymptote $x = -1$; horizontal asymptote $y = 1$ **17** Less and less is remembered as time passes, although the rate of loss diminishes over time

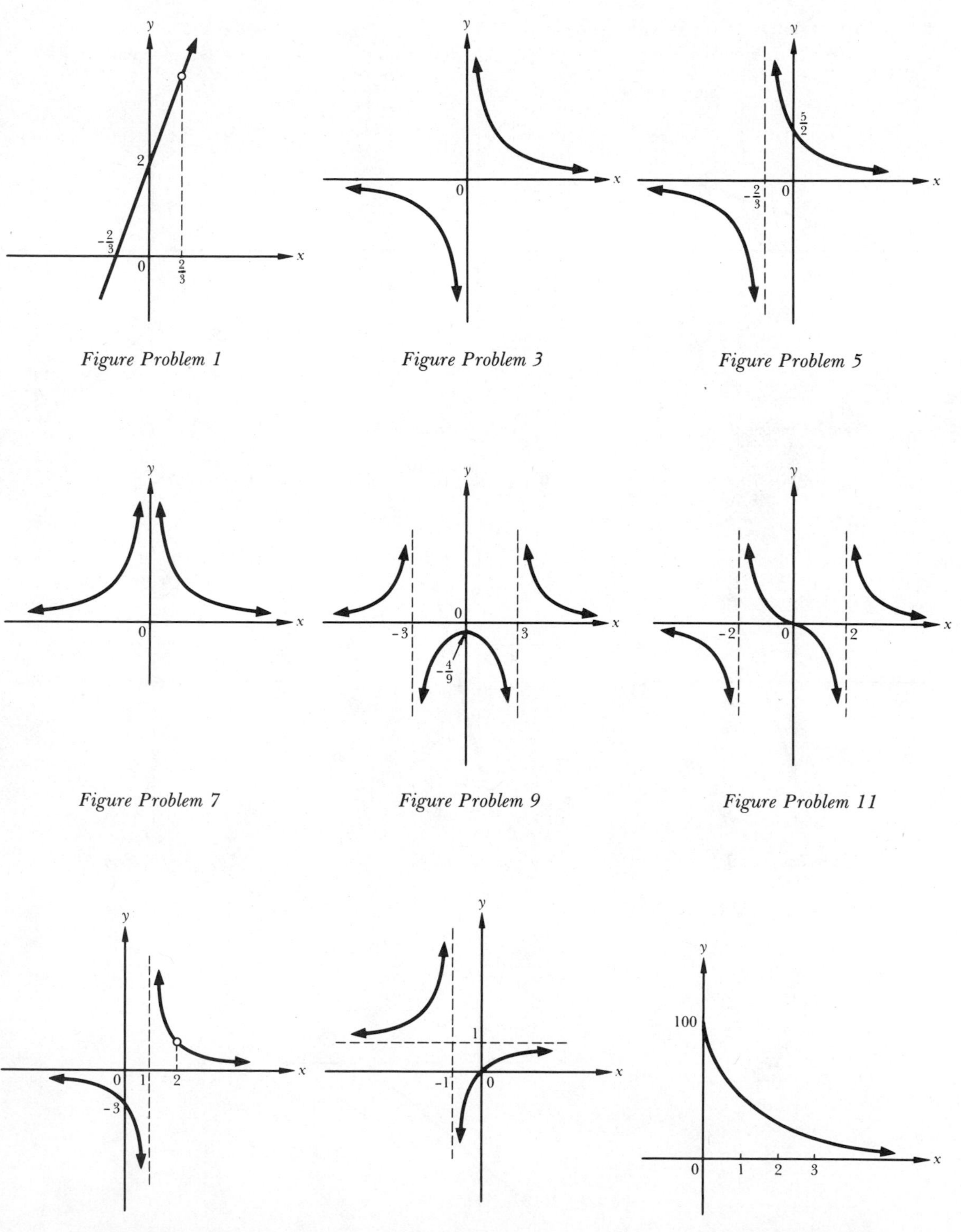

Figure Problem 1

Figure Problem 3

Figure Problem 5

Figure Problem 7

Figure Problem 9

Figure Problem 11

Figure Problem 13

Figure Problem 15

Figure Problem 17

REVIEW PROBLEM SET, page 144

1 $m = \frac{3}{5}$, increasing

3 $m = -\frac{7}{5}$, decreasing

5 $m = -1$, decreasing

7 $y = 3x - 7$

9 $y = 4x - 14$

11 $y = -3x - 3$

13 $y = \frac{7}{8}x - \frac{23}{4}$

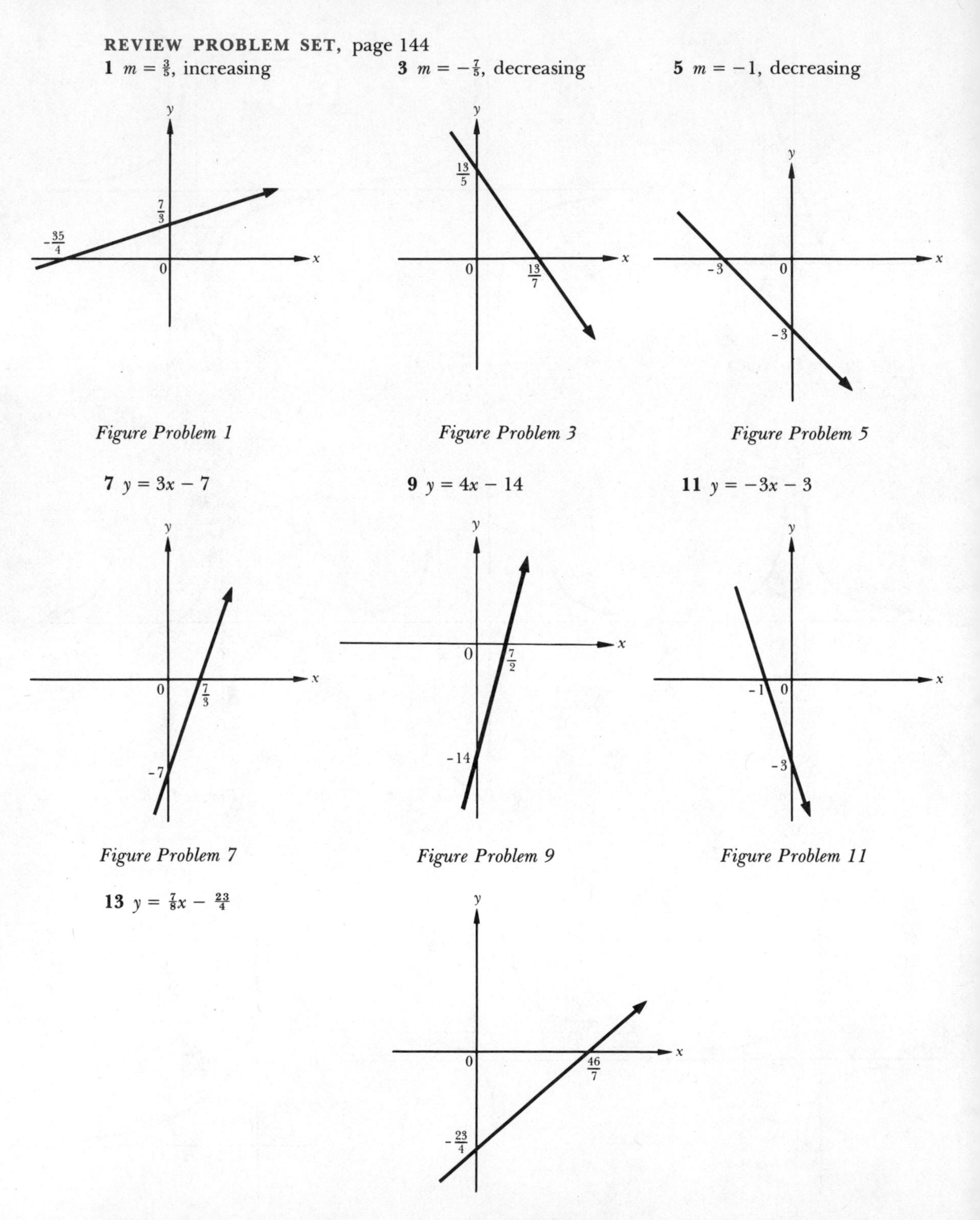

Figure Problem 1

Figure Problem 3

Figure Problem 5

Figure Problem 7

Figure Problem 9

Figure Problem 11

Figure Problem 13

15 (5, 9) **17** $y = 350x - 55$ **19** $\{-1 + \sqrt{7}, -1 - \sqrt{7}\}$ **21** $\left\{-\frac{1}{2}, 1\right\}$

23 $\left\{\frac{-3 - \sqrt{13}}{4}, \frac{-3 + \sqrt{13}}{4}\right\}$ **25** Range = $\{y \mid y \leq 2\}$; inequality = $\{x \mid -\sqrt{2} \leq x \leq \sqrt{2}\}$

27 Range = $\{y \mid y \geq -\frac{121}{24}\}$; inequality = $\{x \mid -\frac{1}{2} \leq x \leq \frac{4}{3}\}$
29 Range = $\{y \mid y \leq 0\}$; inequality = $\{x \mid x \neq 1\}$

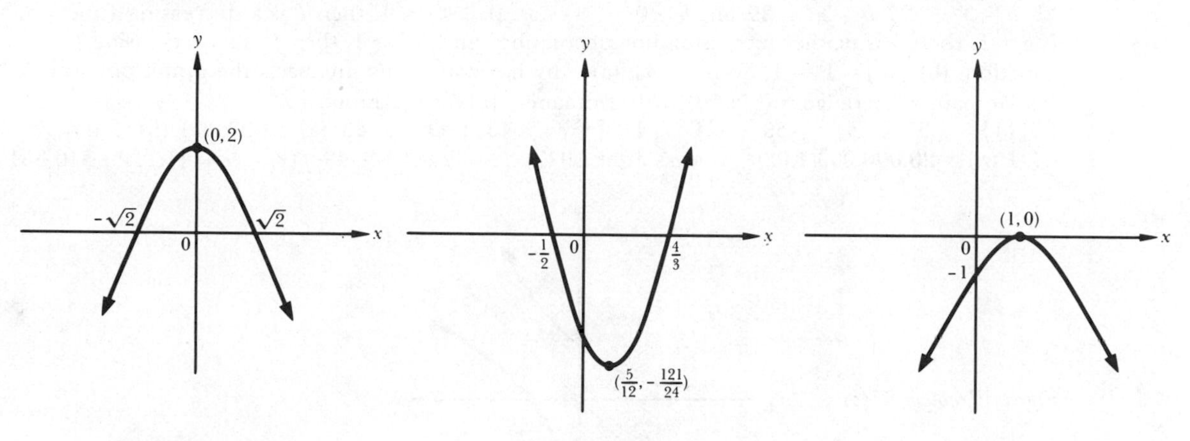

Figure Problem 25 *Figure Problem 27* *Figure Problem 29*

31 600 **33** $-\frac{2}{3}$ **35** 1, 1, 4 **37** 1, −2, 2, 2 **39** $Q(x) = 3x^2 + 11x + 29$; $f(2) = 55$
41 $Q(x) = 2x^2 + 2x + 1$; $f(-1) = -19$ **43** (a) 13 (b) 208 (c) 4 (d) −7 (e) 191/32
45 −1, 2, 3 **47** −3, −1, $\frac{1}{2}$, 3 **49** $\{x \mid x \geq 2 \text{ or } -3 \leq x \leq -1\}$ **51** Domain = $\{x \mid x \neq 5\}$; vertical asymptote $x = 5$; horizontal asymptote $y = 0$ **53** Domain = $\{x \mid x \neq 3 \text{ or } x \neq -3\}$; vertical asymptotes $x = -3$, $x = 3$; horizontal asymptote $y = 0$ **55** Domain = R, horizontal asymptote $y = 1$

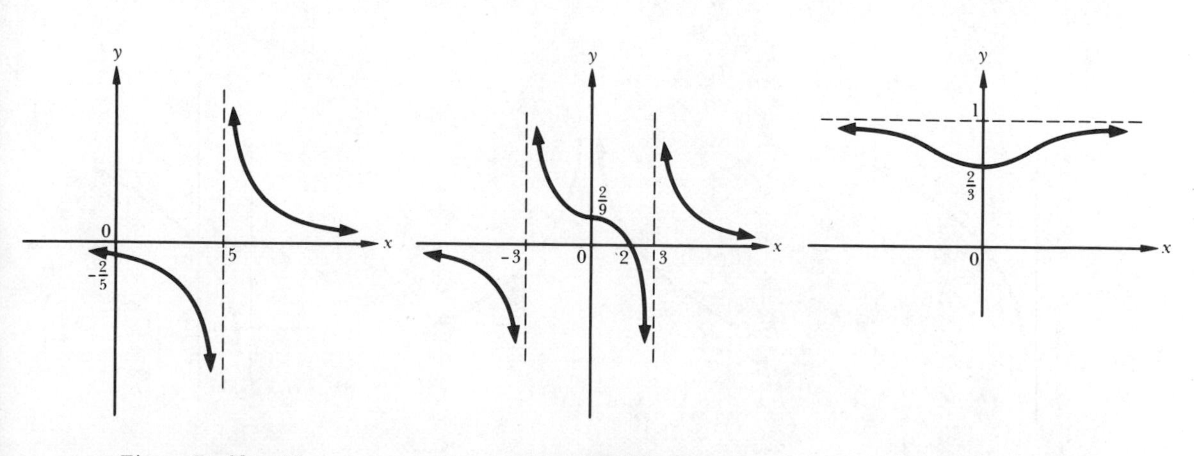

Figure Problem 51 *Figure Problem 53* *Figure Problem 55*

Chapter 4

PROBLEM SET 1, page 151

1 $\frac{3}{2}$ **3** 1 **5** 4 **7** $\frac{4}{9}$ **9** $\frac{1}{2}$ **11** $\frac{1}{8}$ **13** Domain $= R$, range $= \{y|y > 0\}$, increasing **15** Domain $= R$, range $= \{y|y > 0\}$, increasing **17** Domain $= R$, range $= \{y|y > 0\}$, decreasing **19** Domain $= R$, range $= \{y|y > 0\}$, increasing **21** Domain $= R$, range $= \{y|y > 0\}$, increasing **23** Domain $= R$, range $= \{y|y > 0\}$, decreasing **25** $b = 3$ **27** $b = 3$ **29** any $b > 0$ **31** (a) If $0 < b < 1$, then f is a decreasing function; if $b = 1$, then f is neither increasing nor decreasing; and if $b > 1$, then f is an increasing function. (b) $f(x) = 1^x = 1$; even **33** (a) Any horizontal line intersects the graph only once. (c) Domain $= R$; range $= \{y|y > 0\}$ (d) Domain $= \{x|x > 0\}$; range $= R$ **35** $\{1\}$ **37** $\{-4\}$ **39** $\{-\frac{1}{2}\}$ **41** $\{\frac{13}{2}\}$ **43** $\{-1\}$ **45** (a) 2,997,000 (b) 1 day **47** $P(n) = 20{,}000{,}000(1.02)^n$, n years after 1977; 22,082,000 **49** $f(x) = 5000(1.11)^x$; \$10,381

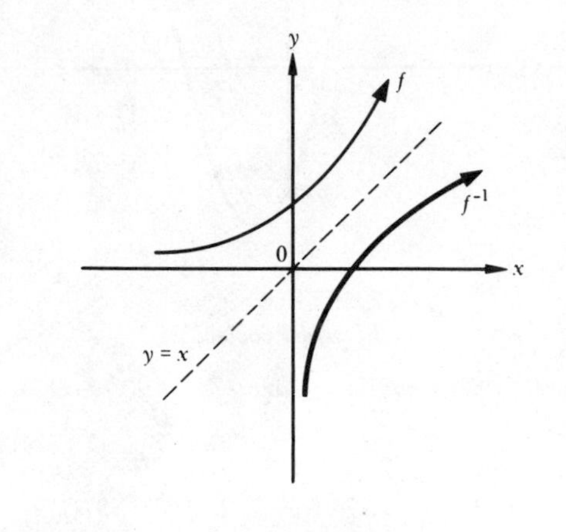

Figure Problem 33(b)

PROBLEM SET 2, page 158

1 $\log_5 125 = 3$ **3** $\log_9 3 = \frac{1}{2}$ **5** $\log_{32} 2 = \frac{1}{5}$ **7** $9^2 = 81$ **9** $(\frac{1}{3})^{-2} = 9$ **11** $36^{3/2} = 216$ **13** 2 **15** -3 **17** -2 **19** 0 **21** $-\frac{17}{2}$ **23** 4 **25** 2 **27** 3 **29** $\{x|x > -\frac{2}{3}\}$ **31** $\{x|x < 1 \text{ or } x > 4\}$ **33** $\{x|x < 1\}$ **35** Domain $= \{x|x > 0\}$; range $= R$; increasing in $(0, \infty)$ **37** Domain $= \{x|x \neq 0\}$; range $= R$; increasing in $(-\infty, 0)$ and decreasing in $(0, \infty)$ **39** Domain $= \{x|x > -1\}$; range $= R$; increasing on R **41** $\log_{10} 3x$; $3 \log_{10} x$ **43** $\log_2(5x - 1)$; $(5 \log_2 x) - 1$

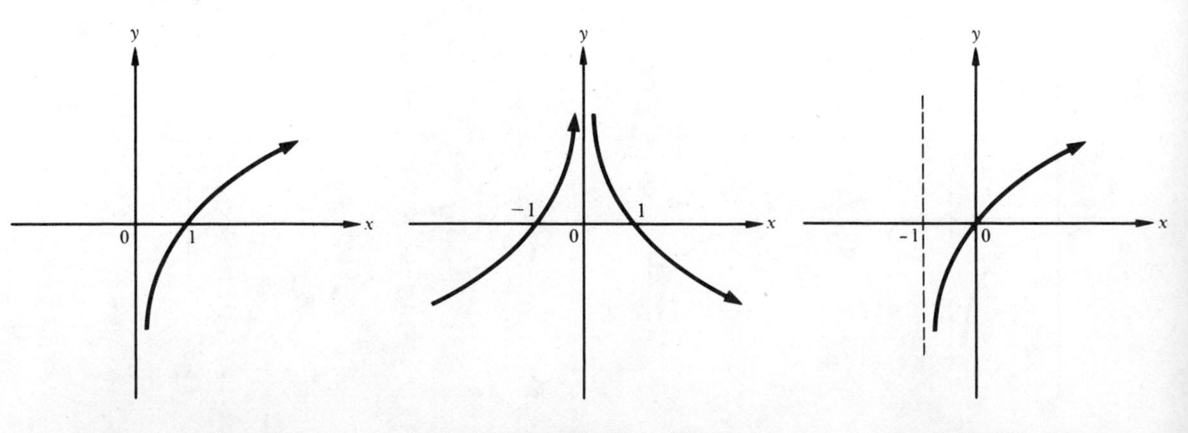

Figure Problem 35 *Figure Problem 37* *Figure Problem 39*

45 As b increases, the graph rises more "slowly."

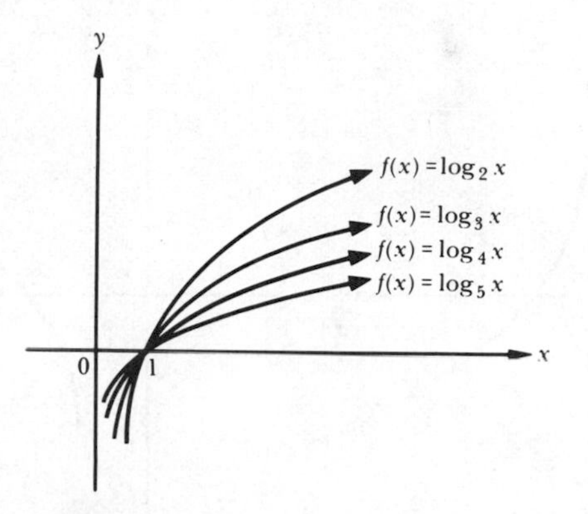

Figure Problem 45

47 For $b < 1$, the graph falls; whereas for $b > 1$, the graph rises.
49 (a) $0 < x < 1$ (b) $x > 1$ (c) $x = 1$ (d) $\log_{10} x_1 < \log_{10} x_2$ **51** \$2742

PROBLEM SET 3, page 162
1 5 **3** $\frac{4}{5}$ **5** $-\frac{1}{6}$ **7** 5 **9** $2 \log_b x + 3 \log_b y$ **11** $\frac{5}{6} \log_b x - \frac{1}{2} \log_b y$
13 $\frac{1}{3} \log_b x + \frac{1}{3} \log_b y$ **15** 0.6020 **17** 0.23855 **19** 3.4771 **21** 1.7781
23 -0.4771 **25** -0.6199 **27** $\log_5 \frac{8}{7}$ **29** $\log_3 2$ **31** $\log_c \frac{a}{7}$ **33** $2 \log_a \left(\frac{a}{\sqrt[3]{x}}\right)$
35 $\{9\}$ **37** $\{-\frac{1}{2}, \frac{1}{2}\}$ **39** $\{x \mid -1 < x < 8\}$ **41** $\{28\}$ **43** $\{\frac{26}{125}\}$ **45** $\{-1, -2\}$
47 $\{7\}$ **49** $\{\frac{1}{3}\}$ **51** $\{22\}$ **53** $B - A - 3C$

PROBLEM SET 4, page 170
1 0.9395 **3** 1.6749 **5** 4.1335 **7** -0.7570 **9** -4.6383 **11** 0.7880 **13** 1.0917
15 -0.2618 **17** 5.6747 **19** 8.20 **21** 55.20 **23** 157 **25** 0.0556 **27** 0.0004
29 3.653 **31** 15.77 **33** 0.1668 **35** 0.007527 **37** 1.6094 **39** 2.4849
41 6.1223 **43** 1.4476 **45** 8.9255 **47** 2.3223 **49** 2.3510

PROBLEM SET 5, page 177
1 56.43 **3** 1.15 **5** 9.01 **7** 0.53 **9** 0.041 **11** $\{1.387\}$ **13** $\{0.609\}$
15 $\{0.241\}$ **17** $\{1.609\}$ **19** $\{-0.277\}$ **21** 123.32 miles
23 (a) 1,808,000 (b) 15.65 minutes **25** (a) 20.48 years (b) 19.88 years (c) 19.88 years (d) 18.07 years **27** (a) \$1271 (b) 11.55 years **29** \$10,650 **31** \$2998 **33** \$40,984

REVIEW PROBLEM SET, page 179
1 Domain $= R$; range $= \{y \mid y > 0\}$; decreasing on R **3** Domain $= R$; range $= \{y \mid y > 0\}$; increasing on R **5** Domain $= R$; range $= \{y \mid y \geq 1\}$; decreasing in $(-\infty, 0]$ and increasing in $[0, \infty)$ **7** $\{-\frac{6}{5}\}$ **9** $\{-6\}$ **11** $\{-4\}$ **13** $\{\frac{5}{3}, -1\}$ **15** Domain $= \{x \mid x > 0\}$; range $= R$; (a) 2 (b) $\frac{3}{2}$ (c) $\frac{5}{4}$ (d) $\frac{1}{20}$ **17** $\{x \mid x \neq 0\}$ **19** R

21 Domain = $\{x \mid x > 0\}$; range = R **23** Domain = $\{x \mid x > 1\}$; range = R

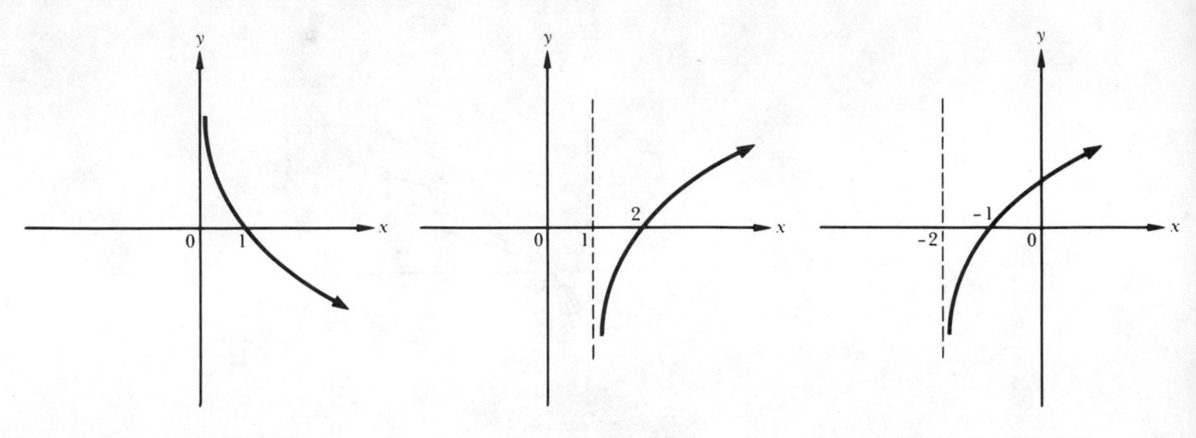

Figure Problem 21 *Figure Problem 23* *Figure Problem 25*

25 Domain = $\{x \mid x > -2\}$; range = R **27** $\frac{3}{2}$ **29** -2 **31** $-\frac{1}{2}$ **33** -8 **35** -10
37 -2 **39** $7 \log_8 5 - 6 \log_8 3$ **41** $\frac{9}{7} \log_2 5$ **43** $6 \log_5 2 + 7 \log_5 3 + 2$ **45** $\log_3 (4/39)$
47 $\log_9 (21/5)$ **49** $9 \log_a x$ **51** 2.524 **53** $\sqrt[3]{2.76} = 1.403$ **55** 0.137 **57** -0.06
59 $\{-\sqrt{5}, \sqrt{5}\}$ **61** $\{x \mid -\frac{8}{3} < x < -2, x \neq -\frac{7}{3}\}$ **63** $\{2\}$ **65** $\{4\}$ **67** $\{80\}$
69 $B - 3C - 2A$ **71** 1.8720 **73** 0.0284 **75** 2.3025 **77** 0.2984 **79** 2.262
81 32.92 **83** 2079 **85** $\{0.778\}$ **87** $\{-1.482, 1.482\}$ **89** $\{-0.524\}$ **91** 7.411
93 0.000908 coulomb **95** \$4503

Chapter 5

PROBLEM SET 1, page 194

1 $\frac{4\pi}{3}$ **3** $\frac{31\pi}{16}$ **5** $-\frac{2\pi}{3}$ **7** $\frac{5\pi}{2}$ **9** None **11** Q_{II} **13** Q_{II} **15** Q_{II}
17 Q_{IV} **19** Q_{III} **21** Q_{IV} **23** Q_{IV} **25** Q_{II} **27** Q_{II} **29** $\left(-\frac{\sqrt{2}}{2}, -\frac{\sqrt{2}}{2}\right)$
31 $\left(\frac{\sqrt{2}}{2}, \frac{\sqrt{2}}{2}\right)$ **33** $\left(\frac{1}{2}, \frac{\sqrt{3}}{2}\right)$ **35** $\left(\frac{\sqrt{3}}{2}, -\frac{1}{2}\right)$ **37** $\frac{\pi}{3}; -\frac{5\pi}{3}$
39 $-\frac{3\pi}{4}; \frac{5\pi}{4}$ **41** 1 **43** 0 **45** Undefined **47** -1 **49** $-\sqrt{2}$ **51** $-\frac{\sqrt{3}}{2}$
53 $-\sqrt{2}$ **55** $\frac{\sqrt{3}}{2}$ **57** $\frac{\sqrt{3}}{2}$ **59** 1

	$\sin t$	$\cos t$	$\tan t$	$\cot t$	$\sec t$	$\csc t$
61	$\frac{1}{2}$	$-\frac{\sqrt{3}}{2}$	$-\frac{\sqrt{3}}{3}$	$-\sqrt{3}$	$-\frac{2\sqrt{3}}{3}$	2
63	$-\frac{12}{13}$	$\frac{5}{13}$	$-\frac{12}{5}$	$-\frac{5}{12}$	$\frac{13}{5}$	$-\frac{13}{12}$
65	$\frac{2\sqrt{13}}{13}$	$-\frac{3\sqrt{13}}{13}$	$-\frac{2}{3}$	$-\frac{3}{2}$	$-\frac{\sqrt{13}}{3}$	$\frac{\sqrt{13}}{2}$

67 (a) sin t increases from 0 to 1; cos t decreases from 1 to 0 (b) sin t decreases from 1 to 0; cos t decreases from 0 to -1 (c) sin t decreases from 0 to -1; cos t increases from -1 to 0 (d) cos t increases from 0 to 1; sin t increases from -1 to 0

PROBLEM SET 2, page 205

1 $P(0.8)$; Q_{I} **3** $P(3.72)$; Q_{III} **5** $P(4.76)$; Q_{IV} **7** $(0, 1)$ **9** $(\sqrt{2}/2, -\sqrt{2}/2)$
11 $(-1, 0)$ **13** $(0, 1)$ **15** $(\sqrt{3}/2, -1/2)$ **17** $(-1/2, \sqrt{3}/2)$ **19** $(\sqrt{2}/2, -\sqrt{2}/2)$
21 $1/2$ **23** $-\sqrt{3}/3$ **25** $\sqrt{2}/2$ **27** $-\sqrt{2}$ **29** $-\sqrt{3}/2$ **31** 1 **33** $-\sqrt{3}/3$
35 $-\sqrt{3}/2$ **37** 1 **39** 2 **41** $1/2$ **43** $\pi/2$ **45** $2\pi/3$ **47** π **49** 4π

51

Quadrant	II	I	IV	III	III	IV	I	IV	I	III
sin t	+	+	−	−	−	−	+	−	+	−
cos t	−	+	+	−	−	+	+	+	+	−
tan t	−	+	−	+	+	−	+	−	+	+
cot t	−	+	−	+	+	−	+	−	+	+
sec t	−	+	+	−	−	+	+	+	+	−
csc t	+	+	−	−	−	−	+	−	+	−

53 $-\frac{1}{2}$; -2 **55** $-\frac{5}{12}$; $-\frac{12}{5}$ **57** $-\frac{13}{5}$; $-\frac{5}{12}$ **59** $\cos(-t) = \cos t$; $\sin(-t) = -\sin t$
61 (a) $\frac{1}{2}$ (b) $\sqrt{3}$ (c) $2\sqrt{3}/3$ **63** $\sin(t^2)$, even; $\sin^2 t$, even

PROBLEM SET 3, page 222

1 Period 2π
Amplitude 10
Phase shift 0

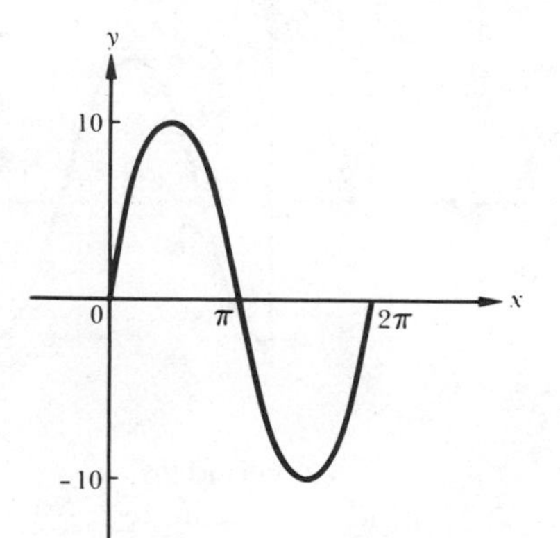

3 Period 2π
Amplitude 2
Phase shift 0

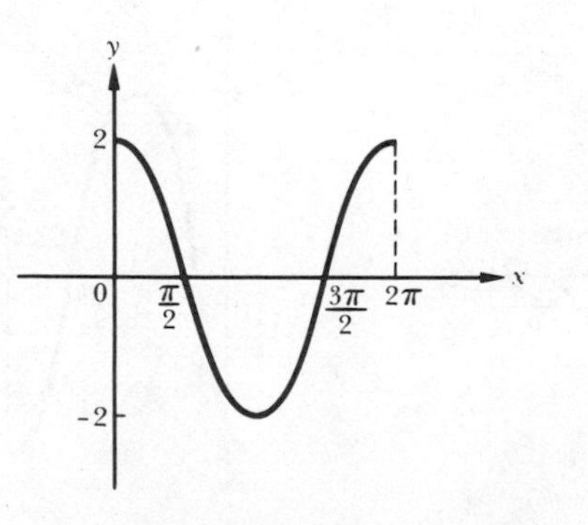

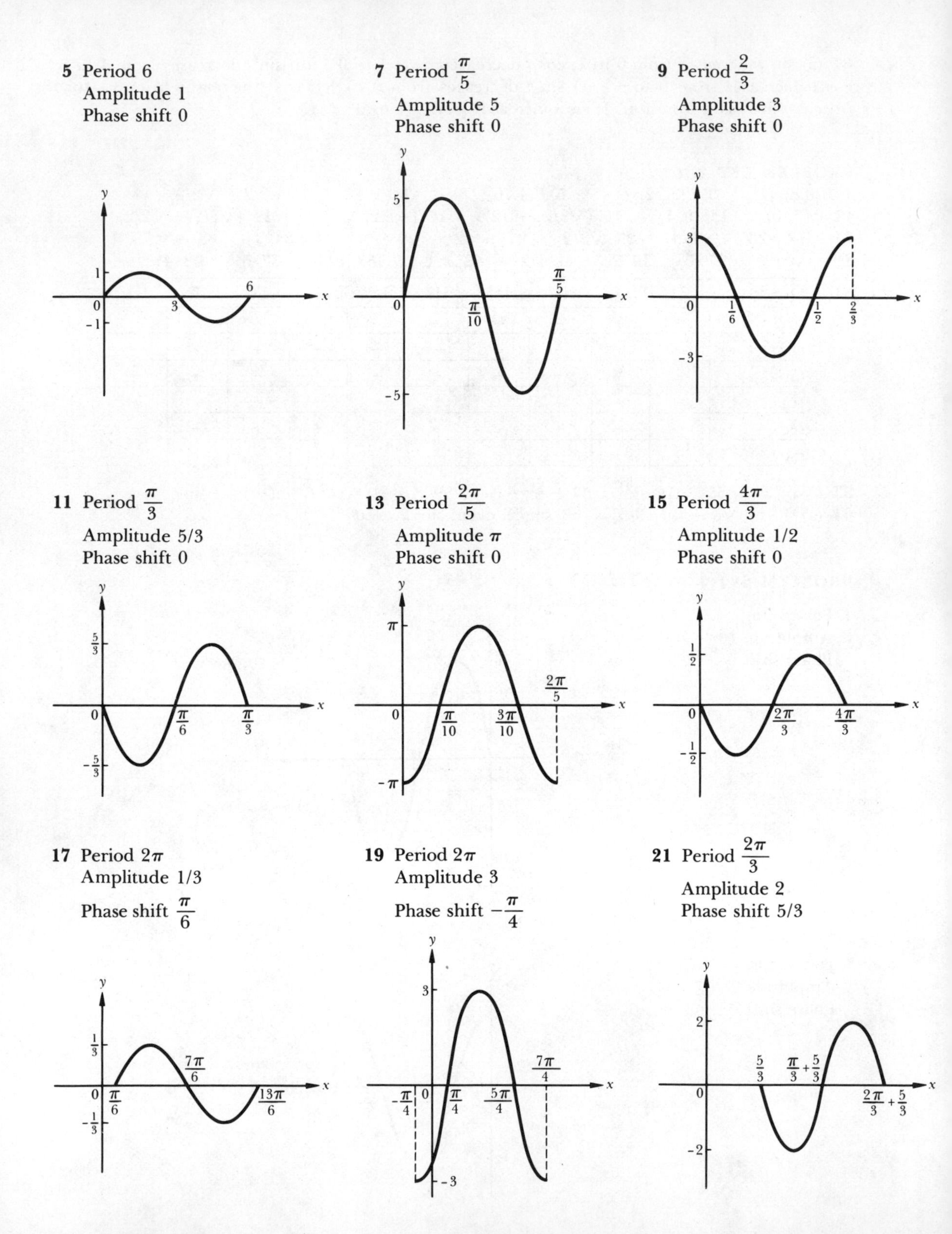
5 Period 6
Amplitude 1
Phase shift 0
7 Period $\frac{\pi}{5}$
Amplitude 5
Phase shift 0
9 Period $\frac{2}{3}$
Amplitude 3
Phase shift 0
11 Period $\frac{\pi}{3}$
Amplitude 5/3
Phase shift 0
13 Period $\frac{2\pi}{5}$
Amplitude π
Phase shift 0
15 Period $\frac{4\pi}{3}$
Amplitude 1/2
Phase shift 0
17 Period 2π
Amplitude 1/3
Phase shift $\frac{\pi}{6}$
19 Period 2π
Amplitude 3
Phase shift $-\frac{\pi}{4}$
21 Period $\frac{2\pi}{3}$
Amplitude 2
Phase shift 5/3

23 Symmetric with respect to y axis

25 (a) Decreases to 0
(b) Decreases from 0
(c) Decreases to 0
(d) Decreases from 0

27

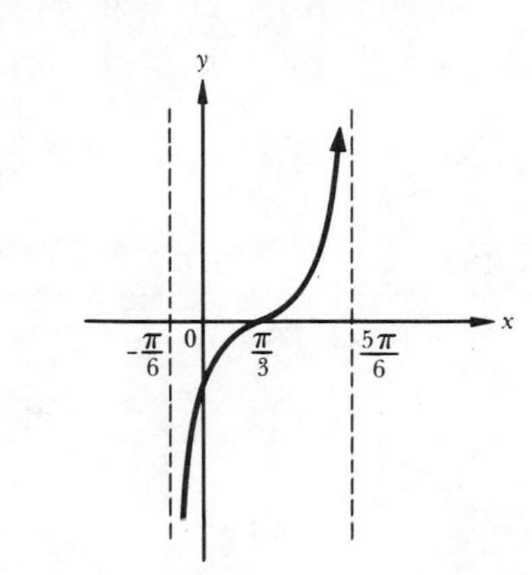

29

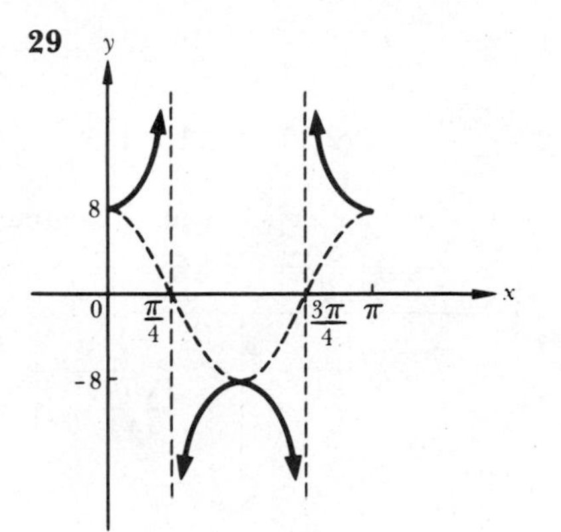

31

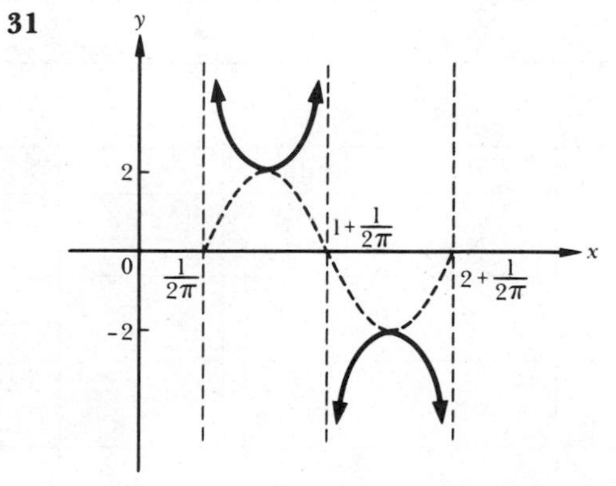

33

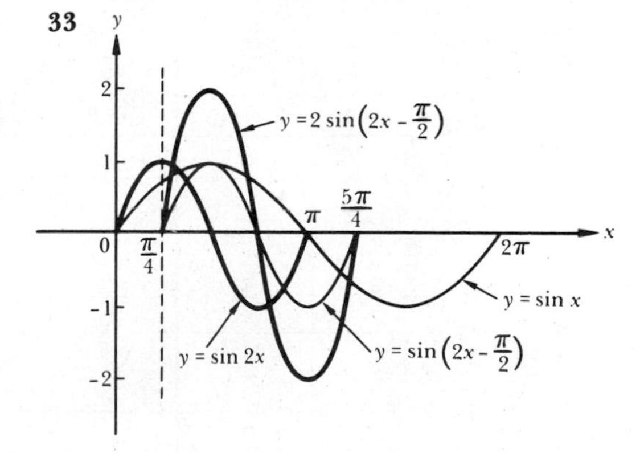

35

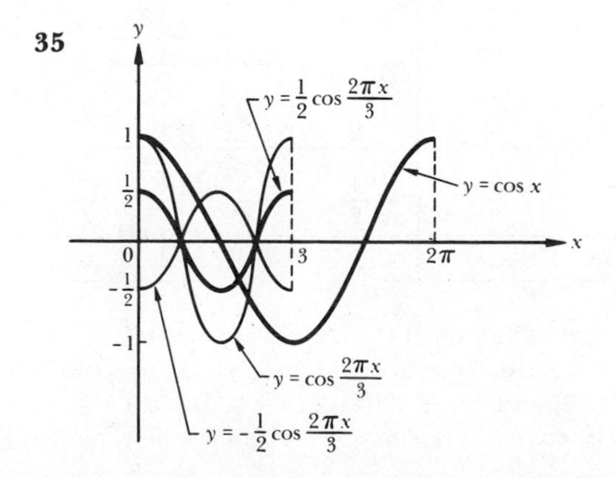

37 (a) $\frac{500}{3\pi} = 53.08$ (b) $\frac{3\pi}{2000} = 0.0047$ (c) $\frac{\pi}{250} = 0.0126$ **39** $\frac{12}{5}; \frac{5}{12}$

PROBLEM SET 4, page 236

1 $87°21'0''$ **3** $-25°33'0''$ **5** $65°22'12''$ **7** $\frac{2\pi}{9}$ **9** $\frac{4\pi}{3}$ **11** $-\frac{19\pi}{36}$ **13** $-\frac{37\pi}{15}$

15 6π **17** $-\frac{2\pi}{5}$ **19** 120° **21** 70° **23** −105° **25** −80° **27** −900°

29 $\frac{720°}{\pi} = 229.30°$ **31** $-135°; -\frac{3\pi}{4}$ **33** $270°; \frac{3\pi}{2}$ **35** $\frac{3\pi}{2}$ inches; $\frac{21\pi}{4}$ square inches

37 $\frac{5\pi}{3}$ centimeters; 5π square centimeters **39** $\frac{7\pi^2}{180}$ centimeters; $\frac{49\pi^2}{360}$ square centimeters

	$\sin\theta$	$\cos\theta$	$\tan\theta$	$\cot\theta$	$\sec\theta$	$\csc\theta$	θ
41	0	−1	0	undef.	−1	undef.	180°
43	$-\frac{10\sqrt{149}}{149}$	$\frac{7\sqrt{149}}{149}$	$-\frac{10}{7}$	$-\frac{7}{10}$	$\frac{\sqrt{149}}{7}$	$-\frac{\sqrt{149}}{10}$	—
45	$\frac{\sqrt{3}}{2}$	$-\frac{1}{2}$	$-\sqrt{3}$	$-\frac{\sqrt{3}}{3}$	−2	$\frac{2\sqrt{3}}{3}$	120°
47	$\pm\frac{\sqrt{10}}{10}$	$\pm\frac{3\sqrt{10}}{10}$	$\frac{1}{3}$	3	$\pm\frac{\sqrt{10}}{3}$	$\pm\sqrt{10}$	—

49

θ Measure							
Degree	Radian	$\sin\theta$	$\cos\theta$	$\tan\theta$	$\cot\theta$	$\sec\theta$	$\csc\theta$
0°	0	0	1	0	undef.	1	undef.
30°	$\frac{\pi}{6}$	$\frac{1}{2}$	$\frac{\sqrt{3}}{2}$	$\frac{\sqrt{3}}{3}$	$\sqrt{3}$	$\frac{2\sqrt{3}}{3}$	2
45°	$\frac{\pi}{4}$	$\frac{\sqrt{2}}{2}$	$\frac{\sqrt{2}}{2}$	1	1	$\sqrt{2}$	$\sqrt{2}$
60°	$\frac{\pi}{3}$	$\frac{\sqrt{3}}{2}$	$\frac{1}{2}$	$\sqrt{3}$	$\frac{\sqrt{3}}{3}$	2	$\frac{2\sqrt{3}}{3}$
90°	$\frac{\pi}{2}$	1	0	undef.	0	undef.	1
120°	$\frac{2\pi}{3}$	$\frac{\sqrt{3}}{2}$	$-\frac{1}{2}$	$-\sqrt{3}$	$-\frac{\sqrt{3}}{3}$	−2	$\frac{2\sqrt{3}}{3}$
135°	$\frac{3\pi}{4}$	$\frac{\sqrt{2}}{2}$	$-\frac{\sqrt{2}}{2}$	−1	−1	$-\sqrt{2}$	$\sqrt{2}$
150°	$\frac{5\pi}{6}$	$\frac{1}{2}$	$-\frac{\sqrt{3}}{2}$	$-\frac{\sqrt{3}}{3}$	$-\sqrt{3}$	$-\frac{2\sqrt{3}}{3}$	2
180°	π	0	−1	0	undef.	−1	undef.

51 390°, −330°, 750°; $\sin 30° = 1/2$, $\cos 30° = \sqrt{3}/2$, $\tan 30° = \sqrt{3}/3$, $\cot 30° = \sqrt{3}$, $\sec 30° = 2\sqrt{3}/3$, $\csc 30° = 2$ **53** $\cos\theta = -\frac{8}{17}$, $\tan\theta = -\frac{15}{8}$, $\cot\theta = -\frac{8}{15}$, $\sec\theta = -\frac{17}{8}$, $\csc\theta = \frac{17}{15}$ **55** $\sin\theta = -\sqrt{2}/2$, $\cos\theta = \sqrt{2}/2$, $\cot\theta = -1$, $\sec\theta = \sqrt{2}$, $\csc\theta = -\sqrt{2}$
57 $\sin\theta = \frac{12}{13}$, $\cos\theta = -\frac{5}{13}$, $\tan\theta = -\frac{12}{5}$, $\cot\theta = -\frac{5}{12}$, $\csc\theta = \frac{13}{12}$

	$\sin\theta$	$\cos\theta$	$\tan\theta$	$\cot\theta$	$\sec\theta$	$\csc\theta$
59	$-\frac{1}{2}$	$\frac{\sqrt{3}}{2}$	$-\frac{\sqrt{3}}{3}$	$-\sqrt{3}$	$\frac{2\sqrt{3}}{3}$	-2
61	$\frac{\sqrt{2}}{2}$	$-\frac{\sqrt{2}}{2}$	-1	-1	$-\sqrt{2}$	$\sqrt{2}$
63	$\frac{\sqrt{3}}{2}$	$-\frac{1}{2}$	$-\sqrt{3}$	$-\frac{\sqrt{3}}{3}$	-2	$\frac{2\sqrt{3}}{3}$
65	$\frac{\sqrt{3}}{2}$	$\frac{1}{2}$	$\sqrt{3}$	$\frac{\sqrt{3}}{3}$	2	$\frac{2\sqrt{3}}{3}$
67	$-\frac{\sqrt{2}}{2}$	$-\frac{\sqrt{2}}{2}$	1	1	$-\sqrt{2}$	$-\sqrt{2}$
69	$\frac{\sqrt{3}}{2}$	$-\frac{1}{2}$	$-\sqrt{3}$	$-\frac{\sqrt{3}}{3}$	-2	$\frac{2\sqrt{3}}{3}$

71 $\frac{20\pi}{3}$ feet **73** $\frac{55}{6}$ units

PROBLEM SET 5, page 249

1 One possibility: sin 2.54 **3** One possibility: tan 0.64 **5** One possibility: sec 330°
7 One possibility: cos 5.58 **9** One possibility: cot 5.87 **11** One possibility: cos (340°52′)
13 0.1790 **15** −3.747 **17** 0.6018 **19** −2.124 **21** −0.9426 **23** −2.971
25 −0.9902 **27** −2.1760 **29** 1.556 **31** −0.5774 **33** −0.9949 **35** 0.7660
37 −0.3335 **39** −1.414 **41** 0.9781 **43** 0.5736 **45** −0.9611 **47** −0.0902
49 −0.0168 **51** 0.8871 **53** −0.6682 **55** 1.165 **57** 1.2024 **59** 0.9870

REVIEW PROBLEM SET, page 250

1 None **3** Q_{II} **5** Q_{II} **7** Q_{I} **9** Q_{III} **11** Q_{III} **13** $(-1, 0)$ **15** $(0, -1)$

17 $\left(\frac{1}{2}, -\frac{\sqrt{3}}{2}\right)$ **19** $\left(\frac{1}{2}, -\frac{\sqrt{3}}{2}\right)$ **21** $\left(-\frac{\sqrt{3}}{2}, \frac{1}{2}\right)$ **23** $\left(\frac{\sqrt{2}}{2}, -\frac{\sqrt{2}}{2}\right)$

25 2 **27** $\frac{\sqrt{3}}{2}$ **29** $-\sqrt{3}$ **31** $-\frac{\sqrt{3}}{2}$ **33** 1 **35** $\frac{3}{2}; -\frac{\sqrt{5}}{2}$ **37** $\frac{1}{5}; \frac{2\sqrt{6}}{5}$

39 Amplitude 1/3
Period 2π
Phase shift 0

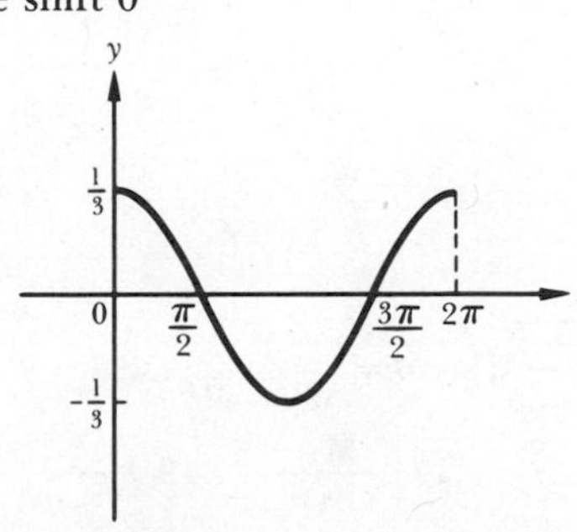

41 Amplitude 2
Period π
Phase shift 0

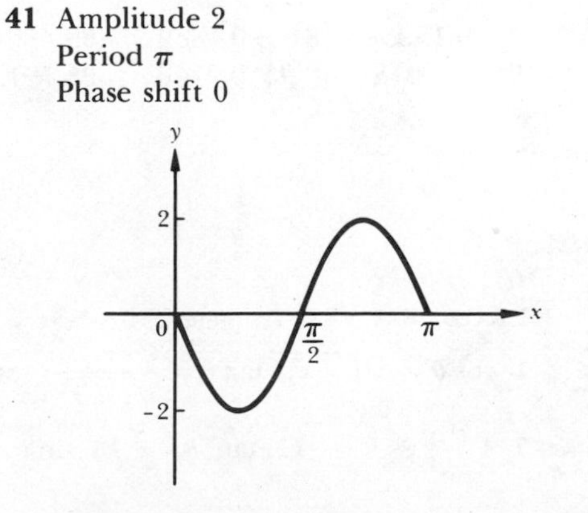

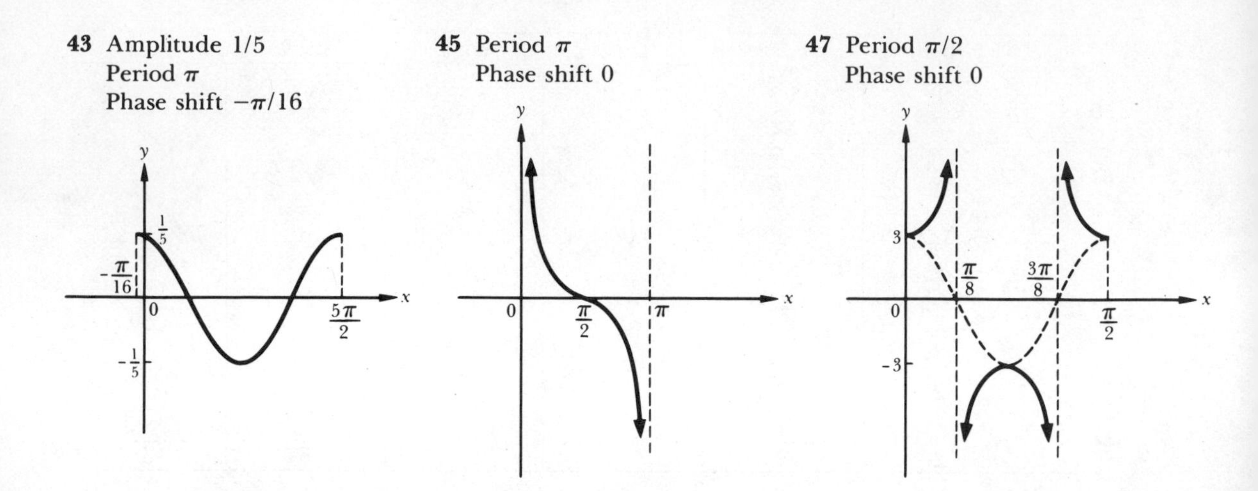

49 $29\pi/18$ **51** $41\pi/9$ **53** $-\pi/15$ **55** 300° **57** −585° **59** −269.43°

61 $\frac{35\pi}{12}$ centimeters; $\frac{245\pi}{24}$ square centimeters **63** 2π inches; 10π square inches

65 5/2; 143.31°; 20 square inches

	$\sin\theta$	$\cos\theta$	$\tan\theta$	$\cot\theta$	$\sec\theta$	$\csc\theta$
67	$-\frac{4}{5}$	$-\frac{3}{5}$	$\frac{4}{3}$	$\frac{3}{4}$	$-\frac{5}{3}$	$-\frac{5}{4}$
69	$-\frac{6\sqrt{61}}{61}$	$\frac{5\sqrt{61}}{61}$	$-\frac{6}{5}$	$-\frac{5}{6}$	$\frac{\sqrt{61}}{5}$	$-\frac{\sqrt{61}}{6}$
71	$-\frac{6\sqrt{61}}{61}$	$-\frac{5\sqrt{61}}{61}$	$\frac{6}{5}$	$\frac{5}{6}$	$-\frac{\sqrt{61}}{5}$	$-\frac{\sqrt{61}}{6}$
73	$\frac{7\sqrt{674}}{674}$	$-\frac{25\sqrt{674}}{674}$	$-\frac{7}{25}$	$-\frac{25}{7}$	$-\frac{\sqrt{674}}{25}$	$\frac{\sqrt{674}}{7}$
75	$\frac{1}{2}$	$\frac{\sqrt{3}}{2}$	$\frac{\sqrt{3}}{3}$	$\sqrt{3}$	$\frac{2\sqrt{3}}{3}$	2
77	$-\frac{1}{2}$	$-\frac{\sqrt{3}}{2}$	$\frac{\sqrt{3}}{3}$	$\sqrt{3}$	$-\frac{2\sqrt{3}}{3}$	−2

79 0.1395 **81** −0.4889 **83** −0.5696 **85** 0.4806 **87** 1.009 **89** −1.054

91 −0.9188 **93** 0.9140 **95** −1.839

Chapter 6

PROBLEM SET 1, page 256

1 $\cos\theta = \sqrt{1-x^2}$; $\tan\theta = \frac{x}{\sqrt{1-x^2}}$; $\cot\theta = \frac{\sqrt{1-x^2}}{x}$; $\sec\theta = \frac{1}{\sqrt{1-x^2}}$; $\csc\theta = \frac{1}{x}$ **3** 0 **5** 1

7 1 **9** 1 **11** $\tan^2\theta$ **13** $\sin t$ **15** $\cos\theta$ **17** $\frac{1}{\cos\theta}$ **19** $\frac{1 + 2\sin t\cos t}{\sin^3 t\cos t}$

PROBLEM SET 2, page 264

1 $\sin 60°$ **3** $\cos 80°$ **5** $\sin t$ **7** $\cos\theta$ **9** $\tan\theta$ **11** $\sin\theta$ **13** $\sec\theta$ **15** $-\cot\theta$
17 $\sin(t+s)$ **19** $-\csc t$ **21** (a) $\dfrac{\sqrt{6}+\sqrt{2}}{4}$ (b) $\dfrac{\sqrt{6}-\sqrt{2}}{4}$ (c) $2+\sqrt{3}$
23 (a) $-\frac{63}{65}$ (b) $-\frac{16}{65}$ (c) $-\frac{33}{65}$ (d) $\frac{56}{65}$ (e) $\frac{63}{16}$ (f) $-\frac{33}{56}$ **29** $2\sin 40°\cos 40°$
31 $\sqrt{\dfrac{1-\cos 20°}{2}}$ **33** $\dfrac{\sqrt{2+\sqrt{2}}}{2}$ **35** $-\left(\dfrac{\sqrt{2+\sqrt{3}}}{2}\right)$ **37** $\dfrac{\sqrt{2+\sqrt{3}}}{2}$ **39** $\sin 6t$
41 $\cos 14t$ **43** $\cos\theta$ **45** (a) $\frac{336}{625}$ (b) $-\frac{527}{625}$ (c) $-\frac{336}{527}$ (d) $\frac{4}{5}$ (e) $-\frac{3}{5}$
47 (a) $\dfrac{\sqrt{6}-\sqrt{2}}{4}$ (b) Same, although they are in different forms

PROBLEM SET 3, page 273

1 $\pi/6$ **3** $3\pi/4$ **5** $-\pi/2$ **7** $-\pi/6$ **9** $\pi/3$ **11** 0.22 **13** 1.36 **15** −0.89
17 2.58 **19** 1.25 **21** $\frac{3}{5}$; $-\frac{7}{25}$ **23** $\sqrt{5}/2$, $-4/3$ **25** $-\frac{24}{7}$

27 (a)

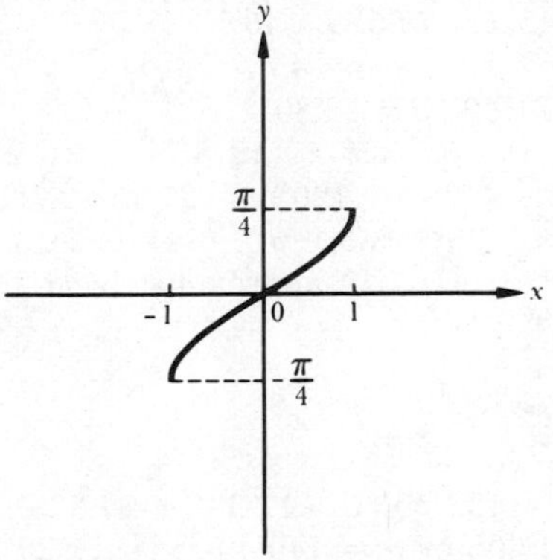

(b)

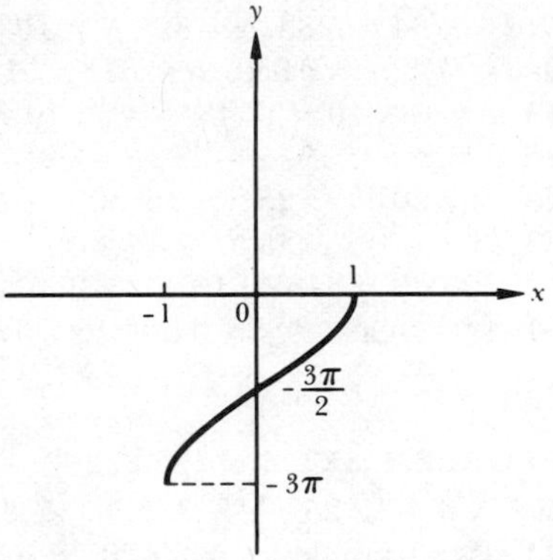

(c)

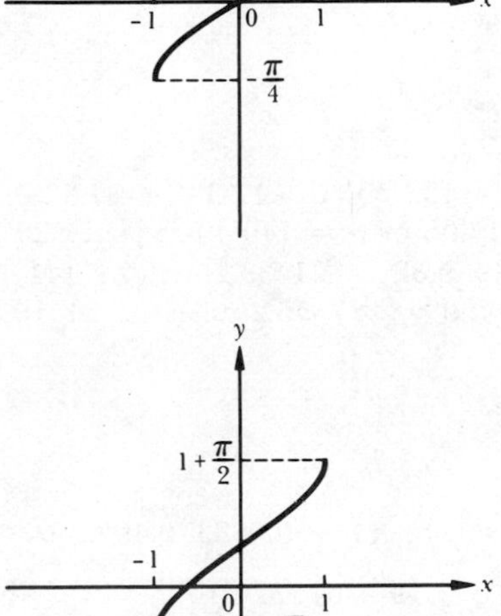

(d)

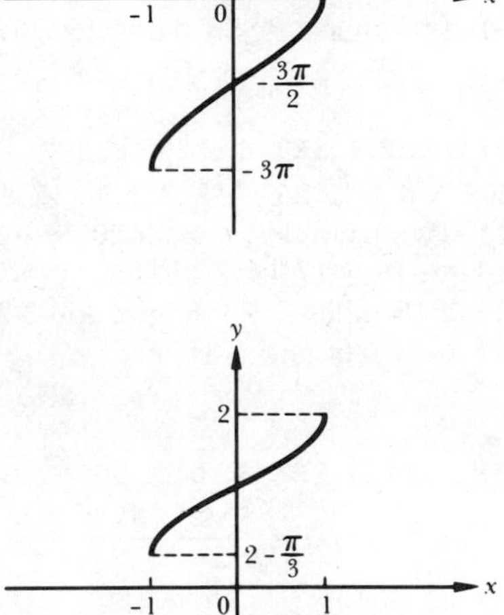

29 $x = \dfrac{1}{2}\cos y$ **31** $x = \dfrac{1}{4}\sin\left(2-\dfrac{y}{2}\right)$ **33** $x = 2\cos\left(\dfrac{3y-1}{4}\right)$
35 (b) $y = \sec^{-1}x$ if and only if $x = \sec y$; domain $= \{x \mid x \geq 1 \text{ or } x \leq -1\}$;
range $= \left\{y \mid 0 \leq y < \dfrac{\pi}{2} \text{ or } \dfrac{\pi}{2} < y \leq \pi\right\}$ **37** $\theta = \dfrac{1}{2}\sin^{-1}\dfrac{M}{k}$

PROBLEM SET 4, page 278

1 $\left\{\frac{\pi}{3}, \frac{2\pi}{3}\right\}$ **3** $\left\{\frac{3\pi}{4}, \frac{7\pi}{4}\right\}$ **5** $\left\{\frac{\pi}{4}, \frac{7\pi}{4}\right\}$ **7** $\{2, 4.28\}$ **9** $\{0.59, 3.73\}$

11 $\left\{\frac{\pi}{6}, \frac{5\pi}{6}\right\}$ **13** $\left\{0, \frac{\pi}{3}, \pi, \frac{5\pi}{3}\right\}$ **15** $\left\{\frac{\pi}{6}, \frac{5\pi}{6}, \frac{7\pi}{6}, \frac{11\pi}{6}\right\}$ **17** $\left\{0, \frac{\pi}{6}, \frac{5\pi}{6}, \pi\right\}$

19 $\{240°, 300°\}$ **21** $\{45°, 75°, 165°, 195°, 285°, 315°\}$ **23** $\{30°, 60°, 210°, 240°\}$

25 $\left\{0.64, \frac{3\pi}{2}\right\}$ **27** $\left\{0, \frac{\pi}{6}, \frac{5\pi}{6}, \pi\right\}$ **29** $\left\{\frac{\pi}{2}, \frac{7\pi}{6}, \frac{11\pi}{6}\right\}$ **31** $\left\{\frac{2\pi}{3}, \frac{4\pi}{3}\right\}$

33 $\{60°, 90°, 270°, 300°\}$ **35** $\{45°, 225°\}$ **37** $\{90°, 120°, 240°, 270°\}$

39 $\{60°, 120°, 240°, 300°\}$ **41** $\left\{t \middle| t = \frac{\pi}{4} + 2n\pi \text{ or } t = \frac{5\pi}{4} + 2n\pi \text{ or } t = n\pi,\ n \in I\right\}$

43 $\{\theta | \theta = n(180°) \text{ or } \theta = 60° + n(360°) \text{ or } \theta = 120° + n(360°),\ n \in I\}$

45 $\{t | t = 0.34 + 2n\pi \text{ or } t = 2.8 + 2n\pi \text{ or } t = 0.73 + 2n\pi \text{ or } t = 2.41 + 2n\pi,\ n \in I\}$

47 12° **49** 62°

PROBLEM SET 5, page 287

5 $c = \sqrt{34} = 5.83, \alpha = 59°, \beta = 31°$ **7** $a = 6.43, b = 7.66, \alpha = 40°$

9 $b = 7.22, c = 18.46, \alpha = 67°$ **11** $c = 5\sqrt{2} = 7.07, \alpha = \beta = 45°$

13 $a = 15\sqrt{10} = 47.43, b = 5\sqrt{10} = 15.81, \alpha = 71°30', \beta = 18°30'$

15 $a = 8, b = 15, \alpha = 28°, \beta = 62°$ **17** $a = 10, \alpha = \beta = 45°$ **19** 4/5 **21** $2\sqrt{85}/85$

23 (2, 120°) **25** (5, 36°50′) **27** $(5\sqrt{2}, 45°)$ **29** (6, 120°) **31** $(5\sqrt{2}, 225°)$

33 $(3\sqrt{3}, 3)$ **35** $(-\frac{7}{2}, 7\sqrt{3}/2)$ **37** (0, −4) **39** $(\pi\sqrt{3}, -\pi)$ **41** $(-2\sqrt{3}, -2)$

43 $200\sqrt{3} = 346.41$ feet **45** 7°30′ **47** 127 feet **49** approximately 41,210 feet

51 4.04 miles **53** 31.44 feet, 7.01 feet

PROBLEM SET 6, page 295

1 $5\sqrt{6} = 12.25$ **3** 3.26 **5** 38.10 **7** $\alpha = 132°30', \beta = 17°30', a = 7.37$ **9** No triangle

11 Two triangles; $\alpha = 39°20', \gamma = 115°40', c = 17.06$ or $\alpha' = 140°40', \gamma' = 14°20', c' = 4.69$

13 $\sqrt{316} = 17.78$ **15** 90° **17** 61.77 **19** 5.82 **21** 33.20 **23** 141°10′

25 2.18 miles at 75° angle, and 3.22 miles at 40°50′ **27** 56.2 miles **29** 10,825 square feet

31 6.76 feet and 1.81 feet

PROBLEM SET 7, page 304

5 $\langle 3, 0\rangle$ **7** $\left\langle \frac{5\sqrt{2}}{2}, \frac{5\sqrt{2}}{2}\right\rangle$ **9** $\left\langle -\frac{\sqrt{3}}{2}, -\frac{1}{2}\right\rangle$ **11** $\langle -0.0883, 0.4922\rangle$

13 $2\sqrt{2}$; 135° **15** 5; 180° **17** 10; 216°50′ **19** $\sqrt{61}$; 320°10′ **21** $\langle -5, -2\rangle; \sqrt{29}$

23 $\langle 5, -11\rangle; \sqrt{146}$ **25** $\langle -6, -6\rangle; 6\sqrt{2}$ **27** $\langle 0, 20\rangle$ **29** $5 + 2\sqrt{5}$ **31** $\langle 8, 16\rangle$

33 $\langle 12, -4\rangle$ **35** 15 **37** (a) (10/3, 2) (b) (1, −1) **39** (a) 3 (b) $-5\sqrt{3}/2$

41 (a) 0 (b) −5 (c) −2 (d) 0 **43** $\langle -12, -3\rangle$; $3\sqrt{17} = 12.37$ miles per hour; 194°

45 $\langle 8, 25\rangle$; $\sqrt{689} = 26.25$ pounds; 72°20′ **47** $\langle 3.1189, -9.5060\rangle$; 10 pounds; 288°10′

49 41°50′ N of E; 391.65 miles per hour

REVIEW PROBLEM SET, page 306

1 $\sin\theta = \frac{x}{\sqrt{4 + x^2}}$; $\cos\theta = \frac{2}{\sqrt{4 + x^2}}$; $\cot\theta = \frac{2}{x}$; $\sec\theta = \frac{\sqrt{4 + x^2}}{2}$; $\csc\theta = \frac{\sqrt{4 + x^2}}{x}$

3 $\sin\theta = \frac{8}{17}$; $\cos\theta = -\frac{15}{17}$; $\tan\theta = -\frac{8}{15}$; $\sec\theta = -\frac{17}{15}$; $\csc\theta = \frac{17}{8}$

5 sin 5° **7** $\tan^2 58°$ **9** tan 6 **11** −cos 2.6

13 $\frac{1}{2}\sin 34°$ **15** $-\frac{240}{289}$ **17** $-\frac{77}{85}$ **19** $-\frac{84}{85}$

21 $\frac{\sqrt{10}}{10}$ **23** $\frac{\sqrt{2-\sqrt{2}}}{2}$ **25** $-\left(\frac{\sqrt{2+\sqrt{3}}}{2}\right)$ **27** $\sqrt{3}+2$ **41** $\frac{\pi}{4}$

43 $-\frac{\pi}{6}$ **45** $\frac{\pi}{4}$ **47** $\frac{3\sqrt{5}}{5}$ **49** 2.25 **51** $x = 2\sin y$ **53** $x = \frac{1}{4}\sec(y-3)$

55 $\{270°\}$ **57** $\{60°, 240°\}$ **59** $\{210°, 330\}$ **61** $\left\{\frac{\pi}{4}, \frac{3\pi}{4}\right\}$ **63** $\left\{0, \frac{2\pi}{3}, \frac{4\pi}{3}\right\}$

65 $\left\{0, \frac{\pi}{6}, \pi, \frac{7\pi}{6}\right\}$ **67** $c = 13, \alpha = 22°40', \beta = 67°20'$ **69** $\alpha = 25°, a = 11.66, c = 27.58$

71 $\alpha = 39°, b = 21, c = 27.01$ **73** $b = 16, c = 20, \alpha = 36°50', \beta = 53°10'$

75 $\left(\frac{5\sqrt{2}}{2}, \frac{5\sqrt{2}}{2}\right)$ **77** $(-1, -1)$ **79** $(0, -3)$ **81** $(3, 180°)$ **83** $(10, 210°)$

85 $(14, 270°)$ **87** $c = 3.66, \beta = 106°50', \alpha = 43°10'$

89 $\gamma = 60°, a = \frac{10\sqrt{6}}{3} = 8.17, b = 11.15$

91 $\gamma = 48°40', \alpha = 101°20'$, and $a = 7.84$, or $\gamma' = 131°20°, \alpha' = 18°40°$, and $a' = 2.56$

93 $\gamma = 54°20', \beta = 10°40', b = 0.98$ **95** $\alpha = 87°, b = 6.7, c = 7.78$ **97** 513.48 feet

99 31.72 feet **101** $\langle 9, 12\rangle$ **103** $\langle -2, -5\rangle$ **105** $\left\langle \frac{5\sqrt{3}}{2}, \frac{5}{2}\right\rangle$

107 $\langle -4\sqrt{3}, 4\rangle$ **109** (a) $\langle 10, -5\rangle; 5\sqrt{5}$ (b) $\langle 10, -7\rangle; \sqrt{149}$ **111** 93°50′

Chapter 7

PROBLEM SET 1, page 316

1 Center (3, 1); radius = 2 **3** Center $(\frac{3}{2}, -2)$; radius = $\frac{3}{2}$ **5** Center $(1, -\frac{3}{2})$; radius = $\frac{7}{2}$

7 Center $(-\frac{3}{2}, -\frac{5}{2})$; radius = $\frac{3}{2}$ **9** $(x-2)^2 + (y-1)^2 = 9$ **11** $(x-5)^2 + (y+4)^2 = 16$

13 $(x-6)^2 + (y+2)^2 = 10$ and $x^2 + (y+4)^2 = 10$ **15** $(x-1)^2 + (y+7)^2 = 49$

17 **i** (a) (2, −1) (b) (1, 5) (c) (4, 10) (d) (2, −2) (e) (−1, 3) (f) (0, 0)
ii (a) (9, −1) (b) (8, 5) (c) (11, 10) (d) (9, −2) (e) (6, 3) (f) (7, 0)

19 **i** (a) (3, −10) (b) (−4, 0) (c) (−2, 1) (d) $(\frac{11}{4}, -\frac{25}{4})$ (e) (−6, −8) (f) (4, −10)
ii (a) (−1, −11) (b) (−8, −1) (c) (−6, 0) (d) $(-\frac{5}{4}, -\frac{29}{4})$ (e) (−10, −9) (f) (0, −11)

21 $\bar{x} = x + \frac{7}{6}, \bar{y} = y - \frac{5}{8}$ **23** $\bar{x} = x - \frac{3}{2}, \bar{y} = y - \frac{7}{8}$

PROBLEM SET 2, page 324

	Vertices	*Foci*
1	(4, 0), (−4, 0), (0, −3), (0, 3)	$(-\sqrt{7}, 0), (\sqrt{7}, 0)$
3	(0, −4), (0, 4), (2, 0), (−2, 0)	$(0, -2\sqrt{3}), (0, 2\sqrt{3})$
5	(−4, 0), (4, 0), (0, −2), (0, 2)	$(-2\sqrt{3}, 0), (2\sqrt{3}, 0)$

	Center	*Vertices*	*Foci*
7	(1, −2)	$(9, -2), (-7, -2), (1, 4\sqrt{3} - 2), (1, -4\sqrt{3} - 2)$	(5, −2), (−3, −2)
9	(−1, 1)	(−3, 1), (1, 1), (−1, 2), (−1, 0)	$(-1-\sqrt{3}, 1), (-1+\sqrt{3}, 1)$
11	(−1, 2)	(−1, 5), (−1, −1), (1, 2), (−3, 2)	$(-1, 2+\sqrt{5}), (-1, 2-\sqrt{5})$
13	(−2, 1)	(−7, 1), (3, 1), (−2, 5), (−2, −3)	(1, 1), (−5, 1)

15 $\frac{(x-3)^2}{4}+\frac{(y+2)^2}{25}=1$ **17** $\frac{(x-3)^2}{4}+\frac{(y-1)^2}{25}=1$ **19** $\frac{(x-2)^2}{4}+\frac{(y-4)^2}{3}=1$

21 $\frac{(x+3)^2}{1}+\frac{(y-1)^2}{25}=1$ **23** (b) $\frac{9}{2}$

25 Circle with center (0, 0) and radius $|a|$

PROBLEM SET 3, page 331

Vertices	Foci	Asymptotes
1 (7, 0), (−7, 0)	$(-\sqrt{85}, 0), (\sqrt{85}, 0)$	$y = \pm \frac{6}{7}x$
3 (0, 3), (0, −3)	$(0, 3\sqrt{2}), (0, -3\sqrt{2})$	$y = \pm x$
5 (3, 0), (−3, 0)	$(\sqrt{10}, 0), (-\sqrt{10}, 0)$	$y = \pm \frac{1}{3}x$

7 $\frac{x^2}{256} - \frac{y^2}{400} = 1$

Center	Vertices	Foci	Asymptotes
9 (5, 4)	(10, 4), (0, 4)	$(5+\sqrt{29}, 4), (5-\sqrt{29}, 4)$	$y-4=\pm\frac{2}{5}(x-5)$
11 (−2, 3)	(2, 3), (−6, 3)	(3, 3), (−7, 3)	$y-3=\pm\frac{3}{4}(x+2)$
13 (4, 1)	(4, 3), (4, −1)	$(4, 1+\sqrt{7}), (4, 1-\sqrt{7})$	$y-1=\pm\frac{2\sqrt{3}}{3}(x-4)$
15 (−4, −2)	(−4, 1), (−4, −5)	$(-4, -2+\sqrt{34}), (-4, -2-\sqrt{34})$	$y+2=\pm\frac{3}{5}(x+4)$

17 $\frac{(y-5)^2}{1}-\frac{(x+1)^2}{3}=1$ **19** $\frac{(y-3)^2}{3}-\frac{(x-2)^2}{1}=1$

21 (a) $a^2 = 7/3$, $b^2 = 35$ (b) $a^2 = 5$, $b^2 = 45/11$

PROBLEM SET 4, page 338

Vertex	Focus	Directrix	Focal Chord
1 (0, 0)	(2, 0)	$x = -2$	8
3 (0, 0)	(0, 1)	$y = -1$	4
5 (0, 0)	$(-\frac{5}{4}, 0)$	$x = \frac{5}{4}$	5
9 (2, 1)	$(2, -\frac{1}{2})$	$y = \frac{5}{2}$	6
11 (−3, −1)	$(-\frac{7}{2}, -1)$	$x = -\frac{5}{2}$	2
13 $(-2, \frac{1}{6})$	$(-2, \frac{2}{3})$	$x = -\frac{1}{3}$	2

15 $(y-2)^2 = 8(x-1)$ **17** $(y-4)^2 = -4(x-1)$ **19** $(x-1)^2 = 4(y-3)$

21 $y = \frac{3}{5}x + \frac{6}{5}$ **23** $\frac{3}{4}$ foot

PROBLEM SET 5, page 344

1 $16\left(y - \frac{15}{4}\right)^2 - 20(x-3)^2 = 125$ **3** $\frac{x^2}{25}+\frac{y^2}{16}=1$ **5** $y^2 = -4(x-3)$

7 $e = \frac{3}{5}$; directrices $y = \pm\frac{25}{3}$ **9** $e = \frac{\sqrt{6}}{3}$; directrices $x = \pm\sqrt{6}$

11 $e = \frac{5}{3}$; directrices $y = \pm\frac{9}{5}$ **13** $e = \frac{\sqrt{6}}{3}$; directrices $y = -11 \pm \frac{1}{4}\sqrt{2978}$

15 Center $= \left(-1, -\frac{3}{2}\right)$; foci: $\left(-1, -\frac{3}{2} \pm \sqrt{29}\right)$; vertices: $\left(-1, -\frac{13}{2}\right), \left(-1, \frac{7}{2}\right)$; asymptotes: $y + \frac{3}{2} = \pm \frac{5}{2}(x+1)$; eccentricity $= \frac{\sqrt{29}}{5}$ **17** (a) $\frac{1}{2}$ (b) $\sqrt{3}a$; $2a$

REVIEW PROBLEM SET, page 345

1 $(x-4)^2 + (y+3)^2 = 25$ **3** $(x-1)^2 + (y-2)^2 = 5$ **5** (a) $(9, 0)$ (b) $(1, 8)$ (c) $(2, -5)$
7 $\frac{x^2}{25} + \frac{y^2}{16} = 1$ **9** $\frac{x^2}{108} + \frac{y^2}{144} = 1$ **11** $\frac{x^2}{9} + \frac{y^2}{5} = 1$ **13** $\frac{x^2}{100} + \frac{y^2}{64} = 1$

Center	*Vertices*	*Foci*	*Eccentricity*	*Directrices*
15 $(0, 0)$	$(2\sqrt{2}, 0), (-2\sqrt{2}, 0), (0, 2\sqrt{3}), (0, -2\sqrt{3})$	$(0, 2), (0, -2)$	$\frac{\sqrt{3}}{3}$	$y = \pm 6$
17 $(-1, 1)$	$(-6, 1), (4, 1), (-1, 4), (-1, -2)$	$(-5, 1), (3, 1)$	$\frac{4}{5}$	$x = \frac{21}{4}$ or $x = -\frac{29}{4}$
19 $(-4, 6)$	$(-8, 6), (0, 6), (-4, 12), (-4, 0)$	$(-4, 6+2\sqrt{5}), (-4, 6-2\sqrt{5})$	$\frac{\sqrt{5}}{3}$	$y = 6 \pm \frac{18\sqrt{5}}{5}$

21 $\frac{x^2}{4} - \frac{y^2}{5} = 1$ **23** $\frac{x^2}{225} - \frac{y^2}{144} = 1$

Center	*Vertices*	*Foci*	*Eccentricity*	*Asymptotes*
25 $(0, 0)$	$(6\sqrt{2}, 0), (-6\sqrt{2}, 0)$	$(4\sqrt{5}, 0), (-4\sqrt{5}, 0)$	$\frac{\sqrt{10}}{3}$	$y = \pm \frac{1}{3}x$
27 $(-2, 3)$	$(-6, 3), (2, 3)$	$(-2+2\sqrt{5}, 3), (-2-2\sqrt{5}, 3)$	$\frac{\sqrt{5}}{2}$	$y - 3 = \pm \frac{1}{2}(x+2)$

29 $(x+4)^2 = -6(y - \frac{3}{2})$ **31** $(x-1)^2 = 3(y + \frac{13}{4})$ **33** $y = \frac{4}{5}x + \frac{8}{5}$ **35** Asymptotes: $y = \pm \frac{3}{2}x$
37 Yes **39** $\frac{(x-5)^2}{25} + \frac{y^2}{9} = 1$ **41** 91.32×10^6 miles **43** $\frac{x^2}{302{,}500} - \frac{y^2}{697{,}500} = 1$

Chapter 8

PROBLEM SET 1, page 355

1 Independent **3** Inconsistent **5** Independent **7** $\{(4, 3)\}$ **9** $\{(1, -1)\}$
11 $\{(\frac{7}{3}, -\frac{4}{3})\}$ **13** $\{(2, 3, -1)\}$ **15** $\{(4, 3, 2)\}$ **17** $\{(1, 1)\}$ **19** $\{(3, 2)\}$
21 $\{(-\frac{3}{2}, \frac{3}{2})\}$ **23** $\{(2, -1)\}$ **25** $\{(1, 1, 1)\}$ **27** $\{(-5, \frac{5}{3}, \frac{28}{3})\}$ **29** $\{(5, 2, 1)\}$
31 7 and 5 **33** 1137 at \$4.25 and 1863 at \$5.50
35 \$1000 at 5 percent and \$3000 at 6 percent **37** 50 miles per hour and 60 miles per hour
39 Car at 40 mph: 5 hours; car at 55 mph: 3 hours
41 \$4000 at 6 percent, \$10,000 at 7 percent, and \$6000 at 8 percent
43 The first is 90 percent pure and the second is 80 percent pure.

PROBLEM SET 2, page 364

1 $\{(3, 5)\}$ **3** $\{(\frac{23}{4}, \frac{13}{2})\}$ **5** $\{(x, y) \mid 3x + y = 4\}$ **7** $\{(1, 2, 3)\}$ **9** $\{(1, 1, 1)\}$
11 $\{(4, 5, 6)\}$ **13** $\{(1, 1)\}$ **15** $\{(3, 0, 5)\}$ **17** $\{(x, 4, 3-x) \mid x \in R\}$
19 $\{(x_2 + 2x_4 + 1, x_2, -3x_4 + 1, x_4, 1) \mid x_2 \in R, x_4 \in R\}$ **21** $\{(2, 3)\}$ **23** $\{(5, 9)\}$
25 $\{(1, 1, 0)\}$ **27** Inconsistent **29** Dependent

PROBLEM SET 3, page 374

1 17 **3** 63 **5** 0 **7** $\frac{1}{4}$ **9** $-\frac{5}{7}$ **11** Theorem 2 **13** Theorem 1 **15** Theorems 1 and 3 **17** 0 **19** -20 **21** 3 **23** $\{(\frac{1}{3}, \frac{2}{3})\}$ **25** $\{(0, 0)\}$ **27** $\{(\frac{19}{4}, -\frac{3}{4}, -\frac{29}{4})\}$ **29** $\{(1, 1, 1)\}$ **31** $\{(3, -1, 2)\}$ **33** 6 and 9 **35** 20 gallons of cream and 10 gallons of milk **37** First is $\frac{5}{3}$, second is $\frac{5}{6}$, third is $\frac{5}{2}$

PROBLEM SET 4, page 381

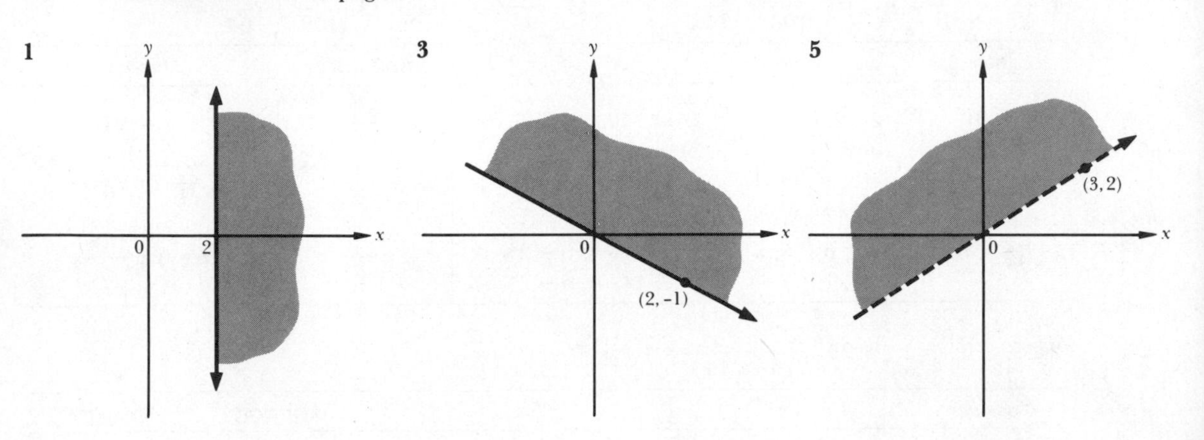

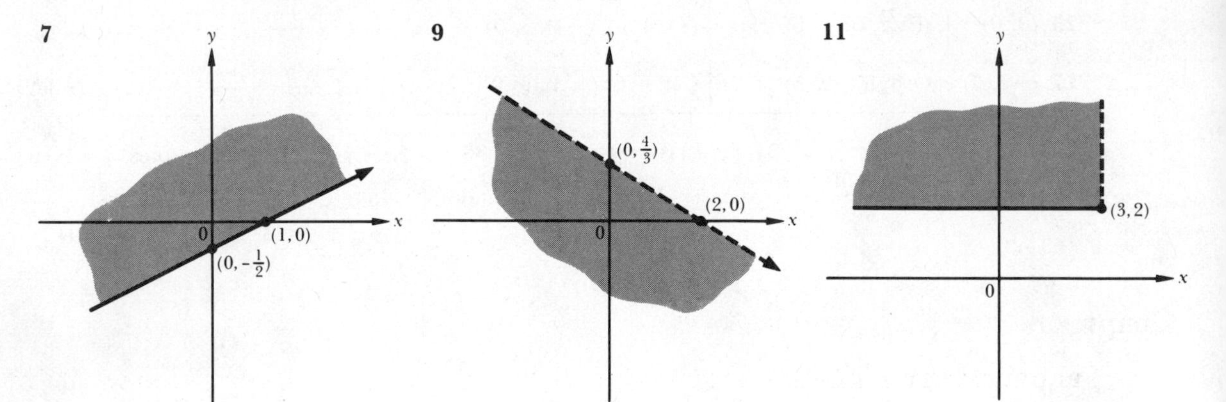

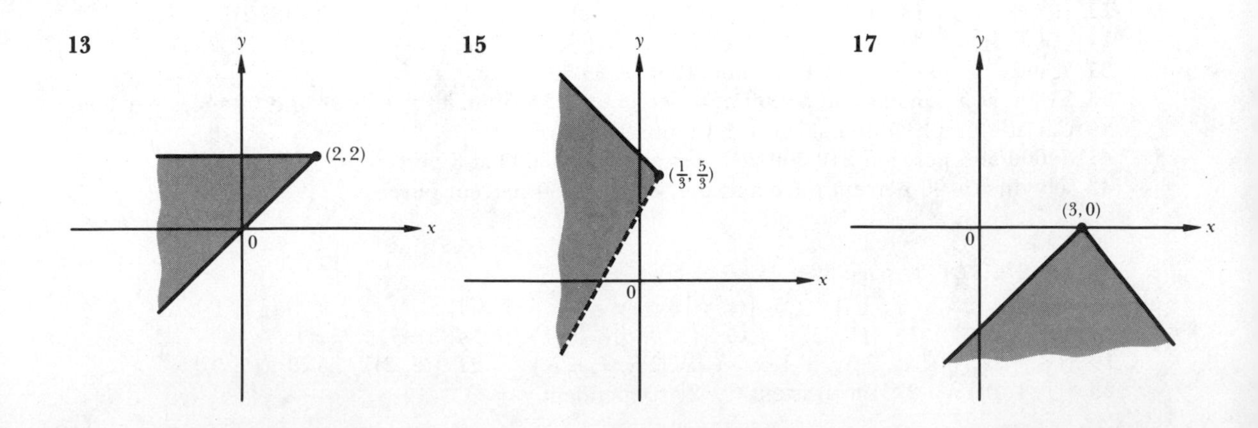

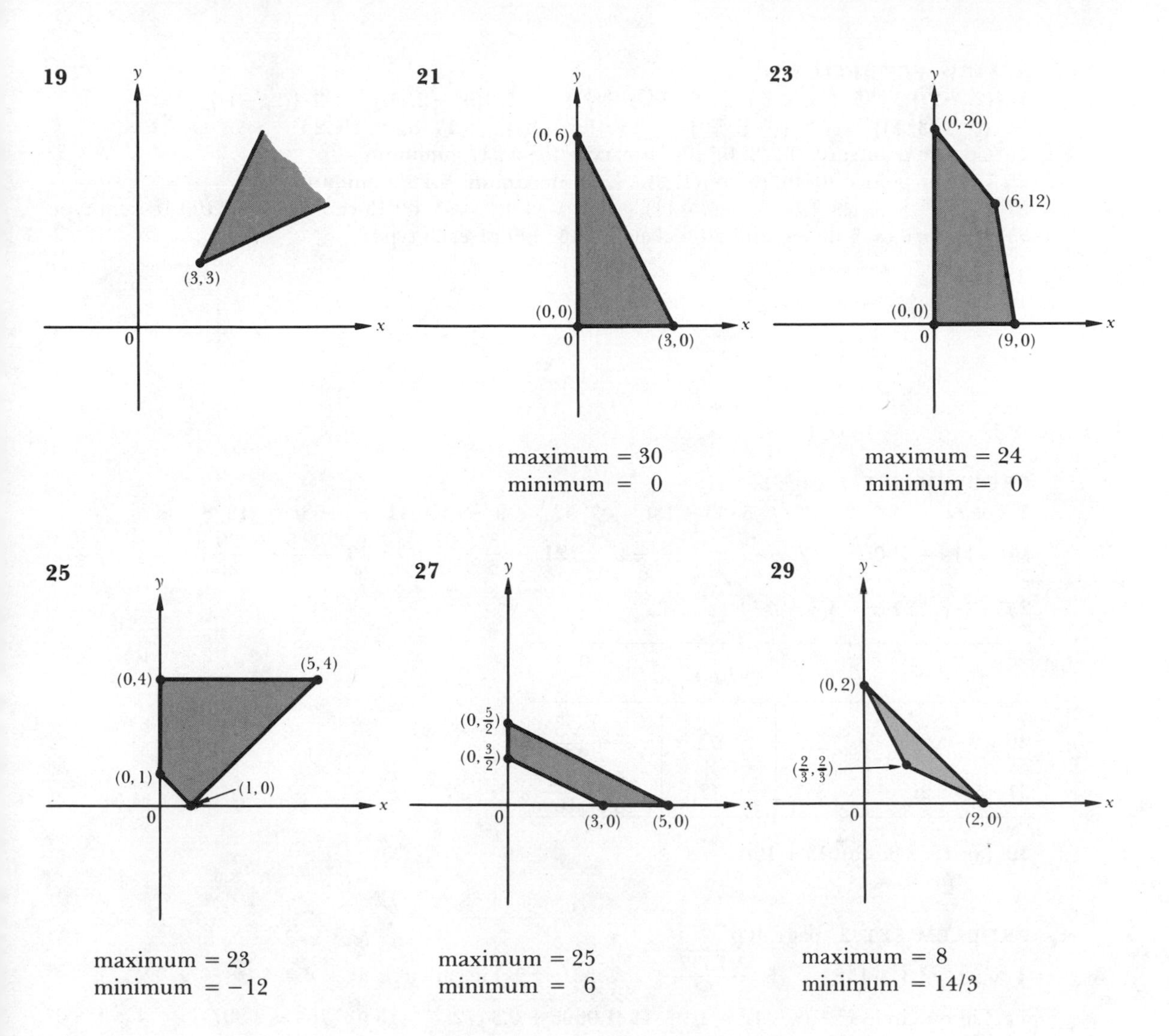

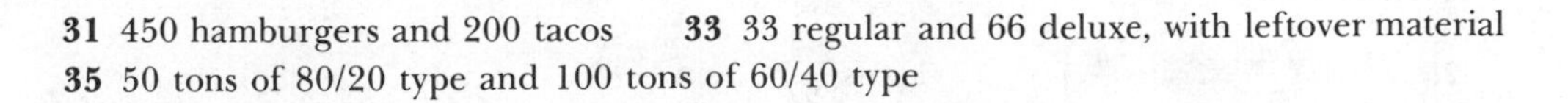

31 450 hamburgers and 200 tacos **33** 33 regular and 66 deluxe, with leftover material
35 50 tons of 80/20 type and 100 tons of 60/40 type

PROBLEM SET 5, page 387

1 $\{(-1, -2), (2, 1)\}$ **3** $\left\{(2, 4), \left(-\frac{4}{5}, -\frac{22}{5}\right)\right\}$ **5** $\varnothing$ **7** $\left\{\left(\frac{7}{2}, \frac{5}{2}\right), \left(\frac{11}{4}, \frac{5}{4}\right)\right\}$ **9** $\varnothing$

11 $\{(5, 4)\}$ **13** $\{(4, 2), (4, -2)\}$ **15** $\{(3, 2), (-3, 2), (3, -2), (-3, -2)\}$

17 $\{(-2\sqrt{6}, 1), (-2\sqrt{6}, -1), (2\sqrt{6}, 1), (2\sqrt{6}, -1)\}$

19 $\{(-2, 2\sqrt{3}), (-2, -2\sqrt{3}), (2, 2\sqrt{3}), (2, -2\sqrt{3})\}$

21 $\left\{(4, 2), (-4, 2), \left(\frac{4\sqrt{15}}{3}, -\frac{10}{3}\right), \left(-\frac{4\sqrt{15}}{3}, -\frac{10}{3}\right)\right\}$ **23** $\{(3, 2), (-3, 2), (-3, -2), (3, -2)\}$

25 $\{(-2, 1), (2, 1)\}$ **27** $\varnothing$ **29** $\left\{\left(\frac{41}{10}, \frac{3\sqrt{91}}{10}\right), \left(\frac{41}{10}, -\frac{3\sqrt{91}}{10}\right)\right\}$

REVIEW PROBLEM SET, page 388

1 $\{(2, -2)\}$ **3** $\{(\frac{1}{2}, -\frac{1}{4})\}$ **5** $\{(\frac{26}{5}, -\frac{36}{5})\}$ **7** $\{(3, -2, 5)\}$ **9** $\{(2, -1)\}$
11 $\{(-1, 3, 2)\}$ **13** $\{(-\frac{2}{9}, \frac{28}{9})\}$ **15** $\{(0, 1, 1)\}$ **17** 5 **19** 26
23 Corner points: (0, 0), (2, 0), (0, 7); maximum = 21, minimum = 0
25 Corner points: (0, 0), (0, 4), (1, 3), $(\frac{7}{4}, 0)$; maximum = 20, minimum = 0
27 $\{(3, -4)\}$ **29** $\{(6, 0), (-5, \sqrt{11}), (-5, -\sqrt{11})\}$ **31** 60 13-cent type and 100 18-cent type
33 10 quarters, 5 dimes, and 20 nickels **35** 300 of each type

Chapter 9

PROBLEM SET 1, page 396

1 $6 + 8i$ **3** $-2 - 2i$ **5** $33 - 13i$ **7** 53 **9** $-i$ **11** $-1 - 3i$ **13** $8 - 6i$
15 $-119 + 120i$ **17** $-\frac{6i}{7}$ **19** $-4i$ **21** $-\frac{3}{5} - \frac{7}{5}i$ **23** $-\frac{3}{25} + \frac{29}{25}i$
25 $-\frac{i}{2}$ **27** $x = 3, y = 4$

	$\bar{z}$	Re z	Im z	$\frac{1}{z}$
29	$2 - \sqrt{3}i$	2	$\sqrt{3}$	$\frac{2}{7} - \frac{\sqrt{3}}{7}i$
31	$5 + 12i$	5	-12	$\frac{5}{169} + \frac{12}{169}i$

39 (a) $13 - 8i$ (b) $13 + 10i$

PROBLEM SET 2, page 406

1 $\sqrt{53}$ **3** $\sqrt{1537}$ **5** $\frac{\sqrt{1537}}{29}$ **7** 53 **9** Lie on circle $x^2 + y^2 = 1$
11 Lie on circle $x^2 + (y - 1)^2 = 1$ **13** $1.9696 + 0.3472i$ **15** $0.7764 - 2.8977i$ **17** $4 + 0i$
19 $\sqrt{2} + \sqrt{2}i$ **21** $\sqrt{2}\left(\cos\frac{5\pi}{4} + i\sin\frac{5\pi}{4}\right)$ **23** $2\left(\cos\frac{3\pi}{2} + i\sin\frac{3\pi}{2}\right)$
25 $2\left(\cos\frac{7\pi}{6} + i\sin\frac{7\pi}{6}\right)$ **27** $1\left(\cos\frac{2\pi}{3} + i\sin\frac{2\pi}{3}\right)$

	z_1z_2	$\frac{z_1}{z_2}$
29	$5(\cos 225° + i\sin 225°) = -\frac{5\sqrt{2}}{2} - \frac{5\sqrt{2}}{2}i$	$5(\cos 115° + i\sin 115°) = -2.1130 + 4.5315i$
31	$6(\cos 90° + i\sin 90°) = 0 + 6i$	$\frac{2}{3}(\cos 10° + i\sin 10°) = 0.6565 + 0.1157i$
33	$8\left(\cos\frac{7\pi}{6} + i\sin\frac{7\pi}{6}\right) = -4\sqrt{3} - 4i$	$2\left(\cos\frac{\pi}{2} + i\sin\frac{\pi}{2}\right) = 0 + 2i$

35 $8\sqrt{2}(\cos 345° + i\sin 345°)$ **37** $\frac{\sqrt{2}}{8}(\cos 105° + i\sin 105°)$

PROBLEM SET 3, page 411

1 $-\frac{\sqrt{3}}{2}-\frac{1}{2}i$ **3** $512-512\sqrt{3}i$ **5** 256 **7** $-125{,}000i$ **9** $-8-8\sqrt{3}i$

11 -2^{30} **13** -1 **15** $\frac{\sqrt{2}}{2}+\frac{\sqrt{2}}{2}i, -\frac{\sqrt{2}}{2}-\frac{\sqrt{2}}{2}i$

17 $2(\cos 0 + i\sin 0), 2\left(\cos\frac{2\pi}{3}+i\sin\frac{2\pi}{3}\right), 2\left(\cos\frac{4\pi}{3}+i\sin\frac{4\pi}{3}\right)$

19 $2\left(\cos\frac{\pi}{4}+i\sin\frac{\pi}{4}\right), 2\left(\cos\frac{3\pi}{4}+i\sin\frac{3\pi}{4}\right), 2\left(\cos\frac{5\pi}{4}+i\sin\frac{5\pi}{4}\right), 2\left(\cos\frac{7\pi}{4}+i\sin\frac{7\pi}{4}\right)$

21 $2\left(\cos\frac{\pi}{5}+i\sin\frac{\pi}{5}\right), 2\left(\cos\frac{3\pi}{5}+i\sin\frac{3\pi}{5}\right), 2(\cos\pi+i\sin\pi),$
$2\left(\cos\frac{7\pi}{5}+i\sin\frac{7\pi}{5}\right), 2\left(\cos\frac{9\pi}{5}+i\sin\frac{9\pi}{5}\right)$

23 $4096\left(\cos\frac{5\pi}{6}+i\sin\frac{5\pi}{6}\right)$ **25** (a) $-\frac{\sqrt{3}}{2}-\frac{1}{2}i$ (b) $\frac{1}{32}-\frac{1}{32}i$

PROBLEM SET 4, page 415

1 $f(x)=\left(x-\frac{1+\sqrt{17}}{4}\right)\left(x-\frac{1-\sqrt{17}}{4}\right)$ **3** $f(x)=x^4-5x^3+7x^2-5x+6$

5 $f(x)=x^4-4x^3+24x^2-40x+100$ **7** $f(x)=x^4-3x^3+x^2+4$

9 $-1-i, \sqrt{2}, -\sqrt{2}$ **11** Yes; 3 **13** Yes; 2

PROBLEM SET 5, page 422

9 3003 **11** 15 **13** 1 **15** 455 **17** 15 **19** $x^5+15x^4+90x^3+270x^2+405x+243$

21 $x^4-8x^3+24x^2-32x+16$ **23** $x^{12}+12x^{11}y+66x^{10}y^2+220x^9y^3$

25 $a^{12}-16a^{21/2}x^2+112a^9x^4-448a^{15/2}x^6$ **27** $\frac{455}{4096}x^{24}a^3$ **29** $\frac{224}{243}x^6a^{12}$ **31** $\frac{28}{243}a^3x^{12}$

PROBLEM SET 6, page 427

1 15 **3** 15 **5** $\frac{5}{6}$ **7** 15 **9** 500 **13** $\sum_{k=0}^{4}(3k+1)$ **15** $\sum_{k=1}^{4}\left(\frac{3}{5}\right)^k$ **17** True

19 False **21** True **23** 1/2 **25** 1/4 **27** 32/99 **29** 46/99 **31** 1186/333

REVIEW PROBLEM SET, page 428

1 $10-5i$ **3** $-1+12i$ **5** $30-16i$ **7** $\frac{5}{7}-\frac{6\sqrt{3}}{7}i$ **9** $x=3, y=-15$

11 $x=-3, y=17/3$ **13** $x=12/13, y=5/13$ **15** $5\sqrt{2}; \pi/4$ **17** $12; \pi/6$

19 $2\sqrt{2}; \pi/4$ **21** $5; 53°10'$

	z_1z_2	$\frac{z_1}{z_2}$
23	$6\left(\cos\frac{3\pi}{2}+i\sin\frac{3\pi}{2}\right)=0-6i$	$\frac{2}{3}\left(\cos\frac{\pi}{2}+i\sin\frac{\pi}{2}\right)=0+\frac{2}{3}i$
25	$12(\cos 322° + i\sin 322°) = 9.4560-7.3884i$	$3(\cos 102° - i\sin 102°) = -0.6237-2.9343i$

27 $1(\cos 300° + i \sin 300°);\ \frac{1}{2} - \frac{\sqrt{3}}{2} i$ **29** $1(\cos 25\pi + i \sin 25\pi);\ -1$

31 $2^{30}(\cos 5\pi + i \sin 5\pi);\ -2^{30}$ **33** $5(\cos 80° + i \sin 80°) = 0.8680 + 4.9240i$

35 $4\left(\cos\frac{\pi}{3} + i \sin\frac{\pi}{3}\right),\ 4(\cos\pi + i\sin\pi),\ 4\left(\cos\frac{5\pi}{3} + i\sin\frac{5\pi}{3}\right)$

37 $2^{1/8}\left(\cos\frac{\pi}{16} + i\sin\frac{\pi}{16}\right),\ 2^{1/8}\left(\cos\frac{9\pi}{16} + i\sin\frac{9\pi}{16}\right),$
$2^{1/8}\left(\cos\frac{17\pi}{16} + i\sin\frac{17\pi}{16}\right),\ 2^{1/8}\left(\cos\frac{25\pi}{16} + i\sin\frac{25\pi}{16}\right)$

39 $f(x) = x^4 + 4x^3 - 5x^2 - 36x - 36$ **41** $f(x) = x^3 - 6x^2 + 13x - 10$

45 $81x^4 + 108x^3y + 54x^2y^2 + 12xy^3 + y^4$ **47** $8x^3 + 12x^2y^{-1} + 6xy^{-2} + y^{-3}$ **49** 95

51 160 **53** 3276 **55** 1/9 **57** 3

Index

Index

Trigonometry

DEFINITION OF TRIGONOMETRIC FUNCTIONS

1 $\sin\theta = \frac{y}{r}$

2 $\cos\theta = \frac{x}{r}$

3 $\tan\theta = \frac{y}{x}, \quad \text{if } x \neq 0$

4 $\csc\theta = \frac{r}{y}, \quad \text{if } y \neq 0$

5 $\sec\theta = \frac{r}{x}, \quad \text{if } x \neq 0$

6 $\cot\theta = \frac{x}{y}, \quad \text{if } y \neq 0$

RIGHT TRIANGLE TRIGONOMETRY

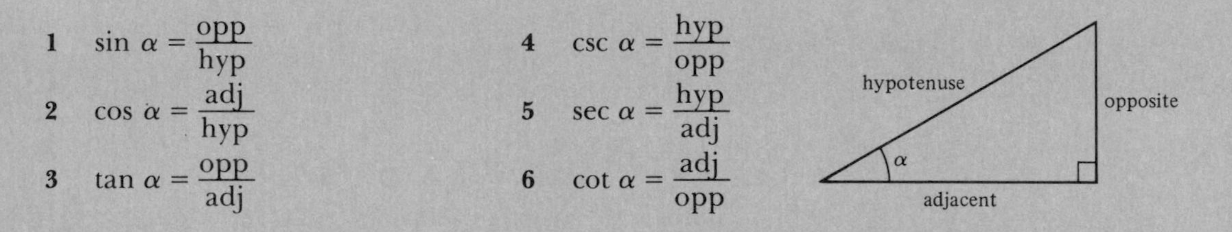

1 $\sin\alpha = \frac{\text{opp}}{\text{hyp}}$

2 $\cos\alpha = \frac{\text{adj}}{\text{hyp}}$

3 $\tan\alpha = \frac{\text{opp}}{\text{adj}}$

4 $\csc\alpha = \frac{\text{hyp}}{\text{opp}}$

5 $\sec\alpha = \frac{\text{hyp}}{\text{adj}}$

6 $\cot\alpha = \frac{\text{adj}}{\text{opp}}$

TRIGONOMETRIC LAWS

1 Law of Sines

$$\frac{\sin\alpha}{a} = \frac{\sin\beta}{b} = \frac{\sin\gamma}{c}$$

2 Law of Cosines

$$c^2 = a^2 + b^2 - 2ab\cos\gamma$$

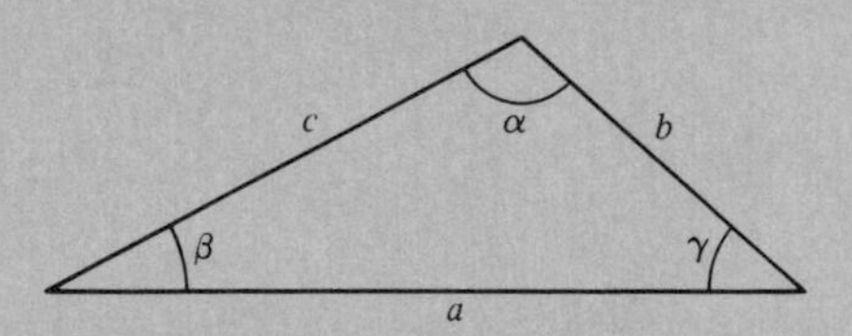